The numerical solution of Volterra equations

CWI Monographs

Centrum voor Wiskunde en Informatica
Centre for Mathematics and Computer Science
P.O. Box 4079, 1009 AB Amsterdam, The Netherlands

The CWI is a research institute of the Stichting Mathematisch Centrum, which was founded on February 11, 1946, as a nonprofit institution aiming at the promotion of mathematics, computer science, and their applications. It is sponsored by the Dutch Government through the Netherlands Organization for the Advancement of Pure Research (Z.W.O.).

CWI Monograph 3

The numerical solution of Volterra equations

H. Brunner

P. J. van der Houwen

1986

North-Holland
Amsterdam · New York · Oxford · Tokyo

ISBN: 0 444 70073 0

Publishers:
Elsevier Science Publishers B.V.
P.O. Box 1991
1000 BZ Amsterdam
The Netherlands

Sole distributors for the U.S.A. and Canada:
Elsevier Science Publishing Company, Inc.
52 Vanderbilt Avenue
New York, N.Y. 10017
U.S.A.

Cover: Tobias Baanders

Library of Congress Cataloging-in-Publication Data

Brunner, H. (Hermann), 1941-
The numerical solution of Volterra equations.

(CWI monograph ; 3)
Bibliography: p.
Includes index.
1. Volterra equations--Numerical solutions.
I. Houwen, P. J. van der (Pieter Jacobus), 1939-
II. Title. III. Series.
QA431.B78 1986 515.4'5 86-12728
ISBN 0-444-70073-0

Printed in the Netherlands

To *Astrid*

To *Aly*

PREFACE

The last fifteen years have witnessed a steadily increasing activity in the development and the analysis of new classes of methods for the numerical solution of integral and integro-differential equations of Volterra type. During the same time, significant advances have been made in the qualitative theory of these equations, and this new insight into the asymptotic behavior of solutions has led to a more frequent use of Volterra equations in the mathematical modeling of various physical and biological phenomena, thus creating a need for reliable and efficient methods for the numerical treatment of the resulting equations.

It is hoped that the present monograph will provide a readable and reasonably comprehensive account of the "state of the art" in the numerical analysis of Volterra integral and integro-differential equations. In order to make the book of interest not only to (upper undergraduate and beginning graduate) mathematics students and researchers, but also to applied mathematicians, physicists, and engineers, we have refrained from introducing any advanced mathematical tools (like abstract functional analytic concepts), even though this would have allowed us to present a number of topics (especially in Chapters 4, 5 and 6) in a more compact manner. Thus, most of the material in this book ought to be readily accessible to a reader possessing no more mathematical background than a good calculus course (including some very basic complex analysis) and a course of elementary linear algebra.

Since a genuine understanding of a numerical method for a Volterra equation can only be acquired if one has a sound grasp of the theory underlying these equations, and since most textbooks on integral equations treat Volterra equations in a very rudimentary way, we have decided to begin the book with a self-contained chapter in which the relevant concepts of, and results from the theory of Volterra integral and integro-differential equations are introduced. This chapter contains also a section on discrete Gronwall inequalities which play a central role in the subsequent convergence analysis of numerical methods. The second chapter is dedicated to a survey of the basic theory of numerical integration; here, we present, in addition to the classical quadrature formulas, the recent theory of (ρ,σ)-reducible quadrature formulas. These formulas will be of crucial importance in the development of linear methods for Volterra equations.

Chapters 3 through 6 deal with the construction and convergence analysis of numerical methods. In Chapter 3 we introduce various classes of linear methods, thus generalizing the well-known classical quadrature methods. Methods of Runge-Kutta type are the subject of Chapter 4. In Chapter 5 we analyze collocation methods (and their discretizations) based on polynomial spline functions. While the Volterra equations occurring in these three chapters are characterized by bounded kernels, Chapter 6 is concerned with the numerical solution of equations possessing weakly singular kernels of Abel type. Here, we study fractional linear multistep methods and spline collocation methods, and give a short introduction to the problem of solving (ill-posed) Abel-type integral equations of the first kind with non-exact (noisy) data.

Chapter 7 focuses on stability results for particular classes of Volterra equations and on the stability of numerical methods applied to these equations; this chapter will also reveal that much research remains to be done in the area of numerical stability, especially as far as equations with weakly singular kernels are concerned. Finally, Chapter 8 contains a survey of computer codes for Volterra integral and integro-differential equations, as well as an extensive collection of test problems which have been used in the literature to assess the performance of numerical methods.

At the end of each chapter the reader will find a section entitled Notes. The role of these notes is two-fold. On the one hand, they are intended to serve as a guide, either to sources of results, to books and papers containing supplementary material on a given topic, or to papers dealing with numerical methods or topics not discussed in the text. It is hoped that they will also restore some of the perspective which has inevitably been lost through the authors' preferences and personal bias. On the other hand, the notes are directed to the actual or prospective researcher: to him or her, they, together with the extensive list of references, should convey a reasonably complete picture of the present state of research on the numerical solution of Volterra equations.

We are grateful to many colleagues for their interest in our work, for the many stimulating conversations, and for their constructive criticism. In particular, we should like to mention P.H.M. Wolkenfelt (Royal Naval Institute, Den Helder), H.J.J. te Riele (CWI), S.P. Nørsett (University of Trondheim), S. McKee and J.A. Scott (University of Oxford), E. Hairer (University of Geneva), Ch. Lubich (University of Innsbruck), I.H. Sloan (University of New South Wales), I.G. Graham (University of Bath), and H. Arndt (University of Bonn). Our special thanks go to Joke Blom (CWI): she not only carried out all the programming work in a most accomplished manner but also suggested many ways for improving the presentation of the material. We are indebted to Paul Beesack (Carleton University, Ottawa) and to Christopher Baker (University of Manchester) for many valuable suggestions.

One of the authors (H.B.) would like to express his gratitude to CWI for the hospitality and the financial support he received during numerous visits and, especially, during his sabbatical leave from Dalhousie University in 1979/80. This and the lively research atmosphere at CWI have been important factors during the writing of the book. Both authors held SERC Senior Fellowships for a total of four weeks in 1980 and 1981; for this they are not only indebted to the SERC but also to their host at the University of Manchester, C.T.H. Baker.

Finally, we would like to thank Mr. R.T. Baanders for drawing the figures, and, in particular, Mrs. J. Kustina of the typing staff at CWI for her patient and expert handling of the many problems in the editing of the manuscripts, and also Joke Blom who assisted in the final editing of the book.

H. Brunner
P.J. van der Houwen
Fribourg/Amsterdam, October 1985.

CONTENTS

CHAPTER 1

AN INTRODUCTION TO THE THEORY OF VOLTERRA EQUATIONS

This opening chapter is devoted to an introduction to the theory of Volterra integral and integro-differential equations. After a brief sketch of the history of these functional equations (beginning essentially with the work of N.H. Abel in 1823) we present those concepts of and results from the quantitative and qualitative theory which will be particularly relevant in the subsequent analysis of numerical methods. A review of certain integral inequalities, including inequalities of Gronwall type, and their discrete counterparts, concludes the chapter.

1.1. Preliminaries

1.1.1. Classification of Volterra equations

An integral equation is a functional equation in which the unknown function appears under one or several integral signs; if, in addition, the equation contains a derivative of this function we call the equation an integro-differential equation. In an integral or integro-differential equation of Volterra type the integrals containing the unknown function are characterized by a variable upper limit of integration. To be more precise, let $I:=[0,T]$ denote a given closed and bounded interval, with $0<T$, and set $S:=\{(t,s):0\leqslant s\leqslant t\leqslant T\}$. An integral equation (for the unknown function y) of the form

$$y(t) = g(t)+\int_0^t k(t,s,y(s))ds, \quad t\in I, \tag{1.1.1}$$

is called a (nonlinear) *Volterra integral equation of the second kind;* here, g and k (the *kernel* of the integral equation) are given real-valued functions. (The function g is occasionally referred to as *forcing function*, or as *initial function.)*

In an integral equation of the first kind the unknown function occurs only under the integral sign; hence, a (nonlinear) *Volterra integral equation of the first kind* is given by

$$\int_0^t k(t,s,y(s))ds = g(t), \quad t\in I. \tag{1.1.2}$$

The functional equation

$$y'(t) = F(t,y(t),z(t)), \quad t\in I, \tag{1.1.3a}$$

with

$$z(t) := \int_0^t k(t,s,y(s))ds, \tag{1.1.3b}$$

is a *first-order Volterra integro-differential equation;* here, one usually looks for a solution which satisfies the initial condition $y(0)=y_0$.

A Volterra integral equation is said to be *linear* if its kernel has the form

$$k(t,s,y) = K(t,s)y. \tag{1.1.4}$$

Accordingly, the linear version of (1.1.3) may be written as

$$y'(t) = p(t)y + q(t) + z(t), \quad t \in I,$$

with (1.1.5)

$$z(t) := \int_0^t K(t,s)y(s)ds.$$

If the kernel of a Volterra equation is the product of a smooth function and a weakly singular (i.e., an unbounded but integrable) function,

$$k(t,s,y) = (t-s)^{-\alpha}\gamma(t,s,y), \quad 0<\alpha<1, \tag{1.1.6}$$

then we speak of an integral (or integro-differential) equation of *Abel type,* since first-kind integral equations whose kernels contain such weak singularities were first studied by Abel (compare Section 1.2).

Let $\theta(t)$ be a continuous function possessing a finite number of zeros in the interval I. The integral equation

$$\theta(t)y(t) = g(t) + \int_0^t k(t,s,y(s))ds, \quad t \in I, \tag{1.1.7}$$

may be viewed as a generalization of (1.1.1): it is called a *Volterra integral equation of the third kind.* There is an intimate connection between linear integral equations of the first and third kind: suppose that the kernel in (1.1.2) is given by (1.1.4), where $\partial K/\partial t \in C(S)$. If g is differentiable on I, then we may differentiate this first-kind equation with respect to t to obtain

$$K(t,t)y(t) = g'(t) - \int_0^t \frac{\partial K(t,s)}{\partial t} y(s)ds, \quad t \in I.$$

This integral equation is of the form (1.1.7) if $K(t,t)$ has a finite number of zeros in the interval I. However, if $K(t,t)$ does not vanish in I then the equation can be written as an integral equation of the second kind, while for $K(t,t) \equiv 0$ it represents again a first-kind equation.

1.1.2. Connections with ordinary differential equations

Let us now consider an important special case of (1.1.1): if the kernel of this equation has the structure

$$k(t,s,y) = \sum_{i=1}^{m} A_i(t) b_i(s,y), \tag{1.1.8}$$

with A_i and b_i being continuous functions, we call k *finitely decomposable.* A second-kind Volterra equation possessing a kernel of this type is equivalent to a system of nonlinear *ordinary differential equations* (ODEs): letting

$$u_i(t) := \int_0^t b_i(s,y(s))ds, \quad i=1, \dots ,m,$$

we have $u_i'(t)=b_i(t,y(t))$, and thus, using the right-hand side of (1.1.1),

$$u_i'(t) = b_i(t,g(t)+\sum_{j=1}^{m} A_j(t)u_j(t)), \quad u_i(0)=0, \quad i=1, \dots ,m. \tag{1.1.9}$$

In the linear case the kernel

$$K(t,s) = \sum_{i=1}^{m} A_i(t) B_i(s) \tag{1.1.10}$$

is said to be *degenerate of rank m* (or: a *Pincherle-Goursat kernel).* Here, the corresponding system of linear first-order differential equations is

$$u_i'(t) = \sum_{j=1}^{m} A_j(t)B_i(t)u_j(t)+B_i(t)g(t), \quad u_i(0)=0, \quad i=1, \dots ,m. \tag{1.1.11}$$

Once the solution of the initial-value problem (1.1.9), or (1.1.11), is known (note that if the functions A_i and B_i in (1.1.10) are continuous on I, then (1.1.11) has a unique, continuously differentiable solution on I), we find that

$$y(t) = g(t)+\sum_{i=1}^{m} u_i(t)A_i(t), \quad t\in I,$$

solves the given integral equation.

We note in passing that the above results are in marked contrast to those for the *Fredholm integral equation of the second kind,*

$$y(t) = g(t)+\lambda\int_0^T K(t,s)y(s)ds, \quad t\in I, \tag{1.1.12}$$

with degenerate kernel (1.1.10): since the limits of integration are now constants we have, setting

$$c_i := \int_0^T B_i(s)y(s)ds, \quad i=1, \dots ,m,$$

the relation

$$c_i = \int_0^T B_i(s)\{g(s)+\lambda\sum_{j=1}^{m} A_j(s)c_j\}ds, \quad i=1, \dots ,m.$$

This represents a system of *linear algebraic equations* for the constants $\{c_i\}$; i.e.,

$$c_i - \lambda \sum_{j=1}^{m} c_j \int_0^T B_i(s)A_j(s)ds = \int_0^T B_i(s)g(s)ds, \quad i=1, \dots, m.$$

Hence, if $\mu := \lambda^{-1}$ is not an eigenvalue of the matrix

$$\left[\int_0^T B_i(s)A_j(s)ds \right], \quad i,j=1, \dots, m,$$

then the unique solution of (1.1.12), (1.1.10) is given by

$$y(t) = g(t) + \lambda \sum_{i=1}^{m} c_i A_i(t), \quad t \in I. \tag{1.1.13}$$

Returning to Volterra integral equations we observe that, for certain special classes of finitely decomposable kernels, there is a more immediate connection with ODEs. Consider, for instance, the initial-value problem,

$$y''(t) = f(t,y(t)), \quad y(0) = y_0, \quad y'(0) = y_0', \tag{1.1.14}$$

where the function f and the initial values y_0, y_0' are prescribed. Repeated integration of (1.1.14) yields

$$y(t) = \int_0^t \int_0^\tau f(s,y(s))ds\, d\tau + y_0' t + y_0, \quad t \in I.$$

This can be written as an integral equation of the second kind,

$$y(t) = y_0 + y_0' t + \int_0^t (t-s) f(s,y(s))ds. \tag{1.1.15}$$

Conversely, assuming that g and k are sufficiently differentiable, and differentiating the second kind equation (1.1.1), we obtain

$$y'(t) = g'(t) + k(t,t,y(t)) + \int_0^t \frac{\partial k}{\partial t}(t,s,y(s))ds, \; y(0) = g(0), \tag{1.1.16}$$

which reduces to an initial-value problem for a *first order ODE* if k does not depend on t, and which is an integro-differential equation otherwise. Differentiation of equation (1.1.16) yields the equation

$$y''(t) = g''(t) + [2\frac{\partial}{\partial t} + \frac{\partial}{\partial s} + y'(t)\frac{\partial}{\partial y}]\, k(t,t,y(t)) + \int_0^t \frac{\partial^2 k}{\partial t^2}(t,s,y(s))ds, \tag{1.1.17}$$

$$y(0) = g(0), \quad y'(0) = g'(0) + k(0,0,g(0)).$$

If $\partial k/\partial t$ does not depend on t, then this equation represents an initial-value problem for a *second-order ODE*. Thus, (1.1.1) can be reduced to an initial-value problem for an n-th order ODE if the n-th derivative with respect to t of

$k(t,s,y)$ vanishes identically. As a consequence, if the kernel in (1.1.1) can be approximated sufficiently closely by a finite Taylor expansion with respect to t, then the solution of (1.1.1) can be approximated by solving an initial-value problem for a higher-order ODE.

For the case where $k(t,s,y)$ is a *polynomial convolution kernel* of the form

$$k(t,s,y) = \sum_{i=0}^{m}(t-s)^i \frac{\phi_i(s)}{i!}y, \tag{1.1.18}$$

yet another relationship can be derived. In Tricomi [1957, p. 18] the following result is given:

Theorem 1.1.1. *The initial-value problem*

$$\begin{aligned} \theta(t)\frac{d^{m+1}z(t)}{dt^{m+1}} &= \sum_{i=0}^{m}\phi_{m-i}(t)\frac{d^iz(t)}{dt^i}+\gamma(t), \quad t\geq 0, \\ \frac{d^iz}{dt^i}(0) &= c_i \quad \text{for} \quad i=0, \dots ,m; \quad \theta(t)\neq 0, \end{aligned} \tag{1.1.19}$$

whose coefficients are assumed to be continuous functions, can be written in the form of the Volterra equation (1.1.7), with k and g given respectively by (1.1.18) and by

$$g(t) := \gamma(t)+\sum_{i=0}^{m}\sum_{j=m-i}^{m} c_j t^{j-m+i}\phi_i(t)/(j-m+i)!, \tag{1.1.20}$$

and with $y(t) := d^{m+1}z(t)/dt^{m+1}$. □

Suppose now that we are given (1.1.7) with k of the form (1.1.18), g and θ arbitrary, $\theta\neq 0$ for $t\geq 0$. It follows from the above theorem that, by solving (1.1.19) with $\gamma(t)=g(t)$ and $c_i=0$ ($i=0, \dots ,m$), we obtain the solution of (1.1.7) from the relation $y(t)=d^{m+1}z(t)/dt^{m+1}$. Specifically, if $\theta(t)\equiv 1$, then we are led to the solution of the second-kind equation (1.1.1).

1.1.3. Connections with partial differential equations

In the preceding subsection we gave examples of ODEs which can be converted into a Volterra equation. In practice, there may be little or no advantage in solving numerically this Volterra equation instead of the original differential equation, because computational methods for ODEs are well developed, and several automatic packages exist (e.g., Shampine & Gordon [1975], NAG FORTRAN Library [1984, Ch. D02]).

In the case of partial differential equations (PDEs) with several independent variables, it may be preferable to solve the equivalent integral equation if it is possible to recast the given problem as an integral equation with fewer independent variables.

In order to illustrate this point we will take a few examples from the literature but omit the details of the conversion which are often complicated and beyond the scope of this book.

Our first example is taken from Noble [1964, p. 225]: it is the problem of an elastic beam being hit by a sphere with velocity V normal to the beam. One wants to know the pressure $p(t)$ produced by the sphere, using for the deflection $\phi(x,t)$ of the beam at the point $x=a$ the representation

$$\phi(a,t) = \int_0^t K(t-s)p(s)ds.$$

Here, a is the point where the sphere hits the beam, and K is a function which can be derived explicitly. The deflection $\phi(x,t)$ of the beam is governed by a PDE in two independent variables,

$$c^4 \frac{\partial^4 \phi}{\partial x^4} = \frac{\partial^2 \phi}{\partial t^2},$$

where c is a known constant; however, the final equation for the pressure $p(t)$ involves only one independent variable and can be shown to assume the form

$$kp^{2/3}(t) = Vt - \int_0^t [\frac{t-s}{m} + K(t-s)]\, p(s)ds, \tag{1.1.21}$$

with m denoting the mass of the sphere, and with k determined by the materials of the beam and sphere, and by the radius of the sphere. By writing $y(t) := p^{2/3}(t)$ this equation becomes a (nonlinear) second-kind Volterra equation with a kernel of convolution type. Once the function $p(t)$ has been determined, the position $z(t)$ of the sphere is given by

$$z(t) = Vt - \frac{1}{m}\int_0^t (t-s)p(s)ds;$$

the quantity $z-\phi$ represents the depth of indentation of the beam by the sphere at time t.

In the reflection of sound pulses by convex parabolic reflectors we are required to solve the equation of sound,

$$\frac{\partial^2 \phi}{\partial t^2} = c^2[\frac{\partial^2 \phi}{\partial x^2} + \frac{1}{w}\frac{\partial \phi}{\partial w} + \frac{\partial^2 \phi}{\partial w^2}],$$

where $w^2 := y^2 + z^2$, (x,y,z) are Cartesian coordinates, t is the time, c the velocity of sound, and ϕ is the velocity potential. Imposing the boundary conditions determined by the geometry of the reflector, Friedlander [1941] showed that this problem can be solved by finding the solution of a Volterra equation of the form

$$y(t) = \frac{1}{(t+2)^\alpha} - \lambda \int_0^t \frac{y(s)}{(t-s+2)^\alpha}\, ds, \tag{1.1.22}$$

where α and λ are constants. For a convex paraboloid Friedlander gives the values $\alpha=\lambda=2$, and for a parabolic cylinder-reflector, the values $\alpha=3/2$, $\lambda=1/\sqrt{2}$.

As a third example we consider the problem of determining the temperature $\phi=\phi(x,t)$ in a nonlinearly radiating semi-infinite solid. It is assumed that energy is radiated away at a rate proportional to $\phi^n(0,t)$, while an external source supplies heat at a rate proportional to a given function of time $f(t)$. Here, n denotes a fixed but arbitrary real number greater than, or equal to, one (with the special case $n=4$ being of particular physical interest: it corresponds to the Stefan-Boltzmann radiation law). As is shown, e.g., in Keller & Olmstead [1972], ϕ is the solution of the following initial-boundary value problem:

$$\frac{\partial\phi}{\partial t} = \frac{\partial^2\phi}{\partial x^2}, \quad x>0, \quad t>0;$$

$$\frac{\partial\phi(0,t)}{\partial x} = c\phi^n(0,t)-f(t), \quad t>0; \quad \phi(x,0) = 0, \quad x\geqslant 0;$$

$$\lim_{x\to\infty} \phi(x,t) = 0, \quad t\geqslant 0.$$

The dimensional constant $c>0$ depends only on properties of the given material. It can be shown that any solution of this initial-boundary value problem must satisfy the equation

$$\phi(x,t) = \pi^{-1/2}\int_0^t (t-s)^{-1/2}\exp\left(\frac{-x^2}{4(t-s)}\right)\{f(s)-c\phi^n(0,s)\}ds, \qquad (1.1.23)$$

$$x\geqslant 0, \quad t\geqslant 0.$$

This means that $\phi(x,t)$ is completely determined once $y(t):=\phi(0,t)$ (the temperature on the boundary) has been found. Thus, upon setting $x=0$ in (1.1.23), we obtain

$$y(t) = \pi^{-1/2}\int_0^t (t-s)^{-1/2}\{f(s)-cy^n(s)\}ds, \quad t\geqslant 0; \qquad (1.1.24)$$

i.e., a nonlinear Volterra integral equation of the second kind with weakly singular kernel.

1.2. A brief history of Volterra integral equations

The origins of the quantitative theory of integral equations with variable (upper) limits of integration go back to the early 19th century. Abel [1823], [1826] considered the problem of determining the equation of a curve in a vertical plane such that the time taken by a mass point to slide, under the influence of gravity, along this curve from a given positive height to the horizontal axis is equal to a prescribed (monotone) function of the height. He showed that this problem can be described by a first-kind integral equation of the form

$$\int_0^t (t-s)^{-\alpha} y(s) ds = g(t), \quad t>0, \tag{1.2.1}$$

with $\alpha=1/2$, and he then proved that, for any $\alpha \in (0,1)$, the solution of (1.2.1) is given by the "inversion formula",

$$y(t) = c_\alpha \frac{d}{dt}\{\int_0^t (t-s)^{\alpha-1} g(s) ds\}, \quad t>0, \tag{1.2.2}$$

with $c_\alpha := sin(\alpha\pi)/\pi$ $(=1/(\Gamma(\alpha)\Gamma(1-\alpha)))$. If $g \in C^1(I)$, with $g(0)=0$, then we have the pair of equations

$$g(t) = \int_0^t (t-s)^{-\alpha} y(s) ds, \quad y(t) = c_\alpha \int_0^t (t-s)^{\alpha-1} g'(s) ds.$$

This, incidentally, is reminiscent of another well-known pair, derived by Fourier in 1811,

$$g(t) = \int_0^\infty \cos(t-s) y(s) ds, \quad y(t) = \frac{2}{\pi} \int_0^\infty \cos(t-s) g(s) ds.$$

Three years after Abel's death the problem of inverting (1.2.1) was also studied by Liouville [1832] (who seems to have been unaware of Abel's work), again in a purely formal manner. In 1884 Sonine extended the inversion formula to cover first-kind integral equations characterized by convolution kernels, $a = a(t-s)$, given by

$$a(z) = z^{-\alpha} \sum_{j=0}^\infty c_j z^j / \Gamma(1+j-\alpha), \quad c_0=1, \quad 0<\alpha<1.$$

Before stepping ahead to 1896, the year when Volterra published his general theory on the inversion of first-kind integral equations, we mention that the discovery of the inversion formula (1.2.2) was the starting point for the systematic development of what is known now as *Fractional Calculus* (whose very origins date back to Leibniz and Euler). The *fractional integral* (of order $\nu>0$) of a function $\phi(t)$ is defined by

$${}_cD_t^{-\nu}\phi(t) := \frac{1}{\Gamma(\nu)} \int_c^t (t-s)^{\nu-1} \phi(s) ds,$$

where c is a given constant. While Liouville (who, in addition to the paper mentioned above, devoted seven more memoirs, totaling close to three hundred pages, to the study of this subject) worked with the value $c=-\infty$, it was Riemann [1847] who suggested $c=0$, the value commonly used today. The fractional derivative is defined in an analogous way. Based on these notions it is thus possible to express the integral equation (1.2.1) and its solution (1.2.2) in terms of these fractional operations. The paper by Davis [1927a] contains a thorough exposition of this theory; in addition, we refer the reader to the monograph by Oldham & Spanier [1974] (which also features a detailed chronological and annotated bibliography on Fractional Calculus).

In order to invert

$$\int_0^t K(t,s)y(s)ds = g(t), \quad t \in I, \quad g(0) = 0, \tag{1.2.3}$$

Volterra [1896] (Nota I) transforms it, by differentiation with respect to t, into an integral equation of the second kind whose kernel and forcing function are, respectively, $\tilde{K}(t,s) := -(\partial K(t,s)/\partial t)/K(t,t)$ and $\tilde{g}(t) := g'(t)/K(t,t)$; if $K(t,t)$ does not vanish on I, and if the derivatives of K and g are continuous, then the (unique) solution of (1.2.3) is given by the "inversion formula"

$$y(t) = g(t) + \int_0^t \tilde{R}(t,s)g(s)ds, \quad t \in I.$$

Here, $\tilde{R}(t,s)$ denotes the so-called *resolvent kernel* of $\tilde{K}(t,s)$; it is defined in terms of the iterated kernels $\tilde{K}_n$ of $\tilde{K}$,

$$\tilde{K}_n(t,s) := \int_s^t \tilde{K}(t,u)\tilde{K}_{n-1}(u,s)du, \quad n \geq 2, \quad \tilde{K}_1(t,s) := \tilde{K}(t,s). \tag{1.2.4}$$

Volterra proved that this series converges absolutely and uniformly on S for any kernel K in (1.2.3) satisfying the conditions stated above (compare Theorem 1.3.1 and Theorem 1.3.5).

Even though Volterra's result was new, his way of attack was not entirely a novel one. In his thesis of 1894, Le Roux had already studied the problem of inverting the "definite integral" (1.2.3) (see Le Roux [1895, pp. 243-246]), using the same approach (without, however, proving the uniform convergence of the infinite series $\Sigma_{n=1}^{\infty} \tilde{K}_n(t,s)$). But second-kind integral equations with variable limit of integration occur already in the work of Liouville [1837] (the name "Volterra integral equation" was coined by Lalesco [1908, p. 126], seemingly following a suggestion by his teacher Picard; the terminology "integral equation of the first (second, third) kind" was first used by Hilbert in connection with his study of Fredholm integral equations, while the name "integral equation" is due to Du Bois-Reymond [1888, p. 228]): he shows that any solution $y(t)$ of $y'' = A_0(t)y$, with $y'(0) - cy(0) = 0$, satisfies the functional equation

$$y(t) = y_0(t) + \int_0^t ds \int_0^s A_0(u)y(u)du, \quad y_0(t) := 1 + ct,$$

which (by Dirichlet's formula) he could have written in the form

$$y(t) = y_0(t) + \int_0^t (t-s)A_0(s)y(s)ds, \quad t \in I.$$

Setting $y(t) := \Sigma_{n=0}^{\infty} y_n(t)$, he determines the terms of this series recursively from

$$y_{n+1}(t) := y_0(t) + \int_0^t ds \int_0^s A_0(u) y_n(u) du, \quad n \geq 0,$$

where pointwise convergence is established by proving that $|y_n(t)| \leq M^n t^{2n}/(2n)!$ for some finite constant M, provided $A_0(t)$ is continuous on I.

The notion of the iterated kernels and the associated "Neumann series" were first used by Caqué [1864, p. 194, 201]: generalizing Liouville's idea, he studied the solution of the $(p+1)$-st order linear differential equation,

$$y^{(p+1)} = \sum_{j=0}^{p} A_j(t) y^{(j)} + A(t),$$

by rewriting the equation as a second-kind integral equation of Volterra type whose kernel is

$$K(t,s) := \sum_{j=0}^{p} A_j(s)(t-s)^{p-j}/(p-j)!.$$

The existence of a solution is then established formally by introducing the iterated kernels and the corresponding Neumann series. At about the same time, Beer [1865, pp. 62-65] uses the same concepts, still in a purely formal way, in connection with the study of second-kind integral equations with fixed limits of integration which arose in the analysis of Poisson's equation in Potential Theory. It was left to Neumann [1877] to furnish the rigorous convergence analysis for the series of iterated kernels (associated with a second-kind integral equation of Fredholm type) now named after him.

In a second paper [1896] (Nota II), Volterra extended his ideas to linear integral equations of the first kind with weakly singular kernels: by using the approach employed by Abel to establish the inversion formula (1.2.2) he shows that

$$\int_0^t (t-s)^{-\alpha} K(t,s) y(s) ds = g(t), \quad t \in I, \quad 0 < \alpha < 1,$$

can be transformed into a first-kind equation with regular kernel, to which the theory of his first Nota applies. The remaining two papers of Volterra [1896] (Nota III, IV) are concerned with the analysis of integral equations of the third kind (see also Volterra [1897] and, especially, Lalesco [1908] for an account and, respectively, an extension of this work).

The next forty years saw mainly a consolidation of Volterra's work; during this time the center stage belonged to the study of Fredholm integral equations and their implications for the development of Functional Analysis. A thorough account of these aspects can be found in Monna [1973] and in Dieudonné [1981]; compare also Davis [1926], Hellinger & Toeplitz [1927], Vivanti [1929], and Elliott [1980].

The last twenty years have witnessed a renewed interest in Volterra equations: the study of qualitative and asymptotic properties of solutions of

Volterra equations (which may be said to have begun with the work of Paley & Wiener [1934]) has experienced a tremendous growth, as indicated by the monograph by Miller [1971] and, especially, the conference proceedings by Londen & Staffans [1979], by Herdman et al. [1981], and by Hannsgen et al. [1982].

1.3. Existence and uniqueness of solutions

If the analysis of numerical methods for Volterra functional equations is to succeed it is imperative to have a grasp of at least the basic quantitative and qualitative theory of such equations. It is the purpose of this section to present, often without detailed proofs, the basic quantitative theory of Volterra equations; in the following section we shall collect some relevant qualitative results. The Notes at the end of Chapter 1 will contain suggestions for further reading, together with a brief guide to the literature.

1.3.1. Linear Volterra integral equations of the second kind

Consider the integral equation

$$y(t) = g(t)+\int_0^t K(t,s)y(s)ds, \quad t\in I, \tag{1.3.1}$$

and assume that the kernel K and the forcing function g are real-valued and continuous on S and I, respectively. (Compare the Notes for a remark on the case where these functions are elements of the space L^2 .) As has already been briefly indicated in the previous section, the so-called iterated kernels and the corresponding resolvent kernel are of fundamental importance in the subsequent analysis.

Definition 1.3.1. Let $K_1(t,s) := K(t,s)$, and set

$$K_n(t,s) := \int_s^t K_1(t,\tau)K_{n-1}(\tau,s)d\tau, \quad (t,s)\in S, \quad n\geq 2. \tag{1.3.2}$$

The functions $\{K_n\colon n=1,2,...\}$ are called the *iterated kernels* associated with the given kernel K in (1.3.1). □

We observe first that, for any r satisfying $1\leq r\leq n-1$,

$$K_n(t,s) = \int_s^t K_r(t,\tau)K_{n-r}(\tau,s)d\tau.$$

Here we have made use of *Dirichlet's formula* which states that

$$\int_0^t\int_0^\tau \Phi(\tau,s)ds\, d\tau = \int_0^t\int_s^t \Phi(\tau,s)d\tau\, ds, \quad (t,s)\in S, \tag{1.3.3}$$

provided the integrals exist.

Using (1.3.2) it is easily verified that, if the given kernel K satisfies $|K(t,s)| \leq M$ for all $(t,s) \in S$, then, for all $n \geq 1$,

$$|K_n(t,s)| \leq M^n \frac{(t-s)^{n-1}}{(n-1)!} \leq M^n \frac{T^{n-1}}{(n-1)!}, \quad (t,s) \in S. \tag{1.3.4}$$

The introduction of the iterated kernels of K is motivated by the *Picard method* for constructing successive approximations $\{y_n(t):n \geq 1\}$ to the exact solution y of (1.3.1) by means of

$$y_n(t) := g(t) + \int_0^t K(t,s) y_{n-1}(s) ds, \quad y_0(t) := g(t) \quad (t \in I). \tag{1.3.5}$$

It follows that the iterate y_n can be expressed in terms of the given function g and the iterated kernels $K_1, \ldots, K_n$, namely,

$$y_n(t) = g(t) + \int_0^t \sum_{m=1}^{n} K_m(t,s) g(s) ds, \quad n \geq 1. \tag{1.3.6}$$

According to the uniform estimate (1.3.4) the infinite series

$$\sum_{m=1}^{\infty} K_m(t,s) := \lim_{n \to \infty} \sum_{m=1}^{n} K_m(t,s)$$

(the *Neumann series* of K) thus converges absolutely and uniformly on S; hence,

$$R(t,s) := \sum_{m=1}^{\infty} K_m(t,s), \quad (t,s) \in S, \tag{1.3.7}$$

is a continuous function on S, called the *resolvent kernel* (or simply: the *resolvent*) of the given kernel K. Due to the uniform convergence of the Neumann series of K equation (1.3.6) yields the classical result, derived first by Volterra in 1896 (compare also Section 1.2), that the expression

$$y(t) = g(t) + \int_0^t R(t,s) g(s) ds, \quad t \in I, \tag{1.3.8}$$

defines a continuous solution of (1.3.1) on the given interval I.

In order to establish uniqueness of this solution we first observe that, by (1.3.7) and (1.3.2),

$$R(t,s) = K_1(t,s) + \sum_{m=2}^{\infty} K_m(t,s) = K(t,s) + \sum_{m=2}^{\infty} \int_s^t K(t,\tau) K_{m-1}(\tau,s) d\tau$$

$$= K(t,s) + \int_s^t K(t,\tau) R(\tau,s) d\tau, \quad (t,s) \in S;$$

in a completely analogous fashion one finds that

$$R(t,s) = K(t,s) + \int_s^t R(t,\tau) K(\tau,s) d\tau, \quad (t,s) \in S.$$

Suppose now that (1.3.1) possesses, in addition to the solution given by (1.3.8), a second solution $z \in C(I)$; i.e.,

$$z(\tau) = g(\tau) + \int_0^\tau K(\tau,s) z(s) ds, \quad \tau \in I.$$

If we multiply this identity by $R(t,\tau)$ and integrate (with respect to τ) over $[0,t]$, we obtain

$$\begin{aligned}\int_0^t R(t,\tau) z(\tau) d\tau &= \int_0^t R(t,\tau) g(\tau) d\tau + \int_0^t \int_0^\tau R(t,\tau) K(\tau,s) z(s) ds\, d\tau \\ &= \int_0^t R(t,\tau) g(\tau) d\tau + \int_0^t z(s) \left[\int_s^t R(t,\tau) K(\tau,s) d\tau \right] ds \\ &= \int_0^t R(t,\tau) g(\tau) d\tau + \int_0^t z(s) [R(t,s) - K(t,s)]\, ds \\ &= \int_0^t R(t,s) z(s) ds + (y(t) - g(t)) - (z(t) - g(t)), \quad t \in I,\end{aligned}$$

and hence $0 = y(t) - z(t)$ for all $t \in I$; in the above, the two principal tools were Dirichlet's formula and the second of the two identities relating K and R. We summarize the results obtained so far in

Theorem 1.3.1. *Let the functions g and K characterizing the integral equation (1.3.1) be continuous on I and S, respectively. Then this equation has a unique solution $y \in C(I)$ given by*

$$y(t) = g(t) + \int_0^t R(t,s) g(s) ds, \quad t \in I, \tag{1.3.9a}$$

where $R \in C(S)$ is the resolvent kernel associated with the given kernel K and defined in (1.3.7). The resolvent kernel satisfies the identities (usually called the Fredholm identities)

$$R(t,s) = K(t,s) + \int_s^t K(t,\tau) R(\tau,s) d\tau, \tag{1.3.9b}$$

and

$$R(t,s) = K(t,s) + \int_s^t R(t,\tau) K(\tau,s) d\tau, \tag{1.3.9c}$$

for all $(t,s) \in S$. □

In certain applications (e.g., in the study of stability of Volterra integral equations of the second kind) it is often convenient to represent the exact solution in a slightly modified form, provided the nonhomogeneous term g is differentiable (Bownds & Cushing [1975a]).

Corollary 1.3.1. *Let* $g \in C^1(I)$*, and let* $U(t,s)$ *denote the (unique) continuous solution of*

$$U(t,s) = 1 + \int_s^t K(t,\tau)U(\tau,s)d\tau, \quad (t,s)\in S. \tag{1.3.10}$$

Then the solution of (1.3.1) is given by

$$y(t) = U(t,0)g(0) + \int_0^t U(t,s)g'(s)ds, \quad t\in I; \tag{1.3.11}$$

furthermore, U *is related to the resolvent of* K *by*

$$\frac{\partial U(t,s)}{\partial s} = -R(t,s), \quad (t,s)\in S. \tag{1.3.12}$$

Proof. For each fixed $s\in I$, the integral equation (1.3.10) possesses a unique continuous solution $U(t,s)$, and we may use Theorem 1.3.1 to write this solution in the form

$$U(t,s) = 1 + \int_s^t R(t,\tau)d\tau, \quad t\in[s,T].$$

Since the resolvent R is continuous on S, differentiation of the above equation with respect to s leads to (1.3.12). The result (1.3.11) is then obtained by substituting (1.3.12) into (1.3.9a) and by employing integration by parts. □

We note that (1.3.10) yields $U(t,t)=1$ for all $t\in I$, and hence, setting $\hat{R}(t,s) := \partial U(t,s)/\partial s$,

$$\hat{R}(t,s) = -K(t,s) + \int_0^t K(t,\tau)\hat{R}(\tau,s)d\tau, \quad (t,s)\in S.$$

Some authors (see, e.g., Miller [1971]) refer to this equation, or to an equivalent one analogous to (1.3.9c), as *resolvent equation.*

Volterra integral equations arising in mathematical modeling processes (e.g. in population dynamics) are often characterized by *convolution kernels;* i.e., (1.3.1) assumes the form

$$y(t) = g(t) + \int_0^t a(t-s)y(s)ds, \quad t\in I. \tag{1.3.13}$$

Since the corresponding iterated kernels inherit this property the resulting resolvent becomes $R(t,s)=\rho(t-s)$.

Corollary 1.3.2. *Under the assumptions of Theorem 1.3.1, and with* $K(t,s) = a(t-s)$*, the unique continuous solution of (1.3.13) is given by*

$$y(t) = g(t) + \int_0^t \rho(t-s)g(s)ds, \quad t\in I, \tag{1.3.14}$$

where the resolvent ρ *satisfies the identities*

$$\rho(t) = a(t)+\int_0^t a(t-\tau)\rho(\tau)d\tau = a(t)+\int_0^t \rho(t-\tau)a(\tau)d\tau, \quad t\in I. \quad \square \tag{1.3.15}$$

Until now the kernel K in (1.3.1) has been assumed to be continuous on S. We turn now to linear equations of the form (1.3.1) where the kernel possesses a *weak singularity of Abel type;* i.e.,

$$G(t,s;\alpha) := (t-s)^{-\alpha}K(t,s), \quad 0<\alpha<1,$$

with $K\in C(S)$. Setting $K_1(t,s;\alpha) := G(t,s;\alpha)$, the iterated kernels are

$$\begin{aligned} K_2(t,s;\alpha) &= \int_s^t \frac{K(t,\tau)K(\tau,s)d\tau}{(t-\tau)^{\alpha}(\tau-s)^{\alpha}} \\ &= (t-s)^{1-2\alpha}\int_0^1 \frac{K(t,s+(t-s)u)K(s+(t-s)u,s)}{u^{\alpha}(1-u)^{\alpha}}du \\ &= (t-s)^{-\alpha}\Psi_2(t,s;\alpha) \end{aligned}$$

with

$$\Psi_2(t,s;\alpha) := (t-s)^{1-\alpha}\Phi_2(t,s;\alpha);$$

in general, we have

$$K_n(t,s;\alpha) = (t-s)^{-\alpha}\Psi_n(t,s;\alpha), \tag{1.3.16}$$

with

$$\Psi_n(t,s;\alpha) = (t-s)^{(n-1)(1-\alpha)}\Phi_n(t,s;\alpha),$$

$$\Phi_n(t,s;\alpha) = \int_0^1 \frac{K(t,s+(t-s)u)\,\Phi_{n-1}(s+(t-s)u,s;\alpha)}{u^{1-(n-1)(1-\alpha)}(1-u)^{\alpha}}du, \quad (t,s)\in S,$$

where $\Phi_1(t,s;\alpha) := K(t,s)$.

Note that K_n is continuous on S whenever $n\geq 1/(1-\alpha)$. Furthermore, if $|K(t,s)|\leq M$ on S, then, for $n\geq 1$,

$$|\Psi_n(t,s;\alpha)|\leq M^n T^{(n-1)(1-\alpha)}(\Gamma(1-\alpha))^n/\Gamma(n(1-\alpha));$$

here we have used that, for $\lambda>0$ and $\mu>0$,

$$\int_0^1 u^{\lambda-1}(1-u)^{\mu-1}du = B(\lambda,\mu) = \frac{\Gamma(\lambda)\Gamma(\mu)}{\Gamma(\lambda+\mu)},$$

with $B(\cdot,\cdot)$ denoting the beta function (Euler's integral of the first kind). (We refer to Henrici [1977], Artin [1964], or Oldham & Spanier [1974] for the definition and properties of the above special functions.) Based on Stirling's formula for the asymptotic behavior of the gamma function,

$$\Gamma(x) = (2\pi/x)^{1/2} x^x e^{-x} e^{J(x)}, \quad x>0,$$

where the so-called Binet function $J(x)$ satisfies $0<J(x)<\frac{1}{12x}$ for all $x>0$ (see Henrici [1977]), one readily verifies that the limit,

$$Q(t,s;\alpha) := \lim_{n\to\infty} \sum_{m=1}^{n} \Psi_m(t,s;\alpha),$$

exists and is continuous for all $(t,s)\in S$. Therefore, in analogy to Theorem 1.3.1, Picard's method of successive approximations can be extended to Volterra equations with weakly singular kernels, and we have

Theorem 1.3.2. *Suppose that the functions g and K are continuous on I and S, respectively, and let* $\alpha\in(0,1)$. *Then the integral equation*

$$y(t) = g(t)+\int_0^t (t-s)^{-\alpha} K(t,s) y(s) ds, \quad t\in I, \tag{1.3.17}$$

has a unique solution $y\in C(I)$ *given by*

$$y(t) = g(t)+\int_0^t R(t,s;\alpha) g(s) ds, \quad t\in I, \tag{1.3.18}$$

where the resolvent kernel $R(t,s;\alpha)$ *has the form*

$$R(t,s;\alpha) = (t-s)^{-\alpha} Q(t,s;\alpha), \quad (t,s)\in S, \tag{1.3.19}$$

with $Q\in C(S)$ *for each* $\alpha\in(0,1)$. *Furthermore, Q satisfies the identities*

$$Q(t,s;\alpha) = K(t,s)+(t-s)^{\alpha}\int_s^t (t-\tau)^{-\alpha}(\tau-s)^{-\alpha} K(t,\tau) Q(\tau,s;\alpha) d\tau \tag{1.3.20}$$

$$= K(t,s)+(t-s)^{\alpha}\int_s^t (t-\tau)^{-\alpha}(\tau-s)^{-\alpha} Q(t,\tau;\alpha) K(\tau,s) d\tau,$$

for all $(t,s)\in S$. □

The special case treated below will serve as an illustration both for Theorem 1.3.2 and for certain results on the asymptotic behavior of solutions of second-kind integral equations with convolution kernels. Before presenting this equation we require some prerequisites. Firstly, we state a result which motivates the use of the special nonhomogeneous term in equation (1.3.21) below (Bellman & Cooke [1963]); note also that, for the case of continuous functions, this is result is a special case of Corollary 1.3.1.

Theorem 1.3.3. *Let* g' *exist and be in* $L^1(I)$, *and suppose that* $a\in L^1(I)$. *If y and z denote, respectively, the solutions of the integral equations*

$$y(t) = g(t)+\int_0^t a(t-s) y(s) ds$$

and

$$z(t) = 1+\int_0^t a(t-s)z(s)ds, \quad t\in I,$$

then these solutions are related by

$$y(t) = g(0)z(t)+\int_0^t z(t-s)g'(s)ds, \quad t\in I. \quad \square$$

Secondly, we introduce the so-called *Mittag-Leffler function* (required in the present context only for real arguments),

$$E_\beta(z) := \sum_{n=0}^{\infty}\frac{z^n}{\Gamma(1+n\beta)}, \quad \beta>0.$$

Note that $E_\beta(0)=1$ for all $\beta>0$; we also have $E_1(z)=\exp(z)$, and $E_{½}(z)=\exp(-z^2)\,\mathrm{erfc}(-z)$, with erfc(.) denoting the complementary error function. Additional properties of $E_\beta(z)$ may be found, e.g., in Erdélyi [1955].

Consider now the integral equation

$$y(t) = 1-\int_0^t \lambda(t-s)^{-\alpha}y(s)ds, \quad t\geqslant 0, \quad \lambda>0, \quad 0\leqslant\alpha<1. \tag{1.3.21}$$

It is easily verified (using (1.3.16)) that the corresponding resolvent kernel $R(t,s;\alpha)$ is of the form $\rho(t-s;\alpha)$ and satisfies

$$1+\int_0^t \rho(t-s;\alpha)ds = E_{1-\alpha}(-\lambda\Gamma(1-\alpha)t^{1-\alpha}).$$

Hence, by Theorem 1.3.2, the exact solution of (1.3.21) is given by

$$y(t) = E_{1-\alpha}(-\lambda\Gamma(1-\alpha)t^{1-\alpha}), \quad t\geqslant 0, \tag{1.3.22}$$

and Theorem 1.3.3 can then be used to find the solution corresponding to a general forcing function in (1.3.21). (Compare also Friedman [1963] for an analysis of (1.3.21) with $\lambda=1$.) The solution corresponding to the special (and practically relevant) value $\alpha=1/2$ can also be found by Laplace transform techniques; here, one has

$$\rho(t-s;\tfrac{1}{2}) = -\lambda(t-s)^{-½}+\lambda^2\pi\exp(-\lambda^2\pi(t-s))\mathrm{erfc}(\lambda(\pi(t-s))^{½})$$

with

$$\mathrm{erfc}(z) := 2\pi^{-½}\int_z^\infty \exp(-t^2)dt,$$

and it then follows that the exact solution can be written as

$$y(t) = \exp(-\lambda^2\pi t)\mathrm{erfc}(\lambda(\pi t)^{½}), \quad t\geqslant 0. \tag{1.3.23}$$

We shall return to these particular Volterra-type integral equations of the second kind when studying smoothness properties of solutions: it will turn out that equations with weakly singular kernels of Abel type and with smooth nonhomogeneous terms g have solutions whose derivatives are unbounded at $t=0$.

1.3.2. Linear Volterra integro-differential equations

In order to establish a result on the representation of the exact solution to the initial-value problem for the linear integro-differential equation,

$$y'(t) = p(t)y(t)+q(t)+\int_0^t K(t,s)y(s)ds, \quad t\in I, \tag{1.3.24}$$

$$y(0) = y_0,$$

with p,q and K assumed to be real-valued and continuous on I and S, respectively, we consider the integrated form of (1.3.24) which, by Dirichlet's formula, may be rewritten as

$$y(t) = g(t)+\int_0^t Q(t,s)y(s)ds, \quad t\in I. \tag{1.3.25}$$

Here, we have introduced the functions

$$g(t) := y_0+\int_0^t q(s)ds, \quad t\in I,$$

and

$$Q(t,s) := p(s)+\int_s^t K(\tau,s)d\tau, \quad (t,s)\in S.$$

Applying Theorem 1.3.1 to the integral equation (1.3.25) and observing the definitions of $g(t)$ and $Q(t,s)$, we find

Theorem 1.3.4. *Let p, q be continuous on I, and suppose that K is continuous on S. Then for every initial value y_0 the integro-differential equation (1.3.24) possesses a unique solution $y\in C^1(I)$ given by*

$$y(t) = R(t,0)y_0+\int_0^t R(t,s)q(s)ds, \quad t\in I, \tag{1.3.26}$$

where the resolvent kernel R is continuously differentiable on S and has the form

$$R(t,s) = 1+\int_s^t r(t,u)du, \quad (t,s)\in S,$$

with r satisfying

$$r(t,s) = Q(t,s)+\int_s^t Q(t,\tau)r(\tau,s)d\tau, \tag{1.3.27}$$

and with

$$Q(t,s) := p(s) + \int_s^t K(u,s)du, \quad (t,s) \in S. \quad \Box$$

If we associate with the integro-differential equation (1.3.24) the so-called *resolvent equation* (see, e.g., Grossman & Miller [1970]),

$$\frac{\partial U(t,s)}{\partial s} = -U(t,s)p(s) - \int_s^t U(t,\tau)K(\tau,s)d\tau, \quad (t,s) \in S,$$

with initial condition $U(t,t)=1$ for $t \in I$, then it follows from the integration by parts formula,

$$\int_0^t \{U(t,s)y'(s) + \frac{\partial U(t,s)}{\partial s} y(s)\}ds = U(t,t)y(t) - U(t,0)y_0,$$

and from the fact that $y(t)$ solves (1.3.24), that this solution has the form

$$y(t) = U(t,0)y_0 + \int_0^t U(t,s)q(s)ds, \quad t \in I.$$

Hence, by the uniqueness of the solution of (1.3.24), the solution $U(t,s)$ of the resolvent equation coincides with $R(t,s)$ defined in Theorem 1.3.4.

Theorem 1.3.4 contains, of course, the analogue of the well-known variation of constants formula for linear first-order differential equations: if $K(t,s) \equiv 0$ on S, then $Q(t,s) = p(s)$, and (1.3.27) reduces to

$$r(t,s) = p(s) + \int_s^t p(\tau)r(\tau,s)d\tau,$$

which yields

$$r(t,s) = p(s)\exp(\int_s^t p(u)du), \quad (t,s) \in S.$$

A straightforward calculation leads to

$$R(t,s) = \exp(\int_s^t p(u)du),$$

and thus to

$$y(t) = \exp(\int_0^t p(u)du)y_0 + \int_0^t \exp(\int_s^t p(u)du)q(s)ds, \quad t \in I.$$

We conclude with the following observation: it has already been seen that a Volterra integro-differential equation can be written as a Volterra integral equation of the second kind (recall (1.3.25)). An alternative to this approach is to view an integro-differential equation as a system of two Volterra integral equations of the second kind. For the linear case (1.3.24), let

$$z(s) := q(s) + \int_0^s K(s,u)y(u)du, \quad s \in I.$$

This allows us to rewrite (1.3.24) in the form

$$y(t) = y_0 + \int_0^t p(s)y(s)ds + \int_0^t [q(s) + \int_0^s K(s,u)y(u)du]\, ds$$

$$= y_0 + \int_0^t p(s)y(s)ds + \int_0^t z(s)ds, \quad t \in I.$$

Thus, (1.3.24) is equivalent to the system

$$\begin{bmatrix} y(t) \\ z(t) \end{bmatrix} = \begin{bmatrix} y_0 \\ q(t) \end{bmatrix} + \int_0^t \begin{bmatrix} p(s) & 1 \\ K(t,s) & 0 \end{bmatrix} \begin{bmatrix} y(s) \\ z(s) \end{bmatrix} ds, \quad t \in I.$$

The extension of this result to nonlinear integro-differential equations, as well as to such equations with weakly singular kernels, is obvious.

1.3.3. Linear Volterra integral equations of the first kind

It has already been indicated in Section 1.2 that the existence of a (unique) solution $y \in C(I)$ of the first-kind integral equation,

$$\int_0^t K(t,s)y(s)ds = g(t), \quad t \in I, \tag{1.3.28}$$

is assured under certain conditions on K and g; obviously, a necessary condition for y to be continuous at $t=0$ is given by $g(0)=0$.

Theorem 1.3.5. *Let K and g satisfy:*

(i) $g \in C^1(I)$, *with* $g(0) = 0$;

(ii) $K \in C(S)$, $\partial K/\partial t \in C(S)$;

(iii) $K(t,t) \neq 0$ *for all* $t \in I$.

Under these conditions the first-kind equation (1.3.28) has a unique solution $y \in C(I)$. □

We have seen that these conditions suffice to transform (1.3.28) into an integral equation of the second kind (to which Theorem 1.3.1 can then be applied). An alternative way of reducing (1.3.28) to a second-kind equation exists if the second part of assumption (ii) is replaced by $\partial K/\partial s \in C(S)$: setting

$$z(t) := \int_0^t y(s)ds$$

and using integration by parts in (1.3.28) (rather than differentiation) we obtain

$$z(t) = \frac{g(t)}{K(t,t)} + \frac{1}{K(t,t)} \int_0^t \frac{\partial K(t,s)}{\partial s} z(s)ds, \quad t \in I.$$

Differentiation of z yields the desired solution y.

We turn now to the general linear Abel integral equation of the first kind,

$$\int_0^t (t-s)^{-\alpha} K(t,s) y(s) ds = g(t), \quad t \in I, \quad 0<\alpha<1. \tag{1.3.29}$$

The following result is due to Volterra [1896] (Nota II).

Theorem 1.3.6. *Let K in (1.3.29) be subject to the assumptions of Theorem 1.3.5, and let $g \in C(I)$, with $g(0)=0$. If, in addition, g is such that the function*

$$G(t) := \int_0^t (t-s)^{\alpha-1} g(s) ds$$

is continuously differentiable on I (this obviously holds if $g \in C^1(I)$), then (1.3.29) possesses a unique solution $y \in C(I)$. Moreover, this solution is given by the solution of another first-kind equation,

$$\int_0^t L(t,s) y(s) ds = G(t), \quad t \in I, \tag{1.3.30}$$

whose kernel,

$$L(t,s) := \int_0^1 \frac{K(s+(t-s)u,s)}{u^\alpha (1-u)^{1-\alpha}} du, \quad (t,s) \in S, \tag{1.3.31}$$

is continuous on S and satisfies $L(t,t)=K(t,t)\pi/\sin(\alpha\pi) \neq 0$ for all $t \in I$; its partial derivative,

$$\frac{\partial L(t,s)}{\partial t} = \int_0^1 \left(\frac{u}{1-u}\right)^{1-\alpha} \frac{\partial K(s+(t-s)u,s)}{\partial t} du,$$

is also continuous on S (note that $\int_0^1 (u/(1-u))^{1-\alpha} du = B(2-\alpha,\alpha)$). □

These results are based on the essential condition that $K(t,t) \neq 0$ for all $t \in I$. If $K(t,s)$ possesses a finite number of zeros along the line $s=t$ of S, then statements on the existence of solutions become more complex. A treatment of this case is beyond the scope of this book (but compare, for example, Lalesco [1908], [1912], and Davis [1927a]). Here, we just cite an example. Consider (1.3.28) with $K(t,s)=2t-3s$ (implying $K(0,0)=0$), and with $g(t)=t^2/2$, $t \in [0,T]$. It is immediately verified that any function of the form $y(t)=1+ct$, $c \in \mathbb{R}$, is a solution. Hence the solution is no longer unique. We are led to some additional insight into this problem by looking at the corresponding second-kind equation: differentiation leads to

$$-ty(t)+2\int_0^t y(s)ds = t,$$

and thus to

$$-ty'(t)+y(t) = 1, \quad t\in[0,T].$$

(Note that this is a special Clairaut differential equation.) Hence we are faced either with an integral equation of the third kind, or with a first-order differential equation where the coefficient of the highest derivative has a zero on the given interval.

On the other hand, if the kernel $K(t,s)$ vanishes identically for $s=t$, then (1.3.28) may still have a unique continuous solution; we illustrate this by two more examples. Consider the two first-kind equations whose kernels are of convolution type,

$$\int_0^t \frac{(t-s)^{n-1}}{(n-1)!}y(s)ds = g(t), \tag{1.3.32}$$

and

$$\int_0^t \frac{(t-s)^{n-\alpha-1}}{\Gamma(n-\alpha)}y(s)ds = g(t), \tag{1.3.33}$$

where $n\geq 1$, $0<\alpha<1$, and $g\in C^n(I)$. If $g^{(j)}(0)=0$, $j=0, \ldots ,n-1$, then both equations possess unique continuous solutions on I: they are, respectively,

$$y(t) = g^{(n)}(t),$$

and

$$y(t) = \frac{1}{\Gamma(\alpha)}\int_0^t (t-s)^{\alpha-1}g^{(n)}(s)ds.$$

Compare also Theorem 1.3.17.

1.3.4. Nonlinear Volterra equations

The special nonlinear Volterra integral equation

$$y(t) = g(t)+\int_0^t K(t,s)\{y(s)+h(s,y(s))\}ds, \quad t\in I, \tag{1.3.34}$$

can be regarded as a natural link between the linear equation (1.3.1) and the general nonlinear equation

$$y(t) = g(t)+\int_0^t k(t,s,y(s))ds, \quad t\in I. \tag{1.3.35}$$

We therefore begin this section by stating a formal result on the relation between the solution of (1.3.1) and that of the perturbed equation (1.3.34) (compare Miller [1971, pp. 191-193]).

Theorem 1.3.7. *Let $K \in C(S)$, $g \in C(I)$, and let R denote the resolvent kernel of K. If h is continuous and such that (1.3.34) has a unique solution $z \in C(I)$ then*

$$z(t)-y(t) = \int_0^t R(t,s)h(s,z(s))ds, \quad t \in I, \tag{1.3.36}$$

where y is the solution of the linear equation (1.3.1). □

The above variation of constants formula represents a generalization of the following analogous result from the theory of ordinary differential equations: if y is the solution of the initial-value problem

$$y' = p(t)y, \quad y(0)=y_0,$$

and if z denotes the solution of the nonlinearly perturbed problem,

$$z' = p(t)z + \Phi(t,z), \quad z(0)=y_0,$$

then

$$z(t)-y(t) = \int_0^t \exp\left(\int_s^t p(u)du\right)\Phi(s,z(s))ds, \quad t \in I,$$

where, as one recalls from Section 1.3.2, the expression $\exp\left(\int_s^t p(u)du\right)$ is the resolvent associated with the kernel of

$$y(t) = y_0 + \int_0^t p(s)y(s)ds.$$

The result of Theorem 1.3.7 is readily extended to the more general nonlinearly perturbed integral equation

$$z(t) = g(t) + \int_0^t K(t,s)z(s)ds + \int_0^t k(t,s,z(s))ds.$$

Setting

$$G(t) := g(t) + \int_0^t k(t,s,z(s))ds,$$

and again denoting by $y(t)$ the solution of the linear equation (1.3.1), we find, by applying Theorem 1.3.1 to

$$z(t) = G(t) + \int_0^t K(t,s)z(s)ds,$$

and by using Dirichlet's formula, that

$$z(t)-y(t) = \int_0^t \hat{R}(t,s,z(s))ds, \quad t\in I,$$

with

$$\hat{R}(t,s,z) := k(t,s,z)+\int_s^t R(t,\tau)k(\tau,s,z)d\tau.$$

The special equation (1.3.34) corresponds to the perturbation

$$k(t,s,z) = h(s,z)K(t,s);$$

hence, the result (1.3.36) derives from the above by recalling the second Fredholm identity (1.3.9c).

We turn now to the general *second-kind integral equation* (1.3.35) for which we give two existence and uniqueness theorems; we refer to the Notes at the end of the chapter for a selection of sources for more general existence theorems.

Theorem 1.3.8. *Let $k(t,s,y)$ be continuous for all $(t,s)\in S$ and all y, and suppose that k satisfies the (uniform) Lipschitz condition,*

$$|k(t,s,y)-k(t,s,z)| \leqslant L|y-z|,$$

where $(t,s)\in S$ and $y,z\in\mathbb{R}$, with Lipschitz constant L being independent of y and z. Then for each $g\in C(I)$ the nonlinear equation (1.3.35) possesses a unique solution $y\in C(I)$. □

Note that, in the linear case $k(t,s,y)=K(t,s)y$, the assumption that $K\in C(S)$ (see Theorem 1.3.1)) implies that $L=\max\{|K(t,s)|:(t,s)\in S\}$.

The proof of Theorem 1.3.8 is based on Picard's method of successive approximations (cf. (1.3.5)) for (1.3.35); we have

$$y_0(t) := g(t),$$

$$y_{n+1}(t) := g(t)+\int_0^t k(t,s,y_n(s))ds \quad (n\geqslant 0).$$

In many cases, however, the conditions imposed on the kernel function $k(t,s,y)$ will not be satisfied for all $y\in\mathbb{R}$ but will hold only in some compact region. As a consequence, the Picard iterates $\{y_n(t)\}$ will, in general, converge uniformly only on some subinterval $[0,T_0]$, with $T_0<T$. (As an illustration, consider the integral equation

$$y(t) = t+\int_0^t (y(s))^2 ds;$$

it has a (unique) continuous solution (namely, $y(t)=\tan(t)$) only if $I\subset[0,\pi/2)$.) The following result makes this more precise (cf. Miller [1971, pp. 26-30]).

Theorem 1.3.9. *Let* $g \in C(I)$, *and assume that* $k(t,s,y)$ *is continuous in the region*

$$\Omega := \{(t,s,y) : (t,s) \in S \quad \text{and} \quad |y-g(t)| \leqslant B\}.$$

In addition, let $k(t,s,y)$ *satisfy a Lipschitz condition of the form*

$$|k(t,s,y)-k(t,s,z)| \leqslant L|y-z|$$

for all triples (t,s,y) *and* (t,s,z) *in* Ω, *and set*

$$M := \max\{|k(t,s,y)| : (t,s,y) \in \Omega\}, \quad T_0 := \min\{T, B/M\}.$$

Then all Picard iterates $\{y_n(t)\}$ *exist, are continuous, and converge, as* $n \to \infty$, *to a continuous solution of (1.3.35) on* $[0,T_0]$. *This solution is the unique continuous solution of (1.3.35) in* $[0,T_0]$. □

A result of the above type is called a *local* existence theorem, as it guarantees the existence of a solution to (1.3.35) only on a sufficiently small interval $[0,T_0]$. However, it is possible to construct a *continuation* of the solution beyond T_0: if we set $z(t) := y(t+T_0)$, then, by (1.3.35),

$$z(t) = g(t+T_0) + \int_0^{T_0} k(t+T_0,s,y(s))ds + \int_{T_0}^{T_0+t} k(t+T_0,s,y(s))ds,$$

or

$$z(t) = g_0(t) + \int_0^t k_0(t,s,z(s))ds,$$

with

$$g_0(t) := g(t+T_0) + \int_0^{T_0} k(t+T_0,s,y(s))ds,$$

and

$$k_0(t,s,z) := k(t+T_0,s+T_0,z), \quad t+T_0 \leqslant T.$$

Since these functions satisfy the hypotheses of Theorem 1.3.9, we may deduce the existence of a (unique) continuous solution $z(t)$ in some interval $[0,T_1]$, with $T_1>0$; this, in turn, implies that $y(t)$ has been continued, as a continuous solution, to the interval $[0,T_0+T_1]$. This process can obviously be repeated. How far the solution can be continued will depend on the constants B and M; details concerning this question may be found in Miller [1971, pp. 30-33; 93-98]; compare also Burton [1983b, pp. 66-89].

The above results can be extended to cover nonlinear Volterra equations of the second kind whose kernels are *weakly singular;* i.e., equations of the type (1.3.35) in which $k(t,s,y)$ is replaced by $p(t,s)k(t,s,y)$ and where the function $p(t,s)$ is subject to the conditions

$$\int_0^t |p(t,s)|ds < \infty \quad \text{for} \quad t \in I, \tag{1.3.37a}$$

and

$$\lim_{v\downarrow 0}\int_{t_1}^{t_1+v}|p(t,s)|ds\to 0,\quad 0\leq t_1<t_1+v\leq t\leq T, \tag{1.3.37b}$$

uniformly in t and t_1 . Examples of such kernel singularities are

$$p(t,s) = (t-s)^{-\alpha},\quad 0<\alpha<1 \quad\text{and}\quad p(t,s)=s(t^2-s^2)^{-1/2};$$

on the other hand, $p(t,s)=(t^2-s^2)^{-1/2}$ does not satisfy (1.3.37b).

As an illustration relating to this last case, consider

$$y(t) = g(t)+\int_0^t(t^2-s^2)^{-1/2}y(s)ds, \tag{1.3.38}$$

with $g(t)\equiv 0$. Obviously, $y(t)\equiv 0$ is a continuous solution. Another continuous solution is given by $y(t)=t$, implying that this equation is not uniquely solvable.

Since a nonlinear Volterra *integro-differential equation* is characterized by two functions, namely, $F(t,y,z)$ (differential part) and $k(t,s,y)$ (integral part) a corresponding existence and uniqueness theorem is an obvious extension of the analogous results for initial-value problems for first-order ordinary differential equations and for integral equations of the form (1.3.35).

Theorem 1.3.10. *In*

$$y'(t) = F(t,y(t),z(t)), \tag{1.3.39}$$

with

$$z(t) := \int_0^t k(t,s,y(s))ds,\quad t\in I,$$

suppose that $F(t,y,z)$ *and* $k(t,s,y)$ *are, respectively, continuous for* $t\in I$ *and* $(t,s)\in S$, *and let the following (uniform) Lipschitz conditions hold:*

(i) $|F(t,y_1,z)-F(t,y_2,z)| \leq L_1|y_1-y_2|$,

(ii) $|F(t,y,z_1)-F(t,y,z_2)| \leq L_2|z_1-z_2|$,

(iii) $|k(t,s,y_1)-k(t,s,y_2)| \leq L_3|y_1-y_2|$,

for all $t\in I$, $(t,s)\in S$, *and* $|y_i|<\infty$, $|z_i|<\infty$ $(i=1,2)$.
Then for each y_0 *there exists exactly one solution* $y\in C^1(I)$ *of (1.3.39) satisfying* $y(0)=y_0$. □

In most applications one encounters nonlinear integro-differential equations of a less general form,

$$y'(t) = f(t,y(t))+\int_0^t k(t,s,y(s))ds,\quad t\in I,\quad y(0)=y_0. \tag{1.3.40}$$

Since here $F(t,y,z)$ is linear in the third argument assumption (ii) holds trivially with $L_2=1$, while (i) becomes the classical Lipschitz condition for the initial-value problem $y'=f(t,y)$, $y(0)=y_0$.

More general (local) existence and uniqueness theorems can be derived on the basis of Theorem 1.3.9, by rewriting the given integro-differential equation as an integral equation (cf. (1.3.25) and the remark at the end of Section 1.3.2). We refrain from stating these results; the reader is referred to Miller [1971] and to Burton [1983b].

Consider now the nonlinear Volterra *integral equation of the first kind* with weakly singular kernel,

$$\int_0^t (t-s)^{-\alpha} k(t,s,y(s))ds = g(t), \quad t\in I \quad (0<\alpha<1). \tag{1.3.41}$$

Theorem 1.3.11. *Suppose that:*

(i) *$k=k(t,s,y)$ satisfies $k\in C^1(S\times\mathbb{R})$;*

(ii) *$k_t:=\partial k/\partial t$ is subject to a (uniform) Lipschitz condition with respect to y, $|k_t(t,s,y)-k_t(t,s,z)|\leqslant L|y-z|$ on $S\times\mathbb{R}$;*

(iii) *for $t=s$, the derivative $k_y:=\partial k/\partial y$ is bounded away from zero: $|k_y(t,t,y)|\geqslant r_0>0$ on $I\times I\times\mathbb{R}$;*

(iv) *g satisfies $g(0)=0$ and is such that the function G defined in Theorem 1.3.6 is in $C^1(I)$.*

Then (1.3.41) possesses a unique solution $y\in C(I)$ for all $\alpha\in(0,1)$. Moreover, $y(0)$ is defined implicitly by

$$k(0,0,y(0)) = \lim_{t\downarrow 0}[(1-\alpha)g(t)/t^{1-\alpha}]. \quad \square \tag{1.3.42}$$

This last statement will gain its importance in connection with certain numerical methods whose recursive character requires that the starting value $y_0=y(0)$ be given.

The above theorem is due to Branca [1978] (but compare also Janikowski [1962]); its proof is based on an existence and uniqueness result for the generalized nonlinear Volterra integral equation

$$y(t) = F(t,z(t)), \quad t\in I,$$

with $z(t)$ as in (1.3.39). As an application of this general result one can then establish an existence and uniqueness theorem for (1.3.41) with $\alpha=0$:

Theorem 1.3.12. *Let (i)-(iii) of Theorem 1.3.11 hold. In addition, assume that $k(t,s,y)$ satisfies a (uniform) Lipschitz condition with respect to y on $S\times\mathbb{R}$, and let $g\in C^1(I)$, with $g(0)=0$. Then (1.3.41) corresponding to $\alpha=0$ has a unique solution $y\in C(I)$.* $\square$

1.3.5. Smoothness of solutions

When analyzing the convergence of a numerical method for a given functional equation one needs information about the smoothness of the exact solution on the domain on which the equation is to be solved. This becomes more significant when one wants to find the maximum order of convergence of a method. In this section we derive a number of smoothness results for Volterra integral equations of the first and second kind whose kernels are weakly singular; analogous results for integro-differential equations may be found by rewriting such equations as (systems of two) integral equations of the second kind, an observation already made in Section 1.3.2. It turns out that while for integral equations (of the first or second kind) with nonsingular kernels smoothness of K and g implies smoothness of the solution y (on the closed interval I), singular first-kind equations with smooth K and g behave quite differently from the analogous second-kind equations (this is not completely surprising in view of the result, already known to Volterra, that a first-kind integral equation with weakly singular kernel can be transformed into an equivalent first-kind equation with regular kernel; see Section 1.3.3).

We begin with two examples. Consider the second-kind equation,

$$y(t) = g(t) - \int_0^t (t-s)^{-\frac{1}{2}}\, y(s)ds, \quad t \geqslant 0:$$

if $g(t) \equiv 1$, then (by (1.3.23))

$$y(t) = \exp(-\pi t)\operatorname{erfc}(\sqrt{\pi t}),$$

and it is easily verified that $y'(t) \sim t^{-\frac{1}{2}}$ as $t \downarrow 0$. Hence y is not smooth at $t=0$. On the other hand, if the smooth nonhomogeneous term $g(t) \equiv 1$ is replaced by the function $g(t) = 1 + 2\sqrt{t}$ which is not smooth for $t=0$, we find (using Laplace transform techniques) that $y(t) \equiv 1$; i.e., the corresponding exact solution is smooth on the closed interval $I := [0, T]$.

The following theorem (Miller & Feldstein [1971]) provides more insight into the above problem.

Theorem 1.3.13. *Let m be a nonnegative integer, let α be a real number with $0 < \alpha < 1$, and suppose that the real-valued function $g(t)$ is analytic in a neighborhood of $I := [0, T]$. Let $h(t,y)$ be real-valued and analytic on an open set containing all real ordered pairs (t,y) with $t \in I$, $|y| < \infty$. Then:*

(a) *the solution of*

$$y(t) = g(t) + \int_0^t (t-s)^{m-\alpha} h(s, y(s))ds \tag{1.3.43}$$

is real analytic in a neighborhood of the left-open interval $0 < t \leqslant T$;

(b) *if α is a rational number, written in lowest terms, $\alpha = p/q$, then $y(t^q)$ is analytic in a neighborhood of $t = 0$.* □

A related result exhibiting the structure of the exact solution of the more general nonlinear second-kind equation (1.3.41), with $\alpha = \frac{1}{2}$,

$$y(t) = g_1(t) + t^{\frac{1}{2}} g_2(t) + \int_0^t (t-s)^{-\frac{1}{2}} k(t,s,y(s))ds, \quad t \in I, \qquad (1.3.44)$$

has been given by de Hoog & Weiss [1973d].

Theorem 1.3.14. *Suppose that the functions characterizing (1.3.44) satisfy the following hypotheses:*

(i) $g_i \in C^m(I)$, $i = 1,2$;

(ii) $k(t,s,y)$ *is continuous with respect to* $(t,s) \in S$ *and satisfies a (uniform) Lipschitz condition with respect to y for all* $y \in \mathbb{R}$;

(iii) $k(t,s,y)$ *has continuous partial derivatives of order m with respect to t and s, and of order* $2m$ *with respect to y, for all* $(t,s) \in S$ *and all y in some open neighborhood of* $y(t)$.

Then the solution of (1.3.44) has the form

$$y(t) = v(t) + t^{\frac{1}{2}} w(t), \quad t \in I, \quad \textit{with} \quad v,w \in C^m(I); \qquad (1.3.45)$$

here, v and w denote the solution of the system

$$v(t) = g_1(t) + \int_0^t p_1(t,s) k_1(t,s,v(s), w(s))ds, \qquad (1.3.46a)$$

$$w(t) = g_2(t) + \int_0^t p_2(t,s) k_2(t,s,v(s), w(s))ds, \quad t \in I. \qquad (1.3.46b)$$

The kernel functions of the above system are given by

$$p_1(t,s) := (s/(t-s))^{\frac{1}{2}}, \quad p_2(t,s) := t^{-\frac{1}{2}},$$

$$k_1(t,s,v,w) := (k(t,s,v+s^{\frac{1}{2}}w) - k(t,s,v-s^{\frac{1}{2}}w))/(2s^{\frac{1}{2}}),$$

$$k_2(t,s,v,w) := (k(t,s,v+s^{\frac{1}{2}}w) + k(t,s,v-s^{\frac{1}{2}}w))/2. \quad \square$$

Note that, by Theorem 1.3.9, (i) and (ii) imply the existence of a continuous solution y to (1.3.44).

The above Theorem 1.3.14. raises the question of whether there exists a result analogous to (1.3.45) for arbitrary values of $\alpha \in (0,1)$. We shall answer this for the linear equation

$$y(t) = g(t) + \int_0^t (t-s)^{-\alpha} K(t,s) y(s)ds, \quad t \in I, \qquad (1.3.47)$$

where

$$g(t) = g_1(t) + t^{\beta} g_2(t), \quad \beta > 0, \quad \beta \notin \mathbb{N}. \qquad (1.3.48)$$

Extensions of the following result to nonlinear Abel-type integral equations of the second kind may be found in Lubich [1983b].

Theorem 1.3.15. *Let $g_i \in C^m(I)$, $i=1,2$, with $m \geq 1$, and let $K \in C^m(S)$. Then the (unique) solution of (1.3.47) is in $C(I) \cap C^m(0,T]$ and has the form*

$$y(t) = g_1(t) + t^\beta g_2(t) + \sum_{n=1}^{\infty} \{\Psi_{n,1}(t;\alpha) + t^\beta \Psi_{n,2}(t;\alpha,\beta)\} t^{n(1-\alpha)}, \quad (1.3.49)$$

where the functions

$$\Psi_{n,1}(t;\alpha) := \int_0^1 (1-\tau)^{n(1-\alpha)-1} \Phi_n(t,\tau t;\alpha) g_1(\tau t) d\tau, \quad (1.3.50a)$$

$$\Psi_{n,2}(t;\alpha,\beta) := \int_0^1 (1-\tau)^{n(1-\alpha)-1} \tau^\beta \Phi_n(t,\tau t;\alpha) g_2(\tau t) d\tau, \quad (1.3.50b)$$

satisfy $\Psi_{n,i} \in C^m(I)$, $i=1,2$ $(n \geq 1)$, with Φ_n defined in the proof below by (1.3.50c). □

Proof. In order to prove this result we first investigate the resolvent kernel associated with the given kernel $G(t,s;\alpha) := (t-s)^{-\alpha} K(t,s)$ in (1.3.47). The iterated kernels $\{K_n\}$ of G have been introduced in Section 1.3.1 (compare (1.3.16)); there, it has been shown that

$$K_n(t,s;\alpha) = (t-s)^{-\alpha} (t-s)^{(n-1)(1-\alpha)} \Phi_n(t,s;\alpha),$$

with Φ_n given by

$$\Phi_n(t,s;\alpha) = \int_0^1 \frac{K(t,s+(t-s)u)\Phi_{n-1}(s+(t-s)u,s;\alpha)du}{u^{1-(n-1)(1-\alpha)}(1-u)^\alpha} \quad (n \geq 2), \quad (1.3.50c)$$

using $\Phi_1(t,s;\alpha) = K(t,s)$. It now follows from our hypothesis on the smoothness of K that $\Phi_n \in C^m(S)$, $n=1,2,\ldots$, for all values of $\alpha \in (0,1)$. Since the resolvent kernel is defined by

$$R(t,s;\alpha) = (t-s)^{-\alpha} \sum_{n=1}^{\infty} (t-s)^{(n-1)(1-\alpha)} \Phi_n(t,s;\alpha),$$

with the infinite series converging absolutely and uniformly on S for $\alpha \in (0,1)$, Theorem 1.3.2 may be invoked to furnish the desired information on the exact solution of (1.3.47): we find that

$$y(t) = g(t) + \int_0^t R(t,s;\alpha) g(s) ds = g_1(t) + t^\beta g_2(t)$$

$$+ \sum_{n=1}^{\infty} \int_0^t (t-s)^{(n-1)(1-\alpha)-\alpha} \Phi_n(t,s;\alpha)[g_1(s) + s^\beta g_2(s)] ds.$$

An obvious substitution leads to the expression

$$y(t) = g_1(t) + t^\beta g_2(t) + \sum_{n=1}^{\infty} \{ \int_0^1 (1-\tau)^{(n-1)(1-\alpha)-\alpha} \Phi_n(t,\tau t;\alpha) g_1(\tau t) d\tau$$

$$+t^{\beta}\int_0^1(1-\tau)^{(n-1)(1-\alpha)-\alpha}\tau^{\beta}\Phi_n(t,\tau t;\alpha)g_2(\tau t)d\tau\}t^{n(1-\alpha)}$$

which holds uniformly on I. The result now follows by recalling the smoothness properties of Φ_n (compare (1.3.50c)) and the smoothness hypotheses for g_1, g_2, and by observing that $(n-1)(1-\alpha)-\alpha=n(1-\alpha)-1$. □

Corollary 1.3.3. *Suppose that the kernel function $K(t,s)$ and the functions g_i in (1.3.47) are constant on their respective domains: $K(t,s)=\lambda$, $(t,s)\in S$, and $g_i(t)=g_i$, $t\in I$. Then the corresponding solution has the form*

$$y(t) = g_1\{1+\sum_{n=1}^{\infty}\lambda^n c_{n,1}(\alpha)t^{n(1-\alpha)}\}+g_2t^{\beta}\{1+\sum_{n=1}^{\infty}\lambda^n c_{n,2}(\alpha,\beta)t^{n(1-\alpha)}\}, \tag{1.3.51}$$

with

$$c_{n,1}(\alpha) := \frac{(\Gamma(1-\alpha))^n}{\Gamma(1+n(1-\alpha))}, \quad c_{n,2}(\alpha,\beta):=\frac{(\Gamma(1-\alpha))^n\Gamma(1+\beta)}{\Gamma(1+\beta+n(1-\alpha))}. \quad \square$$

This result is easily verified by showing that the functions $\{\Phi_n\}$ defined in (1.3.50c) and associated with $K(t,s)=\lambda$ are

$$\Phi_n(t,s;\alpha) = \lambda^n\frac{(\Gamma(1-\alpha))^n}{\Gamma(n(1-\alpha))}, \quad n=1,2,\ldots .$$

The special case given by $g_1=1$, $g_2=0$, has already been treated: its solution can be given in the closed form (1.3.22) (where λ has to be replaced by $-\lambda$).

It is instructive to look briefly at another important special case of Corollary 1.3.3, namely, $\beta=1-\alpha$. Here, we find that

$$c_{n-1,2}(\alpha,1-\alpha) = (1-\alpha)c_{n,1}(\alpha), \quad n=2,3,\ldots ,$$

with $c_{1,1}(\alpha)=1/(1-\alpha)$; we thus obtain

$$y(t) = g_1+(\frac{\lambda}{1-\alpha}g_1+g_2)\,t^{1-\alpha}+\ldots .$$

It follows that the exact solution is given by a power series in t if, and only if, $\lambda g_1/(1-\alpha)+g_2=0$; if this condition holds, then the solution $y(t)$ reduces to the constant g_1. The reader will recall that the special case $\beta=1-\alpha=½$ (i.e. $\alpha=½$) has already been mentioned prior to Theorem 1.3.13.

In equations arising in practical applications the exponent α characterizing the weak singularity of the kernel is usually rational; i.e., $\alpha=p/q$, with p, $q\in\mathbb{N}$; p and q coprime. In view of later applications (especially in connection with the question of the choice of the approximating spaces in collocation methods for (1.3.47)) we present an alternate version of (1.3.49) which, for rational α, sheds more light on the structure of the solution of (1.3.47).

Corollary 1.3.4. *Let the assumptions of Theorem 1.3.15 hold, and suppose that $\alpha=p/q$ (reduced to lowest terms). The solution of (1.3.47) can then be expressed* *in the form*

$$y(t)=v_0(t)+\sum_{s=1}^{q-1} v_s(t)t^{s(1-\alpha)}+t^{\beta}[w_0(t)+\sum_{s=1}^{q-1} w_s(t)t^{s(1-\alpha)}], \qquad (1.3.52)$$

where the functions v_s, w_s *satisfy* $v_s \in C^m(I)$, $w_s \in C^m(I)$, *and are defined by*

$$v_0(t) := g_1(t)+\sum_{l=1}^{\infty} \Psi_{ql,1}(t;\alpha)t^{l(q-p)},$$

$$v_s(t) := \sum_{l=0}^{\infty} \Psi_{ql+s,1}(t;\alpha)t^{l(q-p)}, \quad s=1, \dots ,q-1,$$

$$w_0(t) := g_2(t)+\sum_{l=1}^{\infty} \Psi_{ql,2}(t;\alpha,\beta)t^{l(q-p)},$$

$$w_s(t) := \sum_{l=0}^{\infty} \Psi_{ql+s,2}(t;\alpha,\beta)t^{l(q-p)}, \quad s=1, \dots ,q-1.$$

If $g_2(t)\equiv 0$ *(i.e., if the forcing function in (1.3.47) is smooth on the closed interval* I: $g \in C^m(I)$*), then*

$$y(t) = v_0(t)+\sum_{s=1}^{q-1} v_s(t)t^{s(1-\alpha)}; \qquad (1.3.53)$$

if $g_2(t)\not\equiv 0$, $\beta=1-\alpha$, *then*

$$y(t) = v_0(t)+t^{q-p}w_{q-1}(t)+\sum_{s=1}^{q-1} (v_s(t)+w_{s-1}(t))t^{s(1-\alpha)}. \quad \square \qquad (1.3.54)$$

Illustration: If $\alpha=½$, then (1.3.52) reduces to

$$y(t) = v_0(t)+t^{½}v_1(t)+t^{\beta}[w_0(t)+t^{½}w_1(t)] ; \qquad (1.3.55)$$

for the special choice $\beta=1-\alpha=½$ we have

$$y(t) = v(t)+t^{½}w(t), \quad \text{with} \quad v(t):=v_0(t)+tw_1(t), w(t):=w_0(t)+v_1(t) .$$

This latter result coincides with (1.3.45) given in Theorem 1.3.14.

Proof of Corollary 1.3.4. If $\alpha=p/q$, then $1-\alpha=(q-p)/q$, and hence (i) $l(1-\alpha)\notin\mathbb{N}$ for $l=1, \dots ,q-1$; (ii) for $l=rq+l'$ with $l'<q$, $r\geqslant 1$, we have $l(1-\alpha)=r(q-p)+l'(1-\alpha)$, with $r(q-p)\in\mathbb{N}$. It follows now from (1.3.49) that

$$y(t) = g(t)+\sum_{l=1}^{\infty}(\Psi_{ql,1}(t;\alpha)+t^{\beta}\Psi_{ql,2}(t;\alpha,\beta))(t^{1-\alpha})^{ql}$$

$$+\sum_{l=0}^{\infty}(\Psi_{ql+1,1}(t;\alpha)+t^{\beta}\Psi_{ql+1,2}(t;\alpha,\beta))(t^{1-\alpha})^{ql}t^{1-\alpha}+\dots$$

$$\dots+\sum_{l=0}^{\infty}(\Psi_{ql+q-1,1}(t;\alpha)+t^{\beta}\Psi_{ql+q-1,2}(t;\alpha,\beta))(t^{1-\alpha})^{ql}(t^{1-\alpha})^{q-1}.$$

The assertion (1.3.52) is then immediately established because of the uniform convergence of the infinite series. □

It is clear that if α is irrational, then $n(1-\alpha)\notin\mathbb{N}$, for all $n\in\mathbb{N}$. Hence there does not exist a "finite" representation (like (1.3.52)) for the exact solution; i.e., we have, in this general case, the representation

$$y(t) = \sum_{s=0}^{\infty} v_s(t)t^{s(1-\alpha)} + t^{\beta}\sum_{s=0}^{\infty} w_s(t)t^{s(1-\alpha)}, \tag{1.3.56}$$

with

$$v_0(t) := g_1(t), \quad v_s(t) := \Psi_{s,1}(t;\alpha), \quad s\geq 1,$$
$$w_0(t) := g_2(t), \quad w_s(t) := \Psi_{s,2}(t;\alpha,\beta), \quad s\geq 1,$$

and with $\Psi_{n,1}$ and $\Psi_{n,2}$ given in (1.3.50a), (1.3.50b).

Consider now the *integro-differential equation*,

$$y'(t) = p(t)y(t)+q(t)+\int_0^t (t-s)^{-\alpha}K(t,s)y(s)ds, \quad 0<\alpha<1, \tag{1.3.57}$$

where the function q is of the form

$$q(t) = q_1(t)+t^{\beta}q_2(t), \quad \beta>0, \quad \beta\notin\mathbb{N}. \tag{1.3.58}$$

In analogy to (1.3.25) the initial-value problem for (1.3.57) may be recast as an integral equation of the second-kind,

$$y(t) = g(t)+\int_0^t Q(t,s;\alpha)y(s)ds, \quad t\in I, \tag{1.3.59}$$

where

$$g(t) := y_0 + \int_0^t q(s)ds,$$

and

$$Q(t,s;\alpha) := p(s)+\int_s^t (u-s)^{-\alpha}K(u,s)du, \quad (t,s)\in S.$$

It is now an easy matter to establish the following result.

Theorem 1.3.16. *Let p, K and q_i $(i=1,2)$ in (1.3.57) and (1.3.58) satisfy $p\in C^m(I)$, $K\in C^m(S)$, $q_i\in C^m(I)$, with $m\geq 1$. Then for each y_0 the integro-differential equation (1.3.57) has a unique solution $y\in C^1(I)\cap C^{m+1}(0,T]$ with $y(0)=y_0$. This solution has the form*

$$y(t) = (1+\sum_{n=1}^{\infty}\Psi_{n,0}(t;\alpha)t^{n(2-\alpha)})y_0+\int_0^t [q_1(s)+s^{\beta}q_2(s)]ds +$$

$$+t\sum_{n=1}^{\infty}\Psi_{n,1}(t;\alpha)t^{n(2-\alpha)}+t^{1+\beta}\sum_{n=1}^{\infty}\Psi_{n,2}(t;\alpha,\beta)t^{n(2-\alpha)} \tag{1.3.60}$$

(uniformly on I), where the functions $\Psi_{n,i}$ $(i=0,1,2)$ are in $C^{m+1}(I)$ for all n and all α,β with $\alpha\in(0,1)$, $\beta>0$. □

The representation (1.3.60), whose proof is based on (1.3.59) and on the appropriate generalization of Theorem 1.3.4 to integro-differential equations with weakly singular kernels, is the analogue of (1.3.49) for the solution of a second-kind integral equation with weakly singular kernel; while the latter involves integer powers of $t^{1-\alpha}$, (1.3.60) includes integer powers of $t^{2-\alpha}$.

Corollary 1.3.5. *Let $\alpha=p/q$ (with p, $q\in\mathbb{N}$; p and q coprime), and suppose that the smoothness hypotheses of Theorem 1.3.16 hold. Then there exist functions $v_s\in C^{m+1}(I)$, $w_s\in C^{m+1}(I)$, $s=0, \ldots ,q-1$, such that the solution of the initial-value problem for (1.3.57) may be expressed in the form*

$$y(t) = v_0(t)+\sum_{s=1}^{q-1}v_s(t)t^{s(2-\alpha)}+t^{1+\beta}\{w_0(t)+\sum_{s=1}^{q-1}w_s(t)t^{s(2-\alpha)}\},\quad t\in I. \tag{1.3.61}$$

If $q_2(t)\equiv 0$ (i.e., if the forcing function is smooth on I), then (1.3.61) reduces to

$$y(t) = v_0(t)+\sum_{s=1}^{q-1}v_s(t)t^{s(2-\alpha)},\quad t\in I.\quad \square \tag{1.3.62}$$

This result shows that, in general, the solution of (1.3.57) with smooth functions p, q, and K is not smooth at $t=0$; i.e., we have $y''(t)\sim t^{-\alpha}$ as $t\downarrow 0$.

We shall return to the above results on the behavior of solutions to integral equations of the second kind and to integro-differential equations with weakly singular kernels in Chapter 6 when analyzing the attainable order of convergence of polynomial spline collocation methods for such functional equations.

We now turn our attention to integral *equations of the first kind* with weakly singular kernels; for simplicity we consider only linear equations of the form (1.3.29). Let $g\in C^{m+1}(I)$; it is then easily verified that the function

$$G(t) := \int_0^t (t-s)^{\alpha-1}g(s)ds$$

(see Theorem 1.3.6) is smooth for $0<t\leqslant T$. To be more precise, we have (using repeated integration by parts)

$$G(t) = \frac{1}{\alpha}g(0)t^{\alpha}+\frac{1}{\alpha(\alpha+1)}g'(0)t^{\alpha+1}+\ldots+\frac{1}{(\alpha)_{m+1}}g^{(m)}(0)t^{\alpha+m}$$
$$+\frac{1}{(\alpha)_{m+1}}\int_0^t g^{(m+1)}(s)(t-s)^{\alpha+m}ds,$$

with $(\alpha)_k:=\alpha(\alpha+1)\ldots(\alpha+k-1)$ $(k\geqslant 1)$ denoting the Pochhammer symbol (which can also be expressed as $(\alpha)_k = \Gamma(\alpha+k)/\Gamma(\alpha)$). Based on the results (1.3.30), (1.3.31) of Theorem 1.3.6 we find

Theorem 1.3.17. *Let $K \in C^{m+1}(S)$, with $K(t,t) \neq 0$ on I, and $g \in C^{m+1}(I)$, with $g^{(j)}(0)=0$, $j=0,1,\ldots,m$. Then the unique solution of the first-kind integral equation*

$$\int_0^t (t-s)^{-\alpha} K(t,s) y(s) ds = g(t), \quad t \in I, \quad 0<\alpha<1,$$

has a continuous derivative of order m on the (closed) interval I. □

A special case of this result has already been studied: in (1.3.33) we considered $K(t,s)=(t-s)^{n-\alpha-1}/\Gamma(n-\alpha)$, with $n=m+1$.

1.4. Some results on the asymptotic behavior of Volterra equations

So far we have studied various properties of solutions of Volterra equations on closed and bounded (compact) intervals; in the subsequent discussion of the convergence and the attainable order of convergence of numerical methods it will be these results which will play the central role. However, in the analysis of numerical stability one will require results dealing with the behavior of the solution of a given Volterra equation on infinite intervals of the form $[0,\infty)$. This is, of course, analogous to the situation in the numerical integration of initial-value problems for ordinary differential equations: concepts like A-stability, B-stability, contractivity, etc., are based on classes of, respectively, linear and nonlinear differential equations whose solutions have certain boundedness properties on $[0,\infty)$. It is the purpose of this section to state some first results regarding the above questions.

Before doing so we consider three examples of integral equations of the second kind characterized by convolution kernels $a=a(t-s)$ (see (1.3.13)).

Example 1.4.1. The constant kernel in

$$y(t) = 1-\int_0^t \lambda y(s) ds, \quad t \geq 0, \quad \lambda>0,$$

is not integrable on $\mathbb{R}_+ := [0,\infty)$ (we shall subsequently adopt the notation $a \in L^1(\mathbb{R}_+)$ if a given kernel is integrable on $\mathbb{R}_+$). The resolvent associated with this kernel is

$$\rho = \rho(t-s) = -\lambda \exp(-\lambda(t-s)),$$

with $\rho \in L^1(\mathbb{R}_+)$, since

$$\lim_{t\to\infty} \int_0^t \rho(u) du = -1;$$

we thus find the well-known result

$$y(t) = 1+\int_0^t \rho(u)du \to 0 \quad \text{as} \quad t\to\infty. \quad \square$$

Example 1.4.2. Consider now

$$y(t) = 1-\int_0^t \lambda(t-s)^{-\alpha} y(s)ds, \quad t\geqslant 0, \quad 0<\alpha<1, \quad \lambda>0.$$

This weakly singular equation has already been studied in Section 1.3.1, and it has been shown that the corresponding resolvent $\rho(t-s;\alpha)$ satisfies

$$1+\int_0^t \rho(u;\alpha)du = E_{1-\alpha}(-\lambda\Gamma(1-\alpha)t^{1-\alpha}).$$

Since for $z\in\mathbb{R}$, $z>0$, the Mittag-Leffler function has the property that $\lim_{z\to\infty} E_\beta(-z)=0$ ($\beta\in(0,1]$), it follows that $\rho\in L^1(\mathbb{R}_+)$, and hence

$$y(t) = 1+\int_0^t \rho(u;\alpha)du \to 0 \quad \text{as} \quad t\to\infty,$$

for all $\alpha\in(0,1)$ and all $\lambda>0$. $\square$

Example 1.4.3. The kernel, $a(u):=\frac{1}{2}u^2\exp(-u)$, of the so-called *renewal equation,*

$$y(t) = g(t)+\frac{1}{2}\int_0^t (t-s)^2\exp(s-t)y(s)ds, \quad t\geqslant 0,$$

is in $L^1(\mathbb{R}_+)$; Laplace transform arguments lead to the resolvent being given by

$$\rho(t) = \frac{1}{3}[1-\exp(-3t/2)(\cos(\sqrt{3}t/2)+\sqrt{3}\sin(\sqrt{3}t/2))].$$

Clearly, $\rho\notin L^1(\mathbb{R}_+)$. However, if the above kernel is replaced by $a(t)=-\frac{1}{2}(t-s)^2\exp(s-t)$, then $\rho\in L^1(\mathbb{R}_+)$, since the resolvent is now

$$\rho(t) = -\frac{1}{3}[\exp(-2t)-\exp(-t/2)(\cos(\sqrt{3}t/2)-\sqrt{3}\sin(\sqrt{3}t/2))]. \quad \square$$

For convolution kernels which are integrable on $\mathbb{R}_+$ a necessary and sufficient condition for the associated resolvent kernel to be in $L^1(\mathbb{R}_+)$, too, is given by the celebrated *theorem of Paley and Wiener* (see Paley & Wiener [1934]).

Theorem 1.4.1. *Let $a \in L^1(\mathbb{R}_+)$, and let*

$$A(s) := \int_0^\infty \exp(-st)a(t)dt$$

denote the Laplace transform of a. Then the corresponding resolvent kernel satisfies $\rho \in L^1(\mathbb{R}_+)$ if, and only if, $A(s) \neq 1$ for all s with $\mathrm{Re}(s) \geqslant 0$. □

In Example 1.4.3 above, the kernel $a(t) = \frac{1}{2}(t-s)^2 \exp(s-t)$ leads to $A(s) = 1/(s+1)^3$: the equation $A(s) = 1$ has the solution $s = 0$, hence the resolvent cannot be integrable on $\mathbb{R}_+$. If $a(t) = -\frac{1}{2}(t-s)^2 \exp(s-t)$, then $A(s) = -1/(s+1)^3 = 1$ has the solutions $s_1 = -2$, $s_{2,3} = (-1 \pm i\sqrt{3})/2$. Since all solutions have negative real parts, we have $A(s) \neq 1$ for all s with $\mathrm{Re}(s) \geqslant 0$, implying $\rho \in L^1(\mathbb{R}_+)$.

For integral equations of the second kind satisfying the Paley-Wiener condition one can then derive results on the asymptotic behavior of their solutions. As an example, suppose that the nonhomogeneous term g in (1.3.13) is continuous on $\mathbb{R}_+$ and is such that $g(\infty) := \lim_{t\to\infty} g(t)$ exists. It is not difficult to verify (by making use of (1.3.14) and (1.3.15)) that the exact solution behaves like

$$y(\infty) := \lim_{t\to\infty} y(t) = g(\infty)(1 + \int_0^\infty \rho(u)du) = g(\infty)/(1 - \int_0^\infty a(u)du)$$

(compare also Levin [1977]).

The above theorem of Paley and Wiener has been extended to linear integro-differential equations with convolution kernels,

$$y'(t) = q(t) + \int_0^t a(t-s)y(s)ds, \quad t \in I, \tag{1.4.1}$$

by Grossman & Miller [1973]. According to Theorem 1.3.4 and Corollary 1.3.2, the resolvent kernel associated with (1.4.1) depends only on the difference $t-s$ (since $p(t) \equiv 0$). Denoting this resolvent kernel by $\rho(t-s)$, we have the following result.

Theorem 1.4.2. *Let a in (1.4.1) satisfy $a \in L^1(\mathbb{R}_+)$, and denote by $A(s)$ its Laplace transform. Then the corresponding resolvent kernel ρ is in $L^1(\mathbb{R}_+)$ if, and only if, $A(s) \neq s$ for all s with* $\mathrm{Re}(s) \geqslant 0$. □

We conclude this section with two remarks. First, a very powerful tool for analyzing the asymptotic behavior of Volterra equations is given by the so-called *Tauberian theorems*. Let

$$A(s) := \int_0^\infty \exp(-st)a(t)dt$$

be convergent for $s > 0$. Theorems of *Abelian type* deduce the asymptotic behavior of $A(s)$ as $s\downarrow 0$ from given asymptotic properties of $a(t)$ as $t\to\infty$. The converse problem leads to *Tauberian theorems:* asymptotic properties of $a(t)$ (e.g., the existence of its integral on $\mathbb{R}_+$) are derived from the known behavior of $A(s)$ as $s\downarrow 0$ or $s\to\infty$. A detailed treatment of Tauberian theorems may be found in Widder [1941], Pitt [1958], and Postnikov [1980]; compare also Bellman & Cooke [1963] and Feller [1966].

The second remark concerns the stability of solutions of Volterra integral and integro-differential equations. Unfortunately, a detailed exposition of this topic is beyond the scope of this book, and we shall have to restrict ourselves to quoting a number of basic definitions in Section 7.1.1 where we shall be investigating the stability of numerical methods for these functional equations. The reader interested in further details is referred to the monographs by Miller [1971], Corduneanu [1973], and Burton [1983b]; the papers by Nohel [1964], [1971], Grossman & Miller [1970], [1973], Wagner [1978], Tsalyuk [1979]; and the conference proceedings by Londen & Staffans [1979], Herdman et al. [1981], Hannsgen et al. [1982].

1.5. Gronwall-type inequalities and comparison theorems

1.5.1. Integral inequalities of Gronwall type

Gronwall's original lemma (Gronwall [1918/19]) is concerned with a nonnegative function $y\in C(I)$ satisfying the inequality

$$y(t)\leqslant\int_0^t (My(s)+A)ds,\quad t\in I,\tag{1.5.1}$$

where M and A are nonnegative constants. He showed that the values of y on I can be bounded in terms of A, M, and the length of the given interval; i.e., for $t\in I$ one has

$$y(t)\leqslant AT\exp(MT).$$

Here, we state a more general form of Gronwall's result.

Theorem 1.5.1. *Let g, A and B denote (real-valued) continuous functions on I, and assume that A and B are nonnegative on I. If the continuous function y satisfies*

$$y(t)\leqslant g(t)+\int_0^t A(t)B(s)y(s)ds,\quad t\in I,\tag{1.5.2}$$

then

$$y(t)\leqslant g(t)+\int_0^t A(t)B(s)\exp\Big(\int_s^t A(u)B(u)du\Big)g(s)ds\tag{1.5.3}$$

for all $t\in I$. If $g(t)\leqslant G$ for all $t\in I$, and if $A(t)$ is nondecreasing on I, then (1.5.3) reduces to

$$y(t) \leqslant G \exp(\int_0^t A(t)B(u)du), \quad t \in I.$$

Proof. See Beesack [1975] (Theorem 2.1); compare also Schmidt [1976] where an abstract framework for establishing various Gronwall-type inequalities is given. □

We note in passing that the uniqueness of the solution y to the second-kind Volterra integral equation (1.3.1) could also have been established by employing the above result on Gronwall inequalities. To see this, let y and z in $C(I)$ denote two solutions to (1.3.1). It follows that, for all $t \in I$,

$$|y(t)-z(t)| \leqslant \int_0^t |K(t,s)| \, |y(s)-z(s)| ds \leqslant K_0 \int_0^t |y(s)-z(s)| ds,$$

where $K_0 := \max\{K(t,s)| : (t,s) \in S\}$. This represents a Gronwall inequality (1.5.2) with $g(t) \equiv 0$ (hence $G=0$), $A(t) \equiv 1$, and $B(s) \equiv K_0$, where the role of $y(t)$ is now assumed by $|y(t)-z(t)|$. We thus obtain $|y(t)-z(t)| \leqslant 0$ for all the $t \in I$, and continuity of $y-z$ on I then implies $y(t)-z(t) \equiv 0$ on I.

The expression

$$A(t)B(s)\exp(\int_s^t A(u)B(u)du)$$

in (1.5.3) is recognized to be the resolvent kernel of the degenerate Volterra kernel $K(t,s) = A(t)B(s)$ (cf. (1.1.10)), here assumed to be nonnegative on S. This suggests the following generalization of Theorem 1.5.1 to integral inequalities with general nonnegative kernels (see Chu & Metcalf [1967], Beesack [1969]; also Schmidt [1976]).

Theorem 1.5.2. *Let $g \in C(I)$, $K \in C(S)$, with $K(t,s) \geqslant 0$ for all $(t,s) \in S$. If $y \in C(I)$ obeys the integral inequality*

$$y(t) \leqslant g(t) + \int_0^t K(t,s)y(s)ds, \quad t \in I, \tag{1.5.4}$$

then it can be bounded by

$$y(t) \leqslant g(t) + \int_0^t R(t,s)g(s)ds, \quad t \in I, \tag{1.5.5}$$

where R is the resolvent kernel associated with the kernel K. Furthermore, $R(t,s) \geqslant K(t,s) \geqslant 0$ for all $(t,s) \in S$. □

The last statement is, of course, an immediate consequence of the definition of the iterated kernels of K and the resulting resolvent.

1.5.2. A comparison theorem

The following theorem (due to Beesack [1969]; see also Beesack [1975], Cochran [1972], and Miller [1971] for comparison theorems for nonlinear Volterra integral equations) is representative for a large class of results comparing the solutions of two related second-kind Volterra integral equations.

Theorem 1.5.3. *Let the functions $g_i(t)$ and $K_i(t,s)$, $i=1,2$, be, respectively, continuous on I and S. In addition, suppose that*

(i) $|g_1(t)| \leqslant g_2(t), \quad t \in I;$

(ii) $|K_1(t,s)| \leqslant K_2(t,s), \quad (t,s) \in S.$

Then the (unique) solutions $y_i \in C(I)$ of the integral equations

$$y_i(t) = g_i(t) + \int_0^t K_i(t,s) y_i(s) ds, \quad t \in I, \quad i=1,2,$$

are related by

$$|y_1(t)| \leqslant y_2(t) + |g_1(t)| - g_2(t), \quad t \in I.$$

Moreover, if $0 \leqslant K_1(t,s) \leqslant K_2(t,s)$, $(t,s) \in S$, then

$$y_1(t) \leqslant y_2(t) + g_1(t) - g_2(t), \quad t \in I,$$

and this inequality is valid for arbitrary (continuous) $g_1(t) \leqslant g_2(t)$,

One of the principal applications of comparison theorems is to establish the boundedness of the solution of a given Volterra integral equation. A discussion, together with a number of examples, of this aspect with regard to nonlinear Volterra equations can be found in Miller [1971] which lists additional references on this topic.

1.5.3. Discrete Gronwall-type inequalities

Discretization of Volterra equations by means of quadrature methods or by Runge-Kutta-type methods (to mention just two important classes of numerical methods) and the subsequent convergence analysis leads to discrete inequalities of the form

$$z_n \leqslant g_n + \sum_{i=0}^{n-1} p_i z_i, \quad n=0,1, \dots ,N,$$

where the elements of the sequences $\{g_n\}$, $\{p_n\}$, and $\{z_n\}$ are nonnegative. Moreover, p_n is often bounded by hM, where M will depend on a bound for the given kernel K as well as on properties of the quadrature weights or the Runge-Kutta parameters, and where h is the diameter of the mesh used in the discretization.

In the following we give first a general result from which various special cases will be derived, including one governing the case just mentioned. The results cited are essentially from Jones [1964], Schmidt [1976].

Theorem 1.5.4. *Let $\{z_n\}$, $\{g_n\}$, and $\{p_n\}$ be real-valued sequences defined for $n=0,1,\dots,N$, and assume that $p_n \geqslant 0$ for all n. If, for $n=0,1,\dots,N$,*

$$z_n \leqslant g_n + \sum_{i=0}^{n-1} p_i z_i, \tag{1.5.6}$$

then z_n can be bounded by

$$z_n \leqslant g_n + \sum_{i=0}^{n-1} p_i \Big(\prod_{j=i+1}^{n-1} (1+p_j) \Big) g_i \tag{1.5.7}$$

for all $n=0,1,\dots,N$. (Here, the expression $\prod_{j=i+1}^{n-1}(1+p_j)$ is to be interpreted as unity if $n \leqslant i+1$.) □

We now state explicity two special cases of Theorem 1.5.4.

Corollary 1.5.1. (a) *In Theorem 1.5.4, let $p_n \leqslant c$, $g_n \geqslant 0$, $n=0,1,\dots, N-1$, with $c>0$, and assume that the sequence $\{z_n\}$ satisfies (1.5.6). Then there holds the estimate*

$$z_n \leqslant g_n + c\sum_{i=0}^{n-1} (1+c)^{n-i-1} g_i, \quad n=0,\dots,N. \tag{1.5.8a}$$

If, in addition, $\{g_i\}$ is nondecreasing (so that all $g_i \leqslant g_N$), then we have

$$z_n \leqslant g_N\{1+c\sum_{i=0}^{n-1}(1+c)^{n-i-1}\} = g_n(1+c)^n, \quad n=0,\dots,N. \tag{1.5.8b}$$

(b) *If, in Theorem 1.5.4, $g_n \leqslant g$ for all $n=0,\dots,N$, then (1.5.6) implies (again for $p_n \geqslant 0$) that*

$$z_n \leqslant g\prod_{j=0}^{n-1}(1+p_j), \quad n=0,\dots,N. \quad \square \tag{1.5.9}$$

Finally, in view of later applications, we add an obvious modification of the above results often encountered in the literature on numerical methods for Volterra equations; the result follows from (1.5.9).

Corollary 1.5.2. *Let $z_n \geqslant 0$ for $n=0,1,\dots,N$, and suppose that the sequence $\{z_n\}$ obeys the inequality*

$$z_n \leqslant hC_1\sum_{i=0}^{n-1} z_i + C_2, \quad n=k,\ k+1,\ \cdots, N, \tag{1.5.10}$$

where $k>0$, $C_i>0$, $i=1,2$, and $h>0$. Then the elements of this sequence can be bounded by

$$z_n \leqslant (hC_1 z + C_2)(1+hC_1)^{n-k}, \quad n=k,\dots,N, \tag{1.5.11}$$

provided the starting values $z_0,\dots,z_{k-1}$ satisfy $z_j \leqslant z/k$. □

In the Gronwall-type results for discrete inequalities mentioned so far the coefficients of $\{z_i\}$ on the right-hand side do not depend on n. However, in the discretization of Volterra equations with weakly singular kernels one encounters inequalities of the form

$$z_n \leqslant h^{1-\alpha} C_1 \sum_{i=0}^{n-1} (n-i)^{-\alpha} z_i + C_2, \quad n=0, \ldots, N, \tag{1.5.12}$$

where the coefficient of z_i varies with n. We deal first with the practically important case $\alpha = ½$ (compare McKee [1982a]).

Theorem 1.5.5. *Let $z_n \geqslant 0$, $n=0, \ldots, N$, satisfy the inequality*

$$z_n \leqslant h^{½} C_1 \sum_{i=0}^{n-1} (n-i)^{-½} z_i + C_2, \tag{1.5.13}$$

where $C_1>0$, $C_2>0$, and $h>0$. Then

$$z_n \leqslant C_2(1+C_1 h^{½} + 2C_1(nh)^{½})(1+hC_1^2\pi)^n, \quad n=1, \ldots, N. \tag{1.5.14}$$

Proof. We shall show that the inequality (1.5.13) can be rewritten in the form (1.5.10) (with an appropriate change of C_1 and C_2); application of Corollary 1.5.2 then yields the desired result.

Consider

$$z_k \leqslant h^{½} C_1 \sum_{l=0}^{k-1} (k-l)^{-½} z_l + C_2;$$

multiply this inequality by $h^{½} C_1 (n-k)^{-½}$ (with $n>k$), and sum the resulting inequality from $k=1$ to $k=n-1$ to find

$$h^{½} C_1 \sum_{k=1}^{n-1} (n-k)^{-½} z_k \leqslant hC_1^2 \sum_{k=1}^{n-1} \sum_{l=0}^{k-1} (n-k)^{-½}(k-l)^{-½} z_l + \\ + h^{½} C_1 C_2 \sum_{k=1}^{n-1} (n-k)^{-½}. \tag{1.5.15}$$

If we rewrite (1.5.13) as

$$z_n \leqslant h^{½} C_1 \sum_{k=1}^{n-1} (n-k)^{-½} z_k + h^{½} C_1 n^{-½} z_0 + C_2 \quad (\text{with } z_0 \leqslant C_2),$$

and if we replace the first term on the right-hand side by the result (1.5.15), we are led to

$$z_n \leqslant hC_1^2 \sum_{l=0}^{n-2} [\sum_{k=l+1}^{n-1} (n-k)^{-½}(k-l)^{-½}] z_l \\ + h^{½} C_1 C_2 \sum_{k=1}^{n-1} (n-k)^{-½} + (1+h^{½} C_1) C_2. \tag{1.5.16}$$

Here, we have made use of the discrete version of the Dirichlet formula,

$$\sum_{k=1}^{n-1}\sum_{l=0}^{k-1}(n-k)^{-1/2}(k-l)^{-1/2}z_l = \sum_{l=0}^{n-2}\Big[\sum_{k=l+1}^{n-1}(n-k)^{-1/2}(k-l)^{-1/2}\Big]z_l. \qquad (1.5.17)$$

To achieve our stated aim we have to determine upper bounds for the variable coefficients in the first two terms on the right-hand side of (1.5.16); this will be done by relating these coefficients to Riemann sums of certain improper integrals.

Consider first the function $\Phi(t) := (n-l-t)^{-1/2}t^{-1/2}$, $0<t<n-l$ (where $n-l\geqslant 2$). This function is strictly convex on the given interval and attains its minimum at $t^* := (n-l)/2$. Thus,

$$\begin{aligned}\sum_{k=l+1}^{n-1}(n-k)^{-1/2}(k-l)^{-1/2} &= \sum_{\tau=1}^{n-l-1}(n-l-r)^{-1/2}r^{-1/2}\\ &= \sum_{\tau=1}^{n-l-1}\Phi(r)\leqslant \int_0^{n-l}\Phi(t)dt;\end{aligned}$$

this is an immediate consequence of interpreting the given sum as a (lower) Riemann sum, with the rectangle for the subinterval $[(n-l)/2, (n-l+2)/2]$ (if $n-l$ is even), or $[(n-l-1)/2, (n-l+1)/2]$ (if $n-l$ is odd) missing. But

$$\int_0^{n-l}\Phi(t)dt = \int_0^1(1-\tau)^{-1/2}\tau^{-1/2}d\tau = B(1/2,1/2) = (\sqrt{\pi})^2 = \pi.$$

In an analogous fashion we derive the result

$$\sum_{k=1}^{n-1}(n-k)^{-1/2} \leqslant \int_0^n (n-t)^{-1/2}dt = 2n^{1/2}, \quad n\geqslant 2.$$

Hence, (1.5.16) may be written as

$$z_n \leqslant hC_1^2\pi\sum_{l=0}^{n-1}z_l + (2(nh)^{1/2}C_1 + 1 + h^{1/2}C_1)C_2.$$

This is of the form (1.5.6), where

$$g_n := (2(nh)^{1/2}C_1 + 1 + h^{1/2}C_1)C_2$$

is nondecreasing. Hence, using (1.5.8b) we are led to (1.5.14). □

We note that, in later applications, the product nh will remain uniformly bounded; e.g., we have $nh\leqslant Nh = cT$, where $[0,T]$ denotes the (compact) interval on which the given Volterra equation is to be solved, and where the finite constant $c\geqslant 1$ will depend on the type of partitition for I. Instead of (1.5.14) we may thus write

$$z_n \leqslant C_2(1 + C_1h^{1/2} + C_1^2\pi h + 2C_1(cT)^{1/2})\exp(C_1^2\pi cT), \qquad (1.5.18)$$

where the right-hand side constitutes a uniform bound for $n = 1, \ldots, N$.

The result of Theorem 1.5.5 can be generalized (McKee [1982a], Dixon & McKee [1983]):

Theorem 1.5.6. *Let* $0<\alpha<1$, *and define* ν *to be the smallest positive integer satisfying* $\nu(1-\alpha)\geqslant 1$. *If the nonnegative sequence* $\{z_n\}$ *satisfies the inequality (1.5.12), with* $nh\leqslant Nh=cT$, *then*

$$z_n \leqslant C_2(C_2' + hC_1'(cT)^{\nu(1-\alpha)-1})\exp(C_1'(cT)^{\nu(1-\alpha)}), \quad n=1, \ldots ,N; \quad (1.5.19)$$

here, the constants C_i' *are given by*

$$C_1' := (C_1\Gamma(1-\alpha))^{\nu}/\Gamma(\nu(1-\alpha)),$$

$$C_2' := (1+C_1h^{1-\alpha})\sum_{j=0}^{\nu-2}\gamma^j+\gamma^{\nu-1}, \quad \gamma:=C_1\frac{(cT)^{1-\alpha}}{1-\alpha}. \quad \square$$

The proof uses arguments similar to the ones outlined in the proof of Theorem 1.5.5: here again, the aim is to transform the given inequality (1.5.12) into standard form (1.5.10), where, instead of the integer power h, one will have $(h^{1-\alpha})^{\nu}$, with ν being the smallest (positive) integer for which the product $\nu(1-\alpha)$ is at least equal to one. For $\alpha=\frac{1}{2}$ we obviously have $\nu=2$ (and thus the result contained in (1.5.18)); more generally, if $\alpha=p/q$ (reduced to lowest terms) is rational, then $\nu=[q/(q-p)]$. It is clear that Theorem 1.5.6 contains also the trivial case $\alpha=0$ for which $\nu=1$.

NOTES

Section 1.1.

The terminology "integral equation" was introduced by Du Bois-Reymond [1888]; he writes (p. 228): "... Ich schrieb diese Gleichungen nicht hin, als ob sie etwa das Problem lösten oder doch der Lösung näher führten, sie sollen nur ein Beispiel unter zahllosen sein, dafür, das man bei Randwertproblemen der linearen partiellen Differentialgleichungen beständig vor dieselbe Gattung von Aufgaben gestellt wird, welche jedoch, wie es scheint, für die heutige Analysis im Allgemeinen unüberwindliche Schwierigkeiten darbieten. Ich meine die zweckmässig *Integralgleichungen* zu nennenden Aufgaben, welche darin bestehen, dass die zu bestimmende Function, ausser ihrem sonstigen Vorkommen, in ihnen unter bestimmten Integralen enthalten ist, welche ausser den Integrationsvariablen noch andere enthalten, die eigentlichen Argumente der Gleichungen ..." (An English translation of most of the above can be found in Davis [1926, p. 10].) The more general integral equation with variable limits of integration,

$$y(t) = g(t) + \int_{a(t)}^{b(t)} K(t,s)y(s)ds, \quad t \in I,$$

is often referred to as *Andreoli's integral* equation. Compare, e.g., Andreoli [1914], Davis [1926, pp. 44-45], as well as Hellinger & Toeplitz [1927, pp. 1463-1464], for early references to such integral equations. Ghermanesco [1959] gives an illuminating analysis of the existence of unique or nonunique (e.g., nontrivial) solutions to the homogeneous Andreoli equation with $a(t) = -t$, $b(t) = t$; in Ghermanesco [1961], the analogous integro-differential equation is studied.

In physical modeling problems one frequently encounters Volterra equations of parabolic and hyperbolic type; as up-to-date introductions to these problems we suggest the papers by Tesei in Albrecht & Collatz [1980, pp. 213-227], by Nohel in Londen & Staffans [1979, pp. 220-235] and, by the same author, in Herdman et al. [1981, pp. 3-82] (this paper contains an extensive bibliography). See also the recent paper by Hrusa & Nohel [1984]. The last two books, as well as Hannsgen et al. [1982], are also good sources for results on Volterra equations in more abstract settings.

Due to limitations of space it is not possible to present a description of Volterra integral and integro-differential equations arising in practical applications. Instead, we must refer the reader to the literature: in addition to the bibliography of Noble [1971] (see, in particular, pp. 292-294, pp. 299-300, of the volume "Subject Listing"), we mention the monograph by Miller [1971, pp. 60-74], the papers of Anderssen [1977b], Anderssen & de Hoog [1982], and Gorenflo & Vessella [1984b] on applications of the Abel equation, the contributions by Cushing [1977] and by Diekman [1979] on Volterra equations arising in population dynamics, and the monographs by Bharucha-Reid [1972], Tsokos & Padgett [1974], and Cannon [1984, pp. 82-99] which deal, respectively, with random integral equations and with integral equations occurring in connection with the heat equation. See also Brunner [1982a, pp. 228-229] and the numerous references contained in the bibliography of this book.

Section 1.2.

Gorenflo & Vessella [1984a] contains a survey of the literature on the theory of Abel-type integral equations. A detailed account of Liouville's contributions to the theory of integral equations has been given by Lützen [1982]; see also Dieudonné [1981]. Dirichlet's formula (mentioned in the context of Liouville's work; see also Section 1.3, (1.3.3)) was first given in Dirichlet [1837, p. 45]; its extension to weakly singular integrands is due to Hurwitz [1908]. Picard's method of successive approximations for ODEs is introduced in Picard [1890, pp. 197-200].

Volterra seems to have been led to first-kind integral equations in an early paper of 1884, through a problem in electrostatics. The papers of Volterra [1897] and of Lauricella [1908] contain surveys of the early development of the theory of integral equations with variable (upper) limits of integration. As far as Volterra's work itself is concerned, the first volume of his Opere mathematiche (1955) includes a twelve-page sketch of his scientific work. Compare also the index volume of Acta Mathematica of 1912 which, besides a short description of Volterra's early work, contains brief biographies (as well as photographs) of other mathematicians whose papers appeared in earlier volumes of this journal and who played a role in the development of the theory of integral equations. (We are grateful to Dr. Jan Verwer for bringing this reference to our attention.) The first comprehensive account of the state of the art in Volterra and Fredholm integral equations is due to Bateman [1910]. Evans [1910/1911] may be consulted for an extensive general analysis of singular second-kind Volterra integral equations; see also the recent paper by Reynolds [1984]. The studies of Davis [1926], [1930] focus on the development of the theory after 1910; their bibliographies (reaching back to Abel) list more than 500 papers and books (of which, incidentally, only two deal with the numerical solution of integral equations). The monographs by Lalesco [1912] and by Vivanti [1929] contain also extensive bibliographies (the one in Vivanti is in annotated form and includes Ph.D. theses), as does the survey article by Hellinger & Toeplitz [1927]. Birkhoff [1973, Ch. 13] presents English translations of parts of the papers by Abel [1826] and by Volterra [1896].

As for the more recent development of the theory (and the numerical analysis) of Volterra integral and integro-differential equations, we mention the comprehensive bibliography of Nobel [1971], as well as the survey paper by Tsalyuk [1979] which deals with the over 500 papers on Volterra equations reviewed in Referativnyi Zhurnal "Matematika" between 1966 and 1976; this paper represents a valuable guide especially to work carried out in the U.S.S.R. In addition, compare the bibliography in Fenyö & Stolle [1984b].

Section 1.3.

Of the numerous books treating the classical theory of Volterra integral equations we mention the ones by Lalesco [1912], Bôcher [1913], Vivanti [1929], Kowalewski [1930], Schmeidler [1950], Tricomi [1957], Pogorzelski [1966], Cochran [1972], Krasnov et al. [1977], Burton [1983b], and Fenyö & Stolle [1984a]. In addition, see also the paper of Burton [1983a]. Regarding results on second-kind integral equations with convolution kernels (often referred to as renewal equations), we single out the papers of Feller [1941], of Karlin [1955] (this paper analyzes also the discrete version of the renewal equation), and of Friedman [1963], and the book by Bellman & Cooke [1963, pp. 216-264] (from which Theorem 1.3.3 has been taken). Compare also Feller [1966] for a detailed introduction to renewal theory. Since we have not considered the solution of Volterra equations with convolution kernels by Laplace transform techniques, we refer the interested reader to Widder [1941], Bellman & Cooke [1963], Sneddon [1972], Henrici [1977], and Krasnov et al. [1977]. Existence and uniqueness results in the L^2-setting may be found in Tricomi [1957], Miller [1971], and in Fenyö & Stolle [1984a]. See also the recent contribution of Szufla [1984] on L^p-solutions.

For a treatment of integral operators and integral equations in more abstract settings see, e.g., Atkinson [1976], Linz [1979], Jörgens [1982], and Fenyö & Stolle [1982], [1983]. The monographs by Bharucha-Reid [1972], Tsokos & Padgett [1974], and Hönig [1975] deal with generalizations of Volterra integral equations (equations with random variables) and with Volterra-Stieltjes integral equations.

The study of Davis [1927a] is concerned, in a comprehensive way, with various problems relating to the inversion of Volterra integral equations of the first-kind, including those equations leading to integral equations of the third kind and their connection with differential equations of Fuchsian type. As for third-kind equations, we also mention the paper by Magnitskii [1979]. The question on the existence of explicit inversion formulas for various types of first-kind Abel integral equations is studied in Anderssen [1977b]. The book by Srivastava & Bushman [1977] features extensive tables of explicit inversion formulas for integral equations of the first-kind whose kernels are special functions.

The standard reference for the theory of nonlinear Volterra equations is Miller [1971]; additional references are Davis [1962], Saaty [1967, pp. 262-300], and Burton [1983b, pp. 66-96]. Of the numerous papers dealing with the representation of solutions of nonlinear Volterra equations we cite those of Miller [1968b], Grossman & Miller [1970], [1973], Brauer [1972], Levin [1977], and Bernfeld & Lord [1978]; for additional references the reader is referred to the bibliographies of the relevant papers in London & Staffans [1979], Herdman et al. [1981], and Hannsgen et al. [1982]. Existence and uniqueness results for nonlinear integral equations of the first-kind were given by Janikowski [1962], Branca [1976], [1978], and Dixon, McKee & Jeltsch [1985].

Logan [1976, pp. 90-93] was the first to derive general smoothness results of the form (1.3.49) (with $g_2(t)\equiv 0$) for the solutions of nonlinear integral equations of the second-kind with weakly singular kernels; he also considered kernel singularities of the form $\log(t-s)$. General smoothness results for Abel-type singularities may also be found in Lubich [1983b]; his results are based on formal power series and the implicit function theorem. Theorems 1.3.15 and 1.3.16 are from Brunner [1983]. Atkinson [1974a] studied the behavior of the solution of the general Abel-type integral equation

$$\int_0^t (t^p - s^p)^{-\alpha} K(t,s) y(s) ds = t^\beta g(t), \quad K \text{ and } g \text{ smooth},$$

where $p>0$, $0<\alpha<1$, and $p\alpha+\beta>0$.

Section 1.4.

Theorem 1.4.1 occurs in Paley & Wiener [1934, pp. 58-63]; compare also Miller [1971, pp. 207, 407-409], Corduneanu [1973, pp. 213-219], as well as Bownds & Cushing [1975a], Jordan & Wheeler [1976], Levin [1977], and Gripenberg [1982]. Paley-Wiener-type results for integro-differential equations may be found in Grossman & Miller [1973], Shea & Wainger [1975], and Gripenberg [1982] (see also this last paper for additional references). A discrete version of Theorem 1.4.1 (cf. Chapter 7) has been given by Lubich [1982b], [1983a].

The monograph of Coppel [1965] provides a thorough and lucid introduction to the stability theory for ODEs; see also for extensions to Volterra equations e.g., Miller [1968a], [1971], Bownds & Cushing [1975a], Burton & Mahfoud [1983], and, especially, Burton [1983b]. Nohel's review (Nohel [1976]) of the book by Corduneanu [1973] is more than just that: it gives a detailed description of the state of the art in various areas of qualitative behavior of solutions of Volterra equations, as well as of open problems, and it contains a large number of relevant references. Results on the asymptotic behavior of solutions of Volterra equations may also be found, e.g., in Friedman [1963], Bellman & Cooke [1963], Miller [1971], Ling [1978], Wagner [1978], Burton [1983a], and in Hrusa & Nohel [1984]. The survey paper by Tsalyuk [1979] mentioned before deals with the asymptotic theory of Volterra equations on pp. 725-738. Compare also the Notes on Section 7.1 for additional information.

Section 1.5.

The monograph by Lakshmikantham & Leela [1969] offers a comprehensive study of inequalities and comparison theorems for Volterra integral equations. In addition, compare also Bellman & Cooke [1963], Chu & Metcalf [1967], Beesack [1969], Miller [1971], Cochran [1972], Beesack [1975], Schmidt [1976], Dixon & McKee [1983], and Scott [1984]. For generalizations to several variables consult Corduneanu [1982], [1983], and, especially, Beesack [1984b],

[1984c]; see also Dixon & McKee [1984] for inequalities involving repeated integrals. The survey paper of Agarwal & Thandapani [1981] contains a large number of references on linear and nonlinear generalizations of Gronwall's inequality.

Mikeladze [1935, p. 259] seems to be the first source of a discrete Gronwall inequality. Discrete Gronwall inequalities are discussed in Jones [1964], Beesack [1975], Schmidt [1976], McKee [1982a], Dixon & McKee [1983], Beesack [1984a], and Scott [1984] (the last four papers deal with inequalities of the form (1.5.12) as well as with more general versions). The bound (1.5.19) in Theorem 1.5.6 (and hence also that in Theorem 1.5.5) can be replaced by a bound involving the Mittag-Leffler function (arising from the solution of a majorizing integral equation). Compare Dixon & McKee [1983] and Scott [1984] for details.

CHAPTER 2

NUMERICAL QUADRATURE

This chapter gives a survey of results from the theory of numerical quadrature which we shall need in the construction of integration methods for Volterra equations. For the proofs of classical results we refer to the literature, while for a number of recent (and not yet widely known) results proofs are sketched. In particular, we shall treat the (ρ,σ)-reducible quadrature formulas in some detail.

2.1. General theory

Consider the definite integral

$$\mathbf{I}[\phi] := \int_a^b \phi(t)dt, \quad |a|<\infty, \quad |b|<\infty, \quad a<b, \tag{2.1.1}$$

and a numerical approximation to $\mathbf{I}[\phi]$ of the form

$$\mathbf{I}_n[\phi] := \sum_{j=0}^{n} c_{n,j}\phi(t_{n,j}), \tag{2.1.2}$$

where the $t_{n,j}$, $t_{n,0}<t_{n,1}< \ldots <t_{n,n}$, are *points* or *abscissas* not necessarily lying in the interval $[a,b]$. We will call the $\{c_{n,j}\}$ *coefficients*. The difference $\mathbf{I}[\phi]-\mathbf{I}_n[\phi]$ is the *approximation error* or *quadrature error* and will be denoted by $\mathbf{E}_n[\phi]$. The approximation $\mathbf{I}_n[\phi]$ in the relation

$$\mathbf{I}[\phi] = \mathbf{I}_n[\phi]+\mathbf{E}_n[\phi] \tag{2.1.3}$$

is called an $(n+1)$-point *quadrature formula.*

Usually, the coefficients $\{c_{n,j}\}$ are chosen such that $\mathbf{E}_n[\phi]$ vanishes for all functions ϕ in some test space. For instance, if this space is a certain class of polynomials we obtain the *interpolatory quadrature formulas* (see Section 2.1.2).

2.1.1. The approximation error

Definition 2.1.1. A sequence of quadrature formulas is said to be *convergent* for the function ϕ if $\mathbf{E}_n[\phi]\to 0$ as $n\to\infty$. □

The following general convergence theorem is due to Steklov (1916) and Pólya (1933) (see, e.g., Davis & Rabinowitz [1984, p. 130], Schmeisser & Schirmeier [1976, pp. 189-190]).

Theorem 2.1.1. *Let the points* $\{t_{n,j}\}$ *be contained in the interval* $[a,b]$. *A sequence of quadrature formulas is convergent for all* $\phi \in C[a,b]$ *if, and only if, it converges for all polynomials and satisfies*

$$\sum_{j=0}^{n} |c_{n,j}| < B \quad \text{for all} \quad n \in \mathbb{N},$$

where B is a finite constant independent of n. □

For quadrature formulas with *positive coefficients* we obtain

Corollary 2.1.1. *Consider a sequence of quadrature formulas (2.1.2) with* $t_{n,j} \in [a,b]$ *and with* $c_{n,j} > 0$ $(j=0, \ldots, n;\ n \in \mathbb{N})$. *This sequence converges for all* $\phi \in C[a,b]$ *if, and only if, it converges for all polynomials.*

Proof. If we apply (2.1.2) to (2.1.1) with $\phi(t)=1$, then it follows that

$$\sum_{j=0}^{m} |c_{n,j}| = \sum_{j=0}^{m} c_{n,j} = \mathbf{I}_n[1].$$

Since $\mathbf{I}_n[1] \to \mathbf{I}[1] = b-a$, as $n \to \infty$, $\mathbf{I}_n[1]$ is bounded, and hence the assertion of Corollary 2.1.1 follows from Theorem 2.1.1. □

Polynomials play an important role in the analysis of quadrature formulas and are used to characterize the accuracy of the numerical approximations.

Definition 2.1.2. A quadrature formula has *degree of precision q* if it is exact (i.e., if $\mathbf{E}_n[\phi]=0$) for all polynomials ϕ of degree not exceeding q but if $\mathbf{E}_n[t^{q+1}] \neq 0$. □

In order to give an expression for the approximation error $\mathbf{E}_n[\phi]$ we require the *truncated power functions:* they are defined by

$$x_+^k := \begin{cases} x^k & \text{if } x>0 \\ 0 & \text{if } x \leq 0 \end{cases};\quad k \geq 1. \tag{2.1.4}$$

For $k=0$, we set

$$x_+^0 := \begin{cases} 1 & \text{if } x>0 \\ 0 & \text{if } x<0 \end{cases};$$

the value of this function at the break point $x=0$ will be determined by the context in which it occurs (compare also de Boor [1978, p. 101]). Furthermore, we define the monospline

$$K_m(t) := \frac{(b-t)^m}{m!} - \sum_{j=0}^{n} c_{n,j} \frac{(t_{n,j}-t)_+^{m-1}}{(m-1)!}. \tag{2.1.5}$$

Theorem 2.1.2. *Let the quadrature formula (2.1.2) be of degree of precision q, let all points* $\{t_{n,j}\}$ *belong to* $[a,b]$, *and let* $\phi \in C^m[a,b]$, *with* $1 \leqslant m \leqslant q+1$. *Then*

$$\mathbf{E}_n[\phi] = \int_a^b K_m(t)\phi^{(m)}(t)dt,$$

where $\phi^{(m)}$ *denotes the m-th derivative of* ϕ. *Moreover, if* $K_m(t)$ *does not change its sign on* $[a,b]$, *then there exists a point* $\tilde{t}_n \in [a,b]$ *so that the quadrature error can be expressed in the form*

$$\mathbf{E}_n[\phi] = \phi^{(m)}(\tilde{t}_n)\int_a^b K_m(t)dt.$$

Proof. See, e.g., Stroud [1974, pp. 164-166], Engels [1980, pp. 94-96]. □

This theorem is known as *Peano's kernel theorem* and the function $K_m(t)$ is called *Peano's kernel function.* The theorem can be used to find out how fast a given sequence of quadrature formulas converges as $n \to \infty$. To make this precise we require

Definition 2.1.3. A sequence of quadrature formulas of the form (2.1.2) has *order* r if for all sufficiently smooth ϕ the approximation error satisfies $\mathbf{E}_n[\phi] = O(h^r)$ as $h \downarrow 0$, where $h := \max_{1 \leqslant j \leqslant n} (t_{n,j} - t_{n,j-1})$. □

Example 2.1.1. In order to illustrate Peano's kernel theorem we derive the approximation error for the so-called *repeated trapezoidal rule,* defined by (cf. also Section 2.7)

$$\mathbf{I}_n[\phi] = \frac{h}{2}\phi(a) + h\sum_{j=1}^{n-1}\phi(a+jh) + \frac{h}{2}\phi(b), \quad h := \frac{b-a}{n},$$

when applied to functions $\phi \in C^1[a,b]$. Evidently,

$$\begin{aligned} K_1(t) &= b - t - \tfrac{1}{2}h - h\sum_{j=1}^{n-1}(a+jh-t)_+^0 \\ &= b - t - \tfrac{1}{2}h - h\sum_{j=l+1}^{n-1} 1 \quad \text{for} \quad t \in [a+lh,\ a+lh+h]. \end{aligned}$$

Hence,

$$\begin{aligned} K_1(t) &= b - t - \tfrac{1}{2}h - (n-1-l)h = b - t - \tfrac{1}{2}h - (b-a-h-lh) \\ &= -t + a + (l+\tfrac{1}{2})h \quad \text{for} \quad t \in [a+lh,\ a+lh+h]. \end{aligned}$$

Thus $K_1(t)$, and therefore $\mathbf{E}_n[\phi]$, are $O(h)$, so that the order is $r=1$. □

2.1.2. *Interpolatory quadrature formulas*

In the following, the test space mentioned before will be the space of polynomials. A quadrature formula of degree of precision q satisfies the relations

$$\int_a^b t^l dt = \sum_{j=0}^n c_{n,j} t_{n,j}^l, \quad l=0,1, \ldots ,q.$$

For $n \geqslant q$ this system possesses a solution $\{c_{n,j}\}_{j=0}^n$ since the matrix $(a_{l,j}) := (t_{n,j}^l)$ always has a submatrix of rank $q+1$ whose determinant is a Vandermonde determinant. For $n=q$ there is a unique solution that can be constructed as follows. Let $p_n[t;\phi]$ be the polynomial interpolating the function $\phi(t)$ at the $n+1$ distinct points $t_{n,o}, \ldots ,t_{n,n}$, not necessarily lying in $[a,b]$. Then it can be written in the form

$$p_n[t;\phi] = \sum_{j=0}^n \phi(t_{n,j}) L_{n,j}(t) = \sum_{j=0}^n \phi(t_{n,j}) \frac{\pi_n(t)}{(t-t_{n,j})\pi_n'(t_{n,j})}, \tag{2.1.6}$$

where

$$L_{n,j}(t) := \prod_{\substack{i=0 \\ i \neq j}}^n \frac{t-t_{n,i}}{t_{n,j}-t_{n,i}} \quad \text{and} \quad \pi_n(t) := \prod_{i=0}^n (t-t_{n,i}) \,. \tag{2.1.7}$$

Integration of this interpolating polynomial yields

$$\int_a^b p_n[t;\phi] dt = \sum_{j=0}^n \phi(t_{n,j}) \int_a^b L_{n,j}(t) dt =: \sum_{j=0}^n c_{n,j} \phi(t_{n,j}). \tag{2.1.8}$$

Since we have $p_n[t;\phi] = \phi(t)$ for all polynomials ϕ of degree $\leqslant n$, the coefficients $c_{n,j}$, given by

$$c_{n,j} = \int_a^b L_{n,j}(t) dt, \tag{2.1.9}$$

determine a quadrature formula of degree of precision $q \geqslant n$. This formula is said to be of *interpolatory type.* In Sections 2.3 and 2.4 we will see that the degree of precision of such interpolatory formulas can be made considerably larger than n if the points $\{t_{n,j}\}$ are chosen judiciously.

Since the coefficients of an interpolatory quadrature formula are uniquely defined, we can summarize the preceding considerations in the following theorem.

Theorem 2.1.3. *A quadrature formula based on $n+1$ (distinct) points is of interpolatory type if, and only if, its degree of precision satisfies $q \geqslant n$.* □

In the following we assume that the abscissas $\{t_{n,j}\}$ belong to $[a,b]$; the results, however, can easily be extended to the case where $[a,b]$ does not contain all abscissas.

For sufficiently smooth integrands ϕ the approximation error of an $(n+1)$-

point interpolatory quadrature formula can be derived by using the explicit form of the interpolation error,

$$\phi(t) = p_n[t;\phi]+\frac{1}{(n+1)!}\pi_n(t)\phi^{(n+1)}(\theta(t)), \quad t\in[a,b] \tag{2.1.10}$$

where $\theta(t)\in(a,b)$ (see, e.g., Stroud [1974, p. 52]).

Theorem 2.1.4. *Let $\phi\in C^{n+1}[a,b]$. Then there exists a function $\theta(t)$ with values in the interval (a,b) such that the approximation error of the $(n+1)$-point interpolatory quadrature formula (2.1.2) is given by*

$$\mathbf{E}_n[\phi] = \frac{1}{(n+1)!}\int_a^b \pi_n(t)\phi^{(n+1)}(\theta(t))dt.$$

Proof. Since $\mathbf{I}$ and $\mathbf{I}_n$ are linear functionals we may write, by virtue of (2.1.10),

$$\mathbf{I}[\phi] = \mathbf{I}[p_n]+\frac{1}{(n+1)!}\mathbf{I}[\pi_n(t)\phi^{(n+1)}(\theta(t))].$$

It follows from Theorem 2.1.3 that the interpolating polynomial $p_n[t;\phi]$ is integrated exactly, and hence the first term on the right-hand side can be replaced by $\mathbf{I}_n[p_n]$. The assertion of Theorem 2.1.4 now follows since the interpolating conditions $p_n[t_{n,j};\phi]=\phi(t_{n,j})$ imply $\mathbf{I}_n[p_n]=\mathbf{I}_n[\phi]$. □

Corollary 2.1.2. *Let $\phi\in C^{n+1}[a,b]$. Then the approximation error of the $(n+1)$-point interpolatory quadrature formula (2.1.2) satisfies the inequality*

$$|\mathbf{E}_n[\phi]| \leq \frac{1}{(n+1)!}(b-a)^{n+2}\alpha^{n+1}|\phi^{(n+1)}(\tilde{t}_n)|,$$

where $\tilde{t}_n\in(a,b)$ and $\alpha:=(t_{n,n}-t_{n,o})/(b-a)$. □

The above result is a direct consequence of Peano's kernel theorem 2.1.2, since, for $m = q+1 = n+1$, the Peano kernel of an interpolating quadrature formula is of constant sign on (a,b).

The bound given in this corollary is, in general, rather conservative. However, as suggested by the introduction of the quantity α, it can be quite useful in cases where n and α are fixed and $b-a$ tends to zero. Such cases arise in repeated quadrature formulas (Section 2.7) and in starting formulas for Volterra equations (Section 3.5). Sharper estimates for $\mathbf{E}_n[\phi]$ can be given in special cases; e.g., if the points $\{t_{n,j}\}$ are equally spaced (cf. Theorem 2.2.1), or if $t_{n,j}$ is related to the zeros of certain orthogonal polynomials (cf. the Theorems 2.3.1, 2.4.1 and 2.4.2). In all these cases we find quadrature formulas of the form

$$\int_a^b \phi(t)dt = \mathbf{I}_n[\phi]+e_n(b-a)^{q+2}\phi^{(q+1)}(\tilde{t}_n), \quad \tilde{t}_n\in(a,b) \tag{2.1.11}$$

where q is the degree of precision of the formula and e_n a constant only depending on n. Mapping on the interval [0,1] yields

$$\int_0^1 \phi(a+(b-a)\tau)d\tau = \frac{1}{(b-a)} \mathbf{I}_n[\phi]+e_n(b-a)^{q+1}\phi^{(q+1)}(\tilde{t}_n), \quad (2.1.12)$$

where $\mathbf{I}_n[\phi]$ is the same numerical approximation as in (2.1.11).

In the construction of step-by-step methods for Volterra equations it will often be convenient to write (2.1.12) in the form

$$\int_0^1 \phi(a+\tau h)d\tau = \sum_{i=1}^{m} w_i\phi(a+d_i h)+e_m h^{q+1}\phi^{(q+1)}(\tilde{t}_m), \quad (2.1.13)$$

where $0 \leqslant d_1 < \ldots < d_m \leqslant 1$ and where h denotes the integration step of the method (cf. Table 2.5.1).

2.2. Newton-Cotes formulas

Newton-Cotes type formulas arise if we choose equally spaced points in the interpolatory quadrature formulas defined by (2.1.9). It is straightforward to verify that for

$$t_{n,j} := \text{constant}+jh, \quad j=0,1, \ldots ,n, \quad (2.2.1)$$

the coefficients $c_{n,j}$ can be expressed in the form

$$c_{n,j} = \frac{1}{h^n}\frac{(-1)^{n-j}}{j!(n-j)!}\int_a^b \frac{\pi_n(t)}{t-t_{n,j}}dt. \quad (2.2.2)$$

If $a=t_{n,o}$ and $b=t_{n,n}$, the formulas defined by (2.2.1)-(2.2.2) are called *Newton-Cotes formulas of closed type* because both end points of the integration interval are involved in the set (2.2.1). In Table 2.2.1 the values

$$w_{n,j} := \frac{c_{n,j}}{h}$$

are listed for $n \leqslant 6$; for formulas with more abscissas we refer to Abramowitz & Stegun [1970, p. 886]. The first seven formulas in this sequence have positive weights. The 8-th formula is the first one possessing also negative weights. In the subsequent formulas negative weights may also occur (e.g. for $n=10$). We will see that this phenomenon is related to the divergence of Newton-Cotes sequences (Theorem 2.2.2). In the following the quantities $w_{n,j}$ will be called *weights,* in order to distinguish them from the *coefficients* $c_{n,j}$.

Table 2.2.1. Weights and approximation errors in closed Newton-Cotes formulas ($a = t_{n,o}$, $b = t_{n,n}$, $\tilde{\phi} = \phi(\tilde{t}_n)$)

n	$\frac{1}{d_n}$	$\frac{w_{n,0}}{d_n}$	$\frac{w_{n,1}}{d_n}$	$\frac{w_{n,2}}{d_n}$	$\frac{w_{n,3}}{d_n}$	$\frac{w_{n,4}}{d_n}$	$\frac{w_{n,5}}{d_n}$	$\frac{w_{n,6}}{d_n}$	$\frac{w_{n,7}}{d_n}$	$\mathbf{E}_n[\phi]$
1	2	1	1							$-h^3\tilde{\phi}^{(2)}/12$
2	3	1	4	1						$-h^5\tilde{\phi}^{(4)}/90$
3	8	3	9	9	3					$-3h^5\tilde{\phi}^{(4)}/80$
4	45	14	64	24	64	14				$-8h^7\tilde{\phi}^{(6)}/945$
5	288	95	375	250	250	375	95			$-275h^7\tilde{\phi}^{(6)}/12096$
6	140	41	216	27	272	27	216	41		$-9h^9\tilde{\phi}^{(8)}/1400$

Table 2.2.2. Weights and approximation errors in open Newton-Cotes formulas ($a = t_{n,o} - h$, $b = t_{n,n} + h$, $\tilde{\phi} = \phi(\tilde{t}_n)$)

n	$\frac{1}{d_n}$	$\frac{w_{n,0}}{d_n}$	$\frac{w_{n,1}}{d_n}$	$\frac{w_{n,2}}{d_n}$	$\frac{w_{n,3}}{d_n}$	$\frac{w_{n,4}}{d_n}$	$\frac{w_{n,5}}{d_n}$	$\frac{w_{n,6}}{d_n}$	$\mathbf{E}_n[\phi]$
1	2/3	1	1						$h^3\tilde{\phi}^{(2)}/4$
2	3/4	2	-1	2					$28h^5\tilde{\phi}^{(4)}/90$
3	24/5	11	1	1	11				$95h^5\tilde{\phi}^{(4)}/144$
4	20/6	11	-14	26	-14	11			$41h^7\tilde{\phi}^{(6)}/140$
5	1440/7	611	-453	562	562	-453	611		$5257h^7\tilde{\phi}^{(6)}/8640$
6	945/8	460	-954	2196	-2459	2196	-954	460	$3956h^9\tilde{\psi}^{(8)}/14175$

Table 2.2.3. Weights in half-open Newton-Cotes formulas ($a = t_{n,o}$, $b = t_{n,n} + h$)

n	$\frac{1}{d_n}$	$\frac{w_{n,0}}{d_n}$	$\frac{w_{n,1}}{d_n}$	$\frac{w_{n,2}}{d_n}$	$\frac{w_{n,3}}{d_n}$	$\frac{w_{n,4}}{d_n}$	$\frac{w_{n,5}}{d_n}$	$\frac{w_{n,6}}{d_n}$
2	4/3	1	0	3				
4	144/5	19	-10	120	-70	85		
6	8640/7	751	-840	8547	-11648	14637	-7224	4417

If $a = t_{n,o} - h$ and $b = t_{n,n} + h$ the formulas are said to be of *open type* (see Table 2.2.2).

If either $a = t_{n,o}$, $b = t_{n,n} + h$, or $a = t_{no} - h$, $b = t_{n,n}$, then the formulas are said to be of *half-open type*. For *odd* values of n these half-open formulas coincide with the open formulas corresponding to $n-1$. For *even* values of n a few formulas are listed in Table 2.2.3. The approximation errors are $O(h^{n+2})$. In *explicit* integration methods for Volterra equations one sometimes uses Newton-Cotes formulas which do not involve $\phi(t_{n,n})$. In the following, however, we concentrate on the more important closed-type formulas.

The first three closed Newton-Cotes formulas listed in Table 2.2.1 are the well-known *trapezoidal rule, Simpson's rule,* and *Simpson's 3/8 rule*, respectively. The trapezoidal rule is of first degree of precision, both Simpson rules are of third degree of precision. From the following theorem a general result on the degree of precision of (closed) Newton-Cotes formulas can be derived (the approximation errors listed in Table 2.2.1 can be computed by this theorem).

Theorem 2.2.1. *Let $\phi \in C^{n+2}[a,b]$ if n is even, and $\phi \in C^{n+1}[a,b]$ if n is odd. Then there exists a point $\tilde{t}_n \in (a,b)$ so that the approximation error of the $(n+1)$-point Newton-Cotes formula of closed type is given by*

$$\mathbf{E}_n[\phi] = \frac{1}{(n+2)!}\, \phi^{(n+2)}(\tilde{t}_n) \int_a^b t\pi_n(t)dt \quad \textit{if } n \textit{ is even},$$

$$\mathbf{E}_n[\phi] = \frac{1}{(n+1)!}\, \phi^{(n+1)}(\tilde{t}_n) \int_a^b \pi_n(t)dt \quad \textit{if } n \textit{ is odd}.$$

Proof. See Isaacson & Keller [1966, pp. 310-312]. □

From this theorem it is immediate that an $(n+1)$-point formula has degree of precision $n+1$ if n is even, and degree of precision n if n is odd (cf. (2.1.11)). However, as shown in Theorem 2.2.2, closed Newton-Cotes formulas suffer from a major drawback.

Theorem 2.2.2. *The weights of closed Newton-Cotes formulas satisfy the relation*

$$\lim_{n\to\infty} \frac{1}{n} \sum_{j=0}^{n} |w_{n,j}| = \infty.$$

Proof. See Krylov [1962, p. 86], Brass [1977, pp. 124-127]. □

From this result it follows that the Newton-Cotes coefficients do not satisfy the conditions of Theorem 2.1.1 and therefore a sequence of Newton-Cotes formulas does not necessarily converge to the true value of the integral. Related to this unfavorable behavior is the occurrence of negative weights mentioned before.

2.3. Gauss-Legendre formulas

In Section 2.1 we remarked that the degree of precision of an $(n+1)$-point interpolatory quadrature formula can be made considerably larger than n by choosing the points $\{t_{n,j}\}$ appropriately. The *Gauss-Legendre formulas* are such formulas. They possess the largest possible degree of precision and are defined by the points

$$t_{n,j} = \frac{b-a}{2} z_{n,j} + \frac{b+a}{2}, \quad j=0,1, \ldots, n, \tag{2.3.1}$$

where the $\{z_{n,j}\}$ are the $n+1$ zeros of the Legendre polynomial $P_{n+1}(t)$. Tables of these zeros can be found in Abramowitz & Stegun [1970, p. 916]. The coefficients $\{c_{n,j}\}$ can be expressed as (see also Table 2.3.1)

$$c_{n,j} = \frac{b-a}{(1-z_{n,j})^2 [P'_{n+1}(z_{n,j})]^2} . \tag{2.3.2}$$

Theorem 2.3.1. *Let $\phi \in C^{2n+2}[a,b]$. Then there exists a point $\tilde{t}_n \in (a,b)$ so that the approximation error of the $(n+1)$-point Gauss-Legendre formula is given by*

$$\mathbf{E}_n[\phi] = \frac{(b-a)^{2n+3}[(n+1)!]^4}{(2n+3)[(2n+2)!]^3} \phi^{(2n+2)}(\tilde{t}_n) .$$

Proof. It can be shown that the Peano kernel function $K_{2n+2}(t)$ of the $(n+1)$-point Gauss-Legendre formula does not change its sign on $[a,b]$. We are thus in a position to apply the second part of Theorem 2.1.2: the coefficient of $\phi^{(2n+2)}(\tilde{t}_n)$ in the above expression for $\mathbf{E}_n[\phi]$ is equal to the integral (over $[a,b]$) of this Peano kernel function. (Compare also Stroud [1974, pp. 178-181] for a detailed study of the behavior of the Peano kernel functions for Gauss-Legendre quadrature formulas.) □

It follows from this theorem that an $(n+1)$-point Gauss-Legendre formula has degree of precision $2n+1$. It can be proved that no other interpolatory quadrature formula has a degree of precision as large as $2n+1$ (see e.g. Stroud [1974, p. 137]).

In the preceding section we saw that a sequence of Newton-Cotes formulas does not necessarily converge to $\mathbf{I}[\phi]$. For Gauss-Legendre formulas the following theorem holds (see Davis & Rabinowitz [1984, p. 126]).

Theorem 2.3.2. *A sequence of Gauss-Legendre formulas converges to the true value of $\mathbf{I}[\phi]$ as $n \to \infty$ for all functions $\phi \in C[a,b]$.*

Proof. The above result is based on the fact that the coefficients of the Gauss-Legendre formulas are all positive. Moreover, since these formulas are interpolatory ones, we have

$$\sum_{j=0}^{n} |c_{n,j}| = \sum_{j=0}^{n} c_{n,j} = b-a \quad (n \in \mathbb{N}).$$

The assertion of Theorem 2.3.2 now follows from the Weierstrass approximation theorem and from Corollary 2.1.1. (Compare also Stroud [1974, pp. 142-144] or Davis & Rabinowitz [1984, pp. 126-130] for additional details.) □

Table 2.3.1. Gauss-Legendre points and Gauss-Legendre coefficients on (0,1).

n	$t_{n,0}$	$t_{n,1}$	$t_{n,2}$	$c_{n,0}$	$c_{n,1}$	$c_{n,2}$
0	$\frac{1}{2}$			1		
1	$\frac{3-\sqrt{3}}{6}$	$\frac{3+\sqrt{3}}{6}$		$\frac{1}{2}$	$\frac{1}{2}$	
2	$\frac{5-\sqrt{15}}{10}$	$\frac{1}{2}$	$\frac{5+\sqrt{15}}{10}$	$\frac{5}{18}$	$\frac{8}{18}$	$\frac{5}{18}$

2.4. Radau and Lobatto formulas

The Gauss-Legendre formulas are open formulas because the end points a and b are not involved in the set (2.3.1). However, in the construction of Volterra equations solvers, it is often desirable to include either one or both end points in the set of abscissas $\{t_{n,j}\}$. By choosing the remaining points such that the degree of precision is as large as possible we obtain the families of *Radau* and *Lobatto formulas,* respectively.

These formulas are used in the development of high-order Runge-Kutta methods for Volterra equations.

2.4.1. Radau formulas

Radau formulas are defined by the points

$$t_{n,j} = \frac{b-a}{2} z_{n,j} + \frac{b+a}{2}, \quad j=1, \ldots, n; \quad t_{n,0}=a, \tag{2.4.1}$$

where the $\{z_{n,j}\}$ are the n zeros of the polynomial

$$\frac{P_n(t)+P_{n+1}(t)}{t+1}, \tag{2.4.2}$$

with P_n denoting again the n-th degree Legendre polynomial. Tables of these zeros can be found in Abramowitz & Stegun [1970, p. 916]. The coefficients

$\{c_{n,j}\}$ can be expressed as

$$c_{n,j} = \frac{b-a}{2(1-z_{n,j})}\frac{1}{[P'_n(z_{n,j})]^2}, \quad j=1, \ldots ,n-1; \quad c_{n,0}=\frac{b-a}{(n+1)^2} \tag{2.4.3}$$

(see also Table 2.4.1a).

Theorem 2.4.1a. *Let* $\phi \in C^{2n+1}[a,b]$. *Then there exists a point* $\tilde{t}_n \in (a,b)$ *so that the approximation error of the* $(n+1)$*-point Radau formula corresponding to the abcissas (2.4.1) is given by*

$$\mathbf{E}_n[\phi] = \frac{(b-a)^{2n+2}(n+1)[n!]^4}{2[(2n+1)!]^3}\,\phi^{(2n+1)}(\tilde{t}_n).$$

Proof. See Hildebrand [1974, pp. 406-409]. □

Table 2.4.1a. Radau I points and Radau I coefficients on [0,1)

n	$t_{n,0}$	$t_{n,1}$	$t_{n,2}$	$c_{n,0}$	$c_{n,1}$	$c_{n,2}$
0	0			1		
1	0	$\frac{2}{3}$		$\frac{1}{4}$	$\frac{3}{4}$	
2	0	$\frac{6-\sqrt{6}}{10}$	$\frac{6+\sqrt{6}}{10}$	$\frac{4}{36}$	$\frac{16-\sqrt{6}}{36}$	$\frac{16+\sqrt{6}}{36}$

The Radau formulas defined above will be called *Radau I formulas.* Similar formulas can be derived if instead of a we want to include b in the set of abcissas. These formulas will be called *Radau II formulas;* their abcissas $\bar{t}_{n,j}$ and coefficients $\bar{c}_{n,j}$ are respectively given by $\bar{t}_{n,j} = b+a-t_{n,n-j}$ and $\bar{c}_{n,j} = c_{n,n-j}$, where $t_{n,j}$ and $c_{n,j}$ correspond to the Radau I formulas (see also Table 2.4.1b). The analogue of Theorem 2.4.1 is given below; a proof may be found in Brass [1977, pp. 149-150].

Theorem 2.4.1b. *Let* $\phi \in C^{2n+1}[a,b]$. *Then there exists a point* $\tilde{t}_n \in (a,b)$ *so that the approximation error of the* $(n+1)$*-point Radau II formula is given by*

$$\mathbf{E}_n[\phi] = -\frac{(b-a)^{2n+2}(n+1)[n!]^4}{2[(2n+1)!]^3}\,\phi^{(2n+1)}(\tilde{t}_n). \quad \square$$

Thus, $(n+1)$-point Radau formulas are of degree of precision $2n$.

Table 2.4.1b. Radau II points and Radau II coefficients on (0,1].

n	$t_{n,0}$	$t_{n,1}$	$t_{n,2}$	$c_{n,0}$	$c_{n,1}$	$c_{n,2}$
0	1			1		
1	$\frac{1}{3}$	1		$\frac{3}{4}$	$\frac{1}{4}$	
2	$\frac{4-\sqrt{6}}{10}$	$\frac{4+\sqrt{6}}{10}$	1	$\frac{16+\sqrt{6}}{36}$	$\frac{16-\sqrt{6}}{36}$	$\frac{4}{36}$

2.4.2. Lobatto formulas

Lobatto formulas are defined by the points

$$t_{n,0}=a,\quad t_{n,n}=b,\quad t_{n,j}=\frac{b-a}{2}\,z_{n,j}+\frac{b+a}{2},\quad j=1,2,\ldots,n-1, \tag{2.4.4}$$

where the $\{z_{n,j}\}$ are the $n-1$ zeros of the polynomial $P_n'(t)$, with $P_n(t)$ denoting again the Legendre polynomial. Tables of these zeros may be found in Abramowitz & Stegun [1970, p. 920]. The coefficients $\{c_{n,j}\}$ are given by

$$c_{n,0} = c_{n,n}=\frac{b-a}{n(n+1)},\quad c_{n,j}=\frac{b-a}{n(n+1)}\,\frac{1}{[P_n(z_{n,j})]^2},\quad j=1,2,\ldots,n-1 \tag{2.4.5}$$

(see also Table 2.4.2).

Theorem 2.4.2. *Let $\phi\in C^{2n}[a,b]$. Then there exists a point $\tilde{t}_n\in(a,b)$ so that the approximation error of the $(n+1)$-point Lobatto formula based on the points (2.4.4) is given by*

$$\mathbf{E}_n[\phi] = -\frac{(b-a)^{2n+1}n^3(n+1)[(n-1)!]^4}{(2n+1)[(2n)!]^3}\,\phi^{(2n)}(\tilde{t}_n).$$

Proof. Like the Peano kernels $K_{2n+2}(t)$ and $K_{2n+1}(t)$ for, respectively, the $(n+1)$-point formulas of Gauss-Legendre and of Radau, the Peano kernel function $K_{2n}(t)$ for the $(n+1)$-point Lobatto formula can be shown to be of constant sign over $[a,b]$. Thus, we may again invoke the second part of Theorem 2.1.2 to derive the above form for the approximation error. □

Again, the degree of precision decreases by one to become $2n-1$ for $(n+1)$-point Lobatto formulas.

Table 2.4.2. Lobatto points and Lobatto coefficients on [0,1]

n	$t_{n,0}$	$t_{n,1}$	$t_{n,2}$	$t_{n,3}$	$c_{n,0}$	$c_{n,1}$	$c_{n,2}$	$c_{n,3}$
1	0	1			$\frac{1}{2}$	$\frac{1}{2}$		
2	0	$\frac{1}{2}$	1		$\frac{1}{6}$	$\frac{4}{6}$	$\frac{1}{6}$	
3	0	$\frac{5-\sqrt{5}}{10}$	$\frac{5+\sqrt{5}}{10}$	1	$\frac{1}{12}$	$\frac{5}{12}$	$\frac{5}{12}$	$\frac{1}{12}$

2.5. Summary of error terms

For future reference we summarize, in Table 2.5.1, the error terms of the various interpolatory quadrature formulas discussed in the preceding section when represented in the form (2.1.13).

Table 2.5.1. $\int_0^1 \phi(a+\tau h)d\tau = \sum_{i=1}^{m} w_i\phi(a+d_ih)+e_mh^{q+1}\phi^{(q+1)}(\tilde{t}_m),\ \ 0<\tilde{t}_m<1$

Type of formula		$\phi\in C^{q+1}[a,a+h]$	e_m	Reference
Newton -Cotes	m even	$q=m-1$	$\frac{1}{m!}\int_0^1\prod_{i=1}^{m}(\tau-d_i)d\tau$	Thm. 2.2.1
	m odd	$q=m$	$\frac{1}{(m+1)!}\int_0^1\tau\prod_{i=1}^{m}(\tau-d_i)d\tau$	Thm 2.2.1
Gauss -Legendre		$q=2m-1$	$\frac{(m!)^4}{(2m+1)[(2m)!]^3}$	Thm. 2.3.1
Radau I	$m\geqslant 1$	$q=2m-2$	$\frac{m[(m-1)!]^4}{2[(2m-1)!]^3}$	Thm. 2.4.1a
Radau II	$m\geqslant 1$	$q=2m-2$	$-\frac{m[(m-1)!]^4}{2[(2m-1)!]^3}$	Thm. 2.4.1b
Lobatto	$m\geqslant 2$	$q=2m-3$	$-\frac{m(m-1)^3[(m-2)!]^4}{(2m-1)[(2m-2)!]^3}$	Thm. 2.4.2

2.6. (ρ,σ)-reducible quadrature formulas

Consider the initial value problem for a special ordinary differential equation (ODE):

$$y'(t) = \phi(t), \quad y(a)=0. \tag{2.6.1}$$

Integration of this problem yields

$$y(b) = \mathbf{I}[\phi], \tag{2.6.2}$$

where $\mathbf{I}[\phi]$ is the integral defined in (2.1.1). This relation suggests that we can generate quadrature formulas by considering integration techniques for the initial-value problem (2.6.1). In this section, we will consider *linear multistep* (LM) methods with *uniform* step sizes h. Introducing the *first* and *second characteristic polynomials* (cf. Henrici [1962, p. 218])

$$\rho(z) := \sum_{i=0}^{\kappa} a_i z^{\kappa-i}, \quad \sigma(z) := \sum_{i=0}^{\kappa} b_i z^{\kappa-i}, \tag{2.6.3}$$

the linear κ-step method is defined by

$$\rho(E) y_{n-\kappa} = h\sigma(E)\phi_{n-\kappa}, \quad n \geqslant \kappa, \tag{2.6.4}$$

where y_n denotes the numerical approximation to the solution $y(t)$ of (2.6.1) at $t=t_n$, $\phi_n := \phi(t_n)$, and E is the shift operator defined by

$$Ey_n := y_{n+1}. \tag{2.6.5}$$

Furthermore, $t_0=a$, $y_0=0$, and the additional starting values $y_1, \ldots, y_{\kappa-1}$ are assumed to be prescribed. Evidently, if $b=t_n$, then y_n represents a numerical approximation $\mathbf{I}_n[\phi]$ to $\mathbf{I}[\phi]=y(t_n)$.

In the analysis of quadrature rules based on the recurrence (2.6.4) it is convenient to write $\mathbf{I}_n[\phi]$ in the form (2.1.2). We will assume that the prescribed starting values are of the form

$$y_n = h\sum_{j=0}^{\kappa-1} w_{n,j}\phi_j, \quad n=0,1, \ldots, \kappa-1, \tag{2.6.6a}$$

where the $w_{n,j}$, $j=0,1, \ldots, \kappa-1$ and $n=0,1, \ldots, \kappa-1$, are given starting weights. By virtue of (2.6.4), y_n depends linearly on $\phi_0, \phi_1, \ldots, \phi_n$ for all n, so that we may write

$$y_n = \mathbf{I}_n[\phi] = h\sum_{j=0}^{n} w_{n,j}\phi_j, \quad n \geqslant \kappa. \tag{2.6.6b}$$

Substitution into (2.6.4) reveals that the weights $\{w_{n,j}\}$ and the coefficients $\{a_j, b_j\}$ of the characteristic polynomials are related according to

$$\sum_{i=0}^{\kappa} a_i w_{n-i,j} = \begin{cases} 0 & \text{if } j=0,1, \ldots, n-\kappa-1 \\ b_{n-j} & \text{if } j=n-\kappa, n-\kappa+1, \ldots, n \end{cases}, \quad n \geqslant \kappa, \tag{2.6.7}$$

where $w_{n,j} := 0$ for $j > \max\{n, \ \kappa - 1\}$.

If the starting weights and the characteristic polynomials are given, then the weights $w_{n,j}$ can be generated. We give the following definition (cf. Matthys [1976] and Wolkenfelt [1981a]):

Definition 2.6.1a. The sequence of quadrature formulas in

$$\mathbf{I}[\phi] = h \sum_{j=0}^{n} w_{n,j} \phi(t_j) + \mathbf{E}_n[\phi] \tag{2.6.8}$$

is called (ρ,σ)-reducible if the weights $w_{n,j}$ are generated by (2.6.7). □

This definition can be extended to so-called *block-reducible quadrature formulas* (cf. Wolkenfelt et al. [1981]).

Definition 2.6.1b. The sequence of quadrature formulas in (2.6.8) is called *block-reducible* if the weights $w_{n,j}$ can be partitioned into matrices $V_{n,j}$ (with $V_{n,j} = 0$ for $j > \max\{n, \kappa - 1\}$) such that

$$\sum_{i=0}^{\kappa} A_i V_{n-i,j} = \begin{cases} 0 & \text{if } j = 0, 1, \ldots, n - \kappa - 1 \\ & \\ B_{n-j} & \text{if } j = n - \kappa, n - \kappa + 1, \ldots, n \end{cases}, \quad n \geqslant \kappa, \tag{2.6.9}$$

where A_i and B_i are fixed matrices with $\Sigma_{i=0}^{\kappa} A_i \mathbf{e} = \mathbf{0}$, $\mathbf{e} := [1, \ldots, 1]^T$. □

A block-reducible quadrature formula is also related to an integration method for the initial-value problem (2.6.1), for example, the cyclic linear multistep methods of Donelson & Hansen [1971].

Example 2.6.1. Let us combine the repeated Simpson rule with the 3/8 rule (see Table 2.2.1 for $n = 2$ and $n = 3$, respectively). Then the matrix of weights $(w_{n,j})$ has the form

$$\begin{bmatrix} V_{0,0} & & & \\ V_{1,0} & V_{1,1} & & \\ V_{2,0} & V_{2,1} & V_{2,2} & \\ \vdots & & & \\ V_{n,0} & & \cdots & V_{n,n} \end{bmatrix}$$

$$:=\frac{1}{24}\left[\begin{array}{cccccccccc}
0 & 0 &&&&&&&& \\
x & x &&&&&&&& \\
&&&&&&&&& \\
8 & 32 & 8 & 0 &&&&&& \\
9 & 27 & 27 & 9 &&&&&& \\
&&&&&&&&& \\
8 & 32 & 16 & 32 & 8 & 0 &&&& \\
8 & 32 & 17 & 27 & 27 & 9 &&&& \\
&&&&&&&&& \\
8 & 32 & 16 & 32 & 16 & 32 & 8 & 0 && \\
8 & 32 & 16 & 32 & 17 & 27 & 27 & 9 && \\
&&&&&&&&& \\
8 & 32 & 16 & 32 & 16 & 32 & 16 & 32 & 8 & 0 \\
8 & 32 & 16 & 32 & 16 & 32 & 17 & 27 & 27 & 9 \\
&&&&&&&&& \\
&&&& \cdots &&&&&
\end{array}\right]$$

(compare also Section 2.7 for a discussion of repeated quadrature rules.) It is easily verified that the matrices $V_{n,j}$ satisfy the relations

$$V_{n,j}-V_{n-1,j}+0.V_{n-2,j} = 0 \quad \text{for} \quad j\leqslant n-3, \quad n\geqslant 2,$$

$$V_{n,n}-V_{n-1,n}+0.V_{n-2,n} = \frac{1}{24}\begin{pmatrix} 8 & 0 \\ 27 & 9 \end{pmatrix}, \quad n\geqslant 2,$$

$$V_{n,n-1}-V_{n-1,n-1}+0.V_{n-2,n-1} = \frac{1}{24}\begin{pmatrix} 8 & 32 \\ -10 & 18 \end{pmatrix}, \quad n\geqslant 2,$$

$$V_{n,n-2}-V_{n-1,n-2}+0.V_{n-2,n-2} = \frac{1}{24}\begin{pmatrix} 0 & 0 \\ -1 & 5 \end{pmatrix}, \quad n\geqslant 2.$$

These relations can be written in the form (2.6.9) with

$$\kappa=2, \quad A_0=I, \quad A_1=-I, \quad A_2=0,$$

$$B_0=\frac{1}{24}\begin{pmatrix} 8 & 0 \\ 27 & 9 \end{pmatrix}, \quad B_1=\frac{1}{24}\begin{pmatrix} 8 & 32 \\ -10 & 18 \end{pmatrix}, \quad B_2=\frac{1}{24}\begin{pmatrix} 0 & 0 \\ -1 & 5 \end{pmatrix}.$$

Since the matrices A_i are diagonal, the sequence of quadrature formulas generated by the matrix $(w_{n,j})$ is called *diagonally block-reducible.* □

In the following we will concentrate on (ρ,σ)-reducible quadrature formulas, rather than on block-reducible formulas. In order to analyze such formulas we need some theory of LM methods for ODEs which will be given in the next subsection.

2.6.1. Linear multistep methods

In this section we summarize a number of relevant definitions and theorems from the theory of LM methods for the general initial-value problem

$$y'(t) = f(t,y(t)), \quad y(t_0)=y_0, \quad t\in[t_0,T]$$

(cf., e.g., Henrici [1962], Lambert [1973], or Stetter [1973]).

Definition 2.6.2. (i) A polynomial is said to satisfy the *root condition* if it has all its zeros on the closed unit disk, with those on the unit circle being simple. It is said to satisfy the *strong root condition* if it satisfies the root condition and 1 is its only zero on the unit circle.

(ii) A polynomial is said to be a *simple von Neumann polynomial* if it satisfies the root condition. It is said to be a *Schur polynomial* if it has all zeros in the interior of the unit circle.

(iii) The LM method $\{\rho,\sigma\}$ defined by

$$\rho(E)y_n = h\sigma(E)f(t_n,y_n), \quad n\geqslant 0,$$

is called *zero-stable* if $\rho(z)$ satisfies the root condition.

(iv) The *stability region* of an LM method is the set of points in the (complex) $h\xi$-plane for which the characteristic polynomial $\rho(z)-h\xi\sigma(z)$ is a Schur polynomial.

(v) The LM method is called *of order p* if for all sufficiently differentiable functions y,

$$\mathbf{L}_n[y] := [\rho(E)-h\sigma(E)\frac{d}{dt}]y(t_n)=C_{p+1}h^{p+1}y^{(p+1)}(t_n)+O(h^{p+2}),$$

as $h\downarrow 0$, with $t_n\in[t_0,T]$ and $C_{p+1}\neq 0$. The constant $C_{p+1}/\sigma(1)$ is called the *normalized error constant.*

(vi) If $y(t)$ is the exact solution of the initial-value problem, then $\mathbf{L}_n[y]$ is called the *local truncation error*. If the local truncation error is $O(h^{p+1})$ as $h\downarrow 0$, then the LM method is said to be *consistent of order p with the initial-value problem.*

(vii) The LM method is said to be *convergent* if, for all initial-value problems possessing a unique solution, we have that $y_n\rightarrow y(t_n)$ as $h\downarrow 0$, with t_n fixed, for all $t_n\in[t_0,T]$ and for all convergent starting values y_j, $j=0,1, \ldots ,\kappa-1$. It is called *convergent of order p* if $y_n-y(t_n)=O(h^p)$ as $h\downarrow 0$ with t_n fixed, for all $t_n\in[t_0,T]$. □

Theorem 2.6.1. *The LM method $\{\rho,\sigma\}$ is convergent for all sufficiently smooth right-hand side functions $f(t,y)$ if, and only if, it is zero-stable and consistent of order $p \geqslant 1$.* □

Theorem 2.6.2. *No zero-stable LM method $\{\rho,\sigma\}$ can have an order p higher than $2\lfloor(\kappa+2)/2\rfloor$ (the integer part of a real number will be denoted by $\lfloor\cdot\rfloor$).* □

Theorem 2.6.3. *(a) The LM method $\{\rho,\sigma\}$ is convergent of order p for all sufficiently smooth $f(t,y)$ if, and only if it is consistent of order p and if the starting values satisfy $y_j - y(t_j) = O(h^s)$ for $j=0,1, \ldots ,\kappa-1$ and with $s \geqslant p$, as $h\downarrow 0$.*
(b) Let

$$\rho(e^z) - z\sigma(e^z) = \sum_{q=0}^{\infty} C_q z^q .$$

Then

$$\mathbf{L}_n[y] = \sum_{q=0}^{p+1} C_q \left(h\frac{d}{dt}\right)^q y(t_n) + O(h^{p+2}), \quad y \in C^{p+1}[t_0,T].$$

(c) The coefficients C_q are given by

$$C_0 = \rho(1), \quad C_q = \frac{1}{q!}\sum_{i=0}^{\kappa}(\kappa-i)^{q-1}[(\kappa-i)a_i - qb_i].$$

(d) The coefficients C_q vanish for $q=0,1, \ldots ,p$ if for some value of x the coefficients

$$D_0 = \rho(1), \quad D_q = \frac{1}{q!}\sum_{i=0}^{\kappa}(x-i)^{q-1}[(x-i)a_i - qb_i]$$

vanish for $q=0,1, \ldots ,p$. □

2.6.2. Construction of quadrature weights

In the analysis of the (ρ,σ)-reducible quadrature formulas (2.6.8) we closely follow the analysis of Wolkenfelt [1979], [1981a], [1982].

Let us define the matrix of weights $W_n = (w_{i,j})$ with $i,j = 0,1, \ldots ,n \geqslant \kappa$, and let us partition W_n according to

$$W_n = \begin{bmatrix} w_{0,0} & \cdots & w_{0,\kappa-1} & & & \\ \vdots & & \vdots & & & \\ w_{\kappa-1,0} & \cdots & w_{\kappa-1,\kappa-1} & & & \\ & & & & & \\ w_{\kappa,0} & \cdots & w_{\kappa,\kappa-1} & w_{\kappa,\kappa} & & \\ \vdots & & \vdots & \vdots & & \\ w_{n,0} & \cdots & w_{n,\kappa-1} & w_{n,\kappa} & \cdots & w_{n,n} \end{bmatrix} = \begin{bmatrix} S_\kappa & \\ & \\ \Sigma_{\kappa,n} & \Omega_{\kappa,n} \end{bmatrix},$$

with the partitioning being indicated by dotted lines.

The matrix S_κ

The matrix S_κ contains the prescribed starting weights from (2.6.6a). For future reference we give the first starting matrices based on interpolatory quadrature formulas and the corresponding approximation errors $\mathbf{E}_n[\phi]$:

$$S_2 = \frac{1}{2}\begin{bmatrix} 0 & 0 \\ 1 & 1 \end{bmatrix}, \quad \mathbf{E}_1[\phi] = -\frac{1}{12}h^3\tilde{\phi}^{(2)}\begin{bmatrix} 0 \\ 1 \end{bmatrix}; \tag{2.6.10a}$$

$$S_3 = \frac{1}{12}\begin{bmatrix} 0 & 0 & 0 \\ 5 & 8 & -1 \\ 4 & 16 & 4 \end{bmatrix}, \quad \mathbf{E}_2[\phi] = h^4\begin{bmatrix} 0 \\ \tilde{\phi}^{(3)}/24 \\ -h\tilde{\phi}^{(4)}/90 \end{bmatrix}; \tag{2.6.10b}$$

$$S_4 = \frac{1}{24}\begin{bmatrix} 0 & 0 & 0 & 0 \\ 9 & 19 & -5 & 1 \\ 8 & 32 & 8 & 0 \\ 9 & 27 & 27 & 9 \end{bmatrix}, \quad \mathbf{E}_3[\phi] = -\frac{1}{720}h^5\tilde{\phi}^{(4)}\begin{bmatrix} 0 \\ 19 \\ 8 \\ 27 \end{bmatrix}; \tag{2.6.10c}$$

$$S_5 = \frac{1}{720}\begin{bmatrix} 0 & 0 & 0 & 0 & 0 \\ 251 & 646 & -264 & 106 & -19 \\ 232 & 992 & 192 & 32 & -8 \\ 243 & 918 & 648 & 378 & -27 \\ 224 & 1024 & 384 & 1024 & 224 \end{bmatrix},$$

$$\mathbf{E}_4[\phi] = h^6\begin{bmatrix} 0 \\ 3\tilde{\phi}^{(5)}/160 \\ \tilde{\phi}^{(5)}/90 \\ 3\tilde{\phi}^{(5)}/160 \\ -8h\tilde{\phi}^{(6)}/945 \end{bmatrix}. \tag{2.6.10d}$$

Here, $\tilde{\phi}^{(i)}$ denotes the i-th derivative of $\phi(t)$ at a suitable point $\tilde{t}\in[t_0,t_{\kappa-1}]$. We note that each row in S_κ represents an *interpolatory quadrature formula;* i.e., the j-th row $\mathbf{w}_j$ of weights in S_κ defines a quadrature formula for

$$\int_{t_0}^{t_j}\phi(t)dt.$$

For instance, the last row vector $\mathbf{w}_{\kappa-1}$ defines the (closed) Newton-Cotes formula using κ points (cf. Table 2.2.1).

The matrices $\Sigma_{\kappa,n}$

The weights in the matrix $\Sigma_{\kappa,n}$ can be calculated explicitly from the starting weights by means of (2.6.7). If $\mathbf{w}_0, \ldots, \mathbf{w}_{\kappa-1}$ and $\mathbf{w}_\kappa, \ldots, \mathbf{w}_n$ denote the row vectors of S_κ and $\Sigma_{\kappa,n}$, respectively, then the weights of $\Sigma_{\kappa,n}$ can be computed by the recurrence relation

$$\mathbf{w}_j = -\frac{1}{a_0}\left[\sum_{i=1}^{\kappa} a_i \mathbf{w}_{j-i} - \mathbf{b}_j\right], \quad \mathbf{b}_j := (0, \ldots, 0, b_\kappa, \ldots, b_{j-\kappa+1})^T, \; j \geqslant \kappa, \tag{2.6.11}$$

where $\mathbf{b}_j := \mathbf{0}$ as $j \geqslant 2\kappa$. In the particular case where the LM method is the well-known Adams-Moulton (AM) method defined in Table 2.6.1, we have

$$\mathbf{w}_j = -\frac{a_1}{a_0}\mathbf{w}_{j-1} + \frac{1}{a_0}\mathbf{b}_j = \mathbf{w}_{j-1} + \frac{1}{a_0}\mathbf{b}_j, \quad j \geqslant \kappa. \tag{2.6.11'}$$

Thus, the last row of the starting matrix S_κ generates the weights in the matrix $\Sigma_{\kappa,n}$. For instance, the matrices $\Sigma_{\kappa,n}$ corresponding to (2.6.10) are given by

$$\Sigma_{2,n} = \frac{1}{12}\begin{bmatrix} 5 & 14 \\ 5 & 13 \\ \vdots & \vdots \\ 5 & 13 \end{bmatrix}, \tag{2.6.12a}$$

$$\Sigma_{3,n} = \frac{1}{24}\begin{bmatrix} 9 & 27 & 27 \\ 9 & 28 & 22 \\ 9 & 28 & 23 \\ & \cdots & \\ 9 & 28 & 23 \end{bmatrix}, \tag{2.6.12b}$$

$$\Sigma_{4,n} = \frac{1}{720}\begin{bmatrix} 251 & 916 & 546 & 916 \\ 251 & 897 & 652 & 652 \\ 251 & 897 & 633 & 758 \\ 251 & 897 & 633 & 739 \\ 251 & 897 & 633 & 739 \\ & & \cdots & \\ 251 & 897 & 633 & 739 \end{bmatrix}, \tag{2.6.12c}$$

$$\Sigma_{5,n} = \frac{1}{1440}\begin{bmatrix} 475 & 1875 & 1250 & 1250 & 1875 \\ 475 & 1902 & 1077 & 1732 & 1077 \\ 475 & 1902 & 1104 & 1559 & 1559 \\ 475 & 1902 & 1104 & 1586 & 1386 \\ 475 & 1902 & 1104 & 1586 & 1413 \\ 475 & 1902 & 1104 & 1586 & 1413 \\ & & \cdots & & \\ 475 & 1902 & 1104 & 1586 & 1413 \end{bmatrix}. \tag{2.6.12d}$$

Table 2.6.1. Coefficients of Adams-Moulton formulas ($a_i = 0$, $i > 1$)

κ	a_0	a_1	b_0	b_1	b_2	b_3	b_4	b_5	p	C^*_{p+1}
1	2	-2	1	1					2	-1/12
2	12	-12	5	8	-1				3	-1/24
3	24	-24	9	19	-5	1			4	-19/720
4	720	-720	251	646	-264	106	-19		5	-3/160
5	1440	-1440	475	1427	-798	482	-173	27	6	-863/60480

The matrices $\Omega_{\kappa,n}$

The weights in the matrix $\Omega_{\kappa,n}$ satisfy the equations (cf. (2.6.7))

$$\begin{cases} a_0 w_{j,j} & = b_0, \quad j \geqslant \kappa \\ a_0 w_{j,j-1} + a_1 w_{j-1,j-1} & = b_1, \quad j \geqslant \kappa+1 \\ \quad \cdots & \quad \cdots \\ a_0 w_{j,j-\kappa} + a_1 w_{j-1,j-\kappa} + \ldots + a_\kappa w_{j-\kappa,j-\kappa} & = b_\kappa, \quad j \geqslant 2\kappa \end{cases} \tag{2.6.13a}$$

and, for $j \geqslant 2\kappa+1$,

$$a_0 w_{j,j-\kappa-1} + a_1 w_{j-1,j-\kappa-1} + \ldots + a_\kappa w_{j-\kappa,j-\kappa-1} = 0. \tag{2.6.13b}$$

Thus,

$$w_{j,j} = \frac{b_0}{a_0} \text{ for } j \geqslant \kappa; \quad w_{j,j-1} = \frac{b_1 a_0 - a_1 b_0}{a_0^2} \text{ for } j \geqslant \kappa+1; \text{ etc.}$$

Clearly, the weights in the matrix $\Omega_{\kappa,n}$, satisfy the relation

$$w_{i,j} = w_{i+1,j+1}. \tag{2.6.14}$$

Writing

$$w_{i,j} = \omega_{i-j}, \quad i \geqslant j \geqslant \kappa,$$

the matrix $\Omega_{\kappa,n}$ assumes the *semi-circulant* form

$$\Omega_{\kappa,n} = \begin{bmatrix} \omega_0 & & & & \\ \omega_1 & \omega_0 & & & \\ \omega_2 & \omega_1 & \omega_0 & & \\ & \cdots & & & \\ \omega_{n-\kappa} & \cdots & \omega_2 & \omega_1 & \omega_0 \end{bmatrix}. \tag{2.6.15}$$

It follows from (2.6.13) and (2.6.14) that the weights in $\Omega_{\kappa,n}$ can be found by solving the system

$$\begin{aligned} &a_0\omega_0 &&= b_0 \\ &a_0\omega_1 + a_1\omega_0 &&= b_1 \\ &\cdots &&\cdots \\ &a_0\omega_k + a_1\omega_{k-1} + \ldots + a_k\omega_0 &&= b_\kappa \\ &a_0\omega_j + a_1\omega_{j-1} + \ldots + a_\kappa\omega_{j-\kappa} &&= 0, \quad j = \kappa+1, \kappa+2, \ldots, n-\kappa \end{aligned} \tag{2.6.16}$$

(notice that the ω_j do not depend on the starting weights).

For future reference, we have computed the matrices $\Omega_{\kappa,n}$ for some of the well-known linear multistep methods. Let us again consider the *Adams-Moulton formulas* defined in Table 2.6.1. These linear multistep methods give rise to the system

$$\omega_0 = \frac{b_0}{a_0}, \quad \omega_j = \omega_{j-1} + \frac{b_j}{a_0} \text{ for } j \leqslant \kappa; \quad \omega_j = \omega_{j-1} \text{ for } j > \kappa. \tag{2.6.16'}$$

From this the matrices $\Omega_{\kappa,n}$ can be derived explicitly:

$$\Omega_{2,n} = \frac{1}{12} \begin{bmatrix} 5 & & & & \\ 13 & 5 & & & \\ 12 & 13 & 5 & & \\ & & \cdots & & \\ 12 & \ldots & 12 & 13 & 5 \end{bmatrix}, \tag{2.6.17a}$$

$$\Omega_{3,n} = \frac{1}{24} \begin{bmatrix} 9 & & & & & \\ 28 & 9 & & & & \\ 23 & 28 & 9 & & & \\ 24 & 23 & 28 & 9 & & \\ & & \cdots & & & \\ 24 & \ldots & 24 & 23 & 28 & 9 \end{bmatrix}, \tag{2.6.17b}$$

$$\Omega_{4,n} = \frac{1}{720}\begin{bmatrix} 251 \\ 897 & 251 \\ 633 & 897 & 251 \\ 739 & 633 & 897 & 251 \\ 720 & 739 & 633 & 897 & 251 \\ & & & \cdots \\ 720 & \cdots & 720 & 739 & 633 & 897 & 251 \end{bmatrix}, \tag{2.6.17c}$$

$$\Omega_{5,n} = \frac{1}{1440}\begin{bmatrix} 475 \\ 1902 & 475 \\ 1104 & 1902 & 475 \\ 1586 & 1104 & 1902 & 475 \\ 1413 & 1586 & 1104 & 1902 & 475 \\ 1440 & 1413 & 1586 & 1104 & 1902 & 475 \\ & & & \cdots \\ 1440 & \ldots & 1440 & 1413 & 1586 & 1104 & 1902 & 475 \end{bmatrix}. \tag{2.6.17d}$$

The matrix of weights W_n composed of the starting matrix S_κ defined in (2.6.10), the $\Sigma_{\kappa,n}$ matrix defined in (2.6.12), and the $\Omega_{\kappa,n}$ matrix defined in (2.6.17) generate the so-called *Gregory quadrature rules*. Because of the central role these quadrature formulas play in the numerical solution of Volterra equations, we discuss them in more detail in the next subsection.

2.6.3. The Gregory quadrature rules

Our starting point is the *Euler-Maclaurin summation formula* (see, e.g., Hildebrand [1974], Steinberg [1972], Schmeisser & Schirmeier [1976]). Let $\phi \in C^{2m+2}[a,b]$, $m \geqslant 1$, and let $t_{n,j} = a + jh$, $j = 0, 1, \ldots, n$, with $t_{n,n} = b$. Then

$$\mathbf{I}[\phi] = h[\tfrac{1}{2}\phi(t_{n,0}) + \phi(t_{n,1}) + \ldots + \phi(t_{n,n-1}) + \tfrac{1}{2}\phi(t_{n,n})]$$
$$- \sum_{j=1}^{m} \frac{B_{2j}h^{2j}}{(2j)!}[\phi^{(2j-1)}(b) - \phi^{(2j-1)}(a)] - \mathbf{R}_m[\phi], \tag{2.6.18}$$

where

$$\mathbf{R}_m[\phi] := (b-a)\frac{B_{2m+2}h^{2m+2}}{(2m+2)!}\,\phi^{(2m+2)}(\tilde{t}_n), \quad \tilde{t}_n \in [a,b]\,. \tag{2.6.19}$$

Here, the numbers B_{2j} are the *Bernoulli numbers* defined by

$$t(e^t - 1)^{-1} = \sum_{j=0}^{\infty} \frac{B_j}{j!}\, t^j. \tag{2.6.20}$$

It follows that $B_{2i+1}=0$ $(i \geqslant 1)$; moreover, we have

j	0	1	2	4	6	8
B_j	1	$-\frac{1}{2}$	$\frac{1}{6}$	$-\frac{1}{30}$	$\frac{1}{42}$	$-\frac{1}{30}$

The Gregory quadrature rules result from the Euler-Maclaurin summation formula by suitable discretization of the "end corrections": we have, for $i=1,2, \ldots$,

$$\phi_0^{(i)} := \phi^{(i)}(t_{n,0}) = h^{-i}[\Delta^i - \frac{i}{2}\Delta^{i+1} + \frac{i(3i+5)}{24}\Delta^{i+2} - \ldots]\phi_0$$

$$= h^{-i}[\frac{S_i^{(i)}}{1} + \frac{S_{i+1}^{(i)}}{i+1}\Delta + \frac{S_{i+2}^{(i)}}{(i+1)(i+2)}\Delta^2 + \ldots]\Delta^i\phi_0 \tag{2.6.21a}$$

$$\phi_n^{(i)} := \phi^{(i)}(t_{n,n}) = h^{-i}[\nabla^i + \frac{i}{2}\nabla^{i+1} + \frac{i(3i+5)}{24}\nabla^{i+2} + \ldots]\phi_n$$

$$= h^{-i}[\frac{S_i^{(i)}}{1} - \frac{S_{i+1}^{(i)}}{i+1}\nabla + \frac{S_{i+2}^{(i)}}{(i+1)(i+2)}\nabla^2 - \ldots]\nabla^i\phi_n, \tag{2.6.21b}$$

where Δ^j and ∇^j denote the j-th forward and the j-th backward difference operator, respectively, and where we have set $\phi_0 := \phi(a)$, $\phi_n := \phi(b)$. Here, the numbers $S_j^{(i)}$ are the *Stirling numbers of the first kind* and are defined by the generating function

$$\frac{[\log(1+t)]^i}{i!} = \sum_{j=i}^{\infty} S_j^{(i)} \frac{t^j}{j!}. \tag{2.6.22}$$

The table below lists some of these numbers $S_j^{(i)}$.

j	$i=1$	$i=2$	$i=3$
1	1		
2	-1	1	
3	2	-3	1
4	-6	11	-6

Using (2.6.21) we obtain the following discretization of the "end corrections":

$$\phi^{(2j-1)}(b) - \phi^{(2j-1)}(a) = h^{-2j+1}\left[\frac{S_{2j-1}^{(2j-1)}}{1}(\nabla^{(2j-1)}\phi_n - \Delta^{2j-1}\phi_0) - \frac{S_{2j}^{(2j-1)}}{2j}\right.$$

$$(\nabla^{2j}\phi_n+\Delta^{2j}\phi_0)+\;\dots\;+(-1)^{i-2j+1}\frac{S_i^{(2j-1)}}{\prod_{l=2j}^{i} l}(\nabla^i\phi_n+(-\Delta)^i\phi_0)+\;\dots\;\Bigg].$$

Substitution into (2.6.18) yields

$$\mathbf{I}[\phi] = h[\tfrac{1}{2}\phi(t_{n,0})+\phi(t_{n,1})+\;\dots\;+\phi(t_{n,n-1})+\tfrac{1}{2}\phi(t_{n,n})]$$

$$-h\sum_{i=1}^{q} c_i(\nabla^i\phi_n+(-\Delta)^i\phi_0)+\mathbf{E}_n^q[\phi],\quad q\geq 1, \tag{2.6.23}$$

where the approximation error is given by

$$\mathbf{E}_n^q[\phi] = -h\sum_{i=q+1}^{\infty} c_i(\nabla^i\phi_n+(-\Delta)^i\phi_0)-\mathbf{R}_m[\phi], \tag{2.6.24}$$

and where the coefficients c_i are defined by

$$c_i := \sum_{j=1}^{p}(-1)^{i-2j+1}\frac{B_{2j}S_i^{(2j-1)}}{(2j)!\prod_{l=2j}^{i} l},\quad p:=\lfloor\frac{i+1}{2}\rfloor, \tag{2.6.25}$$

with the convention $\prod_{l=r}^{s} l:=1$ if $s<r$. The first three coefficients are given by

$$c_1 = \frac{1}{2}B_2S_1^{(1)}=\frac{1}{12},\quad c_2=-\frac{1}{2}B_2S_2^{(1)}=\frac{1}{24},$$

$$c_3 = \frac{1}{12}B_2S_3^{(1)}+\frac{1}{24}B_4S_3^{(3)}=\frac{19}{720}.$$

Definition 2.6.3. The quadrature formula defined by (2.6.23),

$$I_n^q[\phi] := h[\tfrac{1}{2}\phi(t_{n,0})+\phi(t_{n,1})+\;\dots\;+\phi(t_{n,n-1})+\tfrac{1}{2}\phi(t_{n,n})]$$

$$-h\sum_{i=1}^{q} c_i(\nabla^i\phi_n+(-\Delta)^i\phi_0),\quad q\geq 1,$$

is called the q-th *Gregory quadrature rule.* □

Theorem 2.6.4. *(a) Let q be even, and assume that $\phi\in C^{q+2}[a,b]$. Then there exists a point $\tilde{t}_n\in[a,b]$ so that the approximation error of the q-th Gregory quadrature formula can be written as*

$$\mathbf{E}_n^q[\phi] = -h^{q+3}[(n-q-1)c_{q+2}+2c_{q+3}]\phi^{(q+2)}(\tilde{t}_n). \tag{2.6.26a}$$

(b) If q is odd, and $\phi\in C^{q+3}[a,b]$, then there exists a $\tilde{t}_n\in[a,b]$ so that the approximation error of the q-th Gregory quadrature formula has the form

$$\mathbf{E}_n^q[\phi] = -c_{q+2}h^{q+2}[\phi^{(q+1)}(a)+\phi^{(q+1)}(b)]+d_qh^{q+3}\phi^{(q+3)}(\tilde{t}_n). \tag{2.6.26b}$$

Here, $d_q\neq 0$ is a known constant.

Proof. Compare Brass [1977, pp. 212-214]. The difference between the two error expressions (2.6.26a) and (2.6.26b) is a consequence of the fact that, while the Peano kernel function $K_{q+2}(t)$ for the q-th Gregory rule with *even* q does not change its sign over $[a,b]$, the Peano kernel function corresponding to *odd* values of q possesses sign changes in $[a,b]$. Hence, in this latter case, the second part of Theorem 2.1.2 is not applicable. □

It follows from this theorem that the sequence of quadrature formulas generated by the q-th Gregory rule has order $q+2$.

Example 2.6.2. The *first Gregory rule* is given by

$$\mathbf{I}_n[\phi] = \frac{1}{12}h[5\phi_0+13\phi_1+12\phi_2+ \ldots +12\phi_{n-2}+13\phi_{n-1}+5\phi_n], \qquad (2.6.27a)$$

the *second* one by

$$\mathbf{I}_n[\phi] = \frac{1}{24}h[9\phi_0+28\phi_1+23\phi_2+24\phi_3+ \ldots +24\phi_{n-3}$$

$$+23\phi_{n-2}+28\phi_{n-1}+9\phi_n]. \qquad (2.6.27b)$$

A comparison with {(2.6.12a), (2.6.17a)} and {(2.6.12b), (2.6.17b)} reveals that they are identical. □

2.6.4. Linear difference equations

We shall frequently need results on the solutions of linear difference equations with constant coefficients, particularly when analyzing the behavior of the quadrature weights constructed in the preceding section. In this section the main properties of linear difference equations are collected.

Homogeneous equations

Consider the *homogeneous equation* with constant coefficients γ_j,

$$\gamma_0 y_n+\gamma_1 y_{n-1}+ \ldots +\gamma_\kappa y_{n-\kappa}=0, \quad n\geqslant\kappa \;\; (\gamma_0\neq 0), \qquad (2.6.28)$$

and define the *characteristic polynomial*

$$G(z) := \sum_{j=0}^{\kappa}\gamma_j z^{\kappa-j}. \qquad (2.6.29)$$

Theorem 2.6.5. *If $G(z)$ has q distinct zeros $z_1, z_2, \ldots, z_q$ with multiplicities $m_1, m_2, \ldots, m_q$ $(m_1+ \ldots +m_q=\kappa)$, then the general solution of (2.6.28) can be written in the form*

$$y_n = \sum_{\nu=1}^{q}\sum_{j=0}^{m_\nu-1} C_{\nu,j}\binom{n}{j}z_\nu^{n-j}, \quad n\geqslant 0, \qquad (2.6.30)$$

where the $C_{\nu,j}$ are constants independent of n, and where we adopt the convention

$$\begin{bmatrix} n \\ j \end{bmatrix} z_\nu^{n-j} := \begin{cases} 0 & \text{for} \quad n=0,1, \ldots ,j-1 \\ 1 & \text{for} \quad n=j \end{cases} .$$

Proof. See e.g. Henrici [1962, p. 210]. □

The κ constants $C_{\nu,j}$ are determined by the starting values $y_0, y_1, \ldots y_{\kappa-1}$. Substitution into (2.6.30) yields a linear, nonsingular system of κ equations for the κ unknowns $C_{\nu,j}$. Hence these constants depend linearly on the starting values.

It follows from Theorem 2.6.5 that the κ different sequences

$$\{n^\mu z_\nu^n\}_{n=0}^\infty, \quad \mu=0,1, \ldots ,m_\nu-1, \quad \nu=1,2, \ldots ,q, \tag{2.6.31}$$

form a *fundamental system* for (2.6.28), since any solution of (2.6.28) can be expressed as a linear combination of the solutions (2.6.31).

In the following theorems explicit expressions for the general solution are given in terms of the starting values for certain special choices of the polynomial $G(z)$. In the proof of these theorems the following lemma plays a crucial role.

Lemma 2.6.1. *The coefficients* $C_{\nu,m_\nu-1}$, $\nu=1,2, \ldots ,q$, *in* (2.6.30) *are given by*

$$C_{\nu,m_\nu-1} = m_\nu! \frac{\Delta_\nu}{G^{(m_\nu)}(z_\nu)},$$

where $G^{(i)}(z)$ *denotes the derivative of order* i *of* $G(z)$, *and where*

$$\Delta_\nu := \sum_{j=0}^{\kappa-1} \gamma_j^{(\nu)} y_{\kappa-1-j}, \quad \gamma_j^{(\nu)} := \sum_{i=0}^{j} \gamma_i z_\nu^{j-i}. \tag{2.6.32}$$

Proof. Along the lines of Henrici [1962, p. 238] (see Wolkenfelt [1979]). □

Theorem 2.6.6. *Let* $G(z)$ *have* r *nonvanishing simple zeros* $z_1, \ldots ,z_r$ *and a zero* $z=0$ *of multiplicity* $m_0=\kappa-r\geqslant 0$, *and let* Δ_ν *be defined by (2.6.32) for* $\nu=1,2, \ldots ,r$. *Then the general solution of (2.6.28) is given by*

$$y_n = \sum_{\nu=1}^{r} \Delta_\nu \frac{z_\nu^n}{G'(z_\nu)}, \quad n\geqslant m_0. \tag{2.6.33}$$

Proof. The result is immediate from Theorem 2.6.5 and Lemma 2.6.1. □

If the number r of nonvanishing simple zeros of G is less than $\kappa-m_0$, then the following generalization of Theorem 2.6.6 can be proved (cf. Wolkenfelt [1979]).

Theorem 2.6.7. *Let $G(z)$ have r nonvanishing simple zeros $z_1, \ldots, z_r$ and $q-r$ zeros $z_{r+1}, \ldots, z_q$ with multiplicities $m_{r+1}, \ldots, m_q$ $(r+m_{r+1}+\ldots+m_q=\kappa)$. Furthermore, let the numbers γ_j^* be defined by*

$$\sum_{j=0}^{\kappa-r} \gamma_j^* z^{\kappa-r-j} = \frac{G(z)}{\prod_{\nu=1}^{r}(z-z_\nu)}. \tag{2.6.34}$$

Then the general solution of (2.6.28) is given by

$$y_n = \sum_{\nu=1}^{r} \Delta_\nu \frac{z_\nu^n}{G'(z_\nu)} + v_n, \quad n \geqslant \kappa, \tag{2.6.35}$$

where Δ_ν is defined by (2.6.32) and where v_n is the solution of

$$\sum_{j=0}^{\kappa-r} \gamma_j^* v_{n-j} = 0, \quad n \geqslant \kappa - r, \tag{2.6.36}$$

corresponding to the starting values $v_n = y_n - \sum_{\nu=1}^{r} \Delta_\nu \frac{z_\nu^n}{G'(z_\nu)}$, $n=0,1,\ldots,\kappa-r-1$. □

Inhomogeneous equations

We next consider the inhomogeneous equation

$$G(E)y_n = g_{n+\kappa}, \quad n \geqslant 0, \tag{2.6.37}$$

where $\{g_n\}$ is a given sequence independent of y_n. Evidently, the general solution of (2.6.37) is given by $\hat{y}_n + p_n$, where $\hat{y}_n$ represents the general solution of the homogeneous equation (2.6.28) and where p_n is a particular solution of (2.6.37).

It is of course possible to express y_n explicitly in terms of the starting values y_j and the inhomogeneities g_n, namely,

$$y_n = \sum_{j=0}^{\kappa-1} a_{j,n} y_j + \sum_{j=\kappa}^{n} a_{j,n} g_j, \quad n \geqslant 0. \tag{2.6.38}$$

Here, the coefficients $a_{j,n}$ depend on the coefficients γ_i of the polynomial $G(z)$. One way of defining the coefficients of the starting values is by the relation

$$\frac{\gamma_0 + \gamma_1 z + \ldots + \gamma_{\kappa-j-1} z^{\kappa-j-1}}{\gamma_0 + \gamma_1 z + \ldots + \gamma_\kappa z^\kappa} = 1 + \sum_{n=\kappa}^{\infty} a_{j,n} z^{n-j}, \quad j=0, \ldots, \kappa-1 \tag{2.6.39}$$

(see e.g. Stetter [1973, p. 204]). From this relation a linear system for the $a_{j,n}$ can be derived,

$$\begin{aligned} \gamma_0 a_{j,\kappa} &= -\bar{\gamma}_{\kappa-j} \\ \gamma_1 a_{j,\kappa} + \gamma_0 a_{j,\kappa+1} &= -\bar{\gamma}_{\kappa-j+1} \\ \gamma_2 a_{j,\kappa} + \gamma_1 a_{j,\kappa+1} + \gamma_0 a_{j,\kappa+2} &= -\bar{\gamma}_{\kappa-j+2} \end{aligned} \tag{2.6.40}$$

. . .

where $\bar{\gamma}_{\kappa-j+i}=0$ for $j<i$ and $\bar{\gamma}_{\kappa-j+i}=\gamma_{\kappa-j+i}$ for $j\geqslant i$. For a discussion of the coefficients $a_{j,n}$ of the inhomogeneous terms in (2.6.38) we refer to, e.g., Stetter [1973, p. 204].

The representation (2.6.38) is useful for deriving estimates of the solution y_n. We mention the following theorems.

Theorem 2.6.8. *The solution of the inhomogeneous equation (2.6.37) satisfies the inequality*

$$|y_n| \leqslant \gamma [\max_{0\leqslant j\leqslant\kappa-1} |y_j| + \max_{\kappa\leqslant j\leqslant n} |g_j|], \quad n\geqslant\kappa,$$

with γ denoting a constant independent of n, if, and only if, all the zeros of $G(z)$ are within the unit circle. □

Theorem 2.6.9. *The solution of the inhomogeneous equation (2.6.37) satisfies the inequality*

$$|y_n|\leqslant\gamma [\max_{0\leqslant j\leqslant\kappa-1} |y_j| + \sum_{j=\kappa}^{n} |g_j|], \quad n\geqslant\kappa,$$

with γ denoting a constant independent of n, if, and only if, $G(z)$ satisfies the root condition. □

We observe that these theorems also hold (and will be applied subsequently) in the more general case where the right-hand side $g_{n+\kappa}$ in (2.6.37) depends on y_n.

2.6.5. Asymptotic behavior of the weights

We are now in a position to study the behavior of the weights $w_{n,j}$ as $n\to\infty$. Firstly, we consider the weights $w_{n,j}$ $(j=0,1, \ldots ,\kappa-1)$ in the matrix $\Sigma_{\kappa,n}$.

Theorem 2.6.10. *If $\rho(z)$ satisfies the strong root condition, then*

$$\bar{\mathbf{w}} := \lim_{n\to\infty} \mathbf{w}_n = \frac{1}{\rho'(1)} \sum_{i=0}^{\kappa-1} a_i \sum_{l=0}^{\kappa-1-i} \mathbf{w}_l,$$

where the weights $\mathbf{w}_l$ occurring on the right-hand side are defined by the starting matrix S_κ.

Proof. We apply to (2.6.11) Theorem 2.6.7, with $G(z)=\rho(z)$, $r=1$, $z_1=1$, $|z_\nu|<1$, $\nu=2, \ldots ,q$, and with starting vectors $\mathbf{w}_0, \ldots , \mathbf{w}_{\kappa-1}$. Then (2.6.11) assumes the form

$$\mathbf{w}_n = \frac{1}{\rho'(1)} \sum_{j=0}^{\kappa-1} \sum_{i=0}^{j} a_i \mathbf{w}_{k-1-j} + \mathbf{v}_n, \quad n\geqslant\kappa, \tag{2.6.41}$$

where $\mathbf{v}_n$ is generated by the recurrence relation

$$\rho^*(E)\mathbf{v}_j=\mathbf{0}, \quad j\geqslant 0, \quad \rho^*(z):=\frac{\rho(z)}{z-1}. \tag{2.6.42}$$

Since the zeros of $\rho^*(z)$ lie in the interior of the unit circle it follows from Theorem 2.6.5 that $\mathbf{v}_n\to\mathbf{0}$ as $n\to\infty$. Hence

$$\lim_{n\to\infty}\mathbf{w}_n = \frac{1}{\rho'(1)}\sum_{j=0}^{\kappa-1}\sum_{i=0}^{j}a_i\mathbf{w}_{\kappa-1-j}.$$

From this expression the assertion of Theorem 2.6.10 follows immediately. □

In a similar way we can derive the asymptotic behavior of the weights ω_j in the matrix $\Omega_{\kappa,n}$.

Theorem 2.6.11. *If $\rho(z)$ satisfies the strong root condition, then*

$$\bar{\omega} := \lim_{n\to\infty}\omega_n = \frac{\sigma(1)}{\rho'(1)}.$$

Proof. We now apply Theorem 2.6.7 to (2.6.16) with starting values $\omega_1, \omega_2, \ldots, \omega_\kappa$ to obtain

$$\omega_n = \frac{1}{\rho'(1)}\sum_{j=0}^{\kappa-1}\sum_{i=0}^{j}a_i\omega_{\kappa-j}+v_n, \quad n\geqslant\kappa, \tag{2.6.43}$$

where v_n is generated by a recurrence relation of the form (2.6.42). Since $\rho^*(z)$ has its zeros within the unit circle we have

$$\lim_{n\to\infty}\omega_n =: \bar{\omega} = \frac{1}{\rho'(1)}\sum_{j=0}^{\kappa-1}\sum_{i=0}^{j}a_i\omega_{\kappa-j} = \frac{1}{\rho'(1)}\sum_{i=0}^{\kappa-1}a_i\sum_{j=1}^{\kappa-i}\omega_j.$$

It follows from (2.6.16) that

$$\begin{aligned}\bar{\omega} &= \frac{1}{\rho'(1)}[(b_1-a_1\omega_0)+(b_2-a_2\omega_0)+\ \ldots\ +(b_\kappa-a_\kappa\omega_0)]\\ &= \frac{1}{\rho'(1)}[\sigma(1)-b_0-(\rho(1)-a_0)\omega_0]=\frac{\sigma(1)}{\rho'(1)},\end{aligned}$$

which proves the theorem. □

2.6.6. *The approximation error*

It was shown by Matthys [1976] that the approximation error of a (ρ,σ)-reducible quadrature formula is at most $O(h^p)$ if p is the order of the generating LM formula. An explicit expression for the approximation error was derived by Wolkenfelt [1981a]. Here, we give a slightly different derivation of this expression.

Theorem 2.6.12. *Let $\phi\in C^p[a,b]$, where $p\geqslant 1$ is the order of the LM formula $\{\rho,\sigma\}$, and let the sequence of (ρ,σ)-reducible quadrature formulas (2.6.8) have* *given approximation errors* $\mathbf{E}_j[\phi]$, $j=0,1,\ldots,\kappa-1$, *with* $\mathbf{E}_0[\phi]=0$. *Then for* $n\geqslant\kappa$ *the approximation error is given by*

$$\mathbf{E}_n[\phi] = C^*_{p+1}[\phi^{(p-1)}(b)-\phi^{(p-1)}(a)]h^p+D_n+O(h^{p+1}),$$

where C^*_{p+1} *is the normalized error constant of the* LM *formula and where* D_n *is given by*

$$\rho(E)D_n = 0, \quad n\geqslant 0, \quad \text{with} \quad D_j:=\mathbf{E}_j[\phi], \ j=0,1,\ldots,\kappa-1. \qquad (2.6.44)$$

Proof. It follows from (2.6.6), (2.6.8) and (2.6.2) that $\mathbf{E}_n[\phi]$ can be expressed as

$$\mathbf{E}_n[\phi] = y(t_n)-y_n,$$

where $t_0=a$, $t_n=b$, and where $y(t)$ is the solution of (2.6.1). Applying the operator $\rho(E)$ and using (2.6.3) yields, since $\phi_n=\phi(t_n)=y'(t_n)$,

$$\begin{aligned}\rho(E)\mathbf{E}_n[\phi] &= \rho(E)y(t_n)-\rho(E)y_n=\rho(E)y(t_n)-h\sigma(E)\phi_n\\ &= \rho(E)y(t_n)-h\sigma(E)\phi(t_n)=\rho(E)y(t_n)-h\sigma(E)y'(t_n).\end{aligned}$$

By virtue of $\{\rho,\sigma\}$ being of order p (see Definition 2.6.2) we find for $\mathbf{E}_n[\phi]$ the recurrence relation

$$\rho(E)\mathbf{E}_n[\phi] = C_{p+1}h^{p+1}\phi^{(p)}(t_n)+O(h^{p+2}), \quad n\geqslant 0, \qquad (2.6.45)$$

where $\mathbf{E}_0[\phi]$, ... , $\mathbf{E}_{\kappa-1}[\phi]$ are the approximation errors of the starting quadrature formulas. From the relation

$$hf'(t) = \frac{1}{\rho'(1)}\rho(E)[f(t)+\text{ constant}]+O(h^2),$$

where $f(t)\in C^1[a,b]$, it follows that we may write, setting $f(t):=\phi^{(p-1)}(t)$,

$$\rho(E)\mathbf{E}_n[\phi] = \frac{C_{p+1}}{\rho'(1)}h^p\rho(E)[\phi^{(p-1)}(t_n)+\text{ constant}]+O(h^{p+2}), \quad n\geqslant 0.$$

It is easily verified that the solution of this difference equation with given starting conditions $\mathbf{E}_0[\phi]=0$, $\mathbf{E}_j[\phi]$, $j=1,2,\ldots,\kappa-1$, may be written as

$$\mathbf{E}_n[\phi] = C^*_{p+1}[\phi^{(p-1)}(t_n)-\phi^{(p-1)}(t_0)]h^p+D_n+D^*_n, \quad n\geqslant 0,$$

where $C^*_{p+1}=C_{p+1}/\rho'(1)$ is the normalized error constant. D_n is given by (2.6.44), and D^*_n is defined by

$$\rho(E)D^*_n = O(h^{p+2}), \quad n\geqslant 0,$$

with starting values

$$D^*_j := -C^*_{p+1}[\phi^{(p-1)}(t_j)-\phi^{(p-1)}(t_0)]h^p, \quad j=0,1,\ldots,\kappa-1.$$

Theorem 2.6.9 implies that $D^*_n=O(h^{p+1})$ as $h\downarrow 0$, and with $t_n=b$ fixed. This proves the theorem. □

Since ρ satisfies the root condition it follows from Theorem 2.6.9 that the solution of (2.6.44) has the same order of magnitude as the starting errors $\mathbf{E}_j[\phi]$, $j=0,1,\ldots,\kappa-1$. This leads to the following corollary of Theorem 2.6.12.

Corollary 2.6.1. *Let p be the order of the LM formula $\{\rho,\sigma\}$, and let the starting errors $\mathbf{E}_j[\phi]$, $j=0,1,\ldots,\kappa-1$ be of order s as $h\downarrow 0$. Then the order of the sequence of $\{\rho,\sigma\}$-reducible quadrature formulas is given by $r:=\min(p,s)$.* □

2.6.7. *Quadrature based on backward differentiation formulas*

A particularly interesting family of quadrature formulas is generated by the backward differentiation (BD) formulas, specified in Table 2.6.2 for $\kappa\leqslant 6$. We will show how the weights can be derived explicitly for $\kappa=2$. For $\kappa\geqslant 3$ they will have to be generated numerically. In actual computation, this is a disadvantage because of a less straightforward implementation than, for instance, that of the Gregory rules. The alternative, the use of the recurrence (2.6.4), shares the same disadvantage. However, their rather good stability properties justify our analyzing them in some detail.

Let us first consider the matrix $\Omega_{\kappa,n}$. Since $b_i=0$ for $i>0$, the system (2.6.16) simplifies to

$$\rho(E)\omega_j = 0, \quad j\geqslant 1-\kappa, \tag{2.6.46}$$

where $\omega_j=0$ if $j<0$ and $\omega_0=b_0/a_0$. This difference equation can be solved explicitly only for $\kappa=2$. We write

$$\omega_j = C_1 z_1^j + C_2 z_2^j,$$

where z_1 and z_2 are the zeros of the characteristic polynomial

$$\rho(z) = a_0 z^2 + a_1 z + a_2.$$

By means of Table 2.6.2 we find that $z_1=1$ and $z_2=1/3$, and by means of the starting values we can determine the coefficients C_1 and C_2, leading to

$$\omega_j = 1-\frac{1}{3}\left(\frac{1}{3}\right)^j.$$

Thus, the matrix $\Omega_{2,n}$ is of the form

$$\Omega_{2,n} = \begin{bmatrix} 2/3 & & & & & & \\ 8/9 & 2/3 & & & & & \\ 26/27 & 8/9 & 2/3 & & & & \\ 80/81 & 26/27 & 8/9 & 2/3 & & & \\ & & \cdots & & & & \\ 1-3^{-n+1} & & \cdots & & 26/27 & 8/9 & 2/3 \end{bmatrix}.$$

Table 2.6.2. Coefficients of backward differentiation formulas ($b_i=0,\ i>0$)

κ	a_0	a_1	a_2	a_3	a_4	a_5	a_6	b_0	p	C^*_{p+1}
2	3	-4	1					2	2	-1/3
3	11	-18	9	-2				6	3	-1/4
4	25	-48	36	-16	3			12	4	-1/5
5	137	-300	300	-200	75	-12		60	5	-1/6
6	147	-360	450	-400	225	-72	10	60	6	-1/7

For $\kappa \geqslant 3$, the zeros of the polynomial $\rho(z)$ are not explicitly available so that the matrix $\Omega_{\kappa,n}$ must be generated numerically. This will be discussed below.

The matrix $\Sigma_{\kappa,n}$ depends on the starting weights. Given a starting matrix S_κ, we can again only find explicitly the matrix $\Sigma_{\kappa,n}$ if $\kappa=2$. Solving the equation

$$\rho(E)\mathbf{w}_j = \mathbf{0}, \quad j \geqslant 2-\kappa \tag{2.6.47}$$

for $\kappa=2$ with starting matrix S_2 defined in Section 2.6.2 yields

$$\mathbf{w}_j = \frac{3}{4}[1-3^{-j}]\mathbf{e}, \quad j \geqslant 2,$$

where $\mathbf{e}=(1,1)^T$. Thus

$$\Sigma_{2,n} = \begin{bmatrix} 2/3 & 2/3 \\ 13/18 & 13/18 \\ 20/27 & 20/27 \\ 121/162 & 121/162 \\ \vdots & \vdots \\ \frac{3}{4}(1-3^{-n}) & \frac{3}{4}(1-3^{-n}) \end{bmatrix}.$$

For $\kappa \geqslant 3$, no explicit expressions for $\mathbf{w}_j$ and ω_j are available. Of course, one can generate them by means of (2.6.11) and (2.6.16), but this is a linearly unstable process in actual computation. To see this, consider the recurrence (2.6.16) in the presence of rounding errors δ_j (the following discussion holds also for (2.6.46)). Thus, instead of the ω_j we obtain values ω_j^* satisfying the inhomogeneous equation

$$\rho(E)\omega_j^* = \delta_{j+\kappa}, \tag{2.6.48}$$

where δ_j is the rounding error introduced during the calculation. Subtracting (2.6.16) and (2.6.48) yields for the numerical error $\Delta\omega_j := \omega_j^* - \omega_j$ the relation

$$\rho(E)\Delta\omega_j = \delta_{j+\kappa},$$

with zero starting values. Applying Theorem 2.6.9 then results in

$$|\Delta\omega_j| \leqslant \gamma \sum_{i=1}^{j} |\delta_j|$$

(note that $\rho(z)$ satisfies the root condition), so that a linear accumulation of rounding errors should be expected.

It is possible, however, to generate the ω_j by a stable process by employing Theorem 2.6.7. Since in the case of BD formulas $\rho(z)$ satisfies the strong root condition, the weights ω_j can also be generated by (cf. (2.6.35))

$$\omega_j = \bar{\omega}+v_j, \quad j\geqslant 0, \tag{2.6.49}$$

where $\bar{\omega}$ is given in Theorem 2.6.11 and where v_j satisfies the recurrence relation

$$\rho^*(E)v_j = 0, \quad j\geqslant 0, \quad \rho^*(z) := \frac{\rho(z)}{z-1}, \tag{2.6.50}$$

with starting vales $v_j=\omega_j-\bar{\omega}$, $j=0,1, \ldots ,\kappa-2$. In practice, the values v_j^* actually obtained now satisfy the strongly stable recursion

$$\rho^*(E)v_j^* = \delta_{j+\kappa},$$

which has a uniformly bounded solution as $n\to\infty$, provided that the rounding errors δ_j are uniformly bounded.

Analogously, the weight vectors $\mathbf{w}_j$ of the matrix $\Sigma_{\kappa,n}$ can be generated in a stable way by means of the recurrence relation

$$\mathbf{w}_j = \bar{\mathbf{w}}+\mathbf{v}_j, \quad j\geqslant 0, \tag{2.6.51}$$

where $\bar{\mathbf{w}}$ is given in Theorem 2.6.10 and where $\mathbf{v}_j$ satisfies relation (2.6.50) with starting values $\mathbf{v}_j=\mathbf{w}_j-\bar{\mathbf{w}}$, $j=0,1, \ldots ,\kappa-2$.

Wolkenfelt [1981a], [1982] showed that the weights ω_j and $\mathbf{w}_j$ reach their asymptotic values $\bar{\omega}$ and $\bar{\mathbf{w}}$ with a precision 10^{-15} for $j\geqslant\bar{n}$, where $\bar{n}=31$, 40, 59, 97 and 227, for $\kappa=2$, 3, 4, 5 and 6, respectively. Therefore, only a finite number of weights have to be generated in actual computation.

Finally, we consider the behavior of the approximation error as $h\downarrow 0$ by applying Theorem 2.6.12 with $p=\kappa$, $C_{p+1}^*=1/(\kappa+1)$, and with starting quadrature rules S_κ as defined in Section 2.6.2; i.e., with $\dot{s}=\kappa+1$. From (2.6.44) it then follows that $D_n=O(h^{\kappa+1})$, hence

$$\mathbf{E}_n[\phi] = \frac{h^\kappa}{\kappa+1}[\phi^{(\kappa-1)}(b)-\phi^{(\kappa-1)}(a)]+O(h^{\kappa+1}), \tag{2.6.52}$$

with $a=t_0$ and $b=t_n$.

2.7 Repeated quadrature formulas

In the preceding sections we have discussed various sequences of quadrature formulas. In the class of interpolatory formulas the Newton-Cotes sequences are unsuitable because convergence is not guaranteed. The Gauss-Legendre, Radau and Lobatto sequences are convergent but require a considerable amount of storage. The (ρ,σ)-reducible rules, when implemented in the form (2.6.4), do not have this drawback (which is advantageous if many points $t_{n,j}$ are involved such as in the integration of Volterra equations), but they often lead to complicated numerical schemes.

However, there is another class of formulas, which also requires only a limited amount of storage, the so-called *repeated quadrature formulas* or *compound formulas*. These formulas are defined by just repeating the application of a *fixed* quadrature formula over an interval $[\alpha,\beta]$ to the ν subintervals of length $\beta-\alpha$ into which the interval $[a,b]$ is divided. Usually, the subintervals are chosen to be of equal length, so that

$$\mathbf{I}_n[\phi] := \sum_{i=1}^{\nu} \sum_{j=0}^{m} c_j \phi(t_j^{(i)}), \quad \nu(\beta-\alpha)=b-a, \tag{2.7.1}$$

where the $\{c_j\}$ are the coefficients of a fixed $(m+1)$-point formula and where $t_0^{(i)}, \ldots, t_m^{(i)}$ are the $m+1$ points in the i-th subinterval.

The following theorem is taken from Davis & Rabinowitz [1984, p. 72].

Theorem 2.7.1. *Let the coefficients $\{c_j\}$ define a fixed $(m+1)$-point quadrature formula over an interval $[\alpha,\beta]$, and let its approximation error be given by an expression of the form*

$$c(\beta-\alpha)^{r+1}\phi^{(r)}(\tilde{t}), \quad \tilde{t}\in[\alpha,\beta], \tag{2.7.2}$$

whenever $\phi\in C^r[\alpha,\beta]$, with c denoting a constant depending on the c_j but not on α,β or ϕ. Then for all $\phi\in C^r[a,b]$ the compound formula (2.7.1) has an approximation error $\mathbf{E}_n[\phi]$ which satisfies

$$\lim_{\nu\to\infty} \nu^r \mathbf{E}_n[\phi] = c(b-a)^r[\phi^{(r-1)}(b)-\phi^{(r-1)}(a)]. \quad \square$$

Corollary 2.7.1. *Let the conditions of Theorem 2.7.1 be satisfied. Then the sequence of repeated quadrature formulas (2.7.1) has order r.*

Proof. Let h be the maximal length of the subintervals. Then $\beta-\alpha\leqslant mh$, and hence

$$\nu = \frac{b-a}{\beta-\alpha} \geqslant \frac{b-a}{mh}.$$

It follows from Theorem 2.7.1 that, for $\nu\to\infty$,

$$\begin{aligned}\mathbf{E}_n[\phi] &= c\nu^{-r}(b-a)^r[\phi^{(r-1)}(b)-\phi^{(r-1)}(a)]+o(\nu^{-r})\\ &\leqslant cm^r h^r[\phi^{(r-1)}(b)-\phi^{(r-1)}(a)]+o(h^r),\end{aligned}$$

so that by Definition 2.1.3 the order is given by r. $\square$

We note in passing that if a given quadrature formula has a Peano kernel function which does not change its sign (e.g., the Newton-Cotes formulas), then the Peano kernel function of the compound quadrature formula based on the given quadrature formula possesses the same property. See also Davis & Rabinowitz [1984, p. 289].

For small values of m the repeated quadrature formula (2.7.1) requires a minimal amount of storage for the coefficients. A disadvantage may be the condition $\nu(\beta-\alpha)=b-a$; that is, the integration interval $[a,b]$ is required to be an exact multiple of the basic subinterval $[\alpha,\beta]$. In numerical methods for Volterra equations we will need approximations $\mathbf{I}_n[\phi]$ for arbitrary intervals. To overcome this difficulty, one usually combines formula (2.7.1) with an additional quadrature formula for integrating over the remaining interval of length $<\beta-\alpha$ (examples of such formulas are given in Examples 2.6.1, 3.2.1 and 3.2.3). Evidently, $\mathbf{E}_n[\phi]$ is bounded by the approximation error $\mathbf{R}[\phi]$ of the $\nu-1$ times repeated basic quadrature formula and by the approximation error $\mathbf{A}[\phi]$ of the additional quadrature formula for covering the last "incomplete" subinterval, i.e.

$$|\mathbf{E}_n[\phi]| \leqslant |\mathbf{R}[\phi]|+|\mathbf{A}[\phi]|. \tag{2.7.3}$$

Example 2.7.1. We determine the order of the sequence of Simpson formulas considered in Example 2.6.1. According to Corollary 2.7.1, $\mathbf{R}[\phi]=O(h^r)$, where r is determined by the approximation error (2.7.2) of the 3-point Simpson formula; it follows from Table 2.2.1 that $r=4$. Table 2.2.1 also implies that $\mathbf{A}[\phi]=O(h^5)$ because $\mathbf{A}[\phi]$ corresponds to the 3/8 rule, i.e., to the 4-point Newton-Cotes formula. Thus, we find $\mathbf{E}_n[\phi]=O(h^4)$ so that the sequence of compound formulas of Example 2.6.1 is of order 4. □

2.8. Product integration

We start with an example. Let $\phi(t)=\sqrt{t}\cos(t)$ and let the quadrature formula be defined by the repeated trapezoidal rule. In Table 2.8.1 the accuracy obtained is listed for a few values of the (constant) step size h. The approximation error roughly behaves as $h^{3/2}$ so that the order r of this sequence of quadrature rule is approximately 1.5, whereas generally the repeated trapezoidal rule is second-order exact. Using the sequence of repeated Simpson rules does not help: the order is still 1.5. The reason is of course that an r-th order behavior is obtained only if $\phi\in C^r[0,1]$. In our example we have $\phi\in C[0,1]$, but $\phi\notin C^1[0,1]$, so that Theorem 2.7.1, and therefore Corollary 2.7.1, cannot be applied for $r\geqslant 1$.

Table 2.8.1. Repeated trapezoidal rule applied to $\int_0^1 \sqrt{t}\cos(t)dt$

h	1/2	1/4	1/8	1/16	1/32	1/64
$\mathbf{E}_n[\phi]$	$8.6_{10^{-2}}$	$2.9_{10^{-2}}$	$9.9_{10^{-3}}$	$3.4_{10^{-3}}$	$1.2_{10^{-3}}$	$4.1_{10^{-4}}$

It is for problems of the form

$$\int_a^b w(t)\psi(t)dt,$$

where ψ is sufficiently differentiable but w is merely integrable on $[a,b]$, that so-called *product integration* is a suitable quadrature technique. Formally, this technique is defined as follows. Let $\psi \in C^{n+1}[a,b]$, and let $\{t_{n,j}\}_{j=0}^n$ be given abscissas in $[a,b]$. Then we write (cf. (2.1.8))

$$\int_a^b w(t)\psi(t)dt = \sum_{j=0}^n c_{n,j}\psi(t_{n,j}) + \mathbf{E}_n[w,\psi], \tag{2.8.1}$$

where

$$c_{n,j} := \int_a^b L_{n,j}(t)w(t)dt, \quad L_{n,j}(t) := \prod_{\substack{i=0\\ i\neq j}}^n \frac{t-t_{n,i}}{t_{n,j}-t_{n,i}}.$$

Theorem 2.8.1. *Let $\psi \in C^{n+1}[a,b]$. If the coefficients $\{c_{n,j}\}$ are bounded, then there exists a function $\theta(t)$ with values in $[a,b]$ such that*

$$\mathbf{E}_n[w,\psi] = \frac{1}{(n+1)!}\int_a^b w(t)\pi_n(t)\phi^{(n+1)}(\theta(t))dt, \tag{2.8.2}$$

where

$$\pi_n(t) := \prod_{j=0}^n (t-t_{n,j}).$$

Proof. The proof is rather similar to that of Theorem 2.1.4. First we observe that the expressions for the $c_{n,j}$ are obtained by requiring that $\mathbf{E}_n[w,\psi]$ vanishes for $\psi = p_n[t;\psi]$, where $p_n[t;\psi]$ is the Lagrange interpolating polynomial for ψ (cf. (2.1.9)). The expression for $\mathbf{E}_n[w,\psi]$ follows from (2.8.1) on substitution of $\psi(t) = p_n[t;\psi] + \pi_n(t)\psi^{(n+1)}(\theta(t))/(n+1)!$. □

As an illustration we derive the quadrature formula for $n=1$ with $t_{1,0}=a$ and $t_{1,1}=b$. We find the two-point formula

$$\int_a^b w(t)\psi(t)dt = \int_a^b \frac{w(t)(t-b)}{a-b}dt\,.\,\psi(a) + \int_a^b \frac{w(t)(t-a)}{b-a}dt\,.\,\psi(b) +$$

$$+\frac{1}{2}\int_a^b w(t)(t-a)(t-b)\psi''(\theta(t))dt. \tag{2.8.3}$$

Note that this formula reduces to the trapezoidal rule if $w(t)\equiv 1$.

Formulas of type (2.8.1) can of course be applied repeatedly as described in the preceding section (compare also Feldstein & Miller [1971]). Thus, dividing the interval $[a,b]$ into n equal subintervals of length h, the repeated application of (2.8.3) yields

$$\begin{aligned}\int_a^b w(t)\psi(t)dt &= h\sum_{i=1}^{n}[\int_0^1(1-\tau)w(t_{i-1}+\tau h)d\tau\,.\,\psi(t_{i-1})\\ &\quad+\int_0^1 \tau w(t_{i-1}+\tau h)d\tau\,.\,\psi(t_i)]+O(h^2),\end{aligned} \tag{2.8.4}$$

where $t_i=a+ih$ and $h=b-a$. In the particular case where $w(t)=\sqrt{t}$, as in our example above, we find

$$\begin{aligned}\int_a^b \sqrt{t}\psi(t)dt &= h\sum_{i=1}^{n}[c_i\psi(t_{i-1})+d_i\psi(t_i)]+O(h^2),\\ c_i &:= \frac{2}{15h^2}[2t_i^{5/2}+3t_{i-1}^{5/2}-5t_it_{i-1}^{3/2}],\\ d_i &:= \frac{2}{15h^2}[3t_i^{5/2}-5t_{i-1}t_i^{3/2}-3t_{i-1}^{5/2}+5t_{i-1}^{5/2}].\end{aligned} \tag{2.8.5}$$

In comparison with the conventional repeated trapezoidal rule, the quadrature weights now depend on the abscissas t_i and have to be computed at each point t_i. This is the price we have to pay for a higher order of convergence. In actual computation one might apply product integration formulas only in regions where the integrand is not smooth and not in the whole integration interval. For example, in the case $\phi(t)=\sqrt{t}\cos(t)$ it is sufficient to restrict the application of (2.8.3) to a *fixed neighborhood* of the origin. In Table 2.8.2 the approximation errors are listed obtained by integrating the first m subintervals using (2.8.3) and by integrating the remaining subintervals using the trapezoidal rule. These results clearly show the second-order behavior if mh is kept fixed while $h\downarrow 0$.

Expressions for the approximation error of a product integration formula which are more general than the one given in Theorem 2.8.1 (i.e., which include functions ψ not contained in the class $C^{n+1}[a,b]$) can be obtained from the following straightforward generalization of Peano's theorem (Theorem 2.1.2):

Table 2.8.2. Product integration for $\int_0^1 \sqrt{t}\cos(t)dt$

h	1/2	1/4	1/8	1/16	1/32	1/64	1/128
$m=1$	$2.3_{10^{-2}}$	$7.8_{10^{-3}}$	$2.5_{10^{-3}}$	$8.3_{10^{-4}}$	$2.7_{10^{-4}}$	$9.1_{10^{-5}}$	$3.1_{10^{-5}}$
$m=2$	$1.1_{10^{-2}}$	$5.6_{10^{-3}}$	$1.9_{10^{-3}}$	$6.4_{10^{-4}}$	$2.1_{10^{-4}}$	$6.8_{10^{-5}}$	$2.2_{10^{-5}}$
$m=3$		$4.1_{10^{-3}}$	$1.6_{10^{-3}}$	$5.5_{10^{-4}}$	$1.8_{10^{-4}}$	$5.8_{10^{-5}}$	$1.9_{10^{-5}}$
$m=4$		$2.8_{10^{-3}}$	$1.4_{10^{-3}}$	$4.9_{10^{-4}}$	$1.6_{10^{-4}}$	$5.1_{10^{-5}}$	$1.6_{10^{-5}}$
$m=5$			$1.2_{10^{-3}}$	$4.4_{10^{-4}}$	$1.5_{10^{-4}}$	$4.7_{10^{-5}}$	$1.5_{10^{-5}}$
$m=6$			$1.0_{10^{-3}}$	$4.1_{10^{-4}}$	$1.4_{10^{-4}}$	$4.4_{10^{-5}}$	$1.4_{10^{-5}}$
$m=7$			$8.5_{10^{-4}}$	$3.8_{10^{-4}}$	$1.3_{10^{-4}}$	$4.1_{10^{-5}}$	$1.3_{10^{-5}}$
$m=8$			$6.9_{10^{-4}}$	$3.5_{10^{-4}}$	$1.2_{10^{-4}}$	$3.9_{10^{-5}}$	$1.2_{10^{-5}}$
⋮			⋮	⋮	⋮	⋮	⋮
$m=1/h$	$1.1_{10^{-2}}$	$2.8_{10^{-3}}$	$6.9_{10^{-4}}$	$1.7_{10^{-4}}$	$4.2_{10^{-5}}$	$1.0_{10^{-5}}$	$1.9_{10^{-5}}$

Theorem 2.8.2. *Assume that the quadrature formula*

$$\mathbf{I}_n[w,\psi] := \sum_{j=0}^{n} c_{n,j}\psi(t_{n,j}), \tag{2.8.6}$$

approximating

$$\mathbf{I}[w,\psi] := \int_a^b w(t)\psi(t)dt,$$

has degree of precision q. If $w(t)$ is nonnegative (with $w(t)\not\equiv 0$) and integrable on $[a,b]$, and if $\psi\in C^m[a,b]$, where $1\leq m\leq q+1$, then the quadrature error $\mathbf{E}_n[w,\psi]:=\mathbf{I}[w,\psi]-\mathbf{I}_n[w,\psi]$ can be expressed in the form

$$\mathbf{E}_n[w,\psi] = \int_a^b K_m(t)\psi^{(m)}(t)dt, \tag{2.8.7}$$

where the Peano kernel function is now defined by

$$K_m(t) := \int_a^b w(x)\frac{(x-t)_+^{m-1}}{(m-1)!}\,dx - \sum_{j=0}^{n} c_{n,j}\frac{(t_{n,j}-t)_+^{m-1}}{(m-1)!}. \quad \square$$

We conclude this chapter by mentioning that the Euler-Maclaurin summation formula (2.6.18) has been generalized to integrals with integrands containing various types of singularities. Due to limitations of space we have to refrain from describing these extensions in detail; instead, the reader is referred to Lyness & Ninham [1967, pp. 166-168], de Hoog & Weiss [1973c, pp. 298-300], Benson [1973], and Schneider [1980].

NOTES

Section 2.1 - 2.5.

A comprehensive treatment of the theory and the application of numerical methods for the evaluation of integrals may be found in the monographs by Krylov [1962], Ghizzetti & Ossicini [1970], Davis & Rabinowitz [1984], Brass [1977], and Engels [1980]. In addition, we also recommend Isaacson & Keller [1966, pp. 300-363], Stroud [1974, pp. 106-205], and Schmeisser & Schirmeier [1976, pp. 188-221], for an introduction to the subject. For special quadrature formulas of Gauss, Radau and Lobatto type see also, e.g., Stroud & Secrest [1966], Ghizzetti & Ossicini [1970, pp. 92-110], and Hildebrand [1974, pp. 379-445]; in addition to the first of these references, tables of weights and abscissas may be found in Glasmacher & Sommer [1966] and in Abramowitz & Stegun [1970, pp. 916-924]. We also mention the recent package of mostly automatic FORTRAN subroutines for one-dimensional integration by Piessens et al. [1983]. Various FORTRAN programs (as well as an extensive bibliography of quadrature routines) may also be found in Davis & Rabinowitz [1984, pp. 480-517].

Theorem 2.1.2 is due to Peano [1913]; compare also Davis & Rabinowitz [1984, pp. 285-295], and Stroud [1974, pp. 163-183], where a slightly more general version and numerous examples are given.

The history of numerical integration begins essentially with the contributions of Gregory (1670), Newton (1676), Cotes (1722), Euler (1738), Maclaurin (1742), and Simpson (1743); the development of special formulas begins with Gauss (1814) (with extensions by Jacobi (1826) and Christoffel (1858)), Lobatto (1852), and Radau (1880). See Gautschi [1981] for a detailed discussion of these historical aspects and for an extensive list of references; compare also Goldstine [1977].

Section 2.6.

The standard references for linear multistep methods for ODEs are Dahlquist [1959] and Henrici [1962]; in addition, compare also Gear [1971], Lambert [1973], Stetter [1973] and Hall & Watt [1976, pp. 20-44]. The notion of (ρ,σ)-reducible quadrature formulas was introduced by Matthys [1976] in connection with the stability analysis of linear multistep methods for Volterra integro-differential equations; a special case (the Gregory formulas, which are reducible to the Adams-Moulton methods) occurs in Brunner & Lambert [1974]. The general theory of (ρ,σ)-reducible quadrature formulas is due to Wolkenfelt [1979], [1981a], [1982]. The starting formulas (2.6.10b), (2.6.10d) occur already in Krylov [1949, p. 43, p. 60].

Gregory formulas are discussed in Steinberg [1972], Baker [1977, pp. 117-123], and, especially, in Brass [1977, pp. 209-214] (where Theorem 2.6.4 can be found). For the Euler-Maclaurin summation formula, see Krylov [1962, p. 216], Hildebrand [1974, pp. 197-203] and Schmeisser & Schirmeier [1976, pp. 228-235].

A concise theory of linear difference equations is given, e.g., in Henrici [1962, pp. 210-215], Isaacson & Keller [1966, pp. 405-410], and in Henrici [1974, pp. 584-590].

Gear [1971, pp. 212-217] and Lambert [1973, p. 242] discuss various properties of backward differentiation formulas, while Hairer & Wanner [1983] present an elegant proof of the instability of BD formulas with $\kappa \geqslant 7$.

Section 2.7.

Convergence properties and error terms for compound quadrature formulas are given, e.g., in Davis & Rabinowitz [1984, pp. 70-74] and in Isaacson & Keller [1966, pp. 336-345]; the Peano kernel for such formulas is analyzed in Davis & Rabinowitz [1984, pp. 289-290].

Section 2.8.

Young [1954a], [1954b] introduced the concept of product integration in the context of the numerical solution of integral equations. See also Baker [1977, pp. 131-144] and Fenyö & Stolle [1984b, pp. 153-156] for an introduction to product integration. Error bounds for compound quadrature of weakly singular integrals are given in Feldstein & Miller [1971]. In addition, compare the papers by Davis & Jacoby [1969], Lyness & Ninham [1967], Benson [1973], de Hoog & Weiss [1973c], and Schneider [1980] for examples and general analyses (including generalizations of the Euler-Maclaurin formula) of product integration formulas.

CHAPTER 3

LINEAR METHODS FOR VOLTERRA EQUATIONS

This chapter treats the convergence analysis of a rather general class of linear multistep methods for Volterra integral equations of the first and second kind and for Volterra integro-differential equations, where kernels are bounded (linear multistep methods for equations with weakly singular kernels are treated in Chapter 6, Section 6.1). These linear methods (which will be called *VLM methods)* include the conventional linear multistep methods as well as a few unconventional families of methods which have recently been proposed and which exhibit an improved stability behavior for problems where large Lipschitz constants are involved (see Chapter 7). We will treat the three types of Volterra equations in separate sections (Sections 3.2, 3.3 and 3.4). Being multistep methods, the VLM methods need starting values which is the subject of Section 3.5. The chapter starts with a brief historical review of early discretization methods.

3.1. Brief survey of early discretization methods

Let $0=t_0<t_1< \dots <t_N=T$, and set $h_l:=t_{l+1}-t_l$. The recurrence relation

$$u(t_n) = g(t_n)+\sum_{l=0}^{n-1} h_l K(t_n,t_l)u(t_l), \quad n=0, \dots ,N, \tag{3.1.1}$$

can be viewed as a special discretization of the Volterra integral equation

$$y(t) = g(t)+\int_0^t K(t,s)y(s)ds, \quad t\in I:=[0,T], \tag{3.1.2}$$

since it is derived from (3.1.2) by restricting the equation to a mesh point $t=t_n$ and by then replacing the corresponding integral by a particular Riemann sum (here, it will be assumed that $K\in C(I)$). The "discrete Volterra equation" (3.1.1) represents a system of linear algebraic equations for the approximations $\mathbf{u}_N:=[u(t_0), \dots ,u(t_N)]^T\in \mathbb{R}^{N+1}$: setting $\mathbf{g}_N:=[g(t_0), \dots ,g(t_N)]^T$ and $A_N:=(a_{n,l})$, with $a_{n,l}:=h_lK(t_n,t_l)$ ($n,l=0, \dots ,N$; $a_{n,l}=0$ whenever $l\geq n$), we may write (3.1.1) in the form

$$(I_N-A_N)\mathbf{u}_N = \mathbf{g}_N. \tag{3.1.3}$$

The lower triangular matrix I_N-A_N reflects, of course, the structure of the Volterra kernel $K(t,s)$ which is defined only for $s\leq t$, and we have $\det(I_N-A_N)=1$ for all values of $N\geq 1$. Hence, the unique solution of (3.1.3) is given by $\mathbf{u}_N=(I_N-A_N)^{-1}\mathbf{g}_N$. In order to derive explicit expressions for the components of the solution vector $\mathbf{u}_N$, let $C_N^{(n)}$ denote the matrix obtained from

$I_N - A_N$ by replacing its n-th column by the vector $\mathbf{g}_N$, and let $C_{n,l}$ be the cofactor associated with the element (l,n) of $C_N^{(n)}$ (i.e., the element g_l). Since it follows by Cramer's rule that $u(t_n) = \det(C_N^{(n)})/\det(I_N - A_N)$ we find, expanding $\det(C_N^{(n)})$ with respect to its n-th column,

$$u(t_n) = g(t_n) + \sum_{l=0}^{n-1} C_{n,l}\, g(t_l), \quad n = 0, \ldots, N.$$

Finally, if C_N is the (strictly) lower triangular matrix formed by the cofactors $C_{n,l}$ with $l < n$ (note that $C_{n,n} = 1$ for all n), then the vector $\mathbf{u}_N$ representing the approximate values at the mesh points $t_0, \ldots, t_N$ is given by

$$\mathbf{u}_N = (I_N + C_N)\mathbf{g}_N. \tag{3.1.4}$$

If we compare (3.1.4) with the expression for the exact solution of (3.1.2),

$$y(t) = g(t) + \int_0^t R(t,s) g(s) ds, \quad t \in I$$

(Theorem 1.3.1), we may interpret C_N as being the discrete resolvent kernel associated with the discrete Volterra equation (3.1.1).

The above approach was employed by Volterra [1896] (Nota I; Opere II, pp. 219-220), and, more explicitly, [1913] (pp. 40-45), to show that the relations holding between corresponding elements of the discrete kernel, A_N, and its discrete resolvent, C_N, yield, by "passing from finiteness to infinity", the continuous analogues (1.3.9) for $K(t,s)$ and its resolvent kernel $R(t,s)$.

We may look at the discretization (3.1.1) from two different points of view:

(i) (3.1.1) has been obtained from (3.1.2) by restricting the given integral equation to the finite mesh $\{t_n : n = 0, \ldots, N\}$ and by then applying a *sequence of quadrature formulas* to the resulting integral terms; in our case, we have employed the (repeated) explicit rectangular rule.

(ii) (3.1.1) results from (3.1.2) by assuming that its exact solution y is approximated, on the interval I rather than just on the mesh $\{t_n\}$, by a *piecewise polynomial function,* such that this approximating function satisfies the integral equation at certain points in I; on each subinterval $[t_n, t_{n+1}]$ the corresponding integral terms are then evaluated by numerical quadrature. In our case, the approximating function is piecewise constant (a step function), and the points at which this function satisfies (3.1.2) are the mesh points $t_0, \ldots, t_{N-1}$; quadrature is based on the explicit rectangular rules for these points.

Let us study these two interpretations of (3.1.1) more closely, with the aim of generalizing them. In (i), the exact solution is approximated by an element $\mathbf{u}_N \in \mathbb{R}^{N+1}$ (which represents the approximate values at the mesh points). Thus, (3.1.1) may be considered to be a particular case of the recursion,

$$y_n = g_n + \sum_{l=0}^{n} h_l w_{n,l} K(t_n, t_l) y_l, \quad n = \kappa, \ldots, N; \quad \kappa \geqslant 1, \tag{3.1.5}$$

where we have set $y_n := u(t_n)$, and where the starting values $y_0 = g(0)$, $y_1, \ldots, y_{\kappa-1}$, have to be prescribed. The construction and analysis of such *direct quadrature methods* (and their generalizations) for Volterra integral and integro-differential equations will be discussed in the present chapter.

In (ii), the full discretization of (3.1.2) is achieved in two stages: first, the given interval I is replaced by a certain finite subset, $X(N)$, and we are looking for a piecewise polynomial u which satisfies the integral equation on this set $X(N)$. This function u will be called the *collocation approximation* to the exact solution y with respect to $X(N)$, the set of collocation points. However, the equation defining u will include integral terms which, in general, cannot be evaluated analytically. Thus, full discretization of the integral equation is obtained after having approximated these integral terms by appropriate quadrature formulas. The resulting approximation $\hat{u}$ will, of course, be different from u but will still be a piecewise polynomial. This approach will be studied in Chapter 5.

Here, we shall give a brief description of the origins of discretization methods for Volterra equations. We first observe that, by the turn of the century, the discretization of initial-value problems for ordinary differential equations had already received a good deal of attention: in addition to the methods of Euler and of Cauchy we encounter the work of Bashforth and Adams (1883), and of Runge [1895], Heun [1900], and Kutta [1901]. Except for Volterra's discretization discussed above (which was not intended for the practical solution of (3.1.2)), the paper by Whittaker [1918] was the first contribution to the numerical solution of Volterra integral equations. It is still written in the classical vein: dealing with the approximate solution of first-kind Abel integral equations of the form (1.3.29), with $K(t,s) = a(t-s)$ given in tabulated form, he essentially adopts the ideas of Sonine [1884] (see Section 1.2) to construct an approximate solution. If the given kernel values are used to define an approximation to the unknown kernel of the form $b(z) := z^{-\alpha} p_n(z)$ (where p_n denotes some appropriate (interpolating) polynomial of degree n), then the given integral equation may be inverted analytically by determining the roots of a certain algebraic equation. An analogous approach works for second-kind integral equations with convolution kernels. The method proposed by Prasad [1924] is much more in the spirit of numerical analysis: here, the ideas of Bashforth and Adams are adapted for the discretization of (3.1.2), yielding a direct quadrature method based on the Gregory formulas (2.6.23) (in the Russian literature, these formulas occur often under the name of Euler-Laplace formulas). Discretization of (3.1.2) by direct quadrature, especially by the Gregory formulas, was also studied by Mikeladze [1935] and by Krylov [1949] (see also Mikeladze [1951]). An account of the "state of the art" in the early 1950s can be found in Fox & Goodwin [1953, pp. 521-534].

A first generalization of Volterra's discretization (3.1.1), viewed as a method for obtaining a piecewise polynomial approximation for the entire interval I, was given by Huber [1939] for first-kind integral equations, including equations with weakly singular kernels: instead of piecewise constant functions he employs continuous piecewise linear functions. This idea was subsequently generalized by Wagner [1954] who used continuous piecewise quadratic polynomials.

The subsequent development of numerical methods will be dealt with in Chapters 3-6. Here, we should like to point out that detailed surveys of the more recent history of numerical methods for Volterra equations can be found in Noble [1964], Kershaw [1974], Baker [1976], te Riele [1979b], Tsalyuk [1979], Linz [1980], Baker [1982a], Brunner [1982a].

3.2. Volterra integral equations of the second kind

Let us now turn to the general Volterra integral equation of the second kind,

$$y(t) = g(t)+\int_0^t k(t,s,y(s))ds, \quad t\in I:=[0,T], \tag{3.2.1}$$

where the functions g and k are given, with k bounded. Throughout this section the conditions for the unique existence of a continuous solution $y(t)$ are assumed to be fulfilled (see Theorems 1.3.8 and 1.3.9).

3.2.1. Direct quadrature methods

Assume that the given interval I is replaced by the mesh points t_n, $n=0, \ldots ,N$ (with $t_N=T$), and let y_n denote a numerical approximation to the exact solution $y(t_n)$ at t_n. Consider the quadrature formula (2.1.3) with $a=t_0=0$ and $b=t_n$, and apply this formula to the integral part of (3.2.1) with abscissas $t_0,t_1, \ldots ,t_n$. Then at $t=t_n$ we obtain the relation

$$y(t_n) = g(t_n)+\sum_{l=0}^{n} c_{n,l}k(t_n,t_l,y(t_l))+\mathbf{E}_n[k]. \tag{3.2.2}$$

If we are given quadrature formulas for $n=\kappa,\kappa+1, \ldots ,N$, and if we write

$$c_{n,l} =: hw_{n,l}, \quad h := \max_{0\leqslant l\leqslant N-1} h_l, \tag{3.2.3}$$

then we find the familiar form of a *direct quadrature method* (*DQ* method) for the general Volterra equation of the second kind (see also the linear form (3.1.5)):

$$y_n = g_n+h\sum_{l=0}^{n} w_{n,l}k(t_n,t_l,y_l), \quad n=\kappa, \ldots ,N, \tag{3.2.4}$$

where $g_n:=g(t_n)$ and where $y_0=g_0, y_1, \ldots , y_{\kappa-1}$ are given *starting values* with κ sufficiently large in order to have small approximation errors $\mathbf{E}_n[k]$ for all $n\geqslant\kappa$.

The system (3.2.4) is a triangular system of $N-\kappa+1$ nonlinear equations. If the $\{w_{n,n}\}$, $n\geqslant\kappa$ vanish, then the y_n can be calculated *explicitly*; otherwise we have to solve successively $N-\kappa+1$ algebraic equations. In this case the direct quadrature method is called *implicit.*

If the starting values are also given by formulas of the form

$$y_n = g_n + h\sum_{l=0}^{\kappa-1} w_{n,l}k(t_n,t_l,y_l), \quad n=0, \dots, \kappa-1, \tag{3.2.5}$$

then the direct quadrature method is completely defined by the points $\{t_l\}_{l=0}^{N}$ and by the matrix of weights

$$W_N = (w_{i,l}), \quad i=0, \dots, N; \quad l=0,1, \dots, N.$$

Note that $w_{i,l}=0$ for $i<\kappa$, $l>\kappa$ and for $i\geqslant\kappa$, $l>i$. Moreover, (3.2.5) can only be used if the kernel function $k(t,s,y)$ can be defined for $s>t$.

Example 3.2.1. Consider quadrature formulas generated by the repeated Simpson rule (if n is even) and by *repeated Simpson* plus *Simpson's 3/8 rule* (if n is odd) (cf. Example 2.7.1). Let the starting values y_0 and y_1 be defined by $y_0=g_0$ and by a 3-point interpolatory formula. Then

$$W_N = \begin{bmatrix} 0 & 0 & 0 & & & & & \\ 5/12 & 8/12 & -1/12 & & & & & \\ 1/3 & 4/3 & 1/3 & & & & & \\ 3/8 & 9/8 & 9/8 & 3/8 & & & & \\ 1/3 & 4/3 & 2/3 & 4/3 & 1/3 & & & \\ 1/3 & 4/3 & 17/24 & 9/8 & 9/8 & 3/8 & & \\ 1/3 & 4/3 & 2/3 & 4/3 & 2/3 & 4/3 & 1/3 & \\ 1/3 & 4/3 & 2/3 & 4/3 & 17/24 & 9/8 & 9/8 & 3/8 \\ & & & & \cdots & & & \end{bmatrix}. \quad \square$$

Example 3.2.2. Consider the quadrature formulas generated by the *BD formulas* (see Section 2.6.7). For the two-step case we derived the weights

$$\{w_{n,l}\}_{l=0}^{n} = \{\frac{3}{4}(1-3^{-n}), \frac{3}{4}(1-3^{-n}), 1-3^{-n+1}, \dots, \frac{26}{27}, \frac{8}{9}, \frac{2}{3}\},$$

where $n\geqslant 2$. Using the trapezoidal rule as starting formula, we find

$$W_N = \begin{bmatrix} 0 & 0 & & & & & \\ 1/2 & 1/2 & & & & & \\ 2/3 & 2/3 & 2/3 & & & & \\ 13/18 & 13/18 & 8/9 & 2/3 & & & \\ 20/27 & 20/27 & 26/27 & 8/9 & 2/3 & & \\ & & \cdots & & & & \\ \frac{3}{4}(1-3^{-N}) & \frac{3}{4}(1-3^{-N}) & 1-3^{-N+1} & & \cdots & 26/27 & 8/9 & 2/3 \end{bmatrix}. \quad \square$$

Example 3.2.3. In the preceding examples the points $\{t_l\}$ are equally spaced. Let us consider the non-equally spaced (scaled) *Lobatto points* (cf. Table 2.4.2),

$$t_1 = t_0+\frac{\sqrt{5}-1}{2}h,\ t_2=t_0+\frac{\sqrt{5}+1}{2}h,\ t_3=t_0+h\sqrt{5}\ ,$$

and define the subsequent points by $t_l=t_{l-3}+h\sqrt{5},\ l=4, \dots ,N.$ For $n=3, 6, 9, \dots$, the weights $\{w_{n,l}\}_{l=0}^{n}$ can be defined by a repeated Lobatto quadrature formula (cf. Section 2.7),

$$\{w_{n,l}\}_{l=0}^{n} = \frac{1}{24}\sqrt{5}\,\{2,\ 5,\ 5,\ 4,\ 5,\ 5,\ 4,\ \dots\ ,\ 4,\ 5,\ 5,\ 2\}. \tag{3.2.6}$$

For $n=4, 7, 10, \dots$, we employ the repeated 4-point Lobatto formula plus an interpolatory formula of the form

$$\int_{t_{n-1}}^{t_n}\phi(t)dt \approx \frac{h}{24}\sqrt{5}\,[v_0\phi_{n-4}+v_1\phi_{n-3}+v_2\phi_{n-2}+v_3\phi_{n-1}+v_4\phi_n], \tag{3.2.7a}$$

where the coefficients $c_j=v_jh\sqrt{5}/24$ can be found by applying (2.1.9). Similarly, for $n=5, 8, 11, \dots$, we use the repeated 4-point Lobatto formula plus the interpolatory formula

$$\int_{t_{n-2}}^{t_n}\phi(t)dt \approx \frac{h}{24}\sqrt{5}\,[w_0\phi_{n-4}+w_1\phi_{n-3}+w_2\phi_{n-2}+w_3\phi_{n-1}+w_4\phi_n]. \tag{3.2.7b}$$

We thus arrive at the matrix of weights

$$W_N=\frac{\sqrt{5}}{24}\left[\begin{array}{cccccccccc}
0 & 0 & 0 & 0 & 0 & & & & & \\
v_4 & v_3 & v_2 & v_1 & v_0 & & & & & \\
w_4 & w_3 & w_2 & w_1 & w_0 & & & & & \\
2 & 10 & 10 & 2 & & & & & & \\
2+v_0 & 10+v_1 & 10+v_2 & 2+v_3 & v_4 & & & & & \\
2 & 10+w_0 & 10+w_1 & 2+w_2 & w_3 & w_4 & & & & \\
2 & 10 & 10 & 4 & 10 & 10 & 2 & & & \\
2 & 10 & 10 & 4+v_0 & 10+v_1 & 10+v_2 & 2+v_3 & v_4 & & \\
2 & 10 & 10 & 4 & 10+w_0 & 10+w_1 & 2+w_2 & w_3 & w_4 & \\
2 & 10 & 10 & 4 & 10 & 10 & 4 & 10 & 10 & 2 \\
 & & & & \cdots & & & & &
\end{array}\right],$$

where the $\{v_j\}$ and $\{w_j\}$ are listed in Table 3.2.1.

Table 3.2.1. Quadrature weights v_j and w_j for the formulas (3.2.7)

j	v_j	w_j
0	-.119 473 775 300 143	-.261 114 561 800 008
1	.449 194 269 498 781	2.636 919 426 949 72
2	-1.682 104 898 799 40	-4.293 250 516 799 28
3	5.607 462 045 101 10	15.945 123 359 449 5
4	2.378 359 213 500 06	3.338 885 438 199 97

In this case, four implicit equations have to be solved simultaneously in order to obtain the starting values $y_1, \dots, y_4$. □

3.2.2. Numerical examples

In this subsection we illustrate numerically the DQ method (3.2.4) where the weights are those given by the weight matrices derived in the preceding examples. As test problem we take the *renewal equation* given in Table 8.2.3a, problem 17 (see also Example 1.4.3),

$$y(t) = \frac{1}{2}t^2e^{-t}+\frac{1}{2}\int_0^t (t-s)^2 e^{s-t} y(s)ds \tag{3.2.8a}$$

with the exact solution

$$y(t) = \frac{1}{3}\,[1-e^{-3t/2}(\cos(\tfrac{1}{2}t\sqrt{3})+\sqrt{3}\sin(\tfrac{1}{2}t\sqrt{3}))]. \tag{3.2.8b}$$

In order to eliminate the effect of possibly less accurate starting values in the quadrature rules, we prescribe the values y_0, y_1, y_2 by using the exact solution.

In Table 3.2.2 the number of correct significant digits obtained at the end point, i.e. the value of

$$sd := -\log_{10}(|y(T)-y_N| \,/\, |y(T)|), \quad T=t_N, \tag{3.2.9}$$

is listed for various values of the number N of integration steps in the interval $[t_0,T]=[0,6]$. The reason for using sd as a measure of accuracy is the following. Let p be the order of the method used. Then we expect $|y(T)-y_N|\approx Ch^p$ as $h\downarrow 0$, where C is some constant independent of h. Hence, by (3.2.9) we obtain

$$sd = sd(h) \approx -\,p\log_{10}(h)-\log_{10}(C\,/\,|y(T)|) \quad \text{as} \quad h\downarrow 0,$$

so that

$$p\approx\frac{sd(h)-sd(2h)}{\log_{10}(2)} \quad \text{as} \quad h\downarrow 0.$$

Hence, from the *sd* measure (3.2.9) the order behavior of the method is easily recognizable. We will define the *effective order* of a numerical method by

$$p_{eff} := \frac{sd(h)-sd(2h)}{\log_{10}(2)}, \tag{3.2.10}$$

where $\log_{10}(2)\approx .3$.

Table 3.2.2. Numbers of correct significant digits for problem (3.2.8) at $t=T=6$

	$N=8$	$N=16$	$N=32$	$N=64$	p_{eff}
DQ(Simpson) method (Ex. 3.2.1)	1.80	2.84	4.02	5.25	4.1
DQ(BD_2) method (Ex. 3.2.2)	1.20	1.80	2.53	3.34	2.7
DQ(Lobatto) method (Ex. 3.2.3)	1.92	3.47	4.78	6.71	6.4

Evidently, p_{eff} is a function of h which converges to the *asymptotic* order p of the method as $h\downarrow 0$. In Table 3.2.2 the effective orders are listed for $h=T/64=3/32$. These figures indicate that the DQ(Lobatto) method has a considerably larger order than the DQ(Simpson) and the DQ(BD_2) methods. In Section 3.2.8, where convergence results are given, it will be shown that the asymptotic orders are, respectively, 4, 2 and 6 (see Example 3.2.10).

3.2.3. (ρ,σ)-*reducible methods*

In the preceding section we assumed that the quadrature weights $w_{n,j}$ are given explicitly. In the class of (ρ,σ)-reducible quadrature formulas this is not always true: for instance, the BD quadrature rules of order $r>2$ do not have explicitly given weights. It has already been remarked in Section 2.6.7 that one can use these rules by employing the recurrence relation (2.6.4). We follow this approach in order to approximate the integral term in (3.2.1). On substituting

$$\phi(t) = k(t_n,t,y(t)), \quad a=t_0, \quad b=t_n,$$

into (2.6.4), the n-th integration step consists in solving the recurrence relation

$$\rho(E)z^{(n)}_{\nu-\kappa} = h\sigma(E)k(t_n,t_{\nu-\kappa},y_{\nu-\kappa}), \quad \nu=\kappa, \ldots ,n, \tag{3.2.11a}$$

and then putting

$$y_n = g(t_n)+z^{(n)}_n. \tag{3.2.11b}$$

Here, $z^{(n)}_l$ is a (ρ,σ)-reducible approximation to the integral term,

$$z^{(n)}_l \approx \int_{t_0}^{t_l} k(t_n,s,y(s))ds,$$

and E is the forward shift operator with respect to the index ν. Evidently, carrying out the process (3.2.11) for $n = \kappa, \ldots, N$ will generally require a significant number of additional operations when compared with the DQ method (3.2.4) employing explicitly given weights. Therefore, a (ρ,σ)-reducible DQ method when implemented according to (3.2.11) is unattractive in actual computation. To overcome this difficulty, Wolkenfelt [1981a], [1983b] proposed a modification of the DQ method (3.2.4) which does not require the repeated use of the recurrence (3.2.11a) and which is therefore easier to program. In the next section this method will be discussed as part of a general class of methods for Volterra equations.

3.2.4. A class of linear methods: VLM methods

Recently, several modifications of the conventional direct quadrature method (3.2.4) have been proposed: the *multilag* and *modified multilag methods* (see Wolkenfelt [1981a], [1983b]), and the *indirect linear multistep methods* (see van der Houwen & te Riele [1981]). These methods, including the DQ method (3.2.4), can be presented in a unified way which enables us to treat consistency and convergence for all these methods at the same time. Therefore, we first define a rather general class of methods of which the DQ method (3.2.4) is a special case.

Let us associate with the integral equation (3.2.1) the function

$$F(t,s) := g(t) + \int_0^s k(t,x,y(x))dx \tag{3.2.12}$$

for $(t,s) \in S := \{(t,s)\colon 0 \leqslant s \leqslant t \leqslant T\}$. Note that $F(t,t) \equiv y(t)$ and $F(t,0) \equiv g(t)$. In addition to y_n (representing the numerical approximation to the exact solution $y(t)$ at $t = t_n$), we introduce the function $\tilde{F}_n(t)$ representing a numerical approximation to $F_n(t) := F(t,t_n)$. Furthermore, we define

$$k_n(t) := k(t,t_n,y_n), \quad n \geqslant 0.$$

Unless otherwise stated, $\tilde{F}_n(t)$ will be given by a quadrature formula of the form

$$\tilde{F}_n(t) = g(t) + h\sum_{l=0}^{n} w_{n,l}k_l(t), \quad n \geqslant \tilde{\kappa}, \tag{3.2.13}$$

where the $w_{n,l}$ are given weights and $\tilde{\kappa}$ is an integer such that $\tilde{F}_n(t)$ is a sufficiently accurate approximation to $F_n(t)$; that is, $\tilde{\kappa}$ should be sufficiently large (we have added the tilde to κ in order to indicate its relation with the lag term $F_n(t)$). We will from now on assume that the mush is *uniform* (i.e., $h_l = h$, $l = 0, \ldots, N-1$), and the quadrature formula is of order r (see Definition 2.1.3):

$$\mathbf{E}_n[k] := \int_0^{t_n} k(t,x,y(x))dx - h\sum_{l=0}^{n} w_{n,l}k(t,t_l,y(t_l)) = O(h^r), \quad n \geqslant \tilde{\kappa}, \tag{3.2.14}$$

as $h\downarrow 0$, with $t\in[0,t_n]$ and $t_n=nh$ fixed. It will be convenient to write

$$\mathbf{E}_n[k] =: E_n(h;t).$$

$F_n(t)$ and $\tilde{F}_n(t)$ are known as the analytical and numerical *lag term*, respectively. (In the literature, one also encounters the names *history term* or *tail term.*) When no confusion arises, both will be referred to as lag terms.

Our general class of methods will consist of two formulas, namely of the *quadrature formula* (3.2.13) for approximating the lag term, and of the *linear multistep formula*

$$\sum_{i=0}^{\kappa}\alpha_i y_{n-i}+\sum_{i=0}^{\kappa}\sum_{j=-\kappa}^{\tilde{\kappa}}\beta_{i,j}\tilde{F}_{n-i}(t_{n+j})= h\sum_{i=0}^{\kappa}\sum_{j=-\kappa}^{\kappa}\gamma_{i,j}k_{n-i}(t_{n+j}), \tag{3.2.15}$$

with $n=\kappa^*, \dots ,N$. The parameters α_i, $\beta_{i,j}$ and $\gamma_{i,j}$ are to be prescribed, and $\kappa^*:=\kappa+\tilde{\kappa}$. Formula (3.2.15) will be called a *Volterra linear multistep formula* (VLM formula). The quadrature formula (3.2.13) and the VLM formula (3.2.15) together are said to constitute a *VLM method.*

A motivation for defining the VLM formula (3.2.15) can be given by first deriving from (3.2.1) and (3.2.12) the equation

$$y'(t)-F_t(t,t) = k(t,t,y(t)),$$

where $F(t,s)$ is assumed to be sufficiently differentiable, and by then applying a linear multistep method to this "differential equation". There results a formula which is linear in $y(t_n)$, $\tilde{F}_n(t_n)$ and $hk_n(t_n)$, and which represents a special case of the VLM formula (3.2.15). In fact, formulas of this special type belong to the family of indirect linear multistep formulas mentioned earlier, and will be discussed in Section 3.2.6. Thus, the VLM formula (3.2.15) is just a generalization of the indirect linear multistep formulas.

Because the kernel $k(t,s,y)$ is not necessarily defined outside S we usually require (cf. Figure 3.2.1)

$$\beta_{i,j} = \gamma_{i,j}=0, \quad j<-i. \tag{3.2.16}$$

Furthermore, if $\beta_{i,j}\neq 0$ and $\gamma_{i,j}\neq 0$ for $j=1, \dots ,\kappa$, we assume that the domain of definition S can be extended to points (t,s) with $T<t\leqslant T+\kappa h$, $0\leqslant s\leqslant T$. This extended domain will be denoted by $\bar{S}$. Throughout this section we only consider $\alpha_0\neq 0$. Finally, it will be assumed that the points $\{t_j\}$ are equally spaced, although most of the analysis can be carried through for non-uniform spacing (compare a similar situation in the analysis of LM methods for ODEs).

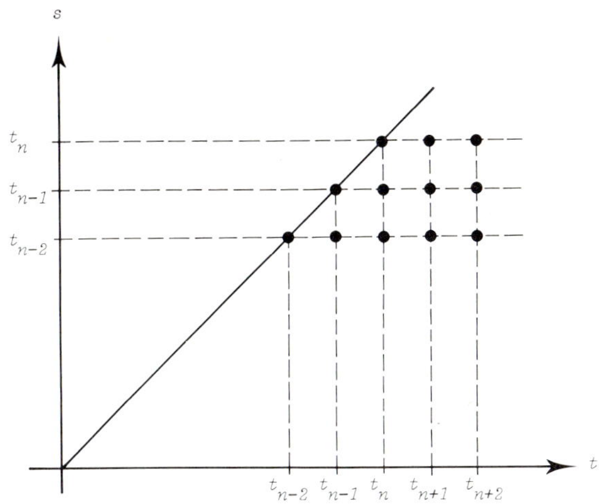

Figure 3.2.1. Domain of definition of (3.2.15) for $\kappa=2$

It is convenient to characterize the VLM formula (3.2.15) by the matrices

$$A=(\alpha_i), \quad B=(\beta_{i,j}), \quad C=(\gamma_{i,j}), \tag{3.2.17}$$

where the row index i and the column index j assume, respectively, the values $0, \dots, \kappa$ and $-\kappa, \dots, \kappa$.

Example 3.2.4. Consider the VLM formula defined by the 1×1 matrices ($\kappa=0$)

$$A=1, \quad B=-1, \quad C=0; \tag{3.2.18}$$

that is, we have the formula

$$y_n = \tilde{F}_n(t_n), \quad n\geqslant\tilde{\kappa}. \tag{3.2.18$'$}$$

Evidently, if $\tilde{F}_n(t)$ is defined by (3.2.13), this method is the *DQ method* (3.2.4). □

Example 3.2.5. Consider the VLM formula

$$3y_n-4y_{n-1}+y_{n-2}+3\tilde{F}_n(t_n)-4\tilde{F}_n(t_{n+1})+\tilde{F}_n(t_{n+2})=2hk_n(t_n), \quad n\geqslant\tilde{\kappa}+2,$$

where $\tilde{F}_n(t)$ is some lag term approximation. In Chapter 7 this formula will turn out to be highly stable for problems whose kernels possess large Lipschitz constants. It belongs to a special class of so-called *indirect backward differentiation (IBD)* formulas to be discussed in Section 3.2.6. □

Example 3.2.6. Another highly stable VLM formula is given by

$$3y_n-4y_{n-1}+y_{n-2}+4(\tilde{F}_{n-1}(t_{n-1})-\tilde{F}_{n-1}(t_n))-(\tilde{F}_{n-2}(t_{n-2})$$
$$-\tilde{F}_{n-2}(t_n))=2hk_n(t_n), \quad n\leqslant\tilde{\kappa}+2.$$

This formula belongs to the class of *modified multilag (MML)* formulas introduced by Wolkenfelt [1981a], [1983b] (see Section 3.2.6). □

We emphasize that the conventional DQ method generated by (3.2.18) involves the lag term at just one point, whereas in the VLM method lag terms and kernel values at several points are involved. In this respect they resemble the Volterra-Runge-Kutta (VRK) methods to be treated in Chapter 4: VRK methods also consist of a *lag term formula* and a *VRK formula* relating the numerical solution, the lag term and the kernel at several points (t_i,t_j) in the (t,s)-plane. However, there the points are not all step points but so-called *non-step points* $(t_i+\theta_i h, t_j+c_j h)$ with θ_i and c_j usually lying in [0,1]. Furthermore, unlike the VLM formula (3.2.15), the VRK formula is a *nonlinear* relation which complicates the theory considerably.

The VLM method {(3.2.13); (3.2.15)} needs the starting values $y_0, y_1, \ldots, y_{\kappa^*-1}$, where $\kappa^*=\tilde{\kappa}+\kappa$. In Section 3.5 we shall discuss methods to obtain these starting values (note that $y_0=g(t_0)$ and $\tilde{F}_0(t)=g(t)$). As soon as the starting values are available, we can compute $y_{\kappa^*}, y_{\kappa^*+1}, \ldots$ by solving successively the equations

$$\alpha_0 y_n + h\sum_{j=-\kappa}^{\kappa}(\beta_{0,j}w_{n,n}-\gamma_{0,j})k(t_{n+j},t_n,y_n) = \Sigma_n, \quad n\geq\kappa^*, \tag{3.2.19}$$

where

$$\Sigma_n := -\sum_{j=-\kappa}^{\kappa}\beta_{0,j}[g(t_{n+j})+h\sum_{l=0}^{n-1}w_{n,l}k_l(t_{n+j})]$$
$$-\sum_{i=1}^{\kappa}[\alpha_i y_{n-i}+\sum_{j=-\kappa}^{\kappa}(\beta_{i,j}\tilde{F}_{n-i}(t_{n+j})-\gamma_{i,j}hk_{n-i}(t_{n+j}))].$$

Here, (3.2.15) is written such that all terms containing y_n occur on the left-hand side and all terms containing the already computed values $y_0, y_1, \ldots, y_{n-1}$ are collected in the right-hand side term Σ_n. The method is *explicit* if

$$\beta_{0,j}w_{n,n}-\gamma_{0,j} = 0, \quad j=-\kappa, \ldots, \kappa, \quad n\geq\kappa^*,$$

and *implicit* otherwise.

Apart from the starting values, the computation of y_n requires $(n+\kappa+1)$ function values of $g(t)$, at least ½$(n+1)(n+2)$ function values of the kernel and, if the method is implicit, $(n-\kappa^*+1)$ times the solution of the equation (3.2.19). However, if the kernel k is of convolution type,

$$k(t,s,y) = K(t-s)f(s,y),$$

the number of kernel evaluations reduces to $(n+1)$ evaluations of K and f.

In practical problems the function $F(t,s)$ may display a rather different behavior with respect to the variables t and s. In such cases it might be more efficient to use alternative formulas for the lag term $\tilde{F}_n(t)$, instead of the one given in (3.2.13). For instance, if we have reasons to assume that $F(t,s)$ varies slowly with respect to s we may use quadrature formulas which do not use all values of $k_l(t)$ (e.g., only those $k_l(t)$ with l even). Alternatively, if $F(t,s)$ changes slowly with respect to t, we may consider computing $\tilde{F}_i(t_j)$ only for even values of i and to interpolate $\tilde{F}_i(t_j)$ for odd values of j. In this way the number of k-evaluations is halved.

In the preceding considerations the Volterra equation was assumed to be a *scalar* equation. If we are given a *system* of Volterra equations, we can formally define a VLM formula of the form (3.2.15) by interpreting y_n, $\tilde{F}_n(t_j)$ and $k_n(t_j)$ as vectors. Thus, each component of the system of VLM formulas is of the form (3.2.15). Of course, these formulas are coupled because the components of $\tilde{F}_n(t_j)$ and $k_n(t_j)$ depend on the whole vector y_n. Usually, the parameter matrices A,B and C are the same for all component VLM formulas. However, one may use different parameter matrices for the successive components. For instance, such formulas might be used in the case of Volterra equations obtained by integrating an integro-differential equation (see Section 3.4.1).

3.2.5. *Consistency of VLM methods*

In analogy to the theory of LM formulas for ODEs, we associate with the VLM formula the linear difference-differential operator $\mathbf{L}_n$: $C^{0,1}(S)\rightarrow\mathbb{R}$, defined by

$$\mathbf{L}_n[X] := \sum_{i=0}^{\kappa}\{\alpha_i X(t_{n-i},t_{n-i}) + \sum_{j=-\kappa}^{\kappa}[\beta_{i,j}-\gamma_{i,j}h\frac{\partial}{\partial s}]X(t_{n+j},t_{n-i})\},\quad n\geqslant\tilde{\kappa}, \tag{3.2.20}$$

where $X\subset C^{0,1}(S)$ with $C^{0,1}(S)$ denoting the space of continuous functions $X(t,s)$ continuously differentiable with respect to s for all $(t,s)\in S$. (In the following, we shall employ the notation $C^r(S):=C^{r,r}(S)$.) However, unlike the ODE case, the relation of the operator $\mathbf{L}_n$ with the VLM formula is not obvious and needs some explanation. Let us substitute the exact solution $y(t)$ of the Volterra integral equation into the VLM method and let us consider the resulting residual. Observing that

$$y(t)=F(t,t),\quad k(t,s,y(s))=\frac{\partial F}{\partial s}(t,s),$$

we obtain the residual term

$$\sum_{i=0}^{\kappa}\{\alpha_i y(t_{n-i})+\sum_{j=-\kappa}^{\kappa}[\beta_{i,j}(g(t_{n+j})+h\sum_{l=0}^{n-i}w_{n-i,l}k(t_{n+j},t_l,y(t_l))) - \gamma_{i,j}hk(t_{n+j},t_{n-i},y(t_{n-i}))]\}=\mathbf{L}_n[F]+\mathbf{Q}_n[y],$$

where

$$\mathbf{Q}_n[y] := -\sum_{i=0}^{\kappa} \sum_{j=-\kappa}^{\kappa} \beta_{i,j} E_{n-i}(h;t_{n+j}),$$

with $E_n(h;t) := \mathbf{E}_n[k]$ defined by (3.2.14). Thus, the exact solution satisfies the numerical scheme apart from the residual term $\mathbf{L}_n[F]$ originating from the VLM formula (3.2.15), and the term $\mathbf{Q}_n[y]$ originating from the quadrature formula (3.2.13). The term $\mathbf{L}_n[F]$ will be called the *local truncation error of the VLM formula* and $\mathbf{Q}_n[y]$ will be called the *lag term error*; together they form the *local truncation error of the VLM method.* This error plays an important role in the convergence analysis of VLM methods; in Section 3.2.8 we shall see that the rate of convergence of the numerical solution to the exact solution is closely related to that of the local truncation error of the VLM method. The convergence of the local truncation error is often called *consistency*:

Definition 3.2.1. (a) The *VLM* and *lag term formulas* are said to be *consistent* if the respective errors $\mathbf{L}_n[F]$ and $\mathbf{Q}_n[y]$ both tend to zero as $h \downarrow 0$, where F and y correspond to the exact solution. They are said to be *consistent of order p* if p is the largest integer so that the errors behave like $O(h^{p+1})$ as $h \downarrow 0$.
(b) The *VLM method* is said to be *consistent of order p with equation (3.2.1) at* $t = t_n$ if both the VLM formula and the lag term formula are consistent of order p. □

For the lag term formula we immediately have that $\mathbf{Q}_n[y] = O(h^r)$ if the quadrature formula defining the lag term is of order r (cf. (3.2.14)).

The order of the truncation error of a VLM formula depends on the parameters α_i, $\beta_{i,j}$ and $\gamma_{i,j}$. By expanding $\mathbf{L}_n[F]$ into a Taylor series about t_n, conditions can be derived for which $\mathbf{L}_n[F] = O(h^{p+1})$ as $h \downarrow 0$. However, such an expansion requires sufficient differentiability of the lag term function $F(t,s)$; that is, the functions g and k, and the solution y must possess continuous derivatives of sufficiently high order. Since the existence and uniqueness conditions only require $g \in C(I)$ and $k \in C(S \times \mathbb{R})$ together with a uniform Lipschitz condition for k (cf. Theorem 1.3.8), we conclude that, generally, we cannot expand $\mathbf{L}_n[F]$ into a Taylor series of sufficiently high order. Therefore, and in complete analogy to the consistency theory for ODEs (see, e.g., Lambert [1973, p. 23]), we allow $\mathbf{L}_n$ to operate on an arbitrary test function X which we may assume to have as many higher derivatives as we require. In this way we can formally define the *order* of the VLM formula.

Definition 3.2.2. The *VLM operator* (3.2.20) and the associated *VLM formula* (3.2.15) are said to be *of (local) order p* at $t = t_n$ if p is the largest integer so that, for all $X \in C^{p+1}(S)$, we have $\mathbf{L}_n[X] = O(h^{p+1})$, with t_n fixed as $h \downarrow 0$. □

The following theorem provides the order conditions in terms of the parameters α_i, $\beta_{i,j}$ and $\gamma_{i,j}$.

Theorem 3.2.1. *The VLM operator (3.2.20) and the associated VLM formula (3.2.15) are of order p if*

$$C_{q,l} := \frac{1}{l!(q-l)!}\sum_{i=0}^{\kappa}[(-i)^q\alpha_i - \sum_{j=-\kappa}^{\kappa} j^{q-l}(-i)^{l-1}(i\beta_{i,j}+l\gamma_{i,j})]=0$$

(with the convention $(-i)^{l-1}l=0$ if $i=l=0$) for $q=0,1,\ldots,p$ and $l=0,1,\ldots,q$, and if $C_{p+1,l}\neq 0$ for at least one value of $l=0,\ldots,p+1$.

Proof. Taylor expansion of $X(t_{n+j},t_{n-i})$ at (t_n,t_n) yields

$$\begin{aligned}\mathbf{L}_n[X] = &\sum_{i=0}^{\kappa}\{\alpha_i\sum_{q=0}^{p}\frac{1}{q!}h^q(-i\frac{\partial}{\partial t}-i\frac{\partial}{\partial s})^q X(t,s)\\ &+\sum_{j=-\kappa}^{\kappa}[\beta_{i,j}-\gamma_{i,j}h\frac{\partial}{\partial s}]\sum_{q=0}^{p}\frac{1}{q!}h^q(j\frac{\partial}{\partial t}-i\frac{\partial}{\partial s})^q X(t,s)\}|_{(t_n,t_n)}\\ &+O(h^{p+1}).\end{aligned}$$

Writing this formula in the form

$$\mathbf{L}_n[X] = \sum_{q=0}^{p}\frac{1}{q!}h^q(D_qX)(t_n,t_n)+O(h^{p+1}),$$

and expanding the differential operator D_q by using the Binomial Theorem, we find

$$\begin{aligned}D_q := &\sum_{i=0}^{\kappa}\{\alpha_i(-i\frac{\partial}{\partial t}-i\frac{\partial}{\partial s})^q\\ &+\sum_{j=-\kappa}^{\kappa}[j\beta_{i,j}\frac{\partial}{\partial t}-(i\beta_{i,j}+q\gamma_{i,j})\frac{\partial}{\partial s}][j\frac{\partial}{\partial t}-i\frac{\partial}{\partial s}]^{q-1}\}\\ &=\sum_{l=0}^{q}\sum_{i=0}^{\kappa}\{(-i)^q\alpha_i-\sum_{j=-\kappa}^{\kappa} j^{q-l}(-i)^{l-1}[i\beta_{i,j}+l\gamma_{i,j}]\}\binom{q}{l}\left(\frac{\partial}{\partial t}\right)^{q-l}\left(\frac{\partial}{\partial s}\right)^{l},\end{aligned}$$

where $(-i)^{l-1}\,l$ is assumed to be zero for $i=l=0$. Thus, for $X\in C^{p+1}(S)$ we have

$$\mathbf{L}_n[X] = \sum_{q=0}^{p}\sum_{l=0}^{q}[C_{q,l}h^q\left(\frac{\partial}{\partial t}\right)^{q-l}\left(\frac{\partial}{\partial s}\right)^{l}X](t_n,t_n)+O(h^{p+1}), \qquad (3.2.21)$$

where the $C_{q,l}$ are defined in the theorem. From this the theorem easily follows. □

The constants $C_{p+1,l}$ will be called the *(principal) error constants* of the VLM formula. The VLM formula (3.2.18), generating the DQ method, satisfies the order equations for arbitrary p so that it has an infinite order of consistency. In this connection, we emphasize that the VLM *formula*, and not the VLM *method*, has an infinite order. The order of the VLM *method* is also determined by the accuracy of the lag term approximation. Furthermore, we should bear in mind that the order conditions for VLM formulas only refer to the VLM formula itself and do not involve the exact solution, whereas the consistency of the VLM formula refers both to the solution and to the formula. A similar situation holds for LM and RK methods for ODEs, and for VRK methods for Volterra equations.

Example 3.2.7. The *IBD* formula given in Example 3.2.5 satisfies the order conditions for $p=2$ and is therefore second-order consistent. This is also true for the MML formula of Example 3.2.6. □

3.2.6. *Special VLM formulas*

Indirect linear multistep formulas

If g and k are differentiable with respect to t, then (3.2.1) can be written as

$$y'(t) = k(t,t,y(t)) + F_t(t,t).$$

Let $\{\rho,\sigma\}$ be an LM formula for ODEs with coefficients $\{a_i,b_i\}$. Its application to the above integro-differential equation yields the scheme

$$\sum_{i=0}^{\kappa} a_i y_{n-i} = h\sum_{i=0}^{\kappa} b_i[k_{n-i}(t_{n-i}) + F_t(t_{n-i},t_{n-i})].$$

Now approximate the derivative F_t by the κ-step *forward differentiation formula* (note that only values of F are used which correspond to points in $\bar{S}$),

$$F_t(t_{n-i},t_{n-i}) \approx -\frac{1}{h}\sum_{l=0}^{\kappa} d_l F(t_{n+l-i},t_{n-i}),$$

where the coefficients d_l are listed in Table 3.2.3 (cf. Abramowitz and Stegun [1970, Table 25.2]). Substitution leads to the VLM formula

$$\sum_{i=0}^{\kappa} a_i y_{n-i} + \sum_{i=0}^{\kappa}\sum_{j=-i}^{\kappa-i} b_i d_{i+j}\tilde{F}_{n-i}(t_{n+j}) = h\sum_{i=0}^{\kappa} b_i k_{n-i}(t_{n-i}),\ n\geqslant\kappa^*, \tag{3.2.22}$$

or equivalently, to the generating matrices

$$A = \begin{bmatrix} a_0 \\ a_1 \\ \vdots \\ a_\kappa \end{bmatrix},\quad B = \begin{bmatrix} \bigcirc & & b_0d_0 & \cdots & b_0d_\kappa \\ & b_1d_0 & b_1d_1 & \cdots & b_1d_\kappa \\ & \dots & \vdots & \dots & \\ b_\kappa d_0 & \cdots & b_\kappa d_\kappa & & \bigcirc \end{bmatrix},$$

$$C = \begin{bmatrix} \bigcirc & & b_0 & 0 & \cdots & 0 \\ & b_1 & & 0 & \cdots & 0 \\ & \dots & \bigcirc & \vdots & & \vdots \\ b_\kappa & & & 0 & \cdots & 0 \end{bmatrix}. \tag{3.2.22$'$}$$

The resulting formulas, although derived by differentiating the Volterra equation, *do not require the differentiability of the functions g and k.* They will be called *indirect linear multistep (ILM) formulas* because they formally arise by applying an LM method for ODEs to the differentiated Volterra equation.

Table 3.2.3. Coefficients of forward differentiation formulas

κ	d_0	d_1	d_2	d_3	d_4	d_5	d_6
2	3/2	-4/2	1/2				
3	11/6	-18/6	9/6	-2/6			
4	25/12	-48/12	36/12	-16/12	3/12		
5	137/60	-300/60	300/60	-200/60	75/60	-12/60	
6	147/60	-360/60	450/60	-400/60	225/60	-72/60	10/60

An interesting family of ILM formulas is obtained by choosing a backward differentiation formula as the generating LM method (see Table 2.6.2). The resulting formulas will be called *IBD formulas.* The generating matrices assume the form

$$A = \begin{bmatrix} a_0 \\ a_1 \\ \vdots \\ a_k \end{bmatrix}, \quad B = \begin{bmatrix} 0 & \cdots & 0 & a_0 & \cdots & a_k \\ 0 & \cdots & 0 & 0 & \cdots & 0 \\ \vdots & & \vdots & \vdots & & \vdots \\ 0 & \cdots & 0 & 0 & \cdots & 0 \end{bmatrix}, \quad C = \begin{bmatrix} 0 & \cdots & 0 & b_0 & 0 & \cdots & 0 \\ 0 & \cdots & 0 & 0 & 0 & \cdots & 0 \\ \vdots & & & & & & \vdots \\ 0 & & & \cdots & & & 0 \end{bmatrix},$$

where we have used the relation $b_0 d_i = a_i$. They have favorable stability properties, as will be seen in Chapter 7.

Theorem 3.2.2. *Let p be the order of the LM formula for ODEs defining the coefficients $\{a_i, b_i\}$. Then the ILM formula (3.2.22) is of order $p^* = \min\{\kappa, p\}$.*

Proof. Using the relations

$$\sum_{i=0}^{\kappa} i^q d_i = \begin{cases} -1 & \text{for } q=1 \\ 0 & \text{for } q=0,\ 2,\ 3,\ \ldots,\kappa \end{cases}, \tag{3.2.23}$$

the constants $C_{q,l}$ given in Theorem 3.2.1 may be written as

$$C_{q,l} = \frac{(-1)^q}{l!(q-l)!} \sum_{i=0}^{\kappa} i^{q-1}(ia_i + qb_i),$$

$0 \leq l \leq q \leq \kappa$. Using Theorem 2.6.3(d) the theorem is readily established. □

Let us assume that $p \geqslant \kappa$. Then $p^* = \kappa$, and the principal error constants turn out to be of the form

$$C_{p^*+1,l} = \frac{(-1)^{p^*-1}}{l!(p^*+1-l)!} \sum_{i=0}^{\kappa} [i^{p^*}(ia_i + (p^*+1)b_i) - R], \quad p^* = \kappa,$$

with $R = \kappa! b_i$ if $l = 0$, and $R = 0$ if $l = 1, \ldots, p^*+1$. In Table 3.2.4 these constants are listed for two popular choices of the coefficients $\{a_i, b_i\}$, viz. the BD and the AM formulas.

Table 3.2.4. Error constants $C_{p^*+1,l} = C^*_{p^*+1,l}\dfrac{1}{l!(p+1-l)!}$, $l = 0, \ldots, p^*+1$, for ILM formulas

Generating LM formula			$\kappa=1$	$\kappa=2$	$\kappa=3$	$\kappa=4$	$\kappa=5$
$BD(p = p^* = \kappa)$	$-C^*_{p^*+1,0}$	=	2	0	$\frac{72}{11}$	0	$\frac{14400}{137}$
	$-C^*_{p^*+1,l>0}$	=	1	$\frac{4}{3}$	$\frac{36}{11}$	$\frac{288}{25}$	$\frac{7200}{137}$
$AM(p = p^* - 1 = \kappa - 1)$	$-C^*_{p^*+1,0}$	=	1	2	6	-24	120
	$-C^*_{p^*+1,l>0}$	=	0	0	0	0	0

Multilag formulas

In Wolkenfelt et al. [1981] and Wolkenfelt [1981a], [1983b] we find formulas which can be characterized by the matrices

$$A = \begin{bmatrix} a_0 \\ 0 \\ \vdots \\ 0 \end{bmatrix}, \quad B = \begin{bmatrix} 0 & \cdots & 0 & 0 & 0 & \cdots & 0 \\ 0 & \cdots & 0 & a_1 & 0 & \cdots & 0 \\ \vdots & & \vdots & \vdots & \vdots & & \vdots \\ 0 & \cdots & 0 & a_\kappa & 0 & \cdots & 0 \end{bmatrix},$$

and

$$C = \begin{bmatrix} 0 & \cdots & 0 & b_0 & 0 & \cdots & 0 \\ \vdots & & \vdots & \vdots & \vdots & & \vdots \\ 0 & \cdots & 0 & b_\kappa & 0 & \cdots & 0 \end{bmatrix}, \tag{3.2.24}$$

where the coefficients $\{a_i, b_i\}$ correspond to those for an LM formula for

ODEs. These formulas were called *multilag* (ML) formulas. He also proposed the *modified multilag* (MML) formulas generated by the parameter matrices

$$A = \begin{bmatrix} a_0 \\ a_1 \\ \vdots \\ a_\kappa \end{bmatrix}, \quad B = \begin{bmatrix} \bigcirc & & & 0 & 0 & \cdots & 0 \\ & & -a_1 & a_1 & 0 & \cdots & 0 \\ & \cdot^{\cdot^{\cdot}} & \bigcirc & \vdots & \vdots & & \vdots \\ -a_\kappa & & & a_\kappa & 0 & \cdots & 0 \end{bmatrix},$$

and

$$C = \begin{bmatrix} 0 & \cdots & 0 & b_0 & 0 & \cdots & 0 \\ \vdots & & \vdots & \vdots & \vdots & & \vdots \\ 0 & \cdots & 0 & b_\kappa & 0 & \cdots & 0 \end{bmatrix}. \tag{3.2.25}$$

The MML formulas were designed to deal with kernels for which the Lipschitz constant with respect to y is large; just as the ILM formulas, they are highly stable if the LM formula $\{a_i,b_i\}$ is chosen to be a backward differentiation formula (see Chapter 7).

Theorem 3.2.3. *Let p be the order of the LM formula for ODEs defining the coefficients $\{a_i,b_i\}$. Then the (M)ML formulas (3.2.24) and (3.2.25) are both of order $p^* = p$.*

Proof. An elementary calculation reveals that

$$C_{q,l} = \begin{cases} 0 & \text{if } \; 0 \leqslant l \leqslant q-1 \\ \dfrac{(-1)^q}{l!(q-l)!} \displaystyle\sum_{i=0}^{\kappa} i^{q-1}[ia_i + qb_i] & \text{if } \; l = q \geqslant 0 \end{cases}.$$

Hence, by virtue of Theorem 2.6.3 (d), the order is $p^* = p$. □

In Table 3.2.5 the principal error constants are listed for (M)ML formulas generated by the BD and the AM formulas. These values suggest that the (M)ML formulas generated by the BD formulas are more accurate than the corresponding ILM formulas (see Table 3.2.4). In the case of the AM formulas the magnitude of the local truncation error in the (M)ML and ILM formulas cannot be compared independently of the problem to be solved (see the numerical experiment in Section 3.2.7).

Table 3.2.5. Error constants $C_{p^*+1,l} = C^*_{p^*+1,l}\dfrac{1}{l!(p^*+1-l)!}$, $l=0, \dots, p^*+1$, for (M)ML formulas

Generating LM formula		$\kappa=1$	$\kappa=2$	$\kappa=3$	$\kappa=4$	$\kappa=5$
$BD(p=p^*=\kappa)$	$C^*_{p^*+1,l<p^*+1}=$	0	0	0	0	0
	$-C^*_{p^*+1,p^*+1}=$	1	$\frac{4}{3}$	$\frac{36}{11}$	$\frac{288}{25}$	$\frac{7200}{137}$
$AM(p=p^*=\kappa+1)$	$C^*_{p^*+1,l<p^*+1}=$	0	0	0	0	0
	$-C^*_{p^*+1,p^*+1}=$	$\frac{1}{2}$	1	$\frac{19}{6}$	$\frac{27}{2}$	$\frac{863}{12}$

The following theorem exhibits a connection between the (M)ML methods and the DQ method.

Theorem 3.2.4. *Let the DQ and (M)ML methods have identical,* (ρ,σ)*-reducible lag terms, and let the (M)ML formulas be generated by the same* $\{\rho,\sigma\}$ *formula. Then the numerical solutions of the three methods are algebraically equivalent provided that the starting values are identical with* $\tilde{F}_n(t_n)=y_n$ *for* $n<\kappa^*$.

Proof. Consider the DQ method in the form (3.2.11). Since

$$g(t_n)+z_j^{(n)} = \tilde{F}_j(t_n),$$

we derive from (3.2.11) the recurrence relation

$$\rho(E)[\tilde{F}_{\nu-\kappa}(t_n)-g(t_n)] = h\sigma(E)k(t_n,t_{\nu-\kappa},y_{\nu-\kappa}).$$

Using the property $\rho(1)=0$, this relation yields for $\nu=n$ the formula

$$a_0\tilde{F}_n(t_n)+\sum_{i=1}^{\kappa}a_i\tilde{F}_{n-i}(t_n) = h\sum_{i=0}^{\kappa}b_ik_{n-i}(t_n).$$

Since $\tilde{F}_n(t_n)=y_n$, this formula is identical with the ML formula (3.2.24). Hence, if the starting values are identical, the numerical solutions obtained by the DQ and ML methods are algebraical equivalent.

Alternatively, we may write

$$\sum_{i=0}^{\kappa}a_i\tilde{F}_{n-i}(t_{n-i})-\sum_{i=0}^{\kappa}a_i[\tilde{F}_{n-i}(t_{n-i})-\tilde{F}_{n-i}(t_n)] = h\sum_{i=0}^{\kappa}b_ik_{n-i}(t_n),$$

and again by observing that $\tilde{F}_n(t_n)=y_n$, we now find algebraical equivalence with the MML method. □

3.2.7. Numerical examples

We will illustrate numerically the effect of the magnitude of the derivatives of F and the error constants on the accuracy of the ILM and (M)ML methods.

First, however, we note that the principal error term in the local truncation error of the (M)ML formulas only involves the (p^*+1)-st derivative of the lag term function $F(t,s)$ with respect to s. Hence, these formulas can be expected to behave better, as this derivative assumes smaller values. Conversely, the ILM formula based on an AM method will behave better as the (p^*+1)-st derivative with respect to t assumes smaller values. Finally, the ILM formula based on BD methods requires that both these derivatives be small in magnitude in order to get accurate results.

Let us consider the following test problem (problem 11a in Table 8.2.3b)

$$y(t) = e^{t^2} + te^{\frac{1}{2}t^2} - t - t\int_0^t s\sqrt{y(s)}\,ds, \tag{3.2.26}$$

whose exact solution is $y(t)=exp(t^2)$. The function $F(t,s)$ is given by

$$F(t,s) = e^{t^2} - t + te^{\frac{1}{2}t^2} - t\int_0^s x\sqrt{y(x)}\,dx.$$

Substitution of the exact solution yields, for $s=t$,

$$F_s = -t^2e^{\frac{1}{2}t^2}, \quad F_{ss} = -t(1+t^2)e^{\frac{1}{2}t^2}, \quad F_{sss} = -t^2(3+t^2)e^{\frac{1}{2}t^2},$$

$$F_t - 2te^{t^2} + t^2e^{\frac{1}{2}t^2}, \quad F_{tt} = 2(1+2t^2)e^{t^2} + t(3+t^2)e^{\frac{1}{2}t^2},$$

$$F_{ttt} = 4t(3+2t^3)e^{t^2} + (3+6t^2+t^4)e^{\frac{1}{2}t^2},$$

indicating that for $t>1$ the derivatives with respect to t are much larger than those with respect to s. Consequently, we expect that the (M)ML methods will produce more accurate results than the ILM methods if the problem is integrated over an interval $[0,T]$ with $T>1$.

We apply, respectively, the DQ, ILM and (M)ML methods. The starting values are taken from the exact solution, and the lag term is evaluated by a Gregory formula defined by the matrix of weights

$$W_N = \begin{bmatrix} S_{r-1} & 0 \\ \\ \Sigma_{r-1,N} & \Omega_{r-1,N} \end{bmatrix},$$

where r is the order of the formula, and S_{r-1}, $\Sigma_{r-1,N}$, $\Omega_{r-1,N}$ are given by (2.6.10), (2.6.12), (2.6.17), respectively. The coefficients $\{a_i,b_i\}$ in the ILM and (M)ML methods correspond to an AM or BD formula (see Section 2.6). The implicit relations which emerge in this nonlinear example are solved by

Newton iteration. The Newton process is stopped as soon as $|y_n^{(i+1)}-y_n^{(i)}|\leq 10^{-12}(1+|y_n^{(i+1)}|)$, where $y_n^{(i)}$ is the i-th Newton approximation to y_n. In the tables of results we denote a particular method by, e.g., $ILM(G_r-AM_p)$, indicating that the ILM method is based on an r-th order Gregory quadrature formula and a p-th order AM formula. The accuracy is expressed in terms of the number of correct significant digits defined by (3.2.9). In Table 3.2.6a we list results obtained by VLM methods whose accuracy is essentially governed by that of the *VLM formula,* since we have chosen $r>p$, i.e., its accuracy is little influenced by the lag term formula.

Table 3.2.6a. Numbers of correct significant digits for problem (3.2.26) at $t=T=4$

h	$DQ(G_6)$	$ILM(G_6-AM_4)$	$ML(G_6-AM_4)$	$MML(G_6-AM_4)$
1/5	5.7	-1.2	5.2	5.0
1/10	7.1	.3	6.4	6.1
1/20	8.8	1.5	7.8	7.2

h	$DQ(G_6)$	$ILM(G_6-BD_4)$	$ML(G_6-BD_4)$	$MML(G_6-BD_4)$
1/5	5.7	*	4.8	4.5
1/10	7.1	.3	6.0	5.4
1/20	8.8	2.0	7.3	6.4

These results clearly demonstrate the rather poor behavior of the ILM methods. In order to show that this behavior is due to the large derivatives of $F(t,s)$ with respect to t we change the initial function $g(t)=e^{t^2}+te^{t^2/2}-t$ to $g(t)=t(1+2t^{5/2}/5)$ which leads to the exact solution $y(t)=t$, yielding lag term derivatives of moderate size (see problem 11b in Table 8.2.3b). The corresponding results are listed in Table 3.2.6b; they clearly confirm our expectations.

In these experiments the DQ method delivers the best results; this is due, of course, to the vanishing truncation error of the VLM formula. The relatively accurate performance of the ML methods is due to the higher order of the *global* error as will be shown in the next section.

Table 3.2.6b. As Table 3.2.6a, with $g(t)=t(1+\frac{2}{5}t^{5/2})$

h	$DQ(G_6)$	$ILM(G_6-AM_4)$	$ML(G_6-AM_4)$	$MML(G_6-AM_4)$
1/5	6.4	4.3	6.2	5.2
1/10	7.2	5.6	7.2	6.2
1/20	7.9	6.0	7.9	7.2
h	$DQ(G_6)$	$ILM(G_6-BD_4)$	$ML(G_6-BD_4)$	$MML(G_6-BD_4)$
1/5	6.4	5.1	4.8	5.1
1/10	7.2	6.0	7.2	6.3
1/20	7.9	6.9	7.9	7.4

3.2.8. Convergence results for VLM methods

Definition 3.2.2. The VLM method is said to be *convergent* if the following statement is true for all functions $g(t)$ and $k(t,s,y)$ satisfying the conditions of Theorem 1.3.8: if $y(t)$ denotes the exact solution to (3.2.1), then

$$\lim_{h\downarrow 0} y_n = y(t), \quad t_n = t \quad \text{fixed},$$

holds for all $t\in I$ and all solutions $\{y_n\}$ of (3.2.15) corresponding to starting values $\{y_j\}$ with $\lim_{h\downarrow 0} y_j = y(t_j)$, $j=0, \ldots, \kappa^*-1$. □

3.2.8.1. Convergence of the VLM method

Before considering the convergence of the VLM solutions we derive the equation the numerical scheme converges to if we substitute a sufficiently differentiable function $y(t)$ (not necessarily the exact solution) and if we then let h tend to zero while keeping the point t fixed. To that end we define the quantities

$$A_q := \sum_{i=0}^{\kappa}(-i)^q \alpha_i. \tag{3.2.27}$$

We observe that $A_0=\alpha(1)$, $A_1=\alpha'(1)-\kappa\alpha(1)$, ..., where

$$\alpha(z) := \sum_{i=0}^{\kappa}\alpha_i z^{\kappa-i}. \tag{3.2.28a}$$

Theorem 3.2.5. *Let the lag term quadrature formula be of order r, let y be a function from* $C^{m+1}(I)$, $m\geq 0$, *and let the corresponding lag term F, defined by (3.2.12), be a function from* $C^{m+1}(S)$. *Then, on substitution of the function y into* *the VLM method* {(3.2.15), (3.2.13)}, *the discrete equation converges to*

$$\sum_{q=0}^{m} h^q \{ \frac{A_q}{q!} \frac{d^q}{dt^q} y(t) + \sum_{l=0}^{q} [C_{q,l} - \frac{A_q}{l!(q-l)!}] \left(\frac{\partial}{\partial t}\right)^{q-l} \left(\frac{\partial}{\partial s}\right)^{l} F(t,t) \}$$

$$= O(h^r + h^{m+1}) \quad \text{as} \quad h \to 0 \quad (\text{with } t \in I \text{ fixed}). \tag{3.2.29}$$

Proof. Let us substitute the function $y(t)$ into the VLM formula (3.2.15). Then, by virtue of (3.2.14), (3.2.20) and (3.2.21), we find

$$\sum_{i=0}^{\kappa} \{ \alpha_i y(t_{n-i}) + \sum_{j=-\kappa}^{\kappa} [\beta_{i,j} \tilde{F}_{n-i}(t_{n+j}) - h\gamma_{i,j} k_{n-i}(t_{n+j})] \}$$

$$= \sum_{i=0}^{\kappa} \{ \alpha_i y(t_{n-i}) + \sum_{j=-\kappa}^{\kappa} [\beta_{i,j} F(t_{n+j}, t_{n-i}) - h\gamma_{i,j} \frac{\partial F}{\partial s}(t_{n+j}, t_{n-i})] \} + O(h^r)$$

$$= \sum_{i=0}^{\kappa} \alpha_i [y(t_{n-i}) - F(t_{n-i}, t_{n-i})] + \mathbf{L}_n[F] + O(h^r)$$

$$= \sum_{i=0}^{\kappa} \alpha_i \sum_{q=0}^{m} \frac{1}{q!} (-ih)^q [\frac{d^q}{dt^q} y(t_n) - \left(\frac{\partial}{\partial t} + \frac{\partial}{\partial s}\right)^{q} F(t_n, t_n)]$$

$$+ \sum_{q=0}^{m} \sum_{l=0}^{q} C_{q,l} h^q \left(\frac{\partial}{\partial t}\right)^{q-l} \left(\frac{\partial}{\partial s}\right)^{l} F(t_n, t_n) + O(h^r + h^{m+1})$$

$$= \sum_{q=0}^{m} \frac{1}{q!} h^q A_q [\frac{d^q}{dt^q} y(t_n) - \sum_{l=0}^{q} \binom{q}{l} \left(\frac{\partial}{\partial t}\right)^{q-l} \left(\frac{\partial}{\partial s}\right)^{l} F(t_n, t_n)]$$

$$+ \sum_{q=0}^{m} \sum_{l=0}^{q} C_{q,l} h^q \left(\frac{\partial}{\partial t}\right)^{q-l} \left(\frac{\partial}{\partial s}\right)^{l} F(t_n, t_n) + O(h^r + h^{m+1}) .$$

From this result the theorem easily follows. □

For future reference we explicitly write down the first terms of (3.2.29). Using the definition of the lag term $F(t,s)$, the first terms up to order h^2 assume the form ($r \geq 3$, $m \geq 2$)

$$\begin{aligned} A_0 y(t) &+ (C_{0,0} - A_0) F(t,t) + h[A_1 y'(t) + (C_{1,0} - A_1) F_t(t,t) \\ &+ (C_{1,1} - A_1) k(t,t,y(t))] + h^2 [\tfrac{1}{2} A_2 y''(t) + (C_{2,0} - \tfrac{1}{2} A_2) F_{tt}(t,t) \\ &+ (C_{2,1} - A_2) k_t(t,t,y(t)) + (C_{2,2} - \tfrac{1}{2} A_2) k_s(t,t,y(t)) \\ &+ (C_{2,2} - \tfrac{1}{2} A_2) y'(t) k_y(t,t,y(t))] + O(h^3) = 0. \end{aligned} \tag{3.2.29′}$$

Example 3.2.8. In the case of the DQ method we have $A_0 = 1$ and $C_{q,l} = 0$. Hence, the numerical scheme will converge to the equation

$$y(t) = F(t,t),$$

which is identical to (3.2.1). Similarly, the ML method generated by (3.2.24) also converges to (3.2.1). □

Example 3.2.9. The ILM method generated by (3.2.22) with $\{a_i,b_i\}$ corresponding to a convergent LM formula for ODEs, leads to $C_{0,0}=C_{1,0}=C_{1,1}=0$ and $A_0=\alpha(1)=0$, $A_1=\alpha'(1)\neq 0$ because the LM formula should be zero-stable; that is, $\alpha(z)$ should have a simple zero at $z=1$ (cf. Theorem 2.6.1). Hence, by Theorem 3.2.5, the numerical scheme will converge to the differentiated Volterra equation

$$y'(t) = k(t,t,y(t))+F_t(t,t).$$

The same limiting equation is obtained for the MML method generated by (3.2.25). □

In analogy to the terminology used in the analysis of numerical schemes for time-dependent partial differential equations we will call equation (3.2.29) the *modified equation* (see e.g. Richtmyer and Morton [1967, p. 331]). This equation provides information about the numerical scheme if sufficiently many terms are taken into account. On the basis of the modified equation one can construct suitable schemes for partial differential equations. To our knowledge, no such approach has been considered for Volterra equations.

It should be realized that the modified equation (3.2.29) is the continuous representation of the VLM method for $h\neq 0$, whereas the limiting equation equals the modified equation if we set $h=0$. As we have seen in the Examples 3.2.8 and 3.2.9, the limiting equation is not unique and depends on the VLM parameters; but the limiting equation is always equivalent with the integral equation $y(t)=F(t,t)$. Of course, the limiting equations derived from (3.2.29) do not present all equations which are equivalent to $y(t)=F(t,t)$. For instance, the equivalent equation

$$y'(t) = k(t,t,F(t,t))+F_t(t,t)$$

cannot be obtained from (3.2.29). Discretization of this equation will lead to numerical methods which do not fit into the VLM class. In Chapter 4, methods are considered which can be interpreted as Runge-Kutta discretizations of the above equation (see the γ-modified methods discussed in Section 4.1.8.3).

In connection with the preceding discussion, we mention yet another equivalent representation of the integral equation $y(t)=F(t,t)$. Consider the equation

$$\Psi(t,s) = g(t)+\int_0^s k(t,x,\Psi(x,x))dx$$

for the function $\Psi(t,s)$. Evidently, this equation contains $y(t)$ as part of its solution, i.e., $y(t)=\Psi(t,t)$. Differentiation with respect to s leads to the initial-value problem

$$\frac{d\Psi}{ds}(t,s) = k(t,s,\Psi(s,s)), \quad \Psi(t,0) = g(t),$$

where t is considered as a parameter (cf. Pouzet [1963]). Linear multistep and Runge-Kutta discretizations of this ordinary differential equation have been considered in Wolkenfelt et al. [1981].

3.2.8.2. Convergence of the VLM solution

In the proof of the convergence theorem presented below it is convenient to use the following notations:

$$\beta(z) := \sum_{i=0}^{\kappa} \beta_i z^{\kappa-i}, \quad \beta_i := \sum_{j=-\kappa}^{\kappa} \beta_{i,j}, \tag{3.2.28b}$$

$$\gamma(z) := \sum_{i=0}^{\kappa} \gamma_i z^{\kappa-i}, \quad \gamma_i := \sum_{j=-\kappa}^{\kappa} \gamma_{i,j}. \tag{3.2.28c}$$

Furthermore, we will write

$$b := \max_{i,j} |\beta_{i,j}|, \quad c := \max_{i,j} |\gamma_{i,j}|, \quad w := \max_{i,j} |w_{i,j}|, \tag{3.2.28d}$$

$$\begin{cases} \Delta k(t,s,y,z) := k(t,s,y) - k(t,s,z) \\ \Delta E(h) := \max\limits_{\substack{i \leqslant j \leqslant N \\ l \leqslant \kappa}} |E_i(h;t_j) - E_i(h;t_{j+l})| \end{cases}, \tag{3.2.28e}$$

$$\begin{cases} E(h) := \max\limits_{\substack{i \leqslant j \leqslant N \\ l \leqslant \kappa}} |E_i(h;t_{j+l})| \\ T(h) := \max\limits_{i \leqslant N} |\mathbf{L}_i[F]| \\ \delta(h) := \max\limits_{1 \leqslant j \leqslant \kappa^* - 1} |y(t_j) - y_j| \end{cases}, \tag{3.2.28f}$$

where $y(t)$ is the exact solution and where $F(t,s)$ and $\tilde{F}_n(t)$ denote the corresponding functions defined in (3.2.12) and (3.2.13). $E(h)$ is the maximal approximation error $\mathbf{E}_n[k]$ arising in the approximation of the lag terms $\tilde{F}_n(t)$ in the interval $[0,t_N]$ (see (3.2.14)), $T(h)$ may be considered as the maximal local truncation error of the VLM formula (3.2.15) in $[0,t_N]$, and $\delta(h)$ is the maximal starting error.

We now prove a general convergence theorem which provides an estimate for the *global error*

$$\epsilon_n := y(t_n) - y_n. \tag{3.2.30}$$

Theorem 3.2.6. *Let the conditions of Theorem 1.3.8 be satisfied.*
(a) *Let $\alpha(z)=\alpha_0 z^\kappa$, $\alpha_0 \neq 0$. Then there exists a constant C independent of h such that for all sufficiently small $h>0$,*

$$|\epsilon_n| \leqslant C[h\delta(h)+E(h)+T(h)], \quad n=\kappa, \ldots ,N.$$

(b) *Let $\beta(z)\equiv 0$, and assume that $\alpha(z)$ and Δk satisfy, respectively, the root condition and the Lipschitz condition*

$$|\Delta k(t,s,y,z) - \Delta k(u,s,y,z)| \leqslant L^* |t-u|\,|y-z|$$

where (t,s,y), (t,s,z), (u,s,y), $(u,s,z) \in S \times \mathbb{R}$ and where the Lipschitz constant L^* is independent of t,u,y and z. Then there exists a constant C not depending on h such that, for all sufficiently small h,

$$|\epsilon_n| \leqslant Ch^{-1}[h\delta(h)+\Delta E(h)+T(h)], \quad n=\kappa, \ldots ,N.$$

Proof. Let $F(t,s)$ correspond to the exact solution $y(t)$. For $n \geqslant \kappa$ we may then write

$$\mathbf{L}_n[F] \equiv \mathbf{L}_n[F] - \sum_{i=0}^{\kappa}[\alpha_i y_{n-i} + \sum_{j=-\kappa}^{\kappa}(\beta_{i,j}\tilde{F}_{n-i}(t_{n+j}) - h\gamma_{i,j}k_{n-i}(t_{n+j}))]$$

$$= \sum_{i=0}^{\kappa}\{\alpha_i[y(t_{n-i})-y_{n-i}] + \sum_{j=-\kappa}^{\kappa}[\beta_{i,j}(F(t_{n+j},t_{n-i}) - \tilde{F}_{n-i}(t_{n+j}))$$

$$-h\gamma_{i,j}(k(t_{n+j},t_{n-i},y(t_{n-i})) - k_{n-i}(t_{n+j}))]\}.$$

Using the notation (3.2.28e) and (3.2.30) we may write

$$\mathbf{L}_n[F] = \sum_{i=0}^{\kappa}\{\alpha_i\epsilon_{n-i} + \sum_{j=-\kappa}^{\kappa}[\beta_{i,j}(F(t_{n+j},t_{n-i}) - \tilde{F}_{n-i}(t_{n+j}))$$

$$-h\gamma_{i,j}\Delta k(t_{n+j},t_{n-i},y(t_{n-i}),y_{n-i})]\}.$$

Substitution of the functions $F(t,s)$ and $\tilde{F}_n(t)$ and using (3.2.14) leads to

$$\mathbf{L}_n[F] = \sum_{i=0}^{\kappa}\{\alpha_i\epsilon_{n-i} + \sum_{j=-\kappa}^{\kappa}[\beta_{i,j}(h\sum_{l=0}^{n-i}w_{n-i,l}\Delta k(t_{n+j},t_l,y(t_l),y_l)$$

$$+E_{n-i}(h;t_{n+j})) - h\gamma_{i,j}\Delta k(t_{n+j},t_{n-i},y(t_{n-i}),y_{n-i})]\}.$$

Thus, we have found for ϵ_n a relation of the form

$$\alpha(E)\epsilon_{n-\kappa} = v_n, \quad n \geqslant \kappa, \tag{3.2.31}$$

where

$$v_n := \mathbf{L}_n[F] - \sum_{i=0}^{\kappa} \sum_{j=-\kappa}^{\kappa} [\beta_{i,j} h \sum_{l=0}^{n} w_{n-i,l} \Delta k(t_{n+j}, t_l, y(t_l), y_l)$$

$$- h\gamma_{i,j} \Delta k(t_{n+j}, t_{n-i}, y(t_{n-i}), y_{n-i}) + \beta_{i,j} E_{n-i}(h; t_{n+j})].$$

(a) $\alpha(z) = \alpha_0 z^{\kappa}$.

In order to derive an upper bound for the solution of this linear difference equation, we want to use the discrete Gronwall inequality stated in Corollary 1.5.2. Therefore, we need an upper bound for $|v_n|$. A straightforward calculation yields

$$|v_n| \leqslant T(h) + \sum_{i=0}^{\kappa} \sum_{j=-\kappa}^{\kappa} [bwLh \sum_{l=0}^{n} |\epsilon_l| + cLh|\epsilon_{n-i}| + bE(h)]$$

$$\leqslant C_0 h \sum_{l=0}^{n} |\epsilon_l| + C_1 E(h) + T(h), \tag{3.2.32}$$

where C_0 and C_1 are constants not depending on h (in the following all constants C_j will be independent of h). It follows from (3.2.31) that

$$|\alpha_0||\epsilon_n| \leqslant C_0 h \sum_{l=0}^{n} |\epsilon_l| + C_1 E(h) + T(h),$$

so that for sufficiently small h,

$$|\epsilon_n| \leqslant \frac{1}{|\alpha_0| - C_0 h}[C_0 h \sum_{l=0}^{n-1} |\epsilon_l| + C_1 E(h) + T(h)]$$

$$\leqslant C_2 h \sum_{l=0}^{n-1} |\epsilon_l| + C_3[E(h) + T(h)].$$

Application of Corollary 1.5.2 yields

$$|\epsilon_n| \leqslant (1 + C_2 h)^{n-\kappa}(\kappa h C_2 \delta(h) + C_3[E(h) + T(h)]), \quad n = \kappa, \ldots, N.$$

Since $nh \leqslant T$, part (a) of the theorem is immediate.

(b) $\beta(z) \equiv 0$, $\alpha(z)$ *satisfies the root condition.*

Instead of directly applying Corollary 1.5.2 to the inequality

$$\sum_{i=0}^{\kappa} |\alpha_i||\epsilon_{n-i}| \leqslant |v_n|$$

(see (3.2.31)), we first apply Theorem 2.6.9 to obtain the "sharper" inequality

$$|\epsilon_n| \leqslant C_0[\delta(h) + \sum_{j=\kappa}^{n} |v_j|], \quad n \geqslant \kappa. \tag{3.2.33}$$

Unfortunately, if we use the upper bound (3.2.32) for $|v_j|$ and then apply Corollary 1.5.2, we cannot prove convergence. However, by using the property that $\beta(z)\equiv 0$, that is, $\beta_i=\Sigma_{j=-\kappa}^{\kappa}\beta_{i,j}=0$, a sharper upper bound than (3.2.32) can be derived. To that end we write

$$|\sum_{j=-\kappa}^{\kappa}\beta_{i,j}\Delta k(t_{n+j},t_l,y(t_l),y_l)| = |\sum_{j=-\kappa}^{\kappa}\beta_{i,j}[\Delta k(t_n,t_l,y(t_l),y_l)$$

$$+\Delta k(t_{n+j},t_l,y(t_l),y_l)-\Delta k(t_n,t_l,y(t_l),y_l)]| \leqslant bL^*h\sum_{j=-\kappa}^{\kappa}|j\epsilon_l|,$$

and similarly,

$$|\sum_{j=-\kappa}^{\kappa}\beta_{i,j}E_{n-i}(h;t_{n+j})| \leqslant b\sum_{j=-\kappa}^{\kappa}\Delta E(h).$$

In this way we obtain instead of (3.2.32) the upper bound

$$|v_n| \leqslant T(h)+\sum_{i=0}^{\kappa}\sum_{j=-\kappa}^{\kappa}[bwL^*h^2\sum_{l=0}^{n}|\epsilon_l|+cL^*h|\epsilon_{n-i}|+b\Delta E(h)]$$

$$\leqslant C_1h\sum_{i=0}^{\kappa}[|\epsilon_{n-i}|+h\sum_{l=0}^{n}|\epsilon_l|]+C_2\Delta E(h)+T(h). \qquad (3.2.32')$$

Substitution into (3.2.33) yields the inequality

$$|\epsilon_n| \leqslant C_3\{\delta(h)+h\sum_{j=\kappa}^{n}[\sum_{i=0}^{\kappa}|\epsilon_{j-i}|+h\sum_{l=0}^{j}|\epsilon_l|+h^{-1}\Delta E(h)+h^{-1}T(h)]\}.$$

It is easily verified that

$$\sum_{j=\kappa}^{n}\sum_{i=0}^{\kappa}|\epsilon_{j-i}| \leqslant (\kappa+1)\sum_{j=0}^{n}|\epsilon_j|.$$

Hence,

$$|\epsilon_n| \leqslant C_4\{\delta(h)+h[(1+nh)\sum_{l=0}^{n}|\epsilon_l|+nh^{-1}\Delta E(h)+nh^{-1}T(h)]\}.$$

Since $nh\leqslant T$, we find for sufficiently small h,

$$|\epsilon_n| \leqslant C_5h\sum_{l=0}^{n-1}|\epsilon_l|+C_6h^{-1}[h\delta(h)+\Delta E(h)+T(h)].$$

Finally, by applying Corollary 1.5.2 we arrive at the estimate

$$|\epsilon_n| \leqslant (1+C_5h)^{n-\kappa}(\kappa hC_5\delta(h)+C_6h^{-1}[h\delta(h)+\Delta E(h)+T(h)]),$$

from which part (b) of the theorem follows. □

The additional Lipschitz condition on k required in part (b) of this theorem is satisfied if, for example, k_t satisfies a Lipschitz condition in y. We then may write

$$|\Delta k(t,s,y,z)-\Delta k(u,s,y,z)| = |k(t,s,y)-k(u,s,y)+k(u,s,z)-k(t,s,z)|$$

$$= \left| \int_u^t k_t(x,s,y)dx - \int_u^t k_t(x,s,z)dx \right| \leq |t-u| L^* |y-z| .$$

In the following corollary we summarize the convergence results obtained in this section.

Corollary 3.2.1. *Let p be the order of the LM formula for ODEs defining the coefficients* $\{a_i, b_i\}$ *employed in the ILM and (M)ML methods, let* $\delta(h)=O(h^s)$, $E(h)=O(h^r)$, *and* $\Delta E(h)=O(h^{r+1})$. *If the functions g and k in (3.2.1) are sufficiently smooth, then the order of convergence* p^* *is given by*

$$p^* = \begin{cases} \min\{s+1, r\} & \text{for the DQ method} \\ \min\{s, r, p, \kappa\} & \text{for the ILM method} \\ \min\{s+1, r, p+1\} & \text{for the ML method} \\ \min\{s, r, p\} & \text{for the MML method} \end{cases} . \quad \square \qquad (3.2.34)$$

The convergence analysis of the DQ methods goes back to Kobayasi [1966], Linz [1968] and Noble [1969]. The (M)ML methods were proved to be convergent in Wolkenfelt [1981a], [1983b].

Example 3.2.10. We illustrate Corollary 3.2.1 by determining the order of convergence of the methods DQ (Simpson), DQ(BD_2) and DQ (Lobatto) tested in Section 3.2.2.

The lag term quadrature formula used in DQ (Simpson) has already been considered in Example 2.7.1 where it was shown that its order is $r=4$. Hence, when exact starting values are used (i.e., $s=\infty$), Corollary 3.2.1 implies that the DQ (Simpson) method is of order 4. This is in agreement with the effective order of convergence indicated in Table 3.2.2.

The DQ(BD_2) method uses a lag term formula whose approximation error is given by (2.6.52). Thus, the order is $r=\kappa=2$, and the order of convergence is therefore 2. In Table 3.2.2, an effective order $p_{eff}=2.7$ was obtained for $h=3/32$. A decrease in h will finally yield $p_{eff}\approx 3$ instead of $p_{eff}=2$. This higher order of convergence is explained by the fact that the coefficients of the h^2-term in the expression for the quadrature error vanish for the particular problem (3.2.8) (see Wolkenfelt [1981b]).

The approximation error of the lag term formula used in the DQ (Lobatto) method can be estimated by (2.7.3). Here, $\mathbf{R}[\phi]$ corresponds to the error of the repeated 4-point Lobatto formula. Using Corollary 2.7.1 and Theorem 2.4.2, we conclude that $\mathbf{R}[\phi]=O(h^6)$. The error $\mathbf{A}[\phi]$ in (2.7.3) corresponds to the 5-point interpolatory quadrature formulas (3.2.7), so that by virtue of Corollary 2.1.1, $\mathbf{A}[\phi]=O(h^6)$. Hence, the sequence of formulas used in DQ (Lobatto) is of order $r=6$, implying that the order of convergence is 6. $\square$

In Section 7.3 it will be shown that VLM methods for which $\beta(z)$ vanishes behave more stably than DQ methods if $k(t,s,y)$ has a large Lipschitz constants with respect to y. Therefore, it is of interest to know the *maximal* order of convergence within the family of VLM methods for which $\beta(z)\equiv 0$. On the basis of Theorem 3.2.6 the following corollary can be proved.

Corollary 3.2.2. *Let $\delta(h)=O(h^s)$, $\Delta E(h)=O(h^{r+1})$, $\beta(z)\equiv 0$, and let $\alpha(z)$ satisfy the root condition. Then the order of convergence p^* of the κ-step VLM method $\{(3.2.13); (3.2.15)\}$ satisfies*

$$p^* \leqslant \begin{cases} \min(s,r,\kappa+1) & \text{for } \kappa \text{ odd} \\ \min(s,r,\kappa+2) & \text{for } \kappa \text{ even} \end{cases}. \tag{3.2.35}$$

Proof. It follows from Theorem 3.2.6(b) that

$$\epsilon_n = O(h^s + h^r + h^{\hat{p}}),$$

where $\hat{p}$ is the order of the VLM formula (3.2.15). By virtue of Theorem 3.2.1, this implies that $C_{q,l}=0$ for $q=0,1, \ldots ,\hat{p}$, $l=0,1, \ldots ,q$. In particular,

$$C_{q,q} = \frac{1}{q!}(-1)^q \sum_{i=0}^{\kappa}[i^q\alpha_i + i^{q-1}q\gamma_i]=0 \quad \text{for} \quad q=0,1, \ldots ,\hat{p},$$

where we have used the notation of (3.2.28). These conditions are just the order conditions of a $\hat{p}$-th order κ-step formula $\{\alpha,\gamma\}$ for ODEs (cf. Theorem 2.6.3(d)). For such formulas it is known that $\hat{p}\leqslant\kappa+1$ for κ *odd*, and $\hat{p}\leqslant\kappa+2$ for κ *even* (see Theorem 2.6.2). This proves the statement of the corollary. □

It follows from this corollary that the MML methods are of maximal attainable order of convergence if we choose the generating LM formula $\{a_i,b_i\}$ optimally, that is, of order $\kappa+1$ for κ *odd*, and of order $\kappa+2$ for κ *even*. We note that the restriction $p\leqslant\kappa$ in the ILM methods is due to the use of a κ-step forward differentiation formula $\{d_i\}$ in the generating matrix B (see (3.2.22′)).

3.2.9. Numerical examples

In this section we shall illustrate the convergence behavior, as predicted by Corollary 3.2.2, by applying the DQ, ILM and (M)ML methods to a number of problems.

As in the preceding experiments the starting values will be taken from the exact solutions. The lag term is evaluated either by a Gregory rule or by a BD quadrature rule (discussed in Section 2.6.7). In both cases the starting weight matrix is defined by interpolatory quadrature formulas of sufficiently high order. The coefficients $\{a_i,b_i\}$ in the ILM and (M)ML methods are taken from Tables 2.6.1 and 2.6.2.

In the tables of results we list the number of correct significant digits *sd* and the effective order p^*_{eff}, defined, respectively, by (3.2.9) and (3.2.10); the effective orders are preceded by >.

Table 3.2.7a. Numbers of correct significant digits and effective orders for problem (3.2.26) at $t = T = 4$

h	$DQ(G_4)$	$ILM(G_4-AM_6)$	$ML(G_4-AM_4)$	$MML(G_4-AM_5)$
1/5	5.0	-2.5	5.0	5.3
	> 3.5	> 9.0	> 3.5	> 4.1
1/10	6.1	0.2	6.1	6.6
	> 3.7	> 6.7	> 3.7	> 4.0
1/20	7.2	2.2	7.2	7.8

h	$DQ(G_5)$	$ILM(G_5-AM_6)$	$ML(G_5-AM_4)$	$MML(G_5-AM_5)$
1/5	5.3	-2.5	5.1	5.3
	> 4.2	> 9.0	> 4.0	> 4.2
1/10	6.6	0.2	6.3	6.6
	> 4.6	> 6.7	> 4.5	> 4.6
1/20	8.0	2.2	7.7	8.0

h	$DQ(G_6)$	$ILM(G_6-AM_6)$	$ML(G_6-AM_4)$	$MML(G_6-AM_5)$
1/5	5.7	-2.5	5.2	5.4
	> 4.9	> 9.0	> 4.2	> 4.2
1/10	7.1	0.2	6.4	6.6
	> 5.4	> 6.7	> 4.6	> 4.6
1/20	8.8	2.2	7.8	8.0

We start with problem (3.2.26) for which results are listed in Table 3.2.7a. We conclude from these values that the order of the DQ(G_r) method tends to $p^* = r$, the order of the ML($G_r - AM_p$) method tends to $p^* = \min(r, p+1)$, and the order of the MML($G_r - AM_p$) method tends to $p^* = \min(r, p)$. This is in agreement with the theoretical results summarized in Corollary 3.2.1. The results obtained by the ILM methods are not (yet) in agreement with the theory for the integration steps listed. This can be explained by the relatively large lag term derivatives. For a discussion of this point we refer to the numerical experiments in Section 3.2.7.

It should be observed that the results furnished by the methods DQ(G_p) and (M)ML($G_p - AM_p$) are not identical because these methods are not algebraically equivalent. This only occurs if the starting values are identical (cf. Theorem 3.2.4) which is not true in the above experiments ((M)ML uses more starting values than DQ). Presumably, the starting errors are relatively small in this experiment so that seemingly identical results are produced.

The solution of problem (3.2.26) is rapidly *increasing*; by changing the equation (3.2.26) to

$$y(t)=(e^{-t}+1)^2-(t^2+t)e^{-t}+t+\frac{t^3}{2}-t\int_0^t s\sqrt{y(s)}\,ds, \tag{3.2.36}$$

(see Table 8.2.3b, problem 11c), we find an exact solution that rapidly reaches a fixed value: $y(t)=[\exp(-t)+1]^2$. Table 3.2.7b gives the analogue of Table 3.2.7a.

Table 3.2.7b. Numbers of correct significant digits and effective order for problem (3.2.36) at $t=T=4$

h	$DQ(G_4)$		$ILM(G_4-AM_6)$		$ML(G_4-AM_4)$		$MML(G_4-AM_5)$	
1/5	6.7		5.3		6.7		5.3	
		> 4.4		> 5.3		> 4.4		> 4.4
1/10	8.0		6.9		8.0		6.6	
		> 4.2		> 5.9		> 4.2		> 4.2
1/20	9.3		8.6		9.3		7.8	
h	$DQ(G_5)$		$ILM(G_5-AM_6)$		$ML(G_5-AM_4)$		$MML(G_5-AM_5)$	
1/5	6.9		6.1		6.5		6.0	
		> 5.3		> 3.1		> 5.4		> 5.5
1/10	8.5		7.1		8.1		7.7	
		> 5.1		> 3.1		> 5.2		> 5.8
1/20	10.1		8.0		9.7		9.4	
h	$DQ(G_6)$		$ILM(G_6-AM_6)$		$ML(G_6-AM_4)$		$MML(G_6-AM_5)$	
1/5	7.4		6.8		6.9		6.4	
		> 6.3		> 1.2		> 5.1		> 4.6
1/10	9.3		7.1		8.4		7.8	
		> 6.1		> 3.1		> 5.1		> 4.9
1/20	11.2		8.0		9.9		9.2	

Next, consider problem 15 of Table 8.2.3a:

$$y(t) = 1-t+\lambda[\tfrac{1}{2}(1-t^2)\log(1+t)+\tfrac{3}{4}t^2-\tfrac{1}{2}t]$$
$$-\lambda\int_0^t \log(1+t-s)y(s)ds, \tag{3.2.37}$$

Table 3.2.8. Numbers of correct significant digits and effective orders for problem (3.2.37) at $t=T=4$ with $\lambda=4$

h	$DQ(G_5)$		$ILM(G_5-AM_6)$		$ML(G_5-AM_4)$		$MML(G_5-AM_5)$	
1/4	4.6		3.4		4.3		6.1	
		> 4.6		> 3.8		> 4.8		> 3.9
1/8	6.0		4.5		5.7		7.3	
		> 5.0		> 4.5		> 4.6		> 3.2
1/16	7.5		5.9		7.1		8.2	
		> 5.0		> 4.7		> 5.0		> 4.0
1/32	9.0		7.3		8.6		9.4	
		> 5.0		> 4.8		> 5.0		> 4.6
1/64	10.5		8.8		10.1		10.8	

whose exact solution is $y(t)=1-t$. The results for $\lambda=4$, given in Table 3.2.8, confirm that the theoretical order is $p^*=5$, that the MML method is the most accurate method, and that the ILM method is the most inaccurate one. However, if the value of λ in (3.2.37) is increased, all methods except the ILM method tend to develop instabilities. In Table 3.2.9 results are listed for $\lambda=100$. In view of the rather smooth functions g and k, one should expect reasonable accuracy for $h=1/4$ or $h=1/8$. For these steps it is now the ILM method which is the more accurate one whereas the DQ method exhibits the lowest accuracy. For a detailed discussion of the stability of VLM methods we refer to Chapter 7.

Table 3.2.9. Numbers of correct significant digits for problem (3.2.37) at $t=T=4$ with $\lambda=100$

h	$DQ(G_5)$	$ILM(G_5-AM_6)$	$ML(G_5-AM_4)$	$MML(G_5-AM_5)$
1/4	-6.5	1.8	-3.7	-2.4
1/8	2.3	4.5	3.7	4.2
1/16	6.3	5.8	6.2	9.0

Our fourth example is the equation of Feller [1941] given in (3.2.8). In Table 3.2.10, results are listed at $t=t_N=T=6$ for DQ methods generated by Gregory rules and by BD formulas. These results show that both families of LM formulas show a comparable behavior of the effective order. Furthermore, the methods based on Gregory quadrature are more accurate than the methods based on BD quadrature; this is probably due to smaller error constants in the Gregory formulas. This difference in accuracy becomes more pronounced as the order increases.

Table 3.2.10. Numbers of correct significant digits and effective orders for problem (3.2.8) at $t=T=6$ for DQ methods

h	$DQ(G_3)$		$DQ(BD_3)$		$DQ(G_4)$		$DQ(BD_4)$		$DQ(G_5)$		$DQ(BD_5)$	
1/4	2.8	> 2.7	2.9	> 1.7	3.1	> 3.7	2.4	> 3.3	3.7	> 4.3	3.2	> 4.0
1/8	3.6	> 3.0	3.4	> 2.7	4.2	> 4.0	3.4	> 4.0	5.0	> 4.7	4.4	> 4.7
1/16	4.5	> 3.0	4.2	> 2.7	5.4	> 4.0	4.6	> 4.0	6.4	> 5.0	5.8	> 5.0
1/32	5.4	> 3.0	5.0	> 3.0	6.6	> 4.0	5.8	> 4.0	7.9	> 5.0	7.3	> 5.0
1/64	6.3		5.9		7.8		6.9		9.4		8.7	

A similar experiment was carried out for the 5-th order ILM and (M)ML methods. The results in Table 3.2.11 show that the MML method is more sensitive to changes in the lag term formula than the ILM and ML methods.

Table 3.2.11. Numbers of correct significant digits and effective orders for problem (3.2.8) at $t=T=6$ for ILM and (M)ML methods

h	ILM (G_5-AM_6)		ILM (BD_5-AM_6)		ML (G_5-AM_4)		ML (BD_5-AM_4)		MML (G_5-AM_5)		MML (BD_5-AM_5)	
1/4	2.4	> 4.0	2.8	> 3.7	3.4	> 4.4	3.2	> 4.0	3.3	> 4.3	2.6	> 4.0
1/8	3.6	> 4.4	3.9	> 4.3	4.7	> 4.7	4.4	> 4.5	4.6	> 4.7	3.8	> 4.5
1/16	4.9	> 4.7	5.2	> 4.6	6.1	> 4.8	5.8	> 4.7	6.0	> 4.8	5.2	> 4.8
1/32	6.3	> 4.8	6.6	> 4.8	7.6	> 4.9	7.2	> 4.9	7.4	> 4.9	6.6	> 4.9
1/64	7.8		8.0		9.1		8.7		8.9		8.1	

Finally, we consider an example of a *nonlinear* convolution equation given in Table 8.2.3a (problem 26)

$$y(t) = e^{-t} + \int_0^t e^{s-t}[y(s)+e^{-y(s)}]ds; \tag{3.2.38}$$

the exact solution is given by $y(t)=\log(t+e)$. In Table 3.2.12 results for fourth-order methods are presented.

Table 3.2.12. Numbers of correct significant digits and effective orders for problem (3.2.38) at $t=T=6$

h	$DQ(G_4)$		$ILM(G_4-AM_5)$		$ML(G_4-AM_3)$		$MML(G_4-AM_4)$	
1/4	3.4		2.4		3.1		3.8	
		$>$ 3.8		$>$ 3.4		$>$ 3.7		$>$ 3.3
1/8	4.6		3.4		4.2		4.8	
		$>$ 3.9		$>$ 3.7		$>$ 3.9		$>$ 3.6
1/16	5.8		4.6		5.4		5.9	

3.2.10. *Extrapolation methods*

Let the approximate solution $y_n=y_n(h)$, $n=0, \ldots ,N$, be generated by a numerical method using step size h, and let $y_n(h/2)$, $n=0, \ldots ,2N$, be obtained by the same numerical method using step size $h/2$. Consider the following linear combination of these numerical solutions:

$$u_n(h;\alpha) := \alpha y_n(h)+(1-\alpha)y_{2n}(\frac{h}{2}), \quad n=0, \ldots ,N. \tag{3.2.39}$$

Here, α is a free parameter that will be chosen so that the numerical solution $u_n(h;\alpha)$ has a higher accuracy than $y_{2n}(h/2)$. The method defined by (3.2.39) is said to be an *extrapolation method.*

In order to determine a suitable value of the extrapolation parameter α we need certain information on the numerical solution $y_n(h)$. Suppose that we can prove that

$$y(nh)-y_n(h) = c_p(nh)h^p+c_q(nh)h^q+O(h^r), \quad r>q>p, \tag{3.2.40}$$

where the functions $c_p(t)$ and $c_q(t)$ do not depend on h. Evidently, by halving the step size, we also have the relation

$$y(nh)-y_{2n}(\frac{h}{2}) = c_p(nh)\left[\frac{h}{2}\right]^p+c_q(nh)\left[\frac{h}{2}\right]^q+O(h^r). \tag{3.2.40$'$}$$

From these relations we derive that the numerical error of $u_n(h;\alpha)$ satisfies

$$y(nh)-u_n(h;\alpha) = (\alpha+\frac{1-\alpha}{2^p})c_p(nh)h^p+(\alpha+\frac{1-\alpha}{2^q})c_q(nh)h^q+O(h^r), \tag{3.2.41}$$

where $n=0, \ldots ,N$. Thus, by setting the coefficient of h^p equal to 0, i.e.,

$$\alpha = \frac{1}{1-2^p}, \tag{3.2.42}$$

we achieve that (3.2.39) defines a method of order q, whereas the original method is of order $p<q$.

The extrapolation process can be continued by defining

$$v_n(h;\beta) := \beta u_n(h;\alpha)+(1-\beta)u_{2n}(\frac{h}{2};\alpha), \quad n=0, \ldots ,N. \tag{3.2.43}$$

It follows from (3.2.41) and (3.2.42) that

$$y(nh)-v_n(h;\beta) = \beta[y(nh)-u_n(h;\alpha)]+(1-\beta)[y(nh)-u_{2n}(\frac{h}{2};\alpha)]$$

$$= (\alpha+\frac{1-\alpha}{2^q})(\beta+\frac{1-\beta}{2^q})c_q(nh)h^q+O(h^r), \tag{3.2.44}$$

for $n=0, \ldots ,N$. We obtain a method by choosing

$$\beta = \frac{1}{1-2^q} \cdot \tag{3.2.45}$$

A rigorous and detailed treatment of the theory of extrapolation methods can be found in the monograph of Marchuk & Shaidurov [1983]. Here, we restrict our considerations to a discussion of the *trapezoidal method* and its extrapolation. The reason for concentrating on the trapezoidal method are: (i) it satisfies a relation of the form (3.2.40) with *even powers* p, q and r, so that every extrapolation step increases the order by 2 (a sketch of the proof of this assertion in given below); (ii) it is of a *simple structure,* so that the resulting extrapolation scheme is easy to implement on a computer.

The trapezoidal method is given by

$$y_n = g(t_n)+h\sum_{l=0}^{n}{}'' k(t_n,t_l,y_l), \tag{3.2.46}$$

where Σ'' means that the first and last term in the summation are divided by 2. It is convenient to introduce a function interpolating the numerical values $y_n(h):=y_n$, $n=0, \ldots ,N$; this function will be denoted by $y(h;t)$.

Theorem 3.2.7. *Let g, k and the solution y of the Volterra equation (3.2.1) be sufficiently smooth. Then, the error of the trapezoidal method (3.2.46) has a Taylor expansion in even powers of h, i.e.,*

$$y(nh)-y_n(h) = c_2(nh)h^2+c_4(nh)h^4+ \ldots ,$$

where the coefficients $c_{2j}(t)$ do not depend on h.

Proof. First we consider the linear case, where

$$k(t,s,y) = K(t,s)y.$$

Let $\phi\in C^{2m+2}(I)$. Then the (repeated) trapezoidal rule $\mathbf{I}_n[\phi]$ satisfies the relation

$$\mathbf{I}[\phi] := \int_0^{t_n}\phi(s)ds = \mathbf{I}_n[\phi]-\alpha_2h^2-\alpha_4h^4- \ldots -\alpha_{2m}h^{2m}-\mathbf{R}_m[\phi],$$

where the coefficients $\{\alpha_{2j}\}$ and the remainder term $\mathbf{R}_m[\phi]$ are given in (2.6.18) and (2.6.19) (notice that α_{2j} and $\mathbf{R}_m[\phi]$ depend on t_n). By means of this relation and by induction it can be shown that

$$y(h;t_n) = g(t_n)+h\sum_{l=0}^{n}{}'' K(t_n,t_l)y(h,t_l)=g(t_n)+\int_0^{t_n} K(t_n,s)y(h,s)ds$$

$$+ \alpha_2 h^2+\alpha_4 h^4+ \ldots +\alpha_{2m}h^{2m}+\mathbf{R}_m[\phi].$$

Hence, at $t=t_n$, the error $\epsilon(h;t):=y(h;t)-y(t)$ satisfies the relation

$$\begin{aligned}\epsilon(h;t) &= \int_0^t K(t,s)\epsilon(h;s)ds+\tilde{g}(t),\\ \tilde{g}(h;t) &:= \alpha_2 h^2+ \ldots +\alpha_{2m}h^{2m}+\mathbf{R}_m[\phi].\end{aligned} \tag{3.2.47}$$

Let $R(t,s)$ be the resolvent kernel associated with K. Then, by Theorem 1.3.1, equation (3.2.47) is solved by

$$\delta(h;t) := \tilde{g}(h;t)+\int_0^t R(t,s)\tilde{g}(h;s)ds.$$

Evidently, at $t=t_n$ the functions $\epsilon(h;t)$ and $\delta(h;t)$ coincide, so that

$$\begin{aligned}\epsilon(h;t_n) &= y(h;t_n)-y(t_n)=y_n(h)-y(nh)\\ &= \tilde{g}(h;t_n)+\int_0^{t_n} R(t_n,s)\tilde{g}(h;s)ds=O(\tilde{g}) \quad \text{as} \quad \tilde{g}\to 0.\end{aligned} \tag{3.2.48}$$

Substitution of $\tilde{g}(h;t)$ yields the result of the theorem for *linear* equations.

Next we consider the *nonlinear* case. Proceeding as in the proof given above, we derive the nonlinear analogue of the error equation (3.2.47)

$$y(h;t)-y(t) = \int_0^t [k(t,s,y(h;s))-k(t,s,y(s))]ds+\tilde{g}(h;t),$$

where $t=t_n$ and $\tilde{g}(h;t)$ is defined as in (3.2.47). From this relation it follows that

$$\epsilon(h;t_n) = \int_0^{t_n} \frac{\partial k}{\partial y}(t_n,s,y(s))\epsilon(h;s)ds+\hat{g}(h;t_n), \tag{3.2.49a}$$

$$\hat{g}(h;t) := \tilde{g}(h;t)+O(\epsilon^2(h;t)). \tag{3.2.49b}$$

Similar to the derivation of (3.2.48), it can be shown that $\epsilon=O(\hat{g})$ as $\hat{g}\to 0$. Since the trapezoidal rule is of order 2 we have from (3.2.49b) that $\hat{g}=O(h^2)+O(h^4)$. Hence $\epsilon=O(h^2)+O(h^4)$. Again using (3.2.49b) we have that $\hat{g}=O(h^2)+O(h^4)+O(h^6)$, so that $\epsilon=O(h^2)+O(h^4)+O(h^6)$. In this way, it is proved that ϵ has a Taylor expansion in *even* powers of h also in the *nonlinear* case. □

Thus, we can construct a method of order $p=2m$ by carrying out $m-1$ extrapolation steps. The total number of kernel evaluations is given by $\frac{1}{2}N^2(1+4+ \ldots +4^{m-1})$, where $N=T/h$ corresponds to the number of basic steps.

3.2.11. Numerical examples

We shall illustrate the efficiency of the extrapolated trapezoidal method by applying this scheme to the same examples as those tested in Section 3.2.9. The basic integration step h will be chosen such that the computational effort involved, in terms of kernel evaluations, is roughly equal to that required by the methods tested in Section 3.2.9 (recall that the extrapolated trapezoidal method of order $p=2m$ requires $\frac{1}{2}N^2(1+4+ \ldots +4^{m-1})$ kernel evaluations). The extrapolated trapezoidal method will be denoted by ETR_p, where p is the order of the method. Thus, $p=2$ indicates no extrapolation, $p=4$ indicates one extrapolation step, etc. The tables below present the ETR analogues of the Tables 3.2.7 - 3.2.12.

Table 3.2.13a. Numbers of correct significant digits and effective orders for problem (3.2.26) at $t=T=4$

h	ETR_2		h	ETR_4		h	ETR_6	
1/5	4.1		4/9	5.3		1	5.5	
		> 2.0			> 3.9			> 5.3
1/10	4.7		4/18	6.4		4/9	7.1	
		> 2.0			> 4.0			> 5.8
1/20	5.3		4/36	7.6		4/18	8.9	

Table 3.2.13b. Numbers of correct significant digits and effective orders for problem (3.2.36) at $t=T=4$

h	ETR_2		h	ETR_4		h	ETR_6	
1/5	4.2		4/9	7.4		1	4.4	
		> 2.0						> 7.9
1/10	4.8		4/18	6.8		4/9	6.8	
		> 2.0			> 3.8			> 6.8
1/20	5.4		4/36	8.0		4/18	8.9	

Table 3.2.14. Numbers of correct significant digits and effective orders for problem (3.2.37) at $t=T=4$ with $\lambda=4$

h	ETR_2		h	ETR_4		h	ETR_6	
1/4	2.8		4/7	3.2		1	2.8	
		> 2.1			> 4.0			
1/8	3.9		4/14	4.4		4/7	6.3	
		> 2.0			> 4.0			> 3.9
1/16	4.1		4/28	5.6		4/14	7.5	
		> 2.0			> 4.0			> 5.8
1/32	4.7		4/56	6.8		4/28	9.3	
		> 2.0			> 4.0			> 6.0
1/64	5.3		4/112	8.0		4/56	11.1	

Table 3.2.15. Numbers of correct significant digits and effective orders for problem (3.2.37) at $t=T=4$ with $\lambda=100$

h	ETR_2		h	ETR_4	h	ETR_6
1/4	-6.9		4/7	-8.4	1	-8.1
1/8	3.3		4/14	-7.7	4/7	-7.9
		> 2.0				
1/16	3.9		4/28	4.9	4/14	-6.5

Table 3.2.16. Numbers of correct significant digits and effective orders for problem (3.2.8) at $t=T=6$

h	ETR_2		h	ETR_4		h	ETR_6	
1/4	4.2		6/11	3.5		1	4.6	
		> 4.0			> 4.0			> 5.3
1/8	5.4		6/22	4.7		6/11	6.2	
		> 4.0			> 4.0			> 6.0
1/16	6.6		6/44	5.9		6/22	8.0	
		> 4.0			> 4.0			> 6.0
1/32	7.8		6/88	7.1		6/44	9.8	
		> 4.0			> 4.0			> 7.1
1/64	9.0		6/176	8.3		6/88	12.0	

Table 3.2.17. Numbers of correct significant digits and effective orders for problem (3.2.38) at $t=T=6$

h	ETR_2		h	ETR_4		h	ETR_6	
1/4	1.7	> 2.0	6/11	2.5	> 4.2	1	3.0	> 5.8
1/8	2.3	> 2.0	6/22	3.8	> 4.0	6/11	4.7	> 6.2
1/16	2.9		6/44	5.0		6/22	6.6	

Except for the stiff example of Table 3.2.15, the ETR_p method reaches its asymptotic order rather quickly. A comparison with the results furnished by the VLM methods (e.g., $DQ(G_4)$ and ETR_4 in the Tables 3.2.7 and 3.2.13, 3.2.10 and 3.2.16, 3.2.12 and 3.2.17, or $DQ(G_6)$ and ETR_6 in the Tables 3.2.7 and 3.2.13) shows that the ETR method is competitive with VLM methods. However, for stiff problems the ETR method is less suitable, particularly because the relatively large basic integration step determines its stability behavior.

3.2.12. Block-by-block methods

The direct quadrature method (3.2.4) represents a quadrature formula for the right-hand side of the Volterra equation (3.2.1) at t_n that uses *only points in the integration interval* $[0,t_n]$. A block-by-block method uses *also points outside the interval* $[0,t_n]$. The generating quadrature formulas of a block-by-block method assume the form

$$y_n = g_n + h\sum_{l=0}^{\nu m} w_{n,l}k(t_n,t_l,y_l),$$
$$n := (\nu-1)m+j; \quad j=1, \dots ,m: \quad \nu=1, \dots , \frac{N}{m}, \tag{3.2.48}$$

where N is assumed to be a multiple of m. For $j<m$ these quadrature formulas use points outside the integration interval $[0,t_n]$. The corresponding matrix of weights, $W_N=(w_{n,l})$, can be partitioned into blocks, and the numerical values $y_{\nu m-m+1}, \dots ,y_{\nu m}$ are computed simultaneously by solving a system of m implicit equations. Obviously, a block-by-block method is self-starting.

Example 3.2.11. Consider the matrix of weights

$$W_N = \frac{1}{12}\begin{bmatrix} 0 & 0 & 0 & & & & & \\ 5 & 8 & -1 & & & & & \\ 4 & 16 & 4 & & & & & \\ 4 & 16 & 9 & 8 & -1 & & & \\ 4 & 16 & 8 & 16 & 4 & & & \\ 4 & 16 & 8 & 16 & 9 & 8 & -1 & \\ 4 & 16 & 8 & 16 & 8 & 16 & 4 & \\ \vdots & & \ddots & & \ddots & & \ddots & \end{bmatrix}.$$

The underlying quadrature rules are the repeated Simpson rule and a three-point, interpolatory formula for integrating over one interval. Each step, from $t_{2\nu}$ to $t_{2\nu+2}$, requires the solution of $m=2$ equations. The order of this method is $p=4$. □

From the starting matrices S_κ given in (2.6.10), block-by-block methods of order $p=\kappa+1$ can be derived in the same way as the method given in Example 3.2.11 is derived from the matrix S_3 specified in (2.6.10c). It is easily verified that all these methods are block-reducible in the sense of Definition 2.6.1b.

A disadvantage of block-by-block methods is that kernel values at points outside the domain of definition S are required. If a smooth extension of the kernel k cannot be obtained, the following modification, suggested by Linz [1967], [1969b], may be considered: Let $p_m[s;y]$ be the polynomial interpolating $y(s)$ with abscissas $t_{(\nu-1)m}, \ldots, t_{\nu m}$. Then, instead of defining a quadrature rule for $k(t_n,s,y(s))$ over the interval $[t_{(\nu-1)m},t_n]$ with abscissas $t_{(\nu-1)m}, \ldots, t_{\nu m}$, a quadrature rule is defined for $k(t_n,s,p_m[s;y])$ only using abscissas in the integration interval $[t_{(\nu-1)m},t_n]$.

Methods of the type (3.2.48) can be recast in the form of an implicit Runge-Kutta method of Pouzet type (*PVRK methods*) with equally spaced abscissas. A detailed treatment of PVRK methods may be found in Chapter 4, and a specification of a modification similar to that of Linz (see also de Hoog & Weiss [1975]) may be found in Chapter 5.

3.3. Volterra integral equations of the first kind

In this section we consider first-kind equations of the form

$$g(t)+\int_0^t k(t,s,y(s))ds = 0, \quad t\in I:=[0,T]. \tag{3.3.1}$$

We assume that the conditions for the existence of a unique solution $y\in C(I)$ hold (cf. Theorem 1.3.12). This implies in particular that, as in Section 3.2, our considerations are restricted to *nonsingular* equations (for the numerical solution of equations with a weakly singular kernel we refer to Chapter 6).

3.3.1. VLM methods

The most simple discretization of (3.3.1) is the *direct quadrature (DQ) method* (cf. (3.2.4)),

$$g_n+h\sum_{l=0}^{n} w_{n,l}k(t_n,t_l,y_l) = 0, \quad n=\tilde{\kappa}, \dots ,N, \tag{3.3.2}$$

or, equivalently,

$$\tilde{F}_n(t_n) = 0, \quad n=\tilde{\kappa}, \dots ,N, \tag{3.3.2$'$}$$

where $\tilde{F}_n(t)$ is the lag term defined by (3.2.13). As before, we added the tilde to κ in order to indicate its relation with $\tilde{F}_n(t)$. The formula (3.3.2′) is of the form (3.2.15) with $\kappa=0$ and

$$A=0, \quad B=1, \quad C=0. \tag{3.3.3}$$

In Section 3.3.3, we shall see that this special linear method will not always converge, in spite of the use of convergent quadrature rules for the lag term. For instance, the higher order Gregory rules were shown by Linz [1967], [1969c] to yield divergent DQ methods. To overcome this difficulty, several other techniques have been proposed in the literature.

If the derivatives $g'(t)$ and $k_t(t,s,y)$ exist and are known explicitly, then it will often be advantageous to generate approximations to the solution of (3.3.1) by applying a DQ method to its differentiated form,

$$k(t,t,y(t)) = -g'(t)-\int_0^t k_t(t,s,y(s))ds. \tag{3.3.4}$$

If $k(t,t,y)\neq 0$ for $t\in I$, then we may consider this equation as an *implicit second-kind equation* with kernel function $k_t(t,s,y)$. As in (3.2.4) one now applies the DQ method to obtain

$$k(t_n,t_n,y_n) = -g_n'-h\sum_{l=0}^{n} w_{n,l}k_t(t_n,t_l,y_l), \quad n=\tilde{\kappa}, \dots , N. \tag{3.3.5}$$

This "indirect" DQ method can be used to generate the numerical approximations y_n. If $k(t,t,y)\equiv 0$ one may try repeated differentiation of (3.3.1).

In actual computation, the derivatives g' and k_t are often approximated by numerical differentiation; e.g., by using the *forward* differentiation formulas (listed in Table 3.2.3) we obtain a scheme which can be written in the form

$$k(t_n,t_n,y_n) = \frac{1}{h}\sum_{j=0}^{\kappa} d_j \tilde{F}_n(t_{n+j}), \tag{3.3.6}$$

where $\tilde{F}_n(t_n)$ is the lag term defined in (3.2.13). The formula (3.3.6) is again a VLM formula of the form (3.2.15). The parameter matrices are now given by

$$A=0, \quad B=\begin{bmatrix} 0 \cdots 0 & d_0 & d_1 & \cdots & d_k \\ 0 \cdots 0 & 0 & 0 & \cdots & 0 \\ \vdots & \vdots & & & \vdots \\ 0 \cdots 0 & 0 & 0 & \cdots & 0 \end{bmatrix}, \quad C=\begin{bmatrix} 0 \cdots 0 & 1 & 0 & \cdots & 0 \\ 0 \cdots 0 & 0 & 0 & \cdots & 0 \\ \vdots & \vdots & \vdots & & \vdots \\ 0 \cdots 0 & 0 & 0 & \cdots & 0 \end{bmatrix}. \tag{3.3.7}$$

The *indirect DQ method* will be denoted by IBD because of the analogy of the matrices B and C in (3.3.7) to those of the IBD formula for second-kind equations (cf. Section 3.2.6).

If k is of the linear form $K(t,s)y$, then equation (3.3.4) can be written as an (explicit) Volterra equation of the second-kind with kernel function

$$k^*(t,s,y) = -\frac{K_t(t,s)}{K(t,t)}y, \quad K(t,t)\neq 0 \tag{3.3.8}$$

to which the linear methods of Section 3.2 are applicable (we observe that the methods then obtained do not include the IBD method just defined because of the new kernel function k^*). In (3.3.8) it is explicitly stated that $K(t,t)$ must not vanish, otherwise the methods of Section 3.2 do not apply. However, in cases where $K(t,t)$ does not vanish but is of small modulus, the methods of Section 3.2 are not all applicable since the kernel k^* then has a large Lipschitz constant with respect to y. Such equations will be called *stiff Volterra equations* of the second-kind. In Chapter 7 we shall return to this type of equations.

It was indicated in Section 1.3.3 that, for linear kernels, yet another second-kind equation can be obtained from (3.3.1) with kernel

$$k^*(t,s,z) = \frac{K_s(t,s)}{K(t,t)}z, \quad K(t,t)\neq 0, \tag{3.3.9}$$

where z is the new unknown function defined in terms of y,

$$z(t) := \int_0^t y(s)ds.$$

Again, the linear methods of Section 3.2 may be used to get numerical approximations to z. However, in order to obtain an approximation to y itself we have to differentiate numerically which is known to be a numerically unstable process as h becomes small.

As for second-kind equations we now define a general class of *VLM methods* for first-kind equations by means of the *quadrature formula* (3.2.13) and the *VLM formula* (3.2.15). Before discussing the basic concepts of consistency and convergence of these methods, we show that the coefficients α_i in (3.2.15) necessarily vanish in the case of first-kind equations. To see this, we substitute the exact solution $y(t)$ into the VLM formula and we require that it satisfies the VLM formula apart from a term $O(h)$. By virtue of Theorem 3.2.5 we find the necessary condition $A_0=0$; that is, $\alpha(1)=0$, so that $\alpha(z)$ must have a zero at $z=1$. Next, apply the VLM method to the problem where $g(t)\equiv 0$ and $k(t,s,y)\equiv y$ (which has the solution $y(t)\equiv 0$). The numerical solution satisfies the recurrence relation

$$\sum_{i=0}^{\kappa}\alpha_i y_{n-i} = -h\sum_{i=0}^{\kappa}\sum_{j=-\kappa}^{\kappa}[\beta_{i,j}\sum_{l=0}^{n-i}w_{n-i,l}y_l-\gamma_{i,j}y_{n-i}], \tag{3.3.10}$$

which for $h=0$ has the trivial solution $y_n=0$ (with t_n fixed) only if $\alpha(z)$ is a Schur polynomial (cf. Theorem 2.6.8). This contradicts the condition $\alpha(1)=0$, unless $\alpha(z)$ *vanishes identically*. From now on it will be assumed that, whenever the VLM formula (3.2.15) is applied to a first-kind equation, the coefficients α_i, and therefore the matrix A, vanish. Notice that the DQ and IBD formulas discussed in this section do have a zero matrix A.

3.3.2. Consistency of VLM methods

Similar to the consistency of VLM formulas for second-kind equations we find that the exact solution of a first-kind equation satisfies the VLM scheme $\{(3.2.13); (3.2.15)\}$, apart from a residual term $\mathbf{L}_n[F]+\mathbf{Q}_n[y]$, where $\mathbf{L}_n$ and $\mathbf{Q}_n$ are defined as before. Consequently, the consistency definition for VLM methods for the first-kind equation (3.3.1) is identical to that for second-kind equations (cf. Definition 3.2.1), provided that all coefficients α_i occurring in the expression defining the operator $\mathbf{L}_n$ equal zero. Therefore, we introduce

Definition 3.3.1. Let $\alpha_i=0$, $i=0, \ldots, \kappa$. The *VLM operator* (3.2.20) and the associated *VLM formula* (3.2.15) are said to be *of (local) order p* at $t=t_n$ if p is the largest integer so that, for all $X\in C^{p+1}(S)$ which vanish on the line $t=s$, we have $\mathbf{L}_n[X]=O(h^{p+1})$, with t_n fixed as $h\downarrow 0$. □

The analogue of Theorem 3.2.1 then reads:

Theorem 3.3.1. *The VLM operator (3.2.20) and the associated VLM formula (3.2.15) are of order p if $\alpha_i=0$, $i=0, \ldots, \kappa$, and if*

$$B_{q,l} := \frac{1}{l!(q-l)!}\sum_{i=0}^{\kappa}\sum_{j=-\kappa}^{\kappa}(j-i)^{q-l-1}(j+i)^{l-1}[\beta_{i,j}(j^2-i^2)$$

$$-\gamma_{i,j}(qj+qi-2lj)]=0$$

for $q=1, \ldots, p$ *and* $l=1, \ldots, q$, *and if* $B_{p+1,l} \neq 0$ *for at least one value of* $l=1, \ldots, p+1$.

Proof. Taylor expansion of $X(t_{n+j}, t_{n-i})$ yields

$$X(t_{n+j},t_{n-i}) = \sum_{q=0}^{p} \frac{1}{q!} h^q [j\frac{\partial}{\partial t} - i\frac{\partial}{\partial s}]^q X(t,s)|_{(t_n,t_n)}\} + O(h^{p+1}) \quad \text{as} \quad h\downarrow 0.$$

In order to exploit the fact that $X(t,t) \equiv 0$ (see Definition 3.3.1), we introduce the variables $\mathbf{u}:=t+s$ and $\mathbf{v}:=t-s$ and write

$$X(t,s) = X(\frac{u+v}{2}, \frac{u-v}{2}) =: Z(u,v).$$

The identity $X(t,t) \equiv 0$ implies that Z and all its derivatives with respect to u vanish for $u=2t$ and $v=0$. In the following we use the notation

$$Z^{(n,m)} := \frac{\partial^n \partial^m Z}{\partial u^n \partial v^m}(2t_n, 0).$$

By means of the Binomial Theorem we have

$$X(t_{n+j},t_{n-i}) = \sum_{q=0}^{p} \frac{1}{q!} h^q [(j-i)\frac{\partial}{\partial u} + (j+i)\frac{\partial}{\partial v}]^q Z(u,v)|_{(2t_n,0)} + O(h^{p+1})$$

$$= \sum_{q=0}^{p} \sum_{l=0}^{q} \frac{1}{q!} h^q \begin{pmatrix} q \\ l \end{pmatrix} (j-i)^{q-l} (j+i)^l Z^{(q-l,l)} + O(h^{p+1}) \tag{3.3.11}$$

and

$$hX_s(t_{n+j},t_{n-i}) = \sum_{q=0}^{p} \sum_{l=0}^{q} \frac{1}{q!} h^{q+1} \begin{pmatrix} q \\ l \end{pmatrix} (j-i)^{q-l} (j+i)^l \cdot$$

$$\cdot [Z^{(q-l+1,l)} - Z^{(q-l,l+1)}] + O(h^{p+1})$$

$$= \sum_{q=0}^{p} \sum_{l=0}^{q} \frac{1}{q!} h^q \begin{pmatrix} q \\ l \end{pmatrix} (j-i)^{q-l-1} (j+i)^{l-1} [qj + qi - 2lj] \cdot$$

$$\cdot Z^{(q-l,l)} + O(h^{p+1}). \tag{3.3.12}$$

If we substitute (3.3.11) and (3.3.12) into $\mathbf{L}_n[X]$ and use the fact that $Z^{(q,0)}=0$, we find

$$\mathbf{L}_n[X] = \sum_{q=1}^{p} \sum_{l=1}^{q} h^q B_{q,l} \left(\frac{\partial^{q-l} \partial^l}{\partial u^{q-l} \partial v^l} Z \right) (2t_n, 0) + O(h^{p+1}) \tag{3.3.13}$$

as $h\downarrow 0$, where $B_{q,l}$ is defined in the theorem. The assertion of Theorem 3.3.1 now follows readily. □

Example 3.3.1. The DQ formula (3.3.3) is easily verified to satisfy the order conditions for all p and has therefore infinite order. □

Example 3.3.2. The IBD formula (3.3.7) yields the coefficients

$$B_{q,l} = \sum_{j=0}^{k} j^q d_j + 0^{q-1}.$$

On the basis of (3.2.23) we conclude that $B_{q,l}=0$ for $q=1, \dots, \kappa$; $l=1, \dots, q$. Hence, the IBD formula is of order κ. □

3.3.3. Special VLM formulas

Indirect linear multistep formulas

Consider the generating matrices (3.2.22′) of the ILM formula for second-kind equations, and replace the matrix A by a zero matrix. The resulting formula will be called an *ILM formula for the first-kind equations* (3.3.1).

Theorem 3.3.2. *The ILM formula (3.2.22′) with $A=0$ is of order $p^*=\kappa$.*

Proof. Substitution of $\beta_{i,j}=b_i d_{j+i}$ and $\gamma_{i,j}=b_i \delta_{i,-j}$ (with δ denoting the Kronecker symbol) yields for the coefficients $B_{q,l}$ the expression

$$B_{q,l} = \frac{1}{l!(q-l)!} \sum_{i=0}^{\kappa} \{ \sum_{j=-i}^{\kappa-i} ((j-i)^{q-l}(j+i)^l b_i d_{i+j}) + l(-2i)^{q-l} 0^{l-1} b_i \}$$

$$= \frac{1}{l!(q-l)!} \sum_{i=0}^{\kappa} b_i \{ \sum_{j=0}^{\kappa} (j^l (j-2i)^{q-l} d_j) + l(-2i)^{q-l} 0^{l-1} \}.$$

Using (3.2.23) it is easily shown that all $B_{q,l}$ with $q \leqslant \kappa$, $l \leqslant q$, vanish. □

We note that Theorem 3.3.2 does not refer to an LM method for ODEs; that is, the coefficients b_i can be chosen arbitrarily. (This is in marked contrast to ILM formulas for second-kind equations (cf. Theorem 3.2.2).) Their values, however, are governed by convergence and stability considerations.

The principal error constant is given by

$$B_{p+1,l} = \frac{(-1)^{\kappa} \kappa!}{l!(p+1-l)!} \sum_{i=0}^{\kappa} b_i, \quad 1 \leqslant l \leqslant p^*+1, \quad p^*=\kappa. \tag{3.3.14}$$

In Table 3.3.1 these constants are given for the ILM (BD) and ILM (AM) formulas.

Table 3.3.1. Error constants $B_{p^*+1,l} = B^*_{p^*+1,l}\dfrac{1}{l!(p^*+1-l)!}$, $l=1, \dots, p^*+1$, for ILM formulas

Generating LM formula			$\kappa=1$	$\kappa=2$	$\kappa=3$	$\kappa=4$	$\kappa=5$
$BD(p^*=\kappa)$	$B^*_{p^*+1,l<0}$	$=$	-1	$\frac{4}{3}$	$-\frac{36}{11}$	$\frac{288}{25}$	$-\frac{7200}{137}$
$AM(p^*=\kappa)$	$B^*_{p^*+1,l>0}$	$=$	-1	2	-6	24	-120

Multilag formulas

The matrix $A:=0$ and the matrices B and C defined in (3.2.24) and (3.2.25) will be said to generate, respectively, an *ML* and *MML formula for the first-kind equations* (3.3.1). These formulas are due to Wolkenfelt [1981a], [1983b].

Theorem 3.3.3. *Let p be the order of the LM formula for ODEs defining the coefficients* $\{a_i,b_i\}$. *Then the (M)ML formulas are both of order* $p^*=p$. □

The proof is analogous to that of Theorem 3.3.2 and is left to the reader. The principal error constants in the (M)ML formulas are given by

$$B_{p^*+1,l} = \frac{(-1)^{p^*+1-l}}{l!(p^*+1-l)!}\sum_{i=0}^{\kappa} i^p[ia_i+(p^*+1)b_i], \quad 1\leq l\leq p^*+1. \quad (3.3.15)$$

For the BD and AM formulas the corresponding $B_{p^*+1,l}$-values are listed in Table 3.3.2. A comparison with Table 3.3.1 reveals that the BD formulas give rise to error constants of the same magnitude in the ILM and (M)ML formulas, whereas the AM formulas lead to slightly smaller error constants in the (M)ML formulas.

Finally, we remark that the (M)ML methods are algebraically equivalent with the DQ method if the lag term approximations and the (M)ML formulas are both based on the same LM formula $\{\tilde{\rho},\tilde{\sigma}\}$. The proof is identical to that of Theorem 3.2.4.

Table 3.3.2. Error constants $B_{p^*+1,l}=B^*_{p^*+1,l}\dfrac{(-1)^{l-p^*-1}}{l!(p^*+1-l)!}$, $l=1, \ldots, p^*+1$, for (M)ML formulas

Generating LM formula		$\kappa=1$	$\kappa=2$	$\kappa=3$	$\kappa=4$	$\kappa=5$
$BD(p^*=\kappa)$	$B^*_{p^*+1,l>0}=$	-1	$\frac{4}{3}$	$-\frac{36}{11}$	$\frac{288}{25}$	$-\frac{7200}{137}$
$AM(p^*=\kappa+1)$	$B^*_{p^*+1,l>0}=$	$\frac{1}{2}$	-1	$\frac{19}{6}$	$-\frac{27}{2}$	$\frac{863}{12}$

3.3.4. Convergence results for VLM methods

Convergence of the VLM method for first kind equations is defined in the same way as in Definition 3.2.2. Before giving convergence results we consider the modified equation corresponding to the VLM method for first-kind Volterra equations. This equation is given by (3.2.29) when we replace A_q by 0 for all q:

$$\sum_{q=0}^{m} h^q \sum_{l=0}^{q} C_{q,l} \left[\frac{\partial}{\partial t}\right]^{q-l} \left[\frac{\partial}{\partial s}\right]^{l} F(t,t) = O(h^r+h^{m+1}), \tag{3.3.16}$$

where $C_{q,l}$ is defined in Theorem 3.2.1, r is the order of accuracy of the lag term formula, and m depends on the differentiability of the substituted lag term $F(t,s)$.

Example 3.3.1. The DQ method has $A_0=0$ and $C_{0,0}=1$, so that the limiting equation is $F(t,t)=0$, i.e. (3.3.1). The same limiting equation is obtained for the ML method. □

Example 3.3.2. For the MML and ILM methods, generated by a convergent LM formula $\{\rho,\sigma\}$, we have

$$C_{q,l} = \sum_{i=0}^{\kappa}(-i)^q a_i.$$

Hence, $C_{0,0}=\rho(1)$ and $C_{1,l}=-\sigma(1)$ lead to the limiting equation

$$k(t,t,y(t))+F_t(t,t) = 0,$$

which is easily recognized as being the differentiated form of (3.3.1). □

We emphasize that the MML and ILM methods, although being discretizations of a differentiated Volterra equation, can be applied to equations in which g and k are *not differentiable* with respect to t.

In the following we summarize the convergence results obtained for the various VLM methods for first-kind Volterra integral equations.

DQ methods

First, we discuss convergence results obtained for DQ methods. The analysis of convergence goes back to Jones [1961] who studied the trapezoidal method in the case of convolution kernels of the linear form

$$k(t,s,y) = K(t-s)y.$$

Jones proved second-order convergence and indicated how to extend his result to systems of convolution equations.

Linz [1967] extended the results of Jones to the class of linear kernels and proved convergence for DQ methods in which the quadrature rule is reducible to the trapezoidal rule ($\tilde{\rho}=z-1, \tilde{\sigma}=(z+1)/2$) or to the midpoint rule ($\tilde{\rho}=z^2-1, \tilde{\sigma}=z$). But he also showed that other DQ methods (e.g., those based on the higher-order Gregory rules, see Section 2.6.3) may be divergent. Gladwin & Jeltsch [1974] too gave "non-convergence" results. They proved that a DQ method based on a $\tilde{\kappa}$-step implicit LM formula of order p derived from interpolatory quadrature (i.e., $\tilde{\rho}=z^{\tilde{\kappa}}-z^{\tilde{\kappa}-r}$) cannot be convergent if $p>\tilde{\kappa}$ (this includes the negative result of Linz because Gregory rules are based on Adams-Moulton methods, for which $\tilde{\rho}=z^{\tilde{\kappa}}-z^{\tilde{\kappa}-1}$).

In order to see the cause of these difficulties we consider the trivial equation

$$g(t)+\int_0^t y(s)ds = 0.$$

Application of a DQ method yields the system of equations

$$g_n+h\sum_{l=0}^{n} w_{n,l}y_l = 0, \quad n=\tilde{\kappa}, \ldots, N.$$

Assuming that the starting values $y_0, \ldots, y_{\tilde{\kappa}-1}$ are given, we can write the solution y_n in the form

$$y_n = \frac{1}{h}\sum_{l=\tilde{\kappa}}^{n} v_{n,l}g_l+\sum_{l=0}^{\tilde{\kappa}-1} v_{n,l}y_l \tag{3.3.17}$$

with suitable coefficients $v_{n,l}$. Suppose that the weights $w_{n,l}$ are all positive. Then it can be expected that many of the coefficients $v_{n,l}$ have alternating signs. The exact solution is given by

$$y(t_n) = g'(t_n),$$

so that the numerical approximation (3.3.17) does not seem to be a very suitable approximation to $y(t_n)$ unless the $v_{n,l}$ decrease sufficiently in magnitude for lower values of l. Let us consider the case where $\tilde{\kappa}=1$; then, by Cramer's rule,

$$v_{n,l} = -\frac{A_{l,n}}{w_{1,1}w_{2,2}\ \cdots\ w_{n,n}}, \quad l=1,\ \dots\ ,n,$$

where $A_{l,n}$ denotes the cofactor of $w_{l,n}$ in the matrix $(w_{i,j})$ with $i,j=1,\ \dots\ ,n$. Thus,

$$v_{n,l} = -\frac{1}{w_{l,l}\ \cdots\ w_{n,n}} \cdot \det \begin{bmatrix} w_{l+1,l} & w_{l+1,l+1} & & \\ \vdots & \vdots & \ddots & \\ w_{n-1,l} & w_{n-1,l+1} & \cdots & w_{n-1,n-1} \\ w_{n,l} & w_{n,l+1} & \cdots & w_{n,n-1} \end{bmatrix}.$$

This expression indicates that the quadrature method will be better as the off-diagonal elements have smaller moduli than the diagonal elements. As an illustration consider the Gregory rules essentially defined by the matrices $\Omega_{\kappa,n}$ given by (2.6.17). Evidently, these matrices have relatively large off-diagonal elements, thus explaining their bad performance in the integration of first-kind equations. In contrast, consider now the matrix of weights generated by the backward differentiation formulas, an example of which is given in Example 3.2.2. Here, the diagonal elements are 2/3, whereas the off-diagonal elements are all less than 1, so that a better performance may be expected. In fact, it can be shown that this method, when applied to kernels of the form $K(y)$, expresses y_n in the form of a *finite* summation of $g(t_l)$-values. Such methods correspond to "local differentiation methods" discussed below.

The above heuristic arguments are quantified by the following, rather general convergence result, due to Wolkenfelt [1981a], [1981c], which holds for $(\tilde{\rho},\tilde{\sigma})$-reducible DQ methods applied to *linear* equations.

Theorem 3.3.4. *Let $A=0$, $B=1$, $C=0$ (DQ formula), let the lag term be defined by a $(\tilde{\rho},\tilde{\sigma})$-reducible quadrature formula of order $r\geqslant 1$ (where $\tilde{\sigma}$ satisfies the root condition), let k be linear in y, and let the conditions of Theorem 1.3.5 be satisfied. Then there exists a constant C, independent of h, such that for all sufficiently small h,*

$$|\epsilon_n| \leqslant C[\delta(h)+E(h)], \quad n=\tilde{\kappa},\ \dots\ ,N,$$

where $\delta(h)$ and $E(h)$ denote, respectively, the maximal starting error and the maximal approximation error. □

The proof of this theorem is rather complicated and space prevents us reproducing it here. We shall only show that $\tilde{\sigma}(z)$ must necessarily satisfy the root condition (Definition 2.6.2 (ii)). Following Wolkenfelt we apply the DQ method to the equation where $g(t)\equiv 0$ and $k(t,s,y)\equiv y$, to obtain the scheme

$$h\sum_{l=0}^{n} w_{n,l}y_l = 0, \quad n\geqslant\tilde{\kappa}. \tag{3.3.2'}$$

Since the quadrature weights correspond to a $(\tilde{\rho},\tilde{\sigma})$-reducible quadrature formula, they satisfy by definition the relations (2.6.7). This suggests multiplying the $(n-i)$-th equation in (3.3.2 ′) by $\tilde{a}_i$ $(i=0,1, \ldots ,\tilde{\kappa})$, summing over i and simplifying by means of (2.6.7), to obtain the equation

$$h\sum_{i=0}^{\tilde{\kappa}}\tilde{a}_i\sum_{l=0}^{n-i}w_{n-i,l}y_l = h\sum_{i=0}^{\tilde{\kappa}}\tilde{b}_i y_{n-i}=0, \quad n\geqslant\tilde{\kappa}.$$

The characteristic polynomial of this linear difference equation is given by $\tilde{\sigma}(z)$, and by virtue of Theorem 2.6.9 we conclude that its solution will become unbounded as $n\to\infty$ if $\tilde{\sigma}(z)$ has roots outside the unit circle or multiple roots on the unit circle. Since $y(t)\equiv 0$, a necessary condition for convergence is, as asserted, the root condition for $\tilde{\sigma}(z)$.

It is because of this condition that the Gregory rules of order $r>2$ do not generate a convergent DQ method. To see this we recall that the polynomials $\tilde{\sigma}(z)$ in the AM formulas (see Table 2.6.1) have a zero to the left of -1 for $\tilde{\kappa}>2$. The BD formulas, on the other hand, possess a polynomial $\tilde{\sigma}$ of the form $\tilde{b}_0 z^{\tilde{\kappa}}$, and are therefore expected to be convergent.

VLM methods

Within the general class of VLM methods a convergence theorem has been given for the MML methods (cf. Wolkenfelt [1981a], [1983b]). Here, we extend this convergence theorem to all methods which have a vanishing polynomial $\beta(z)$ and a polynomial $\gamma(z)$ with all its zeros within the unit circle, that is, $\gamma(z)$ is a Schur polynomial. First, however, we show that $\gamma(z)$ must at least satisfy the root condition (i.e., it must be a simple von Neumann polynomial). Just as in the case of a convergent, $(\tilde{\rho},\tilde{\sigma})$-reducible DQ method (where $\tilde{\sigma}$ is required to satisfy the root condition), we find that the VLM method, when applied to the test equation with $g(t)\equiv 0$ and $k(t,s,y)\equiv y$, reduces to the recurrence relation

$$\gamma(E)y_n = 0, \quad n\geqslant 0.$$

Its solution becomes unbounded as $n\to\infty$ unless γ satisfies the root condition. Thus, whatever we choose for the lag term formula, the (M)ML and ILM methods generated by the AM formulas are not convergent, whereas the BD formulas are still candidates for the generation of convergent VLM methods.

The following theorem provides sufficient conditions for the VLM method to be convergent for first-kind Volterra integral equations.

Theorem 3.3.5. *Let $\alpha(z)\equiv\beta(z)\equiv 0$, let $\gamma(z)$ be a Schur polynomial, let k be linear in y, and let the conditions of Theorem 1.3.5 be satisfied. Then there exists a constant C, independent of h, such that for all sufficiently small h,*

$$|\epsilon_n| \leqslant Ch^{-1}[h\delta(h)+\Delta E(h)+T(h)], \quad n=\kappa, \ldots ,N.$$

Proof. Let

$$k(t,s,y) = K(t,s)y, \quad K_{i,j}=K(t_i,t_j).$$

Then, following the first lines of the proof of Theorem 3.2.6 we obtain from (3.2.31) the relation

$$\sum_{i=0}^{\kappa}\sum_{j=-\kappa}^{\kappa}\gamma_{i,j}K_{n+j,n-i}\epsilon_{n-i} = \sum_{i=0}^{\kappa}\sum_{j=-\kappa}^{\kappa}\beta_{i,j}[\sum_{l=0}^{n}w_{n-i,l}K_{n+j,l}\epsilon_l$$
$$+h^{-1}E_{n-i}(h;t_{n+j})]-h^{-1}\mathbf{L}_n[F], \; n\geqslant\kappa. \tag{3.3.18}$$

Now we write

$$K_{n+j,n-i} = K_{n,n}+(K_{n+j,n-i}-K_{n,n}),$$
$$K_{n+j,l} = K_{n,l}+(K_{n+j,l}-K_{n,l}),$$

and rewrite (3.3.18) to obtain

$$\gamma(E)\epsilon_{n-k} = v_n, \quad n\geqslant\kappa, \tag{3.3.19}$$

where

$$K_{n,n}v_n = \sum_{i=0}^{\kappa}\sum_{j=-\kappa}^{\kappa}[h\gamma_{i,j}\frac{K_{n,n}-K_{n+j,n-i}}{h}\epsilon_{n-i}+\beta_{i,j}\sum_{l=0}^{n}w_{n-i,l}K_{n,l}\epsilon_l$$
$$+ h\beta_{i,j}\sum_{l=0}^{n}\frac{K_{n+j,l}-K_{n,l}}{h}\epsilon_l+h^{-1}\beta_{i,j}E_{n-i}(h;t_{n+j})]-h^{-1}\mathbf{L}_n[F].$$

Since $\gamma(z)$ is a Schur polynomial, that is, all its zeros are within the unit circle, we may apply Theorem 2.6.8 to (3.3.19) to obtain

$$|\epsilon_n| \leqslant C_0[\delta(h)+\max_{\kappa\leqslant j\leqslant n}|v_j|\ ], \; n\geqslant\kappa, \tag{3.3.20}$$

where C_0 (and all subsequent C_i) are independent of h and n. Using the conditions of the theorem we find for v_j the bound

$$|v_j| \leqslant C_1[h\sum_{i=0}^{\kappa}|\epsilon_{n-i}|+h\sum_{l=0}^{j}|\epsilon_l| +h^{-1}\Delta E(h)]+h^{-1}T(h)$$
$$\leqslant C_2[h\sum_{l=0}^{j}|\epsilon_l|+h^{-1}\Delta E_j(h)]+h^{-1}T(h)$$

(cf. the derivation of (3.2.32)). Substitution into (3.3.19) yields, for all sufficiently small h.

$$|\epsilon_n| \leqslant C_3h\sum_{l=0}^{n-1}|\epsilon_l|+C_4h^{-1}[h\delta(h)+\Delta E(h)+T(h)].$$

Applying Corollary 1.5.2 gives

$$|\epsilon_n| \leqslant (1+C_3h)^{n-\kappa}(\kappa hC_3\delta(h)+C_4h^{-1}[h\delta(h)+\Delta E(h)+T(h)]).$$

Since $nh\leqslant T$ we arrive at the result of the theorem. □

Observe that this convergence result is identical to that obtained for VLM methods applied to second-kind equations with $\beta(z)\equiv 0$ (see Theorem 3.2.6 (b)). In the case of the ML method where $\beta(z)\not\equiv 0$, Theorem 3.3.5 does not apply. However, if both the lag term quadrature formula and the ML formula are based on the same LM formula $\{\tilde{\rho},\tilde{\sigma}\}$, the ML method is algebraically equivalent with the $\{\tilde{\rho},\tilde{\sigma}\}$-reducible DQ methods, so that Theorem 3.3.4 can be used. The same is true for MML methods. This last remark indicates that the conditions of Theorem 3.3.5 can be weakened (at least in the case of the MML $(\{\tilde{\rho},\tilde{\sigma}\}-\{\tilde{\rho},\tilde{\sigma}\})$ methods).

In the following corollary we summarize the convergence results obtained in this section.

Corollary 3.3.1. *Let $\tilde{p}$ and p be the orders of the LM formulas $\{\tilde{\rho},\tilde{\sigma}\}$ and $\{\rho,\sigma\}$ employed in the DQ lag term formula and in the MML formula, respectively. Furthermore, let $\delta(h)=O(h^s)$, $E(h)=O(h^r)$, and $\Delta E(h)=O(h^{r+1})$ as $h\downarrow 0$. If the functions g and $k=K(t,s)y$ are sufficiently smooth, then the order of convergence p^* is given by*

$$p^* = \begin{cases} \min\{s,\ \tilde{p}\} & \textit{for the DQ method with } \tilde{\sigma} \textit{ satisfying the root condition} \\ \min\{s,r,\ p\} & \textit{for the MML method with } \gamma \textit{ being a Schur polynomial} \\ \min\{s,r,\kappa\} & \textit{for the ILM method with } \gamma \textit{ being a Schur polynomial.} \end{cases}$$ □

The proof of Theorem 3.3.5 can easily be given for *nonlinear* kernels in the special case where all elements of the matrix C vanish except for those on the first row (compare Wolkenfelt [1983b]).

Theorem 3.3.6. *Let $\alpha(z)\equiv\beta(z)\equiv 0$, let $\gamma(z)=\gamma_0 z^\kappa$ and let the conditions of Theorem 1.3.12 be satisfied. Then there exists a constant C, independent of h, such that for all sufficiently small h,*

$$|\epsilon_n| \leqslant Ch^{-1}[h\delta(h)+\Delta E(h)+T(h)], \quad n=\kappa, \ldots ,N.$$

Proof. From (3.2.31) we derive

$$\sum_{j=-\kappa}^{\kappa} \gamma_{0,j}\Delta k(t_{n+j},t_n,y(t_n),y_n) = u_n,$$

where

$$u_n := -\sum_{i=0}^{\kappa}\sum_{j=-\kappa}^{\kappa} \beta_{i,j}[\sum_{l=0}^{n} w_{n-i,l}\Delta k(t_{n+j},t_l,y(t_l),y_l)$$

$$-h^{-1}E_{n-i}(h;t_{n+j})]-h^{-1}\mathbf{L}_n[F].$$

It then follows from the conditions of Theorem 1.3.12 that

$$\Delta k(t_{n+j},t_n,y(t_n),y_n) = \epsilon_n k_y(t_{n+j},t_n,\bar{y}_n)-\epsilon_n k_y(t_n,t_n,\bar{y}_n)$$

as $h\downarrow 0$ (here, $\bar{y}_n$ is an intermediate point with respect to y_n and $y(t_n)$). Since $k_y(t,t,y)$ is bounded away from zero on $I\times\mathbb{R}$, we have for all sufficiently small h,

$$|\Delta k(t_{n+j},t_n,y(t_n),y_n)| \geqslant L_j|\epsilon_n|$$

so that

$$\sum_{j=-\kappa}^{\kappa} |\gamma_{0,j}|L_j|\epsilon_n| \leqslant |u_n|, \quad n\geqslant\kappa.$$

Proceeding as in the proof of Theorem 3.2.6 (a) we find the convergence result of Theorem 3.3.6. □

3.3.5. Local differentiation methods

The expression (3.3.17) for the numerical solution of the first kind equation with $k(t,s,y)=y$ led Noble [1964, p. 255] to suggest looking for quadrature methods such that only a few coefficients $v_{n,l}$ in (3.3.17) do not vanish. As an example of such methods Noble mentioned the mid-point rule.

Following this suggestion, Holyhead, McKee and Taylor [1975], Holyhead and McKee [1976], and Taylor [1976] describe how to obtain convergent, high-order methods which satisfy the requirement that only a limited number of coefficients in (3.3.17) do not vanish Such methods belong to the so-called 'local differentiation methods' (see Taylor [1976] and Keech [1977b, p. 317]):

Definition 3.3.2. A method will be said to be a *local differentiation method* if, when applied to first-kind equations with $k(t,s,y)=K(y)$, the numerical solution satisfies

$$K(y_n) = \frac{1}{h}\sum_{l=l_1}^{l_2} v_l g_{n+l} = g'(t_n)+O(h^q),$$

for $n\geqslant n_0$, where $\{v_l\}$, l_1, l_2 and $q>0$ are constants independent of n. □

Within the class of $(\tilde{\rho},\tilde{\sigma})$-reducible DQ methods Wolkenfelt [1981c] has given a precise characterization of local differentiation methods.

Theorem 3.3.7. *A convergent $(\tilde{\rho},\tilde{\sigma})$-reducible DQ method applied to first-kind equations with $k(t,s,y)=K(y)$ yields a local differentiation method if, and only if, $\tilde{\sigma}(z)=\tilde{b}_j z^{\tilde{\kappa}-j}$ for some j with $0\leqslant j\leqslant\tilde{\kappa}$.*

Proof. Application of the $(\tilde{\rho},\tilde{\sigma})$-reducible DQ method yields the recurrence relation

$$\tilde{\sigma}(E)K(y_n) = -\frac{1}{h}\tilde{\rho}(E)g(t_n).$$

Suppose that at least two coefficients $\tilde{b}_i$ in $\tilde{\sigma}(z)$ are different from zero. Then this recurrence relation will not lead to a local differentiation method in the sense of Noble. Hence, the necessary part of the theorem is proved. Next, let $\tilde{b}_j$ be the non-zero coefficient in $\tilde{\sigma}(z)$. Then

$$\tilde{b}_j K(y_{n-j}) = -\frac{1}{h}\,\tilde{\rho}(E)g(t_n) = -\tilde{\rho}'(1)g'(t_{n-j}) + O(h)$$

so that

$$K(y_{n-j}) = -\frac{\tilde{\rho}'(1)}{\tilde{\sigma}(1)}\,g'(t_{n-j}) + O(h) = g'(t_{n-j}) + O(h). \quad \square$$

Examples of local differentiation methods within the class of DQ methods are the backward differentiation methods.

Within the class of VLM methods the following characterization of local differentiation methods can be given

Theorem 3.3.8. *A consistent VLM method with $\beta(z)\equiv 0$ applied to first-kind equations with $k(t,s,y)=K(y)$, yields a local differentiation method if, and only if, $\gamma(z)=\gamma_j z^{\kappa-j}$ for some j with $0\leqslant j\leqslant\kappa$.*

Proof. Application of the VLM method yields the recurrence relation

$$\begin{aligned}
\gamma(E)K(y_n) &= \frac{1}{h}E^{\kappa}\sum_{i=0}^{\kappa}\sum_{j=-\kappa}^{\kappa}\beta_{i,j}\tilde{F}_{n-i}(t_{n+j}) \\
&= \frac{1}{h}\sum_{i=0}^{\kappa}\sum_{j=-\kappa}^{\kappa}\beta_{i,j}g(t_{n+j+\kappa}) \\
&= \sum_{i=0}^{\kappa}\sum_{j=-\kappa}^{\kappa} j\beta_{i,j}\frac{dg}{dt}(t_{n+\kappa}) + O(h).
\end{aligned}$$

It follows from the first-order consistency that

$$\begin{aligned}
B_{1,1} := \sum_{i=0}^{\kappa}\sum_{j=-\kappa}^{\kappa}[(j+i)\beta_{i,j}+\gamma_{i,j}] &= \sum_{i=0}^{\kappa}\sum_{j=-\kappa}^{\kappa}[j\beta_{i,j}+\gamma_{i,j}] \\
&= \sum_{i=0}^{\kappa}\sum_{j=-\kappa}^{\kappa} j\beta_{i,j} + \gamma(1) = 0.
\end{aligned}$$

Thus, we find the relation

$$\gamma(E)K(y_n) = -\gamma(1)\frac{dg}{dt}(t_{n+\kappa}) + O(h).$$

The assertion of the theorem can now be proved in a similar way as in the proof of Theorem 3.3.7. $\square$

3.3.6. Numerical examples

We have employed the DQ, ILM and (M)ML methods with a Gregory lag term approximation of order 4 and 5, and with a VLM formula based on a BD formula of order 4 and 5. Starting values were derived from the exact solutions (for further implementational details we refer to Sections 3.2.7 and 3.2.9).

The first two equations to be integrated are problem 3 of Table 8.4.4a:

$$(a^2+1)\int_0^t \cos(t-s)y(s)ds = \sin(t)+a(e^{at}-\cos(t)), \tag{3.3.21}$$

with the exact solution $y(t)=\exp(at)$, and problem 2 of Table 8.4.4a:

$$a\int_0^t e^{a(t-s)}y(s)ds = \sinh(at) \tag{3.3.22}$$

with the exact solution $y(t)=\exp(-at)$. Our third test problem is a *nonlinear* equation, illustrating the performance of VLM methods for nonlinear problems:

$$\int_0^t e^{s-t}[y(s)+e^{-y(s)}]ds = \log(t+e)-e^{-t}; \tag{3.3.23}$$

its exact solution is $y(t)=\log(t+e)$ (cf. problem 7 in Table 8.4.4a). This equation may be considered as the "first-kind" version of equation (3.2.38).

Results for the first test problem are listed in Table 3.3.3; here, an asterisk denotes divergence. The results produced by the ILM and MML methods show that the corresponding asymptotic order is already reached for relatively large values of h. The apparent unstable behavior of the DQ and ML methods is explained by the fact that Gregory quadrature formulas of order $\geqslant 3$ can be traced back to a linear multistep method $\{\tilde{\rho},\tilde{\sigma}\}$ in which $\tilde{\sigma}$ is *not* a simple von Neumann polynomial (cf. Corollary 3.3.1). Notice that here, in contrast to second-kind equations, the ILM method is more accurate than the MML method.

In order to show that the divergence of the DQ and ML method in Table 3.3.3a is due to the lag term formula we have integrated the same problem with BD lag term formulas instead of using Gregory lag term formulas. Results are given in Table 3.3.3b.

Results for problem (3.3.22) are presented in Table 3.3.4. Notice that again the ILM method is superior to the MML method. We also observe that the effective order of the fifth-order ILM method differs considerably from the asymptotic order.

In Table 3.3.5 results for the nonlinear problem (3.3.23) are listed. Here, the ILM and MML method furnish comparable results.

Table 3.3.3a. Numbers of correct significant digits and effective orders for problem (3.3.21) with $a=1$ at $t=T=4$.

h	DQ (G_4)	ILM (G_4-BD_4)		ML (G_4-BD_4)	MML (G_4-BD_4)	
1/10	⋆	4.3	> 4.1	⋆	3.9	> 3.8
1/20	⋆	5.5	> 4.0	⋆	5.1	> 3.9
1/40	⋆	6.7	> 4.0	⋆	6.3	> 4.0
1/80	⋆	7.9		⋆	7.5	
h	DQ (G_5)	ILM (G_5-BD_5)		ML (G_5-BD_5)	MML (G_5-BD_5)	
1/10	⋆	5.6	> 4.9	⋆	4.9	> 4.9
1/20	⋆	7.0	> 5.0	⋆	6.4	> 5.0
1/40	⋆	8.5		⋆	7.9	

Table 3.3.3b. Numbers of correct significant digits and effective orders for problem (3.3.21) with $a=1$ at $t=T=4$

h	DQ (BD_4)		ILM (BD_4-BD_4)		ML (BD_4-BD_4)		MML (BD_4-BD_4)	
1/10	4.3	> 3.7	4.0	> 4.0	4.3	> 3.7	4.3	> 3.7
1/20	5.4	> 3.9	5.2	> 4.0	5.4	> 3.9	5.4	> 3.9
1/40	6.5	> 3.9	6.4	> 4.0	6.5	> 3.9	6.6	> 3.9
1/80	7.7		7.6		7.7		7.8	
h	DQ (BD_5)		ILM (BD_5-BD_5)		ML (BD_5-BD_5)		MML (BD_5-BD_5)	
1/10	5.0	> 4.8	5.8	> 4.2	5.0	> 4.8	5.0	> 4.8
1/20	6.4	> 4.9	7.1	> 4.7	6.4	> 4.9	6.4	> 4.9
1/40	7.9		8.5		7.9		7.9	

Table 3.3.4. Numbers of correct significant digits and effective orders for problem (3.3.22) with $a=1$ at $t=T=4$

h	*ILM* (G_4-BD_4)		*MML* (G_4-BD_4)		*ILM* (G_5-BD_5)		*MML* (G_5-BD_5)	
1/10	2.7		2.0		3.7		2.0	
		> 4.3		> 3.7		> 6.0		> 7.3
1/20	4.0		3.1		5.5		4.1	
		> 4.2		> 3.8		> 5.8		> 4.9
1/40	5.3		4.3		7.2		5.6	
		> 4.2		> 3.9		> 2.5		> 4.1
1/80	6.5		5.5		7.9		6.8	

Table 3.3.5. Numbers of correct significant digits and effective orders for problem (3.3.23) at $t=T=6$.

h	*ILM* (G_4-BD_4)		*MML* (G_4-BD_4)		*ILM* (G_5-BD_5)		*MML* (G_5-BD_5)	
1/4	2.7		2.5		3.4		3.5	
		> 3.5		> 3.5		> 4.4		> 6.4
1/8	3.7		3.6		4.7		5.4	
		> 3.8		> 3.7		> 4.7		> 2.0
1/16	4.8		4.7		6.2		6.0	

3.4. Volterra integro-differential equations

We consider integro-differential equations of the (nonlinear) type

$$\begin{aligned} y'(t) &= f(t,y(t),z(t)) \\ z(t) &:= \int_0^t k(t,s,y(s))ds \end{aligned} \quad , \quad t \in I := [0,T], \tag{3.4.1}$$

where $y(0)=y_0$. Assuming the conditions of Theorem 1.3.10 to be satisfied we are assured of the existence of a unique solution $y \in C^1(I)$.

3.4.1. *VLM methods*

The conventional way to solve (3.4.1) numerically consists in the application of an LM formula for ODEs to the first equation in (3.4.1) and in the approximation of $z(t)$ by a quadrature formula (see Linz [1969d]); i.e.,

$$\begin{aligned} \sum_{i=0}^{\kappa} \hat{\alpha}_i y_{n-i} &= h \sum_{i=0}^{\kappa} \hat{\gamma}_i f_{n-i}, \quad f_n := f(t_n,y_n,z_n), \\ z_n &= \tilde{F}_n(t_n), \end{aligned} \tag{3.4.2}$$

where $\tilde{F}_n(t)$ is again the lag term defined in (3.2.13) (with $g(t)\equiv 0$) and where $\{\hat{\alpha}_i, \hat{\gamma}_i\}$ define the ODE formula. We generalize (3.4.2) in analogy to the VLM formula (3.2.15) and formally define the following VLM formula for integro-differential equations:

$$\sum_{i=0}^{\kappa} \hat{\alpha}_i y_{n-i} = h \sum_{i=0}^{\kappa} \hat{\gamma}_i f_{n-i},$$
$$\sum_{i=0}^{\kappa} \alpha_i z_n + \sum_{i=0}^{\kappa} \sum_{j=-\kappa}^{\kappa} \beta_{i,j} \tilde{F}_{n-i}(t_{n+j}) = h \sum_{i=0}^{\kappa} \sum_{j=-\kappa}^{\kappa} \gamma_{i,j} k_{n-i}(t_{n+j}). \tag{3.4.3}$$

The method $\{(3.2.13); (3.4.3)\}$ will be called a *VLM method* for the integro-differential equation (3.4.1). It differs from (3.4.2) in the formula used for the computation of z_n. This formula can be characterized by the parameter matrices $A=(\alpha_i)$, $B=(\beta_{i,j})$ and $C(\gamma_{i,j})$. Thus, the direct quadrature (DQ) formula (3.4.2) uses $A=1$, $B=-1$ and $C=0$. In Wolkenfelt [1981a], [1983b] we find further examples of VLM formulas which fit into the class (3.4.3). Firstly, he defines the *multilag* (ML) methods: these methods are characterized by those matrices used above to describe the ML methods for Volterra integral equations of the second kind (cf. (3.2.24)). Secondly, he shows that the matrices occurring in the MML methods (cf. (3.2.25)) for second-kind integral equations can be used to define the *modified multilag* (MML) methods for (3.4.1). It seems plausible that further examples are provided by the ILM matrices (3.2.22′).

An alternative way to arrive at the VLM formulas (3.4.3) is to first integrate the differential equation in (3.4.1) resulting in a system of Volterra equations of the second kind,

$$y(t) = y(0) + \int_0^t f(s,y(s),z(s))ds$$
$$, \tag{3.4.1′}$$
$$z(t) = \int_0^t k(t,s,y(s))ds$$

and then to apply the VLM formula (3.2.15) to this system with parameter matrices $(\hat{A},\hat{B},\hat{C})$ and (A,B,C) for the components. Thus,

$$\sum_{i=0}^{\kappa} \hat{\alpha}_i y_{n-i} + \sum_{i=0}^{\kappa} \sum_{j=-\kappa}^{\kappa} \hat{\beta}_{i,j} \tilde{Y}_{n-i}(t_{n+j}) = h \sum_{i=0}^{\kappa} \sum_{j=-\kappa}^{\kappa} \hat{\gamma}_{i,j} f_{n-i}$$
$$, \tag{3.4.3′}$$
$$\sum_{i=0}^{\kappa} \alpha_i y_{n-i} + \sum_{i=0}^{\kappa} \sum_{j=-\kappa}^{\kappa} \beta_{i,j} \tilde{F}_{n-i}(t_{n+j}) = h \sum_{i=0}^{\kappa} \sum_{j=-\kappa}^{\kappa} \gamma_{i,j} k_{n-i}(t_{n+j})$$

where $\tilde{Y}_n(t_j)$ is the lag term approximation to the lag term

$$Y(t,s) := y(0) + \int_0^s f(x,y(x),z(x))dx,$$

corresponding to the first component equation, at $t=t_j$ and $s=t_n$. Here, both Y and f do not depend on t so that by putting $\Sigma_{j=-\kappa}^{\kappa}\hat{\beta}_{i,j}=0$ and by writing $\hat{\gamma}_i:=\Sigma_{j=-\kappa}^{\kappa}\hat{\gamma}_{i,j}$, (3.4.3′) reduces to the form (3.4.3).

3.4.2. Consistency of VLM methods

With the numerical scheme (3.4.3) we associate the operators $\hat{\mathbf{L}}_n$ and $\mathbf{L}_n$ defined by

$$\begin{aligned}\hat{\mathbf{L}}_n[x] &:= \sum_{i=0}^{\kappa}[\hat{\alpha}_i-\hat{\gamma}_i h\frac{d}{dt}]x(t_{n-i}) \\ \mathbf{L}_n[X] &:= \sum_{i=0}^{\kappa}\{\alpha_i X(t_{n-i},t_{n-i})+\sum_{j=-\kappa}^{\kappa}[\beta_{i,j}-\gamma_{i,j}h\frac{\partial}{\partial s}]X(t_{n+j},t_{n-i})\}\end{aligned} \quad ; \qquad (3.4.4)$$

$x(t)$ and $X(t,s)$ are arbitrary functions differentiable with respect to t and s, respectively. The connection with the VLM formulas (3.4.3) is established by observing that substitution of the exact solution $y(t)$ and $z(t)$ into (3.4.3) yields the relation (cf. a similar discussion in Section 3.2.5)

$$\sum_{i=0}^{\kappa}[\hat{\alpha}_i y(t_{n-i})-\hat{\gamma}_i hf(t_{n-i},y(t_{n-i}),z(t_{n-i}))] = \hat{\mathbf{L}}_n[y]$$

$$\begin{aligned}\sum_{i=0}^{\kappa}\{\alpha_i z(t_{n-i})+\sum_{j=-\kappa}^{\kappa}[\beta_{i,j}h\sum_{l=0}^{n-i}w_{n-i,l}k(t_{n+j},t_l,y(t_l)) \\ -\gamma_{i,j}hk(t_{n+j},t_{n-i},y(t_{n-i}))]\} = \mathbf{L}_n[F]+\mathbf{Q}_n[y],\end{aligned}$$

where $\mathbf{Q}_n$ is the operator introduced in Section 3.2.5.

The terms $\hat{\mathbf{L}}_n[y]$ and $\mathbf{L}_n[F]$ will be called the *local truncation errors* of the VLM formulas (3.4.3), and $\mathbf{Q}_n[y]$ will be called the *lag term error.* The convergence of these errors is often called *consistency.*

Definition 3.4.1. (a) The *VLM* and *lag term formulas* are said to be *consistent* if the respective errors $\hat{\mathbf{L}}_n[y]$, $\mathbf{L}_n[F]$ and $\mathbf{Q}_n[y]$ tend to zero as $h\downarrow 0$, where y and F correspond to the exact solution. They are said to be *consistent of order p* if p is the largest integer so that the errors behave like $O(h^{p+1})$ as $h\downarrow 0$.
(b) The *VLM method* is said to be *consistent of order p with equation* (3.4.1) *at* $t=t_n$ if both the VLM formulas and the lag term formula are consistent of order p. □

Definition 3.4.2. The *VLM operators* $\hat{\mathbf{L}}_n$ and $\mathbf{L}_n$, and the associated *VLM formulas* in (3.4.3) are said to be *of (local) order* $\hat{p}$ and p at $t=t_n$ if $\hat{p}$ and p are the largest integers so that, for all $x\in C^{\hat{p}+1}(I)$ and $X\in C^{p+1}(S)$ we have, respectively, $\hat{\mathbf{L}}_n[x]=O(h^{\hat{p}+1})$ and $\mathbf{L}_n[X]=O(h^{p+1})$, with t_n fixed as $h\downarrow 0$. □

Since $\hat{\mathbf{L}}_n$ is of the same form as the linear operator $\mathbf{L}_n$ occurring in ODE theory (compare Definition 2.6.2), the order conditions for $\hat{\mathbf{L}}_n$ are also of the same form. Similarly, $\mathbf{L}_n$ is identical with the operator $\mathbf{L}_n$ defined in (3.2.20), hence they share the same order conditions. From Theorem 2.6.3(d) and Theorem 3.2.1 the following result can be obtained.

Theorem 3.4.1. *The VLM operators $\hat{\mathbf{L}}_n$ and $\mathbf{L}_n$ and the associated VLM formulas in (3.4.3) are of order $\hat{p}$ and p if, respectively,*

$$\hat{C}_q := \sum_{i=0}^{\kappa} i^{q-1}[i\hat{\alpha}_i + q\hat{\gamma}_i] = 0$$

for $q=0,1, \ldots, \hat{p}$, and $C_{q,l}=0$ for $q=0,1, \ldots, p$, $l=0,1, \ldots, q$, where the $C_{q,l}$ are defined in Theorem 3.2.1, and if $\hat{C}_{\hat{p}+1}\neq 0$ and $C_{p+1,l}\neq 0$ for at least one value of $l=0, \ldots, p+1$. □

Evidently, if $\{\hat{a}_i,\hat{\gamma}_i\}$ define a p-th order LM formula for ODEs, then $\hat{p}=p$. Furthermore, in the cases of the DQ, (M)ML and ILM formulas, p is determined by the same expressions derived for the operator $\mathbf{L}_n$ for second-kind Volterra equations (see Corollary 3.2.1).

3.4.3. Convergence results for VLM methods

As in the case of Volterra equations of first and second kind, we first consider the continuous problem to which the linear method {(3.2.13); (3.4.3)} converges as $h\downarrow 0$. We assume that the VLM formulas in (3.4.3) are consistent and that $A_1=\alpha'(1)-\kappa\alpha(1)\neq 0$ (see the discussion in Section 3.2.8). Then, for sufficiently smooth functions k and f the linear method converges to the equations

$$\begin{aligned} y'(t) &= f(t,y(t),z(t)) \\ \alpha(1)[z(t)-F(t,t)]+(\alpha'(1)-\kappa\alpha(1))h[z'(t)-F_t(t,t)-F_s(t,t)] &= 0 \end{aligned} \tag{3.4.5}$$

as $h\downarrow 0$ (see the proof of Theorem 3.2.5). Thus, if $\alpha(1)\neq 0$ (DQ and ML method), then the linear method is a direct discretization of (3.4.1). If $\alpha(1)=0$ (and $\alpha'(1)\neq 0$ by assumption), then the linear method converges to the system

$$\begin{aligned} y'(t) &= f(t,y(t),z(t)) \\ z'(t) &= k(t,t,y(t))+\int_0^t k_t(t,s,y(s))ds \end{aligned} , \tag{3.4.1''}$$

that is, to the system (3.4.1), whose second equation has been differentiated with respect to t. The MML and ILM methods satisfy $\alpha(1)=0$ and are therefore discretizations of the system (3.4.1′′).

Next we prove a general convergence theorem. In the proof it is convenient to use, in addition to the notation introduced in Section 3.2.8, the notation

$$\hat{\alpha}(z) := \sum_{i=0}^{\kappa} \hat{\alpha}_i z^{\kappa-i},$$

$$\eta_n := z(t_n) - z_n,$$

$$\Delta f_n := f(t_n, y(t_n), z(t_n)) - f(t_n, y_n, z_n), \tag{3.4.6}$$

$$\hat{T}(h) := \max_{i \leqslant N} |\hat{\mathbf{L}}_i[y]|, \quad \hat{\delta}(h) := \max_{j \leqslant \kappa-1} |z(t_j) - z_j|,$$

where z and y correspond to the exact solution.

Theorem 3.4.2. *Let the conditions of Theorem 1.3.10 be satisfied. Furthermore, let $\alpha(z)$ and $\hat{\alpha}(z)$ satisfy the root condition.*
(a) *If $\alpha(z) = \alpha_0 z^\kappa$, $\alpha_0 \neq 0$, then there exists a constant C independent of h such that for all sufficiently small h,*

$$|\epsilon_n| \leqslant C[\delta(h) + h\hat{\delta}(h) + E(h) + \Delta E(h) + T(h) + h^{-1}\hat{T}(h)], \quad n = \kappa, \ldots, N.$$

(b) *If $\beta(z) \equiv 0$ then there exists a constant C independent of h such that for all sufficiently small h,*

$$|\epsilon_n| \leqslant C[\delta(h) + \hat{\delta}(h) + h^{-1}\Delta E(h) + h^{-1}T(h) + h^{-1}\hat{T}(h)], \quad n = \kappa, \ldots, N.$$

Proof. Proceeding as in the proof of Theorem 3.2.6 we derive the relations

$$\hat{\mathbf{L}}_n[y] = \sum_{i=0}^{\kappa} [\hat{\alpha}_i \epsilon_{n-i} - h\hat{\gamma}_i \Delta f_{n-i}] \tag{3.4.7a}$$

$$\mathbf{L}_n[F] = \sum_{i=0}^{\kappa} \{\alpha_i \eta_{n-i} + \sum_{j=-\kappa}^{\kappa} [\beta_{i,j}(h \sum_{l=0}^{n-i} w_{n-i,l} \Delta k(t_{n+j}, t_l, y(t_l), y_l)$$

$$+ E_{n-i}(h; t_{n+j})) - h\gamma_{i,j} \Delta k(t_{n+j}, t_{n-i}, y(t_{n-i}), y_{n-i})]\}. \tag{3.4.7b}$$

The first relation is written as (cf. (3.2.31))

$$\sum_{i=0}^{\kappa} \hat{\alpha}_i \epsilon_{n-i} = \hat{v}_n, \tag{3.4.8}$$

where $\hat{v}_n$ satisfies the inequality

$$|\hat{v}_n| := |\hat{\mathbf{L}}_n[y] + h \sum_{i=0}^{\kappa} \hat{\gamma}_i \Delta f_{n-i}|$$

$$\leqslant \hat{T}(h) + h \sum_{i=0}^{\kappa} |\hat{\gamma}_i| [L_1 |\epsilon_{n-i}| + L_2 |\eta_{n-i}|\,],$$

with L_1 and L_2 denoting the Lipschitz constants associated with the function $f(t,y,z)$. Application of Theorem 2.6.8 yields (because $\hat{\alpha}(z)$ satisfies the root condition)

$$|\epsilon_n| \leqslant C_0[h \sum_{j=0}^{n} [|\epsilon_j| + |\eta_j|] + \delta(h) + n\hat{T}(h)], \tag{3.4.9}$$

where C_0 is some constant independent of n and h.

For η_n we derive from (3.4.7b),

$$\sum_{i=0}^{\kappa} \alpha_i \eta_{n-i} = v_n, \tag{3.4.10}$$

where v_n is defined as in (3.2.31).

(a) In the case where $\alpha(z) = \alpha_0 z^\kappa$ we have from (3.2.32),

$$|\eta_n| \leqslant C_1[E(h) + \Delta E(h) + h\sum_{l=0}^{n} |\epsilon_l| + T(h)], \quad n \geqslant \kappa,$$

for some constant C_1. Substitution into (3.4.9) yields

$$|\epsilon_n| \leqslant C_2\{h\sum_{j=\kappa}^{n} [|\epsilon_j| + h\sum_{l=0}^{j} |\epsilon_l| + E(h) + \Delta E(h) + T(h) + h^{-1}\hat{T}(h)] + \delta(h) + h\hat{\delta}(h)\}$$

$$\leqslant C_3\{h\sum_{j=0}^{n} |\epsilon_j| + E(h) + \Delta E(h) + T(h) + h^{-1}\hat{T}(h) + \delta(h) + h\hat{\delta}(h)\},$$

where we have used $nh \leqslant T$. From Corollary 1.5.2, part (a) of the theorem easily follows.

(b) In the case $\beta(z) \equiv 0$ we have from Theorem 2.6.8 and (3.3.32′),

$$|\eta_n| \leqslant C_4\{\hat{\delta}(h) + \sum_{j=\kappa}^{n} [\sum_{i=0}^{\kappa} (h|\epsilon_{j-i}| + h^2 \sum_{l=0}^{j} |\epsilon_l|) \Delta E(h) + T(h)]\}$$

$$\leqslant C_5\{h\sum_{j=0}^{n} |\epsilon_j| + \hat{\delta}(h) + h^{-1}\Delta E(h) + h^{-1}T(h)\}.$$

Substitution into (3.4.9) and applying Corollary 1.5.2 leads to part (b) of the theorem. □

Corollary 3.4.1. *Let $\hat{p}$ and $\tilde{p}$ be the respective orders of the LM formula $\{\hat{\alpha}_i, \hat{\gamma}_i\}$, employed in (3.4.3), and of the LM formula $\{a_i, b_i\}$, employed in the (M)ML and ILM methods; let $\delta(h) = O(h^s)$, $\hat{\delta}(h) = O(h^{\hat{s}})$, $E(h) = O(h^r)$, $\Delta E(h) = O(h^{r+1})$. If the functions f, g and k are sufficiently smooth, then the order of convergence p^* of the VLM method is given by*

$$p^* = \begin{cases} \min(s, \hat{s}+1, r, \hat{p}) & \textit{for the DQ method} \\ \min(s, \hat{s}+1, r, \hat{p}, \tilde{p}+1) & \textit{for the ML method} \\ \min(s, \hat{s}, r, \hat{p}, \tilde{p}) & \textit{for the MML method} \\ \min(s, \hat{s}, r, \hat{p}, \tilde{p}, \kappa) & \textit{for the ILM method} \end{cases} . \quad \square$$

The convergence of the conventional DQ method has already been studied by Linz [1969d] and Mocarsky [1971]. The convergence results for the (M)ML methods were first given in Wolkenfelt [1981a], [1983b].

3.4.4. Numerical examples

The convergence results derived in the preceding section will be illustrated by applying the DQ, ILM and (M)ML methods to a linear and two nonlinear problems. In a similar way as in the Sections 3.2.7, 3.2.9 and 3.3.5, the various methods are denoted by $VLM(G_r - LM_p - LM_{\hat{p}})$, where G_r indicates that the lag term is approximated by a Gregory rule of order r, and where LM_p and $LM_{\hat{p}}$ correspond, respectively, to the $\{a_i, b_i\}$ used in the ILM and (M)ML formulas and to the $\{\hat{\alpha}_i, \hat{\beta}_i\}$ used in the first equation of (3.4.3). For further details on the application of the method we refer to Section 3.2.7.

The problems integrated are given by

$$y'(t) = -t - \frac{1}{(1+t)^2} + \frac{1}{y(t)} \log\left[\frac{1+t}{1+\frac{1}{2}t}\right] + \int_0^t \frac{ds}{1+(1+t)y(s)} \tag{3.4.11}$$

with exact solution $y(t)=1/(1+t)$ (cf. problem 1 of Table 8.3.2c), by

$$y'(t) = 1+y(t)-te^{-t^2} - 2\int_0^t ts\, e^{-y^2(s)} ds \tag{3.4.12}$$

with exact solution $y(t)=t$ (cf. problem 3 of Table 8.3.2b), and by

$$y'(t) = -y(t)+1-2t-\int_0^t t(1-2t)e^{s(s-t)}y(s)ds, \tag{3.4.13}$$

whose exact solution is $y(t)=\exp(-t^2)$ (cf. problem 2 of Table 8.3.2c). The initial values for these equations follow from the exact solutions.

Table 3.4.1 gives results obtained for problem (3.4.11). Since this problem becomes stiff for larger values of t, i.e., $\partial f/\partial y$ becomes negative with large modulus, we have selected a backward differentiation (BD) formula for generating the VLM formula (3.4.3) (cf. Chapter 7). The order of the BD formula was chosen such that the order p^* of the method is just 5 (according to Corollary 3.4.1). The effective order slowly tends to the (asymptotic) order $p^*=5$. We also observe that, in this example, the ILM method is only slightly less accurate than the DQ and (M)ML methods which furnish comparable results.

In Table 3.4.2 results are given for problem (3.4.12) obtained by fourth-order VLM methods. Here, the computed order p^*_{eff} rapidly reaches the asymptotic order in almost all experiments. Notice that the methods lose accuracy when the VLM formulas are based on BD formulas instead of AM formulas (compare a similar behavior in the ODE case). We also observe that the DQ method produces the best results.

Table 3.4.1. Numbers of correct significant digits and effective orders for problem (3.4.11) at $t=T=10$

h	DQ (G_5-BD_5)		ILM $(G_5-BD_5-BD_5)$		ML $(G_5-BD_4-BD_5)$		MML $(G_5-BD_5-BD_5)$	
1/4	6.2		5.9		6.2		6.3	
		> 1.8		> 2.1		> 1.8		> 1.6
1/8	6.7		6.5		6.7		6.8	
		> 2.8		> 3.0		> 2.8		> 2.8
1/16	7.6		7.4		7.6		7.6	
		> 3.7		> 3.8		> 3.7		> 3.7
1/32	8.7		8.6		8.7		8.7	
		> 4.3		> 4.2		> 4.3		> 4.2
1/64	10.0		9.8		10.0		10.0	

Table 3.4.2a. Numbers of correct significant digits and effective orders for problem (3.4.12) at $t=T=2$

h	DQ (G_4-AM_4)		ILM $(G_4-AM_5-AM_4)$		ML $(G_4-AM_3-AM_4)$		MML $(G_4-AM_4-AM_4)$	
1/10	3.9		4.0		3.8		3.9	
		> 3.8		> 4.0		> 4.0		> 3.8
1/20	5.1		5.2		5.0		5.1	
		> 4.0		> 4.1		> 4.0		> 4.0
1/40	6.3		6.5		6.2		6.3	
		> 4.0		> 4.1		> 4.0		> 4.0
1/80	7.5		7.7		7.4		7.5	

Table 3.4.2b. Numbers of correct significant digits and effective orders for problem (3.4.12) at $t=T=2$

h	DQ (G_4-BD_4)		ILM $(G_4-BD_4-BD_4)$		ML $(G_4-BD_3-BD_4)$		MML $(G_4-BD_4-BD_4)$	
1/10	4.0		3.2		3.6		3.6	
		> 3.7		> 4.5		> 4.0		> 3.1
1/20	5.1		4.6		4.8		4.6	
		> 3.9		> 6.2		> 4.1		> 3.9
1/40	6.3		6.4		6.1		5.7	

Finally, results for problem (3.4.13) are listed in Table 3.4.3. Since the exact solution is almost zero at $t=T=4$, we have listed the number of *correct digits* instead of the number of *correct significant* digits, i.e., the value of

$$cd := -\log_{10}(|y(T)-y_N|), \quad T=t_N$$

(cf. (3.2.9)). Here, the ILM method is slightly less accurate than the other VLM methods.

Table 3.4.3. Numbers of correct digits and effective orders for problem (3.4.13) at $t=T=4$

h	DQ (G_4-AM_4)		ILM $(G_4-AM_5-AM_4)$		ML $(G_4-AM_3-AM_4)$		MML $(G_4-AM_4-AM_4)$	
1/10	1.9		1.7		1.9		1.9	
		> 4.2		> 4.0		> 4.2		> 4.2
1/20	3.2		2.9		3.2		3.2	
		> 4.0		> 4.1		> 4.0		> 4.0
1/40	4.4		4.1		4.4		4.4	

3.5. Starting formulas

A plausible approach to finding starting values for VLM methods is the use of self-starting Volterra equations solvers. Such self-starting methods are the (extrapolated) trapezoidal rule discussed in Section 3.2.10, the block-by-block methods discussed in Section 3.2.12, and the one-step methods of Runge-Kutta type to be discussed in the Chapters 4 and 5.

In the literature, however, we find (mainly for second-kind equations) various formulas designed with the purpose to *start* a multistep method and generally *not* intended to be used in the *main computation.* The greater part of these starting formulas can be fitted into the Runge-Kutta type scheme

$$Y_j = g(c_jh)+h\sum_{i=1}^{m} a_{j,i}k(c_jh,c_ih,Y_i), \quad j=1, \dots ,m. \tag{3.5.1}$$

The $\{Y_j\}$ determined by this formula, usually represent approximations to the exact solution $y(t)$ at $t=c_jh$, $j=1, \dots ,m$. In fact, the parameters $\{c_j\}$ and $\{a_{j,i}\}$ are usually chosen in such a way that (3.5.1) represents a direct quadrature formula for the approximation of $y(c_jh)$ using the abscissas $\{c_jh\}$.

If (3.5.1) is to provide the starting values $y_1, \dots ,y_{\kappa-1}$, then $\{c_j\}$ should contain the integers $1, \dots ,\kappa-1$. We then set $y_l:=Y_j$ if $c_j=l$. The order of accuracy of the starting values may affect the overall accuracy of the VLM method (cf. Corollary 3.2.1). For example, if we apply the DQ method based on Gregory rules of order r, then we need the starting values $y_0, \dots ,y_{r-1}$ with order of accuracy $s \geqslant r-1$, in order to obtain an overall accuracy of order r.

In the Sections 3.5.1 and 3.5.2, we give examples of, respectively, *explicit* $(a_{j,i\geqslant j}=0)$ and *implicit* (at least one $a_{j,i\geqslant j}\neq 0$) starting formulas.

3.5.1. Explicit starting formulas

In order to avoid lengthy formulas we represent a particular starting scheme by the diagram

$$\begin{array}{c|c} \mathbf{c} & A \\ \hline \end{array} \quad , \tag{3.5.2}$$

where $\mathbf{c}=(c_j)$ and $A=(a_{j,i})$ (cf. the "Butcher array" for denoting Runge-Kutta methods for ODEs discussed in Section 4.1.3).

We start with an example given by Noble [1964, p. 251]:

$$\begin{array}{c|ccccccc}
0 & 0 \\
1 & 1 \\
1 & 1/2 & 1/2 \\
2 & 0 & 0 & 2 \\
3 & 0 & 0 & 3/2 & 3/2 \\
1 & 9/24 & 0 & 19/24 & -5/24 & 1/24 \\
2 & 1/3 & 0 & 0 & 1/3 & 0 & 4/3 \\
3 & 3/8 & 0 & 0 & 0 & 9/8 & 9/8 & 3/8 \\
\hline
\end{array} \tag{3.5.3}$$

The successive quadrature formulas used are: rectangular rule, trapezoidal rule, midpoint rule, 2-point open Newton-Cotes, 4-point interpolatory rule, Simpson's rule and Simpson's 3/8 rule. The last 3 formulas (stages) provide approximations to $y(h)$, $y(2h)$ and $y(3h)$, respectively, with error of $O(h^4)$. Subsequent starting values of order $\geqslant 4$ can be obtained by open or half-open Newton-Cotes formulas (see Section 2.2). Notice that this starting scheme only employs quadrature rules based on step points (i.e., the $\{c_j\}$ are integers).

An $O(h^5)$ starting scheme, also using *non-step points,* was given by Day [1967a], and is represented by (3.5.4). In this diagram the formulas are indicated that yield the desired approximation to $y(h)$, $y(2h)$ and $y(3h)$. From the 8-th stage on this starting scheme uses open and closed Newton-Cotes formulas in a predictor-corrector-like fashion. Subsequent starting values can be computed in a similar way (as proposed by Day) or, alternatively, by using (half-) open Newton-Cotes formulas.

An alternative 5-th order starting scheme has been given by Iguchi [1972], and is given by (3.5.5). The subsequent starting values y_4, y_5 may be computed as in the preceding starting scheme of Day or, as was done by Iguchi, by using a half-open Newton-Cotes formula as predictor and a closed Newton-Cotes formula as corrector. Iguchi reported that (3.5.5) is superior to (3.5.4) in

0	0													
1/3	1/3													
1/3	1/6	1/6												
2/3	0	0	2/3											
1/2	1/8	0	3/8											
1	1/4	0	0	3/4										
$y(h)\rightarrow$ 1	1/6	0	0	0	2/3	1/6								
3/2	0	0	0	0	3/4	0	3/4							
3/2	3/16	0	0	0	9/16	0	9/16	3/16						
2	0	0	0	0	4/3	0	-2/3	0	4/3					
$y(2h)\rightarrow$ 2	1/6	0	0	0	2/3	0	1/3	0	2/3	1/6				
5/2	0	0	0	0	55/48	0	5/48	0	5/48	0	55/48			
5/2	95/576	0	0	0	375/576	0	250/576	0	250/576	0	375/576	95/576		
3	0	0	0	0	33/20	0	-42/20	0	78/20	0	-42/20	0	33/20	
$y(3h)\rightarrow$ 3	1/6	0	0	0	2/3	0	1/3	0	2/3	0	1/3	0	2/3	1/6

(3.5.4)

0	0									
1/10	1/10									
1/5	0	1/5								
3/10	3/40	0	9/40							
1/2	11/72	0	-15/72	40/72						
7/10	56/360	0	0	49/360	147/360					
$y(h)\rightarrow$ 1	0	0	0	25/24	-26/24	25/24				
2	524/63	0	0	-3875/63	7602/63	-5375/63	1250/63			
$y(2h)\rightarrow$ 2	-39/819	0	0	0	2912/819	-4000/819	2548/819	217/819		
3	273/52	0	0	0	-2964/52	4500/52	-1872/52	0	219/52	
$y(3h)\rightarrow$ 3	15/200	0	0	0	192/200	0	45/200	0	285/200	63/200

(3.5.5)

computational efficiency and accuracy.

A relatively simple algorithm for obtaining starting values is given by (3.5.6) and was proposed by Day [1968]. This scheme is implicit in its fourth stage. In order to "solve" this implicit relation, Day used the predictor formula $Y_4 \approx (Y_0 + Y_3)/2$. The resulting explicit scheme is fourth-order accurate.

Further starting formulas may be found in Campbell & Day [1970].

$$
\begin{array}{r|cccccccc}
0 & 0 \\
1 & 1 \\
1 & 1/2 & 1/2 \\
1/2 & 1/4 & 0 & 0 & 1/4 \\
y(h)\rightarrow 1 & 1/6 & 0 & 1/6 & 2/3 \\
2 & 0 & 0 & 0 & 0 & 2 \\
y(2h)\rightarrow 2 & 1/3 & 0 & 0 & 0 & 4/3 & 1/3 \\
3 & 0 & 0 & 0 & 0 & 3/2 & 0 & 3/2 \\
y(3h)\rightarrow 3 & 3/8 & 0 & 0 & 0 & 9/8 & 0 & 9/8 & 3/8 \\
\hline
\end{array}
\tag{3.5.6}
$$

3.5.2. *Implicit starting formulas*

A fourth-order starting scheme based on closed Newton-Cotes formulas is given by

$$
\begin{array}{c|ccccc}
0 & 0 \\
1/2 & 1/4 & 1/4 \\
1 & 1/6 & 2/3 & 1/6 \\
2 & 1/3 & 0 & 4/3 & 1/3 \\
3 & 3/8 & 0 & 9/8 & 9/8 & 3/8 \\
\hline
\end{array}
\tag{3.5.7}
$$

This scheme only requires the solution of *one* implicit relation at the time. From a computational point of view this may be an advantage, particularly when we are dealing with systems of equations.

Fully implicit starting schemes, requiring the simultaneous solution of $\kappa-1$ implicit relations, are obtained by using the starting matrices S_κ, introduced in Section 2.6.2. We then obtained the diagram

$$
\begin{array}{c|c}
\mathbf{c} & S_\kappa \\
\hline
\end{array}
\quad , \quad \mathbf{c} := (0, 1, \dots, \kappa-1)^T,
\tag{3.5.8}
$$

where S_κ is defined in (2.6.10) for $\kappa=2, \dots, 5$. These starting schemes are of order κ and are based on interpolatory quadrature rules with only step points as abscissas. For $\kappa=2$ and $\kappa=3$ such starting schemes have already been used in the Examples 3.2.3 and 3.2.2, respectively.

NOTES

Section 3.1.

Davis [1926, pp. 21-22] mentions two papers on the numerical solution of integral equations; of these only one (Whittaker [1918]) deals with Volterra equations. In Carson [1926, pp. 145-146] the trapezoidal method is used to solve numerically a convolution type Volterra equation of electric circuit theory. The survey paper by Bernier [1945] refers, in addition, to the method of Huber [1939], while Fox & Goodwin [1953] focus on methods based on the Gregory formulas and on series expansion methods. It is interesting to note that the papers by Prasad [1924], Mikeladze [1935], and Krylov [1949] do not seem to have been noticed until quite recently (they are, for example, not listed in the comprehensive bibliography of Noble [1971]). Mikeladze's paper of 1935 is noteworthy in that it contains, seemingly for the first time, a convergence analysis (using a discrete Gronwall inequality). Among the more recent survey papers are the ones by Mayers [1962], Noble [1964], [1977], Kershaw [1974], van der Houwen [1977c], Brunner [1978c], [1982a] (this reference contains also a list of recent doctoral theses), te Riele [1979b], and Baker [1982a]; Tsalyuk [1979] should be consulted for contributions in Russian between 1966 and 1976.

Numerical methods for first-kind Volterra integral equations are reviewed, e.g., in Baker [1974a], Brunner [1978c], [1982a], te Riele [1979b], Linz [1980], and McKee [1982b], while Cryer [1972] and Baker [1974b], [1976] survey methods for Volterra integro-differential equations.

Of the proceedings of recent conferences reflecting the state of the art in the numerical treatment of Volterra integral equations we cite those edited by Delves & Walsh [1974], Golberg [1979a], te Riele [1979a], Anderssen, de Hoog & Lukas [1980], Albrecht & Collatz [1980], Baker & Miller [1982], and Hämmerlin & Hoffmann [1985]. The monograph by Baker [1977] contains an extensive discussion of Volterra equations; see also Fenyö & Stolle [1984b] for a somewhat shorter survey.

Section 3.2.

Most of the classical DQ methods are based either on the trapezoidal rule, Simpson's rule, or on the Gregory formulas (2.6.23) (in the Russian literature they are occasionally referred to as Euler-Laplace formulas). DQ methods using the former two quadrature rules may be found in Prasad [1924], Mikeladze [1935], [1951], Krylov [1949] (as one of the starting formulas), Collatz [1966], Jones [1961], and Linz [1967], [1968] (see also Noble [1969]); Gregory formulas are employed in Mikeladze [1935], Krylov [1949], Fox & Goodwin [1953], Mayers [1962], Noble [1964], O'Neill [1967], O'Neill & Byrne [1968], Phillips [1971], Steinberg [1972] (see also Rakotch [1973]). Laudet & Oulès [1960] and Pouzet [1962], [1963] base their DQ methods on Adams formulas for ODEs (recall that Adams-Moulton LM methods are related to the Gregory formulas; cf. Section 2.6.3): their approach, which implicitly makes use of the notion of (ρ,σ)-reducibility for a particular class of methods, may be said to constitute the first step towards the generalization to second-kind Volterra integral equations of the Dahlquist-Henrici theory of LM methods for ODEs (see, e.g., Dahlquist [1959], Henrici [1962], Lambert [1973], or Stetter [1973]). Subsequent steps were taken by Spohn [1965], Kobayasi [1966] (compare also Noble [1969] for a brief description of Kobayasi's work), and, especially, Wolkenfelt [1981a], [1982].

Open quadrature formulas yield explicit direct quadrature methods; such methods are studied in Mikeladze [1951] and in Bel'tyukov [1965a]. The use of closed formulas necessitates the solution of nonlinear algebraic equations for the new approximation y_n: in analogy to ODEs, predictor-corrector techniques are almost invariably used to deal with this problem. They occur in Oulès [1960a], Noble [1964, pp. 233-234, 247-252], Garey [1971], [1972a], Garey & Harris [1977], Williams [1980], and Jones & McKee [1985] (see also Chapter 8 for a code based on the theory developed in the last two references). In addition we mention related work by Jain & Sharma [1967] (4-point Lobatto formulas), El-Gendi [1969] (Clenshaw-Curtis quadrature formulas; see also Wolfe [1969] and Phillips [1970] for similar methods based on Chebyshev series), and Evans, Hyslop & Morgan [1981] (Clenshaw-Curtis quadrature formulas).

(ρ,σ)-reducible methods for Volterra integro-differential equations were introduced by Matthys [1976] in the context of deriving A-stable LM methods for such equations. The complete theory of (ρ,σ)-reducible quadrature methods for Volterra integral and integro-differential equations is due to Wolkenfelt [1979], [1981a], [1982] (see also van der Houwen [1982] for the application of such methods to Volterra equations with finitely decomposable kernels). Van der Houwen & te Riele [1977], [1978], [1981] analyzed IBD methods, and Wolkenfelt [1981a], [1983b] introduced ML and MML methods (compare also Wolkenfelt, van der Houwen & Baker [1981]). The general VLM theory was established by van der Houwen & te Riele [1982], [1985].

Extrapolation methods based on the midpoint method have been analyzed in detail by Hock [1977], [1979a], [1979b], [1980], [1981] (compare also Chapter 8 for a computer code implementing Hock's method). Marchuk & Shaidurov [1983, pp. 302-305] present an analysis of extrapolation using the trapezoidal method; this monograph gives a general introduction to the theory and application of extrapolation methods for various types of functional equations.

An abstract convergence analysis of discretization methods for Volterra equations may be found in Scott [1984], [1985], and in Dixon & McKee [1985].

Section 3.3.

Jones [1961] seems to have been the first to give a convergence analysis for the trapezoidal method for first-kind Volterra integral equations. Kobayasi [1967] derived an asymptotic expansion for the error induced by the trapezoidal method; he also showed that it contains an oscillatory component which may lead to numerical instability. Convergence analyses for the trapezoidal methods were also given, e.g., by Linz [1967], [1968], [1969c] (these references contain, in addition, convergence results for the midpoint method and the rectangular (or Euler's) method; a lucid treatment of the latter may also be found in McKee [1984]), Aparcin [1973], and de Hoog [1973] (here, a comparative study of the asymptotic error expansions for the trapezoidal and midpoint methods is presented).

Linz [1968], [1969c] showed that methods for first-kind equations based on Simpson's rule or on Gregory rules with $q \geqslant 1$ are divergent; the same results can also be found in Aparcin [1973]. Krylov [1949, pp. 63-67] applies the Gregory formula with $q=2$ (cf. (2.6.27b)) to a linear first-kind Volterra equation (starting values are generated by means of the starting formula (2.6.10b); since only a few approximations are computed, using a single fixed stepsize, the divergence of the method does not become apparent.

Gladwin [1972], [1973], [1975], [1979] presents a detailed divergence analysis for Gregory methods with $q \geqslant 1$, and he shows how to modify these methods, by introducing certain free parameters, so as to obtain convergent methods of order $p \leqslant 6$ (prior to Gladwin's work, no convergent linear multistep method of order $p>2$ had been known for first-kind Volterra equations). His techniques allow for a simple, explicit construction of higher-order methods (see Gladwin [1975, pp. 44-49], [1979] for details; compare also Gladwin [1982], [1985] for the problem of the optimal choice of the free parameters). It was proved by Gladwin & Jeltsch [1974] that methods which are based on (closed) interpolatory quadrature formulas of order $p>2$ and using equidistant abscissas are necessarily divergent.

Holyhead [1976], Holyhead, McKee & Taylor [1975], and Holyhead & McKee [1976] established a general theory of convergent linear multistep methods, including necessary and sufficient conditions for such methods to be convergent; these conditions involve the eigenvalues of products of certain companion matrices associated with the given method. Although this analysis is more general than Gladwin's, the explicit construction of convergent methods of higher order is quite complex. Compare also the related work by McKee [1979a] (giving a more unified theory), Andrade & McKee [1979] (analyzing methods based on cyclic linear multistep methods for ODEs), and Andrade, Franco & McKee [1981] (extending the results of McKee [1979a] to linear first-kind equations with $K(t,t) \equiv 0$).

A different approach for constructing linear multistep methods of higher order has been proposed by Taylor [1976]: it consists in inverting suitable (local) differentiation formulas satisfying the Dahlquist root condition.

Wolkenfelt [1981a], [1981c] showed that the theory of (ρ,σ)-reducible quadrature formulas yields an elegant way for obtaining convergent DQ methods of arbitrary order; the methods analyzed by Linz [1969c], Gladwin & Jeltsch [1974], Gladwin [1975], [1979], and Taylor [1976] may be viewed as particular cases of this class of methods.

Multilag and modified multilag methods for first-kind Volterra equations were introduced by Wolkenfelt [1981a], [1983b]. The general theory of VLM methods for such integral equations is due to van der Houwen & te Riele [1985].

In contrast to second-kind Volterra equations, the trapezoidal method for first-kind equations is not suitable for extrapolation because of the oscillatory behavior of its error; instead, the midpoint method is used. See Linz [1967], [1968], [1969c], [1971], [1980], and Marchuk & Shaidurov [1983, pp. 305-309].

Finally, we mention the work by Taylor [1982] and Eggermont [1984b] where an abstract (functional-analytic) framework for convergence analyses of quadrature methods for first-kind Volterra equations is given; see also Scott [1984].

Section 3.4.

The use of linear multistep methods for Volterra integro-differential equations seems to date back to Oulès [1960b] (predictor-corrector methods based on the Milne-Simpson formula). Predictor-corrector techniques based on Adams methods were used by Wolfe & Phillips [1968], Linz [1969d], and Day [1970]; see also Makroglou [1977], [1982] for predictor-corrector methods of hybrid type (involving one off-step point). A more comprehensive theory is due to Tavernini [1973]: it encompasses also Volterra functional equations of which integro-differential equations constitute a particular subclass. In this context we cite also the work of Jackiewicz [1979] and Jackiewicz & Kwapisz [1983a]. Smit [1976] established two-sided error bounds for linear multistep methods and for cyclic linear multistep methods. The application of this latter class of methods to integro-differential equations was also investigated by McKee [1979b]. Jackiewicz [1984] used recent defect correction techniques from numerical ODEs to obtain global error estimates for Euler's method.

As has been mentioned before, Matthys [1976] introduced the notion of (ρ,σ)-reducible quadrature formulas and studied their use in the numerical solution of integro-differential equations. This work was subsequently extended by Wolkenfelt [1977], [1979], [1981a], [1982].

Multilag and modified multilag methods for integro-differential equations were suggested by Wolkenfelt [1981a], [1983b]. The general theory of VLM methods for VIDEs is due to van der Houwen & te Riele [1985].

Chang [1982] derived asymptotic error expansions and corresponding extrapolation procedures for the two-step Adams-Moulton method; his work was motivated by earlier work by Linz [1974] where the trapezoidal method and extrapolation are used to solve a first-kind Volterra integro-differential equations (i.e., an equation where both y and y' occur only under integral signs). Extrapolation techniques for certain one-step methods have also been discussed by Feldstein & Sopka [1974].

Section 3.5.

Explicit and implicit (block) starting methods have been proposed by numerous authors. Regarding second-kind Volterra equations we mention Prasad [1924, p. 51], Krylov [1949, p. 43, p. 60], Linz [1967], [1968], [1969b], Day [1967a], [1968], [1969], Campbell [1970], Campbell & Day [1970], [1971], Iguchi [1972], Logan [1976], Kumar [1979], [1981]; in addition see also Baker [1977, pp. 835-849]. Moreover, explicit VRK methods (cf. Chapter 4) can be used as simple starting procedures. Many of the suggested block methods can be viewed either as implicit VRK methods, or as fully discretized collocation methods (cf. Chapter 5).

Krylov [1949, p. 65] seems to be the first source for an (implicit block) starting method for first-kind Volterra equations; it is the one given in (2.6.10b). A semi-explicit third-order starting method is given by the method of Keech [1977b] (see also Keech [1976]).

As for starting methods for integro-differential equations compare, e.g., Day [1967b], [1970], Linz [1969d], Brunner & Lambert [1974], Makroglou [1980]; explicit starting procedures are, of course, also given by explicit VDRK methods (cf. Chapter 4).

CHAPTER 4

RUNGE-KUTTA-TYPE METHODS FOR VOLTERRA EQUATIONS

Runge-Kutta methods for the numerical solution of initial-value problems for ordinary differential equations were introduced at around the turn of the century. The modern theory of this class of methods, due to J.C. Butcher, dates back to the mid-1960s. At about the same time, P. Pouzet and B.A. Bel'tyukov extended the Runge-Kutta idea, in different ways, to Volterra integral equations of the second kind with bounded kernels. In the late 1970s and the early 1980s, a systematic analysis of Runge-Kutta-type methods for second-kind integral equations and for integro-differential equations was given, respectively, by Brunner, Hairer and Nørsett, and by Lubich; the latter also developed the Runge-Kutta theory for second-kind Volterra integral equations with weakly singular kernels.

The present chapter contains an introduction to the theory and the application of Runge-Kutta-type methods for Volterra equations with bounded kernels. It will always be assumed that the conditions ensuring the existence of a unique solution of the Volterra equations considered are satisfied.

4.1. Runge-Kutta methods for integral equations of the second kind

4.1.1. Introduction

Suppose that the given interval $I:=[0,T]$ is again discretized by the uniform mesh $t_n:=nh$ ($n=0, \ldots ,N$: $t_N=T$). The given integral equation (1.1.1) can then be rewritten by relating it to this mesh; i.e.,

$$y(t) = F_n(t)+\int_{t_n}^{t} k(t,s,y(s))ds, \quad t\in[t_n,T], \tag{4.1.1a}$$

where F_n, called the *lag* (or: *tail*) *term*, is given by

$$F_n(t) := g(t)+\int_{0}^{t_n} k(t,s,y(s))ds, \quad n=0, \ldots ,N-1. \tag{4.1.1b}$$

(Note that $F_n(t)$ corresponds to $F(t,t_n)$ introduced in Chapter 3.) Moreover, define

$$h\Phi_n(t) := \int_{t_n}^{t} k(t,s,y(s))ds, \quad t\in[t_n,T], \quad n=0, \ldots ,N-1. \tag{4.1.2}$$

Here, the function $\Phi_n(t)$ will often be referred to as the (exact) *increment function* (with respect to the subinterval $[t_n,t_{n+1}]$).

A Runge-Kutta method for (4.1.1) is based on two approximation processes which, in general, are independent of each other:

(i) An approximation scheme for the increment function $\Phi_n(t)$; i.e., a (local) approximation scheme on the subinterval $[t_n, t_{n+1}]$. The resulting discrete increment function, denoted by $\tilde{\Phi}_n(t)$, will be called a *Volterra-Runge-Kutta formula (VRK formula).*

(ii) An approximation scheme for the lag term $F_n(t)$; i.e., an approximation scheme on the interval $[0, t_n]$. The resulting discrete lag term will be denoted by $\tilde{F}_n(t)$ and will be referred to as *lag term* (or: *tail term*) *formula.*

Hence, the exact equation at $t = t_{n+1} = t_n + h$ will be approximated by

$$y_{n+1} = \tilde{F}_n(t_n + h) + h\tilde{\Phi}_n(t_n + h), \quad n = 0, \ldots, N-1. \tag{4.1.3}$$

In analogy to the *VLM* methods of Chapter 3, we shall call (4.1.3) a *VRK method* if both the VRK formula and the lag term formula have been specified. Also, the two terms on the right-hand side of (4.1.3) will frequently be referred to as, respectively, the *tail approximation* and the *VRK part* of the VRK method.

In the present section we formally define the VRK method, and we then introduce two particular subclasses of VRK methods which have received much attention in the literature, namely, the VRK methods of *Pouzet type,* and the VRK methods of *Bel'tyukov type.* The reader not familiar with Runge-Kutta (RK) methods for ordinary differential equations may first wish to consult Section 4.1.3 where the RK idea is briefly discussed and where references to the relevant literature can be found.

Definition 4.1.1. (a) An *m-stage VRK formula* for (1.1.1) has the form

$$\tilde{\Phi}_n(t) := \sum_{j=1}^{m} b_j k(t + (e_j - 1)h, t_n + c_j h, Y_{n,j}), \tag{4.1.4a}$$

with $\{Y_{n,j}\}$ given by

$$Y_{n,j} = \tilde{F}_n(t_n + \theta_j h) + h \sum_{i=1}^{m} a_{j,i} k(t_n + d_{j,i} h, t_n + c_i h, Y_{n,i}), \tag{4.1.4b}$$

$$j = 1, \ldots, m.$$

Here, the vectors $\boldsymbol{\theta} := (\theta_j)$, $\mathbf{c} := (c_j)$, $\mathbf{e} := (e_j)$, $\mathbf{b} := (b_j)$, and the square matrices $A := (a_{j,i})$, $D := (d_{j,i})$ are given.
(b) The VRK method (4.1.3) is called:

(i) an *extended VRK method* if the lag term formula is given by

$$\tilde{F}_n(t) := g(t) + h \sum_{l=0}^{n-1} \sum_{j=1}^{m} b_j k(t + (e_j - 1)h, t_l + c_j h, Y_{l,j}), \tag{4.1.5a}$$

$$n = 0, \ldots, N-1;$$

(ii) a *mixed VRK method* if $\tilde{F}_n(t)$ has the form (cf. (3.2.13))

$$\tilde{F}_n(t) := g(t)+h\sum_{l=0}^{n} w_{n,l}k(t,t_l,y_l), \quad n=\tilde{\kappa}, \ldots ,N-1; \tag{4.1.5b}$$

(iii) a γ-*modified VRK method* if the lag term formula is defined by

$$\tilde{F}_n^*(t) := \tilde{F}_n(t)+\gamma(t)[y_n-\tilde{F}_n(t_n)], \tag{4.1.6}$$

where $\tilde{F}_n(t)$ is some approximation to $F_n(t)$, and where the function $\gamma(t)$ is such that $\gamma(t_n+\theta_j h)\in[0,1]$. □

Note that in (4.1.5b) the integer $\tilde{\kappa}$ will depend on the desired order of convergence of the quadrature formula with weights $\{w_{n,l}\}$, in analogy to the results of Chapter 3.

In this chapter, the given integral equation (4.1.1) will be a scalar equation. However, just as in the case of VLM methods, the definition of a VRK method is easily extended to systems of Volterra integral equations, and all the results presented below carry over, with obvious modifications, to such systems.

In order to illustrate the location of the kernel values employed in the VRK formula (4.1.4) we consider, in Figure 4.1.1a, the case of a two-stage VRK method ($m=2$). Here, the points indicated by $\times$, $+$, and $\bullet$, correspond, respectively, to those values occurring in the expressions for $Y_{n,1}$, $Y_{n,2}$, and y_{n+1}. In Figure 4.1.1b, we exhibit, respectively, the points used in the lag term approximations for an extended method ($\bullet$) and for a mixed method ($\bigcirc$).

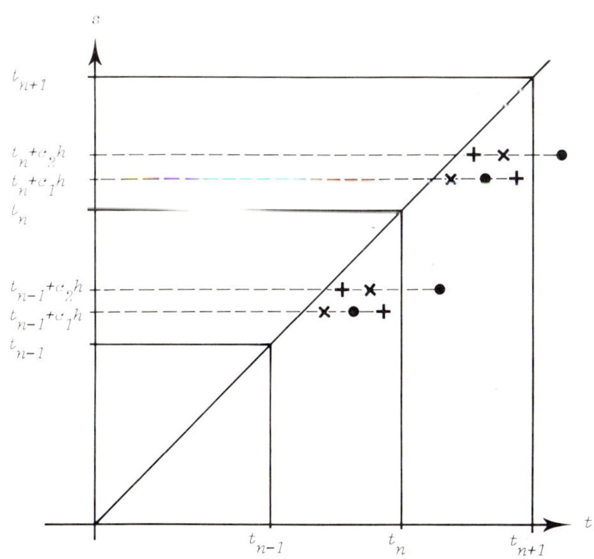

Figure 4.1.1a. VRK part $h\tilde{\Phi}_n(t)$

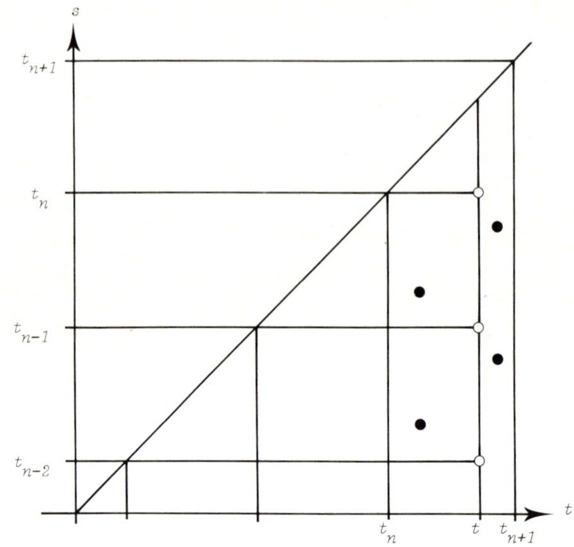

Figure 4.1.1b. Lag term $\tilde{F}_n(t)$

In analogy to the terminology used in RK methods for ODEs we say that the VRK formula is

(i) *explicit* if $a_{j,i}=0$ whenever $i \geqslant j$ (i.e., A is a strictly lower triangular matrix);

(ii) *diagonally implicit* if $a_{j,i}=0$ for all (j,i) with $i>j$ (i.e., A is a lower triangular matrix with nonvanishing diagonal);

(iii) *implicit* if it is neither explicit nor diagonally implicit (i.e., A possesses at least one nonzero element $a_{j,i}$ with $i>j$).

Two particular classes of VRK formulas will play an important role in later applications.

Definition 4.1.2. A VRK formula is of *Pouzet type* (or: a *PVRK formula*) if

$$d_{j,i} = c_j, \quad \theta_j = c_j, \quad e_j = 1, \quad i,j = 1, \ldots, m; \tag{4.1.7}$$

i.e., if its VRK part is completely characterized by the symbolic diagram (the "Butcher array" for ODEs; cf. (4.1.13)),

$$\begin{array}{c|c} \mathbf{c} & A \\ \hline & \mathbf{b}^T \end{array} \quad . \qquad \square$$

Written explicitly, an m-stage PVRK formula is given by

$$Y_{n,j} = \tilde{F}_n(t_n + c_j h) + h \sum_{i=1}^{m} a_{j,i} k(t_n + c_j h, t_n + c_i h, Y_{n,i}), \qquad (4.1.8a)$$

$$j = 1, \dots, m,$$

$$y_{n+1} = \tilde{F}_n(t_n + h) + h \sum_{j=1}^{m} b_j k(t_n + h, t_n + c_j h, Y_{n,j}), \qquad (4.1.8b)$$

$$n = 0, \dots, N-1.$$

In Figure 4.1.2 the analogue of Figure 4.1.1 is given for a PVRK method.

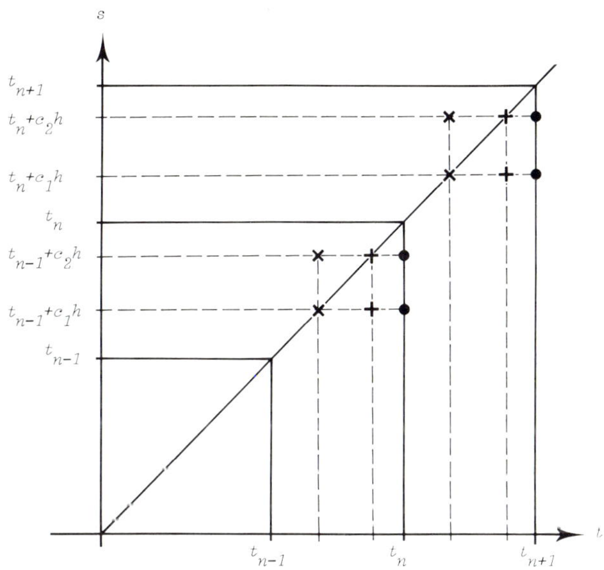

Figure 4.1.2a. PVRK part $h\tilde{\Phi}_n(t)$

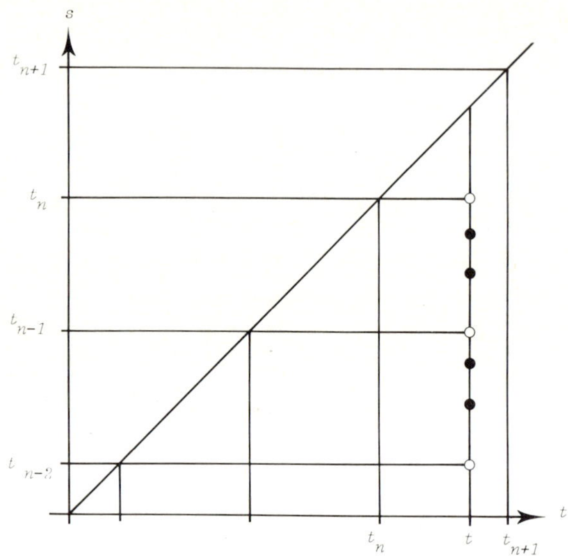

Figure 4.1.2b. Lag term $\tilde{F}_n(t)$

The explicit version of (4.1.8) was introduced by Pouzet [1960], [1962], [1963]. Note that if a PVRK formula is explicit, or diagonally implicit, then we always have $t_n + c_j h \geq t_n + c_i h$ (since $i \leq j$); however, an implicit PVRK formula generally employs kernel values $k(t,s,y)$ with $s > t$ (compare Figure 4.1.2a). In this case the domain of $k(t,s,y)$ must be extended to include the required points (t,s) outside S. This may create a problem concerning the accuracy of the numerical results, since such a continuation of the kernel, even if it is smooth, is necessarily arbitrary and not related to the original equation. The problem can be avoided by an appropriate modification of the VRK part in an implicit Pouzet method; we refer the reader to Section 5.3.4 (especiallv (5.3.43)).

It follows from (4.1.8) that the points (t,s) at which the kernel in the PVRK part has to be evaluated form a two-dimensional array. More precisely, we have

Lemma 4.1.1. *The number $\nu_p(m)$ of kernel evaluations in each step (i.e., for each value $n = 0, \ldots, N-1$) in the VRK part of an m-stage Pouzet method satisfies*

(i) $\nu_p(m) \leq m(m+1)/2$, *if the PVRK formula is explicit;*

(ii) $\nu_p(m) \leq m(m+3)/2$, *if the PVRK formula is diagonally implicit;*

(iii) $\nu_p(m) \leq m(m+1)$, *if the PVRK formula is implicit.* □

A different class of VRK formulas was introduced by Bel'tyukov [1965b]; like Pouzet he considered only explicit methods.

Definition 4.1.3. A VRK formula is of *Bel'tyukov type* (in the following called a *BVRK formula*) if

$$d_{j,i} = d_i, \quad e_j = d_j, \quad \theta_j = c_j, \quad i,j = 1, \dots, m. \tag{4.1.9}$$

Hence, the VRK part of a Bel'tyukov method is characterized by the diagram

$$\begin{array}{c|c|c} \mathbf{d} & \mathbf{c} & A \\ \hline & & \mathbf{b}^T \end{array} \quad . \quad \square$$

An m-stage BVRK formula has thus the form

$$Y_{n,j} = \tilde{F}_n(t_n + c_j h) + h \sum_{i=1}^{m} a_{j,i} k(t_n + d_i h, t_n + c_i h, Y_{n,i}), \tag{4.1.10a}$$

$$j = 1, \dots, m;$$

$$y_{n+1} = \tilde{F}_n(t_n + h) + h \sum_{j=1}^{m} b_j k(t_n + d_j h, t_n + c_j h, Y_{n,j}), \tag{4.1.10b}$$

$$n = 0, \dots, N-1.$$

If the vectors **d** and **c** can be selected so that

$$c_j \leqslant d_j, \quad j = 1, \dots, m, \tag{4.1.11}$$

we say that the corresponding BVRK formula satisfies the *kernel condition:* all points (t,s) at which the kernel has to be evaluated lie then in S, in contrast to implicit PVRK formulas.

The points (t,s) occurring as arguments of the kernel in the VRK part of a Bel'tyukov formula form a one-dimensional array; we thus find

Lemma 4.1.2. *The number $\nu_B(m)$ of kernel evaluations per step in the VRK part of an m-stage Bel'tyukov method satisfies $\nu_B(m) \leqslant m$, regardless of whether the BVRK formula is explicit, diagonally implicit, or implicit* $\square$

A comparison with Lemma 4.1.1 reveals that for a given number of stages, a BVRK formula is expected to require fewer kernel evaluations than a PVRK formula. However, as we shall see in Sections 4.1.5 and 4.1.6, BVRK formulas usually have a lower order of accuracy than PVRK formulas with the same number of stages.

The location of the various kernel values in the (t,s)-plane is shown, for the case $m = 2$, in Figure 4.1.3.

We note in passing that every VRK formula (including the PVRK formulas) can be written as a BVRK formula, with a possible increase in the number of stages. Compare Brunner, Hairer & Nørsett [1982] for details.

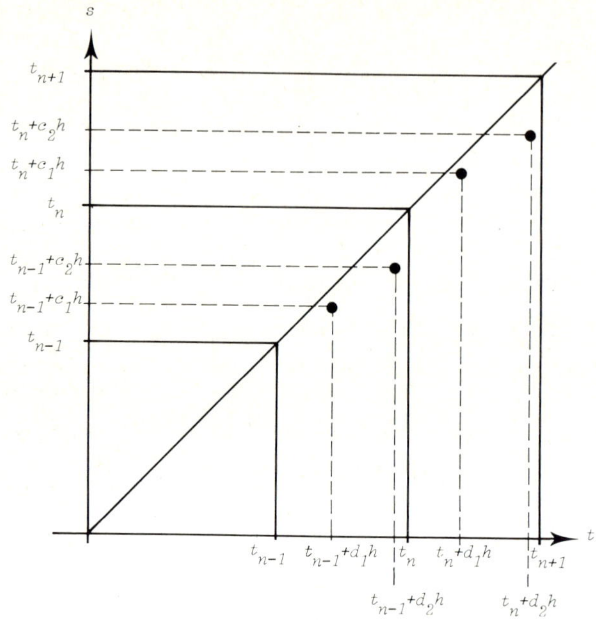

Figure 4.1.3a. BVRK part $h\tilde{\Phi}_n(t)$

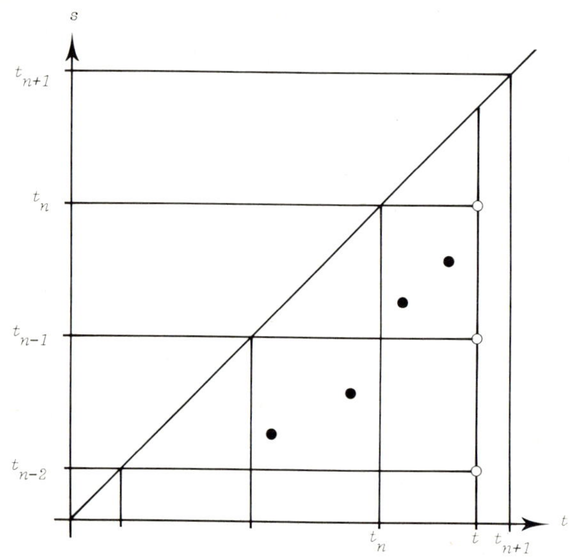

Figure 4.1.3b. Lag term $\tilde{F}_n(t)$

4.1.2. Historical notes

The first paper in which a Runge-Kutta method is proposed for the approximate solution of (linear) second-kind Volterra integral equations is the one by Prasad [1924]; in our terminology, Prasad's method is an extended 4-stage explicit PVRK method of order $p=4$ (compare the second diagram in Example 4.1.3 below). The method is based on Kutta's RK method of 1901 (p. 41). The RK idea was also employed, in a somewhere hidden way, by Krylov [1949 p. 43, p. 60] to generate starting values for direct quadrature methods based on Gregory formulas (compare also Section 3.5).

The first investigations of the convergence order of VRK methods and of the corresponding order conditions for the parameters in the VRK formulas are due to Suyama & Nakamori [1952], to Aparo [1959], and to Oulès [1960a]. Pouzet in his papers of [1960], [1961], [1962] and [1963] undertook a systematic study of the convergence of extended explicit VRK methods (with $p=m\leqslant 4$) now named after him. A similar study, independent of Pouzet's work, was carried out by Gamondi & Italiani [1965]. In 1965, Bel'tyukov introduced and analyzed explicit VRK methods of the form (4.1.10) and of order $p=m\leqslant 3$; these methods were extended to two-dimensional systems of second-kind Volterra integral equations by Bel'tyukov & Kuznechikhina [1976].

The Runge-Kutta-Fehlberg approach (see, e.g., Stetter [1973, pp. 343-347]) of employing embedded RK methods of different orders for ODEs (with the purpose of controlling the local error in the numerical approximations) was adapted for the approximate solution of second-kind Volterra integral equations by Lomakovich & Vasilenko [1968], Lomakovich [1969a], Coroian [1974], and by Lomakovich & Ishchuk [1974] (compare also M. Micula [1981]). Fehlberg's embedding idea served as the basis in the construction of high-order explicit VRK methods (using Hermite-type quadrature formulas) as suggested by Schoedon [1970] and by Filippi & Schoedon [1971]. An account of the state of the art in explicit VRK methods (up to about 1975) may be found in Baker [1977, pp. 848-864].

Implicit VRK methods are of much more recent origin. A special class of implicit PVRK methods was introduced by R. Weiss [1972a] and by de Hoog & Weiss [1975]; they also suggested a modification avoiding the use of kernel values with (t,s) outside S. (It will be shown in Chapter 5 that these modified methods are closely related to, and can in fact be derived from, *collocation methods* in certain polynomial spline spaces.) More general implicit VRK methods were analyzed by van der Houwen [1977b], and by van der Houwen & Blom [1978] (see also van der Houwen [1981]). The general VRK theory dealing with the order conditions for VRK formulas was established by Brunner & Nørsett [1980], and by Brunner, Hairer & Nørsett [1982], while Hairer, Lubich & Nørsett [1983] gave a detailed study of the interplay between the (local) order of the VRK part and that of the tail approximation in an extended or mixed VRK method. As for computer codes based on VRK methods we refer to Chapter 8.

4.1.3. On the order conditions for RK methods in ODEs

The origins of RK methods for the approximate solution of the initial-value problem

$$y' = f(t,y), \quad t \in I, \quad y(0) = y_0,$$

date from the turn of the century when Runge [1895], Heun [1900], and Kutta [1901] introduced the basic idea of avoiding successive differentiation in Taylor series methods by considering instead an appropriate linear combination of different values of $f(t,y)$, chosen such as to preserve the accuracy of the Taylor series method. (We refer the interested reader to the forthcoming monograph by Hairer, Nørsett & Wanner [1986] which contains a detailed history of numerical methods for (4.1.4).) A one-step, m-stage Runge-Kutta method (briefly: an m-stage RK method) for the above initial-value problem is defined by

$$Y_{n,j} = y_n + h\sum_{i=1}^{m} a_{j,i} f(t_n + c_i h, Y_{n,i}), \quad j=1, \ldots ,m, \tag{4.1.12a}$$

$$y_{n+1} = y_n + h\sum_{j=1}^{m} b_j f(t_n + c_j h, Y_{n,j}), \quad n=0, \ldots ,N-1. \tag{4.1.12b}$$

It is usually represented by a symbolic diagram (the "Butcher array") introduced by Butcher [1964b]:

$$\begin{array}{c|ccc} c_1 & a_{1,1} & \cdots & a_{1,m} \\ \vdots & \vdots & & \vdots \\ c_m & a_{m,1} & \cdots & a_{m,m} \\ \hline & b_1 & \cdots & b_m \end{array} \quad =: \quad \begin{array}{c|c} \mathbf{c} & \mathbf{A} \\ \hline & \mathbf{b}^T \end{array} . \tag{4.1.13}$$

Here, the value of c_i is defined by $c_i := \Sigma_{j=1}^{m} a_{i,j}$, $i=1, \ldots ,m$. Even though (4.1.12b) and (4.1.12a) might be viewed as being obtained from the integrated form of the ODE,

$$y(t) = y(t_n) + \int_{t_n}^{t} f(s,y(s))ds, \quad t \in [t_n, T], \tag{4.1.14}$$

by setting $t = t_n + c_j h$ ($j=1, \ldots ,m,m+1$, with $c_{m+1} := 1$) and by then applying m-point quadrature formulas to the integral in (4.1.14), the parameters $a_{j,i}$ and b_j are not necessarily quadrature weights: they are given by a solution of a certain system of (nonlinear) consistency conditions, the so-called *order conditions.* To be somewhat more precise, we wish to obtain conditions ensuring that, for all sufficiently smooth functions f, the value y_n differs from the exact solution at $t = t_n$ by $O(h^{p+1})$ (as $h \downarrow 0$), provided that y_{n-1} is equal to $y(t_{n-1})$ (in other words, we approximate the given differential equation *locally,* namely on $[t_{n-1}, t_n]$, with initial value $y_{n-1} = y(t_{n-1})$). An RK method possessing this property is said to be of *(local) order p.* The algebraic theory of RK methods due to Butcher [1963], [1965], [1972] uses the graph-theoretic notion of rooted trees and thus permits an elegant derivation of the order conditions. (A lucid introduction to this theory may be found in Verner [1979]; compare also Butcher [1976], and Hairer, Nørsett & Wanner [1986] for further references and for a comprehensive presentation of the RK theory for ODEs.)

If n_r and N_p denote, respectively, the number of order conditions attached to a fixed order r $(1 \leqslant r \leqslant p)$ and the total number of order conditions (i.e., $N_p := \Sigma_{r=1}^{p} n_r$), then there emerges the following picture (see Verner [1979]):

Table 4.1.1. Number of order conditions for an RK method

p	1	2	3	4	5	6	7	8	9	10
n_p	1	1	2	4	9	20	48	115	286	719
N_p	1	2	4	8	17	37	85	200	486	1205

Hence, if we want to construct an RK method of order $p \geqslant 1$, the parameters $a_{j,i}$ and b_j in (4.1.12) must satisfy N_p order conditions. In an implicit m-stage RK method the number of parameters is equal to $m(m+1)$. Butcher [1964a] has shown that there exists a (unique) implicit m-stage RK method of order $p=2m$: hence, if $m=3$, the corresponding 12 parameters satisfy $N_6=37$ order conditions. These optimal methods are easily characterized: the parameters are given by

$$a_{j,i} = \int_0^{c_j} L_i(s)ds, \quad b_j = \int_0^1 L_j(s)ds, \quad j,i = 1, \dots ,m,$$

where $L_i(s) := \Pi_{r \neq i}^{m}(s-c_r)/(c_i-c_r)$ denotes the i-th Lagrange fundamental polynomial with respect to the $\{c_j\}$ and where the $\{c_j\}$ are the zeros of the shifted Legendre polynomial $P_m(2s-1)$ (i.e., the Gauss-Legendre points for $(0,1)$).

For explicit RK methods the situation is more complex. Here, an m-stage RK method cannot even attain the order $p=m$ if $m>4$, as shown in Table 4.1.2. (m^* denotes the minimum number of stages needed to achieve order p.)

Table 4.1.2. Attainable order of explicit RK methods

p	1	2	3	4	5	6	7	8	9	10
m^*	1	2	3	4	6	7	9	11	12	13

We conclude with a remark on the diagonally implicit RK methods (frequently also known as semi-implicit, or semi-explicit, RK methods). These methods correspond to $a_{j,i}=0$ for all (j,i) with $i>j$. Among these methods, those with $a_{j,j}=\gamma\neq 0$, $j=1, \ldots ,m$, have been studied by Nørsett [1974] (see also Alexander [1977]). He proved that such methods cannot have an order $p=m+2$ for any m. On the other hand, there exist methods with $p=m+1$. A specific example will be considered in Section 4.1.5.

4.1.4. Order conditions for VRK formulas

In order to construct specific VRK methods of a given order p for the integral equation (1.1.1) we first have to introduce, in analogy to VLM formulas, the concept of the (local) order of consistency of the VRK formula (4.1.4). Let $y(t)$ and $F_n(t)$ denote, respectively, the exact solution and the *exact lag term* of the equation (4.1.1), and suppose that the approximation $\bar{y}_{n+1}$ is defined by

$$\bar{y}_{n+1} = F_n(t_n+h)+h\bar{\Phi}_n(t_n+h), \tag{4.1.15a}$$

where

$$\bar{\Phi}_n(t) := \sum_{j=1}^{m} b_j k(t+(e_j-1)h, t_n+c_jh, \bar{Y}_{n,j}), \tag{4.1.15b}$$

with

$$\bar{Y}_{n,j} = F_n(t_n+\theta_j h)+h\sum_{i=1}^{m} a_{j,i}k(t_n+d_{j,i}h, t_n+c_ih, \bar{Y}_{n,i}), \tag{4.1.15c}$$

$$j=1, \ldots ,m.$$

In other words, we use the VRK formula (4.1.4), with exact lag term, to solve (4.1.1) *locally* on $[t_n,t_{n+1}]$. Here, $\bar{\Phi}_n(t)$ will be called the *local increment function.*

Definition 4.1.4. The VRK formula (4.1.4) is said to be *consistent of (local) order p* if, for all integral equations (1.1.1) with sufficiently differentiable g and k and for all sufficiently small $h>0$, p is the largest integer for which

$$|y(t_n)-\bar{y}_n| \leqslant Ch^{p+1} \quad (t_n=nh=t \text{ fixed, as } h\downarrow 0)$$

holds, with the constant C not depending on n and h. □

We deduce from (4.1.15a) and from (4.1.3) that $\bar{\Phi}_n$ satisfies

$$h(\Phi_n(t_n+h)-\bar{\Phi}_n(t_n+h)) = y(t_{n+1})-\bar{y}_{n+1}=O(h^{p+1}). \tag{4.1.16}$$

Hence, the local increment function $\bar{\Phi}_n$ represents an approximation of order p to the exact increment function, provided that the local approximation (4.1.15) on $[t_n,t_{n+1}]$ employs the exact lag term. The reader may recall that this assumption resembles the "localizing assumption" used in the theory of linear multistep methods for ODEs (cf. Lambert [1973, p. 27]).

As in the theory of RK methods for ODEs, the local order of consistency of a VRK formula (4.1.4) can be completely characterized in terms of the parameters of the formula; the resulting equations will be called the *order conditions for the VRK formula.* As mentioned before, the corresponding VRK theory has been given by Brunner & Nørsett [1980] and by Brunner, Hairer & Nørsett [1982]. Since a detailed description of this theory is beyond the scope of this book, we restrict ourselves to stating the order conditions for VRK formulas with local order $p \leqslant 4$. It will always be assumed that

$$\theta_j = c_j = \sum_{i=1}^{m} a_{j,i}, \quad j=1, \dots ,m. \tag{4.1.17}$$

In Table 4.1.3 we list the order conditions for the VRK formulas (4.1.4) and for the PVRK and BVRK formulas (4.1.8) and (4.1.10); note that all summation indices range from 1 to m.

We observe that the PVRK order conditions are identical with those for the RK method (4.1.12) for ODEs: if, for example, we wish to construct a PVRK formula of local order $p=4$, we have to satisfy eight order conditions (an empty entry in the middle column indicates that the corresponding general VRK order condition leads to a PVRK order condition obtained previously). However, if a BVRK formula is to have the same order $p=4$, the number of order conditions rises to 21. Table 4.1.4 gives a comparison of the number of order conditions for these two classes of VRK formulas; as in Table 4.1.1, n_p denotes the number of order conditions necessary to increase the order from $p-1$ to p, while N_p is the total number of order conditions to attain order p.

Since the PVRK conditions are a subset of the VRK conditions listed in Table 4.1.3, and, as we already observed, the PVRK conditions are just the order conditions for RK methods (compare also Theorem 4.1.1), we conclude that, in any given VRK formula (4.1.4) which is of local order p, the diagram

$$\begin{array}{c|c} \mathbf{c} & \mathrm{A} \\ \hline & \mathbf{b}^T \end{array} \tag{4.1.18}$$

defines a p-th order RK method for ODEs. This RK method will be called the *associated RK method;* it will play a central role in the stability analysis of VRK methods (see Section 7.5).

In the following two tables we list the order conditions for VRK formulas which are either explicit or diagonally implicit. If the formula is explicit, then we have $a_{i,j}=0$ for all (i,j) with $j \geqslant i$, and hence $c_1=0$ (recall (4.1.17) with $i=1$); for a diagonally implicit formula, $a_{i,j}=0$ holds whenever $j>i$.

Table 4.1.3. VRK order conditions

VRK	PVRK ($e_i=1$, $d_{i,j}=c_i$)	BVRK ($e_i=d_i$, $d_{i,j}=d_j$)
p=1:		
$\sum_i b_i=1$	1. $\sum_i b_i=1$	1. $\sum_i b_i=1$
p=2:		
$\sum_i b_i e_i=1$	--	2. $\sum_i b_i d_i=1$
$\sum_i b_i c_i=\frac{1}{2}$	2. $\sum_i b_i c_i=\frac{1}{2}$	3. $\sum_i b_i c_i=\frac{1}{2}$
p=3:		
$\sum_i b_i e_i^2=1$	--	4. $\sum_i b_i d_i^2=1$
$\sum_i b_i c_i e_i=\frac{1}{2}$	--	5. $\sum_i b_i c_i d_i=\frac{1}{2}$
$\sum_i b_i c_i^2=\frac{1}{3}$	3. $\sum_i b_i c_i^2=\frac{1}{3}$	6. $\sum_i b_i c_i^2=\frac{1}{3}$
$\sum_i\sum_j b_i a_{i,j} d_{i,j}=\frac{1}{3}$	--	7. $\sum_i\sum_j b_i a_{i,j} d_j=\frac{1}{3}$
$\sum_i\sum_j b_i a_{i,j} c_j=\frac{1}{6}$	4. $\sum_i\sum_j b_i a_{i,j} c_j=\frac{1}{6}$	8. $\sum_i\sum_j b_i a_{i,j} c_j=\frac{1}{6}$
p=4:		
$\sum_i b_i e_i^3=1$	--	9. $\sum_i b_i d_i^3=1$
$\sum_i b_i c_i e_i^2=\frac{1}{2}$	--	10. $\sum_i b_i c_i d_i^2=\frac{1}{2}$
$\sum_i b_i c_i^2 e_i=\frac{1}{3}$	--	11. $\sum_i b_i c_i^2 d_i=\frac{1}{3}$
$\sum_i b_i c_i^3=\frac{1}{4}$	5. $\sum_i b_i c_i^3=\frac{1}{4}$	12. $\sum_i b_i c_i^3=\frac{1}{4}$
$\sum_i\sum_j b_i a_{i,j} d_{i,j} e_i=\frac{1}{3}$	--	13. $\sum_i\sum_j b_i a_{i,j} d_i d_j=\frac{1}{3}$
$\sum_i\sum_j b_i a_{i,j} d_{i,j} c_i=\frac{1}{4}$	--	12. $\sum_i\sum_j b_i a_{i,j} c_i d_j=\frac{1}{4}$
$\sum_i\sum_j b_i a_{i,j} c_j e_i=\frac{1}{6}$	--	15. $\sum_i\sum_j b_i a_{i,j} d_i c_j=\frac{1}{6}$
$\sum_i\sum_j b_i a_{i,j} c_i c_j=\frac{1}{8}$	6. $\sum_i\sum_j b_i a_{i,j} c_i c_j=\frac{1}{8}$	16. $\sum_i\sum_j b_i a_{i,j} c_i c_j=\frac{1}{8}$
$\sum_i\sum_j b_i a_{i,j} d_{i,j} c_j=\frac{1}{8}$	--	17. $\sum_i\sum_j b_i a_{i,j} c_j d_j=\frac{1}{8}$
$\sum_i\sum_j b_i a_{i,j} c_j^2=\frac{1}{2}$	7. $\sum_i\sum_j b_i a_{i,j} c_j^2=\frac{1}{12}$	18. $\sum_i\sum_j b_i a_{i,j} c_j^2=\frac{1}{12}$
$\sum_i\sum_j b_i a_{i,j} d_{i,j}^2=\frac{1}{4}$	--	19. $\sum_i\sum_j b_i a_{i,j} d_j^2=\frac{1}{4}$
$\sum_i\sum_j\sum_k b_i a_{i,j} a_{j,k} d_{j,k}=\frac{1}{12}$	--	20. $\sum_i\sum_j\sum_k b_i a_{i,j} a_{j,k} d_k=\frac{1}{12}$
$\sum_i\sum_j\sum_k b_i a_{i,j} a_{j,k} c_k=\frac{1}{24}$	8. $\sum_i\sum_j\sum_k b_i a_{i,j} a_{j,k} c_k=\frac{1}{24}$	21. $\sum_i\sum_j\sum_k b_i a_{i,j} a_{j,k} c_k=\frac{1}{24}$

Table 4.1.4. The number of order conditions for PVRK and BVRK formulas

Order p	1	2	3	4	5	6
PVRK n_p	1	1	2	4	9	20
N_p	1	2	4	8	17	37
BVRK n_p	1	2	5	13	37	108
N_p	1	3	8	21	58	166

Table 4.1.5. Order conditions for diagonally implicit and for explicit PVRK formulas

	diagonally implicit	explicit
	p=1:	
1.	$\sum_{i=1}^{m} b_i = 1$	$\sum_{i=1}^{m} b_i = 1$
	p=2:	
2.	$\sum_{i=1}^{m} b_i c_i = \frac{1}{2}$	$\sum_{i=2}^{m} b_i c_i = \frac{1}{2}$
	p=3:	
3.	$\sum_{i=1}^{m} b_i c_i^2 = \frac{1}{3}$	$\sum_{i=2}^{m} b_i c_i^2 = \frac{1}{3}$
4.	$\sum_{i=1}^{m} \sum_{j=1}^{i} b_i a_{i,j} c_j = \frac{1}{6}$	$\sum_{i=3}^{m} \sum_{j=2}^{i-1} b_i a_{i,j} c_j = \frac{1}{6}$
	p=4:	
5.	$\sum_{i=1}^{m} b_i c_i^3 = \frac{1}{4}$	$\sum_{i=2}^{m} b_i c_i^3 = \frac{1}{4}$
6.	$\sum_{i=1}^{m} \sum_{j=1}^{i} b_i a_{i,j} c_i c_j = \frac{1}{8}$	$\sum_{i=3}^{m} \sum_{j=2}^{i-1} b_i a_{i,j} c_i c_j = \frac{1}{8}$
7.	$\sum_{i=1}^{m} \sum_{j=1}^{i} b_i a_{i,j} c_j^2 = \frac{1}{12}$	$\sum_{i=3}^{m} \sum_{j=2}^{i-1} b_i a_{i,j} c_j^2 = \frac{1}{12}$
8.	$\sum_{i=1}^{m} \sum_{j=1}^{i} \sum_{k=1}^{j} b_i a_{i,j} a_{j,k} c_k = \frac{1}{24}$	$\sum_{i=4}^{m} \sum_{j=3}^{i-1} \sum_{k=2}^{j-1} b_i a_{i,j} a_{j,k} c_k = \frac{1}{24}$

Table 4.1.6. Order conditions for diagonally implicit and for explicit BVRK formulas

	diagonally implicit	explicit
	p=1:	
1.	$\sum_{i=1}^{m} b_i = 1$	$\sum_{i=1}^{m} b_i = 1$
	p=2:	
2.	$\sum_{i=1}^{m} b_i d_i = 1$	$\sum_{i=1}^{m} b_i d_i = 1$
3.	$\sum_{i=1}^{m} b_i c_i = 1/2$	$\sum_{i=2}^{m} b_i c_i = 1/2$
	p=3:	
4.	$\sum_{i=1}^{m} b_i d_i^2 = 1$	$\sum_{i=1}^{m} b_i d_i^2 = 1$
5.	$\sum_{i=1}^{m} b_i c_i d_i = 1/2$	$\sum_{i=2}^{m} b_i c_i d_i = 1/2$
6.	$\sum_{i=1}^{m} b_i c_i^2 = 1/3$	$\sum_{i=2}^{m} b_i c_i^2 = 1/3$
7.	$\sum_{i=1}^{m} \sum_{j=1}^{i} b_i a_{i,j} d_j = 1/3$	$\sum_{i=2}^{m} \sum_{j=1}^{i-1} b_i a_{i,j} d_j = 1/3$
8.	$\sum_{i=1}^{m} \sum_{j=1}^{i} b_i a_{i,j} c_j = 1/6$	$\sum_{i=3}^{m} \sum_{j=2}^{i-1} b_i a_{i,j} c_j = 1/6$

In the preceding considerations we assumed that (4.1.17) is always satisfied. This condition does not restrict the class of formulas: no additional degrees of freedom in the order conditions are generated if it is omitted. Moreover, the resulting supplementary order conditions for the $\{\theta_i\}$ match the analogous ones for the $\{c_i\}$. We illustrate this by deriving the "general" order conditions for $p \leqslant 3$. Taylor expansion of $\bar{y}_{n+1}$, $Y_{n,j}$ and $y(t_{n+1})$ yields:

$$\begin{aligned}\bar{y}_{n+1} &= F_n(t_{n+1}) + h\sum_j b_j k + h^2 \sum_j b_j [e_j k_t + c_j k_s + (\bar{Y}_{n,j} - y(t_n))k_y/h] \\ &\quad + h^3 \sum_j b_j [\tfrac{1}{2} e_j^2 k_{tt} + \tfrac{1}{2} c_j^2 k_{ss} + \tfrac{1}{2}(\bar{Y}_{n,j} - y(t_n))^2 k_{yy}/h \\ &\quad + e_j c_j k_{ts} + (\bar{Y}_{n,j} - y(t_n))(e_j k_{ty} + c_j k_{sy})/h] + O(h^4),\end{aligned}$$

$$\begin{aligned}[\bar{Y}_{n,j} - y(t_n)]/h &= \theta_j F_n' + \sum_i a_{j,i} k + h[\tfrac{1}{2}\theta_j^2 F_n'' + \sum_i a_{j,i} d_{j,i} k_t \\ &\quad + \sum_i a_{i,j} c_i k_s + \sum_i a_{j,i} \theta_i k_y F_n'\end{aligned}$$

$$+\sum_{i,l} a_{j,i}a_{i,l}k_y k]+O(h^2),$$

$$y(t_{n+1}) = F_n(t_{n+1})+hk+h^2[k_t+\frac{1}{2}k_s+\frac{1}{2}k_yF_n'+\frac{1}{2}k_yk]$$
$$+\frac{1}{6}h^3[k_yF_n''+2k_yk_t+k_yk_s+k_y^2F_n'+k_y^2k+3k_{tt}+k_{ss}+3+k_{ts}$$
$$+k_{yy}F_n''+k_{yy}k^2+2k_{yy}F_n'k+2k_{sy}F_n'+2k_{sy}k$$
$$+3k_{ty}F_n'+3k_{ty}k]+O(h^4).$$

Here, all summation indices range from 1 to m, and all derivatives are evaluated at $(t_n,t_n,y(t_n))$. The order conditions for general VRK formulas are now obtained by substituting the expansion for $Y_{n,j}$ into the expansion for $\bar{y}_{n+1}$ and by identifying the coefficients of h^j in the expansions for $\bar{y}_{n+1}$ and $y(t_{n+1})$. In Table 4.1.7 these conditions are listed.

Table 4.1.7. General VRK order conditions: $a_i := \sum_j a_{i,j}$

p=1:			
$\sum_i b_i = 1$			
p=2:			
$\sum_i b_i e_i = 1$			
$\sum_i b_i c_i = \frac{1}{2}$	$\sum_i b_i \theta_i = \frac{1}{2}$	$\sum_i b_i a_i = \frac{1}{2}$	
p=3:			
$\sum_i b_i e_i^2 = 1$			
$\sum_i b_i c_i e_i = \frac{1}{2}$	$\sum_i b_i \theta_i e_i = \frac{1}{2}$		
$\sum_i b_i c_i^2 = \frac{1}{3}$	$\sum_i b_i \theta_i^2 = \frac{1}{3}$	$\sum_i b_i c_i \theta_i = \frac{1}{3}$	$\sum_i b_i a_i \theta_i = \frac{1}{3}$
$\sum_i \sum_j b_i a_{i,j} d_{i,j} = \frac{1}{3}$			
$\sum_i \sum_j b_i a_{i,j} c_j = \frac{1}{6}$	$\sum_i \sum_j b_i a_{i,j} \theta_j = \frac{1}{6}$		
	$\sum_i b_i e_i a_i = \frac{1}{2}$	$\sum_i b_i a_i^2 = \frac{1}{3}$	
	$\sum_i b_i c_i a_i = \frac{1}{3}$	$\sum_i \sum_j b_i a_{i,j} a_i = \frac{1}{6}$	

The left-hand column contains the VRK conditions listed in Table 4.1.3. The right-hand conditions are therefore additional conditions. However, the parameters θ_i have to satisfy the same conditions as the parameters c_i.

Therefore, the conditions $c_j=\theta_j$ in (4.1.17) are no restriction. Moreover, on substitution of $\theta_i=c_i$, we see that the $a_i:=\Sigma_j a_{i,j}$ satisfy the same equations as the $\{c_i\}$, so that the condition $c_i=a_i$ does not represent a loss of generality.

4.1.5. PVRK formulas

In his paper of 1963, Pouzet showed that if a given explicit m-stage RK method for the initial-value problem for a first-order ODE possesses the local order $p=m$, then the corresponding (explicit) m-stage VRK formula (4.1.8) has the same order p. This result has been generalized to general m-stage RK methods; its importance lies in the fact that it furnishes an easy way of constructing VRK formulas of a certain order (recall Section 4.1.2).

Theorem 4.1.1. *Let $A=(a_{j,l})$, $\mathbf{b}=(b_j)$ $(j,l=1, \dots ,m)$ define an m-stage RK method of order p for a first-order ODE. Then the corresponding PVRK formula has order p.* □

The proof of this result, based on the algebraic VRK theory, may be found in Brunner, Hairer & Nørsett [1982, Theorem 3.1]; as has already been indicated in Table 4.1.3, the VRK order conditions for Pouzet formulas reduce precisely to those for the RK method for ODEs.

In the following we present a selection of PVRK formulas by giving the corresponding Butcher arrays (4.1.13).

Example 4.1.1. *Explicit PVRK formulas of order $p=m=2$.*

$$\begin{array}{c|cc} 0 & 0 & 0 \\ \frac{1}{2} & \frac{1}{2} & 0 \\ \hline & 0 & 1 \end{array}\,, \qquad \begin{array}{c|cc} 0 & 0 & 0 \\ \frac{2}{3} & \frac{2}{3} & 0 \\ \hline & \frac{1}{4} & \frac{3}{4} \end{array}\,, \qquad \begin{array}{c|cc} 0 & 0 & 0 \\ 1 & 1 & 0 \\ \hline & \frac{1}{2} & \frac{1}{2} \end{array}\,.$$

These formulas are particular cases of

$$\begin{array}{c|cc} 0 & 0 & 0 \\ c_2 & c_2 & 0 \\ \hline & 1-\dfrac{1}{2c_2} & \dfrac{1}{2c_2} \end{array}\,,$$

which represents a formula of order $p=2$ for all $c_2\neq 0$. □

Example 4.1.2. *Explicit PVRK formulas of order $p = m = 3$.*

Here, we have three families of formulas; the first two are characterized by the diagrams

$$
\begin{array}{c|ccc}
0 & 0 & 0 & 0 \\
\frac{2}{3} & \frac{2}{3} & 0 & 0 \\
0 & -\frac{1}{4b} & \frac{1}{4b} & 0 \\
\hline
 & \frac{1}{4}-b & \frac{3}{4} & b
\end{array}
\quad , \quad
\begin{array}{c|ccc}
0 & 0 & 0 & 0 \\
\frac{2}{3} & \frac{2}{3} & 0 & 0 \\
\frac{2}{3} & \frac{2}{3}-\frac{1}{4b} & \frac{1}{4b} & 0 \\
\hline
 & \frac{1}{4} & \frac{3}{4}-b & b
\end{array}
\qquad (b \neq 0).
$$

The third family is given in Example 4.1.12. □

Example 4.1.3. *Explicit PVRK formulas of order $p = m = 4$.*

We list the two formulas based on the classical fourth-order RK methods of Kutta [1901, p. 443, 441]; the second of these VRK formulas was suggested by Prasad [1924] for the numerical solution of linear Volterra integral equations of the second kind (cf. Section 4.1.2).

$$
\begin{array}{c|cccc}
0 & 0 & 0 & 0 & 0 \\
\frac{1}{2} & \frac{1}{2} & 0 & 0 & 0 \\
\frac{1}{2} & 0 & \frac{1}{2} & 0 & 0 \\
1 & 0 & 0 & 1 & 0 \\
\hline
 & \frac{1}{6} & \frac{1}{3} & \frac{1}{3} & \frac{1}{6}
\end{array}
\quad , \quad
\begin{array}{c|cccc}
0 & 0 & 0 & 0 & 0 \\
\frac{1}{3} & \frac{1}{3} & 0 & 0 & 0 \\
\frac{2}{3} & -\frac{1}{3} & 1 & 0 & 0 \\
1 & 1 & -1 & 1 & 0 \\
\hline
 & \frac{1}{8} & \frac{3}{8} & \frac{3}{8} & \frac{1}{8}
\end{array}
\ . \ \square
$$

Example 4.1.4. *Embedded PVRK formulas with $p = 4$ and $p = 5$.*

It is often possible to construct an RK method of, say, order $p+1$ so that it contains an RK method with fewer stages which is of order p; the latter method is then said to be *embedded* in the former. An example is given by the formulas of England (1969) (see, e.g., Lapidus & Seinfeld [1971, pp. 70-71], or van der Houwen [1977a, p. 169]): the number of stages of the two RK methods is, respectively, $m = 4$ and $m = 6$, with corresponding orders $p = 4$ and $p = 5$.

c							
0	0	0	0	0	0	0	$m=4$, $m=6$
$\frac{1}{2}$	$\frac{1}{2}$	0	0	0	0	0	$m=4$, $m=6$
$\frac{1}{2}$	$\frac{1}{4}$	$\frac{1}{4}$	0	0	0	0	$m=4$, $m=6$
1	0	-1	2	0	0	0	$m=4$, $m=6$
$\frac{2}{3}$	$\frac{7}{27}$	$\frac{10}{27}$	0	$\frac{1}{27}$	0	0	$m=6$
$\frac{1}{5}$	$\frac{28}{625}$	$-\frac{1}{5}$	$\frac{546}{625}$	$\frac{54}{625}$	$-\frac{378}{625}$	0	$m=6$
	$\frac{1}{6}$	0	$\frac{2}{3}$	$\frac{1}{6}$	0	0	$(p=4)$
	$\frac{14}{336}$	0	0	$\frac{35}{336}$	$\frac{162}{336}$	$\frac{125}{336}$	$(p=5)$.

The corresponding PVRK method has recently been implemented in a code by Tanfulla & Ribighini [1981]; compare Chapter 8 for further details. □

Example 4.1.5. *A 2-stage diagonally implicit PVRK formula of order $p=3$.*

c		
$\frac{3+\sqrt{3}}{6}$	$\frac{3+\sqrt{3}}{6}$	0
$\frac{3-\sqrt{3}}{6}$	$-\frac{\sqrt{3}}{3}$	$\frac{3+\sqrt{3}}{6}$
	$\frac{1}{2}$	$\frac{1}{2}$

Note that $a_{11}=a_{22}=(3+\sqrt{3})/6$. This PVRK formula is based on a RK method derived by Nørsett [1974]. □

Example 4.1.6. *Implicit PVRK formulas of order $p=2m$ $(m=1,2)$,*

$m=1$ $(p=2)$:

c	
$\frac{1}{2}$	$\frac{1}{2}$
	1

This PVRK formula is called the *implicit Pouzet midpoint formula.*

$m=2$ ($p=4$):

$$\begin{array}{c|cc} \frac{3-\sqrt{3}}{6} & \frac{1}{4} & \frac{3-2\sqrt{3}}{12} \\ \frac{3+\sqrt{3}}{6} & \frac{3+2\sqrt{3}}{12} & \frac{1}{4} \\ \hline & \frac{1}{2} & \frac{1}{2} \end{array}$$

We shall refer to this formula as the *2-stage Gauss-PVRK formula* (compare also Section 5.3.4). □

It will be shown in Chapter 5 that an important subclass of implicit PVRK formulas (including those of optimal order $p=2m$) can be derived by collocation techniques where the exact solution of (1.1.1) is approximated in a certain polynomial spline space. This approach will also suggest a natural way of modifying the Runge-Kutta part of an implicit PVRK method so that the kernel $k(t,s,y)$ need no longer be evaluated at points (t,s) outside the domain S.

4.1.6. BVRK formulas

Tables 4.1.3 and 4.1.5 show that, for a fixed order p, the number of corresponding order conditions is considerably larger for BVRK formulas than for PVRK formulas. This makes the construction of high-order BVRK formulas quite difficult, and it is intuitively clear that it may, in general, not be possible to obtain BVRK formulas which, for given m, possess the same order as a PVRK formula. However, the first example is still quite encouraging.

Example 4.1.7. *The implicit Bel'tyukov midpoint formula: $m=1$, $p=2$.*

This formula (which is the Bel'tyukov analogue of Example 4.1.6 ($m=1$)) is given by

$$\begin{array}{c|c|c} 1 & \frac{1}{2} & \frac{1}{2} \\ \hline & & 1 \end{array} \quad ;$$

written at full length, the formula has the form

$$\begin{aligned} Y_{n,1} &= \tilde{F}_n(t_n+h/2)+\frac{h}{2}\,k(t_n+h,t_n+h/2,Y_{n,1}), \\ y_{n+1} &= \tilde{F}_n(t_n+h)+k(t_n+h,t_n+h/2,Y_{n,1}). \end{aligned}$$

Note that while the midpoint formula of Pouzet type given in Example 4.1.6 requires two kernel evaluations for each n, the Bel'tyukov midpoint formula uses only one such evaluation. □

After this promising start we have the following discouraging result.

Theorem 4.1.2. *There is no 2-stage BVRK formula of order $p=3$.*

Proof. Suppose that $m=2$ and $p=3$, and consider the order conditions 1., 3., and 6. in Table 4.1.3:

$$b_1+b_2 = 1, \quad b_1c_1+b_2c_2=1/2, \quad b_1c_1^2+b_2c_2^2=1/3.$$

These conditions characterize a 2-point quadrature formula whose degree of precision is at least equal to 2. Since there does not exist a quadrature formula possessing this degree of precision but using only one function evaluation (i.e., one quadrature abscissa), we must have

$$c_1 \neq c_2, \quad b_1b_2 \neq 0.$$

If we now subtract, respectively, order condition 1. from 2., and 3. from 5. ($m=2$), we find the equations

$$b_1(d_1-1)+b_2(d_2-1) = 0,$$

$$b_1c_1(d_1-1)+b_2c_2(d_2-1) = 0.$$

According to the above observation, this can only hold if $d_1=d_2=1$. However, by order condition 7., we then obtain

$$b_1(a_{1,1}d_1+a_{1,2}d_2)+b_2(a_{2,1}d_1+a_{2,2}d_2)=b_1(a_{1,1}+a_{1,2})+b_2(a_{2,1}+a_{2,2})=1/3.$$

Here, by (4.1.17), we have $a_{i,1}+a_{i,2}=c_i$ ($i=1,2$). This leads to a contradiction with order condition 3. which says that $b_1c_1+b_2c_2=1/2$. □

In Section 4.1.4 we have seen that, for $1 \leq m \leq 4$, there exist m-stage explicit PVRK formulas of order $p=m$. This is not true for BVRK formulas.

Theorem 4.1.3. *There is no m-stage explicit BVRK formula of order $p=m$ unless $m \leq 3$.*

Proof. (Butcher, private communication, 1981; see also the special case $m=4$ in Brunner, Hairer & Nørsett [1982, p. 160]). Let $m \geq 4$, and define

$$b_i^{(k)} := \sum_{j=1}^{m} b_j^{(k-1)} a_{j,i} \quad (k=1,2, \dots ; \; b_i^{(0)} := b_i).$$

It can be shown that every m-stage explicit RK method for a first-order ODE with order $p=m$ satisfies

$$\sum_{i=1}^{m} b_i^{(m-4)} a_{i,j} = \frac{1}{m-3}\, b_j^{(m-4)}(1-c_j), \quad j=1, \dots ,m.$$

Multiply each side by d_j; then sum over j to obtain

$$\sum_{i=1}^{m} b_i^{(m-4)} \sum_{j=1}^{m} a_{i,j}d_j = \frac{1}{m-3}\Big\{ \sum_{j=1}^{m} b_j^{(m-4)}d_j - \sum_{j=1}^{m} b_j^{(m-4)}c_jd_j \Big\}.$$

On the other hand, the following order conditions hold:

(i) $$\sum_{i=1}^{m}\sum_{j=1}^{m} b_i^{(m-4)} a_{i,j} d_j = 2/(m-1)!,$$

(ii) $$\sum_{j=1}^{m} b_j^{(m-4)} d_j = 2/(m-2)!,$$

(iii) $$\sum_{j=1}^{m} b_j^{(m-4)} c_j d_j = 3/(m-1)!.$$

This leads to $2(m-3)=2m-5$, hence to a contradiction. □

Note that for the case $m=4$, the above order conditions are conditions 7., 2., and 5. in Table 4.1.5 (explicit BVRK).

We add some examples of explicit BVRK formulas; at the time of writting there is still little known on the existence of high-order diagonally implicit and implicit BVRK formulas.

Example 4.1.8. *Explicit BVRK formulas of order $p=m=2$.*

The general form of such a method is

d	**c**	A	
d_1	$c_1=0$	0	0
d_2	c_2	c_2	0
		$1-\frac{1}{2c_2}$	$\frac{1}{2c_2}$

$(c_2 \neq 0)$,

where $d_2=2c_2+(1-2c_2)d_1$. If we choose $d_1=1$, then also $d_2=1$, and we have

d	**c**	A	
1	0	0	0
1	c_2	c_2	0
		$1-\frac{1}{2c_2}$	$\frac{1}{2c_2}$

$(c_2 \neq 0)$.

Two specific second-order BVRK formulas are thus given by

d	**c**	A	
1	0	0	0
1	$\frac{2}{3}$	$\frac{2}{3}$	0
		$\frac{1}{4}$	$\frac{3}{4}$

d	**c**	A	
1	0	0	0
1	$\frac{1}{2}$	$\frac{1}{2}$	0
		0	1

(see Bel'tyukov [1965b, p. 420]). Compare also the first of these two formulas with the analogous PVRK formula shown in Example 4.1.1. □

In the preceding example it has been possible to choose $d_i=1$ $(i=1,2)$. We shall now show that this can no longer be done for $p \geqslant 3$ (this result has already been used implicitly in the proof of Theorem 4.1.2).

Lemma 4.1.3. *If a BVRK formula has order $p \geqslant 3$, then $d_i \neq 1$ for at least one value of $i \in \{1, \ldots, m\}$.*

Proof. Suppose that $p \geqslant 3$ and $d_i=1$ $(i=1, \ldots, m)$. It follows from order condition 7. (Table 4.1.3), using (4.1.17), that

$$\frac{1}{3} = \sum_{i=1}^{m} \sum_{j=1}^{m} b_i a_{i,j} d_j = \sum_{i=1}^{m} b_i \sum_{j=1}^{m} a_{i,j} = \sum_{i=1}^{m} b_i c_i.$$

This contradicts the order condition 3. □

It easily seen that if a BVRK formula has order $p=3$, then $b_3 \neq 0$ (cf. Theorem 4.1.2). Moreover, we have

Lemma 4.1.4. *If a 3-stage BVRK formula possesses the order $p=3$, then*

$$b_i(d_i-1) = 0 \quad (i=1,2,3). \tag{4.1.19}$$

Proof. By subtracting the BVRK order condition 1. from 2., 3. from 5., and 2. from 4. (Table 4.1.3), we obtain the equations

$$\begin{bmatrix} 1 & 1 & 1 \\ c_1 & c_2 & c_3 \\ d_1 & d_2 & d_3 \end{bmatrix} \begin{bmatrix} b_1(d_1-1) \\ b_2(d_2-1) \\ b_3(d_3-1) \end{bmatrix} = 0.$$

Let C denote the matrix occurring on the left-hand side. We wish to show that if the BVRK formula has order $p=3$, then $\det C \neq 0$, implying (4.1.19). Suppose that there exists a vector $\alpha=(\alpha_1,\alpha_2,\alpha_3)^T \neq 0$ such that $\alpha^T C=0$; i.e.

$$\alpha_1+\alpha_2 c_i+\alpha_3 d_i = 0 \quad (i=1,2,3). \tag{4.1.20}$$

Multiplication of this equation by b_i and summation over i yields

$$\alpha_1 \sum_{i=1}^{3} b_i + \alpha_2 \sum_{i=1}^{3} b_i c_i + \alpha_3 \sum_{i=1}^{3} b_i d_i = 0,$$

which, by the order conditions 1., 2., and 3., reduces to

$$\alpha_1 + \frac{1}{2}\alpha_2 + \alpha_3 = 0. \tag{4.1.21}$$

Multiply now (4.1.20) by $b_i c_i$ and sum again over i; taking into account the order conditions 3., 6., and 5., we find

$$\frac{1}{2}\alpha_1 + \frac{1}{3}\alpha_2 + \frac{1}{2}\alpha_3 = 0.$$

Comparison with (4.1.21) shows that we must have $\alpha_2 = 0$, and hence $\alpha_1 + \alpha_3 = 0$. Substitution in (4.1.20) now leads to

$$\alpha_3(d_i - 1) = 0 \quad (i = 1,2,3).$$

Since $d_i \neq 1$ for at least one of the indices $i \in \{1,2,3\}$ (Lemma 4.1.3), we obtain $\alpha_3 = 0$, and thus $\alpha = (\alpha_1, \alpha_2, \alpha_3)^T = 0$, which is contrary to our original hypothesis. □

We now consider those 3-stage BVRK formulas for which two of the $\{d_i\}$ have the value one; by the observation preceding Lemma 4.1.4, we have always $d_3 = 1$.

Case 1: $d_1 = d_3 = 1,\ b_2 = 0.$
It follows from the order conditions that this yields the BVRK diagram given below.

Example 4.1.9. *Explicit BVRK formulas of order* $p = m = 3$ *and with* $d_1 = d_3 = 1$.

d	c	A		
1	0	0	0	0
$d_2 \neq 1$	$1-d_2$	$1-d_2$	0	0
1	$\frac{2}{3}$	$\frac{2(2-3d_2)}{9(1-d_2)}$	$\frac{2}{9(1-d_2)}$	0
		$\frac{1}{4}$	0	$\frac{3}{4}$

Two particular cases, corresponding to the values $d_2 = 1/2$ and $d_2 = 2/3$ (note that the kernel condition is satisfied if $d_2 \geqslant 1/2$) are:

d	c	A			**d**	c	A		
1	0	0	0	0	1	0	0	0	0
$\frac{1}{2}$	$\frac{1}{2}$	$\frac{1}{2}$	0	0	$\frac{2}{3}$	$\frac{1}{3}$	$\frac{1}{3}$	0	0
1	$\frac{2}{3}$	$\frac{2}{9}$	$\frac{4}{9}$	0	1	$\frac{2}{3}$	0	$\frac{2}{3}$	0
		$\frac{1}{4}$	0	$\frac{3}{4}$			$\frac{1}{4}$	0	$\frac{3}{4}$

(the first of these formulas was given by Bel'tyukov [1965b, p. 421]). □

Case 2: $d_2=d_3=1,\ b_1=0.$
Here, the order conditions lead to the following BVRK diagram.

Example 4.1.10. *Explicit BVRK formulas of order* $p=m=3$ *and with* $d_2=d_3=1$.

d	c	A		
$d_1 \neq 1$	0	0	0	0
1	$\frac{1-d_1}{2-3d_1}$	$\frac{1-d_1}{2-3d_1}$	0	0
1	$\frac{3d_1-1}{3d_1}$	$a_{3,1}$	$a_{3,2}$	0
		0	b_2	$1-b_2$

,

where

$$b_2 = \frac{(2-3d_1)^2}{4(1-3d_1+3d_1^2)}, \quad a_{3,2} = \frac{2-3d_1}{6(1-d_1)(1-b_2)}, \quad a_{3,1} = \frac{3d_1-1}{3d_1} - a_{3,2}.$$

The kernel condition will be satisfied if $d_1 \leqslant 1/2$. For $d_1=1/3$ and $d_1=1/2$ we obtain, respectively,

d	c	A		
$\frac{1}{3}$	0	0	0	0
1	$\frac{2}{3}$	$\frac{2}{3}$	0	0
1	0	−1	1	0
		0	$\frac{3}{4}$	$\frac{1}{4}$

d	c	A		
$\frac{1}{2}$	0	0	0	0
1	1	1	0	0
1	$\frac{1}{3}$	$\frac{1}{9}$	$\frac{2}{9}$	0
		0	$\frac{1}{4}$	$\frac{3}{4}$

(The first of these formulas occurs in Bel'tykov [1965b, p. 420]). □

We conclude this section on BVRK formulas with an example of a class of fourth-order formulas possessing five stages.

Example 4.1.11. *A class of explicit BVRK formulas with $m=5$ and $p=4$.*

We have seen (Theorem 4.1.3) that there does not exist an m-stage explicit BVRK formula of order $p=m$ if $m \geqslant 4$. However, there do exist 5-stage explicit BVRK formulas of order $p=4$ (Hairer [1982]). For $c \neq 0$, a class of such methods is given by

d	c	A				
$\frac{3-\sqrt{3}}{4}$	0	0	0	0	0	0
$\frac{3-\sqrt{3}}{4}-c$	c	c	0	0	0	0
1	$\frac{3-\sqrt{3}}{6}$	$\frac{3-\sqrt{3}}{6}+a_{5,2}$	$-a_{5,2}$	0	0	0
$\frac{57+5\sqrt{3}}{92}$	$\frac{9+2\sqrt{3}}{23}$	$\frac{2781-647\sqrt{3}}{6877}-a_{4,2}$	$a_{4,2}$	$a_{4,3}$	0	0
1	$\frac{3+\sqrt{3}}{6}$	$\frac{-3+2\sqrt{3}}{9}-a_{5,2}$	$a_{5,2}$	$\frac{1}{5}$	$\frac{57-5\sqrt{3}}{90}$	0
		0	0	$\frac{1}{2}$	0	$\frac{1}{2}$

,

where $a_{4,2}=(2544-807\sqrt{3})/13754c$, $a_{4,3}=(-90+1245\sqrt{3})/6877$, and $a_{5,2}=(-2+\sqrt{3})/12c$. The kernel condition is satisfied if $0<c \leqslant (3-\sqrt{3})/8$. Note that the underlying quadrature formula is the 2-point Gauss-Legendre formula. □

4.1.7. Lag term formulas

The two important classes of lag term formulas, i.e., the extended and the mixed formulas, have already been mentioned in Definition 4.1.1. They may be viewed as the two extreme special cases of

$$\tilde{F}_n(t) = g(t)+h\sum_{l=-1}^{n-1}\sum_{j=1}^{m+1} w_{n,l,j}k(t+(e_j-1)h,t_l+c_jh,Y_{l,j}) \qquad (4.1.22)$$

(with $w_{n,-1,j}=0$ for $j\leqslant m$, $Y_{l,m+1}:=y_{l+1}$, $c_{m+1}=e_{m+1}:=1$, and $n\geqslant\tilde{\kappa}$), where the quadrature rule constituting the second term on the right-hand side involves both intermediate points and step points. Hence, the *extended lag term formula* (4.1.5a) arises if the weights $\{w_{n,l,j}\}$ satisfy

$$w_{n,l,j} = \begin{cases} b_j, & \text{if } \ 1\leqslant j\leqslant m \quad (0\leqslant l\leqslant n-1),\\ 0\,, & \text{if } \ j=m+1 \ \text{ or } \ l=-1.\end{cases}$$

The *mixed lag term formula* is characterized by

$$w_{n,l,j} = \begin{cases} w_{n,l+1}, & \text{if } \ j=m+1 \quad (-1\leqslant l\leqslant n-1),\\ 0\,, & \text{otherwise.}\end{cases}$$

In other words, the mixed lag term formula employs only step points, while the extended formula is based on the use of intermediate points (recall also Figures 4.1.1b, 4.1.2b, and 4.1.3b).

In analogy to (3.2.14) we shall call the quadrature formula in (4.1.22) *convergent of order r* if, for $1\leqslant n\leqslant N-1$,

$$E_n(h;t) := \int_0^{t_n} k(t,s,y(s))ds - h\sum_{l=-1}^{n-1}\sum_{j=1}^{m+1} w_{n,l,j}k(t+(e_j-1)h,t_l+c_jh,\overline{Y}_{l,j})$$

$$= O(h^r), \quad t\in[t_n,T], \qquad (4.1.23)$$

as $h\downarrow 0$, with $t_n=nh$ fixed. Here, $\overline{Y}_{l,m+1}:=y(t_{l+1})$, while $\overline{Y}_{l,j}$ $(j\leqslant m)$ is defined as in (4.1.15c).

Extended lag term formulas

In the case of an extended lag term formula, the weights $\{w_{n,l,j}\}$ are given by the VRK parameters $\{b_j\}$; as we have mentioned before, these parameters need not correspond to familiar quadrature weights. In this case the determination of the order r in (4.1.23) is more complex than in the case where the $\{b_j\}$ define a classical quadrature formula. However, it follows from the definitions of the increment functions Φ_n and $\overline{\Phi}_n$ (cf. (4.1.2) and (4.1.15b)) that the error $E_n(h;t)$ (recall (3.2.14)) associated with the lag term formula satisfies

$$E_n(h;t) = h\sum_{l=0}^{n-1}(\Phi_l(t)-\overline{\Phi}_l(t)), \quad n=1,\ \dots\ ,N-1. \qquad (4.1.24)$$

Hence, the order of $E_n(h;t)$ depends on how well the local increment function $\overline{\Phi}_n(t)$ approximates the exact increment function $\Phi_n(t)$ on $[t_n,T]$ $(n=1, \ldots ,N-1)$. The following theorem answers this equation for extended VRK methods.

Theorem 4.1.4. *If the VRK formula in an extended VRK method is consistent of order p, then we have*

$$|\Phi_n(t)-\overline{\Phi}_n(t)| \leqslant Ch^p, \quad t\in[t_n,T], \quad n=1, \ldots ,N-1,$$

where the constant $C>0$ is independent of n, h, and t.

Proof. Following Hairer, Lubich & Nørsett [1983, Lemma 2], we consider the system of integral equations,

$$v(t) = \int_a^t k(t-a-h+\xi,s,w(s))ds,$$

$$w(t) = g^*(t)+\int_a^t k(t,s,w(s))ds,$$

where a, $h>0$, and ξ denote parameters, and where $g^*(t)$ is a given, sufficiently smooth function. We approximate $(v(t),w(t))$ at the point $t=a+h$ by executing one step with the extended VRK method, using stepsize h (recall that the definition of a VRK method is extended to systems of Volterra equations in a trivial way). Hence, the numerical approximation to $v(a+h)$ is furnished by

$$v_1 := h\sum_{j=1}^m b_j k(\xi+(e_j-1)h,a+c_jh,W_{0,j}),$$

where $W_{0,j}$ is the j-th intermediate value in the VRK formula. Thus we have, since the VRK formula is of order p,

$$|v(a+h)-v_1| \leqslant Ch^{p+1},$$

with $C=C(\xi,a)$ not depending on h (cf. Definition 4.1.4). Next, we set $a=t_n$ and $g^*(t)=F_n(t)$, where $F_n(t)$ is the exact lag term defined by (4.1.1b). It follows that $w(t)=y(t)$, and $W_{0,j}=Y_{n,j}$, so that

$$v(a+h) = v(t_{n+1}) = \int_{t_n}^{t_{n-1}} k(\xi,s,y(s))ds = h\Phi_n(\xi),$$

and

$$v_1 = h\sum_{j=1}^m b_jk(\xi+(e_j-1)h,t_n+c_jh,\overline{Y}_{n,j}) = h\overline{\Phi}_n(\xi), \quad \xi\in[t_n,T].$$

We are thus led to the assertion of Theorem 4.1.4 by choosing for $C(\xi,a)$ the largest value assumed for $0\leqslant a\leqslant\xi\leqslant T$. □

The following result is a direct consequence of (4.1.24) and of Theorem 4.1.4.

Corollary 4.1.1. *If the VRK formula of an extended VRK method is consistent of order p, then the lag term of the method is also of order p.* □

It is important to notice that extended lag terms are defined for all n with $0 \leqslant n \leqslant N-1$. Therefore, an extended VRK method does not require any starting values, regardless of its order of convergence (which is given by the order of consistency of its VRK formula). This is a strong point in favor of employing extended VRK methods, since, as will be seen below, mixed VRK methods require starting values whose number will depend on the order of the VRK formula and which have to be found by other means.

If an extended VRK method is applied to the second-kind integral equation (1.1.1), both the exact increment function and the lag term in (4.1.1) are discretized by the same method; i.e., once the parameters describing the VRK formula have been selected, the approximation scheme for the lag term is completely determined. On the other hand, the computational effort which goes into the evaluation of the tail approximations $\tilde{F}_n(t_n+\theta_j h)$, $1 \leqslant j \leqslant m$, $0 \leqslant n \leqslant N-1$, becomes very large since the total number of kernel evaluations needed in these evaluations is of the order $N^2m^2/2$ (note that the evaluation of the VRK part involves, in the general case, Nm^2 kernel evaluations; in a BVRK method, this number drops to Nm; see also Lemma 4.1.1 and Lemma 4.1.2). This estimate is, of course, pessimistic, and in many cases the number of distinct kernel values is considerably lower than $N^2m^2/2$. To be precise, let $m_\theta N$ be the number of distinct values in the set $\{n+\theta_j\}$ where $n=0, \ldots N-1$ and $j=1, \ldots ,m$. Then the lag term approximation (4.1.5a) requires about $\Sigma_{n=0}^{N}\Sigma_{j=1}^{m} m_\theta n = \frac{1}{2} m\, m_\theta N^2$ distinct kernel values. Thus, we can reduce the number of distinct kernel values in the lag term by *minimizing the number of distinct, noninteger values* $\{\theta_j\}$.

Example 4.1.12. The three-stage PVRK formulas given in Example 4.1.2 both have $m_\theta=2$ (just as the four-stage method given by the first diagram in Example 4.1.3). However, there exists a three-stage, third-order PVRK family with $m_\theta=1$. Consider the array

$$\begin{array}{c|ccc} 0 & 0 & & \\ c_2 & c_2 & & \\ c_3 & c_3-a_{3,2} & a_{3,2} & 0 \\ \hline & 1-b_2-b_3 & b_2 & b_3 \end{array},$$

where

$$a_{3,2} = \frac{c_3(c_3-c_2)}{c_2(2-3c_2)}, \quad b_2 = \frac{3c_3-2}{6c_2(c_3-c_2)}, \quad b_3 = \frac{3c_3-2}{6c_3(c_3-c_2)},$$

and where c_2 and c_3 are parameters such that $c_2c_3(c_2-2/3)(c_3-c_2)\neq 0$ (compare Butcher [1976, p. 78]). We obtain $m_\theta = 1$ for, e.g., $c_2 = -c_3 = -1$. □

Mixed lag term formulas

In order (given a stepsize h) to reduce the number of kernel evaluations in the tail approximation, there are two other possible avenues open to us:

(i) Instead of a tail approximation employing m intermediate points in each subinterval $[t_l, t_{l+1}]$, $0 \leqslant l \leqslant n-1$, we could resort to a tail approximation which is based on a quadrature formula using as abscissas only the step points $\{t_0, \dots, t_n\}$. This yields, as mentioned above, a mixed VRK method; the total number of kernel evaluations required in the tail approximation is then of the order $N^2m/2$ (as compared to $N^2m^2/2$ in an extended VRK method).

(ii) In addition, we could try to minimize the number of distinct kernel values in (4.1.5b). If $m_\theta \leqslant m$ denotes the number defined above, then the total number of kernel evaluations in the tail approximation will be about $N^2m_\theta/2$ for a mixed VRK method.

The choice of quadrature formulas of appropriate order r will become clear on the basis of Theorem 4.1.6; a popular family of quadrature formulas are the *Gregory rules* (of order $r = p-1$, cf. Theorem 4.1.6) introduced in Section 2.6.3.

The problem of minimizing the number of distinct, noninteger parameters $\{\theta_j\}$ (or, in the case of PVRK or BVRK methods, of distinct, noninteger $\{c_j\}$) is still essentially open. Recall, however, that (a) for an m-stage implicit PVRK formula of (optimal) order $p = 2m$ we have $0 < c_1 < \dots < c_m < 1$ (since the $\{c_j\}$ are the Gauss points for $(0,1)$); and (b) there exist 3-stage explicit BVRK formulas of order $p = 3$ with only two distinct $\{c_j\}$; the case $c_1 = c_2 = c_3$ is not possible (cf. Example 4.1.10).

A disadvantage of employing only step points in the tail approximation is the need for a sufficient number of starting values $\{y_0, \dots, y_{\tilde{\kappa}-1}\}$ in the mixed lag term (4.1.5b), where $\tilde{\kappa}$ is essentially equal to the order of the given quadrature formula (since the abscissas are equally spaced). One possibility of obtaining these starting values is by means of an extended VRK method of appropriate order. Alternatively, suitable starting values may be generated by collocation methods (compare Chapter 5) or by special starting schemes such as those discussed in Section 3.5.

γ-modified lag term formulas

The so-called γ-modification of a given lag term formula is motivated by the requirement that if a VRK method is applied to the integrated form of the

ODE $y'(t)=f(t,y(t))$, then it should be possible to reduce the resulting scheme to an RK method for the given ODE. To be more precise, let us apply the VRK method (4.1.4) to

$$y(t) = y_0 + \int_0^t f(s,y(s))ds, \quad t \in I. \tag{4.1.25}$$

Since the kernel of this special integral equation does not depend on t, we obtain

$$\begin{aligned} Y_{n,j} &= \tilde{F}_n + h \sum_{i=1}^{m} a_{j,i} f(t_n + c_i h, Y_{n,i}), \quad j=1, \ldots, m; \\ y_{n+1} &= \tilde{F}_n + h \sum_{j=1}^{m} b_j f(t_n + c_j h, Y_{n,j}), \quad n=0, \ldots, N-1. \end{aligned} \tag{4.1.26}$$

Here, we have written $\tilde{F}_n$ instead of $\tilde{F}(t_n + \theta_j h)$, as the exact lag term corresponding to (4.1.25) is independent of t. On the other hand, the discretization of the differentiated form of (4.1.25) by an RK method based on the parameters $\{c_j\}$, $\{a_{j,i}\}$, and $\{b_j\}$ yields the scheme (4.1.12). Comparison of (4.1.26) with (4.1.12) reveals that the *two sets of equations are identical if the lag term formula is such that*

$$\tilde{F}_n(t) = y_n, \quad n=1, \ldots, N-1, \tag{4.1.27}$$

for all integral equations of the form (4.1.25). Moreover, it will be shown in Chapter 7 that a lag term possessing this property has a stabilizing effect on the VRK method when it is applied to so-called *stiff problems,* that is, to integral equations where the partial derivative $\partial k/\partial y$ of the kernel $k(t,s,y)$ assumes negative values with large moduli.

Since an extended lag term formula satisfies the recurrence relation

$$\tilde{F}_n(t) = \tilde{F}_{n-1}(t) + h\tilde{\Phi}_{n-1}(t), \quad n=1, \ldots, N-1, \tag{4.1.28}$$

it follows from (4.1.3) that

$$y_n = \tilde{F}_{n-1}(t_n) + h\tilde{\Phi}_{n-1}(t_n) = \tilde{F}_n(t_n), \tag{4.1.29}$$

which, in turn, implies that (4.1.27) holds for all integral equations of the form (4.1.25). However, in the case of a mixed lag term formula, $\tilde{F}_n(t)$ and $\tilde{\Phi}_n(t)$ are not related, and thus (4.1.27) will not be satisfied. Therefore, we introduce the *modified lag term formula* (cf. (4.1.6)),

$$\tilde{F}^*_n(t) := \tilde{F}_n(t) + \gamma(t)[y_n - \tilde{F}_n(t_n)], \quad n=1, \ldots, N-1, \tag{4.1.30}$$

where $\tilde{F}_n(t)$ is an arbitrary lag term approximation. If we choose the function γ so that $\gamma(t) \equiv 1$, then $\tilde{F}^*_n(t)$ satisfies (4.1.27) for all integral equations (4.1.25). An additional advantage of the γ-modification (4.1.30) is that, for any VRK formula with one or several of the $\{\theta_j\}$ vanishing, the choice $\gamma(t_n)=1$, $\gamma(t)=0$ for $t \neq t_n$, yields VRK methods which require no evaluation of the lag term at $t=t_n$. The price we have to pay, however, for the economizing and stabilizing effect of the γ-modified lag term consists in a drop of order, as will be shown in Section 4.1.8.

Lag term formulas for convolution kernels

In practical applications one very frequently encounters integral equations (1.1.1) where the kernel function $k(t,s,y)$ has the form

$$k(t,s,y) = K(t-s)f(s,y).$$

For the linear case we have $f(s,y)=y$: the corresponding kernel $k=K(t-s)y$ is thus a classical *convolution kernel.*

For the nonlinear case the *VRK part* of the VRK method (4.1.3) is thus given by

$$\tilde{\Phi}_n(t_n+h) = \sum_{j=1}^{m} b_j K((e_j-c_j)h)f(t_n+c_jh, Y_{n,j}),$$

with

$$Y_{n,j} = \tilde{F}_n(t_n+\theta_jh)+h\sum_{i=1}^{m} a_{j,i}K((d_{j,i}-c_i)h)f(t_n+c_ih, Y_{n,i}),$$
$$j=1, \dots ,m.$$

We notice that the same K-values are used for all values of $n=0, \dots ,N-1$.

A similar reduction in the number of kernel evaluations occurs in the lag term formulas. The *extended lag term formula* (4.1.5a) at $t=t_n+\theta_jh$ now assumes the form

$$\tilde{F}_n(t_n+\theta_jh) = g(t_n+\theta_jh)$$
$$+h\sum_{l=0}^{n-1}\sum_{\nu=1}^{m} b_\nu K(n-l-1+\theta_j+e_\nu-c_\nu)h)f(t_l+c_\nu h, Y_{l,\nu}),$$

while the *mixed lag term formula* (4.1.5b) becomes

$$\tilde{F}_n(t_n+\theta_jh) = g(t_n+\theta_jh)+h\sum_{l=0}^{n} w_{n,l}K((n-l+\theta_j)h)f(t_l,y_l).$$

For the sake of completeness, we also state the formulas for the PVRK and BVRK methods.

The *PVRK formula* (4.1.8) reduces to

$$Y_{n,j} = \tilde{F}_n(t_n+c_jh)+h\sum_{i=1}^{m} a_{j,i}K((c_j-c_i)h)f(t_n+c_ih, Y_{n,i}),$$
$$j=1, \dots ,m;$$

$$y_{n+1} = \tilde{F}_n(t_n+h)+h\sum_{j=1}^{m} b_jK((1-c_j)h)f(t_n+c_jh, Y_{n,j}),$$
$$n=0, \dots ,N-1.$$

The corresponding *extended lag term formula* is given by

$$\tilde{F}_n(t) = g(t)+h\sum_{l=0}^{n-1}\sum_{j=1}^{m} b_j K(t-t_l-c_jh)f(t_l+c_jh, Y_{l,j}).$$

The *BVRK formula* (4.1.10) now reads

$$Y_{n,j} = \tilde{F}_n(t_n+c_jh)+h\sum_{i=1}^{m} a_{j,i}K((d_i-c_i)h)f(t_n+c_ih, Y_{n,i}),$$

$$j=1, \ldots, m;$$

$$y_{n+1} = \tilde{F}_n(t_n+h)+h\sum_{j=1}^{m} b_j K((d_j-c_j)h)f(t_n+c_jh, Y_{n,j}),$$

$$n=0, \ldots, N-1.$$

The corresponding *extended lag term formula* has the form

$$\tilde{F}_n(t) = g(t)+h\sum_{l=0}^{n-1}\sum_{j=1}^{m} b_j K(t-t_l+(d_j-c_j-1)h)f(t_l+c_jh, Y_{l,j}).$$

The mixed lag term formulas for these two classes of VRK methods are identical; we leave their derivation to the reader.

4.1.8. Convergence results

The general convergence analysis of one-step methods for Volterra integral equations of the second kind (of which the VRK methods represent an important subclass) is due to Hairer, Lubich & Nørsett [1983]. In the present section we exhibit the convergence results for extended, mixed, and γ-modified VRK methods; for the first two classes of methods, the proofs are essentially those of the paper just cited, while for γ-modified VRK methods we follow van der Houwen, Wolkenfelt & Baker [1981].

Extended VRK methods

The subsequent analysis can be rendered into a more concise form by introducing the (discrete) *increment operator* $\mathbf{\Phi}_n$ associated with (4.1.4); it is defined by $\mathbf{\Phi}_n[\tilde{F}_n]$ $(=\mathbf{\Phi}_n[\tilde{F}_n;t,h])\colon=\tilde{\Phi}_n(t)$, $n=0, \ldots, N-1$. As indicated by the notation, this operator depends on the lag term formula $\tilde{F}_n(t)$ characterizing (4.1.46).

Theorem 4.1.5. *Let the following assumptions hold:*

(i) *The VRK formula (4.1.4) is consistent of order p.*

(ii) $\tilde{F}_n(t)$ *is an extended lag term approximation of the form (4.1.5a).*

(iii) *The increment operator* $\mathbf{\Phi}_n$ *satisfies the Lipschitz condition*

$$|\mathbf{\Phi}_n[f_n]-\mathbf{\Phi}_n[g_n]| \leqslant L|f_n-g_n|, \quad t\in[t_n,T], \quad 0\leqslant n\leqslant N-1, \tag{4.1.31}$$

for all real f_n, g_n, with L independent of n and h. Then the approximations $\{y_n\}$ generated by the extended VRK method (4.1.3) satisfy, for all sufficiently differentiable functions g and k,

$$\max_{1\leq n\leq N}|y(t_n)-y_n| \leq Ch^p \quad (\text{as } h\downarrow 0,\ Nh=T),$$

where the constant $C>0$ does not depend on N and h.

Proof. Let F_l be the exact lag term, and define (see Figure 4.1.4 below)

$$\tilde{F}_{l,0} := F_l,$$
$$\tilde{F}_{l,j} := \tilde{F}_{l-1,j-1}+h\Phi_{l-1}[\tilde{F}_{l-1,j-1}], \quad j=1, \ldots, l-1,$$
$$\tilde{F}_{l,l} := \tilde{F}_l,$$

for $l=0, \ldots, N-1$.

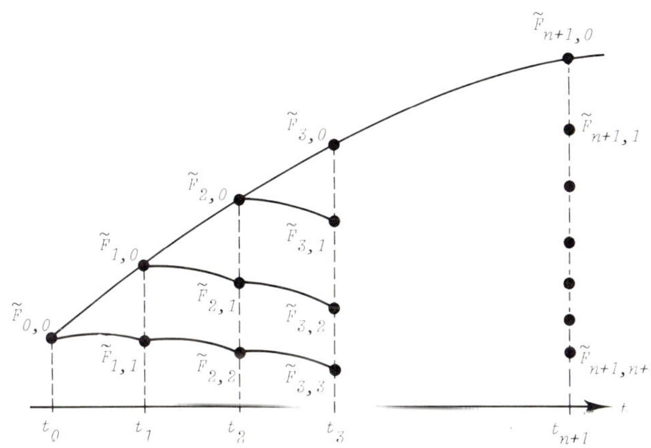

Figure 4.1.4. The auxiliary functions $\tilde{F}_{l,j}$

These auxiliary functions will be used to derive an estimate for $|y(t_{n+1})-y_{n+1}|$. First, we write, by virtue of (4.1.1) and (4.1.29),

$$y(t_{n+1})-y_{n+1} = F_{n+1}(t_{n+1})-\tilde{F}_{n+1}(t_{n+1}).$$

Hence, we obtain, using the above functions $\{\tilde{F}_{l,j}\}$ and observing the assumed Lipschitz condition for Φ_l,

$$|y(t_{n+1})-y_{n+1}| = |\tilde{F}_{n+1,0}-\tilde{F}_{n+1,n+1}|\leq \sum_{j=0}^{n}|\tilde{F}_{n+1,j}-\tilde{F}_{n+1,j+1}|$$

$$= |\tilde{F}_{n+1,0}-\tilde{F}_{n+1,1}|+\sum_{j=1}^{n}|\tilde{F}_{n,j-1}-\tilde{F}_{n,j}+h\Phi_n[\tilde{F}_{n,j-1}]-h\Phi_n[\tilde{F}_{n,j}]|$$

$$\leqslant |\tilde{F}_{n+1,0} - \tilde{F}_{n+1,1}| + (1+hL) \sum_{j=1}^{n} |\tilde{F}_{n,j-1} - \tilde{F}_{n,j}| \leqslant \dots$$

$$\leqslant \sum_{j=0}^{n} (1+hL)^j |\tilde{F}_{n+1-j,0} - \tilde{F}_{n+1-j,1}|$$

$$= \sum_{j=0}^{n} (1+hL)^j |F_{n+1-j} - F_{n-j} - h\mathbf{\Phi}_{n-j}[F_{n-j}]|,$$

where the functions $\tilde{F}_{l,j}$ and F_l are all evaluated at $t = t_{n+1}$. Now, since (4.1.2) and (4.1.15b) imply, for $t \in [t_{n-j}, T]$,

$$F_{n+1-j} - F_{n-j} = h\Phi_{n-j}, \quad \mathbf{\Phi}_{n-j}[F_{n-j}] = h\overline{\Phi}_{n-j},$$

we are led to the estimate

$$|y(t_{n+1}) - y_{n+1}| \leqslant h \sum_{j=0}^{n} (1+hL)^j |\Phi_{n-j}(t_{n+1}) - \overline{\Phi}_{n-j}(t_{n+1})|,$$

$$n = 0, \dots, N-1.$$

The difference between the exact increment function Φ_{n-j} and the local increment function $\overline{\Phi}_{n-j}$ can be estimated by applying Theorem 4.1.4 and by using assumption (i) of the present theorem; we find

$$|y(t_{n+1}) - y_{n+1}| \leqslant Ch^{p+1} \sum_{j=0}^{n} (1+hL)^j \leqslant h^p \frac{C}{L} (\exp(Lt_{n+1}) - 1),$$

where $t_{n+1} = (n+1)h \leqslant T$, $n = 0, \dots, N-1$. This concludes the proof of Theorem 4.1.5. □

It is readily verified that assumption (iii) on the existence of a Lipschitz condition for the increment operator $\mathbf{\Phi}_n$ holds if the kernel function $k(t,s,y)$ itself satisfies a Lipschitz condition, with respect to its third argument y, on a compact neighborhood of $\{(t,s,y(s)): 0 \leqslant s \leqslant t \leqslant T\}$.

Once again, the generalization of Theorem 4.1.5 to systems of integral equations is straightforward; details may be found in Hairer, Lubich & Nørsett [1983].

It has already been mentioned in Section 4.1.2 that a convergence proof for explicit, extended VRK methods with $p = m \leqslant 4$ was given by Pouzet [1963]. Also, R. Weiss [1972a], de Hoog & Weiss [1975], and Brunner & Nørsett [1981] have studied the convergence of certain discretized versions of collocation methods in piecewise polynomial spaces (which, as will be shown in Chapter 5, are closely related to implicit, extended VRK methods).

Mixed VRK methods

The following lemma will play a crucial role in the convergence analysis of mixed VRK methods.

Lemma 4.1.5. *If the elements of the sequence $\{z_n : n=0, \ldots, N\}$ are nonnegative and satisfy the inequality*

$$z_{n+1} \leqslant C_1 z_n + C_2 \sum_{i=0}^{n} z_i + C_3, \quad n=0, \ldots, N-1,$$

with $C_i > 0$ $(i=1,2,3)$, then they are bounded by

$$z_n \leqslant A(n) z_0 + C_3 B(n), \quad n=0, \ldots, N. \tag{4.1.32}$$

Here, $A(n)$, and $B(n)$ are given by

$$A(n) := \frac{(1-R_1)R_1^n + (R_2-1)R_2^n}{R_2-R_1}, \quad B(n) := \frac{R_2^n - R_1^n}{R_2-R_1},$$

with R_1 and R_2 denoting the roots of the equation $R^2-(1+C_1+C_2)R+C_1=0$. □

A proof of this result may be found in Hairer, Lubich & Nørsett [1983]. The estimate (4.1.32) improves a result due to van der Houwen [1980]. Note also that two special cases of Lemma 4.1.5 are well known: the case $C_2=0$ yields a result much used in the convergence analysis of one-step methods for ODEs (see, e.g., Henrici [1962, p. 18]), while the case $C_1=0$ leads to Corollary 1.5.2 (with $k=1$).

We now state the analogue of Theorem 4.1.5 for mixed VRK methods.

Theorem 4.1.6. *Let the following assumptions hold:*

(i) *The VRK formula (4.1.4) is consistent of order p.*

(ii) *$\tilde{F}_n(t)$ is a mixed lag term approximation (4.1.5b) of order r, and its weights are uniformly bounded: $|w_{n,l}| \leqslant w$, $0 \leqslant l \leqslant n$ $(\tilde{\kappa} \leqslant n \leqslant N-1;\ N \in \mathbb{N})$.*

(iii) *The increment operator satisfies the Lipschitz condition (4.1.31).*

(iv) *The starting values $\{y_0, \ldots, y_{\tilde{\kappa}-1}\}$ are of order s: $y(t_j)-y_j=O(h^s)$, $j=0, \ldots, \tilde{\kappa}-1$.*

Then the approximations $\{y_n\}$, $\tilde{\kappa} \leqslant n \leqslant N$, generated by the mixed VRK method (4.1.4) satisfy, for all sufficiently differentiable g and k,

$$\max_{\tilde{\kappa} \leqslant n \leqslant N} |y(t_n)-y_n| \leqslant Ch^{p^*}, \quad (\text{as } h \downarrow 0,\ Nh=T),$$

where $p^* := \min\{p+1, r, s+1\}$, and where the constant $C>0$ is independent of N and h.

Proof. The starting point of the proof is the inequality

$$|\epsilon_{n+1}| := |y(t_{n+1})-y_{n+1}| \leqslant |y(t_{n+1})-\bar{y}_{n+1}| + |\bar{y}_{n+1}-y_{n+1}|,$$

where $\bar{y}_{n+1}$ has been introduced in (4.1.15). Assumption (i) is equivalent to the statement

$$|y(t_{n+1})-\bar{y}_{n+1}| \leqslant C_4 h^{p+1}, \quad n=0, \dots ,N-1,$$

with C_4 denoting a suitable constant. The second term in the above inequality can be shown to satisfy

$$|\bar{y}_{n+1}-y_{n+1}| \leqslant (1+hL)\{C_1|\epsilon_n|+hC_2\sum_{i=0}^{n}|\epsilon_i|+C_3h^r\}.$$

This yields

$$|\epsilon_{n+1}| \leqslant C_1(1+hL)|\epsilon_n|+hC_2(1+hL)\sum_{i=0}^{n}|\epsilon_i|+(1+hL)C_3h^r+C_4h^{p+1},$$

with

$$|\epsilon_n| \leqslant C_0 h^s, \quad n=0, \dots ,\tilde{\kappa}-1.$$

The assertion of Theorem 4.1.6 now follows readily by applying Lemma 4.1.5. Compare also the paper by Hairer, Lubich & Nørsett [1983] for details, as well as for the extension of the convergence result to systems of Volterra integral equations. □

Observe that combining a p-th order consistent VRK formula with a $(p+1)$-st order lag term approximation yields a $(p+1)$-st order mixed VRK method (provided that the starting values are at least of order p). This is in contrast with extended VRK methods, where a p-th order consistent VRK formula always leads to a p-th order method.

γ-modified VRK methods

Before discussing the convergence of γ-modified VRK solutions, we derive the equation to which the method itself converges, as $h \to 0$ (compare a similar derivation for VLM methods in Section 3.2.8).

Let $y(t)$ be a sufficiently smooth function, not necessarily representing the exact solution, and let $F(t,s)$ be the corresponding lag term function defined by (3.2.12). Then, by substituting $y(t)$ into (4.1.3), with $\tilde{F}_n(t)$ replaced by $\tilde{F}^*_n(t)$ defined by (4.1.5c), and observing that, as $h \to 0$,

$$\tilde{F}_n(t) \to F(t,t_n), \quad Y_{n,j} \to F(t_n,t_n),$$

$$\tilde{\Phi}_n(t) \to \sum_{j=1}^{m} b_j k(t_n,t_n,F(t_n,t_n)) = k(t_n,t_n,F(t_n,t_n)),$$

we see that the VRK method converges to the equation

$$[1-\gamma(t)][y(t)-F(t,t)]-h\gamma'(t)[y(t)-F(t,t)]$$
$$+h[y'(t)-F_t(t,t)-k(t,t,F(t,t))] = 0.$$

Thus, if $\gamma(t) \equiv 0$, then the method is a direct discretization of $y(t)=F(t,t)$; i.e., of the integral equation (1.1.1). If $\gamma(t) \equiv 1$, then the method is a discretization of $y'(t)=k(t,t,F(t,t))+F_t(t,t)$. Notice that this limiting equation is not identical with the differentiated form of (1.1.1) which is given by $y'(t)=k(t,t,y(t))+F_t(t,t)$ and which is, in fact, the limiting equation obtained for the ILM and MML methods (cf. Example 3.2.9).

The convergence of γ-modified VRK methods was studied in van der Houwen, Wolkenfelt & Baker [1981]; the convergence result due to these authors can also be deduced by modifying Theorem 4.1.6 (see Hairer, Lubich & Nørsett [1983, p. 575]).

In order to state this result, we introduce (i) the *local truncation error* of the method with respect to the points $\{t_n+c_jh\}$,

$$T_{n,j}(h) := y(t_n+c_jh)-\overline{Y}_{n,j}, \quad j=1, \dots ,m+1, \tag{4.1.33}$$

with $c_{m+1}:=1$, $\overline{Y}_{n,m+1}:=\overline{y}_{n+1}$, and with $\overline{Y}_{n,j}$ $(1\leqslant j\leqslant m)$ given by (4.1.15c); and (ii) the *global error* at these points,

$$\epsilon_{n,j}(h) := y(t_n+c_jh)-Y_{n,j}, \quad j=1, \dots ,m+1, \tag{4.1.34}$$

where $Y_{n,m+1}:=y_{n+1}$. In addition, we define vectors $\mathbf{T}_{n+1}$ and $\boldsymbol{\epsilon}_{n+1}$ whose components are given, respectively, by those values $T_{n,j}$ and $\epsilon_{n,j}$ whose indices j satisfy

$$j\in\{1, \dots ,m+1\}\setminus\{j\colon w_{n,l,j}=0 \quad \text{for all } n \text{ and } l\}.$$

In other words, the components of $\mathbf{T}_{n+1}$ and $\boldsymbol{\epsilon}_{n+1}$ correspond to those stages $Y_{l,j}$ actually occurring in the lag term formula (4.1.22).

Finally, we shall need the *lag term error* for the exact solution $y(t)$; i.e.,

$$E_{n,i}(h) := \int_0^{t_n} k(t_n+\theta_ih,s,y(s))ds$$
$$-h\sum_{l=-1}^{n-1}\sum_{j=1}^{m+1} w_{n,l,j}k(t_n+\theta_ih+(e_j-1)h,t_l+c_jh,y(t_l+c_jh)),$$

and we define

$$\Delta E_{n,i}(h) := E_{n,i}(h)-E_{n,0}(h), \quad i=0, \dots ,m+1, \quad \theta_0:=0, \quad \theta_{m+1}:=1.$$

Theorem 4.1.7. *Let the following assumptions hold:*

(i) *The VRK formula (4.1.4) is consistent of order p.*

(ii) *The lag term is given by the γ-modified form* (4.1.30), *with $\tilde{F}_n(t)$ defined by (4.1.22).*

(iii) *The kernel $k(t,s,y)$ satisfies the Lipschitz condition*

$$|k(t,s,y)-\alpha k(t_n,s,y)-k(t,s,y^*)+\alpha k(t_n,s,y^*)|$$
$$\leqslant L(1-\alpha+\alpha|t-t_n|)|y-y^*|, \quad \alpha\in[0,1],$$

for all (t,s), (t_n,s) *in* S, *and all* y, y^* *in some compact neighborhood of* $\{(t,s,y(s)): 0 \leqslant s \leqslant t \leqslant T\}$.

(iv) *The starting values* $\{y_0, \ldots, y_{\tilde{\kappa}-1}\}$ *are of order s.*

(v) *The function* $\gamma(t)$ *satisfies* $\gamma(t_n + c_j h) =: \gamma_j \in [0,1]$, $j = 1, \ldots, m+1$ $(n = 0, \ldots, N-1)$.

Then the approximations $\{y_n\}$, $n = \tilde{\kappa}, \ldots, N$, *furnished by the* γ*-modified VRK method induce global errors* ϵ_n *for which*

$$\|\epsilon_n\|_\infty \leqslant C_1\{\max_{l,i}|E_{l,i}(h)| + \max_l \|\mathbf{T}_l(h)\|_\infty + h^{s+1}\} +$$

$$+ C_2\{\max_{l,i}|\Delta E_{l,i}(h)| + \max_l |T_{l,m+1}(h)| + h^{s+1}\}/(1 - \gamma_{m+1} + C_3 h),$$

$$n = \tilde{\kappa}, \ldots, N$$

holds, with t_n *fixed,* $h \downarrow 0$. *Here,* C_1, C_2 *and* C_3 *denote positive constants not depending on N and h.* □

For a proof of this result the reader is referred to van der Houwen, Wolkenfelt & Baker [1981].

Let us consider the most important special case of the above theorem, where $\tilde{F}_n(t)$ in the γ-modified lag term (4.1.30) corresponds to a *mixed lag term* of the form (4.1.5b). Since all weights $w_{n,l,j}$ vanish, except those for $j = m+1$, we have

$$\mathbf{T}_{l+1} = T_{l,m+1} = y(t_{l+1}) - \bar{y}_{l+1},$$

and

$$\epsilon_{n+1} = \epsilon_{n,m+1} = y(t_{n+1}) - y_{n+1}.$$

Thus, Theorem 4.1.7 yields the following corollary.

Corollary 4.1.2. *Let the assumptions of Theorem 4.1.7 hold, and assume that the* γ*-modified lag term (4.1.30) is based on a mixed lag term formula (4.1.5b) of order r. If g and k are sufficiently smooth functions, then the approximations* $\{y_n\}$ *generated by the resulting* γ*-modified VRK method satisfy*

$$\max_{\tilde{\kappa} \leqslant n \leqslant N} |y(t_n) - y_n| \leqslant C_1 h^{p_1} + C_2 h^{p_2}/(1 - \gamma_{m+1} + C_3 h), \quad (\text{as } h \downarrow 0,\ Nh = T),$$

where $p_1 := \min\{p+1, r, s+1\}$, $p_2 := \min\{p+1, r+1, s+1\}$, *and where the constants* $C_i > 0$, $i = 1,2,3$, *do not depend on N and h.* □

The effect of the γ-modification becomes apparent: if we choose $\gamma(t)$ so that $\gamma_{m+1} := \gamma(t_{n+1}) < 1$, then the method converges with order $p^* := \min\{p+1, r, s+1\}$ (cf. Theorem 4.1.6). However, if we have $\gamma_{m+1} = 1$, then we may experience a drop of one in the order of convergence; i.e., the order is now given by $p^* := \min\{p, r, s\}$.

We conclude the convergence analysis of γ-modified VRK methods with a comment on the Lipschitz condition (ii) in Theorem 4.1.7. It is not difficult to verify that this condition is fulfilled if both $k(t,s,y)$ and its partial derivative $k_t = \partial k(t,s,y)/\partial t$ satisfy a Lipschitz condition with respect to y, since

$$|k(t,s,y)-\alpha k(t_n,s,y)-k(t,s,y^*)+\alpha k(t_n,s,y^*)|$$

$$= |(1-\alpha)[k(t,s,y)-k(t,s,y^*)]+\alpha\int_{t_n}^{t}[k_t(t,s,y)-k_t(t,s,y^*)]dt|$$

$$\leqslant (1-\alpha)L_1|y-y^*|+\alpha|t-t_n|L_2|y-y^*|.$$

The role of L in assumption (iii) is then taken by $\max\{L_1,L_2\}$.

4.1.9. Summary

We complete our analysis of Runge-Kutta methods for the integral equation

$$y(t) = g(t)+\int_0^t k(t,s,y(s))ds, \quad 0\leqslant t\leqslant T,$$

with a summary of the main types of methods and their orders of convergence.

VRK method:

$$y_{n+1} = \tilde{F}_n(t_n+h)+h\tilde{\Phi}_n(t_n+h), \quad n=0, \ldots ,N-1.$$

VRK formula (m-stages):

$$\tilde{\Phi}_n(t) = \sum_{j=1}^{m} b_j k(t+(e_j-1)h, t_n+c_jh, Y_{n,j}),$$

$$Y_{n,j} = \tilde{F}_n(t_n+\theta_jh)+h\sum_{i=1}^{m} a_{j,i}k(t_n+d_{j,i}h, t_n+c_ih, Y_{n,i}), \quad j=1, \ldots ,m.$$

Extended lag term formula:

$$\tilde{F}_n(t) = g(t)+h\sum_{l=0}^{n-1}\sum_{j=1}^{m} b_j k(t+(e_j-1)h, t_l+c_jh, Y_{l,j}), \quad n=0, \ldots ,N-1.$$

Mixed lag term formula:

$$\tilde{F}_n(t) = g(t)+h\sum_{l=0}^{n} w_{n,l}k(t,t_l,y_l), \quad n=\tilde{\kappa}, \ldots ,N-1.$$

with starting values $\{y_0, \ldots ,y_{\tilde{\kappa}-1}\}$ given.

γ-modified lag term formula:

$$\tilde{F}_n^*(t) = \tilde{F}_n(t)+\gamma(t)[y_n-\tilde{F}_n(t_n)], \quad \gamma(t)\in[0,1].$$

Order of convergence of VRK method:

$$\max_{(n)}|y(t_n)-y_n| \leqslant Ch^{p^*} \quad (h\downarrow 0,\ Nh=T);$$

Extended VRK method: $p^*=p$.

Mixed VRK method: $p^*=\min\{p+1,r,s+1\}$.

γ-modified VRK method based on mixed lag term formula:

$$p^* = \begin{cases}\min\{p+1,r,s+1\}, & \text{if} \quad \gamma(t_n)<1;\\ \min\{p,r,s\}, & \text{if} \quad \gamma(t_n)=1.\end{cases}$$

Here,

p is the order of consistency of the VRK formula;

r is the order of the mixed lag term;

s is the order of the starting values.

4.1.10. Numerical examples

We shall illustrate the convergence results summarized in the preceding section by employing a number of VRK methods which are all of order $p^*=4$.

In our experiments the starting values (if needed) were obtained in the same manner as outlined in Section 3.2.7. The VRK formulas chosen are the standard explicit 4-stage PVRK formula given in Example 4.1.4 (first diagram) and the explicit 5-stage BVRK formula of Hairer given in Example 4.1.11, with $c=(3-\sqrt{3})/8$. We used both the extended, mixed, and γ-modified version of the lag term formula. The mixed lag term formula is based on the fourth-order Gregory formula and will be denoted by G_4; the γ-modified lag term formula is also based on G_4 with $\gamma(t)\equiv 1$ and will be denoted by $(G\gamma)_4$.

In terms of kernel evaluations, the extended PVRK and BVRK methods require, for $N\rightarrow\infty$, $O(8N^2)$ and $O(12.5N^2)$ operations, respectively; in both the mixed and γ-modified case, the orders are $O(2N^2)$ and $O(2.5N^2)$, respectively. In the choice of the integration steps, we have taken into account the computational effort involved: the results listed in the tables below are row-wise comparable, in the sense that, for large N, the numbers of computed kernel values are roughly equal. Moreover, in these tables, each row of results is, in the same sense, comparable with the corresponding row of results listed in the VLM tables of Section 3.2.9 and 3.2.11.

Table 4.1.8a. Numbers of correct significant digits and effective orders for problem (3.2.26) at $t=T=4$

h	Extended $PVRK_4$		h	mixed G_4-PVRK_4		h	γ-modified $(G\gamma)_4-PVRK_4$	
4/5	4.3	> 3.6	2/5	4.8	> 2.1	2/5	5.3	> 3.2
4/10	5.4	> 3.9	2/10	5.4	> 2.9	2/10	6.2	> 3.5
4/20	6.6		2/20	6.3		2/20	7.3	
h	$BVRK_4$		h	G_4-BVRK_4		h	$(G\gamma)_4-BVRK_4$	
1	4.2	> 3.4	4/9	4.9	> 1.4	4/9	5.8	> 2.9
1/2	5.2	> 3.8	4/18	5.3	> 2.7	4/18	6.7	> 2.4
1/4	6.4		4/36	6.1		4/36	7.4	

Table 4.1.8b. Numbers of correct significant digits and effective orders for problem (3.2.36) at $t=T=4$

h	Extended $PVRK_4$		h	mixed G_4-PVRK_4		h	γ-modified $(G\gamma)_4-PVRK_4$	
4/5	*		2/5	1.9	> 2.0	2/5	0.7	> 5.0
4/10	0.7	> 4.9	2/10	2.5	> 5.5	2/10	2.2	> 4.5
4/20	2.2		2/10	4.1		2/20	3.6	
h	$BVRK_4$		h	G_4-BVRK_4		h	$(G\gamma)_4-BVRK_4$	
1	-0.4	> 4.6	4/9	2.2	> 2.7	4/9	1.3	> 4.9
1/2	1.0	> 4.8	4/18	3.0	> 5.4	4/18	2.8	> 4.4
1/4	2.5		4/36	4.6		4/36	4.1	

Table 4.1.9. Numbers of correct significant digits and effective orders for problem (3.2.8) at $t=T=6$

h	Extended $PVRK_4$	order	h	mixed G_4-PVRK_4	order	h	γ-modified $(G\gamma)_4-PVRK_4$	order
1	2.3	> 3.4	1/2	2.5	> 2.7	1/2	1.6	> 3.0
1/2	3.3	> 3.7	1/4	3.3	> 3.4	1/4	2.5	> 3.5
1/4	4.4	> 3.8	1/8	4.3	> 3.7	1/8	3.5	> 3.8
1/8	5.6	> 3.9	1/16	5.4	> 3.8	1/16	4.7	> 3.9
1/16	6.7		1/32	6.6		1/32	5.8	
h	$BVRK_4$		h	G_4-BVRK_4		h	$(G\gamma)_4-BVRK_4$	
6/5	2.3	> 3.3	6/11	2.5	> 2.5	6/11	1.4	> 3.0
6/10	3.3	> 3.7	6/22	3.2	> 3.3	6/22	2.3	> 3.5
6/20	4.4	> 3.8	6/44	4.2	> 3.7	6/44	3.4	> 3.7
6/40	5.6	> 3.9	6/88	5.3	> 3.8	6/88	4.5	> 3.9
6/80	6.7		6/176	6.5		6/176	5.7	

Table 4.1.10. Numbers of correct significant digits and effective orders for problem (3.2.38) at $t=T=6$

h	Extended $PVRK_4$	order	h	mixed G_4-PVRK_4	order	h	γ-modified $(G\gamma)_4-PVRK_4$	order
1	2.3	> 3.5	1/2	2.6	> 2.6	1/2	2.0	> 3.3
1/2	3.3	> 3.8	1/4	3.6	> 3.6	1/4	3.0	> 3.7
1/4	4.5		1/8	4.7		1/8	4.1	
h	BRK_4		h	G_4-BVRK_4		h	$(G\gamma)_4-BVRK_4$	
6/5	1.8	> 3.3	6/11	2.4	> 3.4	6/11	2.0	> 3.2
6/10	2.8	> 3.7	6/22	3.4	> 3.7	6/22	3.0	> 3.6
6/20	3.9		6/44	4.5		6/44	4.1	...

Table 4.1.8a shows the remarkable result that, for the test problem (3.2.26), the γ-modified methods produce the highest accuracies. (As we shall see in the subsequent examples, this is usually not the case.) In this example, the corresponding PVRK and BVRK versions lead to comparable results. A comparison with Table 3.2.7a reveals that the VLM methods furnish more accurate results.

Table 4.1.8b indicates that problem (3.2.26) is a severe test problem for (explicit) VRK methods. The accuracies are low (cf. the corresponding figures in the VLM-table 3.2.7b), and in one case, even a negative numerical solution, denoted by an asterisk, was produced for which the kernel is not defined. Notice that, here, the mixed method is considerably more accurate than the extended method; the PVRK methods are slightly better than the BVRK methods.

In Table 4.1.9 and in Table 4.1.10, we see that the mixed and extended methods, using PVRK or BVRK formulas, all produce comparable accuracies (differing by at most .5 digits). However, these accuracies are again lower than those furnished by, e.g., the fourth-order $DQ(G_4)$ method (see Tables 3.2.10 and 3.2.12).

4.2. Runge-Kutta methods for integro-differential equations

4.2.1. Introduction

In the present section we shall study the application of numerical methods of Runge-Kutta type to Volterra integro-differential equations (VIDEs) of the form

$$y'(t) = f(t,y(t),z(t)), \quad t\in I := [0,T], \ y(0)=y_0, \tag{4.2.1}$$

with

$$z(t) := \int_0^t k(t,s,y(s))ds. \tag{4.2.2}$$

Let the interval I be discretized by the uniform mesh employed also in Section 4.1. We may then put (4.2.1) into the form

$$y'(t) = f(t,y(t),F_n(t)+\int_{t_n}^t k(t,s,y(s))ds), \quad t\in[t_n,T], \ y(0)=y_0, \tag{4.2.3}$$

where F_n, defined now by

$$F_n(t) := \int_0^{t_n} k(t,s,y(s))ds, \quad n=0, \dots ,N-1, \tag{4.2.4}$$

will be referred to as the *lag* (or: *tail*) *term* of the VIDE (4.2.3). As in Section 4.1, we also introduce the corresponding *exact increment function* Φ_n by

$$h\Phi_n(t) := \int_{t_n}^t k(t,s,y(s))ds, \quad t\in[t_n,T], \quad n=0, \dots ,N-1. \tag{4.2.5}$$

(which is, of course, identical with (4.1.2)).

In most applications one encounters VIDEs in a form somewhat less general than (4.2.1), namely,

$$y'(t) = f(t,y(t))+\int_0^t k(t,s,y(s))ds, \quad t\in I, \quad y(0)=y_0; \tag{4.2.6}$$

accordingly, if $t \in [t_n, T]$, then its mesh-related form is

$$y'(t) = f(t,y(t)) + F_n(t) + h\Phi_n(t), \quad n=0, \dots, N-1, \ y(0)=y_0, \quad (4.2.7)$$

with F_n and Φ_n as in (4.2.4) and (4.2.5), respectively. If the kernel function $k(t,s,y)$ vanishes identically, then (4.2.6) reduces to an initial-value problem for the nonlinear ODE $y'=f(t,y)$.

These observations indicate that a Runge-Kutta method for (4.2.1) will be characterized by three approximation processes which, in general, will be independent of each other:

(i) An *RK method* (4.1.12).

(ii) An approximation scheme for the increment function $\Phi_n(t)$; i.e., a (local) approximation scheme on the subinterval $[t_n, t_{n+1}]$. The resulting discrete increment function, denoted by $\tilde{\Phi}_n(t)$, will be called a *VIDE-Runge-Kutta formula (VDRK formula).*

(iii) An approximation scheme for the lag term $F_n(t)$; i.e., an approximation scheme on $[0, t_n]$. This scheme will be referred to as the *lag term formula* (or: *tail term formula*) $\tilde{F}_n(t)$.

In other words, at $t=t_{n+1}(=t_n+h)$, the VIDE (4.2.3) will be discretized by

$$y_{n+1} = y_n + h\sum_{j=1}^{m} b_j f(t_n+c_jh, Y_{n,j}, \tilde{F}_n(t_n+c_jh) + h\tilde{\Phi}_n(t_n+c_jh)), \qquad (4.2.8)$$

$$n=0, \dots, N-1,$$

with given initial value $y_0=y(0)$. Here, the parameters $\mathbf{b}:=(b_j)$ and $\mathbf{c}:=(c_j)$ are those of the RK diagram describing (i) (cf. (4.1.13)).

In analogy to (4.1.3) we shall call (4.2.8) a *VDRK method* if the RK method (4.1.13), the VDRK formula, and the lag term formula have been specified. The formal description of a VDRK method is contained in

Definition 4.2.1. (a) An *m-stage VDRK formula* for (4.2.1) has the form

$$Y_{n,j} = y_n + h\sum_{i=1}^{m} a_{j,i} f(t_n+c_ih, Y_{n,i}, \tilde{F}_n(t_n+c_ih) + h\tilde{\Phi}_n(t_n+c_ih)), \qquad (4.2.9a)$$

$$j=1, \dots, m,$$

with $y_0=y(0)$;

$$\tilde{\Phi}_n(t_n+c_ih) = \sum_{l=1}^{m} \bar{a}_{i,l} k(t_n+d_{i,l}h, t_n+c_lh, Y_{n,l}), \quad i=1, \dots, m. \qquad (4.2.9b)$$

Here, the vectors $\mathbf{c}:=(c_j)$, $\mathbf{b}:=(b_j)$, and the square matrices $A:=(a_{j,i})$, $\bar{A}:=(\bar{a}_{i,l})$, $D:=(d_{i,l})$ are given.

(b) The VDRK method (4.2.8) is called:

(i) an *extended VDRK* method if the lag term formula is given by

$$\tilde{F}_n(t) := h\sum_{l=0}^{n-1}\sum_{j=1}^{m} b_j k(t,t_l+c_jh,Y_{l,j}), \quad n=1,\ldots,N-1; \tag{4.2.10a}$$

(ii) a *mixed VDRK* method if the lag term formula is defined by (cf. (3.2.13))

$$\tilde{F}_n(t) := h\sum_{l=0}^{n} w_{n,l}k(t,t_l,y_l), \quad n=\tilde{\kappa},\ldots,N-1. \quad \square \tag{4.2.10b}$$

In Section 3.4.1 we have shown that VLM methods for integro-differential equations arise naturally by applying a VLM method to the two component equations of the integral equation representation of $\{(4.2.1); (4.2.2)\}$. A similar situation arises in the case of VDRK methods. Let us write (4.2.1) in the integral equation form

$$y(t) = y(0)+\int_0^t f(s,y(s),z(s))ds. \tag{4.2.1'}$$

We now apply a VRK method, as defined in Definition 4.1.1, to the system $\{(4.2.1'), (4.2.2)\}$. However, for the first equation we use parameters $a_{j,i}, b_j$, and for the second equation we use parameters $\bar{a}_{j,i}, \bar{b}_j$. The result of this "composite" VRK method is the scheme

$$\begin{aligned} y_{n+1} &= \tilde{F}_n^{(1)}+h\tilde{\Phi}_n^{(1)} \\ z_{n+1} &= \tilde{F}_n^{(2)}(t_{n+1})+h\tilde{\Phi}_n^{(2)}(t_{n+1}) \end{aligned}, \tag{4.2.11a}$$

$$\begin{aligned} \tilde{\Phi}_n^{(1)} &= \sum_{j=1}^{m} b_j f(t_n+c_jh, Y_{n,j}^{(1)}, Y_{n,j}^{(2)}) \\ \tilde{\Phi}_n^{(2)}(t) &= \sum_{j=1}^{m} \bar{b}_j k(t+(e_j-1)h, t_n+c_jh, Y_{n,j}^{(1)}) \end{aligned}, \tag{4.2.11b}$$

$$\begin{aligned} Y_{n,j}^{(1)} &= \tilde{F}_n^{(1)}+h\sum_{i=1}^{m} a_{j,i}f(t_n+c_ih, Y_{n,i}^{(1)}, Y_{n,i}^{(2)}) \\ Y_{n,j}^{(2)} &= \tilde{F}_n^{(2)}(t_n+\theta_jh)+h\sum_{i=1}^{m} \bar{a}_{j,i}k(t_n+d_{j,i}h, t_n+c_ih, Y_{n,i}^{(1)}) \end{aligned}, \tag{4.2.11c}$$

where the two lag terms $\tilde{F}_n^{(1)}$ and $\tilde{F}_n^{(2)}(t)$ are, respectively, approximations to

$$F_n^{(1)}(t) = F_n^{(1)} = y(0)+\int_0^{t_n} f(s,y(s),z(s))ds = y(t_n),$$

$$F_n^{(2)}(t) = \int_0^{t_n} k(t,s,y(s))ds.$$

It is now easily verified that the approximation scheme for y_{n+1} reduces to that defined by (4.2.8) and (4.2.9) if we set $\tilde{F}_n^{(1)} := y_n$ and $\tilde{F}_n^{(2)}(t) := \tilde{F}_n(t)$.

There exists an *alternative form* of the VDRK formula introduced in Definition 4.2.1. Define

$$U_{n,j} := f(t_n + c_j h, Y_{n,j}, \tilde{F}_n(t_n + c_j h) + h\tilde{\Phi}_n(t_n + c_j h)).$$

Then the VDRK formula (4.2.9) may be written as

$$U_{n,j} = f(t_n + c_j h, y_n + h\sum_{i=1}^{m} a_{j,i} U_{n,i}, \tilde{F}_n(t_n + c_j h) + h\tilde{\Phi}_n(t_n + c_j h)), \qquad (4.2.12a)$$

$$j = 1, \dots, m;$$

$$\tilde{\Phi}_n(t_n + c_i h) = \sum_{l=1}^{m} \bar{a}_{i,l} k(t_n + d_{i,l}h, t_n + c_l h, y_n + h\sum_{j=1}^{m} a_{l,j} U_{l,j}), \qquad (4.2.12b)$$

$$i = 1, \dots, m.$$

Instead of (4.2.8) we now have

$$y_{n+1} = y_n + h\sum_{j=1}^{m} b_j U_{n,j}, \quad n = 0, \dots, N-1, \qquad (4.2.12c)$$

with $y_0 = y(0)$.

Hence, the quantities $\{U_{n,j}\}$ may be interpreted as representing approximations to $y'(t)$ at $t = t_n + c_j h$, while the $\{Y_{n,j}\}$ approximate the solution $y(t)$ itself at these points. We shall see in Chapter 5 that certain fully discretized collocation methods for VIDEs are identical with a special class of VDRK methods based on the form (4.2.12).

In view of applications we also exhibit the VDRK method (4.2.8) and its VDRK formula for the integro-differential equation (4.2.6): the former assumes the form

$$y_{n+1} = y_n + h\sum_{j=1}^{m} b_j f(t_n + c_j h, Y_{n,j}) +$$

$$+ h\sum_{j=1}^{m} b_j \{\tilde{F}_n(t_n + c_j h) + h\tilde{\Phi}_n(t_n + c_j h)\}, \quad n = 0, \dots, N-1, \qquad (4.2.8')$$

while the latter is given by

$$Y_{n,j} = y_n + h\sum_{i=1}^{m} a_{j,i} f(t_n + c_i h, Y_{n,i}) +$$

$$+ h\sum_{i=1}^{m} a_{j,i} \{\tilde{F}_n(t_n + c_i h) + h\tilde{\Phi}_n(t_n + c_i h)\}, \quad j = 1, \dots, m, \qquad (4.2.9'a)$$

and

$$\tilde{\Phi}_n(t_n + c_i h) = \sum_{l=1}^{m} \bar{a}_{i,l} k(t_n + d_{i,l}h, t_n + c_l h, Y_{n,l}), \quad i = 1, \dots, m \qquad (4.2.9'b)$$

(this last equation is, of course, identical with (4.2.9b)).

In the following we shall usually employ the notation

$$Z_{n,j} := \tilde{\Phi}_n(t_n + c_j h)$$

for the values of the discrete increment function at the intermediate points $\{t_n + c_j h\}$.

We look briefly at two examples.

Example 4.2.1. *The* 1-*stage VDRK method.*

For $m = 1$, (4.2.8) and (4.2.9) yield

$$y_{n+1} = y_n + hb_1 f(t_n + c_1 h, Y_{n,1}, \tilde{F}_n(t_n + c_1 h) + hZ_{n,1}),$$

$$n = 0, \ldots, N-1,$$

$$Y_{n,1} = y_n + ha_{1,1} f(t_n + c_1 h, Y_{n,1}, \tilde{F}_n(t_n + c_1 h) + hZ_{n,1}),$$

and

$$Z_{n,1} = \bar{a}_{1,1} k(t_n + d_{1,1} h, t_n + c_1 h, Y_{n,1}).$$

We observe that if $a_{11} = 0$ (i.e., if the RK method (4.1.13) is explicit), then $Y_{n,1}$ and $Z_{n,1}$ can be computed sequentially from the last two formulas; accordingly, we shall call the method an explicit VDRK method (cf. Definition 4.2.2 below). For $a_{1,1} \neq 0$ and $\bar{a}_{1,1} \neq 0$, these formulas constitute a system of two nonlinear equations for $Y_{n,1}$ and $Z_{n,1}$: the corresponding VDRK method is implicit. □

Example 4.2.2. *The* 2-*stage VDRK method.*

For $m = 2$ we obtain the VDRK formula

$$Y_{n,1} = y_n + h\{a_{1,1} f(t_n + c_1 h, Y_{n,1}, \tilde{F}_n(t_n + c_1 h) + hZ_{n,1}) + a_{1,2} f(t_n + c_2 h, Y_{n,2}, \tilde{F}_n(t_n + c_2 h) + hZ_{n,2})\}, \quad (4.2.13a)$$

$$Y_{n,2} = y_n + h\{a_{2,1} f(t_n + c_1 h, Y_{n,1}, \tilde{F}_n(t_n + c_1 h) + hZ_{n,1}) + a_{2,2} f(t_n + c_2 h, Y_{n,2}, \tilde{F}_n(t_n + c_2 h) + hZ_{n,2})\}; \quad (4.2.13b)$$

$$Z_{n,1} = \bar{a}_{1,1} k(t_n + d_{1,1} h, t_n + c_1 h, Y_{n,1}) + \bar{a}_{1,2} k(t_n + d_{1,2} h, t_n + c_2 h, Y_{n,2}); \quad (4.2.13c)$$

$$Z_{n,2} = \bar{a}_{2,1} k(t_n + d_{2,1} h, t_n + c_1 h, Y_{n,1}) + \bar{a}_{2,2} k(t_n + d_{2,2} h, t_n + c_2 h, Y_{n,2}). \quad (4.2.13d)$$

Consider the special case

$$a_{1,1} = a_{1,2} = a_{2,2} = 0, \quad \text{and} \quad \bar{a}_{1,2} = 0:$$

the (appropriately rearranged) equations (4.2.13) are thus

$$\begin{aligned}
Y_{n,1} &= y_n, \quad Z_{n,1} = \bar{a}_{1,1}k(t_n+d_{1,1}h, t_n+c_1h, Y_{n,1}),\\
Y_{n,2} &= y_n + ha_{2,1}f(t_n+c_1h, Y_{n,1}, \tilde{F}_n(t_n+c_1h)+Z_{n,1}),\\
Z_{n,2} &= \bar{a}_{2,1}k(t_n+d_{2,1}h, t_n+c_1h, Y_{n,1}) + \bar{a}_{2,2}k(t_n+d_{2,2}h, t_n+c_2h, Y_{n,2}),
\end{aligned}$$

and the new approximation y_{n+1} is then computed by

$$\begin{aligned}
y_{n+1} = y_n + h\{&b_1f(t_n+c_1h, Y_{n,1}, \tilde{F}_n(t_n+c_1h)+hZ_{n,1})+\\
&+b_2f(t_n+c_2h, Y_{n,2}, \tilde{F}_n(t_n+c_2h)+hZ_{n,2})\}. \square
\end{aligned}$$

Definition 4.2.2. A VDRK method (4.2.8) (or: a VDRK formula (4.2.9)) is said to be

(i) *explicit* if $a_{j,i}=0$, $1 \leqslant j \leqslant i \leqslant m$, and $\bar{a}_{i,l}=0$, $1 \leqslant i < l \leqslant m$ (i.e., if the matrices A and $\bar{A}$ are, respectively, of strictly lower triangular and of lower triangular form);

(ii) *diagonally implicit* if $a_{j,i}=\bar{a}_{j,i}=0$, $1 \leqslant j < i \leqslant m$, with A possessing at least one nonvanishing diagonal element.

(iii) *implicit* if it is neither explicit nor diagonally implicit. □

Two particular classes of VDRK formulas will be of importance in applications.

Definition 4.2.3. A VDRK formula is said to be of *Pouzet type* (a *PVDRK formula*) if

$$a_{i,j} = \bar{a}_{i,j}, \quad d_{j,i} = c_j, \quad j,i=0, \ldots, m; \tag{4.2.14}$$

i.e., if it is completely characterized by the RK diagram (4.1.13). □

Thus, a PVDRK formula has the form

$$Y_{n,j} = y_n + h\sum_{i=1}^{m} a_{j,i}f(t_n+c_ih, Y_{n,i}, \tilde{F}_n(t_n+c_ih)+hZ_{n,i}), \quad j=1, \ldots, m; \tag{4.2.15a}$$

$$Z_{n,i} = \sum_{l=1}^{m} a_{i,l}k(t_n+c_ih, t_n+c_lh, Y_{n,l}), \quad i=1, \ldots, m. \tag{4.2.15b}$$

The location of the kernel values used in (4.1.15b) is reflected in Figure 4.1.2a (compare also (4.2.15b) with (4.1.8a)). Explicit PVDRK methods were introduced by Pouzet [1960], [1962], [1963]. It is clear from (4.2.15b) that implicit PVDRK formulas require that the kernel function $k(t,s,y)$ be known for points (t,s) with $s>t$; hence, the remarks preceding Lemma 4.1.1 apply also in this case.

Definition 4.2.4. A VDRK formula is said to be of *Bel'tyukov type* (a *BVDRK formula*) if

$$d_{i,l} = d_l, \quad i,l = 1, \ldots, m. \tag{4.2.16}$$

It satisfies the *kernel condition* if

$$d_l \geqslant c_l \tag{4.2.17}$$

holds whenever $\bar{a}_{i,l} \neq 0$. A BVDRK formula is therefore characterized by the diagram

$$\begin{array}{c|c|c|c} \mathbf{d} & \bar{A} & \mathbf{c} & A \\ \hline & & & \mathbf{b}^T \end{array} \tag{4.2.18}$$

□

Written explicitly, it becomes

$$Y_{n,j} = y_n + h \sum_{i=1}^{m} a_{j,i} f(t_n + c_i h, Y_{n,i}, \tilde{F}_n(t_n + c_i h) + hZ_{n,i}), \quad j = 1, \ldots, m; \tag{4.2.19a}$$

$$Z_{n,i} = \sum_{l=1}^{m} \bar{a}_{i,l} k(t_n + d_l h, t_n + c_l h, Y_{n,l}), \quad i = 1, \ldots, m. \tag{4.2.19b}$$

Comparison of (4.2.19b) with the BVRK counterpart (4.1.10b) shows that Figure 4.1.3a exhibits the kernel values employed in (4.2.19b). Note also that we can put any VDRK formula into Bel'tyukov form by suitably increasing the number of stages.

VDRK methods of Bel'tyukov type (so named since they are based on the same principle as the BVRK methods for second-kind integral equations) are of very recent origin: they were studied by Lubich [1981], [1982c] in the context of his general theory for VDRK methods; Lubich in these papers also derives a number of explicit BVDRK formulas.

4.2.2. Order conditions for VDRK formulas

As in Section 4.1.3, the (global) order of convergence of a VDRK method for (4.2.1) will be closely related to the (local) order of consistency of its VDRK formula. In order to introduce the latter concept we consider, in analogy to (4.1.15),

$$\bar{y}_{n+1} := y(t_n) + h \sum_{j=1}^{m} b_j f(t_n + c_j h, \bar{Y}_{n,j}, F_n(t_n + c_j h) + h\bar{\Phi}_n(t_n + c_j h)), \tag{4.2.20a}$$

with

$$\bar{Y}_{n,j} := y(t_n) + h \sum_{i=1}^{m} a_{j,i} f(t_n + c_i h, \bar{Y}_{n,i}, F_n(t_n + c_i h) + h\bar{\Phi}_n(t_n + c_i h)), \tag{4.2.20b}$$

$$j = 1, \ldots, m,$$

and

$$\overline{\Phi}_n(t_n+c_ih) := \sum_{l=1}^{m} \overline{a}_{i,l}k(t_n+d_{i,l}h, t_n+c_lh, \overline{Y}_{n,l}), \quad i=1, \dots, m. \tag{4.2.20c}$$

Here, $y(t)$ and $F_n(t)$ denote, respectively, the exact solution and the exact lag term of the VIDE (4.2.1); i.e., we use the VDRK formula (4.2.9), with exact lag term $F_n(t)$, to solve (4.2.1) *locally* on $[t_n, t_{n+1}]$. In analogy to the terminology introduced in Section 4.1.4, we shall call $\overline{\Phi}_n(t)$ the *local increment function.*

Definition 4.2.5. The VDRK formula (4.2.9) is said to be *consistent of (local) order p* if, for all VIDEs (4.2.1) corresponding to sufficiently differentiable f and k and for all sufficiently small $h>0$, p is the largest integer for which

$$|y(t_n)-\overline{y}_n| \leqslant Ch^{p+1} \quad (t_n=nh=t \text{ fixed, as } h\downarrow 0)$$

holds, with the constant $C>0$ not depending on n and h. □

The local order of consistency of VDRK formula (4.2.9) can again be characterized algebraically, by equations involving the parameters defining the formula; these equations will constitute the so-called *order conditions* for the VDRK formula. Since a detailed description of the underlying theory is beyond the scope of this book we shall only list a selection of order conditions and direct the interested reader to Lubich [1982c] for further details.

In the following it will always be assumed that the vector $\mathbf{c}$ and the matrix A are related by

$$c_j = \sum_{i=1}^{m} a_{j,i}, \quad j=1, \dots, m \tag{4.2.21}$$

(cf. Section 4.1.3). Moreover, we introduce parameters $\{\overline{c}_1, \dots, \overline{c}_m\}$ by setting

$$\overline{c}_i := \sum_{l=1}^{m} \overline{a}_{i,l}, \quad i=1, \dots, m. \tag{4.2.22}$$

Table 4.2.1 contains the order conditions for the VDRK formula (4.2.9), with order of consistency $p\in\{1,2,3\}$; note that all summation indices range from 1 to m.

It is a trivial consequence of Definition 4.2.3 that the PVDRK order conditions coincide with the ones for the RK method (4.1.12) (an empty entry in the second column of Table 4.2.1 indicates that the corresponding general order condition in the previous column does not yield a new PVDRK order condition).

The analogues of Table 4.1.5 and 4.1.6, i.e., the order conditions for explicit and diagonally implicit PVDRK and BVDRK formulas, are readily derived from the above table by observing Definition 4.2.2. We refrain from giving them explicitly.

Table 4.2.1. VDRK order conditions

General VDRK (4.2.9)	PVDRK (4.2.15)	BVDRK (4.2.19)
p=1:		
$\sum_i b_i = 1$	1. $\sum_i b_i = 1$	1. $\sum_i b_i = 1$
p=2:		
$\sum_i b_i c_i = 1/2$	2. $\sum_i b_i c_i = 1/2$	2. $\sum_i b_i c_i = 1/2$
$\sum_i b_i \bar{c}_i = 1/2$	--	3. $\sum_i b_i \bar{c}_i = 1/2$
p=3:		
$\sum_i b_i c_i^2 = 1/3$	3. $\sum_i b_i c_i^2 = 1/3$	4. $\sum_i b_i c_i^2 = 1/3$
$\sum_i b_i c_i \bar{c}_i = 1/3$	--	5. $\sum_i b_i c_i \bar{c}_i = 1/3$
$\sum_i b_i \bar{c}_i^2 = 1/3$	--	6. $\sum_i b_i \bar{c}_i^2 = 1/3$
$\sum_i \sum_j b_i a_{i,j} c_j = 1/6$	4. $\sum_i \sum_j b_i a_{i,j} c_j = 1/6$	7. $\sum_i \sum_j b_i a_{i,j} c_j = 1/6$
$\sum_i \sum_j b_i a_{i,j} \bar{c}_j = 1/6$	--	8. $\sum_i \sum_j b_i a_{i,j} \bar{c}_j = 1/6$
$\sum_i \sum_j b_i \bar{a}_{i,j} c_j = 1/6$	--	9. $\sum_i \sum_j b_i \bar{a}_{i,j} c_j = 1/6$
$\sum_i \sum_j b_i \bar{a}_{i,j} d_{i,j} = 1/3$	--	10. $\sum_i \sum_j b_i \bar{a}_{i,j} d_j = 1/3$

4.2.3. PVDRK formulas

Recall first that a VIDE of the form (4.2.1) can be rewritten as a system of two integral equations of the second kind,

$$\begin{bmatrix} y(t) \\ z(t) \end{bmatrix} = \begin{bmatrix} y_0 \\ 0 \end{bmatrix} + \begin{bmatrix} \int_0^t f(s,y(s),z(s))ds \\ \int_0^t k(t,s,y(s))ds \end{bmatrix}, \quad t \in I.$$

If t is contained in the interval $[t_n, T]$, then we may write

$$\begin{bmatrix} y(t) \\ z(t) \end{bmatrix} = \begin{bmatrix} y(t_n) \\ \int_0^{t_n} k(t,s,y(s))ds \end{bmatrix} + \begin{bmatrix} \int_{t_n}^t f(s,y(s),z(s))ds \\ \int_{t_n}^t k(t,s,y(s))ds \end{bmatrix}. \tag{4.2.23}$$

Applying a PVRK method of the form (4.1.8) to the above system, we readily verify that the corresponding PVRK part is identical with the PVDRK formula (4.2.15) (see also (4.2.11)). Hence, we obtain a generalization of a result of Pouzet [1963] derived for m-stage explicit VDRK methods with $m \leqslant 4$; it is due to Lubich [1981], [1982c] and represents the counterpart to Theorem 4.1.1.

Theorem 4.2.1. *The order of consistency of a PVDRK formula (4.2.15) is given by that of the RK method (4.1.13) on which it is based.* □

The diagrams given in Examples 4.1.1 to 4.1.6 may thus be used to generate specific PVDRK formulas. We therefore restrict this discussion to adding an observation concerning m-stage implicit PVDRK formulas of optimal order of consistency, $p=2m$, whose parameters have been given in Section 4.1.3.

Example 4.2.3. *The PVDRK formula with $m=2$ and optimal order $p=4$.*

This method is characterized by the second diagram in Example 4.1.6; according to (4.2.15), we obtain

$$\begin{aligned}
Y_{n,1} &= y_n+\frac{h}{12}\Big\{3f\Big(t_n+\frac{3-\sqrt{3}}{6}h, Y_{n,1}, \tilde{F}_n\Big(t_n+\frac{3-\sqrt{3}}{6}h\Big)+hZ_{n,1}\Big) \\
&\quad +(3-2\sqrt{3})f\Big(t_n+\frac{3+\sqrt{3}}{6}h, Y_{n,2}, \tilde{F}_n\Big(t_n+\frac{3+\sqrt{3}}{6}h\Big)+hZ_{n,2}\Big)\Big\}, \\
Y_{n,2} &= y_n+\frac{h}{12}\Big\{(3+2\sqrt{3})f\Big(t_n+\frac{3-\sqrt{3}}{6}h, Y_{n,1}, \tilde{F}_n\Big(t_n+\frac{3-\sqrt{3}}{6}h\Big)+hZ_{n,1}\Big) \\
&\quad +3f\Big(t_n+\frac{3+\sqrt{3}}{6}h, Y_{n,2}, \tilde{F}_n\Big(t_n+\frac{3+\sqrt{3}}{6}h\Big)+hZ_{n,2}\Big)\Big\}; \\
Z_{n,1} &= \frac{1}{12}\Big\{3k\Big(t_n+\frac{3-\sqrt{3}}{6}h, t_n+\frac{3-\sqrt{3}}{6}h, Y_{n,1}\Big) \\
&\quad +(3-2\sqrt{3})k\Big(t_n+\frac{3-\sqrt{3}}{6}h, t_n+\frac{3+\sqrt{3}}{6}h, Y_{n,2}\Big)\Big\}, \\
Z_{n,2} &= \frac{1}{12}\Big\{(3+2\sqrt{3})k\Big(t_n+\frac{3+\sqrt{3}}{6}h, t_n+\frac{3-\sqrt{3}}{6}h, Y_{n,1}\Big) \\
&\quad +3k\Big(t_n+\frac{3+\sqrt{3}}{6}h, t_n+\frac{3+\sqrt{3}}{6}h, Y_{n,2}\Big)\Big\}; \\
y_{n+1} &= y_n+\frac{h}{2}\Big\{f\Big(t_n+\frac{3-\sqrt{3}}{6}h, Y_{n,1}, \tilde{F}_n\Big(t_n+\frac{3-\sqrt{3}}{6}h\Big)+hZ_{n,1}\Big) \\
&\quad +f\Big(t_n+\frac{3+\sqrt{3}}{6}h, Y_{n,2}, \tilde{F}_n\Big(t_n+\frac{3+\sqrt{3}}{6}h\Big)+hZ_{n,2}\Big)\Big\}.
\end{aligned}$$

The formula involves kernel values $k(t_n+c_ih, t_n+c_lh, Y_{n,l})$ with arguments (t_n+c_ih, t_n+c_lh) not belonging to S if $l>i$. We shall show, in Section 5.4.3 (cf. (5.4.25)) that this difficulty can be avoided by a simple modification of the discrete increment function; see also the examples given in Section 5.4.4. □

4.2.4. BVDRK formulas

Example 4.2.4. *Explicit BVDRK formulas with $p=m=1$.*

Since $c_1=a_{1,1}=0$, $b_1=1$, we have

$$Y_{n,1} = y_n,$$
$$Z_{n,1} = \bar{a}_{1,1}k(t_n+d_1h, t_n, Y_{n,1});$$

the new approximation is then defined by

$$y_{n+1} = y_n+hf(t_n, Y_{n,1}, \tilde{F}_n(t_n)+hZ_{n,1}).$$

For $\bar{a}_{1,1}=0$, $d_1=1$, this yields the *explicit Euler method.* □

Example 4.2.5. *Implicit BVDRK formulas with $m=1$ and $p=2$.*

An implicit one-stage BVDRK formula has local order $p=2$ if the following order conditions hold (cf. Table 4.2.1):

$$b_1 = 1, \quad c_1=a_{1,1}=1/2, \quad \bar{c}_1\ (=\bar{a}_{1,1})=1/2;$$

the kernel condition is satisfied if $d_1 \geqslant c_1=1/2$. We thus obtain:

$$Y_{n,1} = y_n+\frac{h}{2}\,f(t_n+\frac{h}{2}, Y_{n,1}, \tilde{F}_n(t_n+\frac{h}{2})+hZ_{n,1}),$$
$$Z_{n,1} = \frac{h}{2}\,k(t_n+d_1h, t_n+\frac{h}{2}, Y_{n,1}),$$

and

$$y_{n+1} = y_n+hf(t_n+\frac{h}{2}, Y_{n,1}\tilde{F}_n(t_n+\frac{h}{2})+hZ_{n,1})$$

(see also Example 4.2.1). The BVDRK formula corresponding to the choice $d_1=1$ will be called, in analogy to Example 4.1.7, the *implicit Bel'tyukov midpoint formula* for (4.2.1). □

Example 4.2.6. *Explicit BVDRK formula with $m=p=2$.*

These formulas correspond to $c_1=a_{1,1}=0$. We choose $c_2=1$ (thus minimizing the number of $\tilde{F}_n$-evaluations), implying $b_1=b_2=1/2$. We are thus led to

$$Y_{n,1} = y_n,$$
$$Z_{n,1} = \bar{a}_{1,1}k(t_n+d_1h, t_n, Y_{n,1}),$$
$$Y_{n,2} = y_n+ha_{2,1}f(t_n, Y_{n,1}, \tilde{F}_n(t_n)+hZ_{n,1}),$$
$$Z_{n,2} = \bar{a}_{2,1}k(t_n+d_1h, t_n, Y_{n,1})+\bar{a}_{2,2}k(t_n+d_2h, t_n+h, Y_{n,2}),$$

and

$$y_{n+1} = y_n+\frac{h}{2}\,\{f(t_n, Y_{n,1}, \tilde{F}_n(t_n)+hZ_{n,1})+f(t_n+h, Y_{n,2}, \tilde{F}_n(t_n+h)+hZ_{n,2})\}.$$

So far, we have not yet considered the order condition involving the coefficients $\{\bar{a}_{j,i}\}$ (i.e., order condition 3. in Table 4.2.1). In the explicit case, it reads

$$b_1\bar{a}_{1,1}+b_2(\bar{a}_{2,1}+\bar{a}_{2,2}) = \frac{1}{2}(\bar{a}_{1,1}+\bar{a}_{2,1}+\bar{a}_{2,2})=1/2,$$

since $b_1=b_2=1/2$. Hence, in order to minimize the number of kernel evaluations in the expressions for $Z_{n,1}$ and $Z_{n,2}$, we choose either

$$\bar{a}_{1,1} = \bar{a}_{2,1}=0, \quad \bar{a}_{2,2}=1,$$

or

$$\bar{a}_{2,2} = 0, \quad \bar{a}_{2,1}=1-\bar{a}_{1,1}.$$

The two resulting families of 2-stage explicit BVDRK formulas are described by the diagrams

d	$\bar{A}$		**c**	A		
d_1	0	0	0	0	0	
d_2	0	1	1	1	0	
				$\frac{1}{2}$	$\frac{1}{2}$	$\mathbf{b}^T$,

and

d	$\bar{A}$		**c**	A		
d_1	α	0	0	0	0	
d_2	$1-\alpha$	0	1	1	0	
				$\frac{1}{2}$	$\frac{1}{2}$	$\mathbf{b}^T$,

Here, d_1,d_2 and α are free parameters; the kernel conditions (4.2.17) yield, respectively, $d_1 \geqslant 0$, $d_2 \geqslant 1$, and $d_1 \geqslant \alpha$, $d_2 \geqslant 1-\alpha$. Note that these second-order BVDRK formulas require only one kernel evaluation for each n. □

In Section 4.1.6 we have shown that there are no 2-stage BVRK formulas of order $p=3$. For BVDRK formulas, however, the picture is much brighter.

Theorem 4.2.2. *There exist 2-stage BVDRK formulas of order* $p=3$.

Proof. (Lubich [1982c]). Suppose we are given a 2-stage RK method of order $p=3$, characterized by the vectors **b** and **c** and the matrix A. We shall show that we can use this method to generate a BVDRK formula with $\bar{a}_{j,i}=a_{j,i}$.

A glimpse at Table 4.2.1 (BVDRK column) reveals that, due to the above choice, the first nine order conditions are satisfied. Hence we have to show that we can choose d_1 and d_2 satisfying order condition 10.:

$$b_1(a_{1,1}d_1+a_{1,2}d_2)+b_2(a_{2,1}d_1+a_{2,2}d_2) \;=\; 1/3.$$

This is equivalent with

$$d_1(b_1a_{1,1}+b_2a_{2,1})+d_2(b_1a_{1,2}+b_2a_{2,2}) \;=\; 1/3.$$

Since the values of the parameters occurring in the expression between parentheses are given, there clearly exist values for d_1 and d_2 for which this equation will be satisfied. □

Example 4.2.7. *A 2-stage diagonally implicit BVDRK formula with* $p=3$.

Consider the diagonally implicit RK method of Nørsett given in Example 4.1.5. In this case the parameters d_1 and d_2 are subject to

$$\frac{3-\sqrt{3}}{12}\,d_1+\frac{3+\sqrt{3}}{12}\,d_2 \;=\; 1/3.$$

The solution $(d_1,d_2)=(1,\sqrt{3}/3)$ satisfies also the kernel condition. Hence, the resulting diagonally implicit BVDRK method with two stages and order $p=3$ is given by the diagram

$\mathbf{d}$	$\bar{A}=A$		$\mathbf{c}$	A		
1	$\frac{3+\sqrt{3}}{6}$	0	$\frac{3+\sqrt{3}}{6}$	$\frac{3+\sqrt{3}}{6}$	0	
$\frac{\sqrt{3}}{3}$	$-\frac{\sqrt{3}}{3}$	$\frac{3+\sqrt{3}}{6}$	$\frac{3-\sqrt{3}}{6}$	$-\frac{\sqrt{3}}{3}$	$\frac{3+\sqrt{3}}{6}$	
				$\frac{1}{2}$	$\frac{1}{2}$	$\mathbf{b}^T$

Written explicitly, it assumes the form

$$\begin{aligned}
Y_{n,1} &= y_n+\frac{3+\sqrt{3}}{6}hf(t_n+\frac{3+\sqrt{3}}{6}\,h,Y_{n,1},\tilde{F}_n(t_n+\frac{3+\sqrt{3}}{6}\,h)+hZ_{n,1}),\\
Z_{n,1} &= \frac{3+\sqrt{3}}{6}\,k(t_n+h,t_n+\frac{3+\sqrt{3}}{6}\,h,Y_{n,1}),\\
Y_{n,2} &= y_n+\frac{h}{6}\,\{-2\sqrt{3}f(t_n+\frac{3+\sqrt{3}}{6}\,h,Y_{n,1},\tilde{F}_n(t_n+\frac{3+\sqrt{3}}{6}\,h)+hZ_{n,1})\\
&\qquad +(3+\sqrt{3})f(t_n+\frac{3-\sqrt{3}}{6}\,h,Y_{n,2},\tilde{F}_n(t_n+\frac{3-\sqrt{3}}{6}\,h)+hZ_{n,2})\},
\end{aligned}$$

$$Z_{n,2} = \frac{1}{6}\{-2\sqrt{3}k(t_n+h,t_n+\frac{3+\sqrt{3}}{6}h,Y_{n,1})$$

$$+(3+\sqrt{3})k(t_n+\frac{\sqrt{3}}{3}h,t_n+\frac{3-\sqrt{3}}{6}h,Y_{n,2})\};$$

$$y_{n+1} = y_n+\frac{h}{2}\{f(t_n+\frac{3+\sqrt{3}}{6}h,Y_{n,1},\tilde{F}_n(t_n+\frac{3+\sqrt{3}}{6}h)+hZ_{n,1})+$$

$$+f(t_n+\frac{3-\sqrt{3}}{6}h,Y_{n,2},\tilde{F}_n(t_n+\frac{3-\sqrt{3}}{6}h)+hZ_{n,2})\}.$$

The formula requires two kernel evaluations for each step. □

Finally, we consider 4-stage explicit BVDRK formulas. It has been shown by Lubich [1982c] that for such formulas the order $p=m=4$ is attainable; this is in contrast to the result for 4-stage explicit BVRK formulas where the order $p=3$ is best possible (Theorem 4.1.3).

Theorem 4.2.3. *There exist 4-stage explicit BVDRK formulas of order $p=4$.* □

For the proof of this result the reader is referred to Lubich [1982c, pp. 131-133]. We list two of these formulas; they are the only ones satisfying $\bar{a}_{i,j}=a_{i,j}$ and possessing only three distinct values $\{c_i\}$.

Example 4.2.8. *Explicit BVDRK formula with $p=m=4$.*

(i)

d	c	$\bar{A}=A$			
$\frac{1}{2}$	0	0	0	0	0
$\frac{1}{2}$	$\frac{1}{2}$	$\frac{1}{2}$	0	0	0
1	$\frac{1}{2}$	0	$\frac{1}{2}$	0	0
d_4	1	0	0	1	0
		$\frac{1}{6}$	$\frac{1}{3}$	$\frac{1}{3}$	$\frac{1}{6}$	$\mathbf{b}^T$

(ii)

$\mathbf{d}$	$\mathbf{c}$	$\bar{A}=A$				
$\frac{1}{4}$	0	0	0	0	0	
$\frac{3}{4}$	$\frac{1}{2}$	$\frac{1}{2}$	0	0	0	
1	0	$-\frac{3}{2}$	$\frac{3}{2}$	0	0	
d_4	1	$-\frac{5}{6}$	$\frac{3}{2}$	$\frac{1}{3}$	0	
		$\frac{1}{9}$	$\frac{2}{3}$	$\frac{1}{18}$	$\frac{1}{6}$	$\mathbf{b}^T$

Here, d_4 can be chosen arbitrarily: in the numerical algorithm, all terms involving d_4 cancel out. □

We remark that the formula generated by the first array above is algebraically equivalent to the formula generated by the first Butcher array in Example 4.1.3.

At the time of writing, little is known on the attainable order of explicit or implicit BVDRK formulas; in particular, the problem of constructing m-stage implicit BVDRK formulas of optimal order remains unsolved.

4.2.5. Convergence results

In Definition 4.2.1 we have introduced the lag term approximations $\tilde{F}_n(t)$ given, respectively, by the extended lag term formula (4.2.10a) and the mixed lag term formula (4.2.10b). Since the exact lag term $F_n(t)$ in the VIDE (4.2.3) differs from the one in the second-kind integral equation (4.1.1a) only by the (missing) function $g(t)$, the construction and the analysis of the extended and mixed lag term formulas for VIDEs is described completely by the results of Section 4.1.7. As a consequence, the corresponding convergence results (and their proofs) for extended and mixed VDRK methods closely parallel those for VRK methods (cf. Theorems 4.1.5 and 4.1.6).

Extended VDRK methods

In Theorem 4.1.5 we have shown that the global order of convergence of an extended VRK method for a second-kind integral equation equals the (local) order of consistency of its VRK formula. An analogous result holds for extended VDRK methods for the VIDE (4.2.1).

Theorem 4.2.4. *Let the VDRK formula (4.2.9) be consistent of (local) order p, and let $\tilde{F}_n(t)$ be the extended lag term formula (4.2.10a). Moreover, assume that the functions $f(t,y,z)$ and $k(t,s,y)$ characterizing the VIDE (4.2.1) satisfy the* *hypotheses of Theorem 1.3.10 and are such that the corresponding solution* $y(t)$ *is sufficiently smooth on I. Then the approximations* $\{y_n\}$ *generated by the extended VDRK method (4.2.9), (4.2.10a) are convergent of order p; i.e., we have, for* $y_0 = y(0)$,

$$\max_{1 \leq n \leq N} |y(t_n) - y_n| \leq Ch^p \quad (\text{as } h \downarrow 0, \text{ with } Nh = T),$$

where the constant C is independent of N and h. □

The proof of this result may be carried out by suitably adapting the techniques employed in the proof of Theorem 4.1.5; due to limitations of space the details are left to the reader. We refer also to Lubich [1981], [1982c] where a different idea of proof has been used.

A special case of Theorem 4.2.4 (m-stage explicit PVDRK methods of extended form with $p = m \leq 4$) was established by Pouzet [1962, pp. 66-75].

Mixed VDRK methods

The following theorem represents the analogue of Theorem 4.1.6.

Theorem 4.2.5. *Let the following assumptions hold:*

(i) *The VDRK formula (4.2.9) has (local) order of consistency p;*

(ii) $\tilde{F}_n(t)$ *is a mixed lag term approximation of the form (4.2.10b) whose order is r and whose weights* $\{w_{n,l}\}$ *satisfy* $|w_{n,l}| \leq w < \infty$, $0 \leq l \leq n$, $\tilde{\kappa} \leq n \leq N-1$ *(uniformly for all* $N \in \mathbb{N}$ *with* $N > \tilde{\kappa}$*);*

(iii) *the starting values* $\{y_0, \ldots, y_{\tilde{\kappa}-1}\}$ *are such that* $|y(t_j) - y_j| = O(h^s)$, $j = 0, \ldots, \tilde{\kappa} - 1$;

(iv) *the functions* $f(t,y,z)$ *and* $k(t,s,y)$ *characterizing the VIDE (4.2.1) are subject to the hypotheses of Theorem 1.3.10 and are such that the corresponding solution* $y(t)$ *is sufficiently smooth on I.*

If we define $p^* := \min\{p, r, s\}$, *then the approximations* $\{y_n\}$ *generated by the mixed VDRK method (4.2.9), (4.2.10b) are convergent of order* p^**; i.e., we have*

$$\max_{\tilde{\kappa} \leq n \leq N} |y(t_n) - y_n| \leq Ch^{p^*}, \quad n = \tilde{\kappa}, \ldots, N$$

(as $h \downarrow 0$*, with* $Nh = T$*), with the constant C being independent of N and h.* □

The proof of this result can be given along the lines of the one for Theorem 4.1.6; we omit the details. A slightly different version of Theorem 4.2.5 is due to Lubich [1982c, pp. 126-127]; this paper also contains a detailed proof whose techniques are readily adapted to establish Theorem 4.2.5.

4.2.6. Numerical examples

In this section we present the Runge-Kutta analogues of the Tables 3.4.1, 3.4.2 and 3.4.3. The methods tested are the fourth-order PVDRK method, characterized by the first Butcher array in Example 4.1.3, and the fourth-order BVDRK method, defined by the second diagram in Example 4.2.8. In both cases, the mixed methods used the fourth-order Gregory rule, with exact starting values, for approximating the lag term. Notice that both methods are explicit. For experiments with implicit methods we refer to Section 5.4.5.

It has already been observed in Section 3.4.4 that problem (3.4.11) becomes stiff as t increases. For such problems we need a method with a large stability region; just as in the case of ODEs, explicit methods for VIDEs have rather limited stability regions (cf. Chapter 7) so that we should expect an unstable behavior for larger integration steps. In Table 4.2.2, where results for problem (3.4.11) are listed, an asterisk indicates instability. It appears that the BVDRK methods possess larger stability regions than the PVDRK methods.

For the nonstiff problems in the Tables 4.2.3 and 4.2.4, all methods produced more or less comparable results, except for the extended BVDRK in Table 4.2.4. A comparison with the VLM-tables 3.4.2 reveals that the VLM methods are again much more accurate. However, it should be realized that the VLM methods tested are all implicit, whereas the V(D)RK methods tested in this chapter are all explicit.

Table 4.2.2. Numbers of correct significant digits for problem (3.4.11) at $t = T = 10$

h	Extended $PVDRK_4$	h	mixed $PVDRK_4$
1	*	1/2	*
⋮	⋮	⋮	⋮
1/8	*	1/16	*
1/16	*	1/32	5.6
h	$BVDRK_4$	h	$BVDRK_4$
1	*	1/2	*
⋮	⋮	1/4	*
		1/8	*
1/8	*	1/16	1.1
1/16	1.1	1/32	5.5

Table 4.2.3. Numbers of correct significant digits and effective orders for problem (3.4.12) at $t = T = 2$

h	Extended $PVDRK_4$		h	mixed $PVDRK_4$	
4/10	3.0		2/10	2.8	
		> 3.6			> 3.7
4/20	4.1		2/20	3.9	
		> 3.8			> 3.9
4/40	5.2		2/40	5.1	
h	$BVDRK_4$		h	$BVDRK_4$	
4/10	2.7		2/10	2.8	
		> 4.1			> 3.6
4/20	4.0		2/20	3.9	
		> 4.0			> 3.9
4/40	5.2		2/40	5.1	

Table 4.2.4. Numbers of correct significant digits and effective orders for problem (3.4.13) at $t = T = 4$

h	Extended $PVDRK_4$		h	mixed $PVDRK_4$	
4/10	1.0		2/10	0.7	
		> 4.2			> 3.7
4/20	2.3		2/20	1.8	
		> 4.2			> 3.9
4/40	3.5		2/40	3.0	
h	$BVDRK_4$		h	$BVDRK_4$	
4/10	0.3		2/10	0.7	
		> 4.0			> 3.7
4/20	1.5		2/20	1.8	
		> 4.1			> 3.9
4/40	2.7		2/40	3.0	

NOTES

Section 4.1.

A detailed description of explicit PVRK methods may be found in Pouzet's thesis of 1962 (pp. 12-19, 31-44); this work contains also many numerical examples, as well as an analysis of PVRK methods for Volterra integral equations of the form

$$y(t) = g(t) + \int_0^t k(t,s,y(s))ds + \int_0^t \int_0^t h(t,s,\tau,y(s),y(\tau))d\tau\, ds.$$

Van der Houwen [1977b] and van der Houwen & Blom [1978] analyzed implicit VRK methods based on VRK formulas more general than those in Definition 4.1.1: in their work the second argument in $k(t,s,y)$ (cf. (4.1.4)) is given by $t_n + c_{i,j}h$ $(i,j = 1, \ldots, m)$. The first of these references lists the order conditions of such formulas for $p \leqslant 3$.

The thesis by Duncan [1982, pp. 7-32] may be consulted for a very lucid description (and comparison) of (explicit) PVRK and BVRK methods, and of the (implicit) VRK methods of van der Houwen [1977b]. Compare also Hairer [1982] for an analysis of extended VRK methods (this paper contains also a 7-stage explicit PVRK method of order $p = 5$ based on Gaussian quadrature which exhibits certain computational advantages).

High-order implicit PVRK-type based on Gauss, Radau, and Lobatto quadrature formulas have been studied by de Hoog & Weiss [1975] and by Brunner [1980] (compare also Section 5.3.4).

The reader wishing to acquire a deeper understanding of the theory of Runge-Kutta methods of ODEs, including the transformation and embedding techniques of Fehlberg, is referred to the original papers of Butcher [1963], [1964a], [1964b], [1965], [1972] (see also Butcher's survey article [1976] and the recent results on the attainable order of explicit RK methods in Butcher [1985]), to the monographs by Stetter [1973, pp. 107-184], by Dekker & Verwer [1984], and to the forthcoming monograph by Hairer, Nørsett & Wanner [1986]. In addition, compare also Hairer [1981]. Numerous examples of RK methods for ODEs may be found in Lapidus & Seinfeld [1971, pp. 39-74]. Glasmacher & Sommer [1966] present extensive lists of RK parameters for implicit ODE methods based on Gauss, Radau, and Lobatto quadratures. Diagonally implicit RK methods are studied, e.g., in Nørsett [1974], Alexander [1977], and Dekker & Verwer [1984].

The results of Sections 4.1.4, 4.1.5, and 4.1.6 are essentially those given in Brunner & Nørsett [1980] and in Brunner, Hairer & Nørsett [1982]. The concept of a γ-modified lag term formula was introduced by van der Houwen [1980a], [1980b], in the context of improving the stability behavior of VRK methods (see Chapter 7).

A number of authors have adapted the Runge-Kutta-Fehlberg techniques for ODEs to second-kind Volterra integral equations; due to limitations of space we have to confine ourselves to merely pointing out the work of Lomakovich & Vasilenko [1968], Lomakovich [1969a], Lomakovich & Ishchuk [1974], Coroian [1974], Ishchuk [1975], M. Micula [1981], and Pelekh [1984]; see also Schoedon [1970], Filippi & Schoedon [1971], Ishchuk & Lutsev [1981], and Tanfulla & Ribighini [1981] for related ideas, and consult Chapter 8 for remarks on automatic codes based on these methods.

We conclude this section with the following observations: (i) The construction and analysis of diagonally implicit VRK methods with $a_{1,1} = \dots = a_{m,m} \neq 0$ (which, in analogy to such RK methods for ODEs (see Alexander [1977]), may have computational advantages over both implicit and explicit VRK methods, especially when solving systems of integral equations) is essentially an open problem, particularly as far as BVRK methods are concerned. (ii) The VRK theory for Volterra integral equations of the first kind is still to be developed. See, however, the special explicit VRK methods of order $p=3$ derived by Keech [1976], [1977b], and the implicit PVRK-type methods obtained by Brunner [1977], [1978a] (the latter are discussed in Section 5.5.3 of this book).

Section 4.2.

Runge-Kutta methods for ODEs were first extended to Volterra integro-differential equations by Pouzet [1960]; see his thesis of 1962 for details, as well as for examples of m-stage explicit PVDRK methods with $p=m \leq 4$. Wolfe & Phillips [1968] considered 2-stage and 3-stage explicit PVDRK methods of mixed type, with lag term approximations given, respectively, by the Gregory rules (2.6.23) with $q=0$ (trapezoidal rule) and $q=1$. Lihtarnikov [1968] discusses the application of Runge-Kutta methods to Volterra integro-differential equations, either by employing the integrated form of the equation, or by using directly the given equation (compare also MR 42, #4040, for a description of these methods). The VDRK method introduced by Schoedon [1970, pp. 67-81] is of order $p=4$ and is based on a 4-stage explicit RK method and on Simpson's rule, while Lomakovich [1969b] and M.Micula [1973] resorted to Runge-Kutta-Fehlberg techniques to devise numerical methods for VIDEs (cf. Notes for Section 4.1 for relevant references). Tavernini [1971] presented a general theory for one-step methods for Volterra functional differential equations; VIDEs constitute a particular class of such equations (see also Cryer [1972, pp. 72-78] for a review of Tavernini's work). One-step methods for VIDEs were also investigated by Mocarsky [1971].

Using ideas of Henrici [1962, pp. 118-123], Feldstein & Sopka [1974] derived one-step algorithms based on Taylor-expansion techniques (and on extrapolation); in addition, they introduced a related 5-stage explicit VRK method of order $p=4$. See also Goldfine [1973], [1977] for an analysis of Taylor-expansion methods (including the automatic computation of the necessary derivatives) for VIDEs.

Neves [1975a], [1975b] employs the fourth-order explicit RK method of Merson (cf. Lapidus & Seinfeld [1971, pp. 68-69]) to create an automatic package for solving functional differential equations; this method is, as mentioned above, applicable to VIDEs (compare also Chapter 8).

The survey articles by Cryer [1972], Baker [1974b], and Baker [1976] trace the progress made in the development of numerical methods, particularly VRK methods and general one-step methods, up to about 1975.

The modern theory of VDRK methods is due to Lubich [1981], [1982c]; these papers include numerous examples of explicit BVDRK formulas of order $p \leqslant 4$. The paper by Brunner [1984a] gives three classes of m-stage implicit PVDRK-type methods which exhibit the optimal convergence order $p = 2m$ (see also Sections 5.4.3 and 5.4.4).

In addition to the challenging problem of constructing, in a systematic way, BVDRK formulas of order $p \geqslant 5$ (both of explicit and of implicit form), another problem needs to be investigated, namely, the construction and analysis of *diagonally implicit* VDRK formulas of Pouzet and of Bel'tyukov type, where $\bar{a}_{1,1} = \ldots = \bar{a}_{m,m} \neq 0$ (cf. Definition 4.2.2(ii)).

CHAPTER 5

COLLOCATION METHODS FOR VOLTERRA EQUATIONS WITH REGULAR KERNELS

A collocation method is based on the principle of approximating the exact solution of a given functional equation in a suitably chosen finite-dimensional function space (usually, but not always, a subspace of the space containing the solution) such that the approximating element satisfies the functional equation on a certain finite subset (consistent with the dimension of the approximating space) of the interval on which the equation is to be solved. This element will, in general, not satisfy the equation at a point not belonging to this finite subset (called the set of collocation points).

In the present chapter we study collocation methods for Volterra equations with regular kernels: the approximating spaces will be certain polynomial spline spaces. Particular attention will be given to (global and local) convergence results and to the discretization of the collocation equations. Here, we also investigate the connection with certain Volterra-Runge-Kutta methods introduced in Chapter 4.

5.1. The polynomial spline spaces $S_m^{(d)}(Z_N)$

For given $N \in \mathbb{N}$ let

$$\Pi_N : \ 0 = t_0^{(N)} < t_1^{(N)} < \ldots < t_N^{(N)} = T$$

be a partition (or: a mesh) of the interval $I := [0,T]$; unless stated otherwise, we shall assume that the sequence of these partitions is *quasi-uniform*: if we define

$$h_n^{(N)} := t_{n+1}^{(N)} - t_n^{(N)}, \quad h_{\min}^{(N)} := \min_{(n)} h_n^{(N)}, \quad h_{\max}^{(N)} := \max_{(n)} h_n^{(N)},$$

then there exists a finite constant γ independent of N and such that $h_{\max}^{(N)} / h_{\min}^{(N)} \leqslant \gamma$ for all $N \in \mathbb{N}$. This assumption implies that

$$N \cdot h_n^{(N)} \leqslant N \cdot h_{\max}^{(N)} \leqslant \gamma N \cdot h_{\min}^{(N)} \leqslant \gamma T, \quad N \in \mathbb{N}, \tag{5.1.1}$$

a property which will be of crucial importance in the convergence analysis of collocation methods based on polynomial splines. For a uniform partition we have $h_{\min}^{(N)} = h_{\max}^{(N)} (=: h) = T/N$, with $\gamma = 1$; hence, (5.1.1) reduces to the statement $Nh = T$ for all $N \in \mathbb{N}$.

With a given partition Π_N we associate the set of its interior points,

$$Z_N := \{t_n^{(N)} : n = 1, \ldots, N-1\},$$

as well as the subintervals

$$\sigma_0^{(N)} := [t_0^{(N)}, t_1^{(N)}], \quad \sigma_n^{(N)} := (t_n^{(N)}, t_{n+1}^{(N)}], \quad n=1, \dots, N-1.$$

In addition we set $\bar{Z}_N := Z_N \cup \{T\}$. For brevity, we shall usually suppress the superscript N in $t_n^{(N)}$, $h_n^{(N)}$, $h_{\max}^{(N)}$, $h_{\min}^{(N)}$, and $\sigma_n^{(N)}$, and we define $h := h_{\max}$ (called the *diameter* of the mesh Π_N).

Definition 5.1.1. Let m and d be given integers satisfying $-1 \leqslant d \leqslant m-1$, and let π_m denote the space of real polynomials of degree not exceeding m. We call

$$S_m^{(d)}(Z_N) := \{u\colon u(t)|_{t\in\sigma_n} =: u_n(t) \in \pi_m, \quad n=0, \dots, N-1; \tag{5.1.2}$$
$$u_{n-1}^{(j)}(t_n) = u_n^{(j)}(t_n) \text{ for } j=0, \dots, d \text{ and } t_n \in Z_N\}$$

the space of (real) *polynomial splines of degree m* and of *continuity class d*, possessing the *knots* Z_N. If $d=-1$, then the elements of $S_m^{(-1)}(Z_N)$ may have jump discontinuities at the knots Z_N. □

The space $S_m^{(-1)}(Z_N)$ is thus the least smooth of the polynomial spline spaces, while $S_m^{(m-1)}(Z_N)$ (the classical spline space) is the smoothest of these spaces. It is easily seen that $S_m^{(d)}(Z_N)$ is a *linear space*: if u and v denote elements of this space, and if α, β are (real) constants, then $w := \alpha u + \beta v$ is again an element of $S_m^{(d)}(Z_N)$ for any d and m satisfying $-1 \leqslant d \leqslant m-1$. Furthermore, $S_m^{(d)}(Z_N)$ has finite dimension.

Theorem 5.1.1. *The dimension of the piecewise polynomial space $S_m^{(d)}(Z_N)$ is given by*

$$\dim S_m^{(d)}(Z_N) = N(m-d) + d + 1, \quad -1 \leqslant d \leqslant m-1. \quad \square \tag{5.1.3}$$

The proof of this result is straightforward and will be omitted. We note that the integer $\delta := m-d$ is often called the *defect of the polynomial spline space* $S_m^{(d)}(Z_N)$; it varies between $\delta = 1$ and $\delta = m+1$.

The following obvious result will play a role when discussing certain connections between collocation methods for ordinary differential equations and for related Volterra integral equations (Section 5.2).

Corollary 5.1.1. *For $-1 \leqslant d$ and $1 \leqslant m$, we have*

$$\dim S_{m+d}^{(d)}(Z_N) = Nm + d + 1. \quad \square \tag{5.1.4}$$

The following section will reveal that the polynomial spline spaces corresponding to the values $d=0$ and $d=-1$ are of particular importance in many applications.

In order to employ these polynomial spline spaces as approximating spaces in the numerical solution of Volterra equations we need to know a suitable basis for $S_m^{(d)}(Z_N)$, especially when $d \in \{-1, 0\}$. We shall describe two important bases for this polynomial spline space: the *one-sided basis,* and the so-called *B-splines* ("basis splines"). While the former is used mainly for theoretical considerations, the latter will be important in numerical calculations, especially for nonuniform meshes.

Theorem 5.1.2. *A basis for the polynomial spline space* $S_m^{(d)}(Z_N)$ *is given by*

$$\{1, t, \dots, t^m;\ (t-t_1)_+^{m-j}, \dots, (t-t_{N-1})_+^{m-j}\ (j=0, \dots, m-d-1)\};$$

hence, any given element of $S_m^{(d)}(Z_N)$ *possesses a unique representation of the form*

$$u(t) = \sum_{i=0}^{m} \alpha_i t^i + \sum_{l=1}^{N-1} \sum_{j=0}^{m-d-1} \beta_{l,j}(t-t_l)_+^{m-j}. \tag{5.1.5}$$

Here, the functions $(t-t)_+^{m-j}$ *are the truncated power functions introduced in Section 2.1.1 (cf. (2.1.4)).* □

Special cases :

(i) $\delta = m-d = 1$ $(d=m-1)$: *classical splines of degree m.*

$$u(t) = \sum_{i=0}^{m} \alpha_i t^i + \sum_{l=1}^{N-1} \beta_l (t-t_l)_+^m.$$

(ii) $\delta = m$ $(d=0)$: *continuous polynomial splines of degree m.*

$$u(t) = \sum_{i=0}^{m} \alpha_i t^i + \sum_{l=1}^{N-1} \sum_{j=0}^{m-1} \beta_{l,j}(t-t_l)_+^{m-j}.$$

(iii) $\delta = m+1$ $(d=-1)$: *discontinuous polynomial splines of degree m.*

$$u(t) = \sum_{i=0}^{m} \alpha_i t^i + \sum_{l=1}^{N-1} \sum_{j=0}^{m} \beta_{l,j}(t-t_l)_+^{m-j}.$$

The representation of an element of $S_m^{(d)}(Z_N)$ in the form (5.1.5) is generally not well suited for numerical applications, for two reasons: (i) if the given partition is heavily nonuniform then some of the basis functions become almost linearly dependent, leading to a badly conditioned basis; (ii) the evaluation of $u(t)$ near the right endpoint of I will necessitate the evaluation of all, or nearly all, of the basis functions. It is possible to define a more symmetric and more local basis by taking certain linear combinations of the truncated power functions. In order to do so we require the following

Definition 5.1.2. Let m and d be given integers satisfying $-1 \leq d \leq m-1$, and set $\delta := m-d$. For $0=t_0<t_1< \dots <t_{N-1}<t_N=T$ we define the sequence $z_1 \leq \dots \leq z_{2(m+1)+(N-1)\delta}$ by

(i) $z_1 \leq \dots \leq z_{m+1} \leq t_0$, $t_N \leq z_{m+2+(N-1)\delta} \leq \dots \leq z_{2(m+1)+(N-1)\delta}$, and

(ii) $z_{m+2+(n-1)\delta} = \dots = z_{m+n\delta+1} = t_n$, $n=1, \dots, N-1$.

This sequence is called an *extended partition associated with the polynomial spline space* $S_m^{(d)}(Z_N)$; it is, up to the first and last $m+1$ points, uniquely determined. □

Example 5.1.1. Let $d=0$ (defect $\delta=m$). The uniquely determined members of the extended partition associated with $S_m^{(0)}(Z_N)$ are given by:

$$z_{m+2} = \dots = z_{2m+1} = t_1,\ z_{2m+2} = \dots = z_{3m+1} = t_2,\ \dots,$$

$$z_{(N-1)m+2} = \dots = z_{Nm+1} = t_{N-1}.$$

Each interior knot thus occurs with multiplicity $\delta=m$. □

Example 5.1.2. Let $d=-1$ (defect $\delta=m+1$). In this case the uniquely determined points of the extended partition associated with $S_m^{(-1)}(Z_N)$ are:

$$z_{m+2} = \dots = z_{2(m+1)} = t_1,\ z_{2(m+1)+1} = \dots = z_{3(m+1)} = t_2,\ \dots,$$

$$z_{(N-1)(m+1)+1} = \dots = z_{N(m+1)} = t_{N-1}.$$

Here, each interior knot occurs with multiplicity $\delta=m+1$. □

In the following, $[z_1, \dots, z_{r+1}]f$ denotes the *divided difference of order r* of $f(z)$ with respect to the points $z_1, \dots, z_{r+1}$.

Theorem 5.1.3. *Let* $\{z_i : i=1, \dots, 2(m+1)+(N-1)\delta\}$ *be an extended partition associated with* $S_m^{(d)}(Z_N)$, *and suppose that* $z_{2(m+1)+(N-1)\delta} > t_N$. *Define* $m+1+(N-1)\delta$ *functions* $\{B_i(t)\}$ *by setting*

$$B_i(t) := (-1)^{m+1} \cdot (z_{i+m+1} - z_i) \cdot [z_i, \dots, z_{i+m+1}](t-z)_+^m,$$
$$i=1, \dots, m+1+(N-1)\delta, \tag{5.1.6}$$

where the divided difference acts with respect to the variable z in the truncated power function $(t-z)_+^m$. *These functions form a basis for* $S_m^{(d)}(Z_N)$; *they are called B-splines and possess the following properties:*

(i) $B_i(t) \equiv 0$ *for* $t \notin [z_i, z_{i+m+1}]$;

(ii) $B_i(t) > 0$ *for* $t \in [z_i, z_{i+m+1})$;

(iii) $\Sigma_{(i)} B_i(t) \equiv 1$ *for all* $t \in I$. □

We observe that the first and last $m+1$ points of the extended partition associated with $S_m^{(d)}(Z_N)$ are usually chosen to be

$$z_1 = \dots = z_{m+1} = t_0,\ z_{m+2+(N-1)\delta} = \dots = z_{2(m+1)+(N-1)\delta} = t_N.$$

In this case, we have to modify the definition of the last B-spline by altering its value at $t=t_N$ to

$$B_M(t_N) := \lim_{t \uparrow t_N} B_M(t), \tag{5.1.7}$$

with $M := m+1+(N-1)\delta$ (otherwise, this function will not be in $S_m^{(d)}(Z_N)$).

A thorough treatment of B-splines, including stable algorithms for their numerical evaluation, can be found in de Boor [1978] and in Schumaker [1981].

In analogy to (5.1.5), each element $u \in S_m^{(d)}(Z_N)$ has a unique representation of the form

$$u(t) = \sum_{i=1}^{M} b_i B_i(t), \quad t \in I. \tag{5.1.8}$$

For $t \in \sigma_n$ this representation does not, in general, involve all basis functions since the B-splines possess local support (property (i) in Theorem 5.1.3). We illustrate this fact for the case $S_m^{(-1)}(Z_N)$.

Example 5.1.3. Let $d = -1$, and assume that

$$z_1 = \ldots = z_{m+1} = t_0, \quad z_{N(m+1)+1} = \ldots = z_{(N+1)(m+1)} = t_N.$$

The B-spline $B_i(t)$ has the support (z_i, z_{i+m+1}); according to Example 5.1.2, the extended partition associated with $S_m^{(-1)}(Z_N)$ contains each point t_n $(n = 0, \ldots, N)$ with multiplicity $m+1$. Thus, the support of $B_i(t)$ is given by just one of the subintervals (t_n, t_{n+1}); i.e., the B-splines associated with $S_m^{(-1)}(Z_N)$ (and with the above extended partition) have minimal support. This is made precise in the following table.

Table 5.1.1. Local supports of B-splines for $S_m^{(-1)}(Z_N)$.

Interval	(t_0, t_1)	$\ldots$	(t_n, t_{n+1})	$\ldots$	(t_{N-1}, t_N)
	$B_1(t)$		$B_{n(m+1)+1}(t)$		$B_{(N-1)(m+1)+1}(t)$
Support for	$\vdots$		$\vdots$		$\vdots$
	$B_{m+1}(t)$		$B_{(n+1)(m+1)}(t)$		$B_{N(m+1)}(t)$

It then follows that, for $t \in \sigma_n$, $u \in S_m^{(-1)}(Z_N)$ is given by

$$u(t) = \sum_{i=n(m+1)+1}^{(n+1)(m+1)} b_i B_i(t), \tag{5.1.9}$$

where the $m+1$ B-splines involved in this representation are defined by

$$B_{n(m+1)+1}(t) = (-1)^{m+1} \cdot (t_{n+1} - t_n) \cdot [t_n, \ldots, t_n, t_{n+1}](t-z)_+^m,$$

$$B_{n(m+1)+2}(t) = (-1)^{m+1} \cdot (t_{n+1} - t_n) \cdot [t_n, \ldots, t_n, t_{n+1}, t_{n+1}](t-z)_+^m,$$

$$\ldots$$

$$B_{(n+1)(m+1)}(t) = (-1)^{m+1} \cdot (t_{n+1} - t_n) \cdot [t_n, t_{n+1}, \ldots, t_{n+1}](t-z)_+^m,$$

with each divided difference involving $(m+2)$ points. □

5.2. Collocation methods for ordinary differential equations

In this section we review some relevant aspects of collocation methods for the approximate solution of initial-value problems for ODEs. To be specific, we shall consider

$$y'(t) = f(t,y(t)), \quad t \in I := [0,T], \quad y(0) = y_0; \tag{5.2.1}$$

and

$$y''(t) = f(t,y(t)), \quad t \in I, \quad y(0) = y_0, \quad y'(0) = z_0. \tag{5.2.2}$$

It will be shown that collocation using the polynomial spline spaces $S_m^{(0)}(Z_N)$ and $S_{m+1}^{(1)}(Z_N)$ as approximating spaces yields, respectively, implicit Runge-Kutta and Runge-Kutta-Nyström methods for (5.2.1) and (5.2.2).

5.2.1. Connection with Runge-Kutta methods

Let $0 \leq c_1 < \ldots < c_m \leq 1$, and consider the problem of determining an element of $S_m^{(0)}(Z_N)$ which satisfies the differential equation (5.2.1) at the *collocation points* $\{t_n + c_j h\}$ (for ease of exposition, we have chosen the uniform mesh):

$$u_n'(t_n + c_j h) = f(t_n + c_j h, u_n(t_n + c_j h)), \quad j = 1, \ldots, m \quad (n = 0, \ldots, N-1). \tag{5.2.3}$$

The number of these *collocation conditions* equals Nm, while the dimension of $S_m^{(0)}(Z_N)$ is given by $Nm + 1$ (recall (5.1.3)). We thus complement (5.2.3) with the initial condition $u_0(0) = y_0$. Since $u_n' \in \pi_{m-1}$, we may write

$$u_n'(t_n + \tau h) = \sum_{l=1}^{m} u_n'(t_n + c_l h) L_l(\tau), \quad t_n + \tau h \in \sigma_n, \tag{5.2.4}$$

with $L_l(\tau)$ denoting the l-th Lagrange fundamental polynomial associated with the m distinct parameters $\{c_j\}$,

$$L_l(\tau) := \prod_{\substack{r=1 \\ r \neq l}}^{m} (\tau - c_r)/(c_l - c_r).$$

Setting

$$a_{j,l} := \int_0^{c_j} L_l(\tau) d\tau, \quad b_l := \int_0^1 L_l(\tau) d\tau, \quad j,l = 1, \ldots, m,$$

we find

$$u_n(t_n + c_j h) = u_n(t_n) + h \int_0^{c_j} \sum_{l=1}^{m} u_n'(t_n + c_l h) L_l(\tau) d\tau,$$

which, due to (5.2.3) and the above abbreviations, becomes

$$u_n(t_n + c_j h) = u_n(t_n) + h \sum_{l=1}^{m} a_{j,l} f(t_n + c_l h, u_n(t_n + c_l h)), \quad j = 1, \ldots, m. \tag{5.2.5a}$$

In an analogous way we find

$$u_n(t_n+h) = u_n(t_n)+h\sum_{l=1}^{m} b_l f(t_n+c_l h, u_n(t_n+c_l h)),$$
$$n=0, \dots ,N-1. \tag{5.2.5b}$$

If we now identify $u_n(t_n)$ and $u_n(t_n+h)$ with y_n and y_{n+1}, respectively, and if we set $Y_{n,j} := u_n(t_n+c_jh)$, we see that the pair of relations (5.2.5) represents an m-stage *implicit Runge-Kutta method* of the form (4.1.12) for the initial-value problem (5.2.1). We note in passing that the definition of the coefficients $\{a_{j,l}\}$ yields the order condition

$$\sum_{l=1}^{m} a_{j,l} = c_j, \quad j=1, \dots ,m \tag{5.2.6}$$

(since $\Sigma_{l=1}^{m} L_l(\tau) \equiv 1$ for all τ).

5.2.2. Connection with Runge-Kutta-Nyström methods

We now turn our attention to the second-order initial-value problem (5.2.2): let the collocation parameters $\{c_j\}$ be as in Section 5.2.1, and suppose that the approximation $u \in S_{m+1}^{(1)}(Z_N)$ to the solution of (5.2.2) is defined by

$$u_n''(t_n+c_jh) = f(t_n+c_jh, u_n(t_n+c_jh)), \quad j=1, \dots ,m \quad (n=0, \dots ,N-1), \tag{5.2.7}$$

subject to the initial conditions $u_0(0)=y_0$, $u_0'(0)=z_0$. Here, the number of collocation conditions is equal to Nm, while the dimension of the approximating space $S_{m+1}^{(1)}(Z_N)$ is $Nm+2$ (recall (5.1.3)): in analogy to Section 5.2.1, the difference between these two numbers is made up by the two given initial conditions. Since $u_n'' \in \pi_{m-1}$, we may write

$$u_n''(t_n+\tau h) = \sum_{l=1}^{m} u_n''(t_n+c_lh)L_l(\tau), \quad t_n+\tau h \in \sigma_n. \tag{5.2.8}$$

Upon integrating twice and setting

$$\bar{a}_{j,l} := \int_0^{c_j} (c_j-\tau)L_l(\tau)d\tau, \quad \bar{b}_l := \int_0^1 (1-\tau)L_l(\tau)d\tau, \quad j,l=1, \dots ,m,$$

we have also

$$u_n'(t_n+h) = u_n'(t_n)+h\sum_{l=1}^{m} b_l u_n''(t_n+c_lh),$$

and

$$u_n(t_n+h) = u_n(t_n)+hu_n'(t_n)+h^2\sum_{l=1}^{m} \bar{b}_l u_n''(t_n+c_lh).$$

If we now invoke (5.2.7), and if we identify $u_n(t_n)$, $u_n(t_{n+1})$, $u_n'(t_n)$, $u_n'(t_{n+1})$ with y_n, y_{n+1}, y_n', y_{n+1}', respectively, we arrive at

$$Y_{n,j} = y_n + hc_j y'_n + h^2 \sum_{l=1}^{m} \bar{a}_{j,l} f(t_n + c_l h, Y_{n,l}), \quad j=1, \dots, m; \tag{5.2.9a}$$

$$y'_{n+1} = y'_n + h \sum_{l=1}^{m} b_l f(t_n + c_l h, Y_{n,l}); \tag{5.2.9b}$$

$$y_{n+1} = y_n + hy'_n + h^2 \sum_{l=1}^{m} \bar{b}_l f(t_n + c_l h, Y_{n,l}), \quad n=0, \dots, N-1, \tag{5.2.9c}$$

where we have set $Y_{n,j} := u_n(t_n + c_j h)$. These formulas define an *m-stage implicit Runge-Kutta-Nyström method* for the special second-order initial-value problem (5.2.2).

5.2.3. *Approximations in* $S_m^{(-1)}(Z_N)$

The initial-value problems (5.2.1) and (5.2.2) are equivalent, respectively, to the special Volterra integral equations of the second kind,

$$y(t) = y_0 + \int_0^t f(s, y(s))ds, \quad t \in I, \tag{5.2.10}$$

and

$$y(t) = y_0 + z_0 t + \int_0^t (t-s) f(s, y(s))ds, \quad t \in I. \tag{5.2.11}$$

Note that the kernel in (5.2.10) does not depend on the variable t.

In this section we shall focus our attention on certain special cases of (5.2.10) and (5.2.11), in order to establish a first connection between polynomial spline collocation for initial-value problems and for Volterra integral equations and, at the same time, to present a motivation for the use of the nonsmooth space $S_m^{(-1)}(Z_N)$ in collocation methods for Volterra integral equations. These special cases are contained in

$$y(t) = y_0 + \int_0^t (\lambda + \mu(t-s)) y(s) ds; \quad \lambda, \mu \text{ constants.} \tag{5.2.12}$$

The choice $\mu = 0$ leads to $y' = \lambda y$, $y(0) = y_0$, while $\lambda = 0$ yields $y'' = \mu y$, $y(0) = y_0$, $y'(0) = 0$.

It will be natural, in our context, to rewrite (5.2.12) in the form

$$y(t) = F_n(t) + \int_{t_n}^t (\lambda + \mu(t-s)) y(s) ds, \quad t \in \sigma_n, \tag{5.2.13}$$

where

$$F_n(t) := y_0 + \int_0^{t_n} (\lambda + \mu(t-s)) y(s) ds.$$

We wish to approximate the exact solution of (5.2.13) in the piecewise polynomial space $S_{m-1}^{(-1)}(Z_N)$, using the collocation parameters $0 \leqslant c_1 < \ldots < c_m \leqslant 1$ (again, recall Corollary 5.1.1)); i.e., we seek an element $u \in S_{m-1}^{(-1)}(Z_N)$ satisfying

$$u_n(t_n+c_jh) = \bar{F}_n(t_n+c_jh)+h\int_0^{c_j}(\lambda+h\mu(c_j-\tau))u_n(t_n+\tau h)d\tau,$$
$$j=1, \ldots ,m \quad (n=0, \ldots ,N-1), \tag{5.2.14a}$$

with

$$\bar{F}_n(t) := y_0+\int_0^{t_n}(\lambda+\mu(t-s))u(s)ds. \tag{5.2.14b}$$

Theorem 5.2.1. *Let the collocation approximation $u \in S_{m-1}^{(-1)}(Z_n)$ to the integral equation (5.2.13) be given by (5.2.14), with $y_n := u_{n-1}(t_n)$.*
(a) *Let $\mu=0$, and suppose that the initial-value problem $y'=\lambda y$, $y(0)=y_0$, is solved by collocation in $S_m^{(0)}(Z_n)$ (see (5.2.3)), yielding the approximation $\hat{y}_n$ for $t=t_n$ $(n=1, \ldots ,N)$. We then have $\hat{y}_n=y_n$ for $n=1, \ldots ,N$ if, and only if, $c_m=1$.*
(b) *Let $\lambda=0$, and suppose that the initial-value problem $y''=\mu y$, $y(0)=y_0$, $y'(0)=0$, is solved by collocation in $S_{m+1}^{(1)}(Z_N)$ (compare (5.2.7)), furnishing the approximation $\tilde{y}_n$ for $t=t_n$ $(n=1, \ldots ,N)$. Then $\tilde{y}_n=y_n$ holds for $n=1, \ldots ,N$ if, and only if, $c_m=1$.* □

Even though the above results do not remain valid for more general initial-value problems and their associated integral equations, they provide an important indication of why the nonsmooth polynomial spline space $S_{m-1}^{(-1)}(Z_N)$ is a natural choice in collocation methods for Volterra integral equations.

Proof. Let $u \in S_{m-1}^{(-1)}(Z_N)$ satisfy the collocation equation (5.2.14a), and set $Y_{n,j} := u_n(t_n+c_jh)$. Since $u_n \in \pi_{m-1}$, we have

$$u_n(t_n+\tau h) = \sum_{l=1}^{m} Y_{n,l}L_l(\tau), \quad t_n+\tau h \in \sigma_n,$$

and hence (5.2.14a) may be expressed in the form

$$Y_{n,j} = \bar{F}_n(t_n+c_jh)+h\sum_{l=1}^{m}(\lambda a_{j,l}+h\mu\bar{a}_{j,l})Y_{n,l}, \tag{5.2.15a}$$

where $j=1, \ldots ,m$, with $a_{j,l}$ and $\bar{a}_{j,l}$ as in (5.2.5a) and (5.2.9a). The lag term $\bar{F}_n(t_n+c_jh)$ can be written as

$$\bar{F}_n(t_n+c_jh) = y_0+\int_0^{t_n}(\lambda+\mu(t_n-s))u(s)ds+hc_j\mu\int_0^{t_n}u(s)ds$$
$$= \bar{F}_n(t_n)+hc_j\mu\int_0^{t_n}u(s)ds.$$

If $c_m = 1$, it follows from the definition of $\bar{F}_n(t)$ that

$$\bar{F}_n(t_n) = \bar{F}_{n-1}(t_{n-1}+h)+h\int_0^1(\lambda+\mu h(1-\tau))u_{n-1}(t_{n-1}+\tau h)d\tau$$

$$= u_{n-1}(t_{n-1}+h)=u_{n-1}(t_n)=y_n.$$

Furthermore, since

$$\int_{t_l}^{t_{l+1}} u_l(s)ds = h\sum_{j=1}^{m} b_j Y_{l,j},$$

with b_j as in (5.2.5b), we obtain

$$\bar{F}_n(t_n+c_jh) = y_n+h^2c_j\mu\sum_{l=0}^{n-1}<\mathbf{b},\mathbf{Y}^{(l)}>, \tag{5.2.15b}$$

where $\mathbf{b}:=(b_1, \dots ,b_m)^T$ and $\mathbf{Y}^{(l)}:=(Y_{l,1}, \dots ,Y_{l,m})^T$ are in $\mathbb{R}^m$, and where $<\cdot,\cdot>$ denotes the corresponding inner product. In addition, let $\bar{\mathbf{b}}:=(\bar{b}_1, \dots ,\bar{b}_m)^T$, $\mathbf{c}:=(c_1, \dots ,c_m)^T$, $\mathbf{e}:=(1, \dots ,1)^T\in\mathbb{R}^m$, and suppose that A and $\bar{A}$ are matrices of order m defined by the elements $\{a_{j,l}\}$ and $\{\bar{a}_{j,l}\}$ occurring in (5.2.5a) and (5.2.9a). This allows us to rewrite (5.2.15a) as

$$\mathbf{Y}^{(n)} = y_n\mathbf{e}+h^2\mu\sum_{l=0}^{n-1}<\mathbf{b},\mathbf{Y}^{(l)}>\mathbf{c}+h(\lambda A+\mu h\bar{A})\mathbf{Y}^{(n)}, \tag{5.2.16a}$$

while, upon setting $j=m$ in (5.2.15a), we are led to

$$y_{n+1} = u_n(t_{n+1})=y_n+h^2\mu\sum_{l=0}^{n-1}<\mathbf{b},\mathbf{Y}^{(l)}>+h<\lambda\mathbf{b}+h\mu\bar{\mathbf{b}},\mathbf{Y}^{(n)}>. \tag{5.2.16b}$$

For $\mu=0$ we obtain from (5.2.16),

$$\begin{aligned} \mathbf{Y}^{(n)} &= [I-h\lambda A]^{-1}y_n\mathbf{e}\,, \\ y_{n+1} &= y_n+h\lambda<\mathbf{b},\mathbf{Y}^{(n)}>\,, \quad n=0, \dots ,N-1, \end{aligned} \tag{5.2.17}$$

provided h is sufficiently small. To prove part (a) of the above theorem, let $\hat{u}\in S_m^{(0)}(Z_N)$ be the collocation approximation to the initial-value problem for $y'=\lambda y$. Equation (5.2.5) yields

$$\hat{u}_n(t_n+c_jh) = \hat{y}_n+h\lambda\sum_{l=1}^{m}a_{j,l}\hat{u}_n(t_n+c_lh),$$

$$\hat{y}_{n+1} = \hat{y}_n+h\lambda\sum_{l=1}^{m}b_l\hat{u}_n(t_n+c_lh),$$

where we have set $\hat{y}_n:=\hat{u}_n(t_n)$, $\hat{y}_{n+1}:=\hat{u}_n(t_n+h)$. For all sufficiently small $h>0$ we obtain

$$\begin{aligned} \hat{\mathbf{U}}^{(n)} &= [I-h\lambda A]^{-1}\hat{y}_n\mathbf{e}\,, \\ \hat{y}_{n+1} &= \hat{y}_n+h\lambda<\mathbf{b},\hat{\mathbf{U}}^{(n)}>\,, \quad n=0, \dots ,N-1, \end{aligned} \tag{5.2.18}$$

where $\hat{\mathbf{U}}^{(n)}$ is an element of $\mathbb{R}^m$ with components $\hat{u}_n(t_n+c_jh)$. Since $\hat{y}_0$ is equal to the given initial value y_0, it is immediately verified by induction that

$$\hat{\mathbf{U}}^{(n)} = \mathbf{Y}^{(n)}, \quad n=0, \ldots ,N-1,$$

and

$$\hat{y}_n = y_n, \quad n=1, \ldots ,N.$$

Let now $\lambda=0$: in this case, (5.2.16) implies, for all sufficiently small $h>0$,

$$\begin{aligned} \mathbf{Y}^{(n)} &= [I-h^2\mu\bar{A}]^{-1}[y_n\mathbf{e}+h^2\mu\sum_{l=0}^{n-1}<\mathbf{b}, \mathbf{Y}^{(l)}>\mathbf{c}], \\ y_{n+1} &= y_n+h^2\mu\sum_{l=0}^{n-1}<\mathbf{b},\mathbf{Y}^{(l)}>+h^2\mu<\bar{\mathbf{b}}, \mathbf{Y}^{(n)}>. \end{aligned} \tag{5.2.19}$$

Thus, to establish part (b) of Theorem 5.2.1, let $\tilde{u}\in S_{m+1}^{(1)}(Z_N)$ be the collocation approximation to the initial-value problem $y''=\mu y$, $y(0)=y_0$, $y'(0)=\lambda y_0=0$. Arguments similar to those employed above show that (5.2.9) can be written in the form

$$\begin{aligned} \tilde{\mathbf{U}}^{(n)} &= [I-h^2\mu\bar{A}]^{-1}[\tilde{y}_n\mathbf{e}+h\tilde{y}_n'\mathbf{c}], \\ \tilde{y}_{n+1}' &= \tilde{y}_n'+h\mu<\mathbf{b},\tilde{\mathbf{U}}^{(n)}>, \\ \tilde{y}_{n+1} &= \tilde{y}_n+h\tilde{y}_n'+h^2\mu<\bar{\mathbf{b}},\tilde{\mathbf{U}}^{(n)}>. \end{aligned} \tag{5.2.20}$$

The second of these relations can be used to derive the expression

$$\tilde{y}_n' = h\mu\sum_{l=0}^{n-1}<\mathbf{b},\tilde{\mathbf{U}}^{(l)}>$$

(recall that $\tilde{y}_0'=0$); substitution in the remaining equations of (5.2.20) then leads to

$$\begin{aligned} \tilde{\mathbf{U}}^{(n)} &= [I-h^2\mu\bar{A}]^{-1}[\tilde{y}_n\mathbf{e}+h^2\mu\sum_{l=0}^{n-1}<\mathbf{b},\tilde{\mathbf{U}}^{(l)}>\mathbf{c}], \\ \tilde{y}_{n+1} &= \tilde{y}_n+h^2\mu\sum_{l=0}^{n-1}<\mathbf{b},\tilde{\mathbf{U}}^{(l)}>+h^2\mu<\bar{\mathbf{b}},\tilde{\mathbf{U}}^{(n)}>. \end{aligned} \tag{5.2.21}$$

A simple induction argument and comparison with (5.2.19) reveal that

$$\tilde{\mathbf{U}}^{(n)} = \mathbf{Y}^{(n)}, \quad n=0, \ldots ,N-1,$$

and

$$\tilde{y}_n = y_n, \quad n=1, \ldots ,N.$$

This completes the proof of Theorem 5.2.1. □

To conclude this section, let us return to the general initial-value problem (5.2.1) and its integrated form (5.2.10); our aim is to generalize part (a) of

Theorem 5.2.1. We approximate the solution of (5.2.10) in $S_{m-1}^{(-1)}(Z_N)$, employing collocation at the points $\{t_n+c_jh\}$; this approximation u is thus determined from

$$u_n(t_n+c_jh) = \bar{F}_n+h\int_0^{c_j} f(t_n+\tau h, u_n(t_n+\tau h))d\tau, \tag{5.2.22a}$$

where $j=1, \dots, m$, and

$$u_n(t_n+h) = \sum_{l=1}^{m} u_n(t_n+c_lh)L_l(1), \tag{5.2.22b}$$

with

$$\bar{F}_n := y_0+\int_0^{t_n} f(s,u(s))ds.$$

This last expression may be written as

$$\bar{F}_n = \bar{F}_{n-1}+h\int_0^1 f(t_{n-1}+\tau h, u_{n-1}(t_{n-1}+\tau h))d\tau, \quad \bar{F}_0=y_0. \tag{5.2.23}$$

We observe that $\bar{F}_n=u_{n-1}(t_{n-1}+h)=:y_n$ if, and only if, $c_m=1$. Furthermore, the collocation equation (5.2.22a) is not yet fully discretized; i.e., the integrals occurring on the right-hand side of (5.2.22a) have to be approximated by suitable quadrature formulas. In other words, the approximate solution of the integrated form (5.2.1) of the general initial-value problem (5.2.1) by collocation in the polynomial spline space $S_{m-1}^{(-1)}(Z_N)$ does not immediately lead to an implicit Runge-Kutta method (5.2.5) for (5.2.1). However, the following result holds.

Theorem 5.2.2. *Suppose that the special Volterra integral equation (5.2.10) (i.e., the integrated form of the initial-value problem (5.2.1)) is solved by collocation in* $S_{m-1}^{(-1)}(Z_N)$, *using the collocation parameters* $0\leqslant c_1<\dots<c_m\leqslant 1$. *Let the integrals over* $[0,c_j]$ *in the corresponding collocation equation (5.2.22a) be evaluated by interpolatory quadrature based on the m abscissas* $\{t_n+c_lh: l=1, \dots, m\}$. *Then the resulting method coincides with the implicit m-stage Runge-Kutta method (5.2.5) for the initial-value problem (5.2.1) if, and only if,* $c_m=1$.

Proof. Let $\hat{u}\in S_{m-1}^{(-1)}(Z_n)$ denote the solution of the discretized collocation equation (obtained by applying m-point interpolatory quadrature in (5.2.22a)),

$$\hat{u}_n(t_n+c_jh) = \hat{F}_n+h\sum_{l=1}^{m} a_{j,l}f(t_n+c_lh,\hat{u}_n(t_n+c_lh)), \quad j=1, \dots, m \tag{5.2.24a}$$

(with $a_{j,l}$ as in (5.2.5a)). The Lagrange identity,

$$\hat{u}_n(t_n+\tau h) = \sum_{j=1}^{m} \hat{u}_n(t_n+c_jh)L_j(\tau), \; t_n+\tau h\in\sigma_n,$$

yields

$$\hat{y}_{n+1} := \hat{u}_n(t_n+h) = \sum_{j=1}^{m} \hat{u}_n(t_n+c_jh)L_j(1),$$

which, because of (5.2.24a), becomes

$$\hat{y}_{n+1} = \sum_{j=1}^{m} L_j(1)\{\hat{F}_n + h\sum_{l=1}^{m} a_{j,l}f(t_n+c_lh,\hat{u}_n(t_n+c_lh))\},$$

or,

$$\hat{y}_{n+1} = \hat{F}_n + h\sum_{l=1}^{m} f(t_n+c_lh,\hat{u}_n(t_n+c_lh))\sum_{j=1}^{m} L_j(1)a_{j,l}. \qquad (5.2.24b)$$

(Here, we have used the fact that $\Sigma_{j=1}^{m} L_j(\tau) \equiv 1$ for all τ.) It is now obvious that (5.2.24a), (5.2.24b) reduce to (5.2.5a), (5.2.5b) if, and only if, $c_m=1$; in particular, the second sum on the right-hand side of (5.2.24b) then takes the value $a_{m,l}$ which in turn equals b_l. □

An analogous result holds if we solve the integrated form (5.2.11) of the second-order initial-value problem (5.2.2) in $S_{m-1}^{(-1)}(Z_N)$ and then discretize the corresponding collocation equation by interpolatory quadrature based on the abscissas $\{t_n+c_lh\}$: we obtain an implicit Runge-Kutta-Nyström method (5.2.9) if, and only if, $c_m=1$. The details are left to the reader.

The above results may be summarized as follows: if the first-order initial-value problem (5.2.1) is solved by collocation in $S_m^{(0)}(Z_N)$, then this method may be viewed as an implicit m-stage Runge-Kutta method; its order of convergence is known to be at least equal to m (see, for example, Guillou & Soulé [1969], Nørsett & Wanner [1981]). On the other hand, if the integrated form of (5.2.1) (i.e., a Volterra integral equation of the second kind) is solved by collocation in $S_{m-1}^{(-1)}(Z_N)$, using the same collocation points, and if the collocation equation is discretized by interpolatory quadrature (based on the collocation points), then we get identical results, provided $c_m=1$. In the following section we shall generalize the above analysis to integral equations of the form (1.1.1) and to collocation parameters no longer subject to $c_m=1$.

5.3. Polynomial spline collocation for second-kind integral equations

5.3.1. The collocation equation

It has already been pointed out in the previous chapter that, in the context of one-step methods, it is natural to rewrite the integral equation (1.1.1) in the form

$$y(t) = F_n(t) + \int_{t_n}^{t} k(t,s,y(s))ds, \quad t\in\sigma_n, \quad n=0, \dots ,N-1, \qquad (5.3.1)$$

where the lag term $F_n(t)$ is given by

$$F_n(t) := g(t) + \int_0^{t_n} k(t,s,y(s))ds. \tag{5.3.2}$$

According to the remarks made in the previous section the exact solution y of (1.1.1) will be approximated by an element of the polynomial spline space $S_{m-1}^{(-1)}(Z_N)$. Recall that $h_n := t_{n+1} - t_n$ $(n = 0, \dots, N-1)$, and $h := \max_{(n)} h_n$; since the underlying partition is assumed to be quasi-uniform we always have $Nh_n \leq Nh \leq \gamma T$ (see (5.1.1)). Let the finite subset $X(N)$ of I be given by

$$X(N) := \bigcup_{n=0}^{N-1} X_n, \tag{5.3.3}$$

with $X_n := \{t_{n,j} := t_n + c_j h_n \colon 0 \leq c_1 < \dots < c_m \leq 1\}$. The set $X(N)$ will be called the set of *collocation points*: the approximation $u \in S_{m-1}^{(-1)}(Z_N)$ will be required to satisfy the given integral equation (1.1.1) (or, equivalently, (5.3.1)) on this set. More precisely, we are looking for the solution of

$$u_n(t_{n,j}) = \bar{F}_n(t_{n,j}) + h_n \int_0^{c_j} k(t_{n,j}, t_n + vh_n, u_n(t_n + vh_n))dv, \quad j = 1, \dots, m, \tag{5.3.4a}$$

$$u_n(t_n + vh_n) = \sum_{j=1}^{m} L_j(v) u_n(t_{n,j}), \ t_n + vh_n \in \sigma_n, \quad n = 0, \dots, N-1. \tag{5.3.4b}$$

Here, $\bar{F}_n(t)$, the approximation to the exact lag term (5.3.2), is defined by

$$\bar{F}_n(t) := g(t) + \sum_{i=0}^{n-1} h_i \int_0^1 k(t, t_i + vh_i, u_i(t_i + vh_i))dv, \tag{5.3.5}$$

and L_j is the j-th Lagrange fundamental polynomial for the collocation parameters $\{c_j\}$ in (5.3.3).

Before we begin analyzing the collocation equation (5.3.4a), we single out a special case in the choice of the collocation points (5.3.3): it is the case where, for $m \geq 2$, the points t_n and t_{n+1} both belong to X_n.

Theorem 5.3.1. *Let g and k be continuous functions. If the collocation parameters* $\{c_j\}$ *in (5.3.3) are such that* $0 = c_1 < c_2 < \dots < c_{m-1} < c_m = 1$, *then the collocation approximation u defined by (5.3.4) is an element of* $S_{m-1}^{(0)}(Z_N)$. □

The proof of this result is immediate and will thus be omitted. Note that, for the case covered in the above theorem, the cardinality of $X(N)$ is $|X(N)| = N(m-1)+1 = \dim S_{m-1}^{(0)}(Z_N)$, while, for $c_m - c_1 < 1$, we have $|X(N)| = Nm = \dim S_{m-1}^{(-1)}(Z_N)$.

Let us now rewrite the collocation equation (5.3.4a) by introducing

$$\Phi_{n,i}^{(j)}[u_i] := \begin{cases} \int_0^1 k(t_{n,j}, t_i + vh_i, u_i(t_i + vh_i))dv, & \text{if } 0 \leq i \leq n-1, \\ \int_0^{c_j} k(t_{n,j}, t_n + vh_n, u_n(t_n + vh_n))dv, & \text{if } i = n, \end{cases} \tag{5.3.6}$$

where $j=1, \ldots, m$. With this notation (5.3.4a) assumes the form

$$u_n(t_{n,j}) = g(t_{n,j}) + h_n \Phi^{(j)}_{n,n}[u_n] + \sum_{i=0}^{n-1} h_i \Phi^{(j)}_{n,i}[u_i]$$

$$j=1, \ldots, m \quad (n=0, \ldots, N-1). \tag{5.3.7}$$

Example 5.3.1. *Approximation in* $S_1^{(-1)}(Z_N)$.

Here, we have $m=2$, $0 \leqslant c_1 < c_2 \leqslant 1$, and

$$u_i(t_i + vh_i) = \frac{1}{c_2 - c_1}[(c_2 - v)Y_{i,1} + (v - c_1)Y_{i,2}],$$

with $Y_{i,j} := u_i(t_{i,j})$. The collocation equation (5.3.7) is of the form

$$Y_{n,1} = g(t_{n,1}) + h_n \Phi^{(1)}_{n,n}[u_n] + \sum_{i=0}^{n-1} h_i \Phi^{(1)}_{n,i}[u_i],$$

$$Y_{n,2} = g(t_{n,2}) + h_n \Phi^{(2)}_{n,n}[u_n] + \sum_{i=0}^{n-1} h_i \Phi^{(2)}_{n,i}[u_i],$$

and we obtain

$$u_n(t_n + \tau h_n) = \frac{1}{c_2 - c_1}[(c_2 - \tau)Y_{n,1} + (\tau - c_1)Y_{n,2}],$$

$$t_n + \tau h_n \in \sigma_n, \quad n=0, \ldots, N-1.$$

In particular, if $c_2=1$, then $u_n(t_{n+1}) = Y_{n,2}$ for $t_{n+1} \in \bar{Z}_N$. Furthermore, for the special choice of the collocation parameters, $c_1=0$ and $c_2=1$, we generate a continuous piecewise linear approximation, $u \in S_1^{(0)}(Z_N)$: here, we have $\Phi^{(1)}_{n,n}[u_n]=0$ and $Y_{n,1}=u_n(t_n)=u_{n-1}(t_n)$. It is easily verified that the discretization of the integrals $\Phi^{(j)}_{n,i}[u_i]$ (which are all evaluated on $[0,1]$) by two-point interpolatory quadrature using the abscissas $t_{i,l}$ $(l=1,2)$ yields the well-known trapezoidal method. □

Example 5.3.2. *Approximation in* $S_2^{(-1)}(Z_N)$.

This case is described by $m=3$, $0 \leqslant c_1 < c_2 < c_3 \leqslant 1$, and we have

$$u_i(t_i + vh_i) = L_1(v)Y_{i,1} + L_2(v)Y_{i,2} + L_3(v)Y_{i,3}, \quad t_i + vh_i \in \sigma_i,$$

with $Y_{i,j} := u_i(t_{i,j})$, and with

$$L_1(v) = (v - c_2)(v - c_3)/[(c_2 - c_1)(c_3 - c_1)],$$

$$L_2(v) = (v - c_1)(c_3 - v)/[(c_2 - c_1)(c_3 - c_2)],$$

$$L_3(v) = (v - c_1)(v - c_2)/[(c_3 - c_1)(c_3 - c_2)].$$

The collocation equation (5.3.7) can be written as

$$Y_{n,1} = g(t_{n,1}) + h_n \Phi^{(1)}_{n,n}[u_n] + \sum_{i=0}^{n-1} h_i \Phi^{(1)}_{n,i}[u_i],$$

$$Y_{n,2} = g(t_{n,2}) + h_n\Phi^{(2)}_{n,n}[u_n] + \sum_{i=0}^{n-1} h_i\Phi^{(2)}_{n,i}[u_i],$$

$$Y_{n,3} = g(t_{n,3}) + h_n\Phi^{(3)}_{n,n}[u_n] + \sum_{i=0}^{n-1} h_i\Phi^{(3)}_{n,i}[u_i];$$

from which we obtain, upon solving the nonlinear system for the quantities $Y_{n,j}$,

$$u_n(t_n + vh_n) = \sum_{j=1}^{3} L_j(v)Y_{n,j}, \quad t_n + vh_n \in \sigma_n, \quad n = 0, \ldots, N-1.$$

Here, the choice $c_1 = 0$, $c_3 = 1$, implies that the approximation u is an element of $S_2^{(0)}(Z_N)$; as in the previous example, we have $\Phi^{(1)}_{n,n}[u_n] = 0$. If, in addition, we choose $c_2 = ½$, then the discretization of the integrals $\Phi^{(j)}_{n,i}[u_i]$ occurring in the collocation equation by three-point interpolatory quadrature based on the abscissas $t_{i,l}$ ($l = 1,2,3$) leads to the block-by-block method of Simpson. In Section 5.3.5 we shall present additional examples of fully discretized collocation methods. □

For the analysis, as well as for the numerical treatment, of the collocation equation (5.3.7) it is necessary to specify a basis for the approximating space $S^{(-1)}_{m-1}(Z_N)$. It has already been mentioned in Section 5.1 that, unless m is small, the one-sided basis of Theorem 5.1.2 (with $d = -1$) will find its principal role in theoretical considerations; i.e., in the convergence analysis for (5.3.7) and its counterparts associated with equations of the first kind and with integro-differential equations. As for the numerical treatment of (5.3.7), there are essentially two ways open to us: (i) we can try to approximate the integrals (5.3.6) to within a prescribed tolerance by employing suitable adaptive quadrature processes (see, e.g., Davis & Rabinowitz [1984], or Piessens et al. [1983]); or (ii) we can fix the quadrature formulas a priori (e.g., quadrature formulas which are interpolatory and based on the collocation abscissas $\{t_{i,j}\}$). In the first case, it will generally be advantageous to select the B-spline basis (5.1.5) (see also Example 5.1.3); since these basis functions are more symmetric than the one-sided basis functions, they will permit a more uniform sampling of the values of the kernel. According to (5.1.9), the integrals (5.3.6) assume the form

$$\Phi^{(j)}_{n,i}[u_i] = \begin{cases} \int_0^1 k\Big(t_{n,j}, t_i + vh_i, \sum\limits_{l=im+1}^{(i+1)m} b_l B_l(t_i + vh_i)\Big)dv, & 0 \leqslant i \leqslant n-1, \\ \int_0^{c_j} k\Big(t_{n,j}, t_n + vh_n, \sum\limits_{l=nm+1}^{(n+1)m} b_l B_l(t_n + vh_n)\Big)dv, & i = n. \end{cases} \tag{5.3.8}$$

The second case will be studied in detail in Section 5.3.4; it will turn out that the basis given by the (local) Lagrange fundamental polynomials is the natural one.

5.3.2. Global convergence results

In the present section we shall answer the question on the optimal order of global convergence of the collocation method introduced in Section 5.3.1, assuming that the integrals (5.3.6) are evaluated analytically; the effect of their evaluation by certain types of quadrature formulas on the accuracy of the collocation approximation will be investigated in Section 5.3.4. For ease of exposition, we shall formulate and prove our convergence result for the linear Volterra integral equation,

$$y(t) = g(t)+\int_0^t K(t,s)y(s)ds, \quad t\in I; \tag{5.3.9}$$

as will be seen from the remark following the proof of Theorem 5.3.2, this will still allow us to exhibit the essential ideas and techniques of the convergence proof in full generality.

In the following let $e:=y-u$, and denote its restriction to the subinterval σ_n by e_n. This error function will in general have finite jumps on Z_N; we shall use the notation

$$\|e\|_\infty := \sup\{|e_n(t)| : t\in\sigma_n, \quad n=0, \dots ,N-1\}.$$

Recall that $h:=\max_{(n)}(t_{n+1}-t_n)$ denotes the maximum width (i.e., the diameter) of the given mesh.

Theorem 5.3.2. *Suppose that the given functions g and K in (5.3.9) belong, respectively, to $C^m(I)$ and $C^m(S)$. Then there exists an $\bar{h}>0$ such that the collocation equation (5.3.7) defines for each $h\in(0,\bar{h})$ a unique element $u\in S_{m-1}^{(-1)}(Z_N)$. The error induced by this approximation to the exact solution y of (5.3.9) satisfies, for every choice of the collocation parameters $\{c_j\}$ with $0\leqslant c_1< \dots <c_m\leqslant 1$ and for all quasi-uniform mesh sequences,*

$$\|e\|_\infty\leqslant Ch^m, \tag{5.3.10}$$

where C denotes some finite constant independent of h (but depending on the $\{c_j\}$).

Proof. We first outline the basic idea on which the proof of Theorem 5.3.2 will be based; the same idea will be used again, with appropriate modifications, in the proofs of analogous global convergence results for integro-differential equations and integral equations of the first kind, provided their solutions are smooth on I.

The hypotheses for g and K imply that (5.3.9) possesses a unique solution $y\in C^m(I)$. Thus, setting

$$y(t_n+\tau h_n) = \sum_{l=1}^m c_{n,l}\tau^{l-1}+h_n^m R_n(\tau), \quad t_n+\tau h_n\in\sigma_n, \tag{5.3.11}$$

with

$$c_{n,l} := h_n^{l-1} \cdot y^{(l-1)}(t_n)/(l-1)!, \quad R_n(\tau) := y^{(m)}(\xi_n)\tau^m/m!$$

$(t_n < \xi_n < t_n + \tau h_n)$, and writing

$$u_n(t_n + \tau h_n) := \sum_{l=1}^{m} \alpha_{n,l}\tau^{l-1}, \quad t_n + \tau h_n \in \sigma_n, \tag{5.3.12}$$

the restriction of the error function e to the subinterval σ_n is given by

$$e_n(t_n + \tau h_n) = h_n^p \{ \sum_{l=1}^{m} \beta_{n,l}\tau^{l-1} + h_n^{m-p} R_n(\tau) \}, \quad n = 0, \dots, N-1. \tag{5.3.13}$$

Here, $h_n^p \beta_{n,l} := c_{n,l} - \alpha_{n,l}$ $(l = 1, \dots, m)$, and p denotes an integer satisfying $1 \leqslant p \leqslant m$. Our aim now consists in showing that if $u \in S_{m-1}^{(-1)}(Z_N)$ is defined by (5.3.7), then the vectors $\boldsymbol{\beta}_n := (\beta_{n,1}, \dots, \beta_{n,m})^T$ remain uniformly bounded for $n = 0, \dots, N-1$, as $h \downarrow 0$ and $Nh \leqslant \gamma T$ (recall (5.1.1)), whenever $p \leqslant m$. More precisely, $p = m$ is the largest integer for which there exists a finite constant B such that

$$\|\boldsymbol{\beta}_n\|_1 := \sum_{l=1}^{m} |\beta_{n,l}| \leqslant B$$

uniformly in n, as $h \downarrow 0$. This in turn implies, according to (5.3.13), that

$$|e_n(t_n + \tau h_n)| \leqslant h^m (B + M_m), \; t_n + \tau h_n \in \sigma_n, \tag{5.3.14}$$

for $n = 0, \dots, N-1$ $(Nh \leqslant \gamma T)$, } and for all sufficiently small $h > 0$; the constant M_m is given by

$$M_m := \max \{ |y^{(m)}(t)|/m!: t \in I \}.$$

In order to verify (5.3.14) we observe that the restrictions of the error function e to the subintervals σ_i $(0 \leqslant i \leqslant n \leqslant N-1)$ are related by

$$e_n(t_{n,j}) = h_n \int_0^{c_j} K_{n,j}(t_n + \tau h_n) e_n(t_n + \tau h_n) d\tau$$
$$+ \sum_{i=0}^{n-1} h_i \int_0^1 K_{n,j}(t_i + \tau h_i) e_i(t_i + \tau h_i) d\tau, \quad j = 1, \dots, m, \tag{5.3.15}$$

where we have set $K_{n,j}(\cdot) := K(t_{n,j}, \cdot)$. This recurrence relation and (5.3.13) (with $p = m$) can be combined to yield, after division by h_n^m,

$$\sum_{l=1}^{m} \beta_{n,l} (c_j^{l-1} - h_n \int_0^{c_j} K_{n,j}(t_n + \tau h_n) \tau^{l-1} d\tau)$$
$$= \sum_{i=0}^{n-1} h_i (h_i/h_n)^m \sum_{l=1}^{m} \beta_{i,l} \int_0^1 K_{n,j}(t_i + \tau h_i) \tau^{l-1} d\tau + q_{n,j}, \quad j = 1, \dots m, \tag{5.3.16}$$

with

$$q_{n,j} := -R_n(c_j)+h_n\int_0^{c_j} K_{n,j}(t_n+\tau h_n)R_n(\tau)d\tau$$

$$+\sum_{i=0}^{n-1} h_i(h_i/h_n)^m\int_0^1 K_{n,j}(t_i+\tau h_i)R_i(\tau)d\tau. \qquad (5.3.17)$$

In order to make the following analysis more transparent we introduce the square matrices $D_{n,i}$ and V defined by setting

$$D_{n,i} := \begin{cases} (\int_0^1 K_{n,j}(t_i+\tau h_i)\tau^{l-1}d\tau), & \text{if } 0\leqslant i\leqslant n-1, \\ (\int_0^{c_j} K_{n,j}(t_n+\tau h_n)\tau^{l-1}d\tau), & \text{if } i=n \quad (j,l=1, \ldots ,m), \end{cases}$$

and

$$V := (c_j^{l-1}) \quad (j,l=1, \ldots ,m).$$

Furthermore, let $\mathbf{q}_n$ be the vector whose components are $q_{n,1}, \ldots ,q_{n,m}$, and set $K_0:=\max\{|K(t,s)|:(t,s)\in S\}$; since $K\in C(S)$, K_0 is finite. Consider now the left-hand side of (5.3.16): since the Vandermonde matrix V is nonsingular for all collocation parameters $\{c_j\}$ with $0\leqslant c_1< \ldots <c_m\leqslant 1$, there exists an $\bar{h}>0$ such that the matrix $V-h_nD_{n,n}$ possesses a uniformly bounded inverse for all $h_n\in(0,\bar{h})$ ($n=0, \ldots ,N-1$). Hence, under this restriction the vectors $\{\boldsymbol{\beta}_n\}$ are defined by

$$\boldsymbol{\beta}_n = (V-h_nD_{n,n})^{-1}\{\sum_{i=0}^{n-1} h_i(h_i/h_n)^m D_{n,i}\boldsymbol{\beta}_i+\mathbf{q}_n\}, \quad n=0, \ldots ,N-1. \qquad (5.3.18)$$

The components of $\mathbf{q}_n$ can be bounded by

$$|q_{n,j}| \leqslant M_m+hK_0M_m/(m+1)+(N-1)h\gamma^m K_0M_m/(m+1)$$
$$\leqslant M_m(1+\gamma^{m+1}K_0T/(m+1))=:q, \quad j=1, \ldots ,m;$$

here, we have used (5.3.17) and the fact that $h_n\leqslant h$, with $Nh\leqslant\gamma T$, $1\leqslant\gamma<\infty$, for any quasi-uniform mesh sequence. It thus follows that the l_1 norm of $\mathbf{q}_n$ satisfies $\|\mathbf{q}_n\|_1\leqslant mq=:Q$ for some appropriate constant Q not depending on n. Furthermore, since $K(t,s)$ is bounded on S, there exists a constant D_1 such that, for all $h_n\in(0,\bar{h})$ and for $0\leqslant i\leqslant n-1\leqslant N-2$, $\|D_{n,i}\|_1\leqslant D_1$. Thus, if D_0 denotes a uniform upper bound for $\|(V-h_nD_{n,n})^{-1}\|_1$, then (5.3.18) leads to

$$\|\boldsymbol{\beta}_n\|_1 \leqslant h\gamma^m D_0D_1\sum_{i=0}^{n-1}\|\boldsymbol{\beta}_i\|_1+D_0Q, \quad n=0, \ldots ,N-1; \qquad (5.3.19)$$

i.e., we have derived a discrete Gronwall inequality for the l_1-norms of the vectors $\{\boldsymbol{\beta}_n\}$. The uniform boundedness of these norms is established by invoking Corollary 1.5.2, where the role of z_n is taken by $\|\boldsymbol{\beta}_n\|_1$, and where $C_1:=\gamma^m D_0D_1$, $C_2:=D_0Q$. This, together with (5.3.14), completes the proof of Theorem 5.3.2. □

The extension of the above arguments to *nonlinear* Volterra integral equations of the form (1.1.1) is straightforward: after having formed the difference between (1.1.1) (with t replaced by the collocation point $t_{n,j}$) and the collocation equation (5.3.4), we linearize the resulting error equation. Thus, the role of $K_{n,j}(t_i+\tau h_i)$ in (5.3.15) is now assumed by

$$\frac{\partial}{\partial y}k(t_{n,j},t_i+\tau h_i,z_i(\tau)),$$

with $z_i(\tau)$ between $y(t_i+\tau h_i)$ and $u_i(t_i+\tau h_i)$, and the constant K_0 represents an upper bound for $|\partial k(t,s,y)/\partial y|$ on a suitable domain of the form $S\times\{y\in\mathbb{R}:|y|\leq b\}$.

Combining this observation with Theorem 5.3.2 we deduce that, under appropriate smoothness hypotheses, the numerical solution of the integrated form (5.2.10) of the initial-value problem (5.2.1) by collocation in $S_{m-1}^{(-1)}(Z_N)$ yields a method whose global order of convergence is given by $p=m$. On the other hand, if the initial-value problem (5.2.1) is solved in the polynomial spline space $S_m^{(0)}(Z_N)$, using the same collocation parameters, then the global order of convergence of the resulting approximation is also equal to m (which is best possible). We shall return to this last result in Section 5.4.1 where it will be obtained as a special case of the global convergence theorem for integro-differential equations.

We conclude with a remark concerning the smoothness hypotheses for g and K in Theorem 5.3.2; a similar remark applies also to the subsequent convergence results of this chapter. A closer look at the above proof reveals that (5.3.10) remains valid if the kernel $K(t,s)$ is (absolutely) integrable on S (e.g., if it is of the form $(t-s)^{-\alpha}G(t,s)$, $0<\alpha<1$, with $G\in C(S)$) and if the exact solution y is in $C^m(I)$; the nonhomogeneous term g need not be smooth. However, as we have seen in Section 1.3.5 (see also Chapter 6), such a situation represents an exception rather than the rule.

5.3.3. *Local superconvergence on* $\bar{Z}_N$

In the previous section we have shown that the collocation approximation $u\in S_{m-1}^{(-1)}(Z_N)$ to the solution of the integral equation (1.1.1) possesses, under appropriate smoothness hypotheses for g and k, the global order of convergence $p=m$; this order is best possible. Is this value also optimal if we consider only the values of u at the mesh points $\{t_n\}$ (recall from Chapter 3 that direct quadrature methods yield approximate values only at the mesh points)? To be more precise, does there exist an integer $p^*>p=m$ such that, as $h\downarrow 0$ and $Nh\leq\gamma T$ (quasi-uniform mesh sequences),

$$|e(t_n)|\leq C^*h^{p^*}\quad\text{for all}\quad t_n\in\bar{Z}_N:=Z_N\cup\{T\}, \tag{5.3.20}$$

for some suitable choice of the collocation parameters $\{c_j\}$? If such a p^* can be found, then we call the approximation u *locally superconvergent* (on $\bar{Z}_N$).

The following theorem shows that there are various ways of attaining local superconvergence; however, we also have the disappointing result that this cannot be achieved by employing the Gauss-Legendre points (in the following usually referred to as Gauss points). As in the previous section it is assumed that the given mesh sequence is quasi-uniform; in particular, we have $h := \max_{(n)} h_n$, with $h_n := t_{n+1} - t_n$ $(n = 0, \ldots, N-1)$, and $Nh \leqslant \gamma T$ for all $N \in \mathbb{N}$.

Theorem 5.3.3. *Let $u \in S_{m-1}^{(-1)}(Z_N)$ denote the collocation approximation to the solution of the integral equation (5.3.9), and assume that g and K satisfy $g \in C^{2m-\nu}(I)$, $K \in C^{2m-\nu}(S)$, for some $\nu \in \{0, 1, 2\}$ and with $m \geqslant \lfloor \nu/2 \rfloor + 1$.*
(a) *If the collocation parameters $\{c_j\}$ are the zeros of $P_{m-1}(2s-1) - P_m(2s-1)$ (i.e., the Radau II points for $(0,1]$), then we have, for $\nu = 1$,*

$$\max_{t_n \in \underline{Z}_N} |e(t_n)| = O(h^{2m-1}) \quad (\text{as } h \downarrow 0,\ Nh \leqslant \gamma T). \tag{5.3.21}$$

(b) *If the collocation parameters $\{c_j\}$ are the zeros of $s(1-s)P'_{m-1}(2s-1)$ (i.e., the Lobatto points for $[0,1]$), and if $\nu = 2$, then it follows that*

$$\max_{t_n \in \underline{Z}_N} |e(t_n)| = O(h^{2m-2}) \quad (\text{as } h \downarrow 0,\ Nh \leqslant \gamma T). \tag{5.3.22}$$

(c) *If the collocation parameters $\{c_j\}$ are the zeros of $P_m(2s-1)$ (the Gauss points for $(0,1)$), and if $\nu = 0$, then we obtain*

$$\max_{t_n \in \underline{Z}_N} |e(t_n)| = O(h^{m}) \quad (\text{as } h \downarrow 0,\ Nh \leqslant \gamma T); \tag{5.3.23}$$

in other words, collocation at the Gauss abscissas does not lead to local superconvergence on $\overline{Z}_N$.
(d) *If, however, the first $m-1$ of the collocation parameters $\{c_j\}$ are the zeros of $P_{m-1}(2s-1)$, and if $c_m = 1$, then it follows with $\nu = 2$ that (5.3.22) holds; i.e., we encounter the same order of local superconvergence as for the Lobatto points.* □

Proof. The collocation equation for $u \in S_{m-1}^{(-1)}(Z_N)$, which holds only at the collocation points $X(N)$, can be written in the form

$$u(t) = g(t) + \int_0^t K(t,s)u(s)ds - \delta(t), \quad t \in I, \tag{5.3.24}$$

where δ denotes a suitable function, subsequently called the *defect function*, vanishing on $X(N)$. This function is smooth on each subinterval σ_n, with the degree of smoothness given by that of g and K. Moreover, due to the global convergence of u, $\delta(t)$ tends to zero, uniformly on I, as $h \to 0$. Subtraction of (5.3.24) from (5.3.9) yields a second-kind integral equation for the error function $e := y - u$,

$$e(t) = \delta(t) + \int_0^t K(t,s)e(s)ds, \quad t \in I, \tag{5.3.25}$$

whose solution, according to Theorem 1.3.1, is given by

$$e(t) = \delta(t) + \int_0^t R(t,s)\delta(s)ds, \quad t \in I; \tag{5.3.26}$$

here, $R(t,s)$ denotes the resolvent kernel for $K(t,s)$. Let now $t = t_n$ be a point in $\bar{Z}_N$, and write (5.3.26) as

$$e(t_n) = \delta(t_n) + \sum_{i=0}^{n-1} h_i \int_0^1 R(t_n, t_i + \tau h_i)\delta(t_i + \tau h_i)d\tau, \tag{5.3.27}$$

where $t_n \in \bar{Z}_N$. Replacing each of the above integrals by the sum of the corresponding interpolatory m-point quadrature formula based on the abscissas $\{t_{i,l}\}$, and its error term $E_{n,i}$, we are led to

$$e(t_n) = \delta(t_n) + \sum_{i=0}^{n-1} h_i \{ \sum_{l=1}^{m} b_l R(t_n, t_{i,l})\delta(t_{i,l}) + E_{n,i} \},$$

where we have set $b_l := \int_0^1 L_l(v)dv$, with L_l denoting the l-th Lagrange fundamental polynomial for the $\{c_j\}$. Since the defect function vanishes at all collocation points (which now also assume the role of the quadrature abscissas), the above expression for $e(t_n)$ reduces to

$$e(t_n) = \delta(t_n) + \sum_{i=0}^{n-1} h_i E_{n,i}, \quad t_n \in \bar{Z}_N. \tag{5.3.28}$$

This, in turn, implies for $Nh \leqslant \gamma T$ and for all $N \in \mathbb{N}$:

$$|e(t_n)| \leqslant |\delta(t_n)| + Nh \sup\{|E_{n,i}| : 0 \leqslant i \leqslant n-1 \leqslant N-1\}, \quad t_n \in \bar{Z}_N. \tag{5.3.29}$$

Suppose first that $c_m = 1$. Each $t_n \in \bar{Z}_N$ is then a member of $X(N)$, implying $\delta(t_n) = 0$. Thus, according to (5.3.29), the order of $e(t_n)$ coincides with that of the error term $E_{n,i}$ associated with the m-point interpolatory quadrature formula using the abscissas $\{t_i + c_l h_i\}$. A look at Table 2.5.1 shows that, for the Radau II and Lobatto quadrature formulas, $E_{n,i} = O(h_i^{2m-\nu})$, where $\nu = 1$ (Radau II) and $\nu = 2$ (Lobatto); the integrands $R(t,\cdot)\delta(\cdot)$ in (5.3.27) are $(2m-\nu)$-times continuously differentiable on each subinterval σ_i $(0 \leqslant i \leqslant n-1)$, since the resolvent $R(t,s)$ inherits the smoothness properties of the kernel $K(t,s)$ (see (1.3.2), (1.3.7)), and since the defect function is defined by (5.3.24), where $g \in C^{2m-\nu}(I)$ and where u is a polynomial on each σ_i.

If $c_m < 1$, then, in general, $\delta(t_n) \neq 0$; thus, in order to determine the order of $e(t_n)$ we require more precise information on the order of the defect function on the set $\bar{Z}_N$.

Lemma 5.3.1. *Let g and K in (5.3.9) be at least m times continuously differentiable on their respective domains, and let $u \in S_{m-1}^{(-1)}(Z_N)$ denote the collocation approximation defined by (5.3.24). Then:*

$$\delta(t_n + \tau h_n) = O(h^m) \quad \textit{for all} \quad t_n + \tau h_n \notin X(N), \tag{5.3.30}$$

and this order is best possible (i.e., m cannot, in general, be replaced by $m+1$).

Proof. Consider the integral equation

$$y(t) = 1+\int_0^t \lambda y(s)ds, \quad t\in I,$$

where λ is a nonvanishing constant. Let $u\in S_1^{(-1)}(Z_N)$ $(m=2)$. One finds easily that

$$Y_{0,1} := u_0(t_{0,1})=(1-\frac{c_2-c_1}{2}h_0\lambda)/D(h_0\lambda),$$

$$Y_{0,2} := u_0(t_{0,2})=(1+\frac{c_2-c_1}{2}h_0\lambda)/D(h_0\lambda),$$

with

$$D(h_0\lambda) := 1-\frac{c_1+c_2}{2}h_0\lambda+\frac{c_1c_2}{2}(h_0\lambda)^2;$$

this yields, by (5.3.24) and by Example 5.3.1,

$$\delta(t_0+\tau h_0) = (\tau^2-\tau(c_1+c_2)+c_1c_2)(h_0\lambda)^2/(2D(h_0\lambda)), \quad \tau\in[0,1].$$

It is readily verified that $\delta(t_0+\tau h_0)=O(h_0^2)(=O(h_0^m))$, $t_0+\tau h_0\in X_0$, describes the optimal order of the defect function at a point which is not a collocation point. The generalization of the above argument to $u\in S_{m-1}^{(-1)}(Z_N)$ with arbitrary m is obvious. □

Returning to (5.3.29) we thus deduce that $e(t_n)=O(\delta(t_n))=O(h^m)$ for all $t_n\in\overline{Z}_N$, provided that $c_m<1$. This holds in particular for the Gauss-Legendre points since they lie in the open interval $(0,1)$ for any value of m.

Finally, in order to establish part (d) of Theorem 5.3.3 we consider once more (5.3.27): we now replace each of the integrals by the sum of the $(m-1)$-point interpolatory quadrature formula (using the first $(m-1)$ collocation points in each of the subintervals σ_i as abscissas) and the corresponding error term $E'_{n,i}$. Instead of (5.3.28) we obtain

$$\begin{aligned} e(t_n) &= \delta(t_n)+\sum_{i=0}^{n-1} h_i\{\sum_{l=1}^{m-1}{}' b_l R(t_n,t_i+c_lh_i)\delta(t_{i,l})+E'_{n,i}\} \\ &= \delta(t_n)+\sum_{i=0}^{n-1} h_iE'_{n,i}, \quad t_n\in\overline{Z}_N. \end{aligned}$$

Assume now that the first $m-1$ collocation parameters $\{c_1, \ldots ,c_{m-1}\}$ are the zeros of the Legendre polynomial $P_{m-1}(2s-1)$ (i.e., the $m-1$ Gauss points in $(0,1)$). This implies that, under the smoothness hypotheses for g and K and by the above argument used to derive the (piecewise) smoothness properties of the integrand $R(t_n,\cdot)\delta(\cdot)$, $E'_{n,i}=O(h_i^{2(m-1)})$. Since the last collocation parameter is $c_m=1$, we have $\delta(t_n)=0$ for all $t_n\in\overline{Z}_N$. Consequently, there exists a constant C_0' such that $|e(t_n)|\leqslant C_0'h^{2(m-1)}$, $t\in\overline{Z}_N$. This completes the proof of Theorem 5.3.3. □

The arguments employed to establish the above local superconvergence results reveal that the collocation error at each point of $\bar{Z}_N$ is of the order given by the error term associated with the interpolatory quadrature formula whose abscissas are based on the collocation parameters $\{c_j\}$, provided that the last of these parameters satisfies $c_m=1$. This necessary condition for local superconvergence reminds us of Theorem 5.2.2 which states that the suitably discretized collocation equation for the special integral equation (5.2.10) (the integrated form of the initial-value problem (5.2.1)) is identical with an m-stage implicit Runge-Kutta method for the initial-value-problem (5.2.1) if, and only if, $c_m=1$. On the other hand, it is well known that an m-stage implicit Runge-Kutta method for (5.2.1) exhibits the optimal order $p^*=2m$ only if the parameters $\{c_j\}$ are the Gauss points for $(0,1)$. Since this means $c_m<1$, this order of local superconvergence on $\bar{Z}_N$ cannot occur for Volterra integral equations of the second kind. However, it will be shown in Section 5.3.6 that the local order $p^*=2m$ can be attained on $\bar{Z}_N$ by an appropriate *iteration* of the collocation approximation $u \in S_{m-1}^{(-1)}(Z_N)$ employing the Gauss points as the collocation parameters.

It is clear that the list of locally superconvergent collocation methods given in Theorem 5.3.3 is not exhaustive: we need only fix the parameter $c_m=1$, as indicated in part (d). Among the many other possibilities we mention the case where the collocation parameters are *equidistant,* i.e., where $c_j=(j-1)/(m-1)$ $(j=1, \ldots ,m)$. The m-point quadrature formulas based on these parameters are the *Newton-Cotes* formulas whose remainder terms $E_{n,i}$ in (5.3.28) satisfy

$$|E_{n,i}| \leqslant \begin{cases} C_m h_i^m, & \text{if } m \text{ is even,} \\ \\ C_m' h_i^{m+1}, & \text{if } m \text{ is odd,} \end{cases} \tag{5.3.31}$$

provided g and K are sufficiently smooth (compare Table 2.5.1). Therefore, if m is odd, and if the m collocation parameters $\{c_j\}$ are equally spaced (implying in particular $c_1=0$, $c_m=1$), then there results local superconvergence of order $p^*=m+1$ on $\bar{Z}_N$. For example, if $m=3$, then the global order of convergence is given by $p=m=3$, while on $\bar{Z}_N$ we have the order $p^*=m+1=4$. If $c_2=\frac{1}{2}$, and if the integrals $\Phi_{n,i}^{(j)}[u_i]$ in (5.3.7) are approximated by 3-point interpolatory quadrature, the corresponding method reduces to the *block-by-block method of Simpson* in Example 3.2.11 (note that the collocation method with $m=3$ and with $c_1=0$, $c_2=\frac{1}{2}$, $c_3=1$, can also be viewed as a method based on the Lobatto points; according to (5.3.22) of Theorem 5.3.3 we then have $p^*=2m-2=4$).

In *Example 5.3.1.* $(m=2)$ we have the global order $p=m=2$, while the following local orders (on $\bar{Z}_N$) can be attained:

(i) $p^*=2m-1=3$, if $c_1=1/3$, $c_2=1$ (Radau II points);

(ii) $p^*=2m-2=p=2$, if $c_1=0$, $c_2=1$ (Lobatto points);

(iii) $p^*=2(m-1)=p=2$, if $c_1=1/2$ (Gauss point), $c_2=1$.

In *Example 5.3.2.* ($m=3$) the global order is given by $p=m=3$; on $\bar{Z}_N$ we have:

(i) $p^*=2m-1=5$, if $c_1=(4-\sqrt{6})/10$, $c_2=(4+\sqrt{6})/10$, $c_3=1$ (Radau II points);

(ii) $p^*=2m-2=4$, if $c_i=0$, $c_2=1/2$, $c_3=1$ (Lobatto points);

(iii) $p^*=2(m-1)=4$, if $c_1=(3-\sqrt{3})/6$, $c_2=(3+\sqrt{3})/6$ (Gauss points), $c_3=1$.

Here, case (iii) illustrates part (d) of Theorem 5.3.3.

We conclude this section by observing that, in analogy to the results of section 5.3.2, the statements of Theorem 5.3.3 remain valid for the nonlinear Volterra integral equation (1.1.1), provided that $k(t,s,y)$ satisfies the appropriate smoothness and boundedness conditions (compare the remark following the proof of Theorem 5.3.2). We omit the details (see, however, the proof of Theorem 5.3.8 below).

5.3.4. Convergence results for the discretized collocation equation

The convergence results established in the previous section were derived under the assumption that the integrals (5.3.6) occurring in the collocation equation (5.3.7) can be evaluated analytically. Since this is rarely possible in concrete applications, there arises the question as to how to approximate these integrals. The discretized collocation equation (5.3.7) will then yield an approximation $\hat{u}\in S_{m-1}^{(-1)}(Z_N)$ which, in general, is different from u. Setting $\hat{e}(t):=y(t)-\hat{u}(t)$, $\epsilon(t):=u(t)-\hat{u}(t)$, we have

$$|\hat{e}(t)| \leqslant |y(t)-u(t)| + |u(t)-\hat{u}(t)| = |e(t)| + |\epsilon(t)|, \quad t\in I, \tag{5.3.32}$$

where, under the hypotheses of Theorem 5.3.2, $|e(t)|\leqslant Ch^m$ uniformly on I. Thus, we wish to choose quadrature processes for the integrals (5.3.6) for which the perturbation $\epsilon(t)$ has at least the order of the collocation error $e(t)$. To be more specific, let μ_0 and μ_1 be given positive integers, and suppose that the integrals (5.3.6) are approximated by quadrature formulas of the form

$$\hat{\Phi}_{n,i}^{(j)}[u_i] := \begin{cases} \sum_{l=1}^{\mu_1} w_l k(t_{n,j},t_i+d_l h_i,u_i(t_i+d_l h_i)), & \text{if } 0\leqslant i<n, \\ \sum_{l=1}^{\mu_0} w_{j,l} k(t_{n,j},t_n+d_{j,l}h_n,u_n(t_n+d_{j,l}h_n)), & \text{if } i=n, \end{cases} \tag{5.3.33}$$

where $j=1, \ldots ,m$. These quadrature formulas are usually interpolatory ones, with the quadrature parameters $\{d_l\}$ and $\{d_{j,l}\}$ satisfying, respectively, $0\leqslant d_1< \ldots <d_{\mu_1}\leqslant 1$, and $0\leqslant d_{j,1}< \ldots <d_{j,\mu_0}\leqslant c_j$ ($j=1, \ldots ,m$). The quadrature weights are then given by

$$w_l = \int_0^1 \prod_{\substack{r=1\\ r\neq l}}^{\mu_1} (v-d_r)/(d_l-d_r)dv, \quad l=1, \ldots ,\mu_1,$$

and

$$w_{j,l} = \int_0^{c_j} \prod_{\substack{r=1 \\ r\neq l}}^{\mu_0} (v-d_{j,r})/(d_{j,l}-d_{j,r})dv, \quad l=1, \dots ,\mu_0; \quad j=1, \dots ,m.$$

The corresponding error terms are defined by

$$E_{n,i}^{(j)}[u_i] := \Phi_{n,i}^{(j)}[u_i]-\hat{\Phi}_{n,i}^{(j)}[u_i], \quad j=1, \dots ,m \quad (i=0, \dots ,n), \tag{5.3.34}$$

with $\Phi_{n,i}^{(j)}[u_i]$ and $\hat{\Phi}_{n,i}^{(j)}[u_i]$ given by (5.3.6) and (5.3.33).

We now use these quadrature formulas to obtain the *fully discretized version* of the collocation equation (5.3.7). Since the quadrature error terms will be disregarded we generate an approximation $\hat{u} \in S_{m-1}^{(-1)}(Z_N)$ which is defined by

$$\hat{u}_n(t_{n,j}) = g(t_{n,j})+h_n\hat{\Phi}_{n,n}^{(j)}[\hat{u}_n]+\sum_{i=0}^{n-1} h_i\hat{\Phi}_{n,i}^{(j)}[\hat{u}_i],$$
$$j=1, \dots ,m \quad (n=0, \dots ,N-1), \tag{5.3.35}$$

and which will, in general, be different from the approximation $u \in S_{m-1}^{(-1)}(Z_n)$ given by the exact collocation equation (5.3.7). The order of this difference, $\epsilon := u-\hat{u}$, will depend on the choice of the quadrature formulas (5.3.33), as described in the following theorem.

Theorem 5.3.4. *Let g and K in (5.3.9) be at least m times continuously differentiable on I and S, respectively, and let $u \in S_{m-1}^{(-1)}(Z_N)$ denote the approximation to the solution y defined by the (exact) collocation equation (5.3.7). Assume that the quadrature formulas (5.3.33) satisfy*

$$\int_0^1 \phi(t_i+\tau h_i)d\tau-\sum_{l=1}^{\mu_1} w_l\phi(t_i+d_lh_i) = O(h_i^{r_1}), \tag{5.3.36a}$$

and, for $j=1, \dots ,m$,

$$\int_0^{c_j} \phi(t_n+\tau h_n)d\tau-\sum_{l=1}^{\mu_0} w_{j,l}\phi(t_n+d_{j,l}h_n) = O(h_n^{r_0}), \tag{5.3.36b}$$

whenever the integrand is a sufficiently smooth function. If $\hat{u} \in S_{m-1}^{(-1)}(Z_N)$ is the approximation given by the discretized collocation equation (5.3.36), then there exists a finite constant Q such that $\epsilon := u-\hat{u}$ satisfies

$$\|\epsilon\|_\infty \leq Qh^r, \quad \textit{with} \quad r:=\min(r_0+1,r_1), \tag{5.3.37}$$

for all quasi-uniform meshes with sufficiently small $h>0$.

Proof. Let $\epsilon_n(t)$ denote the restriction of $\epsilon(t)$ to the subinterval σ_n. Subtraction of (5.3.35) from (5.3.7) and observing (5.3.34) furnishes, with $k(t,s,y)=K(t,s)y$, the recurrence relation

$$\epsilon_n(t_{n,j}) = h_n\sum_{l=1}^{\mu_0} w_{j,l}K_{n,j}(t_n+d_{j,l}h_n)\epsilon_n(t_n+d_{j,l}h_n)+h_nE_{n,n}^{(j)}[u_n]$$

$$+\sum_{i=0}^{n-1} h_i \sum_{l=1}^{\mu_1} w_l K_{n,j}(t_i+d_l h_i)\epsilon_i(t_i+d_l h_i)+\sum_{i=0}^{n-1} h_i E_{n,i}^{(j)}[u_i], \qquad (5.3.38)$$

$$j=1, \ldots ,m.$$

This equation may be viewed as the discrete analogue of (5.3.15). In order to make the subsequent analysis more lucid, we express (5.3.38) in more compact matrix notation. Let $Q_{n,i}$, $0 \leqslant i \leqslant n \leqslant N-1$, be square matrices of order m whose elements are

$$q_{j,l}^{(n,i)} := \begin{cases} \sum_{s=1}^{\mu_1} w_s K_{n,j}(t_i+d_s h_i)L_l(d_s), & \text{if } 0 \leqslant i \leqslant n-1, \\ \sum_{s=1}^{\mu_0} w_{j,s} K_{n,j}(t_n+d_{j,s} h_n)L_l(d_{j,s}), & \text{if } i=n \end{cases}$$

$$(j,l=1, \ldots ,m);$$

here, we have used the fact that ϵ_n is a polynomial of degree $m-1$, implying

$$\epsilon_n(t_n+\tau h_n) = \sum_{l=1}^{m} \epsilon_n(t_{n,l})L_l(\tau), \quad t_n+\tau h_n \in \sigma_n. \qquad (5.3.39)$$

Furthermore, we introduce the vectors $\mathbf{r}_{n,i} := (E_{n,i}^{(1)}[u_i], \ldots ,E_{n,i}^{(m)}[u_i])^T$ and $\boldsymbol{\eta}_i :=$ $(\epsilon_i(t_{i,1}) \ldots ,\epsilon_i(t_{i,m}))^T$. With this terminology, (5.3.38) becomes

$$(I-h_n Q_{n,n})\boldsymbol{\eta}_n = \sum_{i=0}^{n-1} h_i Q_{n,i}\boldsymbol{\eta}_i + h_n \mathbf{r}_{n,n} + \sum_{i=0}^{n-1} h_i \mathbf{r}_{n,i}, \quad n=0, \ldots ,N-1. \quad (5.3.40)$$

It follows from the assumptions on the order of the quadrature formulas (5.3.34) and from the smoothness hypothesis on the kernel that there exist constants Q_0 and Q_1 such that

$$\|\mathbf{r}_{n,n}\|_1 \leqslant Q_0 h_n^{r_0}, \quad \|\mathbf{r}_{n,i}\|_1 \leqslant Q_1 h_i^{r_1} \quad (0 \leqslant i < n \leqslant N-1).$$

Also, since the quadrature weights are assumed to be bounded, and since the same holds for the kernel $K(t,s)$ on S, there is a constant $\hat{D}_0 < \infty$ with $\|(I-h_n Q_{n,n})^{-1}\|_1 \leqslant \hat{D}_0$, $n=0, \ldots ,N-1$, for all sufficiently small $h_n > 0$. Analogously, there exists a constant $\hat{D}_1$ for which, $\|Q_{n,i}\|_1 \leqslant \hat{D}_1$, $0 \leqslant i < n \leqslant N-1$. Thus, (5.3.40) leads to

$$\|\boldsymbol{\eta}_n\|_1 \leqslant \hat{D}_0\{h\hat{D}_1 \sum_{i=0}^{n-1} \|\boldsymbol{\eta}_i\|_1 + hQ_0 h^{r_0} + NhQ_1 h^{r_1}\}, \quad n=0, \ldots ,N-1$$

(with $h_n \leqslant h$, $Nh \leqslant \gamma T$). This represents a discrete Gronwall inequality for $\|\boldsymbol{\eta}_n\|_1$,

$$\|\boldsymbol{\eta}_n\|_1 \leqslant h\hat{C}_0 \sum_{i=0}^{n-1} \|\boldsymbol{\eta}_i\|_1 + \hat{C}_1 h^r, \quad n=0 ,\ldots ,N-1. \qquad (5.3.41)$$

where

$$\hat{C}_0 := \hat{D}_0\hat{D}_1, \quad \hat{C}_1 := \hat{D}_0(Q_0h^{r_0+1-r}+TQ_1h^{r_1-r}),$$

with $r:=\min(r_0+1,r_1)$. Thus, by Corollary 1.5.2, it follows that $\|\boldsymbol{\eta}_n\|_1 \leqslant Q_2h^r$, $n=0, \ldots, N-1$, with $Nh\leqslant\gamma T$ and $h>0$ sufficiently small. To complete the proof we return to (5.3.39): if

$$\Lambda_m := \max\{\sum_{l=1}^{m} |L_l(\tau)| : 0\leqslant\tau\leqslant1\}$$

denotes the *Lebesgue constant* associated with the collocation parameters $\{c_j\}$, then we have, by (5.3.41),

$$|\epsilon_n(t_n+\tau h_n)| \leqslant \sum_{l=1}^{m} |\epsilon_n(t_{n,l})|\,|L_l(\tau)| \leqslant \|\boldsymbol{\eta}_n\|_1\Lambda_m \leqslant Qh^r,$$

where $t_n+\tau h_n \in \sigma_n$, $n=0, \ldots, N-1$, and with $Q:=Q_2\Lambda_m$. This is equivalent to (5.3.37). □

Theorem 5.3.4 remains valid for nonlinear integral equations of the form (1.1.1); here again, the remark following the proof of Theorem 5.3.2 applies.

Corollary 5.3.1. *Let the assumptions of Theorem 5.3.4 hold. If the quadrature formulas (5.3.33) are of interpolatory type, with $\mu_0=\mu_1=m$, then the approximation $\hat{u}\in S_{m-1}^{(-1)}(Z_N)$ defined by the discretized collocation equation (5.3.35) leads to an error $\hat{e}(t)$ satisfying*

$$\|\hat{e}\|_\infty = O(h^m) \quad (as \quad h\downarrow 0,\ Nh\leqslant\gamma T), \tag{5.3.42}$$

for every choice of the collocation parameters $\{c_j\}$ with $0\leqslant c_1< \ldots <c_m\leqslant 1$.

Proof. According to Theorem 2.1.3 and (2.1.13), an m-point interpolatory quadrature formula is characterized, in the terminology of Theorem 5.3.4, by $\min(r_0,r_1)\geqslant m$. Hence, we have $r\geqslant m$. If we now combine the results of Theorem 5.3.2 and Theorem 5.3.4 in (5.3.32), we obtain

$$|\hat{e}(t)| \leqslant Ch^m + Qh^r = O(h^m)$$

uniformly for $t\in I$. □

The global convergence result (5.3.42) holds in particular for the case where we have $d_l=c_l$, $d_{j,l}=c_jc_l$ ($l=1, \ldots, m$; $j=1, \ldots, m$). The quadrature approximations in the discretized collocation equation (5.3.35) are then given by

$$\hat{\Phi}_{n,i}^{(j)}[\hat{u}_i] = \begin{cases} \sum_{l=1}^{m} b_l k(t_{n,j}, t_i+c_lh_i, \hat{Y}_{i,l}), & \text{if} \quad 0\leqslant i\leqslant n-1, \\ \sum_{l=1}^{m} \bar{a}_{j,l} k(t_{n,j}, t_n+c_jc_lh_n, \sum_{s=1}^{m} L_s(c_jc_l)\hat{Y}_{n,s}), & \text{if} \quad i=n, \end{cases} \tag{5.3.43}$$

where $\bar{a}_{j,l}=c_j b_l$, with $b_l=\int_0^1 L_l(\tau)d\tau$, and where we have set $\hat{u}_n(t_n+\tau h_n)=\sum_{s=1}^m \hat{Y}_{n,s}L_s(\tau)$, $\hat{Y}_{n,s}:=\hat{u}_n(t_{n,s})$. This discretization has already been mentioned in Section 3.2.12, where we discussed the problem of kernel values, outside the domain of definition S, occurring in block-by-block methods. It is easily seen that (5.3.43) does not require kernel values outside S. Runge-Kutta methods employing this discretization, will be called *modified methods of de Hoog and Weiss.* The modified method of de Hoog & Weiss [1975], characterized by $0\leqslant c_1< \ldots <c_m=1$, possesses the global order of convergence $p=m$.

If the kernel $k(t,s,y)$ can be smoothly extended to $S':=\{(t,s)\colon 0\leqslant s\leqslant t+\delta\}\cap I\times I$, for some suitable $\delta>0$, then the integrals $\hat{\Phi}^{(j)}_{n,n}[u_n]$ in (5.3.6) may be approximated by choosing $\mu_0=m$, $d_{j,l}=c_l$:

$$\hat{\Phi}^{(j)}_{n,n}[\hat{u}_n] = \sum_{l=1}^{m} a_{j,l}k(t_{n,j},t_n+c_lh_n,\hat{Y}_{n,l}),$$

with $a_{j,l}=\int_0^{c_j}L_l(v)dv$. The corresponding discretized collocation equation (5.3.35) then represents an implicit, extended *Volterra-Runge-Kutta method of Pouzet type,* provided that $c_m=1$. We have introduced this class of methods in Section 4.1.1 (compare equations (4.1.8a) and (4.1.8b)).

Before turning our attention to the analogue of Theorem 5.3.3 for the fully discretized collocation equation (5.3.35), we summarize the most common discretization procedures for the integrals $\Phi^{(j)}_{n,n}[u_i]$ in (5.3.7). For ease of notation we use the same generic constants $a_{j,l}$ to denote the Runge-Kutta weights.

Table 5.3.1. Discretization of $\Phi^{(j)}_{n,n}[u_n]$

Type	μ_0	$\hat{\Phi}^{(j)}_{n,n}[u_n]$
Explicit	$j-1$	$\sum_{l=1}^{j-1} a_{j,l}k(t_{n,j},t_{n,l},Y_{n,l})$
Diagonally implicit	j	$\sum_{l=1}^{j} a_{j,l}k(t_{n,j},t_{n,l},Y_{n,l})$
Implicit	m	$\sum_{l=1}^{m} a_{j,l}k(t_{n,j},t_{n,l},Y_{n,l})$
Fully implicit	m	$\sum_{l=1}^{m} a_{j,l}k(t_{n,j},t_n+c_jc_lh_n,\sum_{s=1}^{m} L_s(c_jc_l)Y_{n,s})$

Order conditions for explicit and diagonally implicit Volterra-Runge-Kutta methods have been given in Section 4.1.4, in the context of general order results for VRK methods. Recall, in particular, that if the parameters defining an extended VRK method of Pouzet type are those of an m-stage RK method of order r for an ODE, then the PVRK method possesses the same order, provided $c_m = 1$ (compare Theorem 4.1.1).

We now deal with the question of the attainable order of local superconvergence (on $\bar{Z}_N$) in approximations $\hat{u} \in S_{m-1}^{(-1)}(Z_N)$ defined by the fully discretized collocation equation (5.3.35). As before, we state the results for the linear integral equation (5.3.9); their modification for the general case (1.1.1) is once more straightforward. It is again assumed that the underlying mesh sequence is quasi-uniform.

Theorem 5.3.5. *Assume that g and K in (5.3.9) are, respectively, in $C^{2m-\nu}(I)$ and $C^{2m-\nu}(S)$, with $\nu \in \{1,2\}$, and with $m \geq \lfloor \nu/2 \rfloor + 1$. Let $\hat{u} \in S_{m-1}^{(-1)}(Z_N)$ be defined by the fully discretized collocation equation (5.3.35).*
(a) *If the collocation parameters $\{c_j\}$ are the zeros of $P_{m-1}(2s-1) - P_m(2s-1)$ (Radau II points for $(0,1]$), and if the quadrature formulas employed in (5.3.33) are of interpolatory type and characterized by $\mu_0 = \mu_1 = m$, $d_l = c_l$, $d_{j,l} = c_j c_l$ $(l, j = 1, \ldots, m)$, then, for $\nu = 1$,*

$$\max_{t_n \in \bar{Z}_N} |\hat{e}(t_n)| = O(h^{2m-1}) \quad (\text{as } h \downarrow 0,\ Nh \leq \gamma T). \tag{5.3.44}$$

(b) *If the collocation parameters $\{c_j\}$ are given by the zeros of $s(1-s)P'_{m-1}(2s-1)$ (Lobatto points for $[0,1]$), and if the quadrature formulas in (5.3.33) are of the form described under (a), then, for $\nu = 2$,*

$$\max_{t_n \in \bar{Z}_N} |\hat{e}(t_n)| = O(h^{2m-2}) \quad (\text{as } h \downarrow 0,\ Nh \leq \gamma T). \tag{5.3.45}$$

(c) *Let the first $m-1$ collocation parameters $\{c_j\}$ be defined by the zeros of $P_{m-1}(2s-1)$ (i.e., the $m-1$ Gauss points for $(0,1)$), and let $c_m = 1$. Furthermore, suppose that the quadrature formulas used in (5.3.33) are interpolatory $(m-1)$-point formulas characterized by $\mu_0 = \mu_1 = m-1$, $d_l = c_l$ $(l = 1, \ldots, m-1)$, $d_{j,l} = c_j c_l$ $(l = 1, \ldots, m-1;\ j = 1, \ldots, m)$. Then, for $\nu = 2$, (5.3.45) holds.*

Proof. The above assertions follow readily from (5.3.32), using Theorem 5.3.3 (parts (a), (b), and (d)) and Theorem 5.3.4. We omit the details. □

The arguments employed in the proofs of the above global and local convergence results obviously permit the derivation of such convergence results for approximations defined by other discretizations of the collocation equation (5.3.36); e.g., by selecting quadrature parameters $\{d_l\}$ which are different from the collocation parameters $\{c_l\}$, or by choosing $\mu_0 = \mu_1 - 1$ (with $\mu_1 \geq 2$). Since it has been our aim to exhibit the essential ideas behind the interconnections between the exact and the discretized collocation equations and their corresponding approximations, we leave the construction of additional methods to the reader.

5.3.5. Examples

In this section we present a number of specific discretizations of the collocation equation (5.3.7); these methods yield approximations which are locally superconvergent on $\bar{Z}_N$, thus illustrating parts (a), (b), and (c) of Theorem 5.3.5. As before, we write $\hat{u}_n(t_n+\tau h_n)=\Sigma_{l=1}^{m} L_l(\tau)\hat{Y}_{n,l}$, with $\hat{Y}_{n,l}:=\hat{u}_n(t_{n,l})$, and we set $\hat{y}_{n+1}:=\hat{u}_n(t_n+h_n)$.

$\mu_0=\mu_1=m$.

We first consider discretizations based on the quadrature formulas (5.3.7): these formulas are m-point interpolatory formulas using, respectively, the abscissas $\{t_{i,l}\}$ (if $0\leqslant i<n$) and $\{t_n+c_jc_lh_n\}$.

Example 5.3.3. $m=2$; $c_1=1/3$, $c_2=1$ (Radau II points).

The corresponding discretization of (5.3.7) (compare also Example 5.3.1) is found to be

$$\hat{Y}_{n,1} = g(t_{n,1})+\sum_{i=0}^{n-1} h_i\hat{\Phi}_{n,i}^{(1)}[\hat{u}_i]+\frac{h_n}{12}\{3k(t_{n,1},t_n+\frac{h_n}{9},\tfrac{4}{3}\hat{Y}_{n,1}-\tfrac{1}{3}\hat{Y}_{n,2})$$
$$+k(t_{n,1},t_n+\frac{h_n}{3},\hat{Y}_{n,1})\};$$

$$\hat{Y}_{n,2} = g(t_{n,2})+\sum_{i=0}^{n-1} h_i\hat{\Phi}_{n,i}^{(2)}[\hat{u}_i]+\frac{h_n}{4}\{3k(t_{n,2},t_n+\frac{h_n}{3},\hat{Y}_{n,1})$$
$$+k(t_{n,2},t_n+h_n,\hat{Y}_{n,2})\};$$

$$\hat{y}_{n+1} = \hat{Y}_{n,2}, \quad n=0,\ldots,N-1$$

Here, we have

$$\hat{\Phi}_{n,i}^{(j)}[\hat{u}_i] = \tfrac{1}{4}\{3k(t_{n,j},t_{i,1},\hat{Y}_{i,1})+k(t_{n,j},t_{i,2},\hat{Y}_{i,2})\}, \quad j=1,2, \quad i<n.$$

According to Corollary 5.3.1, the global order of convergence of this method is $p=m$, while its local order on $\bar{Z}_N$ is given by $p^*=2m-1=3$ (Theorem 5.3.5, part (a)). □

The next two examples exhibit specific discretizations of the collocation equations treated in Example 5.3.2.

Example 5.3.4. $m=3$, $c_1=0$, $c_2=½$, $c_3=1$ (Lobatto points).

The resulting discretized collocation equation has the form:

$$\hat{Y}_{n,1} = g(t_n)+\sum_{i=0}^{n-1} h_i\hat{\Phi}_{n,i}^{(1)}[\hat{u}_i];$$

$$\hat{Y}_{n,2} = g(t_{n,2})+\sum_{i=0}^{n-1} h_i\hat{\Phi}^{(2)}_{n,i}[\hat{u}_i]+\frac{h_n}{12}\{k(t_{n,2},t_n,\hat{Y}_{n,1})$$
$$+4k(t_{n,2},t_n+\frac{h_n}{4},\frac{3}{8}\hat{Y}_{n,1}+\frac{3}{4}\hat{Y}_{n,2}-\frac{1}{8}\hat{Y}_{n,3})+k(t_{n,2},t_n+\frac{h_n}{2},\hat{Y}_{n,2})\};$$
$$\hat{Y}_{n,3} = g(t_{n,3})+\sum_{i=0}^{n-1} h_i\hat{\Phi}^{(3)}_{n,i}[\hat{u}_i]+\frac{h_n}{6}\{k(t_{n,3},t_n,\hat{Y}_{n,1})$$
$$+4k(t_{n,3},t_n+\frac{h_n}{2},\hat{Y}_{n,2})+k(t_{n,3},t_n+h_n,\hat{Y}_{n,3})\};$$
$$\hat{y}_{n+1} = \hat{Y}_{n,3},\quad n=0,\ \dots\ ,N-1.$$

Here, we have, for $0\leqslant i<n$,

$$\hat{\Phi}^{(j)}_{n,i}[\hat{u}_i] = \frac{1}{6}\{k(t_{n,j},t_i,\hat{Y}_{i,1})+4k(t_{n,j},t_{i,2},\hat{Y}_{i,2})+k(t_{n,j},t_{i,3},\hat{Y}_{i,3})\},\quad j=1,2,3.$$

It follows from Corollary 5.3.1 that this method possesses the global convergence order $p=m=3$; it exhibits local superconvergence of order $p^*=2m-2=4$ on $\bar{Z}_N$ (Theorem 5.3.5, part (b)). □

Example 5.3.5. $m=3$, $c_1=(4-\sqrt{6})/10$, $c_2=(4+\sqrt{6})/10$, $c_3=1$ (Radau II points).

The resulting discretized collocation equation is given by

$$\hat{Y}_{n,1} = g(t_{n,1})+\sum_{i=0}^{n-1} h_i\hat{\Phi}^{(1)}_{n,i}[\hat{u}_i]$$
$$+\frac{h_n}{180}\{(35-10\sqrt{6})k(t_{n,1},t_n+\frac{11-4\sqrt{6}}{50}h_n,\frac{12\sqrt{6}+117}{100}\hat{Y}_{n,1}$$
$$-\frac{108\sqrt{6}+103}{500}\hat{Y}_{n,2}+\frac{24\sqrt{6}+9}{250}\hat{Y}_{n,3})$$
$$+(29-6\sqrt{6})k(t_{n,1},t_n+\frac{h_n}{10},\frac{6\sqrt{6}+9}{20}\hat{Y}_{n,1}-\frac{6\sqrt{6}-9}{20}\hat{Y}_{n,2}+\frac{1}{10}\hat{Y}_{n,3})$$
$$+(8-2\sqrt{6})k(t_{n,1},t_n+\frac{4-\sqrt{6}}{10}h_n,\hat{Y}_{n,1})\};$$

$$\hat{Y}_{n,2} = g(t_{n,2})+\sum_{i=0}^{n-1} h_i\Phi^{(2)}_{n,i}[\hat{u}_i]$$
$$+\frac{h_n}{180}\{29+6\sqrt{6})k(t_{n,2},t_n+\frac{h_n}{10},\frac{6\sqrt{6}+9}{20}\hat{Y}_{n,1}-\frac{6\sqrt{6}-9}{20}\hat{Y}_{n,2}+\frac{1}{10}\hat{Y}_{n,3})$$
$$+(35+10\sqrt{6})k(t_{n,2},t_n+\frac{11+4\sqrt{6}}{50}h_n,\frac{108\sqrt{6}-103}{500}\hat{Y}_{n,1}$$
$$+\frac{117-12\sqrt{6}}{100}\hat{Y}_{n,2}-\frac{24\sqrt{6}-9}{250}\hat{Y}_{n,3})+(8+2\sqrt{6})k(t_{n,2},t_n+\frac{4+\sqrt{6}}{10}h_n,\hat{Y}_{n,2})\};$$

$$\hat{Y}_{n,3} = g(t_{n,3}) + \sum_{i=0}^{n-1} h_i \hat{\Phi}_{n,i}^{(3)}[\hat{u}_i]$$
$$+\frac{h_n}{36}\{(16-\sqrt{6})k(t_{n,3}, t_n + \frac{4-\sqrt{6}}{10}h_n, \hat{Y}_{n,1})$$
$$+(16+\sqrt{6})k(t_{n,3}, t_n + \frac{4+\sqrt{6}}{10}h_n, \hat{Y}_{n,2}) + 4k(t_{n,3}, t_n + h_n, \hat{Y}_{n,3})\};$$

$$\hat{y}_{n+1} = \hat{Y}_{n,3}, \quad n = 0, \ldots, N-1.$$

Here,

$$\hat{\Phi}_{n,i}^{(j)}[\hat{u}_i] = \frac{1}{36}\{(16-\sqrt{6})k(t_{n,j}, t_{i,1}, \hat{Y}_{i,1}) + (16+\sqrt{6})k(t_{n,j}, t_{i,2}, \hat{Y}_{i,2})$$
$$+4k(t_{n,j}, t_{i,3}, \hat{Y}_{i,3})\}, \quad j = 1,2,3 \quad (0 \leqslant i < n).$$

This method possesses the global convergence order $p = m = 3$; its local superconvergence order on $\overline{Z}_N$ is $p^* = 2m-1 = 5$ (Theorem 5.3.5 (a)). It was also derived in de Hoog & Weiss [1975], using product-integration techniques.□

$\mu_0 = \mu_1 = m-1$.

The following two examples illustrate part (c) of Theorem 5.3.5: the first $(m-1)$ collocation parameters are the Gauss points for $(0,1)$, and $c_m = 1$. The quadrature formulas (5.3.33) thus employ $\mu_0 = \mu_1 = m-1$; this implies, in the terminology of Theorem 5.3.4, that $r_0 = r_1 = 2m-2$. Hence, it follows from Theorem 5.3.5 (c) that these discretizations exhibit local superconvergence on $\overline{Z}_N$ of order $p^* = 2m-2$.

Example 5.3.6. $m = 2$; $c_1 = ½$, $c_2 = 1$.

We find:

$$\hat{Y}_{n,1} = g(t_{n,1}) + \sum_{i=0}^{n-1} h_i \hat{\Phi}_{n,i}^{(1)}[\hat{u}_i] + \frac{h_n}{2}k(t_{n,1}, t_n + \frac{h_n}{4}, \frac{3}{2}\hat{Y}_{n,1} - \frac{1}{2}\hat{Y}_{n,2});$$

$$\hat{y}_{n+1} = \hat{Y}_{n,2} = g(t_{n+1}) + \sum_{i=0}^{n-1} h_i \hat{\Phi}_{n,i}^{(2)}[\hat{u}_i] + h_n k(t_{n+1}, t_n + \frac{h_n}{2}, \hat{Y}_{n,1})$$
$$(n = 0, \ldots, N-1),$$

with

$$\hat{\Phi}_{n,i}^{(j)}[\hat{u}_i] = k(t_{n,j}, t_{i,1}, \hat{Y}_{i,1}), \quad j = 1,2 \quad (0 \leqslant i < m).$$

Hence, $p = m = 2 = 2m-2 = p^*$.

We note that in this case ($\hat{u} \in S_1^{(-1)}(Z_N)$) we could also have used the quadrature approximation

$$\hat{\Phi}_{n,n}^{(1)}[\hat{u}_n] = \frac{1}{2}k(t_n + \frac{h_n}{2}, t_n + \frac{h_n}{2}, \hat{Y}_{n,1}),$$

which corresponds to (5.3.34) with $d_{1,1}=c_1$ (instead of $d_{1,1}=c_1c_1$ used above). This yields $r_0=1$ (instead of $r_0=2$) which, according to Theorem 5.3.4, does not reduce the order of convergence. The resulting discretization becomes

$$\hat{Y}_{n,1} = g(t_n+\frac{h_n}{2})+\sum_{i=0}^{n-1} h_i\hat{\Phi}^{(1)}_{n,i}[\hat{u}_i]+\frac{h_n}{2}k(t_n+\frac{h_n}{2},t_n+\frac{h_n}{2},\hat{Y}_{n,1}),$$

with $\hat{Y}_{n,2}$ and $\hat{y}_{n+1}$ as before. This method is called the *Pouzet midpoint method* (see also Brunner, Hairer & Nørsett [1982, p. 117]). □

Example 5.3.7. $m=3$; $c_1=(3-\sqrt{3})/6$, $c_2=(3+\sqrt{3})/6$, $c_3=1$.

The corresponding discretized collocation equation is:

$$\begin{aligned}\hat{Y}_{n,1} = {} & g(t_{n,1})+\sum_{i=0}^{n-1} h_i\hat{\Phi}^{(1)}_{n,i}[\hat{u}_i] \\ & +\frac{3-\sqrt{3}}{12}h_n\{k(t_{n,1},t_n+\frac{2-\sqrt{3}}{6}h_n,\frac{\sqrt{3}+17}{12}\hat{Y}_{n,1} \\ & -\frac{5\sqrt{3}+7}{12}\hat{Y}_{n,2}+\frac{2\sqrt{3}+1}{6}\hat{Y}_{n,3})+k(t_{n,1},t_n+\frac{h_n}{6},\frac{5(1+\sqrt{3})}{12}\hat{Y}_{n,1} \\ & -\frac{5(\sqrt{3}-1)}{12}\hat{Y}_{n,2}+\frac{1}{6}\hat{Y}_{n,3})\};\end{aligned}$$

$$\begin{aligned}\hat{Y}_{n,2} = {} & g(t_{n,2})+\sum_{i=0}^{n-1} h_i\hat{\Phi}^{(2)}_{n,i}[\hat{u}_i] \\ & +\frac{3+\sqrt{3}}{12}h_n\{k(t_{n,2},t_n+\frac{h_n}{6},\frac{5(\sqrt{3}+1)}{12}\hat{Y}_{n,1} \\ & -\frac{5(\sqrt{3}-1)}{12}\hat{Y}_{n,2}+\frac{1}{6}\hat{Y}_{n,3})+k(t_{n,2},t_n+\frac{2+\sqrt{3}}{6}h_n,\frac{5\sqrt{3}-7}{12}\hat{Y}_{n,1} \\ & +\frac{17-\sqrt{3}}{12}\hat{Y}_{n,2}-\frac{2\sqrt{3}-1}{6}\hat{Y}_{n,3})\};\end{aligned}$$

$$\begin{aligned}\hat{Y}_{n,3} = {} & g(t_{n,3})+\sum_{i=0}^{n-1} h_i\hat{\Phi}^{(3)}_{n,i}[\hat{u}_i] \\ & +\frac{h_n}{2}\{k(t_{n,3},t_n+\frac{3-\sqrt{3}}{6}h_n,\hat{Y}_{n,1})+k(t_{n,3},t_n+\frac{3+\sqrt{3}}{6}h_n,\hat{Y}_{n,2})\};\end{aligned}$$

$$\hat{y}_{n+1}=\hat{Y}_{n,3}, \quad n=0, \dots ,N-1.$$

Here, we have

$$\hat{\Phi}^{(j)}_{n,i}[\hat{u}_i] = \frac{1}{2}\{k(t_{n,j},t_{i,1},\hat{Y}_{i,1})+k(t_{n,j},t_{i,2},\hat{Y}_{i,2})\}, \quad j=1,2,3, \quad (0\leqslant i<n).$$

It follows from Theorem 5.3.5 (c) that $p^*=2m-2=4$; the global convergence order is, of course, $p=m=3$. □

5.3.6. Iterated collocation

Suppose that $u \in S_{m-1}^{(-1)}(Z_N)$ is an approximation to the exact solution of the integral equation of the second kind (5.3.9); in our context, u will be the collocation approximation given by the (exact) collocation equation (5.3.7). (More generally, u may represent an approximate solution to y obtained by different means; e.g., by Galerkin techniques.) The *iterated approximation* u^I corresponding to u is then defined by

$$u^I(t) := g(t) + \int_0^t K(t,s)u(s)ds, \quad t \in I. \tag{5.3.46}$$

The notion of iterated approximation was introduced by Sloan [1976] for more general operator equations (and, specifically, for Fredholm integral equations of the second kind); compare, in addition, Sloan [1980], [1984], Chatelin & Lebbar [1981], Chatelin [1983], and Joe [1985a], [1985b], [1985c].

It is immediately clear from (5.3.46) that

(i) u^I is no longer an element of $S_{m-1}^{(-1)}(Z_N)$ (in general, u^I will not even be a polynomial spline); and

(ii) u^I is a continuous function on the entire interval I, provided $g \in C(I)$ and $K \in C(S)$ (i.e., the jump discontinuities on Z_N present in u have been smoothed).

In addition, we have the following important result:

Lemma 5.3.2.

$$u^I(t) = u(t) \quad \textit{for all} \quad t \in X(N); \tag{5.3.47}$$

i.e., the iterated collocation approximation u^I coincides with the original collocation approximation u at all the collocation points.

The proof of this result is a consequence of the fact that $u \in S_{m-1}^{(-1)}(Z_N)$ satisfies

$$u(t) = g(t) + \int_0^t K(t,s)u(s)ds - \delta(t), \quad t \in I,$$

with the defect function $\delta(t)$ vanishing at the collocation points $t \in X(N)$ (see (5.3.24)). Hence, by (5.3.46) we have $u^I(t) - u(t) = \delta(t)$ for all $t \in I$, implying that $u^I(t) - u(t) = 0$ whenever $t \in X(N)$. □

Recall now that if $u \in S_{m-1}^{(-1)}(Z_N)$ is the collocation approximation obtained by employing the m *Gauss points* in each subinterval σ_n as the collocation points, then there does not result local superconvergence at the mesh points: we only have $y(t_n) - u(t_n) = O(h^m)$ for $t_n \in Z_N$ (Theorem 5.3.3, part (c)). However, for the corresponding iterated collocation approximation u^I the rate of convergence on Z_N is doubled. This is made precise in the following theorem.

Theorem 5.3.6. *Let g and K in (5.3.9) be $2m$ times continuously differentiable on their respective domains I and S, and let $u \in S_{m-1}^{(-1)}(Z_N)$ be the collocation approximation defined by (5.3.5), where the collocation parameters $\{c_j\}$ are the zeros of $P_m(2s-1)$ (Gauss points for $(0,1)$) and where the mesh sequence is quasi-uniform. Then the corresponding iterated collocation approximation u^I given by (5.3.46) satisfies*

$$\max_{t_n \in Z_N} |e^I(t_n)| = O(h^{2m}), \tag{5.3.48}$$

with $e^I := y - u^I$, as $h \downarrow 0$ and $Nh \leqslant \gamma T$.

Proof. By (5.3.9) and (5.3.46) the iterated collocation error e^I satisfies

$$e^I(t) = \int_0^t K(t,s) e(s) ds, \quad t \in I. \tag{5.3.49}$$

On the other hand, we have shown in (5.3.26) that the collocation error $e := y - u$ can be expressed in the form

$$e(t) = \delta(t) + \int_0^t R(t,s)\delta(s) ds, \quad t \in I,$$

with $R(t,s)$ being the resolvent kernel of the given kernel $K(t,s)$. If we now employ this expression on the right-hand side of (5.3.49), and if we then make use of the Fredholm identity (1.3.9b), we find that e^I can be written as

$$e^I(t) = \int_0^t R(t,s)\delta(s) ds, \quad t \in I. \tag{5.3.50}$$

Note that here, in contrast to the expression (5.3.26) for the collocation error e, the defect function δ occurs only under the integral sign. Hence, upon setting $t = t_n \in \bar{Z}_N$ in (5.3.50) we obtain, in complete analogy to (5.3.28),

$$e^I(t_n) = \sum_{i=0}^{n-1} h_i E_{n,i}, \quad t_n \in \bar{Z}_N; \tag{5.3.51}$$

here, the terms $E_{n,i}$ are the error terms of m-point Gaussian quadrature formulas with respect to the subintervals σ_i. Since the resolvent R inherits the smoothness of K (and is thus in $C^{2m}(S)$), and since the defect function δ has the same degree of smoothness on each subinterval σ_i, these quadrature errors $E_{n,i}$ $(0 \leqslant i \leqslant n-1 \leqslant N-1)$ are all bounded by

$$|E_{n,i}| \leqslant Qh^{2m} \quad (h_i \leqslant \max_{(i)} h_i =: h),$$

where Q denotes a suitable finite constant not depending on h. This implies, by (5.3.51), that

$$|e^I(t_n)| \leqslant Nh \cdot Qh^{2m} \leqslant \gamma T Q h^{2m} = O(h^{2m}), \quad t_n \in \bar{Z}_N,$$

for all quasi-uniform meshes Π_N. $\square$

We note without proof that the above high-order convergence no longer occurs at points which are not mesh points: we then have

$$e^{I}(t) = O(h^{m}), \quad t_n \in I \setminus \bar{Z}_N.$$

(Compare Brunner [1984b] for details.) This is in marked contrast to iterated collocation for *Fredholm integral equations* of the second kind with smooth kernels where, for collocation at the Gauss points, $e^{I}(t) = O(h^{2m})$ for *all* $t \in I$ (see, e.g., Chatelin & Lebbar [1981]).

Returning to (5.3.46) and its numerical treatment, we first recall from Section 5.3.4 that, in general, one will not be able to compute the collocation approximation $u \in S_{m-1}^{(-1)}(Z_N)$ defined by the exact collocation equation (5.3.7). Instead, an approximation $\hat{u} \in S_{m-1}^{(-1)}(Z_N)$ is determined, namely the solution of the fully discretized collocation equation (5.3.35). Hence, in lieu of (5.3.46) one will consider the iterated approximation $\bar{u}^{I}$ defined by

$$\bar{u}^{I}(t) := g(t) + \int_0^t K(t,s)\hat{u}(s)ds, \quad t \in I. \tag{5.3.52}$$

Since the integral cannot be evaluated analytically, except in special cases, we are forced once more to approximate its value (for a given $t \in I$) by appropriate numerical quadrature. We consider two cases:

(i) $t = t_n \in \bar{Z}_N$: if t is a mesh point, (5.3.52) may be written in the form

$$\bar{u}^{I}(t_n) = g(t_n) + \sum_{i=0}^{n-1} h_i \int_0^1 K(t_n, t_i + vh_i)\hat{u}_i(t_i + vh_i)dv.$$

For the discretization of this equation we choose interpolatory m-point quadrature formulas, with abscissas coinciding with the collocation points; i.e.,

$$\hat{u}^{I}(t_n) := g(t_n) + \sum_{i=0}^{n-1} h_i \sum_{l=1}^{m} b_l K(t_n, t_{i,l})\hat{u}_i(t_{i,l}), \quad t_n \in \bar{Z}_N. \tag{5.3.53}$$

Here, the values $\{\hat{u}_i(t_{i,l})\}$ are defined by (5.3.35), and the quadrature weights are given by

$$b_l := \int_0^1 L_l(v)dv, \quad \text{with} \quad L_l(v) := \prod_{\substack{r=1 \\ r \neq l}}^{m} (v - c_r)/(c_l - c_r)$$

(compare also (5.3.43), $i < n$). Note that (5.3.53) is no longer a recurrence relation: once the collocation approximation $\hat{u}$ has been determined, (5.3.53) may be used to compute specific values of the corresponding iterated approximation (e.g., the value for $t_n = t_N = T$) without first computing its values at preceding mesh points.

(ii) $t = t_n + \tau h_n$, $0 < \tau < 1$ ($t_n \notin \bar{Z}_N$): (5.3.52) can then be rewritten as

$$\overline{u}^I(t) = g(t)+h_n\int_0^{\tau} K(t,t_n+vh_n)\hat{u}_n(t_n+vh_n)dv$$

$$+\sum_{i=0}^{n-1} h_i\int_0^1 K(t,t_i+vh_i)\hat{u}_i(t_i+vh_i)dv.$$

While the integrals occurring under the summation sign may be discretized as in case (i), the remaining integral (for the interval $[0,\tau]$) will be replaced by interpolatory m-point quadrature using the abscissas $\{t_n+\tau c_l h_n\colon l=1, \ldots ,m\}$, in analogy to (5.3.43) ($i=n$). We thus obtain

$$\hat{u}^I(t_n+\tau h_n) := g(t_n+\tau h_n)+h_n\sum_{l=1}^{m} a_l(\tau)K(t_n+\tau h_n,t_n+\tau c_l h_n)\hat{u}_n(t_n+\tau c_l h_n)$$

$$+\sum_{i=0}^{n-1} h_i\sum_{l=1}^{m} b_l K(t_n+\tau h_n,t_{i,l})\hat{u}_i(t_{i,l}), \quad t_n+\tau h_n\in I, \tag{5.3.54}$$

where the quadrature weights $\{a_l(\tau)\}$ are given by $a_l(\tau):=\tau b_l$ ($l=1, \ldots ,m$).

The following result is the discrete analogue of Theorem 5.3.6.

Theorem 5.3.7. *Let g and K in (5.3.9) be subject to the conditions stated in Theorem 5.3.6, and let $\hat{u}\in S_{m-1}^{(-1)}(Z_N)$ denote the collocation approximation obtained from the fully discretized collocation equation (5.3.35), where the collocation parameters $\{c_j\}$ are the zeros of $P_m(2s-1)$ (Gauss points for $(0,1)$), and where the quadrature formulas are the m-point Gauss formulas of the form (5.3.43). If the iterated values $\hat{u}^I(t_n)$ are computed by (5.3.53), then we have, for all quasi-uniform meshes,*

$$\max_{t_n\in \overline{Z}_N} |\hat{e}^I(t_n)| = O(h^{2m}) \tag{5.3.55}$$

(as $h\downarrow 0$, $Nh\leqslant\gamma T$), where $\hat{e}^I := y-\hat{u}^I$.

Proof. The error $\hat{e}^I(t_n)$ can be written as

$$\hat{e}^I(t_n) = (y(t_n)-u^I(t_n))+(u^I(t_n)-\hat{u}^I(t_n)) =: e^I(t_n)+\epsilon^I(t_n),$$

where, for the Gauss collocation points, $e^I(t_n)=O(h^{2m})$ (Theorem 5.3.6). It therefore remains to be shown that the perturbation $\epsilon^I(t_n)$ is of the same order.

It follows from (5.3.46) (setting $t=t_n\in \overline{Z}_N$) and from (5.3.53) that

$$\epsilon^I(t_n) = u^I(t_n)-\hat{u}^I(t_n)$$

$$= \sum_{i=0}^{n-1} h_i\{\int_0^1 K(t_n,t_i+vh_i)u_i(t_i+vh_i)dv - \sum_{l=1}^{m} b_l K(t_n,t_{i,l})\hat{u}_i(t_{i,l})\}$$

$$= \sum_{i=0}^{n-1} h_i\{(\sum_{l=1}^{m} b_l K(t_n,t_{i,l})u_i(t_{i,l})+E_{n,i})-\sum_{l=1}^{m} b_l K(t_n,t_{i,l})\hat{u}_i(t_{i,l})\}, \quad t_n\in \overline{Z}_N,$$

where $E_{n,i}$ is the error term for the m-point Gauss quadrature formula with respect to the subinterval σ_i. This may be written as

$$\epsilon^I(t_n) = \sum_{i=0}^{n-1} h_i \sum_{l=1}^{m} b_l K(t_n,t_{i,l})\epsilon(t_{i,l}) + \sum_{i=0}^{n-1} h_i E_{n,i},$$

with $\epsilon(t_{i,l}) := u_i(t_{i,l}) - \hat{u}_i(t_{i,l})$ (the difference between the exact collocation approximation and the one furnished by the fully discretized collocation equation). However, according to Theorem 5.3.4 (with $r = 2m$; m-point Gauss quadrature), we already know that there exists a finite constant C independent of h and N such that

$$|\epsilon(t_{i,l})| \leq Ch^{2m}$$

for all collocation points $\{t_{i,l}\}$ (which, by assumption, are the Gauss points and coincide with the quadrature abscissas). Moreover, the quadrature errors $E_{n,i}$ satisfy $|E_{n,i}| \leq Qh^{2m}$ $(0 \leq i \leq n-1 \leq N-1)$. These facts can thus be used to derive a uniform upper bound for $\epsilon^I(t_n)$, namely,

$$|\epsilon^I(t_n)| \leq Nh(BK_0C + Q)h^{2m}) \leq \gamma T(BK_0C + Q)h^{2m} = O(h^{2m}), \quad t_n \in \overline{Z}_N,$$

where we have set $B := \max_{(l)} b_l$ (the Gauss quadrature weights are positive and bounded: $b_l \leq 1$), and $K_0 := \max\{|K(t,s)| : (t,s) \in S\}$. This completes the proof of Theorem 5.3.7. □

The above results on the attainable (local) order of convergence of iterated collocation methods can be extended to *nonlinear* Volterra integral equations of the second-kind. If $u \in S_{m-1}^{(-1)}(Z_N)$ is the collocation approximation given by (5.3.7), then its iterate u^I is defined by

$$u^I(t) := g(t) + \int_0^t k(t,s,u(s))ds, \quad t \in I. \tag{5.3.56}$$

Since, however, the concept of the resolvent kernel does not exist for nonlinear Volterra integral equations, we can no longer use the representation (5.3.50) for the error e^I as the basis for the convergence analysis, and hence different techniques will be needed. As we shall see, these techniques make use of a crucial connection between discretized iterated collocation methods and certain extended m-stage implicit PVRK methods.

Let $\hat{u} \in S_{m-1}^{(-1)}(Z_N)$ denote again the solution of the discretized collocation equation (5.3.55), with quadrature formulas (5.3.43) (this corresponds to "fully implicit" discretization of (5.3.7); cf. Table 5.3.1), and let $\tilde{u} \in S_{m-1}^{(-1)}(Z_N)$ be given by the "implicit" discretization of (5.3.7),

$$\tilde{u}_n(t_{n,j}) = g(t_{n,j}) + h_n \tilde{\Phi}_{n,n}^{(j)}[\tilde{u}_n] + \sum_{i=0}^{n-1} h_i \hat{\Phi}_{n,i}^{(j)}[\tilde{u}_i], \tag{5.3.57}$$

$$j = 1, \ldots, m \quad (n = 0, \ldots, N-1),$$

where, for $i < n$, the quadrature approximations $\hat{\Phi}_{n,i}^{(j)}$ are as in (5.3.43), while

$$\tilde{\Phi}_{n,n}^{(j)}[\tilde{u}_n] := \sum_{l=1}^{m} b_l k(t_{n,j}, t_{n,l}, \tilde{u}_n(t_{n,l})), \quad j = 1, \ldots, m. \tag{5.3.58}$$

The values of the corresponding iterated collocation approximations on $\overline{Z}_N$ are then furnished by

$$\hat{u}^I(t_n) := g(t_n)+\sum_{i=0}^{n-1} h_i \sum_{l=1}^{m} b_l k(t_n,t_{i,l},\hat{u}_i(t_{i,l})), \tag{5.3.59}$$

and by

$$\tilde{u}^I(t_n) := g(t_n)+\sum_{i=0}^{n-1} h_i \sum_{l=1}^{m} b_l k(t_n,t_{i,l},\tilde{u}_i(t_{i,l})), \quad t_n\in\overline{Z}_N. \tag{5.3.60}$$

Suppose now that the collocation parameters used in the above discretized collocation equations satisfy $0\leqslant c_1< \ldots <c_m<1$, and set $c_{m+1}:=1$. In addition, we consider the collocation approximation $z\in S_m^{(-1)}(Z_N)$ defined by

$$z_n(t_{n,j}) = g(t_{n,j})+h_n\Phi_{n,n}^{(j)}[z_n]+\sum_{i=0}^{n-1} h_i\Phi_{n,i}^{(j)}[z_i], \tag{5.3.61}$$

$$j=1, \ldots ,m,\ m+1 \quad (n=0, \ldots ,N-1),$$

with integrals $\Phi_{n,i}^{(j)}$ as in (5.3.6). The discretization of (5.3.61) will be achieved by using, respectively, interpolatory *m-point* (instead of $(m+1)$-point) quadrature formulas of the forms (5.3.43), or (5.3.43) (if $i<n$), and (5.3.58). Hence, the resulting collocation approximations $\hat{z}$ and $\tilde{z}$ in $S_m^{(-1)}(Z_N)$ are determined by

$$\hat{z}_n(t_{n,j}) = g(t_{n,j})+h_n\hat{\Phi}_{n,n}^{(j)}[\hat{z}_n]+\sum_{i=0}^{n-1} h_i\hat{\Phi}_{n,i}^{(j)}[\hat{z}_i], \tag{5.3.62}$$

$$j=1, \ldots ,m,\ m+1 \quad (n=0, \ldots ,N-1),$$

and

$$\tilde{z}_n(t_{n,j}) = g(t_{n,j})+h_n\tilde{\Phi}_{n,n}^{(j)}[\tilde{z}_n]+\sum_{i=0}^{n-1} h_i\hat{\Phi}_{n,i}^{(j)}[\tilde{z}_i], \tag{5.3.63}$$

$$j=1, \ldots ,m,\ m+1 \quad (n=0, \ldots ,N-1);$$

on the subinterval σ_n we have, e.g.,

$$\tilde{z}_n(t_n+\tau h_n) = \sum_{l=1}^{m+1} L_l^*(\tau)\tilde{Z}_{n,l}, \tag{5.3.64}$$

where $\tilde{Z}_{n,l}:=\tilde{z}_n(t_{n,l})$, and

$$L_l^*(\tau) := \prod_{k\neq l}^{m+1} (\tau-c_k)/(c_l-c_k).$$

Note that $L_{m+1}^*(c_j)=0$ $(j=1, \ldots ,m)$ implies that the right-hand side of (5.3.63) depends only on the values $\tilde{Z}_{n,1}, \ldots ,\tilde{Z}_{n,m}$, but not on $\tilde{Z}_{n,m+1}$. Thus, in order to obtain $\tilde{z}$ on σ_n it suffices to solve only a nonlinear system in $\mathbb{R}^m$ which is given by the first m equations of (5.3.63). Once the values $\tilde{Z}_{n,1}, \ldots ,\tilde{Z}_{n,m}$ have been found, equation (5.3.63) with $j=m+1$ yields

$$\tilde{Z}_{n,m+1}(=\tilde{z}_n(t_{n+1})) = g(t_{n+1}) + \sum_{i=0}^{n} h_i \sum_{l=1}^{m} b_l k(t_{n+1}, t_{i,l}, \tilde{Z}_{i,l})). \quad (5.3.65)$$

According to Chapter 4, (5.3.63) (with $j=1, \ldots, m$) and (5.3.65) together represent an *extended m-stage implicit PVRK method.* (Note also that the above observations do not apply to (5.3.62) since, by (5.3.43) ($i=n$), the quadrature approximations $\hat{\Phi}^{(j)}_{n,n}[\hat{z}_n]$ depend on all $m+1$ values $\hat{Z}_{n,1}, \ldots, \hat{Z}_{n,m+1}$.)

Theorem 5.3.8. (a) *For all $t_n \in \bar{Z}_N$ the approximations $\{\tilde{z}(t_n)\}$ defined by the extended m-stage implicit PVRK method (5.3.63) ($j=1, \ldots, m$), (5.3.65), coincide with the values $\{\tilde{u}^I(t_n)\}$ given by the discretized iterated collocation method (5.3.60); i.e., we have*

$$\tilde{u}^I(t_n) = \tilde{z}(t_n), \quad t_n \in \bar{Z}_N. \quad (5.3.66)$$

(b) *If the parameters $\{c_j\colon j=1, \ldots, m\}$ are the zeros of $P_m(2s-1)$ (the m Gauss points for* $(0,1)$*), then the iterated approximations (5.3.59) satisfy*

$$\max_{t_n \in Z_N} |y(t_n) - \hat{u}^I(t_n)| = O(h^{2m}) \quad (5.3.67)$$

(as $h \downarrow 0$, $Nh \leqslant \gamma T$), provided g and k in (5.3.1) are sufficiently smooth so that the integral equation has a solution $y \in C^{2m}(I)$.

Proof. (a) It is readily seen that, for all sufficiently small $h>0$, the values $\tilde{Y}_{n,l} := \tilde{u}_n(t_{n,l})$ ($l=1, \ldots, m$) defined by the nonlinear system (5.3.37) are identical with the solution $(\tilde{Z}_{n,1}, \ldots, \tilde{Z}_{n,m})$ of the nonlinear system formed by the first m equations of (5.3.63); that is, $\tilde{Z}_{n,l} = \tilde{Y}_{n,l}$, $l=1, \ldots, m$. Thus, (5.3.60) and (5.3.65) yield

$$\tilde{u}^I(t_{n+1}) = g(t_{n+1}) + \sum_{i=0}^{n} h_i \sum_{l=1}^{m} b_l k(t_{n+1}, t_{i,l}, \tilde{Z}_{i,l}) = \tilde{Z}_{n,m+1} = \tilde{z}_n(t_{n+1}),$$

and this holds for $n=0, \ldots, N-1$.
(b) As has already been remarked, $\hat{u}^I(t_n)$ will not, in general, coincide with $\hat{z}(t_n)$. However, we have by (a),

$$\begin{aligned} |y(t_n) - \hat{u}^I(t_n)| &\leqslant |y(t_n) - \tilde{u}^I(t_n)| + |\tilde{u}^I(t_n) - \hat{u}^I(t_n)| \\ &= |y(t_n) - \tilde{z}(t_n)| + |\tilde{u}^I(t_n) - \hat{u}^I(t_n)|, \quad t_n \in \bar{Z}_N. \end{aligned}$$

If $\{c_1, \ldots, c_m\}$ are the Gauss points for $(0,1)$, then it follows from the proofs of Theorem 5.3.3(d), (c) (with $S_m^{(-1)}(Z_N)$ replacing $S_{m-1}^{(-1)}(Z_N)$) that

$$\max_{t_n \in Z_N} |y(t_n) - \tilde{z}(t_n)| = O(h^{2m}).$$

Moreover, (5.3.59) and (5.3.60) lead to

$$|\tilde{u}^I(t_n) - \hat{u}^I(t_n)| \leqslant h \sum_{i=0}^{n-1} \sum_{l=1}^{m} |k(t_n, t_{i,l}, \tilde{u}_i(t_{i,l})) - k(t_n, t_{i,l}, \hat{u}_i(t_{i,l}))| \leqslant$$

$$\leqslant hL\sum_{i=0}^{n-1}\sum_{l=1}^{m}|\tilde{u}_i(t_{i,l})-\hat{u}_i(t_{i,l})|,$$

where L denotes an upper bound for $|\partial k(t,s,y)/\partial y|$, and where we have used the fact that the Gauss weights $\{b_l\}$ are all positive and satisfy $\Sigma_{l=1}^{m}b_l=1$. Employing similar techniques as in the proofs of Theorems 5.3.4 and 5.3.5 (where the role of $\epsilon_n(t_{n,j})$ is now taken by $\tilde{u}_n(t_{n,j})-\hat{u}_n(t_{n,j})$) we find without any difficulty that

$$\sum_{l=1}^{m}|\tilde{u}_i(t_{i,l})-\hat{u}_i(t_{i,l})| \leqslant Ch^{2m}, \quad i=0, \ldots ,N-1,$$

as $h\downarrow 0$, $Nh\leqslant\gamma T$, where C is a suitable constant not depending on i or h. Hence,

$$|\tilde{u}^I(t_n)-\hat{u}^I(t_n)| \leqslant Nh\,LCh^{2m} \leqslant \gamma TLC\,h^{2m}, \quad t_n\in\bar{Z}_N. \quad \square$$

Finally, we note that the "exact" iterated collocation approximation defined by (5.3.56) satisfies (5.3.48) of Theorem 5.3.6. The proof is straightforward: we have

$$|y(t_n)-u^I(t_n)| \leqslant |y(t_n)-\hat{u}^I(t_n)| + |\hat{u}^I(t_n)-u^I(t_n)|, \quad t_n\in\bar{Z}_N,$$

where, by (5.3.67), the first term on the right-hand side is $O(h^{2m})$. The same estimate holds for the second term; this follows from (5.3.59) and (5.3.56) (with $t=t_n$), using the result that the quadrature processes (5.3.43) are m-point Gauss formulas for the integrals (5.3.6). We leave the details to the reader.

Example 5.3.8. $\hat{u}\in S_1^{(-1)}(Z_N)$: $m=2$; $c_1=(3-\sqrt{3})/6$, $c_2=(3+\sqrt{3})/6$ (Gauss points).

(i) According to (5.3.35) and (5.3.43), the values $\hat{Y}_{n,j}:=\hat{u}_n(t_{n,j})$ $(j=1,2)$ are determined from

$$\{1-\frac{5-\sqrt{3}}{24}h_n\cdot K(t_{n,1},t_n+\frac{2-\sqrt{3}}{6}h_n)-\frac{1+\sqrt{3}}{24}h_n\cdot K(t_{n,1},t_n+\frac{h_n}{6})\}\cdot\hat{Y}_{n,1}$$

$$+\{\frac{\sqrt{3}-1}{24}h_n\cdot K(t_{n,1},t_n+\frac{2-\sqrt{3}}{6}h_n)+\frac{3\sqrt{3}-5}{24}h_n\cdot K(t_{n,1},t_n+\frac{h_n}{6})\}\cdot\hat{Y}_{n,2}$$

$$=g(t_{n,1})+\sum_{i=0}^{n-1}h_i\hat{\Phi}_{n,i}^{(1)}[\hat{u}_i],$$

$$\{-\frac{5+3\sqrt{3}}{24}h_n\cdot K(t_{n,2},t_n+\frac{h_n}{6})-\frac{1+\sqrt{3}}{24}h_n\cdot K(t_{n,2},t_n+\frac{2+\sqrt{3}}{6}h_n)\}\cdot\hat{Y}_{n,1}$$

$$+\{1+\frac{\sqrt{3}-1}{24}h_n\cdot K(t_{n,2},t_n+\frac{h_n}{6})-\frac{5+\sqrt{3}}{24}h_n\cdot K(t_{n,2},t_n+\frac{2+\sqrt{3}}{6}h_n)\}\cdot\hat{Y}_{n,2}$$

$$=g(t_{n,2})+\sum_{i=0}^{n-1}h_i\hat{\Phi}_{n,i}^{(2)}[\hat{u}_i];$$

here,

$$\hat{\Phi}^{(j)}_{n,i}[\hat{u}_i] := \frac{1}{2}\{K(t_{n,j},t_i+c_1h_i)\cdot\hat{Y}_{i,1}+K(t_{n,j},t_i+c_2h_i)\cdot\hat{Y}_{i,2}\},$$

$$0\leqslant i\leqslant n-1 \quad (n=0, \ldots ,N-1).$$

(ii) The corresponding iterated approximations $\hat{u}^I(t)$, $t\in\bar{Z}_N$, are given by (see (5.3.53)):

$$\hat{y}^I_{n+1} := \hat{u}^I(t_{n+1})=g(t_{n+1})+\frac{1}{2}\cdot\sum_{i=0}^{n} h_i\{K(t_{n+1},t_i+c_1h_i)\cdot\hat{Y}_{i,1}$$

$$+K(t_{n+1},t_i+c_2h_i)\cdot\hat{Y}_{i,2}\}, \quad n=0, \ldots ,N-1.$$

Even though the collocation approximation $\hat{u}$ is piecewise linear (with global order of convergence $p=m=2$), the iterated approximations $\{\hat{y}^I_n\}$ on $\bar{Z}_N$ exhibit the (local) order of convergence $p^*=2m=4$.

5.3.7. Numerical examples

We have tested the collocation methods based on:

2 Gauss points and $c_3=1$ (see Example 5.3.7): $PVRK_4(G)$

3 Radau II points (see Example 5.3.5): $PVRK_5(R\ II)$,

and the iterated collocation method based on:

2 Gauss points and iteration at the end point of each step (see Example 5.3.8): $IPVRK_4(G)$.

The $PVRK_4(G)$ method was applied using the quadrature approximations (5.3.43) (fully implicit discretization (cf. Table 5.3.1), alternatively called the de Hoog and Weiss (dHW) modification). In addition, we applied the extended form as defined by Definition 4.1.1 (implicit discretization). The $IPVRK_4(G)$ method was only applied to linear problems.

The integration steps in the tables of results are such that the various experiments require roughly the same amount of computational effort (in terms of kernel evaluations) as the corresponding experiments presented in Sections 3.2.9, 3.2.11 and 4.1.10. In our estimate of the number of kernel values those values required in the Newton process for solving implicit relations are not taken into account.

Table 5.3.2a. Numbers of correct significant digits and effective orders for problem (3.2.26) at $t = T = 4$

h	$PVRK_4(G)$ Extended		dHW mod.		h	$PVRK_5(R\,II)$	
4/8	5.2		5.2		4/7	5.8	
		> 3.8		> 3.8			> 4.8
4/16	6.4		6.4		4/14	7.3	
		> 3.9		> 4.0			> 4.9
4/32	7.6		7.6		4/28	8.8	

Table 5.3.2b. Numbers of correct significant digits and effective orders for problem (3.2.36) at $t = T = 4$

h	$PVRK_4(G)$ Extended		dHW mod.		h	$PVRK_5(R\,II)$	
4/8	3.8		5.4		4/7	4.7	
		> 4.0		> 2.7			> 4.6
4/16	5.0		6.1		4/14	6.1	
		> 4.0		> 3.4			> 4.8
4/32	6.2		7.2		4/28	7.6	

Table 5.3.3. Numbers of correct significant digits and effective orders for problem (3.2.8) at $t = T = 6$

h	$PVRK_4(G)$ Extended		dHW mod.		h	$PVRK_5(R\,II)$		h	$IPVRK_4(G)$	
6/10	3.0		3.6		6/8	4.6		6/12	3.9	
		> 4.0		> 3.9			> 4.6			> 3.9
6/20	4.2		4.7		6/16	6.0		6/24	5.0	
		> 4.0		> 3.9			> 4.8			> 3.9
6/40	5.4		5.9		6/32	7.5		6/48	6.2	
		> 4.0		> 4.0			> 4.9			> 4.0
6/80	6.6		7.1		6/64	9.0		6/96	7.4	
		> 4.0		> 4.0			> 5.0			> 4.0
6/160	7.8		8.3		6/128	10.4		6/192	8.6	

Table 5.3.4. Numbers of correct significant digits and effective orders for problem (3.2.38) at $t = T = 6$

h	$PVRK_4(G)$ Extended		dHW mod.		h	$PVRK_5(R\,II)$	
6/10	3.4		4.0		6/8	4.9	
		> 4.0		> 3.9			> 4.9
6/20	4.6		5.1		6/16	6.4	
		> 4.0		> 3.9			> 4.9
6/40	5.8		6.3		6/32	7.9	

5.4. Polynomial spline collocation for integro-differential equations

5.4.1. Global convergence results

We shall now study the problem of approximating the exact solution of the initial-value problem for an integro-differential equation,

$$y' = f(t,y(t), \int_0^t k(t,s,y(s))ds), \quad t\in I, \quad \text{with} \quad y(0)=y_0, \tag{5.4.1}$$

by polynomial spline functions; our aim is to construct approximations which possess the global convergence order $p=m$ on I. It will turn out that this order is attained if we look for approximations in $S_m^{(0)}(Z_N)$; recall that, by (5.1.4), $dim\ S_m^{(0)}(Z_N)=dim\ S_{m-1}^{(-1)}(Z_N)+1$.

As in the previous chapter, let $\{c_j\}$ denote collocation parameters satisfying $0\leqslant c_1< \ldots <c_m\leqslant 1$. The approximation $u\in S_m^{(0)}(Z_N)$ to the exact solution of (5.4.1) will be determined by the collocation equation,

$$u_n'(t_{n,j}) = f(t_{n,j},u_n(t_{n,j}),h_n\int_0^{c_j} k(t_{n,j},t_n+vh_n,u_n(t_n+vh_n))dv+\overline{F}_n(t_{n,j})), \tag{5.4.2}$$

$$j=1, \ldots ,m \quad (n=0, \ldots ,N-1),$$

with

$$u_n(t_n) = u_{n-1}(t_n), \quad t_n\in Z_N,$$

and with $u_0(0)=y_0$. Here, the expression

$$\overline{F}_n(t) := \sum_{i=0}^{n-1} h_i \int_0^1 k(t,t_i+vh_i,u_i(t_i+vh_i))dv, \quad t\in\sigma_n, \tag{5.4.3}$$

approximates the exact lag term

$$F_n(t) := \int_0^{t_n} k(t,s,y(s))ds.$$

We note in passing that if f and k in (5.4.1) are continuous on their respective domains, then $c_1=0$ and $c_m=1$ imply that the approximation u is not only in $S_m^{(0)}(Z_N)$ but also in $S_m^{(1)}(Z_N)$.

For the linear version of (5.4.1),

$$y' = p(t)y+q(t)+\int_0^t K(t,s)y(s)ds, \quad t\in I, \quad y(0)=y_0, \tag{5.4.4}$$

the collocation equation assumes the form

$$u_n'(t_{n,j}) = p(t_{n,j})u_n(t_{n,j}) + q(t_{n,j})$$

$$+ h_n \int_0^{c_j} K(t_{n,j}, t_n + vh_n) u_n(t_n + vh_n) dv + \bar{F}_n(t_{n,j}), \tag{5.4.5}$$

$$j = 1, \dots, m \quad (n = 0, \dots, N-1),$$

with

$$\bar{F}_n(t_{n,j}) := \sum_{i=0}^{n-1} h_i \int_0^1 K(t_{n,j}, t_i + vh_i) u_i(t_i + vh_i) dv.$$

With the notation introduced in (5.3.6), the collocation equation (5.4.2) can be stated in the form

$$u_n'(t_{n,j}) = f(t_{n,j}, u_n(t_{n,j}), h_n \Phi_{n,n}^{(j)}[u_n] + \sum_{i=0}^{n-1} h_i \Phi_{n,i}^{(j)}[u_i]), \tag{5.4.6}$$

$$j = 1, \dots, m \quad (n = 0, \dots, N-1).$$

This equation is reminiscent, on the one hand, of the collocation equation (5.2.3) for the initial-value problem (5.2.1) (which corresponds, of course, to (5.4.1) with $k(t,s,y) \equiv 0$), and, on the other hand, of the collocation equation (5.3.7) for the second-kind integral equation (5.3.1). Hence, the discretized collocation equation will be based on an implicit Runge-Kutta method of the form (5.3.5) and on quadrature approximations of the type discussed in Section 5.3.4. Before pursuing this aspect further, we analyze the global and local convergence of the approximation furnished by (5.4.6), assuming that the integrals $\Phi_{n,i}^{(j)}[u_i]$ are evaluated analytically. As in the previous sections, we phrase our convergence results for the linear equation (5.4.4); a remark on the extension of these results to the general case (5.4.1) will follow each of the proofs.

Theorem 5.4.1. *Let p, q and K in (5.4.4) be m times continuously differentiable on their respective domains I and S. Then there exists, for a given quasi-uniform mesh sequence* $\{\Pi_N\}$ *with* $h := \max_{(n)}(t_{n+1} - t_n)$ *and* $Nh \leq \gamma T$, *an* $\bar{h} > 0$ *such that the collocation equation (5.4.6) defines for each* $h \in (0, \bar{h})$ *a unique approximation* $u \in S_m^{(0)}(Z_N)$. *For every choice of the collocation parameters* $\{c_j\}$ *with* $0 \leq c_1 < \dots < c_m \leq 1$, *the resulting error function* $e := y - u$ *satisfies*

$$\|e\|_\infty \leq C_0 h^m, \ \|e'\|_\infty \leq C_1 h^m, \tag{5.4.7}$$

with C_0 *and* C_1 *denoting suitable finite constants depending on the* $\{c_j\}$.

We observe that the approximate solution of a Volterra integro-differential equation of the first order in the space of continuous polynomial splines of degree $m \geq 1$ leads to the global order of convergence $p = m$; this is in some contrast to Volterra integral equations of the second kind where the same global order is attained for polynomial splines of degree $m - 1$.

Proof. It is easily verified that the given smoothness hypotheses for p, q and K imply the existence of a (unique) solution $y \in C^{m+1}(I)$ of (5.4.4) (recall Theorem 1.3.4). Hence, in analogy to the proof of Theorem 5.3.2, we set, using Taylor's formula with remainder term,

$$y(t_n+\tau h_n) = \sum_{l=0}^{m} c_{n,l}\tau^l + h_n^{m+1}R_n(\tau), \quad t+\tau h_n \in \sigma_n,$$

with

$$c_{n,l} := h_n^l y^{(l)}(t_n)/l!,$$

$$R_n(\tau) := y^{(m+1)}(\xi_n)\tau^{m+1}/(m+1)! \quad (t_n<\xi_n<t_n+\tau h_n).$$

If the restriction of $u \in S_m^{(0)}(Z_N)$ to the subinterval σ_n is given by

$$u_n(t_n+\tau h_n) = \sum_{l=0}^{m} \alpha_{n,l}\tau^l,$$

then the error function on this subinterval has the form

$$e_n(t_n+\tau h_n) = h_n^p\{\beta_{n,0} + \sum_{l=1}^{m} \beta_{n,l}\tau^l + h_n^{m+1-p}R_n(\tau)\}, \quad t_n+\tau h_n \in \sigma_n. \tag{5.4.8}$$

Here, we have set $h_n^p\beta_{n,l} := c_{n,l} - \alpha_{n,l}$ $(l=0, \ldots, m)$, with the integer p as yet unspecified. We shall show that $p=m$ is the largest possible integer for which the vectors $\boldsymbol{\beta}_n := (\beta_{n,1}, \ldots, \beta_{n,m})^T$ as well as the coefficients $\beta_{n,0}$ remain uniformly bounded as $h := \max_{(n)} h_n$ tends to 0, such that $Nh \leqslant \gamma T$ (recall (5.1.1)).

It follows from (5.4.8) that

$$e_n'(t_n+\tau h_n) = h_n^{p-1}\{\sum_{l=1}^{m} l\beta_{n,l}\tau^{l-1} + h_n^{m+1-p}R_n'(\tau)\}, \quad t_n+\tau h_n \in \sigma_n, \tag{5.4.9}$$

where $R_n'(\tau) := y^{(m+1)}(\eta_n)\tau^m/m!$, with $t_n<\eta_n<t_n+\tau h_n$. If we now subtract the collocation equation (5.4.5) from the integro-differential equation (5.4.4) (with t replaced by $t_{n,j}$), then, setting $p=m$ and employing the expressions (5.4.8) and (5.4.9) for e_n and e_n', we are led to

$$\sum_{l=1}^{m} \{lc_j^{l-1} - h_n p_{n,j} c_j^l - h_n^2 \int_0^{c_j} K_{n,j}(t_n+\tau h_n)\tau^l d\tau\}\cdot\beta_{n,l}$$

$$= \{h_n p_{n,j} + h_n^2 \int_0^{c_j} K_{n,j}(t_n+\tau h_n) d\tau\}\cdot\beta_{n,0}$$

$$+ \sum_{i=0}^{n-1} h_i^2 \gamma_{n,i}^{m-1} \int_0^1 K_{n,j}(t_i+\tau h_i) d\tau \cdot \beta_{i,0} \tag{5.4.10}$$

$$+ \sum_{i=0}^{n-1} h_i^2 \gamma_{n,i}^{m-1} \sum_{l=1}^{m} \int_0^1 K_{n,j}(t_i+\tau h_i)\tau^l d\tau \cdot \beta_{i,l} + r_{n,j},$$

$$j=1, \ldots, m \quad (n=0, \ldots, N-1).$$

Here, we have introduced the abbreviations $p_{n,j}:=p(t_n+c_jh_n)$, $K_{n,j}(\cdot):=K(t_{n,j},\cdot)$, $\gamma_{n,i}:=h_i/h_n$, and

$$r_{n,j} := -h_nR_n'(c_j)+h_n^2p_{n,j}R_n(c_j)+h_n^3\int_0^{c_j}K_{n,j}(t_n+\tau h_n)R_n(\tau)d\tau$$

$$+\sum_{i=0}^{n-1}h_i^3\gamma_{n,i}^{m-1}\int_0^1K_{n,j}(t_i+\tau h_i)R_i(\tau)d\tau. \tag{5.4.11}$$

So far, we have not yet taken into account the continuity of the approximating polynomial spline at the knots Z_N. These continuity conditions yield an additional relationship between $\beta_{n,0}$ and the vectors $\boldsymbol{\beta}_i$ $(i<n)$, namely,

$$\beta_{n,0} = \gamma_{n,n-1}^m\{\beta_{n-1,0}+\sum_{l=1}^m\beta_{n-1,l}+h_{n-1}R_{n-1}(1)\}.$$

It thus follows that

$$\beta_{n,0} = \gamma_{n,0}^m\cdot\beta_{0,0}+\sum_{i=0}^{n-1}\gamma_{n,i}^m\{\sum_{l=1}^m\beta_{i,l}+h_iR_i(1)\},\quad n=1,\ldots,N-1, \tag{5.4.12}$$

where the initial condition $u_0(0)=y_0$ implies $e_0(0)=0$; hence $\beta_{0,0}=0$. Substituting (5.4.12) (and the analogous expressions for $\beta_{i,0}$, $i<n$) in (5.4.10) we are led to the recurrence relation

$$\sum_{l=1}^m\{lc_j^{l-1}-h_np_{n,j}c_j^l-h_n^2\int_0^{c_j}K_{n,j}(t_n+\tau h_n)\tau^ld\tau\}\cdot\beta_{n,l}$$

$$=h_n\cdot\{p_{n,j}+h_n\int_0^{c_j}K_{n,j}(t_n+\tau h_n)d\tau\}\cdot\sum_{i=0}^{n-1}\gamma_{n,i}^m\sum_{l=1}^m\beta_{i,l}$$

$$+\sum_{i=0}^{n-1}h_i^2\gamma_{n,i}^2\sum_{s=0}^{i-1}\gamma_{i,s}^m\sum_{l=1}^m\beta_{s,l}\int_0^1K_{n,j}(t_i+\tau h_i)d\tau$$

$$+\sum_{i=0}^{n-1}h_i^2\cdot\gamma_{n,i}^{m-1}\sum_{l=1}^m\beta_{i,l}\int_0^1K_{n,j}(t_n+\tau h_i)\tau^ld\tau+\tilde{r}_{n,j}, \tag{5.4.13}$$

$$j=1,\ldots,m\quad(n=0,\ldots,N-1),$$

with

$$\tilde{r}_{n,j} := r_{n,j}+h_n\{p_n+h_n\int_0^{c_j}K_{n,j}(t_n+\tau h_n)d\tau\}\sum_{i=0}^{n-1}h_i\cdot\gamma_{n,i}^mR_i(1)$$

$$+\sum_{i=0}^{n-1}h_i\cdot\gamma_{n,i}^{m-1}\int_0^1K_{n,j}(t_i+\tau h_i)d\tau\sum_{s=0}^{i-1}h_s\gamma_{i,s}^mR_s(1) \tag{5.4.14}$$

(the additional terms to $r_{n,j}$ being contributions originating from the remainder terms $R_i(1)$ in (5.4.12)).

Since p and K are bounded on their respective (compact) domains, the inverse of the matrix whose elements are given by the coefficients of $\beta_{n,l}$ on the left-hand side of (5.4.13) exists and is uniformly bounded for all n, provided $h := \max_{(n)} h_n$ is sufficiently small. An appropriate modification of the reasoning used in the proof of Theorem 5.3.2 (following (5.3.17)) thus reveals the existence of an $\bar{h} > 0$ such that, for all $h \in (0, \bar{h})$,

$$\|\boldsymbol{\beta}_n\|_1 \leqslant hD_0 \sum_{i=0}^{n-1} \|\boldsymbol{\beta}_i\|_1 + h^2 D_1 \sum_{i=0}^{n-1} \sum_{s=0}^{i-1} \|\boldsymbol{\beta}_s\|_1 + h^2 D_2 \sum_{i=0}^{n-1} \|\boldsymbol{\beta}_i\|_1 + hR, \quad (5.4.15)$$

where the constants are self-explanatory and where, according to (5.4.14) and (5.4.11), the bound R is given by an expression involving upper bounds for the uniform norms of $p(t)$, $K(t,s)$ and $y^{(m+1)}(t)$. Note also that the assumed quasi-uniformity of the mesh sequence implies that $\gamma_{n,i} \leqslant \max_{(k)} h_k / \min_{(k)} h_k \leqslant \gamma$ for all i and n, with $N \cdot \max_{(n)} h_n = Nh \leqslant \gamma T$.

The above inequality (5.4.15) can be simplified as follows. Since $Nh \leqslant \gamma T$, we find

$$\begin{aligned}\|\boldsymbol{\beta}_n\|_1 &\leqslant h(D_0 + hD_2) \sum_{i=0}^{n-1} \|\boldsymbol{\beta}_i\|_1 + h^2 D_1 \sum_{i=0}^{n-2} (n-i-1) \|\boldsymbol{\beta}_i\|_1 + hR \\ &\leqslant h(D_0 + hD_2) \sum_{i=0}^{n-1} \|\boldsymbol{\beta}_i\|_1 + h^2 N D_1 \|\boldsymbol{\beta}_i\|_1 + hR \\ &\leqslant h(D_0 + hD_2) \sum_{i=0}^{n-1} \|\boldsymbol{\beta}_i\|_1 + hD_3 \sum_{i=0}^{n-1} \|\boldsymbol{\beta}_i\|_1 + hR,\end{aligned}$$

with $D_3 := \gamma T D_1$. In other words, the l_1-norms of the vectors $\boldsymbol{\beta}_n$ are related by the discrete Gronwall inequality

$$\|\boldsymbol{\beta}_n\|_1 \leqslant hD_4 \sum_{i=0}^{n-1} \|\boldsymbol{\beta}_i\|_1 + hR, \quad n = 0, \ldots, N-1,$$

where we have set $D_4 := D_0 + hD_2 + D_3$. Thus, Corollary 1.5.2 allows us to conclude that

$$\|\boldsymbol{\beta}_n\|_1 \leqslant hR(1 + hD_4)\exp(D_4 \gamma T) =: B_0 h, \quad n = 0, \ldots, N-1.$$

We now turn our attention to the coefficients $\beta_{n,0}$: it follows from (5.4.12) that

$$|\beta_{n,0}| \leqslant \gamma^m \sum_{i=0}^{n-1} \left(\sum_{l=1}^{m} |\beta_{i,l}| + hM_{m+1} \right),$$

with

$$M_{m+1} := \max\{ |y^{(m+1)}(t)| / (m+1)!: t \in I \}.$$

Hence,

$$|\beta_{n,0}| \leqslant \gamma^m \sum_{i=0}^{n-1} (\|\beta_i\|_1 + hM_{m+1}) \leqslant \gamma^m N(B_0 h + hM_{m+1})$$

$$\leqslant \gamma^{m+1} T(B_0 + M_{m+1}) =: B_1, \quad n = 1, \ldots, N-1.$$

We see now why the choice $p = m$ in (5.4.8) is best possible: for $p = m+1$, one still finds that $\|\beta_n\|_1 = O(1)$ uniformly as $h \downarrow 0$ and $Nh \leqslant \gamma T$, but this in turn implies that $|\beta_{n,0}|$ does not longer remain bounded as $h \downarrow 0$.

In order to bring the proof to its end, we return to (5.4.8) and (5.4.9): since $p = m$, we have found that, for $t_n + \tau h_n \in \sigma_n$ and $n = 0, \ldots, N-1$,

$$|e_n(t_n + \tau h_n)| \leqslant h^m (B_1 + hB_0 + hM_{m+1}) =: C_0 h^m,$$

$$|e_n'(t_n + \tau h_n)| \leqslant h^{m-1}(mhB_0 + hmM_{m+1}) = h^m(mB_0 + mM_{m+1}) =: C_1 h^m,$$

whenever $h \in (0, \bar{h})$. These two statements are equivalent to (5.4.7). □

There are no difficulties in extending Theorem 5.4.1 to nonlinear integro-differential equations of the form (5.4.1): in the error equation (5.4.10), the roles of $p_{n,j}$ and of $K_{n,j}(t_i + \tau h_i)$ are taken, respectively, by $\partial f(t_n + c_j h_n, z_{n,j})/\partial y$ and $\partial k(t_n + c_j h_n, t_i + \tau h_i, z_i(\tau))/\partial y$, with $z_{n,j}$ and $z_j(\tau)$ denoting suitable intermediate values arising in the application of the Mean-Value Theorem. We leave the details to the reader.

If $K(t,s) \equiv 0$ ($k(t,s,y) \equiv 0$ in the nonlinear case), Theorem 5.4.1 yields the result on the global convergence order of the collocation approximation $u \in S_m^{(0)}(Z_N)$ for the solution of the initial-value problem for $y' = f(t,y)$ (recall (5.2.3) and (5.2.5)). Since it has been known for some time that such a collocation approximation can possess the local order of convergence $p^* = 2m$ on $\bar{Z}_N$, there arises the question as to whether the same superconvergence results hold for the integro-differential equation (5.4.1). This will be answered in the next section.

5.4.2. *Local superconvergence on* $\bar{Z}_N$

Theorem 5.4.2. *Let $u \in S_m^{(0)}(Z_N)$ denote the collocation approximation defined by (5.4.6) where the underlying mesh sequence is assumed to be quasi-uniform. Let the functions p, q and K in (5.4.4) satisfy, respectively, $p \in C^{2m-\nu-1}(I)$, $q \in C^{2m-\nu-1}(I)$, and $K \in C^{2m-\nu-1}(S)$, where the integer ν is in $\{0,1,2\}$ and will be specified below. Finally, let $m \geqslant \lfloor(\nu+1)/2\rfloor + 1$.*
(a) *If the collocation parameters $\{c_j\}$ are the zeros of $P_m(2s-1)$ (Gauss points for $(0,1)$), and if $\nu = 0$, then we have*

$$\max_{t_n \in \bar{Z}_N} |e(t_n)| = O(h^{2m}) \tag{5.4.16}$$

(as $h \downarrow 0$, $Nh \leqslant \gamma T$).

(b) *If the collocation parameters* $\{c_j\}$ *are the zeros of* $P_{m-1}(2s-1)+P_m(2s-1)$ *(Radau I points for* $[0,1)$*), or the zeros of* $P_{m-1}(2s-1)-P_m(2s-1)$ *(Radau II points for* $(0,1]$*), and if* $\nu=1$*, then*

$$\max_{t_n\in\bar{Z}_N}|e(t_n)| = O(h^{2m-1}) \tag{5.4.17}$$

(as $h\downarrow 0$*,* $Nh\leqslant\gamma T$*).*
(c) *If the collocation parameters* $\{c_j\}$ *are the zeros of* $s(1-s)P'_{m-1}(2s-1)$ *(Lobatto points for* $[0,1]$*), and if* $\nu=2$*, then*

$$\max_{t_n\in\bar{Z}_N}|e(t_n)| = O(h^{2m-2}) \tag{5.4.18}$$

(as $h\downarrow 0$*,* $Nh\leqslant\gamma T$*).*

Proof. In analogy to the proof of Theorem 5.3.3 we write the collocation equation (5.4.5) in the form

$$u'(t) = p(t)u(t)+q(t)+\int_0^t K(t,s)u(s)ds-\delta(t), \quad t\in I,$$

where the defect function satisfies $\delta(t)=0$ for $t\in X(N)$, the set of collocation points. Moreover, it follows from the global convergence of u that $\delta(t)\to 0$ uniformly on I, as $h\to 0$. The error function $e:=y-u$ is therefore the (unique) solution of the initial-value problem

$$e'(t) = p(t)e(t)+\delta(t)+\int_0^t K(t,s)e(s)ds, \quad t\in I, \tag{5.4.19}$$

with $e(0)=0$. Let $R(t,s)$ be the resolvent associated with the integro-differential equation (5.4.4) (and hence with (5.4.19); recall Theorem 1.3.4). The solution of (5.4.19) can then be expressed in the form

$$e(t) = R(t,0)e(0)+\int_0^t R(t,s)\delta(s)ds, \quad t\in I,$$

which, by the initial condition $e(0)=0$, reduces to

$$e(t) = \int_0^t R(t,s)\delta(s)ds, \quad t\in I.$$

For $t=t_n\in\bar{Z}_N$ we may write

$$e(t_n) = \sum_{i=0}^{n-1} h_i\int_0^1 R(t_n,t_i+\tau h_i)\delta(t_i+\tau h_i)d\tau.$$

Replace now, in complete analogy to (5.3.28), each integral by the sum of the interpolatory quadrature formula with abscissas $\{t_i+c_l h_i: l=1, \ldots ,m\}$ and the corresponding remainder term $E_{n,i}$: since $\delta(t_i+c_l h_i)=0$, we obtain

$$e(t_n) = \sum_{i=0}^{n-1} h_i E_{n,i}, \quad n=1, \dots, N. \tag{5.4.20}$$

Note now that there is a crucial difference between (5.4.20) and the analogous expression (5.3.28) for second-kind integral equations: (5.4.20) does not contain the value $\delta(t_n)$, in contrast to (5.3.28). (Observe, however, the analogy between (5.4.20) and (5.3.51).) Hence, the condition $c_m = 1$ will not be necessary for attaining local superconvergence on $\bar{Z}_N$. This, in turn, means that the use of the Gauss abscissas as the collocation points leads to local superconvergence on $\bar{Z}_N$ of order $p^* = 2m$, and this is best possible (compare again Table 2.5.1). The statements (b) and (c) are also readily verified; note that (b) holds for *both* types of Radau points, in contrast to the result in part (a) of Theorem 5.3.3. We omit the details but refer to the conclusion of the proof of Theorem 5.3.3, also as regards local superconvergence results for equally spaced collocation parameters corresponding to odd values of m. □

So far, we have only been concerned with superconvergence results for the approximating polynomial spline $u \in S_m^{(0)}(Z_N)$ itself, but not for its derivative $u' \in S_{m-1}^{(-1)}(Z_N)$. Since one is frequently interested in generating collocation approximations u where both u and u' exhibit local superconvergence of the same order, we shall consider the following question: what is the largest integer q^* so that the collocation approximation $u \in S_m^{(0)}(Z_N)$ to the solution y of the integro-differential equation (5.4.4) satisfies

$$y^{(k)}(t_n) - u^{(k)}(t_n) = O(h^{q^*}), \quad k=0,1,$$

whenever $t_n \in \bar{Z}_N$. In other words, we are looking for a *"balanced" optimal collocation approximation* in $S_m^{(0)}(Z_N)$ which yields superconvergence of optimal order q^* on $\bar{Z}_N$ both for u and its derivative u'.

It will be shown that $q^* = 2m - 1$ is best possible; i.e., $q^* = 2m$ cannot occur.

Theorem 5.4.3a. *Let the functions p, q and K in (5.4.4) satisfy the assumptions of Theorem 5.4.2, with $\nu=0$. If $u \in S_m^{(0)}(Z_N)$ denotes the collocation approximation defined by (5.4.6), and if the collocation parameters $\{c_j\}$ are the zeros of $P_m(2s-1)$ (Gauss points for $(0,1)$), then the corresponding collocation error $e := y - u$ behaves like*

$$\max_{t_n \in \bar{Z}_N} |e(t_n)| = O(h^{2m}), \quad \max_{t_n \in \bar{Z}_N} |e'(t_n)| = O(h^m).$$

These orders are best possible and hold for all quasi-uniform meshes.

Theorem 5.4.3b. *Let p, q and K in (5.4.4) be subject to the conditions given in Theorem 5.4.2, with $\nu=1$, and let $u \in S_m^{(0)}(Z_N)$ denote the collocation approximation defined by (5.4.6), where the collocation parameters $\{c_j\}$ are the zeros of $P_{m-1}(2s-1) - P_m(2s-1)$. (Radau II points for $(0,1]$). Then*

$$\max_{t_n \in \overline{Z}_N} |e^{(k)}(t_n)| = O(h^{2m-1}), \quad k=0,1,$$

provided the mesh sequence is quasi-uniform. Here, the exponent $2m-1$ *cannot be replaced by* $2m$*; i.e., balanced local superconvergence of order* $q^* = 2m$ *is not possible.*

Proof. The proofs of Theorem 5.4.3a and 5.4.3b are based on the ideas and techniques employed in the proof of Theorem 5.4.2. Recall first from that proof that the collocation error satisfies

$$e(t) = \int_0^t R(t,s)\delta(s)ds, \quad t \in I,$$

where the defect function $\delta(t)$ is piecewise smooth and vanishes at all collocation points. Moreover, the resolvent $R(t,s)$ has the property that $R(t,t)=1$ for all $t \in I$ (cf. Theorem 1.3.4). Hence,

$$e'(t) = \delta(t) + \int_0^t R_t(t,s)\delta(s)ds, \quad t \in I,$$

where $R_t(t,s) := \partial R(t,s)/\partial t$. For $t = t_n \in \overline{Z}_N$ we may write, in complete analogy to (5.4.20),

$$e(t_n) = \sum_{i=1}^{n-1} h_i E_{n,i},$$

and

$$e'(t_n) = \delta(t_n) + \sum_{i=0}^{n-1} h_i \overline{E}_{n,i},$$

where $E_{n,i}$ and $\overline{E}_{n,i}$ denote, respectively, the quadrature errors associated with the m-point interpolatory quadrature formulas (based on the abscissas $\{t_i + c_l h_i\}$) for the integrals

$$\int_0^t R(t_n, t_i + \tau h_i)\delta(t_i + \tau h_i)d\tau$$

and

$$\int_0^t R_t(t_n, t_i + \tau h_i)\delta(t_i + \tau h_i)d\tau, \quad 0 \leqslant i \leqslant n-1.$$

Suppose first that $c_m < 1$ (i.e., t_n is not a collocation point). It is not difficult to show that this implies $\delta(t_n) = O(h^m)$, where m cannot be replaced by $m+1$ (consider, e.g., the integro-differential equation

$$y'(t) = 1 - \int_0^t y(s)ds, \quad y(0)=0,$$

whose exact solution is $y(t)=\sin(t)$). Consequently, if the collocation parameters are given by the Gauss points for $(0,1)$, then $e'(t_n)=O(h^m)$, thus establishing Theorem 5.4.3a.

Let now $0\leq c_1< \dots <c_m=1$. We obtain, for $t_n\in\overline{Z}_N$,

$$e(t_n) = \sum_{i=0}^{n-1} h_i E_{n,i}$$

and

$$e'(t_n) = \sum_{i=0}^{n-1} h_i \overline{E}_{n,i},$$

since $\delta(t_n)=0$. The (unique) m-point quadrature formula satisfying $c_m=1$ and having optimal order $q^*=2m-1$ is the Radau II formula. The assertion of Theorem 5.4.3b now follows by the arguments employed at the end of the proof of Theorem 5.4.2. □

The statements of Theorems 5.4.2 and 5.4.3 remain valid for nonlinear integro-differential equations (5.4.1), provided the partial derivatives of $f(t,y)$ and of $k(t,s,y)$ with respect to y remain uniformly bounded. We omit the details.

Since the initial-value problem for $y'=f(t,y)$ is a special case of the one for the Volterra integro-differential equation (5.4.1) (corresponding to $k(t,s,y)\equiv 0$), and since collocation in $S_m^{(0)}(Z_N)$ applied to this special problem yields an implicit Runge-Kutta method with m stages (recall (5.2.5)), Theorem 5.4.2 contains also the well-known results on the existence of such *RK* methods whose order of convergence is greater than the number of stages. If $k(t,s,y)\not\equiv 0$, then the collocation equation (5.4.6) is not yet fully discretized. It will be shown in the following section that analogous implicit Volterra-Runge-Kutta methods for (5.4.1) can be derived by a judicious choice of the quadrature formula used in the discretization of (5.4.6). More specifically, there exist *three* m-stage implicit Volterra-Runge-Kutta methods for (5.4.1) of (optimal) order $p^*=2m$, in contrast to ODEs where there is precisely *one* such method.

5.4.3. Convergence results for the fully discretized collocation equation

In Section 5.4.1 we have seen that the collocation equation defining an approximation $u\in S_m^{(0)}(Z_N)$ to the solution of the integro-differential equation (5.4.1) has the form

$$u_n'(t_{n,j}) = f\Big(t_{n,j},u_n(t_{n,j}),h_n\Phi_{n,n}^{(j)}[u_n]+\sum_{i=0}^{n-1} h_i\Phi_{n,i}^{(j)}[u_i]\Big)$$

$$j=1,\dots,m \quad (n=0,\dots,N-1),$$

with $u_0(t_0)=y(0)=y_0$. The integrals $\Phi_{n,i}^{(j)}[u_i]$ are given in (5.3.6). In order to derive the fully discretized version of this collocation equation we proceed as we did in Section 5.3.4, with one important difference concerning the meaning of the values $Y_{n,i}$: since u is now a polynomial spline of *degree m* (as compared to $m-1$ in Section 5.3), its restriction u_n to σ_n is in π_m, and hence $u_n' \in \pi_{m-1}$. We therefore set, using Lagrange's formula,

$$u_n'(t_n+\tau h_n) = \sum_{k=1}^{m} L_k(\tau) Y_{n,k}, \quad t_n+\tau h_n \in \sigma_n,$$

where $Y_{n,k} := u_n'(t_{n,k})$ (recall that in Section 5.3 the meaning of $Y_{n,k}$ was that of the value of the polynomial u_n itself at the collocation point $t_{n,j}$). Integration with respect to τ yields

$$u_n(t_n+vh_n) = y_n + h_n \sum_{k=1}^{m} \lambda_k(v) Y_{n,k},$$

with $y_n := u_n(t_n)$, and with

$$\lambda_k(v) := \int_0^v L_k(\tau)d\tau, \quad L_k(\tau) := \prod_{r\neq k}^{m} (\tau-c_r)/(c_k-c_r).$$

Accordingly, the integrals $\Phi_{n,i}^{(j)}[u_i]$ in (5.4.6) assume the form

$$\Phi_{n,i}^{(j)}[u_i] = \begin{cases} \int_0^1 k(t_{n,j}, t_i+vh_i, y_i+h_i \sum_{k=1}^{m} \lambda_k(v) Y_{i,k}) dv, & \text{if } 0\leqslant i<n, \\ \int_0^{c_j} k(t_{n,j}, t_n+vh_n, y_n+h_n \sum_{k=1}^{m} \lambda_k(v) Y_{n,k}) dv, & \text{if } i=n, \end{cases} \tag{5.4.21}$$

where $j=1, \ldots, m$. As in (5.3.33), the discretization of these integrals will be obtained by quadrature formulas using the abscissas $\{t_i+d_l h_i: 0\leqslant d_1< \ldots <d_{\mu_1}\leqslant 1\}$ (if $i<n$), or $\{t_n+d_{j,l}h_n: 0\leqslant d_{j,1}< \ldots <d_{j,\mu_0}\leqslant c_j\}$, $j=1, \ldots, m$ (if $i=n$); the resulting approximations are then given by

$$\hat{\Phi}_{n,i}^{(j)}[u_i] := \begin{cases} \sum_{l=1}^{\mu_1} w_l k(t_{n,j}, t_i+d_l h_i, y_i+h_i \sum_{k=1}^{m} \lambda_k(d_l) Y_{i,k}), & 0\leqslant i<n, \\ \sum_{l=1}^{\mu_0} w_{j,l} k(t_{n,j}, t_n+d_{j,l}h_n, y_n+h_n \sum_{k=1}^{m} \lambda_k(d_{j,l}) Y_{n,k}), & i=n, \end{cases} \tag{5.4.22}$$

$$j=1, \ldots, m.$$

The *fully discretized* version of the *collocation equation* (5.4.6) is therefore:

$$\hat{u}_n'(t_{n,j}) = f(t_{n,j}, \hat{u}_n(t_{n,j}), h_n\hat{\Phi}_{n,n}^{(j)}[\hat{u}_n] + \sum_{i=0}^{n-1} h_i \hat{\Phi}_{n,i}^{(j)}[\hat{u}_n]), \tag{5.4.23}$$

$$j=1, \ldots, m \quad (n=0, \ldots, N-1),$$

where $\hat{u}_0(t_0)=y(0)=y_0$. For all sufficiently small positive h, (5.4.23) defines an approximation $\hat{u} \in S_m^{(0)}(Z_N)$ which, in general, will be different from the approximation u (in the same space) generated by the exact collocation equation (5.4.6).

For computational purposes it will be convenient to rewrite (5.4.23) by expressing $\hat{u}_n'(t_{n,j})$ and $\hat{u}_n(t_{n,j})$ in terms of $\hat{Y}_{n,k}$; one obtains:

$$\hat{Y}_{n,j} = f(t_{n,j}, \hat{y}_n + h_n \sum_{l=1}^{m} a_{j,l} \hat{Y}_{n,l},\ h_n \hat{\Phi}^{(j)}_{n,n}[\hat{u}_n] + \sum_{i=0}^{n-1} h_i \hat{\Phi}^{(j)}_{n,i}[\hat{u}_i]), \tag{5.4.24a}$$

$$j=1, \ldots ,m \quad (n=0, \ldots ,N-1),$$

with $\hat{y}_n := \hat{u}_n(t_n)$ and $a_{j,l} := \lambda_l(c_j)$; the quadrature approximations $\hat{\Phi}^{(j)}_{n,i}[\hat{u}_i]$ are of the form (5.4.22) where the values $Y_{i,k}$ have now been replaced by $\hat{Y}_{i,k}$. Once the solution $(\hat{Y}_{n,1}, \ldots , \hat{Y}_{n,m})^T$ of the nonlinear system (5.4.24a) has been found, the collocation approximation $\hat{u}$ on σ_n is given by

$$\hat{u}_n(t_n + \tau h_n) = \hat{y}_n + h_n \sum_{k=1}^{m} \lambda_k(\tau) \hat{Y}_{n,k}, \quad t_n + \tau h_n \in \sigma_n. \tag{5.4.24b}$$

More specifically, for $\tau=1$ we have

$$\hat{y}_{n+1} := \hat{u}_n(t_n + h_n) = \hat{y}_n + h_n \sum_{k=1}^{m} b_k \hat{Y}_{n,k}, \tag{5.4.24c}$$

with $b_k := \lambda_k(1)$, and with $\hat{y}_0 = y_0 = y(0)$.

For the particular choice of $\mu_0 = \mu_1 = m$, $d_l = c_l$, $d_{j,l} = c_j c_l$ $(l,j=1, \ldots ,m)$ in (5.4.22), the quadrature expressions $\hat{\Phi}^{(j)}_{n,i}[\hat{u}_i]$ in (5.4.24a) assume the form

$$\hat{\Phi}^{(j)}_{n,i}[\hat{u}_i] = \begin{cases} \sum_{l=1}^{m} b_l k(t_{n,j}, t_{i,l}, \hat{y}_i + h_i \sum_{k=1}^{m} a_{l,k} \hat{Y}_{i,k}), & \text{if } i<n, \\ \sum_{l=1}^{m} \bar{a}_{j,l} k(t_{n,j}, t_n + c_j c_l h_n, \hat{y}_n + h_n \sum_{k=1}^{m} \lambda_k(c_j c_l) \hat{Y}_{n,k}), & \text{if } i=n; \end{cases} \tag{5.4.25}$$

here, we have

$$a_{l,k} := \lambda_k(c_l) = \int_0^{c_l} L_k(v)dv, \quad \bar{a}_{j,l} = c_j b_l, \quad b_l := \int_0^1 L_l(v)dv.$$

If the kernel $k(t,s,y)$ can be smoothly extended to $S' := \{(t,s): 0 \leqslant s \leqslant t+\delta\} \cap I \times I$, for some $\delta > 0$, then one may also resort to the following discretization of the integrals $\Phi^{(j)}_{n,n}[u_n]$ (corresponding to the choice $\mu_0 = m$, $d_{j,l} = c_l$, for all values of j):

$$\hat{\Phi}^{(j)}_{n,n}[\hat{u}_n] = \sum_{l=1}^{m} a_{j,l} k(t_{n,j}, t_n + c_l h_n, \hat{y}_n + h_n \sum_{k=1}^{m} a_{l,k} \hat{Y}_{n,k}), \tag{5.4.26}$$

$$j=1, \ldots ,m,$$

where $a_{l,k}=\lambda_k(c_l)$. In this case, the discretization (5.4.24a) of the collocation equation (5.4.6) is completely characterized by the matrix $A=(a_{j,l})$ and by the vector $\mathbf{b}=(b_1, \dots ,b_m)^T$, which in turn are uniquely determined by the collocation parameters $\{c_1, \dots ,c_m\}$. This discretized collocation method represents an *extended m-stage implicit PVDRK method* (cf. (4.2.12), setting $d_{i,l}=c_i$ and $\bar{a}_{j,l}=a_{j,l}$).

Before presenting a number of specific discretizations of the exact collocation equation (5.4.6) we state

Theorem 5.4.4. *Let p, q and K in (5.4.4) possess continuous derivatives of sufficiently high order on their respective domains. Assume that the quadrature formulas employed in (5.4.22) satisfy, respectively,*

$$\int_0^1 \phi(t_i+vh_i)dv-\sum_{l=1}^{\mu_1} w_l\phi(t_i+d_lh_i) = O(h_i^{r_1}),\quad i<n,$$

and

$$\int_0^{c_j} \phi(t_n+vh_n)dv-\sum_{l=1}^{\mu_0} w_{j,l}\phi(t_n+d_{j,l}h_n) = O(h_n^{r_0}),\quad j=1, \dots ,m,$$

whenever the integrand is a sufficiently smooth function. If $u\in S_m^{(0)}(Z_N)$ and $\hat{u}\in S_m^{(0)}(Z_N)$ denote the approximations to the solution of the integro-differential equation (5.4.4) determined by the collocation equation (5.4.6) and its full discretization (5.4.23), then there exist finite constants Q_0 and Q_1 not depending on h and N such that

$$\|u-\hat{u}\|_\infty\leqslant Q_0h^r,\ \|u'-\hat{u}'\|_\infty\leqslant Q_1h^r,$$

with $r:=\min(r_0+1,\ r_1)$, provided the given mesh sequence is quasi-uniform and $h=\max_{(n)}h_n$ is sufficiently small.

Proof. Let $\epsilon_n:=u_n-\hat{u}_n$, $\epsilon_n':=u_n'-\hat{u}_n'$, and set

$$E_{n,i}^{(j)} := \Phi_{n,i}^{(j)}[u_i]-\hat{\Phi}_{n,i}^{(j)}[u_i],\quad j=1, \dots ,m\quad (0\leqslant i\leqslant n\leqslant N-1),$$

with $\Phi_{n,i}^{(j)}[u_i]$ and $\hat{\Phi}_{n,i}^{(j)}[u_i]$ given by (5.3.6) and (5.4.21). Subtracting (5.4.23) from (5.4.6) we get

$$\begin{aligned}\epsilon_n'(t_{n,j}) &= p_{n,j}\epsilon_n(t_{n,j})+h_n(\Phi_{n,n}^{(j)}[u_n]-\hat{\Phi}_{n,n}^{(j)}[\hat{u}_n])\\ &+\sum_{i=0}^{n-1} h_i(\Phi_{n,i}^{(j)}[u_i]-\hat{\Phi}_{n,i}^{(j)}[\hat{u}_i])= p_{n,j}\epsilon_n(t_{n,j})+h_n(\hat{\Phi}_{n,n}^{(j)}[u_n]-\hat{\Phi}_{n,n}^{(j)}[\hat{u}_n])\\ &+\sum_{i=0}^{n-1} h_i(\hat{\Phi}_{n,i}^{(j)}[u_i]-\hat{\Phi}_{n,i}^{(j)}[\hat{u}_i])+\sum_{i=0}^{n} h_iE_{n,i}^{(j)},\end{aligned}$$

or, using the linearity of the quadrature functionals,

$$\epsilon_n'(t_{n,j}) = p_{n,j}\epsilon_n(t_{n,j}) + h_n\hat{\Phi}_{n,n}^{(j)}[\epsilon_n] + \sum_{i=0}^{n-1} h_i\hat{\Phi}_{n,i}^{(j)}[\epsilon_i] + \sum_{i=0}^{n} h_i E_{n,i}^{(j)}, \tag{5.4.27}$$

$$j=1, \ldots, m \quad (n=0, \ldots, N-1),$$

with $p_{n,j} := p(t_{n,j})$. Since $\epsilon_n \in \pi_m$, we may set

$$\epsilon_n(t_n + \tau h_n) = \sum_{l=0}^{m} \delta_{n,l}\tau^l \quad (\text{with } t_n + \tau h_n \in \sigma_n,\ \tau = (t-t_n)/h_n).$$

By definition, $\epsilon := u - \hat{u}$ is a member of $S_m^{(0)}(Z_N)$; the continuity requirements for its knots Z_N yield, by induction,

$$\delta_{n,0} = \delta_{0,0} + \sum_{i=0}^{n-1}\sum_{l=1}^{m}\delta_{i,l} \quad (n=1, \ldots, N-1), \tag{5.4.28}$$

with $\delta_{0,0} = \epsilon(t_0) = u_0(0) - \hat{u}_0(0) = 0$. Thus, using all the above in (5.4.27), we easily derive the recurrence relation

$$\sum_{l=1}^{m}\{lc_j^{l-1} - h_n p_{n,j} c_j^l - h_n^2 \sum_{s=1}^{\mu_0} w_{j,s} K_{n,j}(t_n + d_{j,s}h_n) d_{j,s}^l\}\delta_{n,l}$$

$$= h_n\{p_{n,j} + h_n \sum_{s=1}^{\mu_0} w_{j,s}K_{n,j}(t_n + d_{j,s}h_n)\}(\sum_{i=0}^{n-1}\sum_{l=1}^{m}\delta_{i,l})$$

$$+ h_n \sum_{i=0}^{n-1} h_i \sum_{l=1}^{m}(\sum_{s=1}^{\mu_1} w_s K_{n,j}(t_i + d_s h_i) d_s^l)\delta_{i,l} \tag{5.4.29}$$

$$+ h_n \sum_{i=0}^{n-1} h_i \sum_{s=1}^{\mu_1} w_s K_{n,j}(t_i + d_s h_i)(\sum_{r=1}^{i-1}\sum_{l=1}^{m}\delta_{r,l}) + h_n \sum_{i=0}^{n} h_i E_{n,i}^{(j)},$$

$$j=1, \ldots, m \quad (n=0, \ldots, N-1).$$

(Observe the analogy between (5.4.29) and (5.4.13).)

Since the above recurrence relation has the same structure as (5.4.13), we may complete the proof of Theorem 5.4.3 by adapting the ideas used in the proof of Theorem 5.4.1 (starting after (5.4.14)): the l_1-norms of the vectors $\boldsymbol{\delta}_n := (\boldsymbol{\delta}_{n,1}, \ldots, \boldsymbol{\delta}_{n,m})^T$ are related by a discrete Gronwall inequality of the type

$$\|\boldsymbol{\delta}_n\|_1 \leqslant hD_0 \sum_{i=0}^{n-1} \|\boldsymbol{\delta}_i\|_1 + h \cdot D_1 h^r \quad (n=1, \ldots, N-1);$$

here, the last term includes an upper bound for $\Sigma_{i=0}^{n} h_i E_{n,i}^{(j)}$ in (5.4.29):

$$|\sum_{i=0}^{n} h_i E_{n,i}^{(j)}| \leqslant h \sum_{i=0}^{n} |E_{n,i}^{(j)}| \leqslant h(Q_0 h^{r_0} + N \cdot Q_1 h^{r_1})$$

$$\leqslant Q_0 h^{r_0+1} + \gamma T \cdot Q_1 h^{r_1} = O(h^r), \quad r := \min(r_0+1, r_0)$$

$$(0 \leqslant n \leqslant N-1;\ Nh \leqslant \gamma T),$$

where we have taken into consideration the hypotheses on the quadrature errors. Hence, according to Corollary 1.5.2, we obtain the estimate

$$\|\delta_n\|_1 \leqslant hD_1h^r(1+hD_0)\exp(D_0\gamma T) =: hDh^r \quad (n=0, \ldots ,N-1;\ Nh \leqslant \gamma T).$$

This in turn implies, by (5.4.28),

$$|\delta_{n,0}| \leqslant \sum_{i=0}^{n-1} \|\delta_i\|_1 \leqslant Nh \cdot Dh^r \leqslant \gamma T \cdot Dh^r \quad (n=1, \ldots ,N-1).$$

Thus, for all $t_n+\tau h_n \in \sigma_n$ and all $n=0, \ldots ,N-1$, with $Nh \leqslant \gamma T$, we find

$$|\epsilon_n(t_n+\tau h_n)| \leqslant |\delta_{n,0}| + \|\delta_n\|_1 \leqslant (\gamma TD+hD)h^r = O(h^r),$$

and

$$|\epsilon_n'(t_n+\tau h_n)| \leqslant h_n^{-1} |\sum_{l=1}^{m} l \cdot \delta_{n,l}| \leqslant (\min_{(n)} h_n)^{-1} m \|\delta_n\|_1$$

$$\leqslant \frac{h}{\min_{(n)} h_n} \cdot mDh^r \leqslant \gamma mDh^r = O(h^r).$$

This is equivalent to the assertion in Theorem 5.4.4. □

Again, the extension of the above arguments to the nonlinear integro-differential equation (5.4.1) is a straightforward matter: the roles of $p_{n,j}$ and of $K_{n,j}$ in (5.4.29) will be assumed by the partial derivatives $f_y(t_{n,j}, \cdot, \cdot)$ and $k_y(t_{n,j}, t_i+d_sh_i, \cdot)$, evaluated at appropriate last arguments. If these partial derivatives are bounded, then the assertions of Theorem 5.4.4 are easily established for the nonlinear case. The details are left to the reader. □

Since the error function corresponding to the approximation $\hat{u} \in S_m^{(0)}(Z_N)$, $\hat{e} := y - \hat{u}$, can be written as $\hat{e} = (y-u)+(u-\hat{u}) = e+\epsilon$, the following result is readily deduced from Theorem 5.4.1, Theorem 5.4.4, and the basic property of m-point interpolatory quadrature formulas (see Section 2.1.2).

Corollary 5.4.1. *Let p, q and K in (5.4.4) be m times continuously differentiable on their respective domains I and S. Assume that $\hat{u} \in S_m^{(0)}(Z_N)$ has been obtained as the solution of the fully discretized collocation equation (5.4.23), where the quadrature approximations $\hat{\Phi}_{n,i}^{(j)}[\hat{u}_i]$ are of interpolatory type of the form (5.4.22), with $\mu_0 = \mu_1 = m$. Then we have*

$$\|\hat{e}^{(k)}\|_\infty = O(h^m), \quad k=0,1 \quad (\textit{as } h \downarrow 0 \textit{ and } Nh \leqslant \gamma T).$$

Furthermore, the discrete analogue of Theorem 5.4.2 (dealing with local superconvergence on $\bar{Z}_N$) holds for the choices of the $\{c_j\}$ listed in that theorem, provided the quadrature approximations are suitably chosen. Instead of stating these results in detail, we shall only focus on the case of optimal local superconvergence.

Theorem 5.4.5. *In (5.4.4), let p, q and K be in $C^{2m}(I)$ and in $C^{2m}(S)$, respectively, where $m \geq 1$. Suppose that the fully discretized collocation equation (5.4.23) is characterized by interpolatory m-point quadrature approximations of the form (5.4.22) (i.e., $\mu_0 = \mu_1 = m$). Then the resulting approximation $\hat{u} \in S_m^{(0)}(Z_N)$ has the property that*

$$\max_{t_n \in Z_N} |\hat{e}(t_n)| = O(h^{2m}), \quad (\text{as } h \downarrow 0 \text{ and } Nh \leq \gamma T)$$

if, and only if, (a), (b), and one of (c), (c′), (c″) hold:

(a) *the collocation parameters $\{c_j\}$ are the zeros of $P_m(2s-1)$ (Gauss points for $(0,1)$);*

(b) $d_l = c_l$ $(l = 1, \ldots, m)$;

(c) $d_{j,l} = c_j c_l$ $(l,j = 1, \ldots, m)$;

(c′) $d_{j,l} = c_j c_l'$ $(l,j = 1, \ldots, m)$, *where the $\{c_l'\}$ denote the zeros of $P_{m-1}(2s-1) + P_m(2s-1)$ (Radau I points for $[0,1)$);*

(c″) $d_{j,l} = c_j c_l''$ $(l,j = 1, \ldots, m)$, *where the $\{c_l''\}$ are the zeros of $P_{m-1}(2s-1) - P_m(2s-1)$ (Radau II points for $(0,1]$).*

Proof. According to Theorem 5.4.4, we have $\epsilon(t_n) = O(h^r)$, with $r = \min(r_0 + 1, r_1)$. Since the choice of the Gauss points as the collocation parameters yields $e(t_n) = O(h^{2m})$ (recall (5.4.16)), and since $\hat{e}(t_n) = e(t_n) + \epsilon(t_n)$, we must have $r_0 \geq 2m-1$ and $r_1 \geq 2m$ (note that, for an m-point quadrature formula considered here, $r_1 = 2m$ is best possible). But:

(i) Assumption (b) implies $r_1 = 2m$ (m-point Gauss quadrature for the integrals $\Phi_{n,i}^{(j)}[u_i]$, where $i < n$);

(ii) assumption (c) implies $r_0 = 2m$ (m-point Gauss quadrature for $\Phi_{n,n}^{(j)}[u_n]$);

(iii) assumptions (c′) and (c″) both imply $r_0 = 2m-1$ (m-point Radau quadratures for $\Phi_{n,n}^{(j)}[u_n]$). □

We refrain from stating explicitly the discrete analogue of Theorem 5.4.3b: it is clear from Theorem 5.4.4 and 5.4.5 that there exist *three balanced optimal collocation methods* in $S_m^{(0)}(Z_N)$ where both $\hat{u}$ and $\hat{u}'$ exhibit local superconvergence of order $q^* = 2m-1$ on $\bar{Z}_N$. (Compare also Example 5.4.4 below.)

We conclude this section by observing that if the kernel $k(t,s,y)$ can be smoothly extended to $S' := \{(t,s): 0 \leq s \leq t + \delta\} \cap I \times I$, for some $\delta > 0$ (such as to be in C^{2m} with respect to t and s), then the use of quadrature approximations of the forms (5.4.26) in the fully discretized collocation equation (5.4.23) permits the construction of three more methods possessing optimal order $p^* = 2m$ on $\bar{Z}_N$. Instead of conditions (c), (c′), and (c″) in Theorem 5.4.4 one has, respectively, $d_{j,l} = c_l$; $d_{j,l} = c_l'$; and $d_{j,l} = c_l''$. As we indicated following (5.4.26), these discretized collocation methods are extended m-stage implicit PVDRK methods, and they possess the optimal order $p^* = 2m$.

5.4.4. Examples of optimal methods

In the present section we list the three optimal, fully discretized collocation methods for $\hat{u} \in S_2^{(0)}(Z_N)$; since $m=2$, the order of convergence of the resulting approximations at $t=t_n \in \overline{Z}_N$ is given by $p^*=2m=4$. In view of practical applications we present the methods for the VIDE

$$y'(t) = f(t,y(t))+\int_0^t k(t,s,y(s))ds, \quad t\in I, \quad y(0)=y_0.$$

These methods are characterized by the collocation parameters $c_1=(3-\sqrt{3})/6$, $c_2=(3+\sqrt{3})/6$ (Gauss points for $(0,1)$), and they have the form

$$\begin{aligned}
\hat{Y}_{n,j} = {} & f(t_{n,j},\hat{y}_n+h_n(a_{j,1}\hat{Y}_{n,1}+a_{j,2}\hat{Y}_{n,2})) \\
& +h_n\{\bar{a}_{j,1}k(t_{n,j},t_n+d_{j,1}h_n,\hat{y}_n+h_n(\lambda_1(d_{j,1})\hat{Y}_{n,1}+\lambda_2(d_{j,1})\hat{Y}_{n,2})) \\
& +\bar{a}_{j,2}k(t_{n,j},t_n+d_{j,2}h_n,\hat{y}_n+h_n(\lambda_1(d_{j,2})\hat{Y}_{n,1}+\lambda_2(d_{j,2})\hat{Y}_{n,2}))\} \\
& +\sum_{i=0}^{n-1} h_i\{b_1k(t_{n,j},t_{i,1},\hat{y}_i+h_i(a_{1,1}\hat{Y}_{i,1}+a_{1,2}\hat{Y}_{i,2})) \\
& +b_2k(t_{n,j},t_{i,2},\hat{y}_i+h_i(a_{2,1}\hat{Y}_{i,1}+a_{2,2}\hat{Y}_{i,2}))\}, \quad j=1,2;
\end{aligned} \tag{5.4.30a}$$

$$\hat{y}_{n+1} := \hat{u}_n(t_{n+1})=\hat{y}_n+h_n(b_1\hat{Y}_{n,1}+b_2\hat{Y}_{n,2}), \quad n=0, \dots ,N-1. \tag{5.4.30b}$$

Here, the weights $\{a_{j,l}: j,l=1,2\}$ and $\{b_l: l=1,2\}$ depend only on the collocation parameters; for the three methods given below these weights are:

$$a_{1,1} = 1/4,\ a_{1,2}=(3-2\sqrt{3})/12,\ a_{2,1}=(3+2\sqrt{3})/12,\ a_{2,2}=1/4;\ b_1=b_2=1/2.$$

Example 5.4.1. $d_{j,l}=c_jc_l$:

$$\begin{aligned}
d_{1,1} &= c_1^2=(2-\sqrt{3})/6,\ d_{1,2}=d_{2,1}=c_1c_2=1/6,\ d_{2,2}=c_2^2=(2+\sqrt{3})/6; \\
\bar{a}_{1,1} &= c_1b_1=(3-\sqrt{3})/12,\ \bar{a}_{1,2}=c_1b_2=(3-\sqrt{3})/12, \\
\bar{a}_{2,1} &=c_2b_1=(3+\sqrt{3})/12,\ \bar{a}_{2,2}=c_2b_2=(3+\sqrt{3})/12; \\
\lambda_1(d_{1,1}) &= (6-\sqrt{3})/72,\ \lambda_2(d_{1,1})=(18-11\sqrt{3})/72, \\
\lambda_1(d_{1,2}) &= (6+5\sqrt{3})/72,\ \lambda_2(d_{1,2})=(6-5\sqrt{3})/72, \\
\lambda_1(d_{2,1}) &= (6+5\sqrt{3})/72,\ \lambda_2(d_{2,1})=(6-5\sqrt{3})/72, \\
\lambda_1(d_{2,2}) &= (18+11\sqrt{3})/72,\ \lambda_2(d_{2,2})=(6+\sqrt{3})/72. \quad \square
\end{aligned}$$

Example 5.4.2. $d_{j,l}=c_jc_l'$, with $c_1'=0$, $c_2'=2/3$ (Radau I points for $[0,1)$):

$$\begin{aligned}
d_{1,1} &= c_1c_1'=0, \quad d_{1,2}=c_1c_2'=(3-\sqrt{3})/9, \\
d_{2,1} &= c_2c_1'=0, \quad d_{2,2}=c_2c_2'=(3+\sqrt{3})/9;
\end{aligned}$$

$$\bar{a}_{1,1} = (3-\sqrt{3})/24, \quad \bar{a}_{1,2}=(3-\sqrt{3})/8,$$
$$\bar{a}_{2,1} = (3+\sqrt{3})/24, \quad \bar{a}_{2,2}=(3+\sqrt{3})/8;$$
$$\lambda_1(d_{1,1}) = 0, \quad \lambda_2(d_{1,1})=0,$$
$$\lambda_1(d_{1,2}) = (3+\sqrt{3})/27, \quad \lambda_2(d_{1,2})=(6-4\sqrt{3})/27,$$
$$\lambda_1(d_{2,1}) = 0, \quad \lambda_2(d_{2,1})=0,$$
$$\lambda_1(d_{2,2}) = (6+4\sqrt{3})/27, \quad \lambda_2(d_{2,2})=(3-\sqrt{3})/27. \quad \square$$

Example 5.4.3. $d_{j,l}=c_jc_l''$, with $c_1''=1/3$, $c_2''=1$ (Radau II points for $(0,1]$):

$$d_{1,1} = c_1c_1''=(3-\sqrt{3})/18, \quad d_{1,2}=c_1c_2''=(3-\sqrt{3})/6,$$
$$d_{2,1} = c_2c_1''=(3+\sqrt{3})/18, \quad d_{2,2}=c_2c_2''=(3+\sqrt{3})/6;$$
$$\bar{a}_{1,1} = (3-\sqrt{3})/18, \quad \bar{a}_{1,2}=(3-\sqrt{3})/24,$$
$$\bar{a}_{2,1} = (3+\sqrt{3})/8, \quad \bar{a}_{2,2}=(3+\sqrt{3})/24;$$
$$\lambda_1(d_{1,1}) = (3+4\sqrt{3})/108, \quad \lambda_2(d_{1,1})=5(3-2\sqrt{3})/108,$$
$$\lambda_1(d_{1,2}) = 1/4, \quad \lambda_2(d_{1,2})=(3-2\sqrt{3})/12,$$
$$\lambda_1(d_{2,1}) = 5(3+2\sqrt{3})/108, \quad \lambda_2(d_{2,1})=(3-4\sqrt{3})/108,$$
$$\lambda_1(d_{2,2}) = (3+2\sqrt{3})/12, \quad \lambda_2(d_{2,2})=1/4. \quad \square$$

Example 5.4.4. We now choose for the collocation parameters the Radau II points: $c_1=1/3$, $c_2=1$. According to Theorem 5.4.3b and the remark following the proof of Theorem 5.4.5, the resulting approximations $\hat{y}_{n+1}$ and $\hat{y}'_{n+1}:=\hat{u}'_n(t_{n+1})$ $(=\hat{Y}_{n,2})$ both converge like $O(h^3)$, while the approximations $\hat{y}'_{n+1}$ defined by the previous three examples converge only quadratically. The parameters characterizing the fully discretized collocation method are:

$$a_{1,1} = 5/1,2, \quad a_{12} = -1/12,$$
$$a_{2,1} = 3/4, \quad a_{2,2} = 1/4;$$
$$b_1 = 3/4, \quad b_2 = 1/4;$$
$$d_{1,1} = c_1^2=1/9, \quad d_{1,2} = d_{2,1}=c_1c_2=1/3, \quad d_{2,2} = c_2^2=1;$$
$$\bar{a}_{1,1} = 1/4, \quad \bar{a}_{1,2} = 1/12,$$
$$\bar{a}_{2,1} = 3/4, \quad \bar{a}_{2,2} = 1/4;$$
$$\lambda_1(d_{1,1}) = 51/324, \quad \lambda_2(d_{1,1}) = -15/324,$$
$$\lambda_1(d_{1,2}) = 15/36, \quad \lambda_2(d_{1,2}) = -3/36,$$
$$\lambda_1(d_{2,2}) = 3/4, \quad \lambda_2(d_{2,2}) = 1/4. \quad \square$$

5.4.5. Numerical examples

The methods specified in the preceding section will be applied to the integro-differential equations used in Section 3.4.4 and 4.2.6 to illustrate the VLM and VDRK methods. As before, the integration steps are chosen in such a way that the results listed below are comparable (with regard to the total number of kernel evaluations) with the corresponding results given in those sections. The methods specified in the Examples 5.4.1-5.4.4 are respectively denoted by $VDRK_4(G-G)$, $VDRK_4(G-RI)$, $VDRK_4(G-RII)$ and $VDRK_3(RII-RII)$. The accuracy of $\hat{y}_N$ as well as $\hat{y}'_N$ are listed (recall that $\hat{y}'_N = \Sigma_{k=1}^m L_k(1)\hat{Y}_{N-1,k}$; if $c_m = 1$ then $\hat{y}'_N = \hat{Y}_{N-1,m}$).

Table 5.4.1. Numbers of correct significant digits and effective orders for problem (3.4.11) at $t=T=10$

	h	$VDRK_4$ $(G-G)$		$VDRK_4$ $(G-RI)$		$VDRK_4$ $(G-RII)$		$VDRK_3$ $(RII-RII)$	
	1/2	4.2	> 4.3	4.2	> 4.3	4.2	> 4.3	4.5	> 2.8
	1/4	5.5	> 4.1	5.5	> 4.1	5.5	> 4.1	5.3	> 2.9
$\hat{y}_N$	1/8	6.8	> 4.0	6.8	> 4.0	6.8	> 4.0	6.2	> 2.9
	1/16	8.0	> 4.0	8.0	> 4.0	8.0	> 4.0	7.1	> 2.9
	1/32	9.2		9.2		9.2		8.0	
	1/2	2.0	> 3.1	2.0	> 3.1	2.0	> 3.1	2.9	> 2.3
	1/4	3.0	> 2.8	3.0	> 2.8	3.0	> 2.8	3.6	> 2.4
$\hat{y}'_N$	1/8	3.8	> 2.5	3.8	> 2.5	3.8	> 2.5	4.3	> 2.5
	1/16	4.5	> 2.4	4.5	> 2.4	4.5	> 2.4	5.0	> 2.7
	1/32	5.3		5.3		5.3		5.8	

Table 5.4.2. Numbers of correct significant digits and effective orders for problem (3.4.12) at $t=T=2$

	h	$VDRK_4$ $(G-G)$		$VDRK_4$ $(G-RI)$		$VDRK_4$ $(G-RII)$		$VDRK_3$ $(RII-RII)$	
	1/5	4.8	> 4.0	4.7	> 3.8	4.9	> 4.2	3.6	> 3.0
$\hat{y}_N$	1/10	5.9	> 4.0	5.8	> 3.9	6.1	> 4.1	4.5	> 3.0
	1/20	7.1		7.0		7.4		5.4	
	1/5	4.2	> 4.0	4.2	> 3.8	4.2	> 4.1	3.1	> 3.0
$\hat{y}'_N$	1/10	5.4	> 4.0	5.3	> 3.9	5.4	> 4.1	4.0	> 3.0
	1/20	6.6		6.5		6.7		4.9	

Table 5.4.3. Numbers of correct digits and effective orders for problem (3.4.13) at $t=T=4$

	h	$VDRK_4$ $(G-G)$		$VDRK_4$ $(G-RI)$		$VDRK_4$ $(G-RII)$		$VDRK_3$ $(RII-RII)$	
$\hat{y}_N$	1/5	2.6		2.6		2.6		0.9	
			> 4.0		> 3.9		> 4.0		> 3.0
	1/10	3.8		3.8		3.8		1.8	
			> 4.0		> 4.0		> 4.0		> 3.0
	1/20	5.0		5.0		5.0		2.7	
$\hat{y}'_N$	1/5	2.1		2.1		2.1		0.4	
			> 4.0		> 3.9		> 4.0		> 3.0
	1/10	3.3		3.3		3.3		1.3	
			> 4.0		> 4.0		> 4.0		> 3.0
	1/20	4.5		4.5		4.5		2.2	

5.5. Polynomial spline collocation for integral equations of the first kind

5.5.1. Global convergence results

In analogy to Section 5.3, the solution of the first-kind Volterra integral equation,

$$g(t)+\int_0^t K(t,s)y(s)ds = 0, \quad t\in I, \tag{5.5.1}$$

will be approximated by a polynomial spline $u\in S_{m-1}^{(-1)}(Z_N)$, with $m\geq 1$. For $t\in\sigma_n$, (5.5.1) may be rewritten as

$$F_n(t)+\int_{t_n}^t K(t,s)y(s)ds = 0, \quad n=0, \dots ,N-1, \tag{5.5.2}$$

where we have set

$$F_n(t) := g(t)+\int_0^{t_n} K(t,s)y(s)ds. \tag{5.5.3}$$

The approximating element u is then determined from the collocation equation corresponding to (5.5.2),

$$\bar{F}_n(t_{n,j})+h_n\int_0^{c_j} K(t_{n,j},t_n+vh_n)u_n(t_n+vh_n)dv = 0, \tag{5.5.4}$$

$$j=1, \dots ,m \quad (n=0, \dots ,N-1),$$

where $\bar{F}_n$ is given by

$$\bar{F}_n(t) := g(t)+\sum_{i=0}^{n-1} h_i\int_0^1 K(t,t_i+vh_i)u_i(t_i+vh_i)dv. \tag{5.5.5}$$

Adopting the notation introduced in Section 5.3.1, the collocation equation

(5.5.4) may be written, in analogy to (5.3.7), in the form

$$h_n \Phi_{n,n}^{(j)}[u_n] = -(g(t_{n,j}) + \sum_{i=0}^{n-1} h_i \Phi_{n,i}^{(j)}[u_i]), \tag{5.5.6}$$

$$j=1, \dots, m \quad (n=0, \dots, N-1),$$

where the integrals $\Phi_{n,i}^{(j)}[u_i]$ are defined as in (5.3.6) (and where $k(t,s,y)$ has now been replaced by $K(t,s)y$).

The convergence behavior of the resulting approximation is described by the following theorem.

Theorem 5.5.1a. *Let g and K in (5.5.1) possess continuous derivatives of order $m+1$ on their respective domains, and suppose that $g(0)=0$, $|K(t,t)| \geq \rho > 0$ for all $t \in I$. Then the collocation equation (5.5.6) defines, for all quasi-uniform mesh sequences with sufficiently small $h := \max_{(n)} h_n$, a unique approximation $u \in S_{m-1}^{(-1)}(Z_N)$. This collocation approximation yields an error with*

$$\|e\|_\infty = O(h^m) \tag{5.5.7}$$

(as $h \downarrow 0$ and $Nh \leq \gamma T$) , for all collocation parameters $\{c_j\}$ with $0 < c_1 < \dots < c_m = 1$. If $0 < c_1 < \dots c_m < 1$, then (5.5.7) holds for all $m \geq 1$ if, and only if,

$$\prod_{j=1}^{m} (1-c_j)/c_j < 1. \tag{5.5.8}$$

Remark. The last assertion may be slightly weakened: if we consider only *odd* values of m, then (5.5.7) remains valid if $\Pi_{j=1}^m (1-c_j)/c_j \leq 1$. However, in the case of equality, the corresponding approximation u exhibits a behavior often called *weak instability* (or: weak stability). We refer to Brunner [1978a] for details.

Proof. We give the proof for the case $c_m = 1$; for the case $c_m < 1$, the reader is referred to Brunner [1978a]. It follows from (5.5.2) and (5.5.4) that the error function $e := y - u$, with $e_n(t) := y(t) - u_n(t)$, $t \in \sigma_n$, satisfies the recurrence relation

$$h_n \int_0^{c_j} K_{n,j}(t_n + vh_n) e_n(t_n + vh_n) dv = -\sum_{i=0}^{n-1} h_i \int_0^1 K_{n,j}(t_i + vh_i) e_i(t_i + vh_i) dv, \tag{5.5.9}$$

$$j=1, \dots, m \quad (n=0, \dots, N-1),$$

where we have again set $K_{n,j}(\cdot) := K(t_{n,j}, \cdot)$. Since the assumptions for g and K imply the existence of a unique solution $y \in C^m(I)$ for (5.5.1), we may write, using the notation introduced in the proof of Theorem 5.3.2,

$$e_n(t_n + vh_n) = h_n^m \{ \sum_{l=1}^{m} \beta_{n,l} v^{l-1} + R_n(v) \}, \quad t_n + vh_n \in \sigma_n.$$

The recurrence relation (5.5.9) can thus be converted into a recurrence relation for the components of the vectors $\boldsymbol{\beta}_n := (\beta_{n,1}, \ldots, \beta_{n,m})^T$:

$$h_n^{m+1} \sum_{l=1}^{m} \beta_{n,l} \int_0^{c_j} K_{n,j}(t_n + vh_n) v^{l-1} dv \tag{5.5.10}$$

$$= -\sum_{i=0}^{n-1} h_i^{m+1} \sum_{l=1}^{m} \beta_{i,l} \int_0^1 K_{n,j}(t_i + vh_i) v^{l-1} dv + q_{n,j},$$

$$j = 1, \ldots, m \quad (n = 0, \ldots, N-1),$$

with

$$q_{n,j} := -\{h_n^{m+1} \int_0^{c_j} K_{n,j}(t_n + vh_n) R_n(v) dv \tag{5.5.11}$$

$$+ \sum_{i=0}^{n-1} h_i^{m+1} \int_0^1 K_{n,j}(t_i + vh_i) R_i(v) dv\}.$$

Replace now n by $n-1$ and j by m, and subtract the so modified equation from the original equation (5.5.10) to obtain (using the assumption that $c_m = 1$)

$$h_n^{m+1} \sum_{l=1}^{m} \beta_{n,l} \int_0^{c_j} K_{n,j}(t_n + vh_n) v^{l-1} dv$$

$$= -\sum_{i=0}^{n-1} h_i^{m+1} \sum_{l=1}^{m} \beta_{i,l} \int_0^1 [K_{n,j}(t_i + vh_i) - K_{n-1,m}(t_i + vh_i)] v^{l-1} dv \tag{5.5.12}$$

$$+ (q_{n,j} - q_{n-1,m}), \quad j = 1, \ldots, m \quad (n = 1, \ldots, N-1).$$

Because of the smoothness hypothesis for K we may apply the Mean-Value Theorem on the right-hand side of (5.5.12); it yields (since $t_{n-1} + c_m h_{n-1} = t_n$)

$$K_{n,j}(t_i + vh_i) - K_{n-1,m}(t_i + vh_i) \;=\; c_j h_n \cdot K_t(\cdot\, , t_i + vh_i),$$

where we have set $K_t(\cdot\, , t_i + vh_i) := (\partial/\partial t) K(t_n + \theta_{n,j} c_j h_n, t_i + vh_i)$, with $0 < \theta_{n,j} < 1$. If an analogous step is carried out in the remainder terms $(q_{n,j} - q_{n-1,m})$, we may express (5.5.12) in the form

$$\sum_{l=1}^{m} \beta_{n,l} \int_0^{c_j} K_{n,j}(t_n + vh_n) v^{l-1} dv$$

$$= -c_j h_n \sum_{i=0}^{n-1} \gamma_{n,i}^{m+1} \sum_{l=1}^{m} \beta_{i,l} \int_0^1 K_t(\cdot\, , t_i + vh_i) v^{l-1} dv + r_{n,j}, \tag{5.5.13}$$

$$j = 1, \ldots, m \quad (n = 1, \ldots, N-1),$$

with $\gamma_{n,i} := h_i / h_n$, and with

$$r_{n,j} := -\{\int_0^{c_j} K_{n,j}(t_n+vh_n)R_n(v)dv$$

$$+c_jh_n\sum_{i=0}^{n-1}\gamma_{n,i}^{m+1}\int_0^1 K_t(\cdot,t_i+vh_i)R_i(v)dv\}. \qquad (5.5.14)$$

With matrices $D_{n,i}$ and vectors $\mathbf{r}_n$ defined in an obvious way, we rewrite (5.5.13) more compactly as

$$D_{n,n}\boldsymbol{\beta}_n = h_n\sum_{i=0}^{n-1}\gamma_{n,i}^{m+1}\cdot D_{n,i}\boldsymbol{\beta}_i+r_n \quad (n=1, \dots ,N-1), \qquad (5.5.15)$$

where the vector $\boldsymbol{\beta}_0$ is the solution of the system

$$\sum_{l=1}^{m}\beta_{0,l}\int_0^{c_j}K_{0,j}(t_0+vh_0)v^{l-1}dv = -\int_0^{c_j}K_{0,j}(t_0+vh_0)R_0(v)dv \quad (j=1, \dots ,m),$$

The assumption that $K(t,t)$ be bounded away from 0 on I implies that the matrix $D_{n,n}$ has a bounded inverse for all n and all quasi-uniform meshes with sufficiently small $h>0$. Moreover, since $K\in C^{m+1}(S)$, K_t is bounded on S, and thus the matrices $D_{n,i}$ $(i<n)$ in (5.5.15) have bounded norms (recall that $Nh\leqslant\gamma T$ for any quasi-uniform mesh sequence, and that $\gamma_{n,i}\leqslant\gamma$). Hence, there exist finite constants C_0 and C_1 such that the l_1-norms of the vectors $\boldsymbol{\beta}_n$ defined by (5.5.15) satisfy

$$\|\boldsymbol{\beta}_n\|_1 \leqslant hC_0\sum_{i=0}^{n-1}\|\boldsymbol{\beta}_i\|_1+C_1 \quad (n=0, \dots ,N-1).$$

This is again a discrete Gronwall inequality; the by now well-known argument leads to the result that $\|\boldsymbol{\beta}_n\|_1\leqslant B$ uniformly for $n=0, \dots ,N-1$, as $h\downarrow 0$ and $Nh\leqslant\gamma T$. Recalling the expression for the errors $e_n(t_n+vh_n)$, we readily deduce (5.5.7). □

We note in passing that the differencing procedure which led from (5.5.10) to (5.5.12) may be viewed as the discrete analogue of the differentiation step which was used to transform a first-kind Volterra integral equation into an integral equation of the second kind (cf. Theorem 1.3.5).

Theorem 5.5.1a states that if the last of the collocation parameters equals one (i.e., if the points $\overline{Z}_N$ are collocation points), then the (discontinuous) collocation approximation $u\in S_{m-1}^{(-1)}(Z_N)$ is convergent of order $p=m$, regardless of how one chooses the remaining collocation parameters. This is no longer true for collocation approximations in the continuous polynomial spline space $S_{m-1}^{(0)}(Z_N)$.

Theorem 5.5.1b. *Let g and K in (5.5.1) be subject to the conditions stated in Theorem 5.5.1a. Assume that* $u \in S_{m-1}^{(0)}(Z_N)$ $(m \geq 2)$ *denotes the collocation approximation corresponding to collocation parameters* $\{c_j\}$ *with* $0 < c_1 < \dots < c_{m-1} = 1$, *with the underlying mesh sequence being quasi-uniform. Then the resulting collocation error behaves like*

$$\|e\|_\infty = O(h^m)$$

(as $h \downarrow 0$, $Nh \leq \gamma T$*) for all* $m \geq 2$ *if, and only if, the parameters* $\{c_j\}$ *satisfy*

$$\prod_{j=1}^{m-2} (1-c_j)/c_j < 1.$$

Proof. Recall first that

$$\dim S_{m-1}^{(0)}(Z_N) = N(m-1)+1 = \dim S_{m-1}^{(-1)}(Z_N) - (N-1).$$

Hence, collocation is now characterized by $(m-1)$ collocation parameters but, in contrast to the previous case, we now require an initial value $u(0)$. The collocation equation defining $u \in S_{m-1}^{(0)}(Z_N)$ has again the form (5.5.4) but with the index j now ranging from one to $m-1$ only. Additional equations are given by

$$u_n(t_n) = u_{n-1}(t_n), \quad t \in Z_N$$

(*continuity conditions* for Z_N), and

$$u_0(t_0) = y(0)$$

(*initial condition*). Thus, the error equation (5.5.9) holds with $j = 1, \dots, m-1$, and we have

$$e_n(t_n) = e_{n-1}(t_n), \quad n = 1, \dots, N-1,$$

and

$$e_0(t_0) = 0.$$

Proceeding along the lines of the proof of Theorem 5.5.1a, we are led to a recurrence relation of the form (5.5.12), where again $j = 1, \dots, m-1$; in addition, there holds the relation

$$h_n^m \beta_{n,1} = h_{n-1}^m \{ \sum_{l=1}^{m} \beta_{n-1,l} + R_{n-1}(1) \} \quad n = 1, \dots, N-1,$$

with

$$\beta_{0,1} = 0.$$

Let now, for ease of exposition, $h_n = h$ for all n (uniform mesh), and set $K(t,s) \equiv 1$ (the case of a general kernel $K(t,s)$ can be reduced to this special case by appropriate Taylor-expansion techniques, along the lines of Brunner [1978a]). Hence, it follows from (5.5.13) (with $j = 1, \dots, m-1$) and (5.5.14) that the vectors $\boldsymbol{\beta}_n := (\beta_{n,1}, \dots, \beta_{n,m})^T$ satisfy the recurrence relation

$$P\boldsymbol{\beta}_n = Q\boldsymbol{\beta}_{n-1} + \mathbf{r}_n, \quad n=1, \dots, N-1,$$

where P and Q denote m by m matrices given by

$$P := \begin{bmatrix} 1 \quad 0 \quad \cdots \quad 0 \\ \\ \int_0^{c_j} v^{l-1} dv \\ (j=1, \dots, m-1; \\ l=1, \dots, m) \end{bmatrix}, \quad Q := \begin{bmatrix} 1 & 1 & \cdots & 1 \\ & & 0 & \\ & & & \end{bmatrix},$$

and where $\mathbf{r}_n \in \mathbb{R}^m$ has the components

$$[R_{n-1}(1), -\int_0^{c_j} R_n(v)dv \quad (j=1, \dots, m-1)]^T.$$

It is easily verified that the nontrivial eigenvalue of the rank-one matrix $P^{-1}Q$ (which is independent of n) is given by

$$\lambda_1 := (-1)^{m-1} \prod_{k=1}^{m-2} (1-c_k)/c_k$$

(compare also de Hoog & Weiss [1973b]). Hence, the sequence $\|\boldsymbol{\beta}_n\|_1$ is uniformly bounded, as $h \downarrow 0$ and $Nh \leqslant \gamma T$, for all integers $m \geqslant 2$ if, and only if, $|\lambda_1| < 1$. □

Note that for even values of m, the assertion of Theorem 5.5.1b can be shown to remain valid if $\lambda_1 = -1$, in analogy to the remark following Theorem 5.5.1a.

In order to prepare the ground for the discretization of the exact collocation equation (5.5.6), suppose that we are given, in addition to the collocation parameters $\{c_j\}$, another set of m parameters $\{v_j\}$, with $0 \leqslant v_1 < \dots < v_m \leqslant 1$. With this new set we associate the Lagrange fundamental polynomials

$$\bar{L}_l(v) := \prod_{r \neq l}^{m} (v - v_r)/(v_l - v_r) \quad (l=1, \dots, m).$$

Setting $U_{n,l} := u_n(t_n + v_l h_n)$, we may write $u_n(t_n + vh_n) = \sum_{l=1}^{m} \bar{L}_l(v) U_{n,l}$ $(t_n + vh_n \in \sigma_n)$. If we have $v_j = c_j$, $j=1, \dots, m$, then $U_{n,l} = Y_{n,l}$ and $\bar{L}_l(v) \equiv L_l(v)$. The integrals $\Phi_{n,i}^{(j)}[u_i]$ occurring in the collocation equation (5.5.6) may now be expressed in the form

$$\Phi_{n,i}^{(j)}[u_i] = \begin{cases} \sum_{l=1}^{m} \int_0^1 K(t_{n,j}, t_i + vh_i) \bar{L}_l(v) dv \cdot U_{i,l}, & \text{if } 0 \leqslant i \leqslant n-1, \\ \sum_{l=1}^{m} \int_0^{c_j} K(t_{n,j}, t_n + vh_n) \bar{L}_l(v) dv \cdot U_{n,l}, & \text{if } i = n, \end{cases} \tag{5.5.16}$$

where $j=1, \ldots, m$. With (5.5.16), the collocation equation (5.5.6) represents a recurrence relation for the values $\{U_{n,j}\}$; i.e., for the values of the approximating polynomial spline at the points $\{t_n+v_jh_n\}$, while collocation takes place at the points $\{t_{n,j}\}$ characterized by the parameters $\{c_j\}$. It will be shown in the following section that a judicious choice of these two set of parameters, $\{c_j\}$ and $\{v_j\}$, leads to certain superconvergence results which, however, are no longer as spectacular as those for integral equations of the second kind.

5.5.2. *Local differentiation formulas based on collocation*

Before studying the discretization of the general collocation equation (5.5.6) we treat the special case corresponding to $K(t,s)\equiv 1$. If $g\in C^1(I)$, and if $g(0)=0$, then the exact solution of (5.5.1) is given by $y(t)=-g'(t)$. Moreover, since the integrands in (5.5.16) are now polynomials of degree $(m-1)$, the approximation of these integrals by interpolatory m-point quadrature formulas would reproduce the exact value (which can, of course, be found analytically). Thus, setting

$$\lambda_{j,l} := \int_0^{c_j} \bar{L}_l(v)dv, \quad \bar{b}_l := \int_0^1 \bar{L}_l(v)dv \quad (j,l=1, \ldots, m),$$

(5.5.6) reduces to

$$h_n \sum_{l=1}^{m} \lambda_{j,l} U_{n,l} = -g(t_{n,j}) - \sum_{i=0}^{n-1} h_i \sum_{l=1}^{m} \bar{b}_l U_{i,l}, \tag{5.5.17}$$

$$j=1, \ldots, m \quad (n=0, \ldots, N-1).$$

Let now $n\geqslant 1$, then replace n by $n-1$, j by m, and subtract this new equation from the original equation (5.5.17). Assuming $c_m=1$, we obtain

$$\sum_{l=1}^{m} \lambda_{j,l} U_{n,l} = -h_n^{-1}(g(t_{n,j})-g(t_n)), \quad j=1, \ldots, m. \tag{5.5.18}$$

In other words, the values $\{U_{n,l}\}$, which approximate the exact values $\{y(t_n+v_lh_n)=-g'(t_n+v_lh_n)\}$, depend only on the $m+1$ values $\{g(t_{n,j})$: $j=0, \ldots, m\}$ (with $c_0:=0$) of the given function g in the subinterval $[t_n, t_{n+1}]$. Because of this local dependence, (5.5.18) represents a family of so-called *local differentiation formulas* (compare Section 3.3.5). Note that the local dependence is lost if $c_m<1$; furthermore, collocation in $S_m^{(0)}(Z_N)$ (i.e., the use of continuous polynomial splines) does not lead to local differentiation formulas.

Example 5.5.1. $u\in S_1^{(-1)}(Z_N)$; $m=2$; $0<c_1<c_2=1$, $0\leqslant v_1<v_2\leqslant 1$. It is easily verified that (5.5.18) yields the solution

$$U_{n,1} = \frac{-1}{h_n c_1(1-c_1)} \cdot \{c_1(2v_1-c_1)g(t_n+h_n)+(1-2v_1)g(t_n+c_1h_n)$$
$$+(c_1^2+2v_1(1-c_1)-1)g(t_n)\},$$

$$U_{n,2} = \frac{-1}{h_n c_1(1-c_1)} \cdot \{c_1(2v_2-c_1)g(t_n+h_n)+(1-2v_2)g(t_n+c_1h_n)$$
$$+(c_1^2+2v_2(1-c_1)-1)g(t_n)\}.$$

Recall that these values approximate $y(t_n+v_lh_n)=-g'(t_n+v_lh_n)$, $l=1,2$. For $c_1=1/2$, $c_2=1$, we have the approximations

$$U_{n,1} = -h_n^{-1}\cdot\{(4v_1-1)g(t_n+h_n)+4(1-2v_1)g(t_n+\frac{h_n}{2})+(4v_1-3)g(t_n)\},$$

$$U_{n,2} = -h_n^{-1}\cdot\{(4v_2-1)g(t_n+h_n)+4(1-2v_2)g(t_n+\frac{h_n}{2})+(4v_2-3)g(t_n)\}.$$

(i) Let $v_1=c_1=1/2$, $v_2=c_2=1$. We find

$$U_{n,1}=Y_{n,1}=-h_n^{-1}\cdot\{g(t_n+h_n)-g(t_n)\}$$

(central differentiation formula),

$$U_{n,2} = Y_{n,2}=-h_n^{-1}\cdot\{3g(t_n+h_n)-4g(t_n+h_n/2)+g(t_n)\}$$

(backward differentiation formula).

It is well known that both approximations satisfy $y(t_n+v_lh_n) - U_{n,l}= O(h_n^2)$; according to Theorem 5.5.1, this agrees with the global order of convergence.

(ii) Let $v_1=(3-\sqrt{3})/6$ ($\neq c_1$), $v_2=(3+\sqrt{3})/6$ ($\neq c_2$): Gauss points for $(0,1)$.

The corresponding approximations for $g'(t_n+v_lh_n)$ have the form

$$U_{n,1} = -(3h_n)^{-1}\cdot\{(3-2\sqrt{3})g(t_n+h_n)+4\sqrt{3}g(t_n+h_n/2) -(3+2\sqrt{3})g(t_n)\}, \tag{5.5.19a}$$

$$U_{n,2} = -(3h_n)^{-1}\cdot\{(3+2\sqrt{3})g(t_n+h_n)-4\sqrt{3}g(t_n+h_n/2) -(3-2\sqrt{3})g(t_n)\}. \tag{5.5.19b}$$

It will be shown below (Theorem 5.5.2) that $y(t_n+v_lh_n)-U_{n,l}=O(h_n^3)$ $(=O(h_n^{m+1}))$: there has resulted an increase in the order of convergence of one over the global convergence order. Since this will be shown to be best possible, the expressions (5.5.19) represent *optimal differentiation formulas* using three function evaluations. (Compare Brunner [1978a]; see also Salzer [1960] and Ash & Jones [1981] where such optimal formulas have been found by a different approach.)

Theorem 5.5.2. *In (5.5.1), assume that $K(t,s)\equiv 1$ on S, and let $g\in C^{m+2}(I)$, with $g(0)=0$. Suppose that $u\in S_{m-1}^{(-1)}(Z_N)$ is the approximation defined by the collocation equation (5.5.17).*
(a) *If the collocation parameters $\{c_j\}$ are the zeros of*

$$q_m(s) := (P_{m+1}(2s-1)-P_{m-1}(2s-1))/s \tag{5.5.20}$$

(i.e., those m of the $m+1$ Lobatto points which are contained in the interval $(0,1]$), and if the parameters $\{v_j\}$ are given by the zeros of $P_m(2s-1)$ (the Gauss points for $(0,1)$), then

$$y(t_n+v_lh_n)-U_{n,l} = O(h^{m+1}), \quad l=1, \ldots ,m \quad (n=0, \ldots ,N-1) \tag{5.5.21}$$

(as $h\downarrow 0$, $Nh\leqslant\gamma T$), where $U_{n,l}:=u_n(t_n+v_lh_n)$. This order $p^*=m+1$ is best possible.
(b) *At the mesh points $t_n\in\bar{Z}_N$, we can do not better than $y(t_n)-u(t_n)=O(h^m)$ for any $\{c_j\}$ with $0<c_1< \ldots <c_m=1$.*

Hence, a family of optimal local differentiation formulas (5.5.18) for $[t_n,t_{n+1}]$ (based on collocation in $S_{m-1}^{(-1)}(Z_N)$) consists of linear combinations of the values of g at the $m+1$ Lobatto points for that interval; at the m Gauss points for (t_n,t_{n+1}), the error is of order $p^*=m+1$.

Illustration:

m	p^*	Collocation at $\{t_n+c_jh_n\}$: $\{c_j\}$	Evaluation at $\{t_n+v_lh_n\}$: $\{v_l\}$
1	2	1	1/2
2	3	1/2, 1	$(3-\sqrt{3})/6$, $(3+\sqrt{3})/6$
3	4	$(5-\sqrt{5})/10$, $(5+\sqrt{5})/10$, 1	$(5-\sqrt{15})/10$, 1/2, $(5+\sqrt{15})/10$
4	5	$(7-\sqrt{21})/14$, 1/2, $(7+\sqrt{21})/14$, 1	$(1\mp[15\pm2\sqrt{30})/35]^{1/2})/2$

Proof. According to (5.5.9), the error $e:=y-u$ satisfies

$$\int_0^{c_j} e_n(t_n+vh_n)dv = -\sum_{i=0}^{n-1}\int_0^1 e_i(t_i+vh_i)dv, \quad j=1, \ldots ,m \quad (n=0, \ldots ,N-1).$$

Using $c_m=1$, this can be reduced to

$$\int_0^{c_j} e_n(t_n+vh_n)dv=0, \quad j=1, \ldots ,m \quad (n=0, \ldots ,N-1). \tag{5.5.22}$$

Since the exact solution y is in $C^{m+1}(I)$, we may employ Taylor's formula to write

$$y(t_n+vh_n) = \sum_{l=1}^{m} \bar{L}_l(v)y(t_n+v_lh_n)+h_n^m y^{(m)}(t_n+\theta_n vh_n)w_m(v)/m!$$

$$(0<\theta_n<1),$$

with $w_m(v):=\Pi_{j=1}^m(v-v_j)$; an obvious modification then leads to

$$y(t_n+vh_n) = \sum_{l=1}^{m} \bar{L}_l(v)y(t_n+v_lh_n)$$

$$+\frac{h_n}{m!}\cdot\{y^{(m)}(t_n)+\theta_n vh_n y^{(m+1)}(t_n+\theta_n' vh_n)\}\cdot w_m(v),$$

where $0<\theta_n'<1$. Substitution in (5.5.22) yields

$$\sum_{l=1}^{m} \int_0^{c_j} \bar{L}_l(v)dv\cdot e_n(t_n+v_lh_n)$$

$$=-(1/m!)h_n^m y^{(m)}(t_n)\int_0^{c_j} w_m(v)dv+O(h_n^{m+1}),\quad j=1,\ldots,m. \tag{5.5.23}$$

The above represents a (nonsingular) linear system for the values of the error function at the (distinct) points $\{t_n+v_lh_n\}$: thus, these values will satisfy

$$e_n(t_n+v_lh_n) = O(h_n^{m+1}),\quad l=1,\ldots,m, \tag{5.5.24}$$

for all $y\in C^{m+1}(I)$ (where $y(t)=-g'(t)$) if, and only if, the parameters $\{v_j\}$ are such that

$$\int_0^{c_j} w_m(v)dv = 0 \tag{5.5.25}$$

holds for $j-1,\ldots,m$. Do two sets $\{c_j\}$ and $\{v_l\}$ exist for which (5.5.25) can be satisfied? The answer is in the affirmative; in fact, we have

Lemma 5.5.1. *Let* $w_m(s):=P_m(2s-1)$. *Then:*

$$Q_{m+1}(s) := \int_0^s w_m(v)dv=\text{const.}\{P_{m-1}(2s-1)-P_{m+1}(2s-1)\} \tag{5.5.26}$$

This result can be obtained as a special case of a more general identity for Jacobi polynomials; see, e.g., Szegö [1939, p.72], or Ghizzetti & Ossicini [1970, pp. 62-63]. □

Since the zeros of the above polynomial $Q_{m+1}(s)$ are the $m+1$ Lobatto points for $[0,1]$, comparison of (5.5.26) with (5.5.25) shows that the *m positive* Lobatto points will have to assume the role of the collocation parameters $\{c_j\}$.

The zeros of $w_m(s)$ (which are the m Gauss points for $(0,1)$) then form the set $\{v_l\}$. It is easily seen that this is not the only way of selecting parameters $\{c_j\}$ and $\{v_j\}$ satisfying (5.5.25): just choose m distinct parameters $\{c_j\}$ in $(0,1]$ and define $\bar{Q}_{m+1}(s):=s\cdot\Pi_{j=1}^{m}(s-c_j)$. By Rolle's Theorem, the polynomial $\bar{w}_m(s):=\bar{Q}'_{m+1}(s)$ has m distinct zeros in $(0,1)$; these zeros will then be the parameters $\{v_j\}$. The reason for using the sets described in Lemma 5.5.1 is that they are given by well-known quadrature abscissas; note also that, for *odd* values of m, we have $v_{(m+1)/2}=\frac{1}{2}$ as one of the Gauss points, implying that local superconvergence of order $p^*=m+1$ occurs at the midpoint of each subinterval $[t_n,t_{n+1}]$.

The result that $p^*=m+1$ is the optimal order of a differentiation formula based on $m+1$ function values has been established by Salzer [1960]; it can also be derived by an obvious modification of the above arguments.

It remains to be shown that local superconvergence is not possible at the mesh points $\bar{Z}_N$. In order to do so, assume that $v_m=1$ and that $e_n(t_n+h_n)=O(h_n^{m+1})$ for $n=0, \ldots ,N-1$. Since we now have $w_m(v)=(v-1)\Pi_{j=1}^{m-1}(v-v_j)$, we must satisfy the condition (5.5.25) for this polynomial w_m. Let $c_1, \ldots ,c_r$ be the zeros of $Q_{m+1}(s):=\int_0^s w_m(v)dv$ in $(0,1]$. We have $r\leqslant m$. If $r=m$, then these zeros necessarily interlace with those of $w_m(v)$; i.e., $0\leqslant v_1<c_1\leqslant \ldots \leqslant c_{m-1}\leqslant v_m<c_m$. Since $v_m=1$, $Q_{m+1}(s)$ cannot have m distinct zeros in $(0,1]$. □

Theorem 5.5.2 remains valid for first-kind integral equations of the form (5.5.1), provided $K\in C^{m+2}(S)$, with $K(t,t)$ bounded away from zero. Details may be found in Brunner [1978b], [1979].

We conclude this section by observing that if the functions g and K in the first-kind integral equation (5.5.1) satisfy the assumptions of Theorem 1.3.5, then it will be advantageous to transform it into an integral equation of the second-kind; while collocation in $S_{m-1}^{(-1)}(Z_N)$ for the former equation yields local superconvergence of order at most $m+1$ (at the knots Z_N, the order cannot even exceed m), collocation in the same polynomial spline space for the corresponding second-kind equation leads to local superconvergence of order $2m-1$ on $\bar{Z}_N$ if the Radau II points serve as the collocation parameter, or of order $2m$ if one employs iterated collocation using the Gauss points.

5.5.3. *Convergence results for the fully discretized collocation equation*

We now return to the general first-kind integral equation (5.5.1) and the corresponding exact collocation equation (5.5.6). Let $\mu\geqslant 1$, and assume that $0\leqslant d_1< \ldots <d_\mu\leqslant 1$ and $d_{j,s}:=c_jd_s$ $(s=1, \ldots ,\mu;\ j=1, \ldots ,m)$ are the parameters defining the (interpolatory) μ-point quadrature approximations for the integrals occurring in (5.5.6); i.e., we employ (5.3.33), where $k(t,s,u)$ has been replaced by $K(t,s)u$, and where $\mu_0=\mu_1=\mu$:

$$\hat{\Phi}_{n,i}^{(j)}[u_i] := \begin{cases} \sum_{s=1}^{\mu} w_s K(t_{n,j}, t_i + d_s h_i) u_i(t_i + d_s h_i), & \text{if } 0 \leqslant i < n, \\ \sum_{s=1}^{\mu} w_{j,s} K(t_{n,j}, t_n + d_{j,s} h_n) u_n(t_n + d_{j,s} h_n), & \text{if } i = n, \end{cases} \tag{5.5.27a}$$

$$j = 1, \dots, m.$$

Here, the quadrature weights are given by

$$w_s := \left(\prod_{r \neq s}^{\mu} (d_s - d_r)\right)^{-1} \cdot \int_0^1 \prod_{r \neq s}^{\mu} (v - d_r) dv, \tag{5.5.27b}$$

$$w_{j,s} := c_j w_s, \quad s = 1, \dots, \mu \quad (j = 1, \dots, m); \tag{5.5.27c}$$

we assume that, for all sufficiently smooth γ,

$$\int_0^1 \gamma(t_i + vh_i) dv - \sum_{s=1}^{\mu} w_s \gamma(t_i + d_s h_i) = O(h_i^r) \quad (r \geqslant \mu). \tag{5.5.27d}$$

Instead of $u \in S_{m-1}^{(-1)}(Z_N)$, we compute an approximation $\hat{u} \in S_{m-1}^{(-1)}(Z_N)$ by solving the *fully discretized* version of the collocation equation (5.5.6),

$$h_n \hat{\Phi}_{n,n}^{(j)}[\hat{u}_n] = -(g(t_{n,j}) + \sum_{i=0}^{n-1} h_i \hat{\Phi}_{n,i}^{(j)}[\hat{u}_i]), \tag{5.5.28}$$

$$j = 1, \dots, m \quad (n = 0, \dots, N-1).$$

Setting

$$\hat{u}_n(t_n + vh_n) = \sum_{l=1}^{m} L_l(v) \hat{Y}_{n,l}, \quad t_n + vh_n \in \sigma_n, \tag{5.5.29}$$

with $L_l(v) = \Pi_{r \neq l}^m (v - c_r)/(c_l - c_r)$, we see that (5.5.28) represents, for each n, a system of m linear equations for $\hat{Y}_{n,l} := \hat{u}_n(t_{n,l})$ $(l = 1, \dots, m)$. This is reminiscent of (5.3.35), except that now the equation does not contain a "free" term $\hat{Y}_{n,j}$.

Theorem 5.5.3. *Let K and g in (5.5.1) be subject to the conditions stated in Theorem 5.5.1. Assume that the exact collocation equation (5.5.6) has been discretized by interpolatory μ-point quadrature of the form (5.5.27). Then the fully discretized collocation equation (5.5.28) defines, for any quasi-uniform mesh sequence with sufficiently small $h := \max_{(n)} h_n > 0$ and for all collocation parameters $\{c_j\}$ with $0 < c_1 < \dots < c_m = 1$, a unique approximation $\hat{u} \in S_{m-1}^{(-1)}(Z_N)$. The error $\hat{e} := y - \hat{u}$ induced by this approximation satisfies*

$$\|\hat{e}\|_\infty = O(h^\nu) \quad (\text{as } h \downarrow 0 \text{ and } Nh \leqslant \gamma T),$$

where $\nu := \min(m, r)$, with r defined in (5.5.27d).

Proof. Set

$$E_{n,i}^{(j)} := \Phi_{n,i}^{(j)}[u_i] - \hat{\Phi}_{n,i}^{(j)}[u_i],$$

with $\Phi_{n,i}^{(j)}[u_i]$ and $\hat{\Phi}_{n,i}^{(j)}[u_i]$ given by (5.3.6) and (5.5.27). The exact collocation equation (5.5.6) can thus be written as

$$h_n(\hat{\Phi}_{n,n}^{(j)}[u_i] + E_{n,n}^{(j)}) = -(g(t_{n,j}) + \sum_{i=0}^{n-1} h_i(\hat{\Phi}_{n,i}^{(j)}[u_i] + E_{n,i}^{(j)}));$$

since $|K(t,t)| \geqslant \rho > 0$ for all $t \in I$, it possesses a unique solution $u \in S_{m-1}^{(-1)}(Z_N)$ whenever $h = \max_{(n)} h_n > 0$ is sufficiently small. The fully discretized equation (5.5.28) is obtained by deleting the error terms $E_{n,i}^{(j)}$ and by replacing u by $\hat{u}$ in the above equation. Since the quadrature approximations are based on interpolatory μ-point formulas, the corresponding quadrature errors satisfy, for all sufficiently smooth integrands, $E_{n,i}^{(j)} = O(h^r)$, with $r \geqslant \mu \geqslant 1$, $j = 1, \ldots, m$ $(0 \leqslant i \leqslant n \leqslant N-1)$. This holds in particular for $i = n$: thus, by the Banach Lemma, (5.5.28) is uniquely solvable for all sufficiently small $h > 0$.

Consider now the perturbation $\epsilon := u - \hat{u}$, with $\epsilon_n(t) := u_n(t) - \hat{u}_n(t)$ $(t \in \sigma_n)$. If we subtract (5.5.28) from (5.5.6), and if we use the linearity of the quadrature functionals $\hat{\Phi}_{n,i}^{(j)}$, we obtain

$$h_n \hat{\Phi}_{n,n}^{(j)}[\epsilon_n] = -\sum_{i=0}^{n-1} h_i \hat{\Phi}_{n,i}^{(j)}[\epsilon_i] - \sum_{i=0}^{n} h_i E_{n,i}^{(j)}, \tag{5.5.30}$$

$$j = 1, \ldots, m \quad (n = 0, \ldots, N-1).$$

We wish to show that ϵ satisfies $\|\epsilon\|_\infty = O(h^r)$; due to the estimate

$$\|\hat{e}\|_\infty = \|(y-u) + (u - \hat{u})\|_\infty \leqslant \|e\|_\infty + \|\epsilon\|_\infty,$$

the second assertion of Theorem 5.5.3 then follows since $\|e\|_\infty = O(h^m)$ (Theorem 5.5.1a).

In order to establish this asymptotic result, modify (5.5.30) by replacing n and j by $n-1$ and m, respectively, and subtract the resulting equation from the original one. We find, after dividing by h_n,

$$\begin{aligned} \hat{\Phi}_{n,n}^{(j)}[\epsilon_n] &= -\sum_{i=0}^{n-1} (h_i/h_n)\{\hat{\Phi}_{n,i}^{(j)}[\epsilon_i] - \hat{\Phi}_{n-1,i}^{(m)}[\epsilon_i]\} \\ &\quad - E_{n,n}^{(j)} - \sum_{i=0}^{n-1} (h_i/h_n)\{E_{n,i}^{(j)} - E_{n-1,i}^{(m)}\}. \end{aligned} \tag{5.5.31}$$

Here, we have used the assumption that $c_m = 1$, implying $t_{n-1} + c_m h_{n-1} = t_n$. The smoothness conditions imposed on the kernel K yield, for $i < n$,

$$\begin{aligned} \hat{\Phi}_{n,i}^{(j)}[\epsilon_i] - \hat{\Phi}_{n-1,i}^{(m)}[\epsilon_i] &= \sum_{s=1}^{\mu} w_s\{K(t_{n,j}, t_i + d_s h_i) - K(t_n, t_i + d_s h_i)\}\epsilon_i(t_i + d_s h_i) \\ &= c_j h_n \sum_{s=1}^{\mu} w_s K_t(\cdot\,, t_i + d_s h_i)\epsilon_i(t_i + d_s h_i), \end{aligned}$$

where the partial derivative $K_t := \partial K/\partial t$ is to be evaluated at a suitable first argument determined by the Mean-Value Theorem, in complete analogy to (5.5.13). In a similar way we deduce that

$$\sum_{i=0}^{n-1} |E_{n,i}^{(j)} - E_{n-1,i}^{(m)}| \leq Nh \cdot Ch^r, \quad n=1, \ldots, N-1 \text{ (with } Nh \leq \gamma T),$$

for some finite constant C not depending on h or N. Since, as shown in the first part of the proof, (5.5.31) has a unique solution $\eta_n := (\epsilon_n(t_{n,1}), \ldots, \epsilon_n(t_{n,m}))^T$ for all sufficiently small $h>0$ (the matrix characterizing the left-hand side of the system (5.5.31) is identical with that in (5.5.28), except for the factor h_n), we conclude, using an argument analogous to the one which led to (5.3.41), that there exist finite constants $\hat{C}_0$ and $\hat{C}_1$ such that

$$\|\eta_n\|_1 \leq h\hat{C}_0 \sum_{i=0}^{n-1} \|\eta_i\|_1 + \hat{C}_1 h^r \quad (n=0, \ldots, N-1).$$

Thus, the quantities $\|\eta_n\|_1$ governed by this discrete Gronwall inequality satisfy $\|\eta_n\|_1 = O(h^r)$, $n=0, \ldots, N-1$ (as $h \downarrow 0$ and $Nh \leq \gamma T$). The proof is brought to its end by observing that ϵ_n is a polynomial of degree not exceeding $m-1$ and that, therefore,

$$|\epsilon_n(t_n + vh_n)| \leq \sum_{l=1}^{m} |L_l(v)\epsilon_n(t_n + c_l h_n)| \leq \text{const.} \|\eta_n\|_1 = O(h^r),$$

which holds uniformly in v and n (cf. the conclusion of the proof of Theorem 5.3.4). □

Theorem 5.5.3 is readily modified to deal with the collocation approximation $\hat{u} \in S_{m-1}^{(0)}(Z_N)$ corresponding to the discretization of the collocation equation analyzed in Theorem 5.5.1b. We refrain from presenting the details.

It has been mentioned at the end of Section 5.5.2 that the approximation $u \in S_{m-1}^{(-1)}(Z_N)$ defined by the exact collocation equation (5.5.6) exhibits local superconvergence of order $p^* = m+1$ at the Gauss points in each subinterval σ_n, provided the collocation parameters $\{c_j\}$ are the zeros of the polynomial $(P_{m+1}(2s-1) - P_{m-1}(2s-1))/s$ (recall (5.5.20)). How do we have to choose the quadrature parameters $\{d_l\}$ in (5.5.27a) if the approximation $\hat{u}$, obtained from the fully discretized collocation equation (5.5.28), is to inherit this superconvergence behavior? If $\mu = m$, and if $d_l = c_l$, then this will not hold: the quadrature errors associated with these special collocation parameters will satisfy only $E_{n,i}^{(j)} = O(h^m)$. A glimpse at the proof of Theorem 5.5.3 shows that we must have $E_{n,i}^{(j)} = O(h^r)$ with $r \geq m+1$, since this will yield

$$\epsilon_n(t_{n,j}) = O(h^r), \quad j=1, \ldots, m \quad (n=0, \ldots, N-1),$$

and thus

$$|\hat{e}_n(t_{n,j})| \leq |e_n(t_{n,j})| + |\epsilon(t_{n,j})| = O(h^{m+1}).$$

Interpolatory μ-point quadrature formulas with $r \geqslant m+1$ include (cf. Chapter 2):

(i) the Gauss formulas (with $\mu = \lfloor (m+2)/2 \rfloor$);

(ii) the Radau formulas (with $\mu = \lfloor (m+3)/2 \rfloor$);

(iii) the Lobatto formulas (with $\mu = \lfloor (m+4)/2 \rfloor$);

(iv) the Newton-Cotes formulas (with $\mu = m$, if m is odd; hence $r = m+1$).

We leave the details to the reader.

5.5.4. Examples

In this section we present two examples corresponding to the choice $\mu = m$ in (5.5.27a). For this case it is convenient to set

$$\hat{u}_n(t_n + vh_n) = \sum_{l=1}^{m} L_l(v)\hat{U}_{n,l}, \quad t_n + vh_n \in \sigma_n, \tag{5.5.32}$$

where now $\hat{U}_{n,l} := \hat{u}_n(t_n + d_l h_n)$ and $L_l(v) := \Pi_{r \neq l}^{m}(v - d_r)/(d_l - d_r)$. Thus, the fully discretized collocation equation assumes the form

$$\sum_{l=1}^{m}\Big(\sum_{s=1}^{m} w_{j,s} L_l(c_j d_s) K(t_{n,j}, t_n + c_j d_s h_n)\Big)\hat{U}_{n,l} =$$

$$= -h_n^{-1} g(t_{n,j}) - \sum_{i=0}^{n-1}(h_i/h_n)\sum_{s=1}^{m} w_s K(t_{n,j}, t_i + d_s h_i)\hat{U}_{i,s} \tag{5.5.33}$$

$$j = 1, \dots, m \quad (n = 0, \dots, N-1).$$

Example 5.5.2. Approximation in $S_1^{(-1)}(Z_N)$: $m = 2$; $c_1 = ½$, $c_2 = 1$; $0 \leqslant d_1 < d_2 \leqslant 1$.

	(i): $d_1 = 0$, $d_2 = 2/3$	(ii): $d_1 = 1/3$, $d_2 = 1$	(iii): $d_l = c_l$
$w_{1,1}L_1(c_1 d_1)$	1/8	15/32	3/4
$w_{1,2}L_1(c_1 d_2)$	3/16	3/32	0
$w_{1,1}L_2(c_1 d_1)$	0	−3/32	−1/4
$w_{1,2}L_2(c_1 d_2)$	3/16	1/32	0
$w_{2,1}L_1(c_2 d_1)$	1/4	3/4	1
$w_{2,2}L_1(c_2 d_2)$	0	0	0
$w_{2,1}L_2(c_2 d_1)$	0	0	0
$w_{2,2}L_2(c_2 d_2)$	3/4	1/4	0
w_1	1/4	3/4	0
w_2	3/4	1/4	0

In the cases (i) and (ii), quadrature is based on the *Radau points* for $[0,1)$ and $(0,1]$, respectively; hence $r=2m-1=3$. Consequently, there occurs local superconvergence of order $p^*=m+1=3$ at the Gauss points $t_n+(3\mp\sqrt{3})h_n/6$. The corresponding approximations are given by

$$(i) \quad \hat{u}_n(t_n+\frac{3\mp\sqrt{3}}{6}h_n)=\frac{1}{4}\{(1\pm\sqrt{3})\hat{U}_{n,1}+(3\mp\sqrt{3})\hat{U}_{n,2}\}$$

$$(ii) \quad \hat{u}_n(t_n+\frac{3\mp\sqrt{3}}{6}h_n)=\frac{1}{4}\{(3\pm\sqrt{3})\hat{U}_{n,1}+(1\mp\sqrt{3})\hat{U}_{n,2}\}.$$

For the case (iii), the convergence order at the Gauss points equals only $p=m=2$, since the quadrature errors $E_{n,i}^{(j)}$ now satisfy $E_{n,i}^{(j)}=O(h^m)=O(h^2)$. Note, however, that (iii) defines a *diagonally implicit Runge-Kutta method* if we write the collocation equations in (5.5.33) in the reverse order ($j=2$, $j=1$):

$$K(t_n+h_n,t_n+h_n/2)\hat{Y}_{n,1}$$
$$= -h_n^{-1}g(t_n+h_n)-\sum_{i=0}^{n-1}(h_i/h_n)K(t_n+h_n,t_n+h_i/2)\hat{Y}_{i,1}, \tag{5.5.34a}$$
$$\frac{3}{4}K(t_n+h_n/2,t_n+h_n/4)\hat{Y}_{n,1}-\frac{1}{4}K(t_n+h_n/2,t_n+h_n/4)\hat{Y}_{n,2}$$
$$=-h_n^{-1}g(t_n+h_n/2)-\sum_{i=0}^{n-1}(h_i/h_n)K(t_n+h_n/2,t_i+h_i/2)\hat{Y}_{i,1}. \tag{5.5.34b}$$

(Since the quadrature parameters $\{d_l\}$ coincide with the collocation parameters $\{c_j\}$, we have $\hat{U}_{n,l}=\hat{Y}_{n,l}$.)

Example 5.5.3. Approximation in $S_2^{(-1)}(Z_N)$: $m=3$; $c_1=(5-\sqrt{5})/10$, $c_2=(5+\sqrt{5})/10$, $c_3=1$ (zeros of $(P_4(2s-1)-P_2(2s-1))/s$); $d_1=0$, $d_2=½$, $d_3=1$ (Lobatto points: $r=2m-2=4$).

		$s=1$	$s=2$
$j=1$	$l=1$	$w_{1,1}L_1(c_1d_1)=(5-\sqrt{5})/60$	$w_{1,2}L_1(c_1d_1)=(15+\sqrt{5})/150$
	$l=2$	$w_{1,1}L_2(c_1d_1)=0$	$w_{1,2}L_2(c_1d_2)=(20-6\sqrt{5})/75$
	$l=3$	$w_{1,1}L_3(c_1d_1)=0$	$w_{1,2}L_3(c_1d_1)=-(5-\sqrt{5})/150$
$j=2$	$l=1$	$w_{2,1}L_1(c_1d_1)=(5+\sqrt{5})/60$	$w_{2,2}L_1(c_2d_2)=(15-\sqrt{5})/150$
	$l=2$	$w_{2,1}L_2(c_2d_1)=0$	$w_{2,2}L_2(c_2d_2)=(20+6\sqrt{5})/75$
	$l=3$	$w_{2,1}L_3(c_2d_1=0$	$w_{2,2}L_3(c_2d_2)=-(5+\sqrt{5})/150$
$j=3$	$l=1$	$w_{3,1}L_1(c_3d_1)=1/6$	$w_{3,2}L_1(c_3d_2)=0$
	$l=2$	$w_{3,1}L_2(c_3d_1)=0$	$w_{3,2}L_2(c_3d_2)=2/3$
	$l=3$	$w_{3,1}L_3(c_3d_1)=0$	$w_{3,2}L_3(c_3d_2)=0$
		$w_1=1/6$	$w_2=2/3$

		$s=3$
$j=1$	$l=1$	$w_{1,3}L_1(c_1d_3)=\sqrt{5}/150$
	$l=2$	$w_{1,3}L_2(c_1d_3)=(5-\sqrt{5})/75$
	$l=3$	$w_{1,3}L_3(c_1d_3)=(5-3\sqrt{5})/300$
$j=2$	$l=1$	$w_{2,3}L_1(c_2d_3)=-\sqrt{5}/150$
	$l=2$	$w_{2,3}L_2(c_2d_3)=(5+\sqrt{5})/75$
	$l=3$	$w_{2,3}L_3(c_2d_3)=(5+3\sqrt{5})/300$
$j=3$	$l=1$	$w_{3,3}L_1(c_3d_3)=0$
	$l=2$	$w_{3,3}L_2(c_3d_3)=0$
	$l=3$	$w_{3,3}L_3(c_3d_3)=1/6$
		$w_3=1/6$

It follows by the special choice of the parameters $\{c_j\}$ and $\{d_l\}$ that at the Gauss points in each subinterval σ_n (characterized by the zeros of $P_3(2s-1)$) we get local superconvergence of order $p^*=m+1=4$. In particular, the midpoint $t_{n+1/2}:=t_n+h_n/2$ is one of these Gauss points; here, the approximation is given by

$$\hat{u}_n(t_{n+\frac{1}{2}}) = \hat{U}_{n,2}, \tag{5.5.35}$$

since $t_{n+\frac{1}{2}}$ is also one of the quadrature abscissas in $[t_n,t_{n+1}]$.

Note that, in order to decrease the number of kernel evaluations, it might be advantageous to replace the above three-point Lobatto quadrature formula (which is, of course, Simpson's rule) by the two-point Gauss formula: in this case we would have $\mu=2$, $d_1=(3-\sqrt{3})/6$, $d_2=(3+\sqrt{3})/6$. We omit the details.

5.5.5. Numerical examples

Our discussion of collocation methods for Volterra equations of the first-kind is concluded by presenting a few experiments with the generalized midpoint method, using Lobatto points, as specified in Example 5.5.3. Results are given for $\hat{U}_{N-1,2}$ and $\hat{U}_{N-1,3}$.

Table 5.5.1. Numbers of correct significant digits and effective orders for problem (3.3.21) with $a=1$

h	$t=t_{N-1/2}=4-h/2$		$t=t_N=T=4$	
4/13	4.9		3.9	
		> 3.9		> 2.6
4/26	6.0		4.6	
		> 3.9		> 2.8
4/52	7.2		5.5	
		> 4.0		> 2.9
4/104	8.4		6.4	

Table 5.5.2. Numbers of correct significant digits and effective orders for problem (3.3.22) with $a=1$

h	$t=t_{N-1/2}=4-h/2$		$t=t_N=T=4$	
4/13	4.8		4.1	
		> 4.1		> 2.0
4/26	6.0		4.7	
		> 4.0		> 2.7
4/52	7.2		5.5	
		> 3.7		> 2.9
4/104	8.3		6.4	

Table 5.5.3. Numbers of correct significant digits and effective orders for problem (3.3.23) with $a=1$

h	$t=t_{N-1/2}=6-h/2$		$t=t_N=T=6$	
6/8	4.4		3.3	
		> 3.9		> 3.8
6/16	5.6		4.4	
		> 3.9		> 3.9
6/32	6.8		5.6	

NOTES

Section 5.1.

The books by de Boor [1978] and by Schumaker [1981] present a comprehensive treatment of the theory and numerical analysis of polynomial spline functions. In addition, see also Prenter [1975]. Historical notes on splines in general, and on B-splines in particular, may be found in Schumaker [1981, pp. 9-11, 181-185]. Theorem 5.1.3 is taken from Schumaker [1981, pp. 116-117]. This monograph contains also a thorough introduction to divided differences.

Drawbacks of one-sided polynomial spline bases are discussed in de Boor [1978, pp. 104-106]. Readers interested in sources on the motivation for using polynomial spline functions, rather than classical polynomials, in various types of approximation, especially in interpolation problems, are referred to de Boor [1978, Ch. II], Schumaker [1981, pp. 101-107], and to the illuminating paper of Elliott [1979].

Section 5.2.

Loscalzo [1969] and Loscalzo & Talbot [1967] introduced collocation methods in classical spline spaces $S_m^{(m-1)}(Z_N)$ for initial-value problems for first-order ODEs; see also Hung [1970]. Callender [1971] employed $S_m^{(1)}(Z_N)$ as the approximating space for collocation. A general analysis of polynomial spline collocation (including multiple collocation points) is due to Mülthei; see Mülthei [1980] and the references (by the same author) listed there. Recently, Keller [1982] studied collocation methods in certain nonpolynomial spline spaces.

Polynomial spline collocation methods in $S_m^{(m-1)}(Z_N)$ for initial-value problems for second-order ODEs were analyzed by Harvey [1971], G. Micula [1973] (compare also G. Micula [1978, pp. 184-200]). A general analysis of projection methods (of which collocation is a special case) for the solution of initial-value problems for kth order ODEs was given by Kramarz [1978].

So-called perturbed collocation methods (which furnish an elegant framework for the study of order results for Runge-Kutta methods for ODEs) were introduced by Nørsett [1980] and Nørsett & Wanner [1981]. This work generalizes earlier results by Guillou & Soulé [1969] and by Wright [1970] who showed that certain implicit Runge-Kutta methods for ODEs are in fact collocation methods, thus linking the attainable order of such a Runge-Kutta method with that of the underlying quadrature formula.

Certain aspects of collocation as applied to the integrated form (5.2.10) of the initial-value problem (5.2.1), particularly the question of superconvergence on $\bar{Z}_N$, was studied by Axelsson [1969].

Section 5.3.

The idea of using elements from the polynomial spline space $S_1^{(0)}(Z_N)$ (piecewise linear polynomials) to obtain numerical approximations to solutions of integral equations is apparently due to Huber [1939]; compare also Wagner [1954]. Kadner [1967] showed that Huber's method can be obtained as a special case of his collocation approach. Bel'tyukov [1961] generated piecewise solutions by approximating g and k in (5.3.9) by step functions. Hung [1968], [1970] proved that while collocation in $S_1^{(0)}(Z_N)$ and $S_2^{(1)}(Z_N)$ yields convergent methods, collocation using classical cubic splines of higher degree (i.e., $S_3^{(2)}(Z_N)$) leads to divergent approximations. If, however, approximations in this space are constructed by Hermite-type collocation (where $t=0$ and $t=T$ are collocation points of multiplicity two), then the resulting process converges (Netravali [1973]; the method was extended to second-kind Volterra integral equations in $\mathbb{R}^2$ by Singh [1976]). Convergence results for the space $S_m^{(0)}(Z_N)$ (using equidistant collocation points) were given by Brunner [1971], while convergence and stability results for various polynomial spline spaces were obtained by El Tom [1971], [1973], [1975] (see also the review-type paper by El Tom [1976]). In addition, compare M. Micula & G. Micula [1973], G. Micula [1979], G. Micula [1978, pp. 243-249], and Malina [1975a], [1975b]. A general convergence analysis of Hermite-type collocation methods for $S_{2m-1}^{(m-1)}(Z_N)$ is due to Esser [1976], [1978]; he considered generalized second-kind Volterra integral equations whose kernels are of the form $k(t,s,y(s),y(a(t,s)))$, with $0 \leqslant a(t,s) \leqslant \max\{\theta t, t-b\}$ ($\theta \in [0,1)$, $b>0$).

Global convergence and local superconvergence of collocation approximations in $S_{m-1}^{(0)}(Z_N)$ and $S_{m-1}^{(-1)}(Z_N)$ to solutions of second-kind Volterra integral equations with regular kernels was studied by Brunner [1980], Brunner & Nørsett [1981]; in addition, compare Brunner [1978c] and, especially, Brunner [1981] where a general framework for deriving local superconvergence results for integral and integro-differential equations may be found. Iterated collocation approximations for Volterra integral equations are considered in Brunner [1984b]; the general concept of iterated approximations for operator equations (with applications to Fredholm integral equations of the second-kind) is due to Sloan [1976]; see also Chandler [1979], Graham [1980], [1982b], Chatelin & Lebbar [1981], Chatelin [1983], Sloan [1984] and Joe [1985a], [1985b], [1985c] (here, an in-depth analysis of the discretized collocation and iterated collocation equation may be found). A different philosophy for improving the accuracy of approximate solutions is based on the notion of so-called defect correction (introduced originally for initial-value problems in ODEs); see the paper by Sathiaraj & Sankar [1982].

Another, widely analyzed and employed class of one-step methods for second-kind Volterra integral equations are the block-by-block methods; as shown, e.g., in Brunner [1977], [1980], these methods can be viewed as particular discretizations of collocation methods in polynomial spline spaces. Initially, they were derived by means of product-integration techniques (Young [1954a], [1954b]) and used mainly to generate starting values for LM methods (see Day

[1967a], [1968], [1969], [1977], Campbell [1970], Campbell & Day [1970], [1971], O'Neill [1967], O'Neill & Byrne [1968], Linz [1967], [1969b]). Weiss [1972a] and de Hoog & Weiss [1975] gave a detailed convergence and stability analysis of block-by-block methods (called here "implicit Runge-Kutta methods"); an illuminating discussion of various practical aspects of these methods (such as the distribution of the kernel values occurring in the method in S, and motivations for selecting the collocation parameters) was given by Schlichte [1984, pp. 44-56] (see also section 8.2 for the automatic routine of Schlichte [1984] which is based on the method of de Hoog & Weiss [1975]). In addition, we refer to Garey [1971], [1972b], [1975a], [1975b], Kumar [1979], [1980a], [1981], and Malina [1975a] for related methods.

Of the papers dealing with various other methods for obtaining global approximations to solutions of second-kind Volterra integral equations we cite those of El-Gendi [1969], Bownds & Wood [1977], Porath & Tabbert [1979], Kumar [1980b] (solutions based on Chebyshev series); Goldfine [1973], [1977] (Taylor expansion methods; see also section 5.4); Bownds [1978a], P. Weiss [1978], MacCamy & Weiss [1979a], [1979b] (Galerkin-type methods; the last two of these papers deal with Galerkin approximations to more general Volterra equations in Hilbert space). Finally, we mention the methods for equations with degenerate (finitely decomposable) kernels (obtained by suitably approximating the given kernel) introduced by Porath [1975a] (who presents a general theoretical framework), Bownds & Wood [1976], Bownds [1978b], [1979], [1981], [1982a] (see also Section 8.2), and Golberg [1978].

Section 5.4.

Hung [1970] extended the spline method of Loscalzo (cf. Notes to Section 5.2) to integro-differential equations: obviously, collocation in classical polynomial spline spaces $S_m^{(m-1)}(Z_N)$ leads to divergent approximations if $m \geqslant 4$. However, Hermite-type collocation in $S_4^{(2)}(Z_N)$ (with the knots being double collocation points) yields a convergent fourth-order method (see also Cryer [1972, pp. 68 69]). Results similar to those derived by Hung were also given by Guzek & Kemper [1973] (for $S_3^{(2)}(Z_N)$) and by G. Micula [1975], [1978, pp. 254-266] (for $S_m^{(m-1)}(Z_N)$). Brunner [1973a] analyzes the convergence of collocation approximations in $S_m^{(1)}(Z_N)$, and Brunner & Lambert [1974] employed $S_m^{(0)}(Z_N)$ to obtain starting values for LM methods for Volterra integro-differential equations. A complete convergence theory of collocation approximations in $S_m^{(0)}(Z_N)$, including local superconvergence results and the discretization of the collocation equations, may be found in Brunner [1984a] (see also Brunner [1981]). Wahr [1977] studied the convergence and application of collocation methods to higher-order (linear) VIDEs.

Day [1967b], [1970] derived block-by-block methods to generate starting values; see also Linz [1969d] for a block-by-block method based on Simpson's rule (this method is also reviewed in Cryer [1972, pp. 66-67]). Makroglou [1977], [1980] uses the techniques of Weiss [1972a] and of de Hoog & Weiss

[1975] to construct block-by-block (or: implicit Runge-Kutta) methods for Volterra integro-differential equations; as shown in Brunner [1984a], these methods correspond to particular discretizations of collocation methods using $S_m^{(0)}(Z_N)$.

A general theory of one-step methods for functional differential equations of retarded type is due to Tavernini [1971]; this theory is applicable to Volterra integro-differential equations. One-step methods were also studied by Mocarsky [1971] and Garey [1975c]. Goldfine [1973], [1977] introduced Taylor expansion methods for Volterra integro-differential equations of higher order; symbolic differentiation is performed by a PL/l-FORMAC program. The recent paper by Bellen [1985] contains, among other things, a concise survey of recent advances in the numerical solution of VIDEs by collocation and related methods.

Section 5.5.

Huber [1939] seems to have been the first to use piecewise (linear) polynomials in the numerical solution of first-kind Volterra integral equations. The first rigorous convergence analysis for collocation approximations in $S_1^{(0)}(Z_N)$ is due to Hung [1970, pp. 124-145]. He also showed that while Richardson extrapolation can be applied to increase the order of convergence, the use of the space $S_2^{(1)}(Z_N)$ leads to divergent collocation approximations. Brunner [1971], [1973a] showed that collocation in $S_m^{(0)}(Z_N)$, employing equidistant collocation points, yields convergent approximations of order $m+1$. El Tom [1974b], [1976] generalized the results of Hung to less smooth polynomial spline spaces, using a discretized form of the collocation equation. The general convergence theory of collocation methods in $S_m^{(-1)}(Z_N)$ was established in Brunner [1978a], while the local superconvergence results summarized in Theorem 5.5.2 may be found, for general kernels $K(t,s)$, in Brunner [1978b], [1979]. Eggermont [1982], [1983] gave a general function-analytic framework for proving the results of Brunner just mentioned (in addition, he also considers equations with a special class of kernels satisfying $K(t,t)\equiv 0$ for all $t\in I$). Compare also Eggermont [1985] for additional superconvergence results.

The development of block-by-block methods (based again on product-integration techniques) dates back to the work of Linz [1971] and of Anderssen & White [1971]. The most comprehensive analyses of block-by-block methods for first-kind Volterra integral equations with regular kernels (which, as shown in Brunner [1977], [1978a], correspond to particular discretizations of collocation methods in $S_m^{(0)}(Z_N)$ and $S_{m-1}^{(-1)}(Z_N)$) is due to Weiss [1972a], de Hoog & Weiss [1973a], [1973b]; see also Smarzewski [1976]. Block-by-block methods in $\mathbb{R}^2$ were considered by Ten Men Yan [1979a].

Certain Volterra integral operators have the property that they map Haar subspaces into (weak) Haar subspaces: this fact forms the basis for a class of global methods where the solution of a first-kind Volterra integral equation is found in a given Haar subspace (e.g., π_m) so that the uniform norm of the difference between the left-hand and the right-hand side (i.e., the residual) is minimized, see Brunner [1974b], [1976].

Due to limitations of space and scope we have to refrain from discussing methods (based on solving first-kind Volterra integral equations with $K(t,s)\equiv 1$) for the numerical differentiation of non-exact (experimental) data. Instead we point out some of the principal references treating various aspects of this problem (which is ill-posed) in detail. The application of Tikhonov's regularization procedure is lucidly analyzed in Cullum [1971] (see also Tikhonov & Arsenin [1977], Groetsch [1984], and Morozov [1984] for background material on regularization methods). Anderssen & Bloomfield [1974] showed that Cullum's regularization method is equivalent to a certain spectrum analysis procedure; this observation is used to derive an automatic computational procedure for numerical differentiation (involving an estimation of the optimum value of the regularization parameter, and using fast Fourier transform techniques). This paper contains also an excellent review of other techniques for numerical differentiation; its lecture should be followed by that of the paper by Lyness [1977]. In addition, see Anderssen & de Hoog [1984].

Regular first-kind integral equations with non-exact data were studied by, e.g., Zheludev [1975], [1981], Radziuk [1977], Linz [1982], Jakeman & Young [1980], [1984]. (Compare also Section 6.4 and its Notes for equations with weakly singular kernels.)

CHAPTER 6

VOLTERRA EQUATIONS WITH WEAKLY SINGULAR KERNELS

So far we have only considered Volterra equations with regular (bounded) kernels where the smoothness of the solution is determined by the smoothness of the given kernel and forcing function. Once we admit weakly singular kernels then, as we have seen in Section 1.3.5, the resulting solutions are typically nonsmooth near the initial point of the interval of integration (i.e., their derivatives become unbounded). This means that if a numerical method is to possess a high order of convergence one has to take into account, in some way, the singular behavior of the exact solution near this point. In linear multistep methods this will be reflected in the special construction of the starting weights, while in polynomial spline collocation methods we shall be forced to use graded meshes, instead of quasi-uniform ones. If one wants to retain quasi-uniform meshes, then one will have to abandon polynomial spline spaces in favour of certain nonpolynomial spline spaces whose dimension will be considerably larger.

6.1. Fractional linear multistep methods

6.1.1. Introduction

Consider the second-kind integral equation with weakly singular (Abel-type) kernel,

$$y(t) = g(t)+\int_0^t (t-s)^{-\alpha}k(t,s,y(s))ds, \quad t\in I, \quad 0<\alpha<1, \tag{6.1.1}$$

where $k(t,s,y)$ is a bounded (and usually smooth) function. For a given uniform mesh $t_n=nh$, $n=0, \ldots ,N$; $N\geqslant 1$, of I, (6.1.1) will be discretized, in analogy to (3.2.4), by direct quadrature methods of the form

$$y_n = g(t_n)+h^{1-\alpha}\sum_{j=0}^{n} w_{n,j}(\alpha)k(t_n,t_j,y_j), \quad \kappa\leqslant n\leqslant N, \tag{6.1.2}$$

where the quadrature weights $\{w_{n,j}(\alpha)\}$ will now depend on α. It is obvious that the direct quadrature methods based on (ρ,σ)-reducible quadrature formulas (cf. Section 3.2.3) are not applicable to (6.1.1) since, for $t=t_n$, they employ kernel values with $(t,s)=(t_n,t_n)$; in the presence of a weak singularity $(t-s)^{-\alpha}$, these are not bounded. It will be shown, however, that the ideas underlying the (ρ,σ)-reducible formulas can be generalized to derive so-called fractional convolution quadratures for Riemann-Liouville integrals.

Before presenting these fractional convolution quadratures (introduced by Lubich [1983c], [1985b]), we briefly look at the discrete Volterra equation satisfied by the errors $e_n := y(t_n) - y_n$ induced by (6.1.2); this will exhibit certain properties which these quadratures are to possess. Setting $t = t_n$ in (6.1.1) and subtracting (6.1.2) we find

$$e_n = \int_0^{t_n} (t_n - s)^{-\alpha} k(t_n, s, y(s)) ds - h^{1-\alpha} \sum_{j=0}^{n} w_{n,j}(\alpha) k(t_n, t_j, y_j).$$

This may be written as

$$e_n = h^{1-\alpha} \sum_{j=0}^{n} w_{n,j}(\alpha) k_y(t_n, t_j, z_j) e_j + \mathbf{E}_n^\alpha[\phi_n], \tag{6.1.3}$$

where, by the Mean-Value Theorem, z_j denotes an appropriate value between $y(t_j)$ and y_j, and where

$$\mathbf{E}_n^\alpha[\phi_n] := \int_0^{t_n} (t_n - s)^{-\alpha} \phi_n(s) ds - h^{1-\alpha} \sum_{j=0}^{n} w_{n,j}(\alpha) \phi_n(t_j), \tag{6.1.4}$$

with $\phi_n(s) := k(t_n, s, y(s))$, denotes the quadrature error associated with the given quadrature formula. If the partial derivative $k_y(t,s,y)$ is bounded, i.e., if $|k_y(t,s,y)| \leqslant K_0 < \infty$ for all $(t,s) \in S$ and all y with $|y(s) - y| \leqslant b$ (see also Theorem 1.3.9), then (6.1.3) yields, by

$$(1 - h^{1-\alpha} w_{n,n}(\alpha) k_y(t_n, t_n, z_n)) e_n = h^{1-\alpha} \sum_{j=0}^{n-1} w_{n,j}(\alpha) k_y(t_n, t_j, z_j) e_j + \mathbf{E}_n^\alpha[\phi_n],$$

the generalized discrete Gronwall inequality

$$|e_n| \leqslant C_0 h^{1-\alpha} \sum_{j=0}^{n-1} |w_{n,j}(\alpha)| \, |e_j| + C_1 |\mathbf{E}_n^\alpha[\phi_n]|, \tag{6.1.5}$$

provided h is sufficiently small. Here, the finite constants C_0 and C_1 depend on K_0 and on α.

Recall now Theorem 1.5.6: it states that if the quadrature weights $w_{n,j}(\alpha)$ are such that, for some finite constant W_0,

$$|w_{n,j}(\alpha)| \leqslant W_0 (n-j)^{-\alpha}, \quad 0 \leqslant j \leqslant n-1 \leqslant N-1, \tag{6.1.6}$$

then, as $N \to \infty$, $Nh = T$, (6.1.5) implies

$$|e_n| \leqslant C \cdot \sup\{|\mathbf{E}_n^\alpha[\phi_n]| : 1 \leqslant n \leqslant N\}, \quad n = 1, \ldots, N. \tag{6.1.7}$$

In other words, the order of convergence of the approximations $\{y_n\}$ generated by (6.1.2) is governed by the order of the quadrature errors $\{\mathbf{E}_n^\alpha[\phi_n]\}$, provided the quadrature weights are suitably bounded, as described in (6.1.6).

Thus, there arises the following problem: construct a sequence of quadrature formulas using equidistant abscissas $\{t_j\}$ and possessing weights $w_{n,j} = w_{n,j}(\alpha)$, such that:

(i) for $j < n$, the weights $w_{n,j}(\alpha)$ satisfy (6.1.6), while $w_{n,n}(\alpha)$ is bounded;

(ii) the quadrature errors $\{\mathbf{E}_n^\alpha[\phi]\}$ can be bounded by

$$|\mathbf{E}_n^\alpha[\phi]| \leqslant Qh^p, \quad n=1, \ldots ,N, \tag{6.1.8}$$

where $p \in \mathbb{N}$ is prescribed, and where ϕ comes from a suitable class of functions. As we have seen in Section 1.3.5, the derivative of the solution of (6.1.1) behaves like $y'(t) \sim t^{-\alpha}$ near $t=0$; hence, this class of functions will have to include functions of the form $\phi(t)=t^{1-\alpha}\psi(t)$, with ψ being smooth on I.

The following section is dedicated to a detailed study of this problem.

6.1.2. Convolution quadratures for integrals of fractional order

In Section 2.6 we derived sequences of so-called (ρ,σ)-reducible quadrature formulas for the approximate evaluation of integrals

$$\mathbf{J}[\phi](t) := \int_0^t \phi(s)ds, \quad t \in I, \tag{6.1.9}$$

where $\phi \in C(I)$, and where $t=t_n=nh$ is a mesh point. These quadrature formulas were then used in Chapter 3 to obtain linear multistep methods for the numerical solution of Volterra equations with bounded kernels. The underlying ideas used by Matthys [1976] and by Wolkenfelt [1981a], [1982] have recently been generalized by Lubich [1983c], [1985b] to obtain *fractional convolution quadratures* for the approximation of *fractional integrals* of the form

$$\mathbf{J}^\alpha[\phi](t) := \frac{1}{\Gamma(1-\alpha)} \int_0^t (t-s)^{-\alpha}\phi(s)ds, \quad t \in I, \ 0<\alpha<1. \tag{6.1.10}$$

(In order to be consistent with Section 2.6, we have changed the notation introduced in Section 1.2: $\mathbf{J}^\alpha[\phi](t)$ corresponds to ${}_0D_t^{-(1-\alpha)}\phi(t)$, the *Riemann-Liouville integral of order* $1-\alpha$.) Note that the integral in (6.1.10) may be viewed as the *convolution* of the functions $f(t):=t^{-\alpha}$ and $\phi(t)$.

In order to see better the motivation for the form of the fractional convolution quadratures for (6.1.10) (cf. Definition 6.1.1 below) we look again, from a somewhat different viewpoint, at the case $\alpha=0$ (for which (6.1.10) reduces to (6.1.9)). Let us first rewrite the (ρ,σ)-reducible quadrature formula (2.6.8) (with $\kappa-1$ replaced by $s \geqslant 0$) as

$$\mathbf{J}_n[\phi] = h\sum_{j=0}^{s} \bar{w}_{n,j}\phi_j + h\sum_{j=0}^{n} \omega_{n-j}\phi_j, \quad n=0, \ldots ,N. \tag{6.1.11}$$

Here, we have set $\phi_j := \phi(t_j)$ and

$$\bar{w}_{n,j} := w_{n,j} - \omega_{n-j}, \quad j=0, \ldots ,s, \tag{6.1.12}$$

with the weights $\{\omega_l\}$ satisfying $\omega_l=0$ if $l<0$. Hence, instead of describing the (ρ,σ)-reducible quadrature formula by means of the matrices S_κ, $\Sigma_{\kappa,n}$, and $\Omega_{\kappa,n}$ (see Section 2.6.2), we may characterize it by the matrix

$$W_N := S_{s,N} + \Omega_N, \tag{6.1.13a}$$

where the $(N+1)$ by $(N+1)$ matrices $S_{s,N}$ (the starting matrix) and Ω_N (the convolution matrix) are defined by

$$S_{s,N} := \begin{bmatrix} \bar{w}_{0,0} & \cdots & \bar{w}_{0,s} & & \\ \vdots & & \vdots & & \\ \bar{w}_{n,0} & \cdots & \bar{w}_{n,s} & O & \\ \vdots & & \vdots & & \\ \bar{w}_{N,0} & \cdots & \bar{w}_{N,s} & & \end{bmatrix}, \tag{6.1.13b}$$

and by

$$\Omega_N := \begin{bmatrix} \omega_0 & & & & & \\ \omega_1 & \omega_0 & & O & & \\ \vdots & & \ddots & & & \\ \omega_n & \omega_{n-1} & \cdots & \omega_0 & & \\ \vdots & & & & \ddots & \\ \omega_N & \omega_{N-1} & \cdots & & \cdots & \omega_0 \end{bmatrix}. \tag{6.1.13c}$$

It follows from the recurrence relation (2.6.7) that the elements of Ω_N (which we shall refer to as the *convolution weights*) are given by the coefficients of the (formal) power series

$$\omega(z) := \frac{\sigma(z^{-1})}{\rho(z^{-1})} = \sum_{n=0}^{\infty} \omega_n z^n, \tag{6.1.14}$$

where ρ and σ denote the characteristic polynomials introduced in (2.6.3).

Suppose now that we are given an implicit linear κ-step method (ρ,σ) of order $p \geqslant 1$. By (6.1.14), the matrix Ω_N of convolution weights is thus completely determined. There remains the problem of finding a matrix $S_{s,N}$ (with suitably chosen s) such that the corresponding quadrature error satisfies

$$\mathbf{E}_n[\phi] := \mathbf{J}[\phi](t_n) - \mathbf{J}_n[\phi] = O(h^p), \quad n = 1, \ldots, N, \tag{6.1.15}$$

whenever ϕ is a sufficiently smooth function. (Recall that the order p of a convergent linear κ-step method (ρ,σ) cannot exceed $\kappa+2$; for the practically important families of the Adams-Moulton methods and the BD methods, we have $p=\kappa$. As will be seen below, (6.1.15) will hold if we choose $s=p-1$.)

Let

$$\mathbf{\Delta}_n[\phi] := h\sum_{j=0}^{n} \omega_{n-j}\phi_j \tag{6.1.16}$$

denote the convolution part (characterized by the matrix (6.1.13c)) of the (ρ,σ)-reducible quadrature formula (6.1.11), and let

$$\mathbf{\Delta}_n[\phi] := \mathbf{J}[\phi](t_n) - \mathbf{\Omega}_n[\phi] \tag{6.1.17}$$

be the error associated with $\mathbf{\Omega}_n[\phi]$. It follows from (6.1.15) and (6.1.11) that

$$\mathbf{E}_n[\phi] = \mathbf{\Delta}_n[\phi] - h\sum_{j=0}^{n} \overline{w}_{n,j}\phi_j. \tag{6.1.18}$$

Thus, given Ω_n and $p \in \mathbb{N}$, we wish to find starting weights $\{\overline{w}_{n,j}: 0 \leqslant j \leqslant s\}$ such that, for all sufficiently smooth functions ϕ,

$$h\sum_{j=0}^{s} \overline{w}_{n,j}\phi_j = \mathbf{\Delta}_n[\phi] + O(h^p), \quad 1 \leqslant n \leqslant N \quad (Nh = T). \tag{6.1.19}$$

The following result (which is in essence a rephrasing of Theorem 2.6.12 for the present approach) shows how to choose these starting weights.

Lemma 6.1.1. *Let the linear* κ*-step method* (ρ, σ) *be zero-stable and consistent of order* $p \geqslant 1$. *If* $s = p - 1$, *and if the starting weights* $\{\overline{w}_{n,j}: 0 \leqslant j \leqslant s\}$ *are given, for each* $n = 1, \ldots, N$, *by the solution of the linear system*

$$\sum_{j=0}^{s} \overline{w}_{n,j} j^q = n^{q+1}/(q+1) - \sum_{j=0}^{n} \omega_{n-j} j^q, \quad q = 0, \ldots, s, \tag{6.1.20}$$

then the quadrature errors $\mathbf{E}_n[\phi]$ *defined by (6.1.15) satisfy* $\mathbf{E}_n[\phi] = O(h^p)$ *for all* $\phi \in C^p(I)$.

Proof. If we use $\phi(t) = t^q$ in (6.1.18), and if we then neglect the $O(h^p)$-term on its right-hand side, we find (after division by h^{q+1}) the linear system (6.1.20). It has a unique solution for each $n \geqslant 1$, since it is characterized by a Vandermonde matrix (which does not depend on n). This construction of the starting weights implies, by (6.1.18), that $\mathbf{E}_n[\phi] = 0$ for all $\phi \in \pi_s$, where $s = p - 1$. In other words, the degree of precision of the quadrature formula $\mathbf{I}_n[\phi]$ equals $p - 1$. The assertion of Lemma 6.1.1 now follows from the Peano kernel theorem (Theorem 2.1.2). □

To illustrate the connection between the present approach and the one employed in Section 2.6.7, as well as to prepare the ground for its generalization to fractional integrals (6.1.10), we return to an example considered already in Section 2.6.7.

Example. 6.1.1. Let $\rho(z) = 3z^2 - 4z + 1$, $\sigma(z) = 2z^2$. The corresponding linear 2-step method is a BD method; its order is $p = \kappa = 2$. The power series (6.1.14) associated with this method may be written as

$$\omega(z) = \frac{2}{z^2 - 4z + 3} =: \tfrac{2}{3}(1 + Q(z))^{-1} = \sum_{n=0}^{\infty} \omega_n z^n,$$

where $Q(z) := -4z/3 + z^2/3$ (compare also Section 6.1.3, where the motivation for introducing $Q(z)$ is given). It is readily verified that the coefficients $\{\omega_n\}$

are given by

$$\omega_n = 1-3^{-(n+1)}, \quad n=0,1, \ldots;$$

these numbers coincide, of course, with the elements of the matrix $\Omega_{2,n}$ in Section 2.6.7. Consider now (6.1.20): setting $s=p-1=1$, this linear system for the starting weights $\{\bar{w}_{n,j}: j=0,1\}$ reduces to

$$\bar{w}_{n,0}+\bar{w}_{n,1} = n-\sum_{j=0}^{n}\omega_{n-j} \quad (\text{if } q=0),$$

$$\bar{w}_{n,1} = n^2/2-\sum_{j=1}^{n} j\omega_{n-j} \quad (\text{if } q=1).$$

We find

$$\bar{w}_{n,0}+\omega_n = \frac{n(2-n)}{2}+\sum_{j=2}^{n}(j-1)\omega_{n-j}, \tag{6.1.21a}$$

and

$$\bar{w}_{n,1}+\omega_{n-1} = \frac{n^2}{2}-\sum_{j=2}^{n} j\omega_{n-j}; \tag{6.1.21b}$$

by (6.1.12), these values are equal, respectively, to the weights $w_{n,0}$ and $w_{n,1}$ in (2.6.8), and it easily seen that (6.1.21) leads to

$$w_{n,0}+w_{n,1} = \tfrac{3}{2}(1-3^{-n}), \quad w_{n,0}-w_{n,1}=0,$$

and hence to

$$w_{n,0} = w_{n,1}=\tfrac{3}{4}(1-3^{-n}).$$

These are the weights characterizing the matrices S_2 and $\Sigma_{2,n}$ given in (2.6.10a) and Section 2.6.7, respectively. □

Following Lubich [1985b], we shall now show that the above ideas carry over, in a natural manner, to the construction of sequences of so-called fractional convolution quadratures for the approximation of fractional integrals of the form (6.1.10) (as well as to more general fractional integrals; we refer the interested reader to Lubich [1985b] for a comprehensive analysis). First, we state

Definition 6.1.1. An approximation to the fractional integral of order $1-\alpha$,

$$\mathbf{J}^{\alpha}[\phi](t_n) = \frac{1}{\Gamma(1-\alpha)}\int_0^{t_n}(t_n-s)^{-\alpha}\phi(s)ds, \quad 0<\alpha<1,$$

of the form

$$\mathbf{J}_n^{\alpha}[\phi] := h^{1-\alpha}\sum_{j=0}^{s}\bar{w}_{n,j}(\alpha)\phi_j+h^{1-\alpha}\sum_{j=0}^{n}\omega_{n-j}(\alpha)\phi_j, \tag{6.1.22}$$

with fixed $s \geqslant 0$, with starting weights $\{\overline{w}_{n,j}(\alpha)\}$ and convolution weights $\{\omega_{n-j}(\alpha)\}$ not depending on h, is called a *fractional convolution quadrature* for $\mathbf{J}^\alpha[\phi](t)$. The term

$$\Omega_n^\alpha[\phi] := h^{1-\alpha} \sum_{j=0}^{n} \omega_{n-j}(\alpha)\phi_j \tag{6.1.23}$$

will be referred to as its *convolution part.* □

In analogy to (6.1.13), a fractional convolution quadrature (6.1.22) is thus completely characterized by the matrix

$$W_N(\alpha) := S_{s,N}(\alpha) + \Omega_N(\alpha), \tag{6.1.24a}$$

where $S_{s,N}(\alpha)$ and $\Omega_N(\alpha)$ are $(N+1)$ by $(N+1)$ matrices (with $N \geqslant s$) of the form

$$S_{s,N}(\alpha) := \begin{bmatrix} \overline{w}_{0,0}(\alpha) & \cdots & \overline{w}_{0,s}(\alpha) & \\ \vdots & & \vdots & \\ \overline{w}_{n,0}(\alpha) & \cdots & \overline{w}_{n,s}(\alpha) & O \\ \vdots & & \vdots & \\ \overline{w}_{N,0}(\alpha) & \cdots & \overline{w}_{N,s}(\alpha) & \end{bmatrix} \tag{6.1.24b}$$

$$\Omega_N(\alpha) := \begin{bmatrix} \omega_0(\alpha) & & & & & \\ \omega_1(\alpha) & \omega_0(\alpha) & & & O & \\ \vdots & & \ddots & & & \\ \omega_n(\alpha) & \omega_{n-1}(\alpha) & \cdots & \omega_0(\alpha) & & \\ \vdots & & & & \ddots & \\ \omega_N(\alpha) & \omega_{N-1}(\alpha) & \cdots & & & \omega_0(\alpha) \end{bmatrix} \tag{6.1.24c}$$

Alternatively, we may express (6.1.22) in a form analogous to (2.6.8); i.e.,

$$\mathbf{J}_n^\alpha[\phi] = h^{1-\alpha} \sum_{j=0}^{n} w_{n,j}(\alpha)\phi_j, \quad n = s+1, \ldots, N, \tag{6.1.25a}$$

with quadrature weights

$$w_{n,j}(\alpha) := \begin{cases} \overline{w}_{n,j}(\alpha) + \omega_{n-j}(\alpha), & \text{if} \quad 0 \leqslant j \leqslant s; \\ \\ \omega_{n-j}(\alpha), & \text{if} \quad s+1 \leqslant j \leqslant n, \end{cases} \tag{6.1.25b}$$

where we set $\omega_l(\alpha) = 0$ if $l < 0$. For $n = 0, \ldots, s$, we have

$$\mathbf{J}_n^\alpha[\phi] = h^{1-\alpha} \sum_{j=0}^{s} w_{n,j}(\alpha). \tag{6.1.25c}$$

Suppose that the weights $\{\omega_n(\alpha)\}$ of the convolution part $\Omega_n^\alpha[\phi]$ of (6.1.22) are generated by the (formal) power series

$$\omega(z;\alpha) := \sum_{n=0}^{\infty} \omega_n(\alpha) z^n. \tag{6.1.26}$$

It is clear that stability, consistency, and convergence of a fractional convolution quadrature (6.1.22) will be governed by the corresponding properties of its convolution part (6.1.23), in analogy to the regular case $\alpha=0$. (Compare also Lubich [1983a] for an analysis of the case $\alpha=0$.)

Definition 6.1.2. The convolution part $\Omega_n^\alpha[\phi]$ of (6.1.22) is said to be *consistent of order p* (with respect to the fractional integral $\mathbf{J}^\alpha[\phi]$) if its generating power series $\omega(z;\alpha)$ defined in (6.1.26) satisfies

$$h^{1-\alpha}\omega(e^{-h};\alpha) = 1+O(h^p). \quad \square \tag{6.1.27}$$

For $\alpha=0$, (6.1.27) reduces to $h\omega(e^{-h})=1+O(h^p)$, with $\omega(z)$ given in (6.1.14): this is known to be equivalent to the condition for consistency of order p of the method (ρ,σ) stated in Definition 2.6.2 (compare Henrici [1962, p. 226]).

Definition 6.1.3. The convolution part $\Omega_n^\alpha[\phi]$ of (6.1.22) is said to be *stable* (with respect to $\mathbf{J}^\alpha[\phi]$) if its weights satisfy

$$\omega_n(\alpha) = O(n^{-\alpha}), \quad n>0. \quad \square \tag{6.1.28}$$

Note that (6.1.28 corresponds essentially to the desired property (6.1.6) of a direct quadrature method for (6.1.1), except that we have not yet specified the starting weights $\{\overline{w}_{n,j}(\alpha)\}$ in (6.1.25b).

In the following we shall restrict our analysis to that class of fractional convolution quadratures which may be viewed as the direct analogues of (ρ,σ)-reducible quadrature formulas for $\alpha=0$. Let

$$\rho(z) := \sum_{j=0}^{\kappa} a_j z^{\kappa-j}, \quad \sigma(z) := \sum_{j=0}^{\kappa} b_j z^{\kappa-j} \quad (\kappa\geq 1;\ a_0 b_0\neq 0),$$

denote the characteristic polynomials of an implicit linear κ-step method, and assume that they have no common factors. The following result is a special case of a more general theorem due to Lubich [1985b].

Theorem 6.1.1. *Assume that the implicit linear κ-step method (ρ,σ) is zero-stable and consistent of order $p\geq 1$ (i.e., convergent of order p), and let $\alpha\in(0,1)$. If the weights $\{\omega_n(\alpha)\}$ of the convolution part $\Omega_n^\alpha[\phi]$ of (6.1.22) are generated by the (formal) power series*

$$\omega(z;\alpha) := (\omega(z))^{1-\alpha} = \sum_{n=0}^{\infty} \omega_n(\alpha) z^n, \tag{6.1.29}$$

with $\omega(z) := \sigma(z^{-1})/\rho(z^{-1})$ *(cf. (6.1.14)), then* $\Omega_n^\alpha[\phi]$ *is both stable and consistent of order p.* □

Before sketching the proof of Theorem 6.1.1, we illustrate it by three examples; the second and third of these will reveal the basic problem we are facing when computing the convolution weights $\{\omega_n(\alpha)\}$ and which will be discussed in more detail at the end of Section 6.1.3.

Example 6.1.2. Let $\rho(z) = z-1$, $\sigma(z) = z$ (implicit Euler method: $\kappa = p = 1$). We have

$$\omega(z) = (1-z)^{-1},$$

and thus, by (6.1.29) and by the binominal series,

$$\omega(z;\alpha) = (1-z)^{-(1-\alpha)} = \sum_{n=0}^{\infty} (-1)^n \binom{-(1-\alpha)}{n} z^n. \tag{6.1.30}$$

Here, the binomial coefficient may be expressed as

$$\binom{-(1-\alpha)}{n} = (-1)^n \frac{(1-\alpha)_n}{n!} = (-1)^n \frac{\Gamma(n+1-\alpha)}{\Gamma(1-\alpha)\Gamma(n+1)}.$$

Moreover,

$$\frac{\Gamma(z+\beta)}{\Gamma(z+\gamma)} = z^{\beta-\gamma}(1+\tfrac{1}{2}(\beta-\gamma)(\beta+\gamma-1)z^{-1}+O(z^{-2}))$$

(cf. Erdélyi [1953, p. 47]), implying

$$\frac{\Gamma(n+1-\alpha)}{\Gamma(n+1)} = n^{-\alpha}(1-\tfrac{1}{2}\alpha(1-\alpha)n^{-1}+O(n^{-2})).$$

The convolution weights $\{\omega_n(\alpha)\}$ defined by (6.1.30) are thus given by

$$\omega_n(\alpha) = \frac{1}{n!}\,\frac{\Gamma(n+1-\alpha)}{\Gamma(1-\alpha)}, \tag{6.1.31}$$

and they satisfy

$$\omega_n(\alpha) = n^{-\alpha}(1-\tfrac{1}{2}\alpha(1-\alpha)n^{-1}+O(n^{-2}))/\Gamma(1-\alpha) = O(n^{-\alpha}), \quad n \geq 1.$$

In summary, the convolution part $\Omega_n^\alpha[\phi]$ has the form

$$\Omega_n^\alpha[\phi] = \frac{h^{1-\alpha}}{\Gamma(1-\alpha)} \sum_{j=0}^{n} \frac{\Gamma(n-j+1-\alpha)}{(n-j)!}\,\phi_j; \tag{6.1.32}$$

it is stable and consistent of order $p = 1$. For $\alpha = 1/2$, we find

$$\Omega_n^{1/2}[\phi] = h^{1/2} \sum_{j=0}^{n} \frac{\prod_{l=1}^{n-j}(2l-1)}{2^{n-j}(n-j)!}\,\phi_j. \quad \square$$

Example 6.1.3. Let $\rho(z)=z-1$, $\sigma(z)=\frac{1}{2}(z+1)$ (trapezoidal method: $\kappa=p=2$). From

$$\omega(z)=\frac{1}{2}\frac{1+z}{1-z}=\frac{1}{2}(1+Q(z)),\quad Q(z):=\frac{2z}{1-z},$$

we obtain, by (6.1.29),

$$\omega(z;\alpha)=2^{-(1-\alpha)}(1+Q(z))^{1-\alpha}. \tag{6.1.33}$$

Here, $Q(z)=2z+2z^2+\ldots$ is an example of a (formal) power series called a *nonunit*; i.e., the constant term of the power series vanishes (compare Henrici [1974, p. 12]. As will be shown in detail in Section 6.1.3, (6.1.33) represents the central problem in the numerical construction of the convolution weights $\{\omega_n(\alpha)\}$: given a formal power series which is a nonunit, evaluate $(1+Q(z))^\beta$, where $\beta=1-\alpha$, or $\beta=-(1-\alpha)$ (cf. Example 6.1.4). □

Example 6.1.4. BD methods with $\kappa\leqslant 6$ are zero-stable and consistent of order $p=\kappa$; they are given by

$$\rho(z)=a_0z^\kappa+\ldots+a_\kappa,\quad \sigma(z)=b_0z^\kappa$$

(compare Table 2.6.2). Hence, the corresponding $\omega(z)$ has the form

$$\omega(z)=\frac{b_0}{a_0}(1+Q(z))^{-1},$$

where

$$Q(z):=\frac{a_1}{a_0}z+\ldots+\frac{a_\kappa}{a_0}z^\kappa$$

is a nonunit. In order to compute the weights $\{\omega_n(\alpha)\}$ we need to find $(1+Q(z))^{-(1-\alpha)}$ in

$$\omega(z;\alpha)=(b_0/a_0)^{1-\alpha}(1+Q(z))^{-(1-\alpha)}. \tag{6.1.34}$$

The resulting convolution part $\Omega_n^\alpha[\phi]$ is stable and consistent of order p. □

We now return to Theorem 6.1.1: instead of furnishing a detailed proof (this may be found in Lubich [1985b]), we outline the principal steps leading to the assertion of Theorem 6.1.1.

First, write $\omega(z;\alpha)$ (defined by (6.1.29)) in the form

$$\omega(z;\alpha)=(1-z)^{-\mu}\tilde{\omega}(z;\alpha),$$

where $\tilde{\omega}(z;\alpha)$ is holomorphic at $z=1$, with $\tilde{\omega}(1;\alpha)\neq 0$. Since $\omega(z;\alpha)=(\omega(z))^{1-\alpha}$, with $\omega(z)=\sigma(z^{-1})/\rho(z^{-1})$, and with $\rho(z)$ possessing a simple zero at $z=1$, we must have $\mu=1-\alpha$, and $\tilde{\omega}(1;\alpha)=1$ (since, by consistency, $\sigma(1)=\rho'(1)$). Hence,

$$\omega(z;\alpha)=(1-z)^{-(1-\alpha)}\{\sum_{j=0}^{M-1}c_j(\alpha)(1-z)^j+(1-z)^M\tilde{r}(z;\alpha)\}, \tag{6.1.35}$$

with $\tilde{r}(z;\alpha)$ denoting a suitable function holomorphic at $z=1$. Consistency of order p can be characterized in terms of the coefficients $\{c_j(\alpha)\}$ in (6.1.35). To see this, consider the term $h^{1-\alpha}\omega(e^{-h};\alpha)$ in the consistency condition (6.1.27): according to (6.1.35) (with z replaced by e^{-h}) we obtain

$$h^{1-\alpha}\omega(e^{-h};\alpha) = (h/(1-e^{-h}))^{1-\alpha}\tilde{\omega}(e^{-h};\alpha)$$
$$=(h/(1-e^{-h}))^{1-\alpha}\{\sum_{j=0}^{M-1} c_j(\alpha)(1-e^{-h})^j+(1-e^{-h})^M\tilde{r}(e^{-h};\alpha)\},$$

where the right-hand side is $O(h^p)$ if, and only if,

$$\sum_{j=0}^{M-1} c_j(\alpha)(1-e^{-h})^j+(1-e^{-h})^M\tilde{r}(e^{-h};\alpha) = (h/(1-e^{-h}))^{-(1-\alpha)}+O(h^p).$$

But this can hold only if the coefficients $\{c_j(\alpha)\}$ coincide with the first p coefficients $\{\gamma_j(\alpha)\}$ in

$$(-\log(z)/(1-z))^{-(1-\alpha)} = \sum_{j=0}^{\infty}\gamma_j(\alpha)(1-z)^j; \tag{6.1.36}$$

i.e., if

$$\tilde{\omega}(z;\alpha) = (-\log(z)/(1-z))^{-(1-\alpha)}+O((1-z)^p). \tag{6.1.37}$$

However, since the method (ρ,σ) is, by assumption, consistent of order p, we have

$$h\omega(e^{-h}) = 1+O(h^p), \tag{6.1.38}$$

$$h\omega(e^{-h}) = h\frac{\sigma(e^h)}{\rho(e^h)} = \frac{h}{1-e^{-h}}\frac{\sigma(e^h)}{\tilde{\rho}(e^h)} =: \frac{h}{1-e^{-h}}\tilde{\omega}(e^{-h}).$$

Hence, by (6.1.38) and letting $z:=e^{-h}$, we must have

$$\tilde{\omega}(z) = -\frac{1-z}{\log(z)}+O((1-z)^p).$$

But we also find, by (6.1.29),

$$\tilde{\omega}(z;\alpha) = (\tilde{\omega}(z))^{1-\alpha} = \left(-\frac{1-z}{\log(z)}\right)^{1-\alpha}(1+O((1-z)^p),$$

which leads to (6.1.37).

The stability condition $\omega_n(\alpha)=O(n^{-\alpha})$ can be established by using (6.1.35), and by observing that the coefficients in the binomial series for $(1-z)^{-(1-\alpha)}$ possess the asymptotic expansion

$$(-1)^n\binom{-(1-\alpha)}{n} = \frac{n^{-\alpha}}{\Gamma(1-\alpha)}(1-\tfrac{1}{2}\alpha(1-\alpha)n^{-1}+O(n^{-2}))$$

(cf. Example 6.1.2). The reader is referred to Lubich [1985b] for additional details.

We now define what we mean by the order of convergence of the convolution part $\Omega_n^\alpha[\phi]$ of (6.1.22); in contrast to the regular case $\alpha=0$, we shall have to augment the class of test functions so as to admit certain nonsmooth functions.

Definition 6.1.4. The convolution part $\Omega_n^\alpha[\phi]$ $(0<\alpha<1)$ of the convolution quadrature (6.1.22) is said to be *convergent of order p* if, for all $\phi(t)=t^{\lambda-1}$ with $\lambda>0$, the corresponding error $\Delta_n^\alpha[\phi]:=\mathbf{J}^\alpha[\phi](t_n)-\mathbf{W}_n^\alpha[\phi]$ satisfies

$$\mathbf{D}_n^\alpha[\phi] = O(h^\lambda)+O(h^p), \quad t_n=nh\in[a,T], \quad a>0 \quad \text{fixed.} \quad \square \tag{6.1.39}$$

Note that, for $t>0$, we have

$$\mathbf{J}^\alpha[\phi](t) = \frac{1}{\Gamma(1-\alpha)}\int_0^t (t-s)^{-\alpha}\phi(s)ds = \frac{t^{1-\alpha}}{\Gamma(1-\alpha)}\int_0^1 (1-\tau)^{-\alpha}\phi(\tau t)d\tau.$$

Hence, for $\phi(t)=t^{\lambda-1}$ $(\lambda>0)$, and for $0<a\leqslant t$, we obtain

$$\mathbf{J}^\alpha[\phi](t) = \frac{t^{\lambda-\alpha}}{\Gamma(1-\alpha}\int_0^1 (1-\tau)^{-\alpha}\tau^{\lambda-1}d\tau = \frac{t^{\lambda-\alpha}}{\Gamma(1-\alpha)}B(1-\alpha,\lambda),$$

or, recalling that Euler's beta function is related to the gamma function by $B(1-\alpha,\lambda)=\Gamma(1-\alpha)\Gamma(\lambda)/\Gamma(\lambda+1-\alpha)$,

$$\mathbf{J}^\alpha[\phi](t) = \frac{\Gamma(\lambda)}{\Gamma(\lambda+1-\alpha)}\, t^{\lambda-\alpha}.$$

Also, if $t_n\geqslant a>0$, then (6.1.23) can be written as

$$\Omega_n^\alpha[\phi] = t_n^{1-\alpha}\left(\frac{h}{t_n}\right)^{1-\alpha}\sum_{j=0}^n \omega_{n-j}(\alpha)\phi(\frac{j}{n}t_n).$$

For $\phi(t)=t^{\lambda-1}$, $\lambda>0$, this yields

$$\Omega_n^\alpha[\phi] = n^{\alpha-1}\sum_{j=0}^n \omega_{n-j}(\alpha)\left(\frac{j}{n}\right)^{\lambda-1} t_n^{\lambda-\alpha},$$

and thus

$$\Delta_n^\alpha[\phi] = t_n^{\lambda-\alpha}\left\{\frac{\Gamma(\lambda)}{\Gamma(\lambda+1-\alpha)} - n^{\alpha-1}\sum_{j=0}^n \omega_{n-j}(\alpha)\left(\frac{j}{n}\right)^{\lambda-1}\right\}.$$

This shows that, for fixed $t_n=nh=:t>0$, we could have stated condition (6.1.39) for the normalized interval of integration $[0,1]$, by setting $t=1$. Observe, incidentally, that both the fractional integral $\mathbf{J}^\alpha[\phi](t)$ and the convolution part $\Omega_n^\alpha[\phi]$ in the fractional convolution quadrature (6.1.22) possess analogous homogeneity properties.

The following theorem, again due to Lubich [1985b], relates the notions of stability, consistency, and convergence for $\Omega_n^\alpha[\phi]$, in analogy to the regular case $\alpha=0$.

Theorem 6.1.2. *Let the implicit linear κ-step method (ρ,σ) be zero-stable and consistent of order $p \geqslant 1$, and assume that $\alpha \in (0,1)$. If the weights $\{\omega_n(\alpha)\}$ of the convolution part $\Omega_n^\alpha[\phi]$ of (6.1.22) are generated as in Theorem 6.1.1, namely by the formal power series*

$$\omega(z;\alpha) := (\omega(z))^{1-\alpha} = (\sigma(z^{-1})/\rho(z^{-1}))^{1-\alpha},$$

then $\Omega_n^\alpha[\phi]$ is convergent of order p. In other words, if $\Omega_n^\alpha[\phi]$ is stable and consistent of order p, then it is convergent of order p. □

This result represents a particular case of a more general theorem (Lubich [1985b] which states that $\Omega_n^\alpha[\phi]$ is convergent of order p if, and only if, it is stable and consistent of order p; there, it is assumed that the generating power series for the convolution weights $\Omega_N(\alpha)$ (cf. (6.1.24c)) is given by

$$\omega(z;\alpha) = (R_1(z))^{1-\alpha} R_2(z),$$

where R_1 and R_2 are rational functions. We refer the reader to the original paper for the proof of this result.

The foregoing analysis gives rise to the following questions:

(i) Suppose we are given the convolution part $\Omega_n^\alpha[\phi]$ of (6.1.22), with convergence order equal to some $p \geqslant 1$. How can we construct a starting quadrature $S_{s,N}(\alpha)$ (defined in (6.1.24b)) with weights $\{\overline{w}_{n,j}(\alpha)\colon 0 \leqslant j \leqslant s\}$ so that the convolution quadrature error $\mathbf{E}_n^\alpha[\phi] := \mathbf{J}^\alpha[\phi](t_n) - \mathbf{J}_n^\alpha[\phi]$ satisfies, for all $\phi(t) = t^{\lambda-1}\tilde{\phi}(t)$, with $\tilde{\phi} \in C^p(I)$, and with $\lambda > 0$,

$$\mathbf{E}_n^\alpha[\phi] = O(h^p) \tag{6.1.40}$$

(uniformly for $t_n \in [a,T]$, $a > 0$) ?

(ii) Let (ρ,σ) be a given implicit linear κ-step method which is subject to the hypotheses stated in Theorem 6.1.1. How can we compute the convolution weights $\{\omega_n(\alpha)\}$ and the starting weights $\{\overline{w}_{n,j}(\alpha)\}$ in an efficient manner?

These questions will be studied in detail in the next section.

6.1.3. Computation of the fractional convolution quadrature weights

The fractional convolution quadrature error,

$$\mathbf{E}_n^\alpha[\phi] := \mathbf{J}^\alpha[\phi](t_n) - \mathbf{J}_n^\alpha[\phi],$$

can be written as

$$\mathbf{E}_n^\alpha[\phi] = \mathbf{\Delta}_n^\alpha[\phi] - h^{1-\alpha}\sum_{j=0}^{s} \overline{w}_{n,j}(\alpha)\phi_j, \tag{6.1.41}$$

where $\mathbf{\Delta}_n^\alpha[\phi]$ is the error induced by the convolution part $\Omega_n^\alpha[\phi]$ of $\mathbf{J}_n^\alpha[\phi]$. In the case $\alpha = 0$, we determined the starting weights $\{\overline{w}_{n,j}\colon 0 \leqslant j \leqslant s\}$ by requiring that the quadrature formula $\mathbf{J}_n[\phi]$ have degree of precision $p-1$ (cf. Lemma 6.1.1). In other words, $\mathbf{E}_n[\phi]$ was to vanish on the polynomial space π_{p-1}. For the singular case, $0<\alpha<1$, this test space will no longer suffice: as an example, recall that the solution of the weakly singular Volterra integral equation of the second-kind (6.1.1) near $t=0$ behaves typically like the (nonsmooth) function $\phi(t)=c_0+c_1t^{1-\alpha}$, with $c_1\neq 0$.

In order to prepare the ground for the general case we first show that, for any function of the form $\phi(t)=t^{\lambda-1}\tilde{\phi}(t)$, with $\lambda>0$, and with $\tilde{\phi}$ smooth on I, there exists a starting quadrature so that $\mathbf{E}_n^\alpha[\phi]=O(h^p)$ and $\bar{w}_{n,j}(\alpha)=O(n^{-\alpha})$.

Theorem 6.1.3. (Lubich [1985b]). *Let the convolution part $\Omega_n^\alpha[\phi]$ of (6.1.22) be generated by $\omega(z;\alpha)=(\sigma(z^{-1})/\rho(z^{-1}))^{1-\alpha}$, and assume that it is convergent of order $p\geq 1$. Suppose that the starting weights $\{\bar{w}_{n,j}(\alpha)\}$ are given by the solution of the linear system*

$$\sum_{j=1}^{s}\bar{w}_{n,j}(\alpha)j^{\mu+\lambda-1} = \frac{\Gamma(\mu+\lambda)}{\Gamma(\mu+\lambda+1-\alpha)}n^{\mu+\lambda-\alpha} - \sum_{j=1}^{n}\omega_{n-j}(\alpha)j^{\mu+\lambda-1}, \tag{6.1.42}$$

$\mu=0,\ldots,s-1$ $(n\geq 1)$, *where $\lambda>0$ is fixed. If $s\in\mathbb{N}$ is such that*

$$s+\lambda-1<p\leq s+\lambda, \tag{6.1.43}$$

then:

(i) $\bar{w}_{n,j}(\alpha)=O(n^{-\alpha})$, $j=1,\ldots,s$; $n\geq 1$;

(ii) *for any function $\phi(t)=t^{\lambda-1}\tilde{\phi}(t)$, with $\tilde{\phi}\in C^p(I)$, $\mathbf{E}_n^\alpha[\phi]=O(h^p)$, uniformly for $t_n=nh=:t\in[a,T]$ $(a>0)$.*

Proof. For $\tilde{\phi}\in C^p(I)$ we may write, by Taylor's formula with integral representation of the remainder term,

$$\tilde{\phi}(t) = \sum_{\mu=0}^{p-1}\frac{\tilde{\phi}^{(\mu)}(0)}{\mu!}t^\mu + \frac{1}{(p-1)!}\int_0^t (t-\tau)^{p-1}\tilde{\phi}^{(p)}(\tau)d\tau.$$

This yields, for $t>0$,

$$\phi(t) = \sum_{\mu=0}^{p-1}\frac{\tilde{\phi}^{(\mu)}(0)}{\mu!}t^{\mu+\lambda-1} + \frac{t^{p+\lambda-1}}{(p-1)!}\int_0^1 (1-v)^{p-1}\tilde{\phi}^{(p)}(tv)dv. \tag{6.1.44}$$

In direct analogy to the case $\alpha=0$, we now wish to determine starting weights $\{\bar{w}_{n,j}(\alpha)\}$ such that, for $\psi(t):=t^{\mu+\lambda-1}$ ($\mu=0,\ldots,s-1$; $\lambda>0$ fixed),

$$\mathbf{E}_n^\alpha[\psi] = \Delta_n^\alpha[\psi]-h^{1-\alpha}\sum_{j=1}^{s}\bar{w}_{n,j}(\alpha)\psi_j = 0.$$

This is equivalent to

$$h^{1-\alpha}\sum_{j=1}^{s}\bar{w}_{n,j}(\alpha)(jh)^{\mu+\lambda-1} = \frac{\Gamma(\mu+\lambda)}{\Gamma(\mu+\lambda+1-\alpha)}(nh)^{\mu+\lambda-\alpha}$$
$$- h^{1-\alpha}\sum_{j=1}^{n}\omega_{n-j}(\alpha)(jh)^{\mu+\lambda-1},\quad \mu=0,\ldots,s-1, \tag{6.1.45}$$

where the right-hand side equals $\mathbf{\Delta}_n^\alpha[\psi]$. Hence, by (6.1.39),

$$\mathbf{\Delta}_n^\alpha[\phi] = O(h^{\mu+\lambda})+O(h^p).$$

After dividing both sides of (6.1.45) by $h^{\mu+\lambda-\alpha}$ we obtain the system (6.1.42); i.e., we have, for $\mu=0, \dots ,s-1$,

$$\sum_{j=1}^{s} \overline{w}_{n,j}(\alpha)j^{\mu+\lambda-1} = \mathbf{\Delta}_n^\alpha[\psi]/h^{\mu+\lambda-\alpha} = O(h^\alpha)+O(h^{p-\mu-\lambda+\alpha}),$$

where, for fixed $t_n=nh>0$, $h=O(n^{-1})$.

If we choose s so that (6.1.43) holds, then it follows that

$$p-\mu-\lambda+\alpha \geqslant p+\alpha-(s+\lambda-1)>\alpha,$$

(since $\mu\leqslant s-1$), implying

$$\sum_{j=1}^{s} \overline{w}_{n,j}(\alpha)j^{\mu+\lambda-1} = O(n^{-\alpha}), \quad \mu=0, \dots ,s-1,$$

and so

$$\overline{w}_{n,j}(\alpha) = O(n^{-\alpha}), \quad j=1, \dots ,s \quad (n\geqslant 1).$$

Moreover, for $\psi(t)=t^{s+\lambda-1}$ we find

$$\mathbf{E}_n^\alpha[\psi] = \mathbf{\Delta}_n^\alpha[\psi]-h^{1-\alpha}\sum_{j=1}^{s}\overline{w}_{n,j}(\alpha)(jh)^{s+\lambda-1}=(O(h^{s+\lambda})+O(h^p))+O(h^{s+\lambda})$$

(since $\overline{w}_{n,j}(\alpha)=O(h^\alpha)$), or, by (6.1.43),

$$\mathbf{E}_n^\alpha[\psi] = O(h^p).$$

Assertion (ii) of Theorem 6.1.3 now follows readily from (6.1.44), using the linearity property of $\mathbf{E}_n^\alpha[\phi]$ and of $\mathbf{\Delta}_n^\alpha[\phi]$. □

Theorem 6.1.3 shows that the starting weights $\{\overline{w}_{n,j}(\alpha)\}$ depend on the choice of λ; that is, if we change the value of λ, we obtain a different set of starting weights. This is, of course, in contrast to the convolution weights $\{\omega_n(\alpha)\}$ which are determined by the choice of the power series $\omega(z;\alpha)$; thus, they only depend on α, and (6.1.39) holds for any $\lambda>0$.

So far, we have only considered functions ϕ which are either undefined at $t=0$ (for $0<\lambda<1$), or which vanish at $t=0$ (if $\lambda\geqslant 1$); in the former case we have set $\phi(0):=0$. In view of subsequent applications we shall now analyze the general case where ϕ contains smooth as well as nonsmooth components. The relevant example comes from Corollary 1.3.4 where we have shown that the solution of a linear Volterra integral equation of the second kind with weakly singular kernel, but with smooth functions $g(t)$ and $K(t,s)$, is of the form

$$y(t) = \sum_{j=0}^{q-1} v_j(t)t^{j(1-\alpha)}, \quad v_j \quad \text{smooth on } I,$$

provided α is rational, $\alpha=m/q$, with m and q coprime. This solution is thus generated by functions of the form

$$\phi_{\mu,l}(t) := t^{\mu+l(1-\alpha)}, \quad 0<\alpha<1, \tag{6.1.46}$$

where μ and l are nonnegative integers. Accordingly, our task will consist in determining a starting quadrature for which

$$\mathbf{E}_n^\alpha[\phi_{\mu,l}] = \mathbf{\Delta}_n^\alpha[\phi_{\mu,l}] - h^{1-\alpha}\sum_{j=0}^{s}\overline{w}_{n,j}(\alpha)\phi_{\mu,l}(jh) = 0,$$

for all pairs (μ,l) contained in a suitable subset of $\mathbb{N}_0\times\mathbb{N}_0$ (depending on α). In order to describe this subset we introduce the following notation. For given $p\in\mathbb{N}$ and $\alpha\in(0,1)$ we set

$$l_p(\alpha) := \lfloor (p-1)/(1-\alpha) \rfloor ,$$

and we associate with each $l\in\mathbb{N}_0$ satisfying $l\leqslant l_p(\alpha)$ the integer $\mu_p(\alpha;l)$, defined to be the largest nonnegative integer so that

$$\mu_p(\alpha;l) \leqslant p-1-l(1-\alpha). \tag{6.1.47}$$

Finally, define the index sets

$$M_p(\alpha) := \{(\mu,l)\in\mathbb{N}_0\times\mathbb{N}_0 \colon l\leqslant l_p(\alpha),\ \mu\leqslant\mu_p(\alpha;l)\}, \tag{6.1.48a}$$

and, if α is rational, $\alpha=m/q$ (with m and q coprime),

$$\overline{M}_p(\alpha) := \{(\mu,l)\in\mathbb{N}_0\times\mathbb{N}_0 \colon l\leqslant\overline{l}_p(\alpha),\ \mu\leqslant\mu_p(\alpha;l)\}, \tag{6.1.48b}$$

where $\overline{l}_p(\alpha) := \min\{q-1, l_p(\alpha)\}$. Let $|M_p(\alpha)|$ denote the cardinality of $M_p(\alpha)$.

The introduction of $\overline{M}_p(\alpha)$ is motivated by the following observation. If $\alpha=m/q$, and if $q\leqslant l_p(\alpha)$, then powers $t^{\mu+l(1-\alpha)}$, with $(\mu,l)\in M_p(\alpha)$ but $(\mu,l)\notin\overline{M}_p(\alpha)$, coincide with powers $t^{\mu'+l'(1-\alpha)}$ for some $(\mu',l')\in\overline{M}_p(\alpha)$. To see this, assume $q\leqslant l\leqslant l_p(\alpha)$, and set $l=\nu q+l'$, where $\nu\in\mathbb{N}_0$, $l'\leqslant q-1$. We then find that

$$\mu+l(1-\alpha) = \mu+(\nu q+l')(q-m)/q = \mu+\nu(q-m)+l'(1-\alpha) =: \mu'+l'(1-\alpha),$$

and it is easily seen that $(\mu',l')\in\overline{M}_p(\alpha)$, and

$$t^{\mu+l(1-\alpha)} = t^{\mu'+l'(1-\alpha)}.$$

Theorem 6.1.4. *Let the convolution part* $\mathbf{\Omega}_n^\alpha[\phi]$ *of (6.1.22) be generated by* $\omega(z;\alpha)=(\sigma(z^{-1})/\rho(z^{-1}))^{1-\alpha}$, *and assume that it is convergent of order* $p\geqslant 1$. *Suppose that the starting weights* $\{\overline{w}_{n,j}(\alpha):0\leqslant j\leqslant s\}$ *are given by the solution of the linear system*

$$\sum_{j=0}^{s}\overline{w}_{n,j}(\alpha)j^{\mu+l(1-\alpha)} = C_{\mu,l}(\alpha)n^{\mu+(l+1)(1-\alpha)} - \sum_{j=0}^{n}\omega_{n-j}(\alpha)j^{\mu+l(1-\alpha)}, \tag{6.1.49}$$

with $C_{\mu,l}(\alpha) := \Gamma(\mu+l(1-\alpha)+1)/\Gamma(\mu+(l+1)(1-\alpha)+1)$, *and where*

$$(\mu,l)\in\begin{cases}M_p(\alpha), & \textit{if } \alpha \textit{ is irrational};\\ \overline{M}_p(\alpha), & \textit{if } \alpha=m/q, \textit{ with } m \textit{ and } q \textit{ coprime};\end{cases}$$

$$s = \begin{cases}|M_p(\alpha)|-1, & \textit{if } \alpha \textit{ is irrational};\\ |\overline{M}_p(\alpha)|-1, & \textit{if } \alpha=m/q, \textit{ with } m \textit{ and } q \textit{ coprime};\end{cases}$$

Then:

(i) $\overline{w}_{n,j}(\alpha) = O(n^{-\alpha}),\ j=0,\ \dots\ ,s\ \ (n\geqslant 1)$;

(ii) *for any function* $\phi(t) := \sum_{l=0}^{M}\tilde{\phi}_l(t)t^{l(1-\alpha)}$, *with* $\tilde{\phi}_l\in C^p(I)$, *the fractional convolution quadrature error satisfies*

$$\mathbf{E}_n^\alpha[\phi] = O(h^p),$$

uniformly for any fixed $t_n=nh=:t\in[a,T]$ $(a>0)$.

Proof. The proof of Theorem 6.1.4 proceeds along the lines of the one for Theorem 6.1.3: we shall construct a starting quadrature so that the fractional convolution quadrature error vanishes for certain functions of the form (6.1.46); i.e.,

$$\mathbf{E}_n^\alpha[\phi_{\mu,l}] = 0 \quad \text{for } all \quad (\mu,l)\in M_p(\alpha).$$

By (6.1.41), this is equivalent to

$$h^{1-\alpha}\sum_{j=0}^{s}\overline{w}_{n,j}(\alpha)(jh)^{\mu+l(1-\alpha)} = \mathbf{\Delta}_n^\alpha[\phi_{\mu,l}],\quad (\mu,l)\in M_p(\alpha),$$

where, by our hypothesis on the order of convergence of $\mathbf{\Omega}_n^\alpha[\phi]$,

$$\mathbf{\Delta}_n^\alpha[\phi_{\mu,l}] = O(h^{\mu+l(1-\alpha)+1})+O(h^p).$$

(This construction may be viewed as a generalization of the one encountered in Theorem 6.1.3: instead of employing a single value of $\lambda>0$, we now use values $\lambda_l:=l(1-\alpha)+1$, where $l\leqslant l_p(\alpha)$.)

It follows that the starting weights satisfy

$$\sum_{j=0}^{s}\overline{w}_{n,j}(\alpha)j^{\mu+l(1-\alpha)} = O(h^\alpha)+O(h^{p-\mu-(l+1)(1-\alpha)}),\quad (\mu,l)\in M_p(\alpha).$$

Since $\mu\leqslant\mu_p(\alpha;l)\leqslant p-1-l(1-\alpha)$, we have

$$p-\mu-(l+1)(1-\alpha)\geqslant p-(p-1-l(1-\alpha))-(l+1)(1-\alpha) = \alpha,$$

provided $l\leqslant l_p(\alpha)$. Hence,

$$\sum_{j=0}^{s}\overline{w}_{n,j}(\alpha)j^{\mu+l(1-\alpha)} = O(h^\alpha),\quad (\mu,l)\in M_p(\alpha),$$

and this leads to $\bar{w}_{n,j}(\alpha)=O(h^{\alpha})=O(n^{-\alpha})$, whenever $t_n=nh>0$.

Consider now a function $\phi_{\mu,l}(t)$ (as defined in (6.1.46)) with $(\mu,l)\notin M_p(\alpha)$: by (6.1.47) we have $\mu+l(1-\alpha)>p-1$, and this yields, by (6.1.41),

$$\mathbf{E}_n^{\alpha}[\phi_{\mu,l}] = (O(h^{\mu+l(1-\alpha)+1})+O(h^p))-h^{1-\alpha}\sum_{j=0}^{s}\bar{w}_{n,j}(\alpha)(jh)^{\mu+l(1-\alpha)}=O(h^p)$$

(note again that $\bar{w}_{n,j}(\alpha)=O(h^{\alpha})$). Assertion (ii) of Theorem 6.1.4 then follows, since the functions ϕ under consideration have, by the assumed smoothness for $\tilde{\phi}_l$, and thus by Taylor's formula, the form

$$\phi(t) = \sum_{l=0}^{M}\Big(\sum_{\mu=0}^{p-1}\tilde{\phi}_l^{(\mu)}(0)t^{\mu}+R_l(t)\Big)t^{l(1-\alpha)} =: \sum_{l=0}^{M}\sum_{\mu=0}^{p-1}a_{\mu,l}\phi_{\mu,l}(t)+R(t;\alpha).$$

As we already observed, before stating Theorem 6.1.4, rational values of α ($\alpha=m/q$, with m and q coprime) lead to a linear system (6.1.49) of lower dimension, provided $q-1<l_p(\alpha)$. In this case, $\bar{M}_p(\alpha)$ is a proper subset of $M_p(\alpha)$; however, it follows from the observation just mentioned that we have $\mathbf{E}_n^{\alpha}[\phi_{\mu,l}]=0$ not only for $(\mu,l)\in\bar{M}_p(\alpha)$, but for all $(\mu,l)\in M_p(\alpha)$. Thus, the proof is readily adapted to the case of rational α. □

Example 6.1.5. Let $\alpha=1/2, p\geqslant 2$. The index set $\bar{M}_p(\alpha)$ is given by

$$\bar{M}_p(\tfrac{1}{2}) = \{(0,0),\ (1,0),\ \dots\ ,(p-1,0);\quad (0,1),\ (1,1),\ \dots\ ,(p-2,1)\},$$

since we have $\bar{l}_p(\tfrac{1}{2})=1$, $\mu_p(\tfrac{1}{2};0)=p-1$, and $\mu_p(\tfrac{1}{2};1)=p-2$. Hence, $|\bar{M}_p(\tfrac{1}{2})|=2p-1$, leading to $s=2p-2$. Accordingly, the starting weights $\{\bar{w}_{n,j}(\tfrac{1}{2}): 0\leqslant j\leqslant 2p-2\}$ are found by solving the linear system

$$\sum_{j=0}^{2p-2}\bar{w}_{n,j}(\tfrac{1}{2})j^{\mu+l/2} = C_{\mu,l}(\tfrac{1}{2})n^{\mu+(l+1)/2} - \sum_{j=0}^{n}\omega_{n-j}(\tfrac{1}{2})j^{\mu+l/2}, \tag{6.1.50}$$

$(\mu,l)\in\bar{M}_p(\tfrac{1}{2})$, with

$$C_{\mu,l}(\tfrac{1}{2}) = \Gamma(\mu+l/2+1)/\Gamma(\mu+l/2+3/2).\ \square$$

In order to set up the linear system (6.1.48) for the starting weights $\{\bar{w}_{n,j}(\alpha)\}$ we have to evaluate finite sums of the form

$$\sum_{j=0}^{n}\omega_{n-j}(\alpha)j^{\beta},\quad \text{with}\quad \beta := \mu+l(1-\alpha),$$

and where $0\leqslant n\leqslant N$. These sums may be viewed as *convolutions* of the sequences $\{\omega_j(\alpha)\}$ and $\{j^{\beta}\}$. If N is large, then summation is most efficiently carried out by Fast Fourier Transform (FFT) techniques. An analogous remark applies, of course, also to the evaluation of the convolution part $\Omega_n^{\alpha}[\phi]$ in (6.1.22). A lucid introduction to FFT techniques, including numerous applications, can be found in Henrici [1979].

We conclude this section with a remark on the computation of the coefficients $\{\omega_n(\alpha)\}$ of the power series $\omega(z;\alpha)$ introduced in (6.1.29). For large N one will again employ FFT techniques for their computation (compare Henrici [1979], pp. 520-521). However, there exists an alternative method that can be used if N assumes only moderately large values (which will be the case in many practical applications: here, values in the range $50 \leqslant N \leqslant 200$ will often be sufficient). The corresponding algorithm is based on the composition of two (formal) power series,

$$B_\beta(z) := 1 + \binom{\beta}{1} z + \binom{\beta}{2} z^2 + \ldots \quad \text{(binomial series)},$$

and

$$Q(z) := q_1 z + q_2 z^2 + \ldots\,.$$

This latter series is a so-called *nonunit* (i.e. $q_0 = 0$). We wish to find the recurrence relation for the coefficients in the composition

$$B_\beta \circ Q =: c_0(\beta) + c_1(\beta) z + \ldots\,, \tag{6.1.51}$$

where

$$B_\beta \circ Q := B_\beta(Q(z))$$

("raising $1+Q(z)$ to the power β").

Theorem 6.1.5. (Henrici [1974, pp. 41-42]) *The coefficients $\{c_n(\beta)\}$ of the composition (6.1.51) of the binomial series B_β with the nonunit Q can be computed recursively by*

$$\begin{aligned} c_0(\beta) &= 1, \\ c_n(\beta) &= \frac{1}{n} \sum_{j=0}^{n-1} [\beta(n-j) - j] c_j(\beta) q_{n-j} \\ &= \frac{1}{n} \sum_{j=1}^{n} [(\beta+1)j - n] c_{n-j}(\beta) q_j, \quad n = 1,2, \ldots\,. \end{aligned}$$

The computation of the N coefficients $c_1(\beta), \ldots, c_N(\beta)$ can be carried out in $N^2 + 2N - 2$ multiplications and $N-1$ divisions. □

We note that the computation of the N coefficients $c_1(\beta), \ldots, c_N(\beta)$ by means of FFT techniques would require only $O(N \log(N))$ additions and multiplications.

According to (6.1.29), we must determine the power series $(\omega(z))^{1-\alpha}$, where $\omega(z) := \sigma(z^{-1})/\rho(z^{-1})$, with ρ and σ given by

$$\rho(z) := \sum_{j=0}^{\kappa} a_j z^{\kappa-j}, \quad \sigma(z) := \sum_{j=0}^{\kappa} b_j z^{\kappa-j}, \quad \kappa \geqslant 1;\ a_0 b_0 \neq 0.$$

Thus,

$$\omega(z) = \frac{b_0}{a_0}(1+\frac{b_1}{b_0}z^2 + \dots + \frac{b_\kappa}{b_0}z^\kappa)/(1+\frac{a_1}{a_0}z + \dots + \frac{a_\kappa}{a_0}z^\kappa)$$

$$=: \frac{b_0}{a_0}(1+Q(z)),$$

where $Q(z) := q_1 z + q_2 z^2 + \dots$ is a nonunit. Setting

$$\alpha_j := a_j/a_0, \quad \beta_j := b_j/b_0, \quad j=0, \dots, \kappa,$$

we find that the coefficients of $Q(z)$ are given by the recurrence relations

$$q_n = \begin{cases} \beta_n - \sum_{j=1}^{n} \alpha_j q_{n-j}, & \text{if } 1 \leqslant n \leqslant \kappa, \\ -\sum_{j=1}^{\kappa} \alpha_j q_{n-j}, & \text{if } \kappa < n. \end{cases}$$

Thus the power series

$$\omega(z;\alpha) = \sum_{n=0}^{\infty} \omega_n(\alpha) z^n = \left(\frac{b_0}{a_0}\right)^{1-\alpha} (1+Q(z))^{1-\alpha}$$

may be computed by applying Theorem 6.1.5 with $\beta := 1-\alpha$, and where $\omega_n(\alpha) := c_n(\beta)$.

6.1.4. Fractional linear multistep methods for second-kind equations

If we apply a fractional convolution quadrature (6.1.22) to the integral term in (6.1.1) (with t replaced by $t_n = nh$, $n > 0$), we obtain

$$y(t_n) = g(t_n) + h^{1-\alpha} \sum_{j=0}^{s} \tilde{w}_{n,j}(\alpha) k(t_n, t_j, y(t_j))$$
$$+ h^{1-\alpha} \sum_{j=0}^{n} \tilde{\omega}_{n-j}(\alpha) k(t_n, t_j, y(t_j)) + \tilde{\mathbf{E}}_n^\alpha[\phi], \tag{6.1.52}$$

where

$$\tilde{w}_{n,j}(\alpha) := \Gamma(1-\alpha)\overline{w}_{n,j}(\alpha), \quad \tilde{\omega}_{n-j}(\alpha) := \Gamma(1-\alpha)\omega_{n-j}(\alpha), \tag{6.1.53}$$

and

$$\tilde{\mathbf{E}}_n^\alpha[\phi_n] := \Gamma(1-\alpha)\mathbf{E}_n^\alpha[\phi_n]$$

(note that we now approximate an integral of the form $\int_0^t (t-\tau)^{-\alpha}\phi(\tau)d\tau$, rather than $(1/\Gamma(1-\alpha))\int_0^t (t-\tau)^{-\alpha}\phi(\tau)d\tau$, as in (6.1.10)). We have also set

$$\phi_n(\tau) := k(t_n,\tau,y(\tau)), \tag{6.1.54}$$

where $y(t)$ is the exact solution of (6.1.1).

The following definition is motivated by (6.1.52).

Definition 6.1.5. A *fractional linear multistep method* for the second-kind integral equation (6.1.1) has the form

$$y_n = g(t_n)+h^{1-\alpha}\sum_{j=0}^{s}\tilde{w}_{n,j}(\alpha)k(t_n,t_j,y_j)$$
$$+h^{1-\alpha}\sum_{j=0}^{n}\tilde{\omega}_{n-j}(\alpha)k(t_n,t_j,y_j), \quad 1\leqslant n\leqslant N, \tag{6.1.55}$$

with s fixed. The quadrature weights $\{\tilde{w}_{n,j}(\alpha)\}$ and $\{\tilde{\omega}_{n-j}(\alpha)\}$ are as in (6.1.53), with $\{\overline{w}_{n,j}(\alpha)\}$ and $\{\omega_{n-j}(\alpha)\}$ denoting, respectively, the starting weights and the convolution weights of the fractional convolution quadrature (6.1.22). □

Let us look more closely at (6.1.55). Observe first that $y_0=y(t_0)=g(0)$ is known (provided g and k are continuous functions). For $n=1, \ldots ,s$, we have

$$y_n = h^{1-\alpha}\sum_{j=1}^{s}(\tilde{w}_{n,j}(\alpha)+\tilde{\omega}_{n-j}(\alpha))k(t_n,t_j,y_j)$$
$$+\{g(t_n)+h^{1-\alpha}(\tilde{w}_{n,0}(\alpha)+\tilde{\omega}_n(\alpha))k(t_n,0,y_0)\}, \quad n=1, \ldots ,s, \tag{6.1.56}$$

with $\tilde{\omega}_l(\alpha)=0$ if $l<0$. This represents a system of s nonlinear equations for the approximations $\{y_1, \ldots ,y_s\}$; note that, for $s>1$, it involves values of $k(t,\tau,y)$ with $\tau>t$. If $k(t,\tau,y)=K(t,\tau)y$, then the corresponding linear system in $\mathbb{R}^s$ has the form

$$\left[a_{n,j}(\alpha)\right]\begin{bmatrix}y_1\\ \vdots\\ y_s\end{bmatrix} - \begin{bmatrix}g_1(\alpha)\\ \vdots\\ g_s(\alpha)\end{bmatrix}, \tag{6.1.57a}$$

with matrix elements

$$a_{n,j}(\alpha) := \begin{cases} -h^{1-\alpha}(\tilde{w}_{n,j}(\alpha)+\tilde{\omega}_{n-j}(\alpha))K(t_n,t_j), & \text{if } j<n,\\ 1-h^{1-\alpha}(\tilde{w}_{n,n}(\alpha)+\tilde{\omega}_0(\alpha))K(t_n,t_n), & \text{if } j=n,\\ -h^{1-\alpha}\tilde{w}_{n,j}(\alpha)K(t_n,t_j), & \text{if } j>n \end{cases} \tag{6.1.57b}$$

$$(n,j = 1, \ldots ,s),$$

and

$$g_n(\alpha) := g(t_n)+h^{1-\alpha}(\tilde{w}_{n,0}(\alpha)+\tilde{\omega}_n(\alpha))K(t_n,0)y_0. \tag{6.1.57c}$$

Suppose now that "starting values" $\{y_1, \ldots, y_s\}$ have been found, and let $s+1 \leqslant n \leqslant N$. The corresponding approximations $\{y_{s+1}, \ldots, y_N\}$ are then determined recursively from

$$y_n = h^{1-\alpha}\tilde{\omega}_0(\alpha)k(t_n,t_n,y_n)+Q_n, \tag{6.1.58a}$$

with

$$Q_n := g(t_n)+h^{1-\alpha}\sum_{j=0}^{n-1}\tilde{\omega}_{n-j}(\alpha)k(t_n,t_j,y_j)+h^{1-\alpha}\sum_{j=0}^{s}\tilde{w}_{n,j}(\alpha)k(t_n,t_j,y_j). \tag{6.1.58b}$$

Hence, for each $n>s$, we have to solve a nonlinear equation for y_n.

Example 6.1.6. Let $\alpha=1/2$, and assume that

(i) the convolution part $\Omega_n^\alpha[\phi]$ of the convolution quadrature (6.1.22), generated by $\omega(z;\alpha)=(\sigma(z^{-1})/\rho(z^{-1}))^{1-\alpha}$, is stable and consistent of order $p\geqslant 1$;

(ii) the starting quadrature is given by (6.1.50) (Example 6.1.5).

We thus have $s=2p-2$. For $p\geqslant 2$, the required starting values $\{y_1, \ldots, y_{2p-2}\}$ are found by solving the nonlinear system

$$y_n = (h\pi)^{1/2}\sum_{j=1}^{2p-2}(\bar{w}_{n,j}(\tfrac{1}{2})+\omega_{n-j}(\tfrac{1}{2}))k(t_n,t_j,y_j) \tag{6.1.59a}$$

$$+\{g(t_n)+(h\pi)^{1/2}(\bar{w}_{n,0}(\tfrac{1}{2})+\omega_n(\tfrac{1}{2}))k(t_n,0,g(0))\}, \quad n=1, \ldots, 2p-2,$$

since, by (6.1.53),

$$\tilde{w}_{n,j}(\tfrac{1}{2}) = \Gamma(\tfrac{1}{2})\bar{w}_{n,j}(\tfrac{1}{2}) = \pi^{1/2}\bar{w}_{n,j}(\tfrac{1}{2}), \quad \tilde{\omega}_{n-j}(\tfrac{1}{2}) = \pi^{1/2}\omega_{n-j}(\tfrac{1}{2}).$$

For $2p-1 \leqslant n \leqslant N$, the recurrence relation for the corresponding approximations $\{y_n\}$ reads:

$$y_n = (h\pi)^{1/2}\omega_0(\tfrac{1}{2})k(t_n,t_n,y_n) \tag{6.1.59b}$$

$$+ \{g(t_n)+(h\pi)^{1/2}(\sum_{j=0}^{n-1}\omega_{n-j}(\tfrac{1}{2})k(t_n,t_j,y_j)+\sum_{j=0}^{2p-2}\bar{w}_{n,j}(\tfrac{1}{2})k(t_n,t_j,y_j))\}.$$

As will be shown below (cf. (6.1.60b)), the approximations $\{y_n\}$ determined by (6.1.59) converge with order p; i.e.,

$$\max_{0\leqslant n\leqslant N}|y(t_n)-y_n| = O(h^p) \quad (\text{as } h\downarrow 0,\ Nh=T).$$

Theorem 6.1.6. (Lubich [1985c]) *Let the following assumptions hold:*

(i) *The implicit linear* κ*-step method* (ρ,σ) *is zero-stable and consistent of order* $p\geqslant 1$*;*

(ii) *the convolution part* $\Omega_n^\alpha[\phi]$ *of the fractional convolution quadrature (6.1.22) is generated by* $\omega(z;\alpha)=(\sigma(z^{-1})/\rho(z^{-1}))^{1-\alpha}$;

(iii) *the starting weights* $\{\bar{w}_{n,j}(\alpha):0\leqslant j\leqslant s\}$ *in (6.1.22) are given by the solution of the linear system (6.1.49).*

If the second-kind Volterra integral equation (6.1.1) (with $0<\alpha<1$*) is discretized by the fractional linear multistep method (6.1.55) based on the fractional convolution quadrature (6.1.22), and if* $g(t)$ *and* $k(t,\tau,y)$ *are sufficiently smooth, then the approximations* $\{y_n\}$ *to the exact solution* $y(t)$ *of (6.1.1) satisfy*

$$\max_{0\leqslant n\leqslant N}|y(t_n)-y_n| = O(h^{p+\nu(\alpha)}) \quad (\text{as } h\downarrow 0,\ Nh=T), \tag{6.1.60a}$$

with $\nu(\alpha)\in(-\alpha,0]$ *defined in (6.1.68) below. For rational* α *of the form* $\alpha=1/q$ $(q\geqslant 2)$, *the above estimate can be replaced by*

$$\max_{0\leqslant n\leqslant N}|y(t_n)-y_n| = O(h^{p}) \quad (\text{as } h\downarrow 0,\ Nh=T). \tag{6.1.60b}$$

Proof. For ease of notation we carry out the proof for the linear version of (6.1.1), for which $k(t,\tau,y)=K(t,\tau)y$. Let $e_n:=y(t_n)-y_n$. By (6.1.52) and (6.1.55), e_n is subject to the recurrence relation

$$e_n = h^{1-\alpha}\sum_{j=0}^{s}(\tilde{w}_{n,j}(\alpha)+\tilde{\omega}_{n-j}(\alpha))K(t_n,t_j)e_j$$
$$+h^{1-\alpha}\sum_{j=s+1}^{n}\tilde{\omega}_{n-j}(\alpha)K(t_n,t_j)e_j+\tilde{\mathbf{E}}_n^{\alpha}[\phi_n],\quad n=1,\ldots,N, \tag{6.1.61}$$

with $e_0=y(0)-g(0)=0$.

Suppose first that $1\leqslant n\leqslant s$. For these values of n, (6.1.61) represents a linear system in $\mathbb{R}^s$ for the errors $e_1,\ldots,e_s$; i.e., we have

$$\left[a_{n,j}(\alpha)\right]\begin{bmatrix}e_1\\ \vdots\\ e_s\end{bmatrix} = \begin{bmatrix}\tilde{\mathbf{E}}_1^{\alpha}[\phi_1]\\ \vdots\\ \tilde{\mathbf{E}}_s^{\alpha}[\phi_s]\end{bmatrix}, \tag{6.1.62}$$

with the matrix elements $\{a_{n,j}(\alpha)\}$ defined in (6.1.57b), and with

$$\phi_n(\tau) := K(t_n,\tau)y(\tau). \tag{6.1.63}$$

If $h>0$ is sufficiently small, then, by Theorem 6.1.1 and 6.1.4, the above system (6.1.62) is nonsingular; the order of the "starting errors" $\{e_1,\ldots,e_s\}$ is thus given by that of the quadrature errors $\tilde{\mathbf{E}}_n^{\alpha}[\phi_n]$ $(1\leqslant n\leqslant s)$. Before estimating these errors (note that now $t_n=nh$, $1\leqslant n\leqslant s$, does not remain bounded away from 0, as $h\downarrow 0$, since s is a fixed integer), we consider the case where $s+1\leqslant n\leqslant N$. Here, (6.1.61) yields

$$(1-h^{1-\alpha}\tilde{\omega}_0(\alpha)K(t_n,t_n))e_n = h^{1-\alpha}\sum_{j=s+1}^{n-1}\tilde{\omega}_{n-j}(\alpha)K(t_n,t_j)e_j$$
$$+h^{1-\alpha}\sum_{j=1}^{s}(\tilde{w}_{n,j}(\alpha)+\tilde{\omega}_{n-j}(\alpha))K(t_n,t_j)e_j+\tilde{\mathbf{E}}_n^{\alpha}[\phi_n]. \tag{6.1.64}$$

By Theorem 6.1.1 and 6.1.4, and by (6.1.53), there exist finite constants $W_0(\alpha)$ and $W_1(\alpha)$ such that

$$|\tilde{w}_{n,j}(\alpha)| \leqslant W_0(\alpha)n^{-\alpha}, \quad j=1, \ldots ,s \quad (n \geqslant 1),$$

and

$$|\tilde{\omega}_{n-j}(\alpha)| \leqslant W_1(\alpha)(n-j)^{-\alpha}, \quad 0 \leqslant j < n \leqslant N.$$

Moreover, $|(1-h^{1-\alpha}\tilde{\omega}_0(\alpha)K(t_n,t_n))^{-1}| \leqslant C_1(\alpha) < \infty$, for all sufficiently small $h>0$. Since $|K(t,\tau)| \leqslant K_0 < \infty$ on S, (6.1.64) leads to

$$|e_n| \leqslant C_1(\alpha)\{h^{1-\alpha}W_1(\alpha)K_0 \sum_{j=s+1}^{n-1} (n-j)^{-\alpha}|e_j|$$
$$+h^{1-\alpha}K_0(W_0(\alpha)\sum_{j=1}^{s} n^{-\alpha}|e_j| + W_1(\alpha)\sum_{j=1}^{s}(n-j)^{-\alpha}|e_j|) + |\tilde{\mathbf{E}}_n^{\alpha}[\phi_n]|\},$$
$$n=s+1, \ldots ,N.$$

Letting $W(\alpha):=\max\{W_0(\alpha), W_1(\alpha)\}$, and observing that, for $n \geqslant s+1$, $n^{-\alpha} < (n-j)^{-\alpha}$ $(j=1, \ldots ,s)$, we find

$$|e_n| \leqslant h^{1-\alpha}C_0(\alpha)\sum_{j=1}^{n-1}(n-j)^{-\alpha}|e_j| + C_1(\alpha)|\tilde{\mathbf{E}}_n^{\alpha}[\phi_n]|, \quad n=s+1, \ldots ,N, \tag{6.1.65}$$

with

$$C_0(\alpha) := 2C_1(\alpha)K_0W(\alpha).$$

This represents a generalized discrete Gronwall inequality for $\{|e_{s+1}|, \ldots ,|e_N|\}$, with starting values $\{|e_1|, \ldots ,|e_s|\}$ given by (6.1.62).

The remaining principal task consists in determining the asymptotic behavior (as $h \downarrow 0$, $Nh=T$) of the fractional convolution quadrature errors $\{\tilde{\mathbf{E}}_n^{\alpha}[\phi_n]\}$. Consider a function of the form $\phi_{\mu,l}(t):=t^{\mu+l(1-\alpha)}$, where (μ,l) is not in the set $M_p(\alpha)$ (cf. (6.1.48a)). By (6.1.47), this implies that

$$\mu+l(1-\alpha) > p-1.$$

Recalling the homogeneity property of $\mathbf{\Omega}_n^{\alpha}[\phi]$ mentioned before (cf. the remark following Definition 6.1.4), we deduce that

$$\mathbf{\Delta}_n^{\alpha}[\phi_{\mu,l}] = O(t_n^{\mu+(l+1)(1-\alpha)-p}h^p), \quad n=1, \ldots ,N. \tag{6.1.66}$$

Hence, by the construction of the starting weights $\{\bar{w}_{n,j}(\alpha): 0 \leqslant j \leqslant s\}$, with $s=|M_p(\alpha)|-1$ in the case of general $\alpha \in (0,1)$, it is readily seen that (6.1.66) yields

$$\tilde{\mathbf{E}}_n^{\alpha}[\phi_{\mu,l}] = O(t_n^{\mu+(l+1)(1-\alpha)-p}h^p), \quad n=1, \ldots ,N, \quad (\mu,l) \notin M_p(\alpha). \tag{6.1.67}$$

But, by (6.1.47),

$$\mu+(l+1)(1-\alpha)-p = \mu+l(1-\alpha)-p+1-\alpha > p-1-p+1-\alpha = -\alpha,$$

implying

$$\tilde{\mathbf{E}}_n^\alpha[\phi_{\mu,l}] = O(t_n^{\nu(\alpha)}h^p), \quad n=1,\dots,N, \quad (\mu,l)\notin M_p(\alpha),$$

with

$$\nu(\alpha) := \min\{\mu+l(1-\alpha)-p\colon (\mu,l)\notin M_p(\alpha)\}>-1. \tag{6.1.68}$$

Recall now that the exact solution y of the linear version of (6.1.1) is of the form

$$y(\tau) = \sum_{l=0}^{\infty}\tilde{\phi}_l(\tau)\tau^{l(1-\alpha)},$$

with $\tilde{\phi}_l$ denoting suitable smooth functions (see Theorem 1.3.5, with $g_2(t)\equiv 0$). Thus, by Theorem 6.1.4, and by (6.1.68), the above leads to

$$\tilde{\mathbf{E}}_n^\alpha[\phi_n] = O(t_n^{\nu(\alpha)}h^p), \quad n=1,\dots,N. \tag{6.1.69}$$

Returning to the discrete Gronwall inequality (6.1.65), and combining the statements of Theorem 1.5.6 and of Corollary 1.5.2, we deduce on the basis of (6.1.65) and (6.1.69) that

$$e_n = O(t_n^{\nu(\alpha)}h^p), \quad n=1,\dots,N \quad (\text{as } h\downarrow 0,\ Nh=T).$$

This is equivalent to (6.1.60a).

Assume then that $\alpha=1/q$, with $q\in\mathbb{N}$, $q\geqslant 2$. By (6.1.47), $\mu_p(\alpha;l)$ denotes the largest nonnegative integer for which $\mu_p(\alpha;l)\leqslant p-1-l(1-\alpha)$, with $l\leqslant l_p(\alpha)$. Hence, if $(\mu,l)\notin M_p(\alpha)$, we have $\mu\geqslant\mu_p(\alpha;l)+1$. Since α is rational, it suffices to restrict our discussion to the set $\overline{M}_p(\alpha)$ (cf (6.1.48b)). It is easily seen that, for $\alpha=1/q$, we have

$$\mu_p(q^{-1};l) = p-(l+1), \quad l=0,\dots,q-1,$$

and hence $\mu_p(q^{-1};l)+1=p-l$. It thus follows for $(\mu,l)\notin\overline{M}_p(\alpha)$ that

$$\begin{aligned}\mu+(l+1)(1-\alpha)-p &= \mu+(l+1)(1-1/q)-p\\ &\geqslant (p-l)+(l+1)-(l+1)/q-p\geqslant 1-1=0.\end{aligned}$$

Hence, by (6.1.67),

$$\tilde{\mathbf{E}}_n^\alpha[\phi_n] = O(h^p), \quad n=1,\dots,N \quad (Nh=T).$$

Assertion (6.1.60b) now follows immediately. □

6.2. Polynomial spline collocation for second-kind equations

In Chapter 5 we used polynomial spline functions from $S_{m-1}^{(d)}(Z_N)$, with $d\in\{-1,0\}$, to approximate the solutions of second-kind Volterra integral equations with bounded kernels. If the given functions $g(t)$ and $K(t,s)$ are m times continuously differentiable on their respective domains it follows that the corresponding solution has the same degree of smoothness on I. This, in turn, implies that the collocation approximation from the above spline space exhibits the (optimal) order of global convergence $p=m$. However, if we admit kernels containing a weak singularity of the form $(t-s)^{-\alpha}$, with $0<\alpha<1$, and if we employ quasi-uniform mesh sequences, then the global order of convergence of the collocation approximation drops to $p=1-\alpha$, regardless of how we choose its degree $m-1$. This is due to the fact that smooth $g(t)$ and $K(t,s)$ now yield a solution whose first derivative near $t=0$ behaves like $y'(t) \sim t^{-\alpha}$. In order to restore the optimal order of convergence the underlying mesh has to be graded so as to reflect the structure of $y(t)$ near $t=0$.

6.2.1. The approximation power of polynomial spline functions

We use this section to review a number of classical results dealing with the approximation power of polynomials and of polynomial spline functions with respect to nonsmooth functions (with the typical representative being $f(t)=t^{\beta}$, $o<\beta<1$, $t\in[0,1]$). We begin with

Definition 6.2.1. A function f defined on the interval I is said to satisfy a *Hölder condition on I* with *Hölder exponent* β, $0<\beta\leqslant 1$, if there exists a finite constant L such that

$$|f(t_1)-f(t_2)| \leqslant L|t_1-t_2|^{\beta} \tag{6.2.1}$$

for all t_1, $t_2\in I$. For $\beta=1$ we obtain the usual *Lipschitz condition.* The space of functions satisfying (6.2.1) on I will be denoted by $\mathrm{Lip}_{\beta}(I)$.

It is easily verified that the function $f(t)=t^{\beta}$, $0<\beta<1$, lies in $\mathrm{Lip}_{\beta}[0,1]$.

The celebrated results of Jackson describe how well functions with a given degree of smoothness can be approximated by polynomials with respect to the uniform norm.. The following theorem contains those results which will be relevant in the subsequent analysis; see e.g., Timan [1963], de Boor [1978], and Schumaker [1981] for additional details. In the following, π_N will again denote the space of (real) polynomials of degree not exceeding N.

Theorem 6.2.1. *For a given function $f\in C(I)$ let $p_N^*(t)$ denote its best uniform approximation from π_N on a compact interval I, and define $E_N^*(f):=\|f-p_N^*\|_{\infty}$.*

(i) *If $f\in\mathrm{Lip}_{\beta}(I)$, with $0<\beta\leqslant 1$, then $E_N^*(f)\leqslant C_0N^{-\beta}$, $N\geqslant 1$.*

(ii) *If $f\in C^k(I)$ and $f^{(k)}\in\mathrm{Lip}_{\beta}(I)$, with $0<\beta\leqslant 1$, then, for $N\geqslant k+1$,*

$$E_N^*(f) \leqslant C_kN^{-(k+\beta)}.$$

Here, the constant C_k is essentially given by the constant L occurring in the Hölder condition for $f^{(k)}(t)$. □

This result tells us that the best uniform approximation to a function $f\in C^m(I)$ by polynomials of degree N cannot convergence faster than $O(N^{-m})$. We recall also that if f is approximated, not in the uniform sense but by interpolation in π_N, then it may happen that the sequence of interpolating polynomials does not even converge uniformly to f on I, as the degree N tends to infinity.

In the approximation of a given function f by polynomial spline functions (piecewise polynomials) a different strategy is followed: here, the degree of the piecewise polynomials is kept fixed while the number of subintervals is increased. In order to illustrate the approximation power of polynomial spline functions we first consider an example.

Let $f(t)=t^{\beta}$, $0<\beta<1$, $t\in[0,1]$, and assume that $u\in S_1^{(0)}(Z_N)$ interpolates f at the points $0=t_0<t_1<\ldots<t_N=1$. The approximation is a (continuous) broken line, and hence the interpolation error, $e(t):=f(t)-u(t)$, on the subinterval $[t_i,t_{i+1}]$ is given by

$$e(t) = t^{\beta}-[t_i^{\beta}(t_{i+1}-t)+t_{i+1}^{\beta}(t-t_i)]/h_i, \tag{6.2.2}$$

with $h_i:=t_{i+1}-t_i$. It is easily verified that $e(t)$ assumes its maximum in (t_i,t_{i+1}) at the point

$$\xi_i := (\beta h_i/(t_{i+1}^{\beta}-t_i^{\beta}))^{1/(1-\beta)}. \tag{6.2.3}$$

Since geometrical considerations show that for equally spaced abscissas, $\|e\|_{\infty}=|e(\xi_0)|$, we find, by (6.2.2) and (6.2.3), the following result.

Theorem 6.2.2. *Let $f(t)=t^{\beta}$, $0<\beta<1$, and let $u\in S_1^{(0)}(Z_N)$ denote the (unique) interpolant to f with respect to the points $0=t_0<t_1<\ldots<t_n=1$. If $h_i:=t_{i+1}-t_i=1/N$ for all $i=0,\ldots,N-1$ (uniform partition), then the interpolation error satisfies*

$$\|e\|_{\infty} = c_{1,0}(\beta)N^{-\beta}, \tag{6.2.4}$$

with $c_{1,0}(\beta):=\beta^{1/(1-\beta)}(\beta^{-1}-1)$.

We note in passing that the above order of convergence, $O(N^{-\beta})$, remains the same if the points $\{t_i\}$ correspond to quasi-uniform partitions of $[0,1]$.

As has been mentioned above, the function $f(t)=t^{\beta}$ of Theorem 6.2.2 belongs to the space $\mathrm{Lip}_{\beta}[0,1]$. Hence the order of convergence attained by interpolation in $S_1^{(0)}(Z_N)$ is equal to that of the best uniform approximation to f from π_N (Theorem 6.2.1(i)).

It is readily seen that if $f\in C^2[0,1]$, then interpolation in $S_1^{(0)}(Z_N)$ at the points $\{t_i\}$ induces an error which satisfies $\|e\|_{\infty}=O(N^{-2})$. The following result shows that this optimal order of convergence can also be obtained for the function $f(t)=t^{\beta}$, $0<\beta<1$, provided that the underlying partition of $[0,1]$ is no longer uniform (or quasi-uniform).

Theorem 6.2.3. *Let $\Pi_N^{(r)}$ denote the graded partition of $[0,1]$ defined by*

$$t_i := (i/N)^r, \quad i=0,\ldots,N \quad (\textit{with } N\geq 2), \tag{6.2.5}$$

where the so-called grading exponent r is given by

$$r = 2/\beta. \tag{6.2.6}$$

If $u \in S_1^{(0)}(Z_N)$ *interpolates* $f(t)=t^\beta$, $0<\beta<1$, *at the points* $\{t_i\}$ *defined by* (6.2.5), *then we obtain*

$$\|e\|_\infty = O(N^{-2}). \quad \square \tag{6.2.7}$$

The proof of this result can be found by direct calculation, using (6.2.2) with $t=\xi_i$ given in (6.2.3). (A detailed analysis for the case $\beta=1/2$ may be found in Powell [1981, pp. 254-255]. Theorem 6.2.3 can, however, also be obtained as a special case of the results given below (see, e.g., de Boor [1978, pp. 46; 189-191], and Schumaker [1981, pp. 268-296], as well as the Notes for this section). We begin with a result relating to the space $S_1^{(0)}(Z_N)$ (cf. de Boor [1978, p. 46]).

Theorem 6.2.4. *Let* $f \in C^2(0,1)$ *be such that* $|f''(t)|$ *is monotone near* $t=0$ *and* $t=1$, *with* $\int_0^1 |f''(t)|^{1/2}dt<\infty$. *If* $u \in S_1^{(0)}(Z_N)$ *interpolates* f *at the points* $\{t_i\colon i=0, \ldots ,N\}$ *given by*

$$\int_0^{t_i} |f''(t)|^{1/2}dt = \frac{i}{N}\int_0^1 |f''(t)|^{1/2}dt, \quad i=1, \ldots ,N-1, \tag{6.2.8}$$

with $t_0=0$ *and* $t_N=1$, *then the resulting error* $e(t):=f(t)-u(t)$ *satisfies*

$$\|e\|_\infty = O(N^{-2});$$

i.e., we obtain the optimal order of convergence with respect to $S_1^{(0)}(Z_n)$. $\square$

The function $f(t)=t^\beta$, $0<\beta<1$, clearly satisfies the hypotheses of Theorem 6.2.4, and it readily seen that condition (6.2.8) becomes

$$\int_0^{t_i} t^{\beta/2-1}dt = \frac{i}{N}\int_0^1 t^{\beta/2-1}dt, \quad i=1, \ldots ,N-1,$$

implying $t_i^{\beta/2}=i/N$. This is equivalent to (6.2.5), (6.2.6).

So far we have only considered approximation by means of interpolation with polynomial splines. Let now u^* be a best uniform approximation from $S_m^{(0)}(Z_N)$ to $f(t)=t^\beta$, with $0<\beta<0$, on $[0,1]$. If the knots Z_N are equally spaced, then we obtain, for all $m\geqslant 1$, $\|f-u^*\|_\infty=O(N^{-\beta})$ (cf. de Boor [1978, p. 170]. Hence, the rate of convergence is no better than the one found in Theorem 6.2.2, regardless of the choice of the degree m. However, if the knots are placed in a judicious manner, then we can recover the optimal order of convergence, as the following result shows (compare, e.g., de Boor [1978, pp. 189-190], and Schumaker [1981, pp. 286-294], where the theorem is given for more general polynomial spline spaces).

Theorem 6.2.5. *Let $f \in C[0,1]$ have the following properties:*
(i) $f^{(m+1)}$ continuously at all but finitely many points of $[0,1]$;
(ii) $\int_0^1 |f^{(m+1)}(t)|^{1/(m+1)} dt < \infty$.
If the knots Z_N are chosen such that

$$\int_0^{t_i} |f^{(m+1)}(t)|^{1/(m+1)} dt = \frac{i}{N} \int_0^1 |f^{(m+1)}(t)|^{1/(m+1)} dt, \quad i=1, \dots, N-1, \tag{6.2.9}$$

and if $u^* \in S_m^{(0)}(Z_N)$ denotes a best uniform approximation to f on I (i.e., $\|f-u^*\|_\infty \leqslant \|f-u\|_\infty$ for all $u \in S_m^{(0)}(Z_N)$), then

$$\|f-u^*\|_\infty \leqslant C_0 N^{-(m+1)},$$

where the constant C_0 depends on the integral given in (ii) above. □

Example 6.2.1. Let $f(t)=t^\beta$, with $0<\beta<1$. For this function, condition (6.2.9) becomes, after simplifying,

$$\int_0^{t_i} t^{\beta/(m+1)-1} dt = \frac{i}{N} \int_0^1 t^{\beta/(m+1)-1} dt, \quad i=1, \dots, N-1,$$

and this leads to $t_i^{\beta/(m+1)} = i/N$, hence to

$$t_i = (i/N)^{(m+1)/\beta}, \quad i=1, \dots, N-1. \quad \square \tag{6.2.10}$$

Example 6.2.2. Let $f(t)=t\log(t)$. Since, for $m \geqslant 1$, $f^{(m+1)}(t)=(-1)^{m+1}(m-1)!t^{-m}$, condition (6.2.9) yields the knots given by

$$t_i = (i/N)^{m+1}, \quad i=1, \dots, N-1.$$

Note that this result corresponds formally to (6.2.10) with $\beta=1$. □

In Section 1.3.5 we have shown that the solution of the integral equation

$$y(t) - g(t) + \int_0^t (t-s)^{-\alpha} K(t,s) y(s) ds, \quad 0<\alpha<1,$$

where g and K are smooth functions, is of the form $y \in \mathrm{Lip}_\alpha(I)$. Hence, according to the above results on the approximation power of polynomial spline functions, we can only hope to construct collocation approximations with optimal order of convergence on $[0,T]$ if the underlying mesh (i.e., the knots Z_N of the approximating spline) is a graded mesh of the type (6.2.10), where the role of β will be taken by $1-\alpha$. More precisely, it will be shown (Theorem 6.2.7) that collocation approximations in $S_{m-1}^{(-1)}(Z_N)$, or in $S_{m-1}^{(0)}(Z_N)$, will exhibit global convergence of order $p=m$ if the knots Z_N are given by

$$t_i = (i/N)^r T, \quad i=1, \dots, N-1, \tag{6.2.11}$$

with grading exponent $r=m/(1-\alpha)$. We shall also see that the condition (6.2.11) is obtained in a straightforward manner and without resorting to any of the above results. We have stated these theorems merely to indicate what orders of global convergence may be anticipated if we use polynomial spline functions to approximate the solutions of second-kind Volterra integral equations with weakly singular kernels.

6.2.2. *Polynomial spline collocation on quasi-uniform meshes*

As in Section 5.3 the polynomial spline space $S_{m-1}^{(-1)}(Z_N)$ will be used to obtain collocation approximations to the solution of the integral equation

$$y(t) = g(t)+\int_0^t (t-s)^{-\alpha}k(t,s,y(s))ds, \quad t\in I := [0,T], \tag{6.2.12}$$

where $0<\alpha<1$, and where $k(t,s,y)$ is a smooth function. Given a partition Π_N of the interval I, we rewrite this equation as

$$y(t) = F_n(t)+\int_{t_n}^t (t-s)^{-\alpha}k(t,s,y(s))ds, \quad t\in\sigma_n, \tag{6.2.13}$$

where the lag term $F_n(t)$ is defined by

$$F_n(t) := g(t)+\int_0^{t_n} (t-s)^{-\alpha}k(t,s,y(s))ds, \quad n=0, \dots ,N-1. \tag{6.2.14}$$

Employing again the collocation points introduced in (5.3.3),

$$X(N) := \bigcup_{n=0}^{N-1} X_n, \text{ with } X_n := \{t_{n,j}:=t_n+c_jh_n:\ 0\leqslant c_1< \dots <c_m\leqslant 1\},$$

we seek an element $u\in S_{m-1}^{(-1)}(Z_N)$ which satisfies (6.2.13) at each point of X_n; i.e.,

$$u_n(t_{n,j}) = \bar{F}_n(t_{n,j})+h_n^{1-\alpha}\int_0^{c_j}(c_j-v)^{-\alpha}k(t_{n,j},t_n+vh_n,u_n(t_n+vh_n))dv, \tag{6.2.15a}$$

$j=1, \dots ,m$ $(n=0, \dots ,N-1)$, with the lag term approximation $\bar{F}_n(t_{n,j})$ given by

$$\bar{F}_n(t_{n,j}) := g(t_{n,j}) \tag{6.2.15b}$$
$$+\sum_{i=0}^{n-1} h_i^{1-\alpha}\int_0^1\left[\frac{t_{n,j}-t_i}{h_i}-v\right]^{-\alpha} k(t_{n,j},t_i+vh_i,u_i(t_i+vh_i))dv.$$

Observe that if g and k are continuous functions, then the special choice of the collocation parameters, $0=c_1<c_2< \dots <c_{m-1}<c_m=1$, yields an approximation u which is continuous on I; i.e., u belongs then to $S_{m-1}^{(0)}(Z_N)$. This is, of course, in complete analogy with the result of Theorem 5.3.1.

For the subsequent analysis we introduce the integrals

$$\Phi_{n,i}^{(j)}[u_i;\alpha] := \begin{cases} \int_0^1 \left[\dfrac{t_{n,j}-t_i}{h_i} - v\right]^{-\alpha} k(t_{n,j},t_i+vh_i,u_i(t_i+vh_i))dv, & \text{if } 0\leqslant i\leqslant n-1, \\ \\ \int_0^{c_j} (c_j-v)^{-\alpha} k(t_{n,j},t_n+vh_n,u_n(t_n+vh_n))dv, & \text{if } i=n \end{cases} \tag{6.2.16}$$

$$(j=1, \ldots ,m).$$

With this notation the collocation equation (6.2.15a) assumes the form

$$u_n(t_{n,j}) = g(t_{n,j})+h_n^{1-\alpha}\Phi_{n,n}^{(j)}[u_n;\alpha]+\sum_{i=0}^{n-1} h_i^{1-\alpha}\Phi_{n,i}^{(j)}[u_i;\alpha], \tag{6.2.17}$$

$$j=1, \ldots ,m \quad (n=0, \ldots ,N-1).$$

Thus, if on the subinterval σ_n the approximation $u\in S_{m-1}^{(-1)}(Z_N)$ is given by its Lagrange form,

$$u_n(t_n+vh_n) := \sum_{j=1}^{m} L_j(v)Y_{n,j}, \quad t_n+vh_n\in\sigma_n, \tag{6.2.18}$$

where $Y_{n,j}:=u_n(t_{n,j})$, then (6.2.17) represents a sequence of N nonlinear systems in $\mathbb{R}^m$ for the components of the vectors $\mathbf{Y}_n:=(Y_{n,1}, \ldots ,Y_{n,m})^T$ $(n=0, \ldots ,N-1)$. Once $\mathbf{Y}_n$ has been computed, then the values of u on σ_n can be found by means of (6.2.18).

How well will u approximate the exact solution of (6.2.12)? As we have shown in Section 1.3.5, the solution of (6.2.12) corresponding to smooth g and k will possess unbounded derivatives at $t=0$; more precisely, y' behaves like $y'(t)\sim t^{-\alpha}$ near $t=0$. Hence, according to the results of the previous Section 6.2.1, all we can hope for if we use uniform (or quasi-uniform) meshes Π_N is $\|e\|_\infty:=\|y-u\|_\infty=O(N^{-(1-\alpha)})$, regardless of how we choose m. This is made precise in the following theorem which we state, in accordance with the philosophy outlined in Chapter 5, for the linear version of (6.2.12),

$$y(t) = g(t)+\int_0^t (t-s)^{-\alpha}K(t,s)y(s)ds, \quad t\in I, \quad 0<\alpha<1. \tag{6.2.19}$$

Theorem 6.2.6. *Let the functions g and K in (6.2.19) belong to $C^m(I)$ and $C^m(S)$, respectively, with $m\geqslant 1$, and assume that neither function vanishes identically. Then there exists an $\bar{h}>0$ such that the collocation equation (6.2.17) defines for each $h\in(0,\bar{h})$ a unique approximation $u\in S_{m-1}^{(-1)}(Z_N)$. The error induced by this approximation satisfies, for every choice of the collocation parameters $\{c_j\}$ with $0\leqslant c_1< \ldots <c_m\leqslant 1$, and for all quasi-uniform mesh sequences,*

$$\|e\|_\infty = O(N^{-(1-\alpha)}). \tag{6.2.20}$$

The exponent $1-\alpha$ is best possible, in the sense that it cannot be replaced by any number greater than $1-\alpha$.

Proof. Set $t=t_{n,j}$ in (6.2.19), and subtract the collocation equation (6.2.17) (with $k(t,s,y)$ replaced by $K(t,s)y$) from the resulting equation. Denoting by $e_n:=y-u_n$ the restriction of the collocation error to the subinterval σ_n we obtain

$$e_n(t_{n,j}) = h_n^{1-\alpha}\int_0^{c_j}(c_j-v)^{-\alpha}K_{n,j}(t_n+vh_n)e_n(t_n+vh_n)dv$$

$$+\sum_{i=0}^{n-1}h_i^{1-\alpha}\int_0^1\left(\frac{t_{n,j}-t_i}{h_i}-v\right)^{-\alpha}K_{n,j}(t_i+vh_i)e_i(t_i+vh_i)dv. \qquad (6.2.21)$$

Here, we have again set $K_{n,j}(s):=K(t_{n,j},s)$.

Suppose, for ease of notation, that α is rational: $\alpha=p/q$, with p and q coprime. Thus, by (1.3.53), the solution of (6.2.19) on σ_n has the form

$$y(t_n+vh_n) = \sum_{s=0}^{q-1}v_s(t_n+vh_n)(t_n+vh_n)^{s(1-\alpha)}, \qquad (6.2.22)$$

where $v_s\in C^m(I)$ $(s=0, \ldots ,q-1)$. For $n=0$ we therefore set

$$v_s(t_0+vh_0) = \sum_{l=1}^{m}c_{0,l}^{(s)}v^{l-1}+h_0^mR_{0,s}(v), \quad t_0+vh_0\in\sigma_0, \qquad (6.2.23)$$

with

$$c_{0,l}^{(s)} := v_s^{(l-1)}(t_0)h_0^{l-1}/(l-1)!,$$

and

$$R_{0,s}(v) := v_s^{(m)}(\xi_{0,s})v^m/m! \quad (t_0<\xi_{0,s}<t_0+vh_0).$$

Substituting (6.2.23) in (6.2.22), and setting

$$(t_0+vh_0)^{s(1-\alpha)} = h_0^{s(1-\alpha)}(1+(v^{s(1-\alpha)}-1))$$

(note that $t_0=0$), we are led to

$$y(t_0+vh_0) = \sum_{l=1}^{m}c_{0,l}v^{l-1}+h_0^{1-\alpha}C_0(v)+h_0^mR_0(v), \quad v\in[0,1], \qquad (6.2.24)$$

where

$$c_{0,l} := \sum_{s=0}^{q-1}h_0^{s(1-\alpha)}c_{0,l}^{(s)}, \quad C_0(v) := \sum_{s=1}^{q-1}h_0^{s(1-\alpha)}(v^{s(1-\alpha)}-1)\sum_{l=1}^{m}c_{0,l}^{(s)}v^{l-1},$$

and

$$R_0(v) := \sum_{s=0}^{q-1}h_0^{s(1-\alpha)}R_{0,s}(v)v^{s(1-\alpha)}.$$

On the other hand, if $n \geq 1$, then $t_n > 0$, and thus the exact solution y of (6.2.19) is m times continuously differentiable for all $t \in [t_n, T]$. For $t_n + vh_n \in \sigma_n$ we may then write

$$y(t_n + vh_n) = \sum_{l=1}^{m} c_{n,l} v^{l-1} + h_n^m R_n(v), \tag{6.2.25}$$

with

$$c_{n,l} := y^{(l-1)}(t_n) h_n^{l-1}/(l-1)!, \quad R_n(v) := y^{(m)}(t_n + \theta_n v h_n) v^m / m! \quad (0 < \theta_n < 1).$$

Let u_n be given in the form

$$u_n(t_n + vh_n) = \sum_{l=1}^{m} \alpha_{n,l} v^{l-1}, \quad t_n + vh_n \in \sigma_n.$$

Thus, collecting the above results, we obtain

$$e_n(t_n + vh_n) = \begin{cases} \sum_{l=1}^{m} \beta_{0,l} v^{l-1} + h_0^{1-\alpha} C_0(v) + h_0^m R_0(v), & \text{if } n = 0, \\ \sum_{l=1}^{m} \beta_{n,l} v^{l-1} + h_n^m R_n(v), & \text{if } 1 \leq n \leq N-1, \end{cases} \tag{6.2.26}$$

with $\beta_{n,l} := c_{n,l} - \alpha_{n,l}$.

Note that (6.2.26) is the analogue of (5.3.13): in the present situation the expression for $e_0(t_0 + vh_0)$ contains not only the remainder term $h_0^m R_0(v)$ but also the lower-order term $h_0^{1-\alpha} C_0(v)$, reflecting the fact that e_0 will have unbounded derivatives at $t = 0$.

We now return to the error equation (6.2.21): with the above expressions (6.2.26) it becomes a recurrence relation for the vectors $\beta_n := (\beta_{n,1}, \ldots, \beta_{n,m})^T \in \mathbb{R}^m$, namely,

$$\begin{aligned} &\sum_{l=1}^{m} \beta_{n,l} \{ c_j^{l-1} - h_n^{1-\alpha} \int_0^{c_j} (c_j - v)^{-\alpha} K_{n,j}(t_n + vh_n) v^{l-1} dv \} \\ &\quad = \sum_{i=0}^{n-1} h_i^{1-\alpha} \sum_{l=1}^{m} \beta_{i,l} \int_0^1 \left[\frac{t_{n,j} - t_i}{h_i} - v \right]^{-\alpha} K_{n,j}(t_i + vh_i) v^{l-1} dv + q_{n,j}, \end{aligned} \tag{6.2.27}$$

where the remainder terms $q_{n,j}$ are given by

$$\begin{aligned} q_{n,j} := &-h_n^m R_n(c_j) + h_n^{1-\alpha} \int_0^{c_j} (c_j - v)^{-\alpha} K_{n,j}(t_n + vh_n)(h_n^m R_n(v)) dv \\ &+ \sum_{i=1}^{n-1} h_i^{1-\alpha} \int_0^1 \left[\frac{t_{n,j} - t_i}{h_i} - v \right]^{-\alpha} K_{n,j}(t_i + vh_i)(h_i^m R_i(v)) dv \qquad (6.2.28a) \\ &+ h_0^{1-\alpha} \int_0^1 \left[\frac{t_{n,j} - t_0}{h_0} - v \right]^{-\alpha} K_{n,j}(t_0 + vh_0)(h_0^{1-\alpha} C_0(v) + h_0^m R_0(v)) dv; \end{aligned}$$

for the initial interval σ_0 we have

$$q_{0,j} := h_0^{1-\alpha}(-C_0(c_j)-h_0^{m-1-\alpha}R_0(c_j)+ \tag{6.2.28b}$$
$$h_0^{1-\alpha}\int_0^{c_j}(c_j-v)^{-\alpha}K_{0,j}(t_0+vh_0)(C_0(v)+h_0^{m-1-\alpha}R_0(v))dv).$$

We shall now show that the norms $\|\boldsymbol{\beta}_n\|_1 := \Sigma_{l=1}^{m}|\beta_{n,l}|$ satisfy a generalized discrete Gronwall inequality of the form (1.5.12),

$$\|\boldsymbol{\beta}_n\|_1 \leqslant C_0 h^{1-\alpha}\sum_{i=0}^{n-1}(n-i)^{-\alpha}\|\boldsymbol{\beta}_i\|_1 + C_1\|\mathbf{q}_n\|_1, \quad n=0, \dots, N-1, \tag{6.2.29}$$

where $\mathbf{q}_n := (q_{n,1}, \dots, q_{n,m})^T$ is the vector whose components are defined by (6.2.28). According to Theorem 1.5.6, this inequality will imply that

$$\|\boldsymbol{\beta}_n\|_1 = O(\|\mathbf{q}_n\|_1), \quad n=0, \dots, N-1, \tag{6.2.30}$$

provided the constants C_0 and C_1 are finite, as $h\downarrow 0$, with Nh bounded.

In order to establish (6.2.29) we convert (6.2.27) into a more compact form, using vector notation (compare also the analysis following (5.3.17)). For $0\leqslant i\leqslant n-1$, define the m by m matrices $D_{n,i}(\alpha)$ by

$$D_{n,i}(\alpha) := (\int_0^1 \left[\frac{t_{n,j}-t_i}{h_i} - v\right]^{-\alpha} K_{n,j}(t_i+vh_i)v^{l-1}dv) \tag{6.2.31}$$

(where both j and l vary between 1 and m). Moreover, let

$$D_{n,n}(\alpha) := (\int_0^{c_j}(c_j-v)^{-\alpha}K_{n,j}(t_n+vh_n)v^{l-1}dv), \tag{6.2.32}$$

and let $V := (c_j^{l-1})$ denote the Vandermonde matrix of order m associated with the collocation parameters $\{c_j\}$. The recurrence relation (6.2.27) can thus be written as

$$(V-h_n^{1-\alpha}D_{n,n}(\alpha))\boldsymbol{\beta}_n = \sum_{i=0}^{n-1}h_i^{1-\alpha}D_{n,i}(\alpha)\boldsymbol{\beta}_i + \mathbf{q}_n, \quad n=0, \dots, N-1. \tag{6.2.33}$$

Here, the matrix $V-h_n^{1-\alpha}D_{n,n}(\alpha)$ possesses a uniformly bounded inverse for all meshes Π_N with sufficiently small diameter h since the kernel $K(t,s)$ is continuous and hence bounded on S. Thus, there exists a finite constant C_1, independent of h and N, such that

$$\|(V-h_n^{1-\alpha}D_{n,n}(\alpha))^{-1}\|_1 \leqslant C_1, \quad n=0, \dots, N-1, \tag{6.2.34}$$

whenever h is smaller than some positive $\bar{h}$. For these values of h, (6.2.33) defines a unique sequence of vectors $\{\boldsymbol{\beta}_n\}$ which are given by

$$\boldsymbol{\beta}_n = (V-h_n^{1-\alpha}D_{n,n}(\alpha))^{-1}\{\sum_{i=0}^{n-1}h_i^{1-\alpha}D_{n,i}(\alpha)\boldsymbol{\beta}_i + \mathbf{q}_n\}. \quad \square \tag{6.2.35}$$

We observe, for later reference, that until now we have not yet made use of the assumption that the mesh sequence $\{\Pi_N\}$ be quasi-uniform: the only condition we have imposed so far is that the mesh diameter h tend to zero as N (the number of subintervals defined by Π_N) tends to infinity. Thus, the above portion of the proof for Theorem 6.2.6 will carry over to the case of graded meshes (Theorem 6.2.7).

We now require some auxiliary results.

Lemma 6.2.1. *Consider a quasi-uniform sequence of meshes Π_N for I, and let $0<\alpha<1$. If $0\leqslant i\leqslant n-1$ $(n\leqslant N-1)$, then*

$$\int_0^1 \left[\frac{t_{n,j}-t_i}{h_i}-v\right]^{-\alpha} v^{l-1}dv < \frac{\gamma^\alpha(1+\gamma)^\alpha}{1-\alpha}(n-i)^{-\alpha}, \quad j,l=1, \dots ,m, \tag{6.2.36}$$

for all $\{c_j\}$ with $c_j\in[0,1]$. Here, $\gamma\geqslant 1$ denotes the parameter characterizing these quasi-uniform meshes (cf. Section 5.1).

Proof. For $i=n-1$ the result is obviously true. Thus, let $i\leqslant n-2$. We have, using $t_{n,j}\geqslant t_n$,

$$\begin{aligned} I_{n,i}(\alpha) :&= \int_0^1 \left[\frac{t_{n,j}-t_i}{h_i}-v\right]^{-\alpha} v^{l-1}dv \leqslant \int_0^1 \left[\frac{t_{n,j}-t_i}{h_i}-v\right]^{-\alpha} dv \\ &\leqslant \frac{1}{1-\alpha}\left\{\left[\frac{t_n-t_i}{h_i}\right]^{1-\alpha}-\left[\frac{t_n-t_i}{h_i}-1\right]^{1-\alpha}\right\} \\ &= \frac{1}{1-\alpha}\left[\frac{t_n-t_i}{h_i}\right]^{1-\alpha}\left\{1-\left(1-\left[\frac{t_n-t_i}{h_i}\right]^{-1}\right)^{1-\alpha}\right\}. \end{aligned}$$

Application of the Mean-Value Theorem yields

$$\left(1-\left[\frac{t_n-t_i}{h_i}\right]^{-1}\right)^{1-\alpha} = 1-(1-\alpha)\left[\frac{t_n-t_i}{h_i}\right]^{-1}\left(1-\theta_{n,i}\left[\frac{t_n-t_i}{h_i}\right]^{-1}\right)^{-\alpha},$$

for some $\theta_{n,i}$ with $0<\theta_{n,i}<1$, and this leads to the estimate

$$I_{n,i}(\alpha) \leqslant \left[\frac{t_n-t_i}{h_i}\right]^{-\alpha}\left(1-\theta_{n,i}\left[\frac{t_n-t_i}{h_i}\right]^{-1}\right)^{-\alpha}. \tag{6.2.37}$$

For quasi-uniform mesh sequences it holds that

$$(t_n-t_i)/h_i \geqslant (n-i)h'/h \geqslant \gamma^{-1}(n-i),$$

with $h':=h_{\min}$. Moreover, since $i\leqslant n-2$, we have

$$1-\theta_{n,i}\left[\frac{t_n-t_i}{h_i}\right]^{-1} = 1-\theta_{n,i}\left[\frac{h_i+\dots+h_{n-1}}{h_i}\right]^{-1} \geqslant 1-\left[\frac{h_i+h_{i+1}}{h_i}\right]^{-1}$$

$$= 1-(1+h_{i+1}/h_i)^{-1} \geqslant 1-(1+h'/h)^{-1} \geqslant 1-(1+1/\gamma)^{-1}=(1+\gamma)^{-1}.$$

With these results, (6.2.37) may be replaced by

$$I_{n,i}(\alpha) \leqslant \gamma^\alpha(1+\gamma)^\alpha(n-i)^{-\alpha} < \frac{\gamma^\alpha(1+\gamma)^\alpha}{1-\alpha}(n-i)^{-\alpha}. \quad \square$$

Lemma 6.2.2. *Let the assumptions of Lemma 6.2.1 hold. Then*

$$\sum_{i=0}^{n-1} h_i^{1-\alpha}\int_0^1 \left[\frac{t_{n,j}-t_i}{h_i}-v\right]^{-\alpha} v^{l-1}dv \leqslant T^{1-\alpha}/(1-\alpha), \quad 1\leqslant n\leqslant N-1. \tag{6.2.38}$$

Proof. Since $c_j \in [0,1]$, we have

$$\begin{aligned}\sum_{i=0}^{n-1} h_i^{1-\alpha}\int_0^1 \left[\frac{t_{n,j}-t_i}{h_i}-v\right]^{-\alpha} v^{l-1}dv &\leqslant \sum_{i=0}^{n-1} h_i^{1-\alpha}\int_0^1 \left[\frac{t_n-t_i}{h_i}-v\right]^{-\alpha} dv \\ &= \frac{1}{1-\alpha}\sum_{i=0}^{n-1} h_i^{1-\alpha}\left\{\left[\frac{t_n-t_i}{h_i}\right]^{1-\alpha} - \left[\frac{t_n-t_{i+1}}{h_i}\right]^{1-\alpha}\right\} \\ &= \frac{1}{1-\alpha}\sum_{i=0}^{n-1}\{(t_n-t_i)^{1-\alpha}-(t_n-t_{i+1})^{1-\alpha}\} = t_n^{1-\alpha}/(1-\alpha).\end{aligned}$$

As $t_n \leqslant T$ for all $n \leqslant N$, the result follows. $\square$

Lemma 6.2.1 will allow us to derive upper bounds for the l_1-norms of the matrices $D_{n,i}(\alpha)$ introduced in (6.2.31): if K_0 is such that $|K(t,s)|\leqslant K_0$ for all $(t,s)\in S$, then we find

$$\|D_{n,i}(\alpha)\|_1 \leqslant C(\alpha)(n-i)^{-\alpha}, \quad i=0, \ldots, n-1, \tag{6.2.39}$$

with $C(\alpha) := mK_0\gamma^\alpha(1+\gamma)^\alpha/(1-\alpha)$. Hence, it follows from (6.2.35) and from the estimates (6.2.34) and (6.2.39) that the discrete Gronwall inequality (6.2.29) holds with $C_0 := C_1 C(\alpha)$.

Our final task will consist in determining upper bounds for the norms $\|\mathbf{q}_n\|_1$. Consider first (6.2.28b): since the functions v_s in (6.2.22) are all in $C^m(I)$, equations (6.2.23) and (6.2.24) yield $\|\mathbf{q}_0\|_1 = O(h_0^{1-\alpha}) = O(N^{-(1-\alpha)})$, since $h_0 \leqslant h \leqslant \gamma TN^{-1}$. In order to obtain a similar estimate for $\|\mathbf{q}_n\|_1$ with $n>0$ (cf. (6.2.28a)) we need to know upper bounds for the remainder terms $h_i^m R_i(v)$, since they will govern the order of $\|q_n\|_1$. Note first that, by (6.2.25) and (6.2.22) we have (using Leibniz's product rule)

$$h_i^m|R_i(v)| \leqslant h_i^m\{M_{0,m} + \sum_{s=1}^{q-1}\sum_{k=0}^{m}|\gamma_{m,k}^{(s)}|M_{s,m-k}(t_i+\theta_i vh_i)^{s(1-\alpha)-k}\}, \tag{6.2.40}$$

$i \geqslant 1$, $v\in[0,1]$, where we have set $M_{s,\mu} := \max\{|v_s^{(\mu)}(t)| : t\in I\}$,

$$\gamma_{m,k}^{(s)} := \binom{m}{k}\binom{s(1-\alpha)}{k}k!,$$

and where $\theta_i \in (0,1)$. Since

$$(t_i+\theta_i v h_i)^{s(1-\alpha)-k} = t_i^{s(1-\alpha)-k} + O(h_i), \quad i \geq 1,$$

we see that the order of $h_i^m R_i(v)$ will be determined by that of $h_i^m t_i^{s(1-\alpha)-k}$, $k=0, \dots, m$.

Lemma 6.2.3. *Consider any quasi-uniform mesh sequence* $\{\Pi_N\}$ *for the interval* I, *and assume that* $1 \leq k \leq m$. *Then, for* $0<\alpha<1$ *and* $s \geq 1$,

$$h_i^m t_i^{s(1-\alpha)-k} = \begin{cases} O(N^{-m}), & \text{if } s(1-\alpha)-k \geq 0, \\ O(N^{-(1-\alpha)}), & \text{if } s(1-\alpha)-k<0, \end{cases} \tag{6.2.41}$$

$i=1, \dots, N-1$.

Proof. If s and k are such that $s(1-\alpha)-k \geq 0$, then the corresponding assertion in (6.2.41) is trivially satisfied since $t_i \leq T$ and $h_i \leq \gamma T N^{-1}$ (cf. (5.1.1)).

Let now s and k be such that $s(1-\alpha)-k<0$ (there clearly exist such values: take, e.g., $s=k=1$). Since the mesh sequence is quasi-uniform we find, for $i>1$,

$$t_i \geq t_1 = h_0 \geq h_{\min} \geq h/\gamma \geq \gamma^{-1} T N^{-1},$$

and hence

$$\begin{aligned} h_i^m t_i^{s(1-\alpha)-k} &\leq (\gamma T)^m N^{-m} (\gamma^{-1} T)^{s(1-\alpha)-k} N^{-(s(1-\alpha)-k)} \\ &= (\gamma T)^m (\gamma^{-1} T)^{s(1-\alpha)-k} N^{-(m+s(1-\alpha)-k)} \\ &\leq (\gamma T)^m (\gamma^{-1} T)^{s(1-\alpha)-k} N^{-(1-\alpha)}; \end{aligned}$$

here, the last step is a consequence of $m+s(1-\alpha)-k \geq s(1-\alpha) \geq 1-\alpha>0$, as k and s satisfy, respectively, $1 \leq k \leq m$ and $s \geq 1$. □

Returning to (6.2.40) we deduce that there exists a constant Q_m (depending on α) such that

$$h_i^m |R_i(v)| \leq Q_m N^{-(1-\alpha)}, \quad v \in [0,1], \quad i=1, \dots, N-1.$$

Thus, (6.2.28a) yields the estimate

$$|q_{n,j}| = O(N^{-(1-\alpha)}), \quad j=1, \dots, m \quad (n=1, \dots, N-1),$$

which then leads to

$$\|\mathbf{q}_n\|_1 = O(N^{-(1-\alpha)}), \quad n=1, \dots, N-1 \quad (\text{as } h \downarrow 0,\ Nh \leq \gamma T). \tag{6.2.42}$$

The proof of Theorem 6.2.6 is brought to its end by recalling (6.2.30) (which implies $\|\boldsymbol{\beta}_n\|_1 = O(N^{-(1-\alpha)})$) and by using (6.2.26). □

The above proof is readily extended to cover the case where α is not rational: the theoretical basis is now provided by Theorem 1.3.15 (with

$g_2(t)\equiv 0$) which states that the exact solution of (6.2.19) is of the form

$$y(t) = g(t) + \sum_{n=1}^{\infty} \psi_n(t;\alpha) t^{n(1-\alpha)}, \quad t \in I, \tag{6.2.43}$$

where, by the hypotheses of Theorem 6.2.6, the functions ψ_n are all in $C^m(I)$, and where the series converges absolutely and uniformly. We leave the details to the reader.

6.2.3. *Polynomial spline collocation on graded meshes*

Theorem 6.2.7. *Let the functions g and k in (6.2.19) belong to $C^m(I)$ and $C^m(S)$, respectively, with $m \geq 1$, and assume that neither function vanishes identically. If we choose the mesh $\Pi_N^{(r)}$ to be the graded mesh given by*

$$t_i = (i/N)^r T, \quad i=0, \ldots ,N \quad (N \geq 2), \tag{6.2.44a}$$

with grading exponent

$$r = m/(1-\alpha), \tag{6.2.44b}$$

then the corresponding collocation equation (6.2.17) defines, for all sufficiently large N, a unique approximation $u \in S_{m-1}^{(-1)}(Z_N)$. The error induced by this approximation satisfies

$$\|e\|_\infty = O(N^{-m}), \tag{6.2.45}$$

and this holds for all collocation parameters $\{c_j\}$ with $0 \leq c_1 < \ldots < c_m \leq 1$. □

This result tells us that if we base polynomial spline collocation for a second-kind integral equation with weakly singular kernel (but with smooth g and K) on specially *graded meshed* of the form (6.2.44), then the resulting approximation possesses no longer the low order of convergence of Theorem 6.2.6 but converges optimally with the order derived in Example 6.2.1, except that now we have $\beta = 1-\alpha$, and m instead of $m+1$.

Proof. We begin by stating three relevant properties of graded meshes of the type (6.2.44a) with grading exponent $r > 1$.

$$t_i = i^r t_1, \quad i=1, \ldots ,N, \quad \text{with} \quad t_1 = h_0 \ (=h_{\min}) = TN^{-r}; \tag{6.2.46}$$

$$h = h_{N-1} = rTN^{-1}(1-\theta N^{-1})^{r-1} \leq rTN^{-1} \quad (0<\theta<1); \tag{6.2.47}$$

$$h/h_{\min} = h_{N-1}/h_0 = rN^{r-1}(1-\theta N^{-1})^{r-1} \quad (0<\theta<1). \tag{6.2.48}$$

The last property shows that a sequence of graded meshes (6.2.44a) with $r > 1$ is not quasi-uniform, since $h/h_{\min} \to \infty$, as $N \to \infty$. However, the crucial property in the following convergence arguments will be (6.2.47): it shows that, as $N \to \infty$, the mesh diameter tends to zero. According to the observations made in the previous section prior to Lemma 6.2.1, the first part of the proof for Theorem 6.2.6, up to (and including) (6.2.35), is also valid for graded meshes of the above form since, as we have just noted, $h\downarrow 0$ as $N\to\infty$. What will change drastically in the present situation is the estimate for the norms $\|\mathbf{q}_n\|_1$ in (6.2.29): it will be shown (Lemma 6.2.5) that choosing r by (6.2.44b) implies $\|\mathbf{q}_n\|_1=O(N^{-m})$. Before deriving this important result we present the analogue of Lemma 6.2.1 for graded meshes.

Lemma 6.2.4. *Consider a graded mesh $\Pi_N^{(r)}$ of the form (6.2.44a), with grading exponent $r>1$. If the collocation parameters $\{c_j\}$ are in $[0,1]$, then, for $0<\alpha<1$ and $i=0,\ldots,n-1$ (with $n\leqslant N-1$),*

$$\int_0^1 \left[\frac{t_{n,j}-t_i}{h_i}-v\right]^{-\alpha} v^{l-1}dv<\frac{2^\alpha}{1-\alpha}(n-i)^{-\alpha},\quad j,l=1,\ldots,m. \tag{6.2.49}$$

Proof. It is easy to see that inequality (6.2.37) in the proof for Lemma 6.2.1 remains valid for graded meshes. But, in contrast to uniform meshes, we now have

$$\frac{t_n-t_i}{h_i}=\frac{h_{n-1}+\ldots+h_i}{h_i}\geqslant\frac{(n-1)h_i}{h_i}=n-i,$$

since $r>1$ implies that the stepsize sequence is strictly monotone increasing: $0<h_0<\ldots<h_{N-1}$. For $i\geqslant n-2$ we find

$$1-\theta_{n,i}\left[\frac{t_n-t_i}{h_i}\right]^{-1}>1-\left[\frac{t_n-t_i}{h_i}\right]^{-1}>1-\left[\frac{h_i+h_{i+1}}{h_i}\right]^{-1}$$

$$=1-(1+h_{i+1}/h_i)^{-1}>1-1/2=1/2,$$

and this leads to

$$\left(1-\theta_{n,i}\left[\frac{t_n-t_i}{h_i}\right]^{-1}\right)^{-\alpha}<2^\alpha<2^\alpha/(1-\alpha).$$

Hence, by (6.2.37),

$$I_{n,i}(\alpha)<\frac{2^\alpha}{1-\alpha}(n-i)^{-\alpha},\quad 0\leqslant i\leqslant n-1.\quad\square$$

Lemma 6.2.5. *Let $\Pi_N^{(r)}$ be a graded mesh of the form (6.2.44a), and assume that the grading exponent r is the one given by (6.2.44b). Then for, all integers k,v and s satisfying $1\leqslant k\leqslant v\leqslant m$, and $s\geqslant 1$, we have*

$$h_i^v t_i^{s(1-\alpha)-k}=O(N^{-v}),\quad i=1,\ldots,N-1. \tag{6.2.50}$$

Proof. The stepsize $h_i=t_{i+1}-t_i$ can be written as

$$h_i=((i+1)^r-i^r)t_1=i^r((1+i^{-1})^r-1)TN^{-r}$$

(cf. (6.2.46)), or, upon application of the Mean-Value Theorem, as

$$h_i = ri^{r-1}(1+\theta_i i^{-1})^{r-1}TN^{-r}, \quad \text{with} \quad \theta_i \in (0,1).$$

This implies

$$h_i \leqslant r2^{r-1}Ti^{r-1}N^{-r}, \quad i=1, \dots, N-1. \tag{6.2.51}$$

If we combine this last result with (6.2.46) we are led to

$$h_i^\nu t_i^{s(1-\alpha)-k} \leqslant ci^{r(\nu-k)+rs(1-\alpha)-\nu}N^{-r(\nu-k)-rs(1-\alpha)},$$

where the constant c is defined by

$$c := r^\nu 2^{\nu(r-1)}T^{\nu-k+s(1-\alpha)}.$$

Let now $r=m/(1-\alpha)$. We obtain

$$h_i^\nu t_i^{s(1-\alpha)-k} \leqslant c(i/N)^{r(\nu-k)}i^{ms-\nu}N^{-ms},$$

where the exponent of i satisfies $ms-\nu \geqslant ms-m=m(s-1)\geqslant 0$, since $\nu \leqslant m$ and $s \geqslant 1$. It thus follows that, for all i with $1 \leqslant i \leqslant N-1$, and for $1 \leqslant k \leqslant \nu \leqslant m$,

$$h_i^\nu t_i^{s(1-\alpha)-k} \leqslant cN^{ms-\nu}N^{-ms} = cN^{-\nu}. \quad \square$$

Consider now the bound for the terms $h_i^m|R_i(v)|$ derived in (6.2.40): according to Lemma 6.2.5 with $\nu=m$, we deduce that there exists a constant Q_m (depending on α) so that

$$h_i^m|R_i(v)| \leqslant Q_m N^{-m}, \quad v\in[0,1], \quad i=1, \dots, N-1.$$

This result can then be used in (6.2.28a) to establish the estimate

$$\|\mathbf{q}_n\|_1 = O(N^{-m}), \quad n=1, \dots, N-1 \quad (\text{as } N\to\infty).$$

Moreover, (6.2.28b) yields $\|\mathbf{q}_0\|_1=O(h_0^{1-\alpha})$: since, by (6.2.46), we have $h_0=TN^{-r}$, with $r=m/(1-\alpha)$, this is equivalent to $\|\mathbf{q}_0\|_1 = O(N^{-r(1-\alpha)})=O(N^{-m})$. Using (6.2.30) and arguments analogous to those in the proof of Theorem 6.2.6 we find

$$\|\boldsymbol{\beta}_n\|_1 = O(N^{-m}), \quad n=0, \dots, N-1 \quad (\text{as } N\to\infty),$$

and hence, by (6.2.26),

$$|e_n(t_n+vh_n)| = O(N^{-m}), \tag{6.2.52}$$

uniformly for $v\in[0,1]$ and $n=0, \dots, N-1$. This is, of course, equivalent to the assertion (6.2.45) of Theorem 6.2.7. $\square$

So far it has been assumed that the nonhomogeneous term g in the given integral equation (6.2.19) is in $C^m(I)$. However, the results of Theorem 6.2.6 and 6.2.7 (as well as those of Section 6.2.4 below) remain valid if g has the form

$$g(t) := g_1(t)+t^{1-\alpha}g_2(t),$$

with $g_1, g_2 \in C^m(I)$: this is a consequence of Corollary 1.3.4 which shows that the corresponding exact solution will still be of the form (6.2.22).

The use of graded meshes of the form (6.2.44) has a severe drawback which will restrict their applicability. Recall from (6.2.46) that the initial stepsize h_0 of the graded mesh $\Pi_N^{(r)}$ is given by $h_0 = N^{-r}T$, where $r = m/(1-\alpha)$. For the practical value $\alpha = 1/2$ we find $h_0 = N^{-2m}T$. Hence, if we work with cubic polynomial splines ($m = 4$), we have to start the collocation method on a subinterval whose length is, for $T = 1$, $h_0 = N^{-8}$: it is obvious that, even for rather moderate values of N, this may create serious rounding error problems, contaminating the subsequent recursive procedure, unless one is prepared to compute in multiple precision arithmetic.

There are two possible ways to avoid this problem: (i) If one is interested in generating an approximate solution to y which exhibits a high order of convergence on the *entire* interval I, then one may resort to collocation in certain *nonpolynomial spline spaces* (reflecting the singular behavior of y near $t = 0$) but on *uniform meshes.* This approach will be briefly discussed in Section 6.2.5; compare also te Riele [1982] and Brunner & te Riele [1984]. (ii)) If the given physical problem calls for an accurate approximation only near the *right end-point* $t = T$ of I, then we suggest to employ *polynomial spline* collocation in $S_{m-1}^{(-1)}(Z_N)$ on *uniform meshes:* numerical experience has shown that the corresponding spline approximation is "self-correcting" in the sense that it exhibits not only the optimal convergence order $p = m$ near $t = T$ but is, as a rule, more accurate than the spline approximation employing optimally graded meshes.

We conclude this section with a result which, in practical applications, may be used as a compromise between uniform meshes and graded meshes with optimal grading exponent $r = m/(1-\alpha)$: instead of employing a fully graded mesh (6.2.44) we choose a "partial grading" corresponding to $r = \mu/(1-\alpha)$, where μ lies between $1-\alpha$ and m.

Theorem 6.2.8. *Let g and K in (6.2.19) be subject to the hypotheses stated in Theorem 6.2.7. If $\Pi_N^{(r)}$ is a partition defined by*

$$t_i = (i/N)^r T, \quad i = 0, \ldots, N \quad (N \geq 2), \tag{6.2.53a}$$

with grading exponent given by

$$r = \mu/(1-\alpha), \quad 1-\alpha \leq \mu \leq m, \tag{6.2.53b}$$

then the resulting collocation approximation $u \in S_{m-1}^{(-1)}(Z_N)$ determined by the collocation equation (6.2.17) induces an error which behaves like

$$\|e\|_\infty = O(N^{-\mu}). \tag{6.2.54}$$

This holds for all collocation parameters $\{c_j\}$ with $0 \leq c_1 < \ldots < c_m \leq 1$. □

The choice $\mu = 1$ (i.e., $r = 1/(1-\alpha)$) leads to *linear convergence* for all $m \geq 1$ and all $\alpha \in (0,1)$.

The proof for this theorem is readily carried out by an obvious modification of the proof for Theorem 6.2.7: assertion (6.2.50) in Lemma 6.2.5 is now replaced by

$$h_i^m t_i^{s(1-\alpha)-k} = O(N^{-\mu}), \quad i=1, \ldots, N-1.$$

Details are left to the reader.

6.2.4. *Discretization of the collocation equation*

As in Section 5.3.4 we are again faced with the problem of rendering the collocation equation (6.2.17) into a form which is amenable to numerical computation, by approximating the integrals (6.2.16) by means of suitable quadrature processes. Now, however, the problem is complicated by the fact that the integrands in $\Phi_{n,n}^{(j)}[u_n;\alpha]$ and in $\Phi_{n,n-1}^{(1)}[u_{n-1};\alpha]$ (if $c_1=0$) are unbounded on the intervals of integration $[0,c_j]$ and $[0,1]$, respectively. We shall therefore approximate the integrals (6.2.16) by *product integration formulas* of the form

$$\hat{\Phi}_{n,i}^{(j)}[u_i;\alpha] := \begin{cases} \sum_{l=1}^{\mu_1} w_{j,l}^{(n,i)}(\alpha)k(t_{n,j},t_i+d_lh_i,u_i(t_i+d_lh_i)), & \text{if } 0\leqslant i\leqslant n-1, \\ \sum_{l=1}^{\mu_0} w_{j,l}(\alpha)k(t_{n,j},t_n+d_{j,l}h_n,u_n(t_n+d_{j,l}h_n)), & \text{if } i=n. \end{cases} \tag{6.2.55}$$

Here, the parameters characterizing the quadrature abscissas are of the form

$$0\leqslant d_1< \ldots <d_{\mu_1}\leqslant 1,$$

and

$$0\leqslant d_{j,1} <\ldots< d_{j,\mu_0}\leqslant c_j, \quad j=1, \ldots ,m,$$

with μ_0 and μ_1 denoting given positive integers (cf. (5.3.33)). The weights are given by

$$w_{j,l}^{(n,i)}(\alpha) = \int_0^1 \left[\frac{t_{n,j}-t_i}{h_i} - v\right]^{-\alpha} \lambda_l(v)dv, \quad i=0, \ldots ,n-1 \tag{6.2.56a}$$

$(j=1, \ldots ,m;\ l=1, \ldots ,\mu_1)$, and by

$$w_{j,l}(\alpha) := \int_0^{c_j} (c_j-v)^{-\alpha}\lambda_{j,l}(v)dv \tag{6.2.56b}$$

$(j=1, \ldots ,m;\ l=1, \ldots ,\mu_0)$, where $\lambda_l(v)$ and $\lambda_{j,l}(v)$ are, respectively, the Lagrange fundamental polynomials associated with the quadrature parameters:

$$\lambda_l(v) := \prod_{\substack{k=1\\k\neq l}}^{m} (v-d_k)/(d_l-d_k), \tag{6.2.57a}$$

and

$$\lambda_{j,l}(v) := \prod_{\substack{k=1 \\ k \neq l}}^{m} (v - d_{j,k})/(d_{j,l} - d_{j,k}). \tag{6.2.57b}$$

Using these quadrature approximations in the collocation equation (6.2.17) we obtain its fully discretized version,

$$\hat{u}_n(t_{n,j}) = g(t_{n,j}) + h_n^{1-\alpha}\hat{\Phi}_{n,n}^{(j)}[\hat{u}_n;\alpha] + \sum_{i=0}^{n-1} h_i^{1-\alpha}\hat{\Phi}_{n,i}^{(j)}[\hat{u}_i;\alpha], \tag{6.2.58}$$

$$j = 1, \ldots, m \quad (n = 0, \ldots, N-1).$$

It determines, for all meshes with sufficiently small mesh diameter, a unique approximation $\hat{u} \in S_{m-1}^{(-1)}(Z_N)$ which, due to the errors introduced in the approximation of the integrals (6.2.16) by (6.2.55), will generally be different from the solution u of (6.2.17). Since on the subinterval σ_n this approximation $\hat{u}$ is given by

$$\hat{u}_n(t_n + vh_n) = \sum_{j=1}^{m} L_j(v)\hat{Y}_{n,j}, \; t_n + vh_n \in \sigma_n, \tag{6.2.59}$$

with $\hat{Y}_{n,j} := \hat{u}_n(t_{n,j})$, (6.2.58) constitutes a sequence of N nonlinear systems for the vectors $\hat{\mathbf{Y}}_n := (\hat{Y}_{n,1}, \ldots, \hat{Y}_{n,m})^T \in \mathbb{R}^m$, $n = 0, \ldots, N-1$. Once $\hat{\mathbf{Y}}_n$ has been computed, (6.2.59) will allow the evaluation of $\hat{u}$ on σ_n.

In order to investigate the order of convergence of the collocation approximation $\hat{u}$, let

$$\hat{e} := y - \hat{u}, \quad \epsilon := u - \hat{u},$$

where u and $\hat{u}$ denote, respectively, the solutions of (6.2.17) and (6.2.58). Since $e = (y-u) + (u - \hat{u})$, we have

$$\|\hat{e}\|_\infty \leqslant \|e\|_\infty + \|\epsilon\|_\infty. \tag{6.2.60}$$

Here, the order of $\|e\|_\infty$ is given by Theorem 6.2.6 and Theorem 6.2.7. The order of the perturbation ϵ due to the full discretization of (6.2.17) by numerical integration will essentially depend on the choice of μ_0 and μ_1 in (6.2.55), as shown in the following result.

Theorem 6.2.9. *Let g and K in (6.2.19) be m times continuously differentiable on their respective domains. Assume that u, $\hat{u} \in S_{m-1}^{(-1)}(Z_N)$ denote the solutions of (6.2.17) and its fully discretized counterpart (6.2.58), where the quadrature formulas (6.2.55), (6.2.56) have been used. Then $\epsilon := u - \hat{u}$ satisfies*

$$\|\epsilon\|_\infty = O(N^{-\mu}), \tag{6.2.61}$$

where $\mu := \min(\mu_0 + 1 - \alpha, \mu_1)$. This holds for quasi-uniform and for graded mesh sequences (6.2.44a) with $r > 1$.

Proof. Let

$$E_{n,i}^{(j)}[u_i;\alpha] := \Phi_{n,i}^{(j)}[u_i;\alpha] - \hat{\Phi}_{n,i}^{(j)}[u_i;\alpha]. \tag{6.2.62}$$

Subtracting (6.2.58) from (6.2.17), and setting $k(t,s,y) = K(t,s)y$, we obtain

$$\epsilon_n(t_{n,j}) = h_n^{1-\alpha}\hat{\Phi}_{n,n}^{(j)}[\epsilon_n;\alpha] + \sum_{i=0}^{n-1} h_i^{1-\alpha}\hat{\Phi}_{n,i}^{(j)}[\epsilon_i;\alpha], \tag{6.2.63}$$

where ϵ_n is the restriction of ϵ to the subinterval σ_n. Since $\epsilon_n \in \pi_{m-1}$ we may write

$$\epsilon_n(t_n + vh_n) = \sum_{l=1}^{m} L_l(v)\epsilon_n(t_{n,l}), \quad t_n + vh_n \in \sigma_n, \tag{6.2.64}$$

with $L_l(v)$ representing the l-th Lagrange fundamental polynomial for the collocation parameters. The terms $\hat{\Phi}_{n,i}^{(j)}[\epsilon_i;\alpha]$ in (6.2.63) are thus of the form

$$\hat{\Phi}_{n,i}^{(j)}[\epsilon_i;\alpha] = \begin{cases} \sum_{l=1}^{m}\left(\sum_{s=1}^{\mu_1} w_{j,s}^{(n,i)}(\alpha)K_{n,j}(t_i + d_s h_i)L_l(d_s)\right)\epsilon_i(t_{i,l}), & \text{if } i < n, \\ \sum_{l=1}^{m}\left(\sum_{s=1}^{\mu_0} w_{j,s}(\alpha)K_{n,j}(t_n + d_{j,s}h_n)L_l(d_{j,s})\right)\epsilon_n(t_{n,l}), & \text{if } i = n. \end{cases}$$

Let now $Q_{n,i}(\alpha)$ denote the square matrix of order m whose entry (j,l) is given by

$$q_{j,l}^{(n,i)}(\alpha) := \begin{cases} \sum_{s=1}^{\mu_1} w_{j,s}^{(n,i)}(\alpha)K_{n,j}(t_i + d_s h_i)L_l(d_s), & \text{if } 0 \leqslant i \leqslant n-1, \\ \sum_{s=1}^{\mu_0} w_{j,s}(\alpha)K_{n,j}(t_n + d_{j,s}h_n)L_l(d_{j,s}), & \text{if } i = n, \end{cases} \tag{6.2.65}$$

and introduce the vectors

$$\mathbf{r}_{n,i} := (E_{n,i}^{(1)}[u_i;\alpha], \dots, E_{n,i}^{(m)}[u_i;\alpha])^T, \quad \boldsymbol{\eta}_i := (\epsilon_i(t_{i,1}), \dots, \epsilon_i(t_{i,m}))^T.$$

With this notation equation (6.2.63) assumes the form

$$(I - h_n^{1-\alpha}Q_{n,n}(\alpha))\boldsymbol{\eta}_n = \sum_{i=0}^{n-1} h_i^{1-\alpha}Q_{n,i}(\alpha)\boldsymbol{\eta}_i + \sum_{i=0}^{n} h_i^{1-\alpha}\mathbf{r}_{n,i}, \tag{6.2.66}$$

where I is the identity matrix. Consider first the matrix multiplying $\boldsymbol{\eta}_n$: since the elements of the matrix $Q_{n,n}(\alpha)$ are bounded (this follows from the boundedness of the kernel $K(t,s)$ and that of the quadrature weights (6.2.56b)), and since $h_n \leqslant h = O(N^{-1})$, $n = 0, \dots, N-1$, both for quasi-uniform and for graded mesh sequences, there exists a finite constant Q_0' not depending on h, such that, for all sufficiently large N,

$$\|(I - h_n^{1-\alpha}Q_{n,n}(\alpha))^{-1}\|_1 \leqslant Q_0', \quad n = 0, \dots, N-1. \tag{6.2.67}$$

We shall show that the norms $\|\boldsymbol{\eta}_n\|_1$ are governed, in analogy to the norms $\|\boldsymbol{\beta}_n\|_1$ encountered in the previous section (cf. (6.2.29)), by a discrete Gronwall inequality. In order to do so we require the following

Lemma 6.2.6. *The quadrature weights* $\{w_{j,l}^{(n,i)}(\alpha)\}$ *defined by (6.2.56a) satisfy*

$$|w_{j,l}^{(n,i)}(\alpha)| \leqslant w(\alpha)(n-i)^{-\alpha}, \quad j=1, \ldots, m; \; l=1, \ldots, \mu_1 \quad (i<n), \tag{6.2.68}$$

where the constant $w(\alpha)$ *is given by*

$$w(\alpha) := \begin{cases} \dfrac{\gamma^\alpha(1+\gamma)^\alpha}{1-\alpha}\,\Lambda_1, & \textit{if the mesh sequence is quasi-uniform,} \\[2ex] \dfrac{2^\alpha}{1-\alpha}\,\Lambda_1, & \textit{if the mesh sequence is given by (6.2.44a), } r>1. \end{cases}$$

Here, $\Lambda_1 := \max\{\sum_{l=1}^{m} |\lambda_l(v)| : v\in[0,1]\}$ *denotes the Lebesgue constant associated with the quadrature parameters* $\{d_1, \ldots, d_{\mu_1}\}$. □

This result is an immediate consequence of Lemma 6.2.1 (for quasi-uniform meshes), or of Lemma 6.2.4 (for graded meshes); we omit the proof.

We now use Lemma 6.2.6 to derive upper bounds for the norms $\|Q_{n,i}(\alpha)\|_1$. By (6.2.65) we find

$$\begin{aligned} \|Q_{n,i}(\alpha)\|_1 &:= \max\{\sum_{j=1}^{m} |q_{j,l}^{(n,i)}(\alpha)| : l=1, \ldots, m\} \\ &\leqslant Q(\alpha)(n-i)^{-\alpha}, \quad i=0, \ldots, n-1 \quad (n\leqslant N-1), \end{aligned}$$

with the constant $Q(\alpha)$ depending on the upper bound K_0 for $|K(t,s)|$ and on the Lebesgue constant Λ_1. Using (6.2.67) and the above estimate in (6.2.66) we obtain

$$\|\boldsymbol{\eta}_n\|_1 \leqslant Q_0 h^{1-\alpha} \sum_{i=0}^{n-1} (n-i)^{-\alpha} \|\boldsymbol{\eta}_i\|_1 + Q_0' \sum_{i=0}^{n} h_i^{1-\alpha} \|\mathbf{r}_{n,i}\|_1, \tag{6.2.69}$$

with $Q_0 := Q_0' Q(\alpha)$. This is the desired discrete Gronwall inequality. Hence, in analogy to (6.2.30), the order of the norms $\|\boldsymbol{\eta}_n\|_1$ will be given by the order of the expressions $\Sigma_{i=0}^{n} h_i^{1-\alpha} \|\mathbf{r}_{n,i}\|_1$.

Lemma 6.2.7. *Let the assumptions of Theorem 6.2.7 hold. Then we have*

$$\sum_{i=0}^{n} h_i^{1-\alpha} \|\mathbf{r}_{n,i}\|_1 = O(N^{-\mu}), \quad n=0, \ldots, N-1, \tag{6.2.70}$$

with $\mu := \min(\mu_0+1-\alpha, \mu_1)$*, regardless of whether we consider quasi-uniform or graded mesh sequences of the form (6.2.44a).*

Proof. Recall that the components of the vectors $\mathbf{r}_{n,i}$ are the quadrature errors introduced in (6.2.62). According to the hypotheses imposed on the quadrature formulas (cf. (6.2.56)) and on the kernel $K(t,s)$ these quadrature errors are bounded. More precisely, we have

$$|E_{n,i}^{(j)}[u_i;\alpha]| \leqslant \begin{cases} C_1 h_i^{\mu_1} \int_0^1 \left[\frac{t_{n,j}-t_i}{h_i} - v\right]^{-\alpha} dv, & \text{if } 0 \leqslant i \leqslant n-1, \\ C_0 h_n^{\mu_0} \int_0^{c_j} (c_j - v)^{-\alpha} dv, & \text{if } i = n, \end{cases} \tag{6.2.71}$$

where C_0 and C_1 are suitable finite constants. To see this, recall that the quadrature formulas (6.2.55), (6.2.56) have been obtained by replacing the smooth part of the integrand in (6.2.16) by an interpolating polynomial of degree $\mu_1 - 1$ (if $i \leqslant n-1$), or of degree $\mu_0 - 1$ (if $i = n$). Since the resulting interpolation errors contain, respectively, the derivatives of order μ_0 (for $i \leqslant n-1$), or of order μ_0 (for $i = n$) of the interpolated function, the result contained in (6.2.71) follows readily.

For $i \leqslant n-1$ we then obtain, using $t_{n,j} - t_i \geqslant t_n - t_i$,

$$\begin{aligned} \|\mathbf{r}_{n,i}\|_1 &= \sum_{j=1}^{m} |E_{n,i}^{(j)}[u_i;\alpha]| \leqslant mC_1 h_i^{\mu_1} \int_0^1 \left[\frac{t_n - t_i}{h_i} - v\right]^{-\alpha} dv \\ &= \frac{mC_1 h_i^{\mu_1}}{1-\alpha} \left\{ \left[\frac{t_n - t_i}{h_i}\right]^{1-\alpha} - \left[\frac{t_n - t_{i+1}}{h_i}\right]^{1-\alpha} \right\}. \end{aligned}$$

If $i = n$, then

$$\|\mathbf{r}_{n,n}\|_1 \leqslant \frac{mC_0 h_n^{\mu_0}}{1-\alpha}.$$

Thus, it follows that

$$\begin{aligned} \sum_{i=0}^{n} h_i^{1-\alpha} \|\mathbf{r}_{n,i}\|_1 &= h_n^{1-\alpha} \|\mathbf{r}_{n,n}\|_1 + \sum_{i=0}^{n-1} h_i^{1-\alpha} \|\mathbf{r}_{n,i}\|_1 \\ &\leqslant C(\alpha)\{h^{\mu_0+1-\alpha} + h^{\mu_1} \sum_{i=0}^{n-1} \{(t_n - t_i)^{1-\alpha} - (t_n - t_{i+1})^{1-\alpha}\} \\ &= C(\alpha)\{h^{\mu_0+1-\alpha} + h^{\mu_1} t_n^{1-\alpha}\} \leqslant C(\alpha) h^{\mu} \{h^{\mu_0+1-\alpha-\mu} + T^{1-\alpha} h^{\mu_1-\mu}\}, \end{aligned}$$

with $C(\alpha) := \max(mC_0/(1-\alpha),\ mC_1/(1-\alpha))$, and with $\mu := \min(\mu_0 + 1 - \alpha,\ \mu_1)$. Since the coefficient of h is uniformly bounded we have established (6.2.70). □

We return now to the discrete Gronwall inequality (6.2.69). Since for the two types of mesh sequences considered the diameter h satisfies $h = O(N^{-1})$, it follows that $\|\boldsymbol{\eta}_n\|_1 = O(N^{-\mu})$, on the basis of the estimates given in Lemma

6.2.7. Moreover, (6.2.64) implies

$$|\epsilon_n(t_n+vh_n)| \leqslant \sum_{l=1}^{m}|L_l(v)|\,|\epsilon_n(t_{n,l})| \leqslant \Lambda_1\|\eta_n\|_1 = O(N^{-\mu}),$$

uniformly for $t_n+vh_n\in\sigma_n$ and for $n=0, \ldots ,N-1$ (as $h\downarrow 0$, with $Nh\leqslant\gamma T$ or $Nh<rT$, depending on whether quasi-uniform or graded mesh sequences are used). This is equivalent to (6.2.61). □

We are now in a position to derive a result which is the analogue of Corollary 5.3.1.

Corollary 6.2.1. *Let g and K in (6.2.19) satisfy the conditions stated in Theorem 6.2.9. Suppose that the quadrature formulas (6.2.55) used to obtain the fully discretized collocation equation (6.2.58) correspond to*

$$\mu_0=\mu_1=m, \quad d_l=c_l, \quad d_{j,l}=c_{j,l}=c_jc_l \quad (j,l=1, \ldots ,m), \tag{6.2.72}$$

with weights given by (6.2.56) (see also (6.2.74) below). Then the resulting collocation approximation $\hat{u}\in S_{m-1}^{(-1)}(Z_N)$ yields an error $\hat{e}:=y-\hat{u}$ with

$$\|\hat{e}\|_\infty = O(N^{-m}),$$

provided we employ graded meshes $\Pi_N^{(r)}$ with

$$t_i = (i/N)^rT, \quad i=0, \ldots ,N, \quad \textit{where} \quad r=m/(1-\alpha). \quad \square$$

The proof of this result is straightforward: we use the results given in Theorem 6.2.7 and Theorem 6.2.9 (with $\mu=m$), together with (6.2.60).

Before presenting an example of a collocation method of the type described in Corollary 6.2.1 we rewrite the integrals (6.2.55) corresponding to (6.2.72). Using (6.2.59) these integrals assume the form

$$\hat{\Phi}_{n,i}^{(j)}[u_i;\alpha] = \begin{cases} \sum_{l=1}^{m} w_{j,l}^{(n,i)}(\alpha)k(t_{n,j},t_{i,l},\hat{Y}_{i,l}), & \text{if } 0\leqslant i\leqslant n-1, \\ \sum_{l=1}^{m} w_{j,l}(\alpha)k\left(t_{n,j},t_n+c_jc_lh_n, \sum_{s=1}^{m} L_s(c_jc_l)\hat{Y}_{n,s}\right), & \text{if } i=n, \end{cases} \tag{6.2.73}$$

and the weights are given by

$$w_{j,l}^{(n,i)}(\alpha) = \int_0^1 \left[\frac{t_{n,j}-t_i}{h_i}-v\right]^{-\alpha} L_l(v)dv, \quad (j,l=1, \ldots ,m), \tag{6.2.74a}$$

and by

$$w_{j,l}(\alpha) = c_j^{1-\alpha}\int_0^1(1-v)^{-\alpha}L_l(v)dv \quad (j,l=1, \ldots ,m), \tag{6.2.74b}$$

since (6.2.57b) reduces to

$$\lambda_{j,l}(v) = c_j^{-(m-1)} \prod_{k \neq l}^{m} (v - c_j c_k)/(c_l - c_k).$$

An obvious change of variable in (6.2.56b) then yields (6.2.74b).

Example 6.2.3. For the case $\hat{u} \in S_1^{(-1)}(Z_N)$ $(m=2)$ the quadrature weights (6.2.74) are:

$$w_{j,1}^{(n,i)}(\alpha) = b(\alpha)\{\left(\frac{t_{n,j}-t_i}{h_i}\right)^{-\alpha}\left((2-\alpha)c_2 - \frac{t_{n,j}-t_i}{h_i}\right)$$
$$- \left(\frac{t_{n,j}-t_{i+1}}{h_i}\right)^{1-\alpha}\left((2-\alpha)(c_2-1) - \frac{t_{n,j}-t_{i+1}}{h_i}\right)\},$$
$$w_{j,2}^{(n,i)}(\alpha) = b(\alpha)\{\left(\frac{t_{n,j}-t_i}{h_i}\right)^{1-\alpha}\left(\frac{t_{n,j}-t_i}{h_i} - (2-\alpha)c_1\right)$$
$$- \left(\frac{t_{n,j}-t_{i+1}}{h_i}\right)^{1-\alpha}\left(\frac{t_{n,j}-t_{i+1}}{h_i} - (2-\alpha)(c_1-1)\right)\}, \quad j=1,2$$
$$(0 \leqslant i \leqslant n-1),$$

and

$$w_{j,1}(\alpha) = b(\alpha)c_j^{1-\alpha}((2-\alpha)c_2 - 1),$$
$$w_{j,2}(\alpha) = b(\alpha)c_j^{1-\alpha}(1-(2-\alpha)c_1), \quad j=1,2,$$

with

$$b(\alpha) := ((1-\alpha)(2-\alpha)(c_2-c_1))^{-1}.$$

The corresponding fully discretized collocation equation (6.2.58) reads:

$$\hat{Y}_{n,j} = \tilde{F}_n(t_{n,j}) + h_n^{1-\alpha}\{w_{j,1}(\alpha)k(t_{n,j}, t_n + c_j c_1 h_n, L_1(c_j c_1)\hat{Y}_{n,1} + L_2(c_j c_1)\hat{Y}_{n,2})$$
$$+ w_{j,2}(\alpha)k(t_{n,j}, t_n + c_j c_2 h_n, L_1(c_j c_2)\hat{Y}_{n,1} + L_2(c_j c_2)\hat{Y}_{n,2})\}, \quad j=1,2,$$

with

$$\tilde{F}_n(t_{n,j}) := g(t_{n,j}) + \sum_{i=0}^{n-1} h_i^{1-\alpha}\{w_{j,1}^{(n,i)}(\alpha)k(t_{n,j}, t_{i,1}, \hat{Y}_{i,1}) + w_{j,2}^{(n,i)}(\alpha)k(t_{n,j}, t_{i,2}, \hat{Y}_{i,2})\},$$

and with

$$L_1(v) = (c_2 - v)/(c_2 - c_1), \quad L_2(v) = (v - c_1)/(c_2 - c_1).$$

According to Corollary 6.2.1 and to Theorem 6.2.8, the convergence behavior of the resulting collocation approximation is described by

$$\|y - \hat{u}\|_\infty = \begin{cases} O(N^{-2}), & \text{for graded meshes with } r = 2/(1-\alpha), \\ O(N^{-1}), & \text{for graded meshes with } r = 1/(1-\alpha), \\ O(N^{-(1-\alpha)}), & \text{for uniform meshes } \quad (r=1). \end{cases}$$

6.2.5. Collocation in nonpolynomial spline spaces

The theoretical and numerical results derived in the previous sections raise the following question: suppose we want to construct a collocation approximation u to the solution of the integral equation (6.2.12) (or (6.2.19)), employing a *uniform* mesh sequence $\{\Pi_N\}$, so that, for a given $m \in \mathbb{N}$, we obtain $\|y-u\|_\infty = O(N^{-m})$ as $N \to \infty$, $Nh = T$. How do we have to choose the approximating space containing u? According to Theorem 6.2.6 and the results of Section 6.2.1, a polynomial spline space will not do since the nonsmooth behavior of the solution y near $t=0$ implies that in such a space we can do no better than $\|y-u\|_\infty = O(N^{-(1-\alpha)})$, regardless of the degree of the polynomial spline.

A glimpse at the proof of Theorem 6.2.6 (especially at (6.2.26) and (6.2.41)) furnishes a first indication as to the answer to the above question: the approximating space will have to contain functions which "match", to an appropriate extent, the fractional powers of t in the exact solution,

$$y(t) = v_0(t) + \sum_{s=1}^{q-1} v_s(t) t^{s(1-\alpha)}, \tag{6.2.75}$$

especially near $t=0$. (For ease of exposition we consider only the case where α is rational: $\alpha = p/q$, with p and q coprime.) Under the hypotheses that g and K in (6.2.19) be elements of $C^m(I)$ and $C^m(S)$, respectively, we may write, using $h = TN^{-1}$,

$$v_s(t_n + vh) = \sum_{l=1}^{m} c_{n,l}^{(s)} v^{l-1} + h^m R_{s,n}(v), \quad s = 0, \ldots, q-1, \tag{6.2.76a}$$

with

$$c_{n,l}^{(s)} := v_s^{(l-1)}(t_n) h^{l-1}/(l-1)!,$$

and

$$R_{s,n}(v) := v_s^{(m)}(\xi_{s,n}) v^m/m!, \quad t_n < \xi_{s,n} < t_n + vh \in \sigma_n. \tag{6.2.76b}$$

For the following analysis we rewrite the terms $(t_n + vh)^{s(1-\alpha)}$ in "scaled" form,

$$(t_n + vh)^{s(1-\alpha)} = t_{n+1}^{s(1-\alpha)} \left(1 - \frac{1-v}{n+1}\right)^{s(1-\alpha)}$$

(this follows from $t_n + vh = t_{n+1} - (1-v)h$, where $t_{n+1} = (n+1)h$), and we define the functions

$$\phi_{s,n}(v;\alpha) := \left(1 - \frac{1-v}{n+1}\right)^{s(1-\alpha)}, \quad v \in [0,1] \ (s = 0, \ldots, q-1). \tag{6.2.77}$$

Thus, for $t = t_n + vh \in \sigma_n$, the exact solution (6.2.75) assumes the form

$$y(t_n + vh) = \sum_{s=0}^{q-1} \left(\sum_{l=1}^{m} d_{n,l}^{(s)} v^{l-1}\right) \phi_{s,n}(v;\alpha) + h^m R_n(v), \tag{6.2.78}$$

where we have set

$$d_{n,l}^{(s)} := t_{n+1}^{s(1-\alpha)} c_{n,l}^{(s)},$$

and

$$R_n(v) := \sum_{s=0}^{q-1} t_{n+1}^{s(1-\alpha)} R_{s,n}(v)\phi_{s,n}(v;\alpha), \quad n=0, \ldots, N-1,$$

with $c_{n,l}^{(s)}$ and $R_{s,n}(v)$ coming from (6.2.76a). For $\alpha=1/2$ we have

$$y(t_n+vh) = \sum_{l=1}^{m} d_{n,l}^{(0)} v^{l-1} + \sum_{l=1}^{m} d_{n,l}^{(1)} v^{l-1} \phi_{n,1}(v;\tfrac{1}{2}) + h^m R_n(v), \quad t_n+vh \in \sigma_n, \tag{6.2.79a}$$

with

$$\phi_{n,1}(v;\tfrac{1}{2}) = \Big(1-\frac{1-v}{n+1}\Big)^{\frac{1}{2}}.$$

On the first subinterval $[0,h]$ this becomes

$$y(t_0+vh) = \sum_{l=1}^{m} d_{0,l}^{(0)} v^{l-1} + \sum_{l=1}^{m} d_{0,l}^{(1)} v^{l-\frac{1}{2}} + h^m R_0(v). \tag{6.2.79b}$$

The above observations motivate the following definition.

Definition 6.2.2. Let $\alpha=p/q$, with p and q coprime, and assume that Π_N is the uniform partition of I, with $h=TN^{-1}$. Then for given $m \in \mathbb{N}$ we define the *nonpolynomial spline space*

$$V_{m-1}^{(-1)}(Z_N;\alpha) := \{u \colon u|_{\sigma_n} = u_n(t_n+vh) := \sum_{s=0}^{q-1} u_{s,n}(v)\phi_{s,n}(v;\alpha), \tag{6.2.80}$$

$$\text{with } u_{s,n} \in \pi_{m-1}, \ n=0, \ldots, N-1\}.$$

Here, the functions $\phi_{s,n}(v;\alpha)$ are given in (6.2.77), and the knots Z_N are again the interior points of the mesh Π_N. □

Thus, the nonpolynomial spline space corresponding to $\alpha=1/2$ is defined by

$$V_{m-1}^{(-1)}(Z_N;\tfrac{1}{2}) = \{u \colon u|_{\sigma_n} = u_n(t_n+vh) = u_{0,n}(v) + u_{1,n}(v)\phi_{1,n}(v;\tfrac{1}{2}), \tag{6.2.81}$$

$$\text{with } u_{0,n}, u_{1,n} \text{ in } \pi_{m-1}, \quad n=0, \ldots, N-1\}.$$

It is clear that $V_{m-1}^{(-1)}(Z_N;\alpha)$ is a linear space whose dimension is given by

$$\dim V_{m-1}^{(-1)}(Z_N;\alpha) = qNm = q \cdot \dim S_{m-1}^{(-1)}(Z_N). \tag{6.2.82}$$

In particular, the dimension of $V_{m-1}^{(-1)}(Z_N;1/2)$ is twice that of the polynomial spline space $S_{m-1}^{(-1)}(Z_N)$ employed in the previous sections.

The extension of Definition 6.2.2 to smoother nonpolynomial spline spaces $V_{m-1}^{(d)}(Z_N;\alpha)$, with $0 \leq d \leq m-2$, is obvious, and we thus omit the detailed definition. We note also that Definition 6.2.2 exists in a more general form, where the degrees of the polynomials $u_{s,n}$ are not equal to the fixed value $m-1$ but only satisfy $0 \leq m_s \leq m-1$, where m_s is chosen suitably. We refer to

Brunner [1982b], [1983] for details.

In view of practical applications, and due to limitations of space we shall from now on restrict the analysis to the case $\alpha=1/2$. Setting

$$u_{s,n}(t_n+vh) = \sum_{l=1}^{m} \alpha_{n,l}^{(s)} v^{l-1}, \quad s=0,1,$$

we have, for $n=0, \ldots, N-1$,

$$u_n(t_n+vh) = \sum_{l=1}^{m} \alpha_{n,l}^{(0)} v^{l-1} + \sum_{l=1}^{m} \alpha_{n,l}^{(1)} v^{l-1} \phi_{1,n}(v;\tfrac{1}{2}), \quad t_n+vh \in \sigma_n, \qquad (6.2.83)$$

where

$$\phi_{1,n}(v;\tfrac{1}{2}) = (1-\frac{1-v}{n+1})^{½}$$

(compare also (6.2.79a), (6.2.79b)).

We now wish to determine, by collocation, an element $u \in V_{m-1}^{(-1)}(Z_N;1/2)$ whose restriction to the subinterval σ_n is given by (6.2.83) and which satisfies the integral equation (6.2.12) (with $\alpha=1/2$) on an appropriate set of collocation points. According to the dimension of the approximating space $V_{m-1}^{(-1)}(Z_N;1/2)$, the number of collocation points must be $2Nm$, and they will be given by

$$X(N;\tfrac{1}{2}) := \bigcup_{n=0}^{N-1} X_n(\tfrac{1}{2}),$$

with

$$X_n(\tfrac{1}{2}) := \{t_{n,j} := t_n + c_j h \colon 0 \leqslant c_1 < \ldots < c_{2m} \leqslant 1\}.$$

Hence we formally obtain a collocation equation of the familiar form (6.2.17) (with $\alpha=1/2$), where the integrals $\Phi_{n,i}^{(j)}[u_i;1/2]$ are given by (6.2.16). However, for each valuc of $n=0, \ldots, N-1$, the number of collocation conditions is now equal to $2m$ (instead of m), and u_n is of the form (6.2.83).

Consider the resulting collocation error $e := y - u$: it satisfies an equation of the form (6.2.21) (with $\alpha=1/2$, and with $j=1, \ldots, 2m$), and its representation on σ_n is given by, using (6.2.79a) and (6.2.83),

$$e_n(t_n+vh) = \sum_{l=1}^{m} \beta_{n,l}^{(0)} v^{l-1} + \sum_{l=1}^{m} \beta_{n,l}^{(1)} v^{l-1} \phi_{1,n}(v;\tfrac{1}{2}) + h^m R_n(v),$$

where

$$\beta_{n,l}^{(s)} := d_{n,l}^{(s)} - \alpha_{n,l}^{(s)}, \quad l=1, \ldots, m \quad (s=0,1).$$

The remainder terms $R_n(v)$ introduced in (6.2.78) and (6.2.76b) are all bounded,

$$|R_n(v)| \leqslant T^{\frac{1}{2}} \sum_{s=0}^{1} |R_{s,n}(v)||\phi_{s,n}(v;\tfrac{1}{2})|$$

$$\leqslant T^{\frac{1}{2}} \sum_{s=0}^{1} M_{m,s} =: M_m, \quad v \in [0,1] \quad (n=0, \dots, N-1),$$

since $v_s \in C^m(I)$, $t_{n+1} \leqslant T$, and $|\phi_{s,n}(v;1/2)| \leqslant 1$ for $v \in [0,1]$ $(s=0,1;$ $n=0, \dots, N-1)$. Consequently,

$$|e_n(t_n+vh_n)| \leqslant \sum_{l=1}^{m} |\beta_{n,l}^{(0)}| + \sum_{l=1}^{m} |\beta_{n,l}^{(1)}| + h^m M_m \tag{6.2.84}$$

$$= \|\boldsymbol{\beta}_n^{(0)}\|_1 + \|\boldsymbol{\beta}_n^{(1)}\|_1 + h^m M_m, \quad t_n + vh \in \sigma_n \quad (n=0, \dots, N-1),$$

where the vectors $\boldsymbol{\beta}_n^{(s)}$ are given by $\boldsymbol{\beta}_n^{(s)} := (\beta_{n,1}^{(s)}, \dots, \beta_{n,m}^{(s)})^T$.

If we can show that the l_1-norms of these vectors satisfy

$$\|\boldsymbol{\beta}_n^{(s)}\|_1 \leqslant B_s h^m, \quad s=0,1,$$

uniformly for $n=0, \dots, N-1$, as $n \to \infty$, $Nh=T$, then it follows from (6.2.84) that

$$\|e\|_\infty = O(N^{-m});$$

i.e., we obtain the same order of convergence as for polynomial spline collocation in $S_{m-1}^{(-1)}(Z_N)$ using graded meshes of the form (6.2.44). For this we have, of course, paid a price: since we now employ an approximating space whose dimension is twice that of $S_{m-1}^{(-1)}(Z_N)$ we have to solve N systems in $\mathbb{R}^{2m}$, instead of N systems in $\mathbb{R}^m$.

In order to establish uniform boundedness for $\|\boldsymbol{\beta}_n^{(s)}\|_1$ one proceeds as in the proof of Theorem 6.2.6: one shows that the vectors $\boldsymbol{\beta}_n := (\beta_{n,1}^{(0)}, \dots, \beta_{n,m}^{(0)};$ $\beta_{n,1}^{(1)}, \dots, \beta_{n,m}^{(1)})^T \in \mathbb{R}^{2m}$ satisfy a discrete Gronwall inequality of the form (6.2.29), where $\mathbf{q}_n$ is $O(N^{-m})$: it depends on the remainder terms $h^m R_n(v)$ which are all $O(h^m)$, as indicated above. We omit the details.

Nonpolynomial spline collocation methods for (6.2.12) with $\alpha = 1/2$ were introduced by te Riele [1982]: he used nonpolynomial splines of the form (6.2.81) to approximate the exact solution on the first few subintervals $\sigma_0, \dots, \sigma_k$ (i.e., close to $t=0$ where the solution is not smooth). For $t \geqslant t_{k+1}$, the solution is approximated in the polynomial spline space $S_{m-1}^{(-1)}(Z_N)$. This paper deals also with the problem of designing suitable quadrature formulas (exact for certain powers of $v^{1/2}$) for the full discretization of the collocation equation on $[0, t_{k+1}]$. However, the question of how to choose k (i.e., when to switch from nonpolynomial splines to polynomial splines) remains essentially open (compare also Brunner & te Riele [1984]).

6.2.6. Product integration and iterated collocation methods

So far we have only investigated the attainable order p of *global* convergence (on I) of spline collocation approximations to the solution y of

$$y(t) = g(t)+\int_0^t(t-s)^{-\alpha}k(t,s,y(s))ds, \quad t\in I. \tag{6.2.85}$$

We have seen that $p=m$ is best possible in $S_{m-1}^{(-1)}(Z_N)$: this value is attained if one employs knots Z_N which are graded according to (6.2.44) (cf. Theorem 6.2.7 and Corollary 6.2.1). As in Section 5.3.3 there arises the question as to whether a judicious choice of the collocation parameters $\{c_j\}$ yields locally (on $\overline{Z}_N$) a higher order p^* of convergence (*local superconvergence* on $\overline{Z}_N$). We recall that, for $\alpha=0$, $p^*=2m-1$ is possible if collocation is based on the Radau II points, while iterated collocation using the Gauss points yields even $p^*=2m$. In the present section we shall briefly indicate that, for $0<\alpha<1$ in (6.2.85) and for suitable graded meshes, we always have $m\leqslant p^*\leqslant m+1-\alpha$: the attainable order of local superconvergence cannot exceed $m+1-\alpha$. We shall also show that the same result holds for a class of so-called product integration methods for (6.2.85).

In a *product integration method* the bounded part of the integrand in (6.2.85), $k(t,s,y(s))$, is approximated by polynomial spline interpolation (recall that this is in contrast to collocation where it is the solution y that is approximated in a spline space). The resulting integrals can then be evaluated analytically; i.e., product integration techniques yield a full discretization of the given integral equation. In our context, let $t=t_n+vh_n$ be a point in the subinterval σ_n, and rewrite (6.2.85) as

$$y(t) = g(t)+h_n^{1-\alpha}\int_0^v(v-\tau)^{-\alpha}k(t,t_n+\tau h_n,y(t_n+\tau h_n))d\tau \tag{6.2.86}$$
$$+\sum_{i=0}^{n-1}h_i^{1-\alpha}\int_0^1\left[\frac{t-t_i}{h_i}-\tau\right]^{-\alpha}k(t,t_i+\tau h_i,y(t_i+\tau h_i))d\tau, \quad n=0,\ldots,N-1.$$

On each subinterval σ_i, interpolation will be by polynomials of degree $m-1$: if $i\leqslant n-1$, then the interpolation abscissas will be the points $t_{i,l}:=t_i+c_lh_i$ $(0\leqslant c_1<\ldots<c_m\leqslant 1)$, while for $i=n$ (i.e., on the subinterval $[t_n,t_n+vh_n]$, with $v\leqslant 1$) we interpolate at the points $\{t_n+vc_lh_n\colon l=1,\ldots,m\}$; as in Section 6.2.4, this latter choice implies that only kernel values $k(t,s,\cdot)$ with $(t,s)\in S$ are being used. The corresponding interpolants for the terms $k(t,t_i+\tau h_i,y(t_i+\tau h_i))$ in (6.2.86) are thus in π_{m-1}, and substitution in (6.2.86) yields the following equation for the approximation z to the solution y of (6.2.85):

$$z(t) = g(t)+h_n^{1-\alpha}\sum_{l=1}^{m}w_l(v;\alpha)k(t,t_n+vc_lh_n,z(t_n+vc_lh_n))$$
$$+\sum_{i=0}^{n-1}h_i^{1-\alpha}\sum_{l=1}^{m}w_l^{(n,i)}(v;\alpha)k(t,t_{i,l},Z_{i,l}), \quad t=t_n+vh_n\in\sigma_n. \tag{6.2.87}$$

Here, we have set $Z_{i,l}:=z(t_{i,l})$. The quadrature weights are given by

$$w_l^{(n,i)}(v;\alpha) := \int_0^1 \left[\frac{t-t_i}{h_i} - \tau\right]^{-\alpha} L_l(\tau)d\tau, \quad i \leq n-1, \tag{6.2.88a}$$

and by

$$w_l(v;\alpha) := v^{1-\alpha}\int_0^1 (1-\tau)^{-\alpha} L_l(\tau)d\tau, \tag{6.2.88b}$$

with $t=t_n+vh_n \in \sigma_n$; they can be evaluated analytically.

Suppose for a moment that we know the values $\{Z_{i,l}\colon\ l=1, \ldots ,m;\ i=0, \ldots ,n\}$. Then (6.2.87) allows us to compute the approximation $z(t)$ at $t=t_{n+1}$: since $v=1$, we have

$$\begin{aligned} z(t_{n+1}) &= g(t_{n+1})+h_n^{1-\alpha}\sum_{l=1}^{m} w_l(1;\alpha)k(t_{n+1},t_{n,l},Z_{n,l}) \\ &\quad + \sum_{i=0}^{n-1} h_i^{1-\alpha}\sum_{l=1}^{m} w_l^{(n,i)}(1;\alpha)k(t_{n+1},t_{i,l},Z_{i,l}). \end{aligned} \tag{6.2.89}$$

If, however, we wanted to compute $z(t_n+vh_n)$ for some $v<1$ (e.g., for $v=c_j$), then we face a problem: since the restriction of z to the subinterval σ_n is not a polynomial (in contrast to polynomial spline collocation), it cannot be expressed exactly by a linear combination of the values $\{Z_{n,l}\}$. Hence, (6.2.87) cannot be used to determine the desired approximations $\{Z_{n,1}, \ldots ,Z_{n,m}\}$ (which correspond to $v=c_j$, $j=1, \ldots ,m$). Instead, we employ

$$\begin{aligned} Z_{n,j} &= g(t_{n,j})+h_n^{1-\alpha}\sum_{l=1}^{m} w_{j,l}(\alpha)k\Big(t_{n,j},t_n+c_jc_lh_n, \sum_{k=1}^{m} L_k(c_jc_l)Z_{n,k}\Big) \\ &\quad + \sum_{i=0}^{n-1} h_i^{1-\alpha}\sum_{l=1}^{m} w_{j,l}^{(n,i)}(\alpha)k(t_{n,j},t_{i,l},Z_{i,l}), \quad j=1, \ldots ,m; \end{aligned} \tag{6.2.90}$$

here, we have $w_{j,l}^{(n,i)}(\alpha):=w_l^{(n,i)}(c_j;\alpha)$ $(i\leq n-1)$, and $w_{j,l}(\alpha):=w_l(c_j;\alpha)$: these weights agree, of course, with the ones introduced in (6.2.74a) and (6.2.74b).

Therefore, the *product integration method* for (6.2.85) consists of the equations (6.2.90) and (6.2.89): once the solution $(Z_{n,1}, \ldots ,Z_{n,m})^T$ of the nonlinear system (6.2.90) has been found, (6.2.89) furnishes the approximation $z(t)$ at $t=t_{n+1}\in\overline{Z}_N$. (Note that, for $c_m=1$, we have $z(t_{n+1})=Z_{n,m}$.)

We observe immediately that the equation (6.2.90) is identical with the fully discretized collocation equation (6.2.58) (where the integrals $\hat{\Phi}_{n,i}^{(j)}$ are those in (6.2.73)). Hence, *the product integration approximation $z(t_{n+1})$ given by (6.2.89) can be viewed as the iterated collocation approximation $\hat{u}^I(t_{n+1})$* corresponding to $\hat{u}\in S_{m-1}^{(-1)}(Z_N)$ defined by (6.2.58); i.e.,

$$\hat{u}^I(t_{n+1}) := g(t_{n+1})+\sum_{i=0}^{n} h_i^{1-\alpha} w_l^{(n,i)}(1;\alpha)k(t_{n+1},t_{i,l},\hat{Y}_{i,l}), \tag{6.2.91}$$

where $w_l^{(n,n)}(1;\alpha) := w_l(1;\alpha)$, and where the approximations $\hat{Y}_{i,l} := \hat{u}_i(t_{i,l})$ are determined by (6.2.58), (6.2.73): since, as noted above, we have $Z_{i,l} = Y_{i,l}$ for all l and i, this implies that $z(t_{n+1}) = \hat{u}^I(t_{n+1})$ for $t_{n+1} \in \bar{Z}_N$. In other words, the order of convergence for the product integration approximations is given by that of the iterated collocation approximations on $\bar{Z}_N$.

Theorem 6.2.10. *Let $\hat{u} \in S_{m_{(j)}-1}^{(-1)}(Z_N)$ denote the collocation approximation defined by (6.2.58), with integrals $\Phi_{n,i}^{(j)}$ given by (6.2.73), and let $\hat{u}^I(t_n)$ be the corresponding iterated collocation approximation determined according to (6.2.91) (or, equivalently, the product integration approximation at $t = t_n$ given by (6.2.90), (6.2.89)). Then, for $0 < \alpha < 1$, the exponent $p(\alpha)$ in*

$$\max_{t_n \in \bar{Z}_N} |y(t_n) - \hat{u}^I(t_n)| = O(N^{-p(\alpha)}) \quad (\text{as } N \to \infty,\ Nh \text{ fixed}) \qquad (6.2.92)$$

cannot exceed the value $m+1-\alpha$. The optimal value $p^*(\alpha) := m+1-\alpha$ is attained, for sufficiently smooth g and k in (6.2.85), if
(i) the mesh Π_N is graded and given by

$$t_n = (n/N)^r T \quad (n = 0, \ldots, N), \quad \text{with} \quad r = m/(1-\alpha); \qquad (6.2.93)$$

and
(ii) the collocation parameters $\{c_j\}$ satisfy

$$\int_0^1 \prod_{j=1}^m (\tau - c_j)\, d\tau = 0. \quad \square \qquad (6.2.94)$$

Note that we have $\hat{u}^I(t_n) = \hat{u}(t_n)$ $(= \hat{Y}_{n,m})$ if $c_m = 1$. Hence, if the collocation parameters $\{c_j\}$ are given by the Radau II points or by the Lobatto points, then the above theorem describes the local convergence behavior of the collocation approximation $\hat{u} \in S_{m-1}^{(-1)}(Z_N)$ on $\bar{Z}_N$.

Example 6.2.4. The *Gauss points* satisfy (6.2.94) for all m (recall Lemma 5.5.1: we have $\int_0^s P_m(2\tau - 1)d\tau = \text{const.}\ (P_{m-1}(2s-1) - P_{m+1}(2s-1))$, and the Legendre polynomials P_{m-1} and P_{m+1} have the same values at the endpoints of $[-1,1]$). Hence, the discretized iterated collocation approximation defined by (6.2.91) and based on collocation in $S_{m-1}^{(-1)}(Z_N)$ at the Gauss points yields

$$\max_{t_n \in \bar{Z}_N} |y(t_n) - \hat{u}^I(t_n)| = O(N^{-(m+1-\alpha)}) \quad (0 < \alpha < 1), \qquad (6.2.95)$$

provided the underlying mesh is graded as in (6.2.93). $\square$

Example 6.2.5. Let the collocation parameters be the *Radau II points* (i.e., the zero of $P_{m-1}(2s-1) - P_m(2s-1)$). By the foregoing example, these points also satisfy (6.2.94), and hence the local superconvergence results (6.2.95) holds, too, with $\hat{u}^I(t_n) = \hat{u}(t_n)$. $\square$

Example 6.2.6. Let $m=3$; $c_1=0$, $c_2=1/2$, $c_3=1$ (*Lobatto points*). These points $\{c_j\}$ satisfy (6.2.94). The corresponding product integration method (6.2.90), (6.2.89), is known as the *product Simpson method* (see, e.g., Logan [1976]). On graded meshes (6.2.93), with grading exponent $r=3/(1-\alpha)$, we have

$$\max_{t_n\in \bar{Z}_N}|y(t_n)-\hat{u}(t_n)| = O(N^{-(4-\alpha)}) \quad (0<\alpha<1).$$

If $\alpha=1/2$, then the approximations converge like $O(N^{-7/2})$. □

The *proof* of Theorem 6.2.10 can be carried out on the basis of the ideas introduced in Section 5.3.6 for the case $\alpha=0$. We sketch it for the linear version of (6.2.85). Let $e^I(t):=y(t)-u^I(t)$ be the error associated with the exact iterated collocation approximation,

$$u^I(t) := g(t)+\int_0^t(t-s)^{-\alpha}K(t,s)u(s)ds, \quad t\in I,$$

where $u\in S_{m-1}^{(-1)}(Z_N)$ is given by (6.2.17), (6.2.18). This error satisfies, in complete analogy to (5.3.50),

$$e^I(t) = \int_0^t R(t,s;\alpha)\delta(s)ds, \quad t\in I, \tag{6.2.96a}$$

where δ is the defect function defined by

$$u(t) = g(t)-\delta(t)+\int_0^t(t-s)^{-\alpha}K(t,s)u(s)ds, \quad \text{with} \quad \delta|_{X(N)}=0$$

(cf (5.3.24); compare also Brunner [1984c]), and where the resolvent kernel $R(t,s;\alpha)$ of $(t-s)^{-\alpha}K(t,s)$ has the form

$$R(t,s;\alpha) = (t-s)^{-\alpha}Q(t,s;\alpha), \quad \text{with} \quad Q\in C(S) \tag{6.2.96b}$$

(cf. Theorem 1.3.2). Proceeding as in (5.3.51), it follows that the order of the values $e^I(t_n)$ is given by the order of the error in the m-point product integration formula based on the abscissas $\{c_j\}$. Since the functions $Q(t,s;\alpha)$ and $\delta(s)$ forming the bounded part of the integrand in (6.2.96a) are not smooth, but only Hölder continuous on S and I, respectively, it follows from results by Schneider [1980], [1981] that the order of $e^I(t_n)$, $t_n\in\bar{Z}_n$, attains $p^*(\alpha)=m+1-\alpha$ only if (6.2.93) and (6.2.94) hold. If these conditions are not satisfied, then the local convergence order on $\bar{Z}_N$ coincides with the global convergence order on I. On the basis of these results for the exact collocation approximation u one now readily derives the results of Theorem 6.2.10 for the discretized collocation approximations $\hat{u}$ and $\hat{u}^I$, by employing the techniques introduced in Section 5.3.6. We leave the details to the reader.

6.3. Polynomial spline collocation for integro-differential equations

In this section we extend the analysis of Section 5.4 to VIDEs of the form

$$y'(t) = f(t,y(t)) + \int_0^t (t-s)^{-\alpha} k(t,s,y(s))ds, \quad t \in I := [0,T], \quad 0<\alpha<1, \tag{6.3.1}$$

with given initial value $y(0)=y_0$. For ease of exposition we shall, as in Section 5.4, often resort to its linear counterpart,

$$y'(t) = p(t)y(t) + q(t) + \int_0^t (t-s)^{-\alpha} K(t,s)y(s)ds, \quad t \in I, \quad 0<\alpha<1. \tag{6.3.2}$$

For given collocation parameters $\{c_j\}$ with $0 \leq c_1 < \dots < c_m \leq 1$ we determine a collocation approximation $u \in S_m^{(0)}(Z_N)$ from

$$u_n'(t_{n,j}) = f(t_{n,j}, u_n(t_{n,j})) + h_n^{1-\alpha} \Phi_{n,n}^{(j)}[u_n;\alpha] + \sum_{i=0}^{n-1} h_i^{1-\alpha} \Phi_{n,i}^{(j)}[u_i;\alpha], \tag{6.3.3}$$

$j=1, \dots ,m$ $(n=0, \dots ,N-1)$, with $u_0(0)=y_0$. Here, the integrals $\Phi_{n,i}^{(j)}[u_i;\alpha]$ are those introduced in (6.2.16), except that now we have $u_i \in \pi_m$, instead of π_{m-1}. It is again easily seen that if the collocation parameters are such that $0=c_1< \dots <c_m=1$, then the resulting u is an element of the smoother polynomial spline space $S_m^{(1)}(Z_N)$, provided f and k are continuous functions.

We have shown in Theorem 1.3.16 and in Corollary 1.3.5 that, for smooth functions p, q, and K, the solution of (6.3.2) behaves like $y(t) = \text{const.}\ t^{2-\alpha} +$ smoother functions near $t=0$; this also holds for smooth f and k in (6.3.1). Hence, by the results of Section 6.2.1 on the approximation power of polynomial spline functions, the order of convergence will depend on the type of mesh sequence we employ; as in Section 6.2, we can only hope to achieve convergence of optimal order if we work with suitably graded meshes.

6.3.1. *Polynomial spline collocation on quasi uniform meshes*

Theorem 6.3.1. *Let p, q, and K in (6.3.2) be m times continuously differentiable on their respective domains I and S, and assume that q and K do not vanish identically. If $u \in S_m^{(0)}(Z_N)$ denotes the collocation approximation determined by (6.3.3), and if the underlying mesh sequence is quasi-uniform, then, for all collocation parameters $\{c_j\}$ with $0 \leq c_1 < \dots < c_m \leq 1$, the error $e := y - u$ satisfies*

$$\|e^{(k)}\|_\infty = O(N^{-(1-\alpha)}), \quad k=0,1. \quad \square \tag{6.3.4}$$

We observe that, for quasi-uniform meshes, the order of convergence of u is governed by the degree of smoothness of the derivative of the exact solution, and not by that of the solution itself. This is, of course, not surprising since the collocation equation defining u involves also u' (which is an element of $S_{m-1}^{(-1)}(Z_N)$) as an approximation to y'.

Proof. The collocation error e satisfies

$$e_n'(t_{n,j}) = p(t_{n,j})e_n(t_{n,j})+h_n^{1-\alpha}\int_0^{c_j}(c_j-v)^{-\alpha}K_{n,j}(t_n+vh_n)e_n(t_n+vh_n)dv$$
$$+\sum_{i=0}^{n-1}h_i^{1-\alpha}\int_0^1\left[\frac{t_{n,j}-t_i}{h_i}-v\right]^{-\alpha}K_{n,j}(t_i+vh_i)e_i(t_i+vh_i)dv, \tag{6.3.5}$$
$$j=1,\ \dots\ ,m \quad (n=0,\ \dots\ ,N-1),$$

with $e_0(t_0)=y(0)-u_0(0)=0$, and where e_n denotes again the restriction of e to the subinterval σ_n.

For ease of notation we assume, as in Section 6.2, that α is rational: $\alpha=p/q$, with p and q coprime. Hence, according to (1.3.62), the solution of (6.3.2) has the form

$$y(t) = \sum_{s=0}^{q-1}w_s(t)t^{s(2-\alpha)}, \quad w_s\in C^{m+1}(I) \quad (s=0,\ \dots\ ,q-1).$$

On the first subinterval σ_0 we have, in complete analogy to (6.2.24),

$$y(t_0+vh_0) = \sum_{l=0}^{m}c_{0,l}v^l+h_0^{2-\alpha}C_0(v)+h_0^{m+1}R_0(v), \quad v\in[0,1], \tag{6.3.6}$$

where we have set

$$c_{0,l} := \sum_{s=0}^{q-1}h_0^{s(2-\alpha)}c_{0,l}^{(s)}, \quad R_0(v) := \sum_{s=0}^{q-1}h_0^{s(2-\alpha)}R_{0,s}(v)v^{s(2-\alpha)},$$

and

$$C_0(v) := \sum_{s=1}^{q-1}h_0^{(s-1)(2-\alpha)}(v^{s(2-\alpha)}-1)\sum_{l=0}^{m}c_{0,l}^{(s)}v^l.$$

This follows from the application of Taylor's formula to the functions $w_s(t_0+vh_0)$ in the expression for $y(t_0+vh_0)$: we have

$$w_s(t_0+vh_0) = \sum_{l=0}^{m}c_{0,l}^{(s)}v^l+h_0^{m+1}R_{0,s}(v),$$

with

$$c_{0,l}^{(s)} := w_s^{(l)}(t_0)h_0^l/l!,$$

and

$$R_{0,s}(v) := \frac{1}{m!}\int_0^v(v-z)^m w_s^{(m+1)}(t_0+zh_0)dz, \quad s=0,\ \dots\ ,q-1.$$

If $n\geq 1$, then $t_n>0$, and we may write

$$y(t_n+vh_n) = \sum_{l=0}^{m}c_{n,l}v^l+h_n^{m+1}R_n(v), \quad v\in[0,1], \tag{6.3.7}$$

where

$$c_{n,l} := y^{(l)}(t_n)h_n^l/l!,$$

and

$$R_n(v) := \frac{1}{m!}\int_0^v (v-z)^m y^{(m+1)}(t_n+zh_n)dz.$$

The corresponding expressions for the first derivative of the exact solution on the subintervals σ_0 and σ_n, $n=1, \ldots, N-1$, are, respectively,

$$y'(t_0+vh_0) = h_0^{-1}\{\sum_{l=1}^m lc_{0,l}v^{l-1} + h_0^{2-\alpha}C_0'(v) + h_0^{m+1}R_0'(v)\}, \tag{6.3.8}$$

and

$$y'(t_n+vh_n) = h_n^{-1}\{\sum_{l=1}^m lc_{n,l}v^{l-1} + h_n^{m+1}R_n'(v)\}, \quad n\geqslant 1. \tag{6.3.9}$$

Thus, setting

$$u_n(t_n+vh_n) = \sum_{l=0}^m \alpha_{n,l}v^l, \quad v\in[0,1], \quad n=0, \ldots, N-1,$$

we obtain the following expressions for he restrictions of the collocation error and its derivative to the subintervals $\{\sigma_n\}$:

$$e_n(t_n+vh_n) = \begin{cases} \sum_{l=0}^m \beta_{0,l}v^l + h_0^{2-\alpha}C_0(v) + h_0^{m+1}R_0(v), & \text{if} \quad n=0, \\ \sum_{l=0}^m \beta_{n,l}v^l + h_n^{m+1}R_n(v), & \text{if} \quad 1\leqslant n\leqslant N-1; \end{cases} \tag{6.3.10}$$

and

$$e_n'(t_n+vh_n) = \begin{cases} h_0^{-1}\{\sum_{l=1}^m l\beta_{0,l}v^{l-1} + h_0^{2-\alpha}C_0'(v) + h_0^{m+1}R_0'(v)\}, & \text{if } n=0, \\ h_n^{-1}\{\sum_{l=1}^m l\beta_{n,l}v^{l-1} + h_n^{m+1}R_n'(v)\}, & \text{if } 1\leqslant n\leqslant N-1, \end{cases} \tag{6.3.11}$$

with $\beta_{n,l} := c_{n,l} - \alpha_{n,l}$.

If we now employ (6.3.10) and (6.3.11) in the error equation (6.3.5), then, setting $p_{n,j} := p(t_{n,j})$, we find

$$\sum_{l=1}^m \{lc_j^{l-1} - h_n p_{n,j} c_j^l - h_n^{2-\alpha}\int_0^{c_j} (c_j-v)^{-\alpha}K_{n,j}(t_n+vh_n)v^l dv\}\beta_{n,l}$$

$$= h_n\{p_{n,j} + h_n^{1-\alpha}\int_0^{c_j}(c_j-v)^{-\alpha}K_{n,j}(t_n+vh_n)dv\}\beta_{n,0}$$

$$+h_n\sum_{i=0}^{n-1}h_i^{1-\alpha}\beta_{i,0}\int_0^1\left[\frac{t_{n,j}-t_i}{h_i}-v\right]^{-\alpha}K_{n,j}(t_i+vh_i)dv$$

$$+h_n\sum_{i=0}^{n-1}h_i^{1-\alpha}\sum_{l=1}^{m}\beta_{i,l}\int_0^1\left[\frac{t_{n,j}-t_i}{h_i}-v\right]^{-\alpha}K_{n,j}(t_i+vh_i)v^l dv+h_n r_{n,j}, \tag{6.3.12}$$

with

$$r_{n,j}:=-h_n^m R_n'(c_j)+h_n^{m+1}p_{n,j}R_n(c_j)$$

$$+h_n^{1-\alpha}\int_0^{c_j}(c_j-v)^{-\alpha}K_{n,j}(t_n+vh_n)(h_n^{m+1}R_n(v))dv \tag{6.3.13}$$

$$+\sum_{i=1}^{n-1}h_i^{1-\alpha}\int_0^1\left[\frac{t_{n,j}-t_i}{h_i}-v\right]^{-\alpha}K_{n,j}(t_i+vh_i)(h_i^{m+1}R_i(v))dv$$

$$+h_0^{1-\alpha}\int_0^1\left[\frac{t_{n,j}-t_0}{h_0}-v\right]^{-\alpha}K_{n,j}(t_0+vh_0)(h_0^{2-\alpha}C_0(v)+h_0^{m+1}R_0(v))dv.$$

On the first subinterval, (6.3.12) and (6.3.13) become, respectively,

$$\sum_{l=1}^{m}\{lc_j^{l-1}-h_0p_{0,j}c_j^l-h_0^{2-\alpha}\int_0^{c_j}(c_j-v)^{-\alpha}K_{0,j}(t_0+vh_0)v^l dv\}\beta_{0,l}$$

$$=\{h_0p_{0,j}+h_0^{2-\alpha}\int_0^{c_j}(c_j-v)^{-\alpha}K_{0,j}(t_0+vh_0)dv\}\beta_{0,0}+h_0r_{0,j}, \tag{6.3.14}$$

and

$$r_{0,j}:=h_0^{1-\alpha}\{-C_0'(c_j)-h_0^{m+\alpha-1}R_0'(c_j)+h_0p_{0,j}(C_0(c_j)+h_0^{m+\alpha-1}R_0(c_j))$$

$$+\int_0^{c_j}(c_j-v)^{-\alpha}K_{0,j}(t_0+vh_0)(h_0^{2-\alpha}C_0(v)+h_0^{m+1}R_0(v))dv\}. \tag{6.3.15}$$

Since the error function is continuous at the knots Z_N, we have $e_n(t_n)=e_{n-1}(t_n)$ for all $t_n\in Z_N$, implying, by (6.3.10) and using $C_0(1)=0$, that

$$\beta_{n,0}=\beta_{0,0}+\sum_{i=0}^{n-1}\sum_{l=1}^{m}\beta_{i,l}+\sum_{i=0}^{n-1}h_i^{m+1}R_i(1). \tag{6.3.16}$$

the given initial value $e_0(t_0)=y(0)-u_0(0)=0$ furnishes the additional equation

$$\beta_{0,0}=h_0^{2-\alpha}\sum_{s=1}^{q-1}h_0^{(s-1)(2-\alpha)}w_s(0) \tag{6.3.17}$$

(note that $c_{0,0}^{(s)}=w_s(0)$). Hence, it follows from $w_s\in C^{m+1}(I)$ that there exist finite constants B_0 so that

$$|\beta_{0,0}|\leqslant B_0h_0^{2-\alpha}. \tag{6.3.18}$$

Combining these results with (6.3.16), we obtain the estimate

$$|\beta_{n,0}| \leqslant \sum_{i=0}^{n-1} \|\boldsymbol{\beta}_i\|_1 + h_0^{2-\alpha} B_0 + \sum_{i=0}^{n-1} h_i^{m+1} |R_i(1)|, \qquad (6.3.19)$$

$$n = 1, \dots, N-1,$$

where the vector $\boldsymbol{\beta}_i \in \mathbb{R}^m$ is given by $\boldsymbol{\beta}_i := (\beta_{i,1}, \dots, \beta_{i,m})^T$.

We now rewrite (6.3.12) as a recurrence relation for these vectors, by introducing the matrices

$$D_{n,n}(\alpha) := (lc_j^{l-1} - h_n p_{n,j} c_j^l - h_n^{2-\alpha} \int_0^{c_j} (c_j - v)^{-\alpha} K_{n,j}(t_n + vh_n) v^l dv) \qquad (6.3.20a)$$

and

$$D_{n,i}(\alpha) := (\int_0^1 \left[\frac{t_{n,j} - t_i}{h_i} - v\right]^{-\alpha} K_{n,j}(t_i + vh_i) v^l dv), \quad i < n \qquad (6.3.20b)$$

$$(j, l = 1, \dots, m),$$

and the m-vectors

$$\mathbf{p}_n := (p_{n,1}, \dots, p_{n,m})^T,$$

$$\mathbf{d}_{n,n}(\alpha) := (\int_0^{c_j} (c_j - v)^{-\alpha} K_{n,j}(t_n + vh_n) dv)^T, \qquad (6.3.21a)$$

$$\mathbf{d}_{n,i}(\alpha) := (\int_0^1 \left[\frac{t_{n,j} - t_i}{h_i} - v\right]^{-\alpha} K_{n,j}(t_i + vh_i) dv)^T, \quad i < n, \qquad (6.3.21b)$$

and $\mathbf{r}_n := (r_{n,1}, \dots, r_{n,m})^T$, with components given by (6.3.13) and (6.3.15). The desired recurrence relation then assumes the form

$$D_{n,n}(\alpha)\boldsymbol{\beta}_n = h_n \sum_{i=0}^{n-1} h_i^{1-\alpha} D_{n,i}(\alpha)\boldsymbol{\beta}_i$$

$$+ h_n \{(\mathbf{p}_n + h_n^{1-\alpha} \mathbf{d}_{n,n}(\alpha))\beta_{n,0} + \sum_{i=0}^{n-1} h_i^{1-\alpha} \mathbf{d}_{n,i}(\alpha)\beta_{i,0}\} + h_n \mathbf{r}_n, \qquad (6.3.22)$$

$$n = 0, \dots, N-1,$$

where the scalars $\beta_{i,0}$ are subject to (6.3.18) and (6.3.19).

Observe first that for all mesh sequences $\{\Pi_N\}$ possessing mesh diameters which tend to zero, as N tends to infinity, the matrices $D_{n,n}(\alpha)$ defined by (6.3.20a) have uniformly bounded inverses for all sufficiently large N, since $p(t)$ and $K(t,s)$ are continuous, and since

$$\det(lc_j^{l-1}) = m! \det(V) \neq 0,$$

where V denotes the Vandermonde matrix associated with the collocation parameters $\{c_j\}$. Thus, (6.3.22) defines a unique sequence of vectors $\{\boldsymbol{\beta}_n\}$ for all sufficiently small mesh diameters, and this holds both for quasi-uniform and graded mesh sequences.

We shall now show that the l_1-norms of these vectors satisfy a discrete Gronwall inequality of the form (1.5.10). The arguments for achieving this coincide to a large extent with those used in Section 6.2.2: since $|K(t,s)| \leqslant K_0$ for all $(t,s) \in S$ it follows from Lemma 6.2.1 and from (6.3.20b) that

$$\|D_{n,i}(\alpha)\|_1 \leqslant d_0(\alpha)(n-i)^{-\alpha}, \quad 0 \leqslant i < n \leqslant N-1,$$

with $d_0(\alpha) := mK_0\gamma^\alpha(1+\gamma)^\alpha/(1-\alpha)$; we also find

$$\|\mathbf{d}_{n,i}(\alpha)\|_1 \leqslant d_0(\alpha)(n-i)^{-\alpha}, \quad 0 \leqslant i < n \leqslant N-1.$$

Hence, since $D_{n,n}^{-1}(\alpha)$ is uniformly bounded for all sufficiently large N; i.e., $\|D_{n,n}^{-1}(\alpha)\|_1 \leqslant C_1 < \infty$, $n=0, \ldots, N-1$, there exist finite constants C_0', C_1', and C_2' so that (6.3.22), (6.3.18), and (6.3.19) imply

$$\begin{aligned}
\|\boldsymbol{\beta}_n\|_1 \leqslant\; & h\{C_0' h^{1-\alpha}\sum_{i=0}^{n-1}(n-i)^{-\alpha}\|\boldsymbol{\beta}_i\|_1 + C_1'\sum_{i=0}^{n-1}\|\boldsymbol{\beta}_i\|_1 \\
& + C_1' h^{1-\alpha}\sum_{i=1}^{n-1}\sum_{k=0}^{i-1}(n-i)^{-\alpha}\|\boldsymbol{\beta}_k\|_1\} \\
& + h\{h_0^{2-\alpha}B_0(C_0' + C_2' h^{1-\alpha}\sum_{i=0}^{n-1}(n-i)^{-\alpha}) \\
& + C_1'\sum_{i=0}^{n-1}h_i^{m+1}|R_i(1)| + C_2' h^{1-\alpha}\sum_{i=1}^{n-1}\sum_{k=0}^{i-1}(n-i)^{-\alpha}h_k^{m+1}|R_k(1)|\} + hC_1\|\mathbf{r}_n\|_1.
\end{aligned}$$

Here, the right-hand side may be simplified by employing the discrete form of the Dirichlet formula for the double sums; e.g.,

$$\sum_{i=1}^{n-1}\sum_{k=0}^{i-1}(n-i)^{-\alpha}\|\boldsymbol{\beta}_k\|_1 = \sum_{i=0}^{n-2}\Big(\sum_{k=i+1}^{n-1}(n-k)^{-\alpha}\Big)\|\boldsymbol{\beta}_i\|_1.$$

Setting $C_0 := \max(C_0', C_1', C_2')$, and using

$$\begin{aligned}
1 + h^{1-\alpha}\sum_{k=i+1}^{n-2}(n-k)^{-\alpha} &< 1 + h^{1-\alpha}\sum_{k=0}^{n-1}(n-k)^{-\alpha} < 1 + h^{1-\alpha}\int_0^n v^{-\alpha}\,dv \\
&< 1 + (Nh)^{1-\alpha}/(1-\alpha), \quad n \leqslant N-1,
\end{aligned}$$

we obtain the desired discrete Gronwall inequality:

$$\|\boldsymbol{\beta}_n\|_1 \leqslant hC_0(1 + (Nh)^{1-\alpha}/(1-\alpha))\sum_{i=0}^{n-1}\|\boldsymbol{\beta}_i\|_1 + hz_n, \tag{6.3.23a}$$

with

$$\begin{aligned}
z_n := \;& h_0^{2-\alpha}C_0B_0(1 + (Nh)^{1-\alpha}/(1-\alpha)) \\
& + C_0(1 + (Nh)^{1-\alpha}/(1-\alpha))\sum_{i=0}^{n-1}h_i^{m+1}|R_i(1)| + C_1\|\mathbf{r}_n\|_1,
\end{aligned} \tag{6.3.23b}$$

where the components of $\mathbf{r}_n$ are given in (6.3.13), (6.3.14). Note that, in (6.3.23a), the factor following h and preceding the sum is bounded for all N since $Nh \leqslant \gamma T$ for any quasi-uniform mesh sequence.

It follows from Corollary 1.5.2 that the order of $\|\boldsymbol{\beta}_n\|_1$ is determined by the orders of the terms hz_n in (6.3.23a). These, in turn, will be governed by the orders of the remainder terms $h_i^{m+1}R_i(v)$, $v \in [0,1]$, and $h_n^m R_n'(c_j)$, as is easily seen by recalling (6.3.23b), (6.3.13), and (6.3.15). These terms all contain the $(m+1)$st derivative of the exact solution of (6.3.2) which, by Leibniz's product rule, has the form

$$y^{(m+1)}(t) = w_0^{(m+1)}(t) + \sum_{s=1}^{q-1} \sum_{k=0}^{m+1} \gamma_{m,k}^{(s)} w_s^{(m+1-k)}(t) t^{s(2-\alpha)-k}, \quad t>0,$$

with

$$\gamma_{m,k}^{(s)} := \binom{m+1}{k} \binom{s(2-\alpha)}{k} k!.$$

Thus, in complete analogy to the proof of Theorem 6.2.7, the orders of these remainder terms will hinge on the orders of the products $h_i^{m+\lambda} t_i^{s(2-\alpha)-k}$ $(i=1, \ldots, N-1)$, where $\lambda \in \{0,1\}$ and $k=0, \ldots, m+1$.

Lemma 6.3.1. *Consider any quasi-uniform sequence of meshes* $\{\Pi_N\}$ *for* I, *and assume that* $0<\alpha<1$, $\lambda \in \{0,1\}$, *and* $0 \leqslant k \leqslant m+1$, *Then, for* $s \geqslant 1$ *and* $i=1, \ldots, N-1$,

$$h_i^{m+\lambda} t_i^{s(2-\alpha)-k} = \begin{cases} O(N^{-(m+\lambda)}), & \text{if } s(2-\alpha)-k \geqslant 0, \\ O(N^{-(1+\lambda-\alpha)}), & \text{if } s(2-\alpha)-k<0. \quad \square \end{cases} \tag{6.3.24}$$

The proof of Lemma 6.2.3 is readily adapted to the present situation; we omit the details.

It is now easily verified that (6.3.24) implies $h_i^{m+1}R_i(v) = O(N^{-(2-\alpha)})$ and $h_i^m R_i'(v) = O(N^{-(1-\alpha)})$, $v \in [0,1]$ $(i=1, \ldots, N-1)$. These estimates may then be used in (6.3.13), (6.3.15), and (6.3.23b) to show that

$$hz_n = O(N^{-(2-\alpha)}), \quad n=0, \ldots, N-1,$$

and this, as mentioned above, leads to

$$\|\boldsymbol{\beta}_n\|_1 = O(hz_n) = O(N^{-(2-\alpha)}), \quad n=0, \ldots, N-1 \text{ (as } h \downarrow 0,\ Nh \leqslant \gamma T).$$

Using (6.3.19), and observing that $Nh \leqslant \gamma T$, we also find

$$|\beta_{n,0}| = N \cdot O(N^{-(2-\alpha)}) = O(N^{-(1-\alpha)}), \quad n=1, \ldots, N-1,$$

and, from (6.3.18),

$$\beta_{0,0} = O(h_0^{2-\alpha}) = O(N^{-(2-\alpha)}),$$

since $h_0 \leqslant h \leqslant \gamma T N^{-1}$. The proof of Theorem 6.3.1 is concluded by using the above estimates in the error expressions (6.3.10) and (6.3.11). □

6.3.2. Polynomial spline collocation on graded meshes

Theorem 6.3.2. *Let p, q, and K in (6.3.2) be m times continuously differentiable on I and S, respectively, and assume that q and K do not vanish identically. If $u \in S_m^{(0)}(Z_N)$ is the collocation approximation defined by (6.3.3), and if the underlying mesh sequence consists of graded meshes $\Pi_N^{(r)}$ of the form*

$$t_i = (i/N)^r T, \quad i=0, \dots, N, \tag{6.3.25a}$$

with grading exponent

$$r = m/(1-\alpha), \tag{6.3.25b}$$

then the corresponding collocation error $e := y - u$ satisfies

$$\|e^{(k)}\|_\infty = O(N^{-m}), \quad k=0,1; \tag{6.3.26}$$

this holds for all collocation parameters $\{c_j\}$ with $0 \leqslant c_1 < \dots < c_m \leqslant 1$.

Proof. The arguments leading to the discrete Gronwall inequality (6.3.23) in the proof of Theorem 6.3.1 remain valid for graded meshes with $r>1$, since we have $h < rTN^{-1}$ (cf. (6.2.47)); the only change concerns the constant $d_0(\alpha)$ used in the bounds for $\|D_{n,i}(\alpha)\|_1$ and $\|\mathbf{d}_{n,i}(\alpha)\|_1$: it is now given, according to Lemma 6.2.4, by $d_0(\alpha) = mK_0 2^\alpha/(1-\alpha)$.

However, instead of Lemma 6.3.1 we now have

Lemma 6.3.2. *Consider the sequence of graded meshes defined by (6.3.25), and assume that $0<\alpha<1$, $\lambda \in \{0,1\}$, and $0 \leqslant k \leqslant m+1$. Then, for $s \geqslant 1$ and $i=1, \dots, N-1$,*

$$h_i^{m+\lambda} t_i^{s(2-\alpha)-k} = O(N^{-(m+\lambda)}), \tag{6.3.27}$$

regardless of whether $s(2-\alpha)-k$ is positive or negative.

Proof. By (6.2.51) and (6.3.25a) we have, setting $c := (r2^{r-1}T)^{m+\lambda} T^{s(2-\alpha)-k}$,

$$\begin{aligned} h_i^{m+\lambda} t_i^{s(2-\alpha)-k} &\leqslant c i^{(r-1)(m+\lambda)+rs(2-\alpha)-rk} N^{-r(m+\lambda)-rs(2-\alpha)+rk} \\ &= c i^{r(m+1-k)+r[s(2-\alpha)-(1-\lambda)]-(m+\lambda)} N^{-r(m+1-k)-r[s(2-\alpha)-(1-\lambda)]} \\ &\leqslant c i^{r[s(2-\alpha)-(1-\lambda)]-(m+\lambda)} N^{-r[s(2-\alpha)-(1-\lambda)]}, \end{aligned}$$

$$k=0, \dots, m+1.$$

If, for some real number $\mu \geqslant 0$, the exponent of i satisfies

$$r(2-\alpha-(1-\lambda))-(m+\lambda) \geqslant -\mu,$$

or, equivalently,

$$r(1+\lambda-\alpha)-(m+\lambda-\mu) \geqslant 0, \tag{6.3.28a}$$

then it follows that, for all $s \geqslant 1$ and $k \leqslant m+1$,

$$h_i^{m+\lambda} t_i^{s(2-\alpha)-k} = O(N^{-(m+\lambda-\mu)}), \quad i=1, \dots, N-1. \tag{6.3.28b}$$

These results contain the desired information on how to choose the grading exponent r for the graded meshes (6.3.25a): since optimal order, $O(N^{-(m+\lambda)})$, in (6.3.28b) is achieved if, and only if, $\mu=0$, (6.3.28a) yields

$$r \geqslant (m+\lambda)/(1+\lambda-\alpha),$$

where $\lambda \in \{0,1\}$ is given. Since this is to hold for $\lambda=0$ as well as for $\lambda=1$, we must have

$$r \geqslant m/(1-\alpha).$$

Hence, the smallest grading exponent r for which (6.3.27) is valid is the one given in (6.3.25b). □

Recall now (6.3.13), (6.3.15), and (6.3.23b): since the mesh diameter h of the graded mesh (6.3.25) satisfies $h=O(N^{-1})$, and since $h_0=O(N^{-r})$ (cf. (6.2.46)), the application of Lemma 6.3.2 leads to

$$hz_n = O(N^{-(m+1)}), \quad n=0, \dots, N-1,$$

and hence, by Corollary 1.5.2, to

$$\|\beta_n\|_1 = O(hz_n)=O(N^{-(m+1)}), \quad n=0, \dots, N-1 \ \ (\text{as } h\downarrow 0, \text{ with } Nh<rT).$$

Moreover, (6.3.19) yields

$$|\beta_{n,0}| = N \cdot O(N^{-(m+1)}) = O(N^{-m}), \quad n=0, \dots, N-1.$$

Thus, assertion (6.3.26) in Theorem 6.3.2 follows readily from the error expressions (6.3.10) and (6.3.11). □

What happens if we choose for the grading exponent r in (6.3.25a) the value

$$r = (m+1)/(2-\alpha)$$

(which, according to Theorem 6.2.5, is the correct grading exponent when approximating the function $f(t)=t^{2-\alpha}$ $(0<\alpha<1)$ in $S_m^{(0)}(Z_N)$, but which is smaller than the optimal grading exponent $r=m/(1-\alpha)$ given in Theorem 6.3.2)? It is easily seen that (6.3.28a) holds with

$$\mu = \frac{m+1}{2-\alpha} - 1$$

for both $\lambda=0$ and $\lambda=1$. Hence (6.3.28b) yields, for $\lambda=0$,

$$h_i^m t_i^{s(2-\alpha)-k} = O(N^{-\nu}), \quad s \geqslant 1, \ k \leqslant m+1,$$

where

$$\nu := m - \mu = \frac{(m+1)(1-\alpha)}{2-\alpha} .$$

The argument used in the above proof then reveals that

$$\|e^{(k)}\|_\infty = O(N^{-\nu}), \quad k=0,1.$$

Thus, if $\alpha = \frac{1}{2}$, the global convergence order corresponding to graded meshes with grading exponent $r=(m+1)/(2-\alpha)$ is given by

$$\|e^{(k)}\|_\infty = O(N^{-(m+1)/3}), \quad k=0,1.$$

Finally, if we choose $r=1$, then (6.3.28a) yields $\mu=m-(1-\alpha)$, and hence $\nu=1-\alpha$ (both for $\lambda=0$ and $\lambda=1$): since $r=1$ yields the sequence of uniform meshes, the resulting order of convergence,

$$\|e^{(k)}\|_\infty = O(N^{-(1-\alpha)}), \quad k=0,1,$$

agrees, of course with the one established in Theorem 6.3.1.

6.3.3. Discretization of the collocation equation

The fully discretized form of the collocation equation (6.3.3) is obtained by approximating the integrals $\Phi_{n,i}^{(j)}[u_i;\alpha]$ (cf. (6.2.16)) by means of suitable product integration techniques, in analogy to Section 6.2.4. Denoting these quadrature approximations by $\hat{\Phi}_{n,i}^{(j)}[u_i;\alpha]$, the resulting discrete version of (6.3.3),

$$\hat{u}_n'(t_{n,j}) = f(t_{n,j},\hat{u}_n(t_{n,j})) + h_n^{1-\alpha}\hat{\Phi}_{n,n}^{(j)}[\hat{u}_n;\alpha] + \sum_{i=0}^{n-1} h_i^{1-\alpha}\hat{\Phi}_{n,i}^{(j)}[\hat{u}_i;\alpha], \tag{6.3.29}$$

$$j=1, \ldots ,m \quad (n=0, \ldots ,N-1),$$

with $\hat{u}_0(t_0)=y_0$, defines, for all sufficiently small mesh diameters h, a polynomial spline $\hat{u}\in S_m^{(0)}(Z_N)$ which, due to the errors induced by numerical integration, will generally be different from the collocation approximation $u\in S_m^{(0)}(Z_N)$ furnished by the exact collocation equation (6.3.3).

In the following we restrict our analysis to the case where the approximations $\hat{\Phi}_{n,i}^{(j)}[u_i;\alpha]$ are of the form

$$\hat{\Phi}_{n,i}^{(j)}[u_i;\alpha] := \begin{cases} \sum_{l=1}^{m} w_{j,l}^{(n,i)}(\alpha)k(t_{n,j},t_{i,l},u_i(t_{i,l})), & \text{if } 0\leq i\leq n-1, \\ \sum_{l=1}^{m} w_{j,l}(\alpha)k(t_{n,j},t_n+c_jc_lh_n,u_n(t_n+c_jc_lh_n)), & \text{if } i=n; \end{cases} \tag{6.3.30}$$

more general approximations may be treated along the lines of Section 6.2.4. According to (6.2.74), the quadrature weights in (6.3.30) are given by

$$w_{j,l}^{(n,i)}(\alpha) := \int_0^1 \left[\frac{t_{n,i}-t_i}{h_i} - v\right]^{-\alpha} L_l(v)dv, \quad i\leq n-1, \tag{6.3.31a}$$

and by

$$w_{j,l}(\alpha) := c_j^{1-\alpha} \int_0^1 (1-v)^{-\alpha} L_l(v)dv \quad (j,l=1, \ldots ,m), \tag{6.3.31b}$$

with

$$L_l(v) := \prod_{\substack{k=1 \\ k\neq l}}^{m} (v-c_k)/(c_l-c_k).$$

Setting

$$\lambda_k(v) := \int_0^v L_k(\tau)d\tau,$$

we have, as in (5.4.24b),

$$\hat{u}_n(t_n+vh_n) = \hat{y}_n + h_n \sum_{k=1}^{m} \lambda_k(v)\hat{Y}_{n,k}, \quad v\in[0,1], \tag{6.3.32}$$

with $\hat{y}_n := \hat{u}_n(t_n)$, and with $\hat{Y}_{n,k} := \hat{u}_n{}'(t_{n,k})$. Thus, (6.3.29) represents, for each $n=0, \ldots ,N-1$, a system of m nonlinear equations for $(\hat{Y}_{n,1}, \ldots ,\hat{Y}_{n,m})^T$; once these values have been found, the approximating spline $\hat{u}\in S_m^{(0)}(Z_N)$ on σ_n is given by (6.3.32).

The analysis of the global order of convergence of $\hat{u}$ is a synthesis of the techniques used, respectively, in Section 5.4.3 and Section 6.2.4. We therefore leave the proof of the following theorem to the reader.

Theorem 6.3.3. *Let p, q, and K in (6.3.2) be m times continuously differentiable on their respective domains I and S, and assume that q and K do not vanish identically. Moreover, let $\hat{u}\in S_m^{(0)}(Z_N)$ denote the solution of the fully discretized collocation equation (6.3.29), with quadrature approximations $\hat{\Phi}_{n,i}^{(j)}[u_i;\alpha]$ given by (6.3.30), (6.3.31).*

(i) *If the underlying mesh sequence $\{\Pi_N\}$ is quasi-uniform, then the error $\hat{e} := y-\hat{u}$ satisfies*

$$\|\hat{e}^{(k)}\|_\infty = O(N^{-(1-\alpha)}), \quad k=0,1. \tag{6.3.33a}$$

(ii) *If $\{\Pi_N^{(r)}\}$ is the sequence of graded meshes (6.3.25a), and if the grading exponent is given by $r=m/(1-\alpha)$, then we have*

$$\|\hat{e}^{(k)}\|_\infty = O(N^{-m}), \quad k=0,1. \tag{6.3.33b}$$

These convergence results hold for all collocation parameters $\{c_j\}$ with $0\leqslant c_1 < \ldots < c_m \leqslant 1$. □

Example 6.3.1. Let the approximating spline space be $S_2^{(0)}(Z_N)$ $(m=2)$. Setting $Y_{n,j} = \hat{u}_n{}'(t_{n,j})$, $j=1,2$, with $0\leqslant c_1 < c_2 \leqslant 1$, the corresponding fully discretized collocation equation (6.3.3) can be written as

$$\hat{Y}_{n,j} = f(t_{n,j},\hat{u}_n(t_{n,j}))+h_n^{1-\alpha}\hat{\Phi}_{n,n}^{(j)}[\hat{u}_n;\alpha]+\sum_{i=0}^{n-1} h_i^{1-\alpha}\hat{\Phi}_{n,i}^{(j)}[\hat{u}_i;\alpha], \tag{6.3.34a}$$

$$j=1,2 \quad (n=0, \dots ,N-1),$$

with

$$\hat{u}_n(t_n+vh_n) = \hat{y}_n+h_n\sum_{k=1}^{2}\lambda_k(v)\hat{Y}_{n,k}, \quad v\in[0,1], \tag{6.3.34b}$$

where

$$\lambda_1(v) = \frac{1}{c_2-c_1}\int_0^v(c_2-\tau)d\tau, \quad \lambda_2(v) = \frac{1}{c_2-c_1}\int_0^v(\tau-c_1)d\tau. \tag{6.3.34c}$$

The quadrature approximations are given by

$$\hat{\Phi}_{n,i}^{(j)}[\hat{u}_i;\alpha] = \sum_{l=1}^{2} w_{j,l}^{(n,i)}(\alpha)k(t_{n,j},t_{i,l},\hat{u}_i(t_{i,l})), \quad j=1,2, \tag{6.3.35a}$$

and by

$$\hat{\Phi}_{n,n}^{(j)}[\hat{u}_n;\alpha] = \sum_{l=1}^{2} w_{j,l}(\alpha)k(t_{n,j},t_n+c_jc_lh_n,\hat{u}_n(t_n+c_jc_lh_n)), \tag{6.3.35b}$$

with weights

$$\begin{aligned} w_{j,1}^{(n,i)}(\alpha) &= \frac{1}{c_2-c_1}\int_0^1\left[\frac{t_{n,j}-t_i}{h_i}-v\right]^{-\alpha}(c_2-v)dv,\\ w_{j,2}^{(n,i)}(\alpha) &= \frac{1}{c_2-c_1}\int_0^1\left[\frac{t_{n,j}-t_i}{h_i}-v\right]^{-\alpha}(v-c_1)dv, \quad j=1,2, \end{aligned} \tag{6.3.35c}$$

and

$$\begin{aligned} w_{j,1}(\alpha) &= \frac{c_j^{1-\alpha}}{c_2-c_1}\int_0^1(1-v)^{-\alpha}(c_2-v)dv,\\ w_{j,2}(\alpha) &= \frac{c_j^{1-\alpha}}{c_2-c_1}\int_0^1(1-v)^{-\alpha}(v-c_1)dv; \end{aligned} \tag{6.3.35d}$$

they are easily computed analytically.

Once the values $\hat{Y}_{n,1}$ and $\hat{Y}_{n,2}$ have been determined from (6.3.34a), the collocation approximation $\hat{u}\in S_2^{(0)}(Z_N)$ on σ_n is given by (6.3.34b), while its derivative (which is an element of $S_1^{(-1)}(Z_N)$) on σ_n has the form

$$\hat{u}_n'(t_n+vh_n) = \frac{1}{c_2-c_1}\{(c_2-v)\hat{Y}_{n,1}+(v-c_1)\hat{Y}_{n,2}\}, \quad t_n+vh_n\in\sigma_n.$$

If we employ *graded meshes* with $r=2/(1-\alpha)$, then, by Theorem 6.3.3,

$$\|y^{(k)}-\hat{u}^{(k)}\|_\infty = O(N^{-2}), \quad k=0,1,$$

while for *uniform* (or quasi-uniform) *meshes* we obtain only

$$\|y^{(k)}-\hat{u}^{(k)}\|_\infty = O(N^{-(1-\alpha)}), \quad k=0,1. \quad \square$$

6.4. First-kind integral equations with weakly singular kernels

Formally, the collocation techniques of Section 6.2 are readily extended to first-kind Volterra integral equations with weakly singular kernels, as well as to more general Abel-type equations. However, there are, as we shall see, two problems connected with such an approach: while the first one concerns the still open question on how to choose the collocation parameters so as to obtain a convergent method (and which will eventually be settled), the second one relates to the *ill-posed* nature of the given integral equation and will be particularly serious when the function $g(t)$ is known only in the form of experimental (noisy) data, thus severely limiting the practical importance of collocation methods.

6.4.1. Polynomial spline collocation and product integration methods

In this section we consider collocation approximations in the polynomial spline spaces $S_{m-1}^{(-1)}(Z_N)$ and $S_m^{(0)}(Z_N)$ to the solution of the (linear) first-kind integral equation

$$g(t)+\int_0^t (t-s)^{-\alpha}K(t,s)y(s)ds = 0, \quad t\in I:=[0,T] \;\; (0<\alpha<1), \tag{6.4.1}$$

where the functions $K(t,s)$ and $g(t)$ are supposed to be known analytically (i.e., with no error present). We shall also assume that they are subject to the hypotheses of Theorem 1.3.6. If Π_N denotes a partition of the given interval I, we relate (6.4.1) to Π_N by rewriting it as

$$F_n(t)+\int_{t_n}^t (t-s)^{-\alpha}K(t,s)y(s)ds = 0, \quad t\in\sigma_n, \tag{6.4.2}$$

where the lag term $F_n(t)$ is defined by

$$F_n(t) := g(t)+\int_0^{t_n} (t-s)^{-\alpha}K(t,s)y(s)ds, \quad n=0, \dots ,N-1, \tag{6.4.3}$$

(see also (5.5.2), (6.2.14)).

In analogy to Section 5.5.1 we seek an element $u\in S_{m-1}^{(-1)}(Z_N)$ which satisfies (6.4.2) at the collocation points

$$X_n := \{t_{n,j}:=t_n+c_jh_n\colon 0<c_1< \dots <c_m\leq 1\} \;\; (n=0, \dots ,N-1).$$

The collocation equation defining u thus reads

$$\bar{F}_n(t_{n,j})+h_n^{1-\alpha}\int_0^{c_j} (c_j-v)^{-\alpha}K_{n,j}(t_n+vh_n)u_n(t_n+vh_n)dv = 0, \tag{6.4.4}$$

$$j=1, \dots ,m \quad (n=0, \dots ,N-1);$$

here, we have set $K_{n,j}(s):=K(t_{n,j},s)$, and

$$\bar{F}_n(t_{n,j}) := g(t_{n,j})+\sum_{i=0}^{n-1} h_i^{1-\alpha}\int_0^1 \left[\frac{t_{n,j}-t_i}{h_i}-v\right]^{-\alpha} K_{n,j}(t_i+vh_i)u_i(t_i+vh_i)dv. \quad (6.4.5)$$

Alternatively, the collocation equation (6.4.4) may be written in the form

$$\Phi_{n,n}^{(j)}[u_n;\alpha] = -(h_n^{\alpha-1}g(t_{n,j})+\sum_{i=0}^{n-1}(h_i/h_n)^{1-\alpha}\Phi_{n,i}^{(j)}[u_i;\alpha]), \quad (6.4.6)$$

where (cf. (6.2.16))

$$\Phi_{n,i}^{(j)}[u_i;\alpha] := \begin{cases} \int_0^1 \left[\frac{t_{n,j}-t_i}{h_i}-v\right]^{-\alpha} K_{n,j}(t_i+vh_i)u_i(t_i+vh_i)dv, & \text{if } 0\leqslant i\leqslant n-1, \\ \int_0^{c_j}(c_j-v)^{-\alpha}K_{n,j}(t_n+vh_n)u_n(t_n+vh_n)dv, & \text{if } i=n. \end{cases} \quad (6.4.7)$$

Suppose now that we are given a set of parameters $\{v_j\}$ satisfying $0\leqslant v_1< \dots <v_m\leqslant 1$ (this set may or may not coincide with that of the collocation parameters $\{c_j\}$). Setting $U_{n,s}:=u_n(t_n+v_sh_n)$, and denoting by $\bar{L}_s(v)$ the Lagrange fundamental polynomials associated with the m parameters $\{v_j\}$, the collocation approximation u on the subinterval σ_n is given by

$$u_n(t_n+vh_n) = \sum_{s=1}^{m}\bar{L}_s(v)U_{n,s}, \quad t_n+vh_n\in\sigma_n. \quad (6.4.8)$$

Hence, by (6.4.7) and (6.4.8), (6.4.6) represents, for each $n=0, \dots ,N-1$, a linear system in $\mathbb{R}^m$ for the vector $\mathbf{U}_n:=(U_{n,1}, \dots ,U_{n,m})^T$. Once $\mathbf{U}_n$ has been found, we know the values of u on σ_n by means of (6.4.8).

Before dealing with the fully discretized version of the collocation equation (6.4.6) we look briefly at collocation in the continuous polynomial spline space $S_m^{(0)}(Z_N)$ (recall that the dimension of this space exceeds the one of $S_{m-1}^{(-1)}(Z_N)$ by one; cf. Corollary 5.1.1). Employing the same set of collocation points X_n $(n=0, \dots ,N-1)$ we see that the collocation approximation $u\in S_m^{(0)}(Z_N)$ must satisfy (6.4.4), too; now, however, we have $u_n\in\pi_m$. The continuity requirements for $t_n\in Z_N$ furnish, in addition, the conditions

$$u_n(t_n) = u_{n-1}(t_n), \quad t_n\in Z_N. \quad (6.4.9a)$$

Moreover, the starting value $u(0)=y(0)$ must be known: it follows from Theorem 1.3.11 that this value is defined by

$$y(0) = \frac{1-\alpha}{K(0,0)}\lim_{t\downarrow 0}(g(t)/t^{1-\alpha}). \quad (6.4.9b)$$

As before, it will again be convenient to introduce parameters $\{v_j\}$ satisfying $0=v_0<v_1< \dots <v_m=1$; instead of (6.4.8) we now have the representation

$$u_n(t_n+vh_n) = \sum_{s=0}^{m} \tilde{L}_s(v)U_{n,s}, \quad t_n+vh_n \in \sigma_n, \tag{6.4.10a}$$

with

$$\tilde{L}_s(v) := \prod_{\substack{r=0 \\ r\neq s}}^{m} (v-v_r)/(v_s-v_r)$$

denoting the s-th Lagrange fundamental polynomial with respect to the $m+1$ parameters $\{v_j\}$. The continuity conditions (6.4.9a) translate into

$$U_{n,0} = U_{n-1,m}, \quad n=1, \ldots ,N-1, \tag{6.4.10b}$$

with $U_{0,0}=u_n(0)=y(0)$ given by (6.4.9a).

Hence, by (6.4.10) and (6.4.7), (6.4.6) constitutes a linear system in $\mathbb{R}^m$ for $\mathbf{U}_n:=(U_{n,1}, \ldots ,U_{n,m})^T$ (note that $U_{n,0}$ is given by (6.4.10b)), and the interpolation formula (6.4.10a) can be used to evaluate u on σ_n.

If (6.4.1) represents the classical Abel integral equation (i.e., if $K(t,s)=1$ for all $(t,s)\in S$), then the integrals (6.4.7) can be found analytically. However, in more general problems it will usually be necessary to approximate these integrals by suitable numerical quadrature processes. Since we have discussed this problem in some detail in Section 6.2.4 we shall now only mention the case where the integrals (6.4.7) are evaluated by μ-point product quadrature formulas of the form

$$\hat{\Phi}^{(j)}_{n,i}[u_i;\alpha] := \begin{cases} \sum_{l=1}^{\mu} w^{(n,i)}_{j,l}(\alpha)K_{n,j}(t_i+d_l h_i)u_i(t_i+d_l h_i), & \text{if } 0\leqslant i\leqslant n-1, \\ \sum_{l=1}^{\mu} w_{j,l}(\alpha)K_{n,j}(t_n+d_{j,l}h_n)u_n(t_n+d_{j,l}h_n), & \text{if } i=n, \end{cases} \tag{6.4.11}$$

with $0\leqslant d_1< \ldots <d_\mu\leqslant 1$, $d_{j,l}:=c_j d_l$, and with weights given, respectively, by

$$w^{(n,i)}_{j,l}(\alpha) := \prod_{\substack{r=1 \\ r\neq l}}^{\mu} (d_l-d_r)^{-1} \int_0^1 \left[\frac{t_{n,j}-t_i}{h_i} - v\right]^{-\alpha} \prod_{\substack{r-1 \\ r\neq l}}^{\mu} (v-d_r)dv, \tag{6.4.12a}$$

and

$$w_{j,l}(\alpha) := c_j^{1-\alpha} \prod_{\substack{r=1 \\ r\neq l}}^{\mu} (d_l-d_r)^{-1} \int_0^1 (1-v)^{-\alpha} \prod_{\substack{r=1 \\ r\neq l}}^{\mu} (v-d_r)dv \tag{6.4.12b}$$

(compare also (6.2.56), (6.2.57)). Generalizations in the sense of (6.2.55) are obviously possible.

The fully discretized collocation equation corresponding to (6.4.6) is thus of the form

$$\hat{\Phi}^{(j)}_{n,n}[\hat{u}_n;\alpha] = -(h_n^{\alpha-1}g(t_{n,j})+\sum_{i=0}^{n-1} (h_i/h_n)^{1-\alpha}\hat{\Phi}^{(j)}_{n,i}[\hat{u}_i;\alpha]); \tag{6.4.13}$$

it defines an element $\hat{u} \in S_m^{(-1)}(Z_N)$ which, in general, will not coincide with the approximation u determined by (6.4.6). For computational purposes one will again replace $\hat{u}$ by the Lagrange representation analogous to (6.4.8), where the role of $U_{n,s}$ will now be assumed by $\hat{U}_{n,s} := \hat{u}_n(t_n + v_s h_n)$.

The case $\hat{u} \in S_m^{(0)}(Z_N)$ is treated in an almost identical manner. We leave the details to the reader (compare, however, Example 6.4.2 below).

Example 6.4.1. Approximation in $S_0^{(-1)}(Z_N)$, with $0 < c_1 \leq 1$. Since we have $L_1(v) \equiv 1$, and $U_{n,1} = u_n(t_n + v_1 h_n)$, with $v_1 \in [0,1]$, the resulting collocation equation (6.4.6) has the form

$$U_{n,1} \int_0^{c_1} (c_1 - v)^{-\alpha} K_{n,1}(t_n + vh_n)dv = -(h_n^{\alpha-1} g(t_{n,1}) + \sum_{i=0}^{n-1} (h_i/h_n)^{1-\alpha} U_{i,1} \int_0^1 \left(\frac{t_{n,1} - t_i}{h_i} - v\right)^{-\alpha} K_{n,1}(t_i + vh_i)dv). \tag{6.4.14}$$

For $c_1 = 1$, $v_1 = 1/2$, this yields a recurrence relation for the approximation of $y_{n+1/2} := u_n(t_n + h_n/2) = U_{n,1}$, namely,

$$y_{n+1/2} \int_0^1 (1-v)^{-\alpha} K(t_{n+1}, t_n + vh_n)dv = -(h_n^{\alpha-1} g(t_{n+1}) + \sum_{i=0}^{n-1} (h_i/h_n)^{1-\alpha} y_{i+1/2} \int_0^1 \left(\frac{t_{n+1} - t_i}{h_i} - v\right)^{-\alpha} K(t_{n+1}, t_i + vh_i)dv). \tag{6.4.15}$$

For the full discretization of (6.4.14) we choose $\mu = 1$, and $d_1 = v_1$. Hence, the weights (6.4.12) become

$$w_{1,1}^{(n,i)}(\alpha) = \frac{1}{1-\alpha} \left\{ \left(\frac{t_{n,1} - t_i}{h_i}\right)^{1-\alpha} - \left(\frac{t_{n,1} - t_{i+1}}{h_i}\right)^{1-\alpha} \right\}, \quad 0 \leq i \leq n-1,$$

and

$$w_{1,1}(\alpha) = c_1^{1-\alpha}/(1-\alpha).$$

This leads to the fully discretized version of (6.4.14),

$$w_{1,1}(\alpha) K_{n,1}(t_n + c_1 d_1 h_n) \hat{U}_{n,1} = -(h_n^{\alpha-1} g(t_{n,1}) + \sum_{i=0}^{n-1} (h_i/h_n)^{1-\alpha} w_{1,1}^{(n,i)}(\alpha) K_{n,1}(t_i + d_1 h_i) \hat{U}_{i,1}). \tag{6.4.16}$$

If, in particular, we work with values $c_1 = 1$ and $d_1 = 1/2$ *(product midpoint rule)*, then the following discretization of (6.4.15) is obtained:

$$\frac{1}{1-\alpha} K(t_{n+1}, t_n + h_n/2) \hat{y}_{n+1/2} = -h_n^{\alpha-1} g(t_{n+1}) - \frac{1}{1-\alpha} \sum_{i=0}^{n-1} (h_i/h_n)^{1-\alpha} \left\{ \left(\frac{t_{n+1} - t_i}{h_i}\right)^{1-\alpha} \right.$$

$$-\left(\frac{t_{n+1}-t_{i+1}}{h_i}\right)^{1-\alpha}\}K(t_{n+1},t_i+h_i/2)\hat{y}_{i+1/2}, \tag{6.4.17}$$

with $\hat{y}_{n+1/2}:=\hat{u}_n(t_n+h_n/2)$. This discretization (6.4.17) represents the so-called *product midpoint method* for (6.4.1) which, for uniform partitions Π_N, was introduced and analyzed by Weiss & Anderssen [1972] (see also Eggermont [1979]). □

Example 6.4.2. Approximation in $S_1^{(0)}(Z_N)$, with $0<c_1\leqslant 1$. Let $v_0=0$, $v_1=1$, implying that (6.4.10a) reduces to

$$u_n(t_n+vh_n) = (1-v)U_{n,0}+vU_{n,1}, \quad v\in[0,1].$$

Setting $y_n:=U_{n-1,1}(=U_{n,0})$ the collocation equation defining u may be written in the form

$$y_{n+1}\int_0^{c_1}(c_1-v)^{-\alpha}K_{n,1}(t_n+vh_n)v\,dv$$

$$=-(h_n^{\alpha-1}g(t_{n,1})+\{y_n\int_0^{c_1}(c_1-v)^{-\alpha}K_{n,1}(t_n+vh_n)(1-v)dv \tag{6.4.18}$$

$$+\sum_{i=0}^{n-1}(h_i/h_n)^{1-\alpha}\int_0^1\left(\frac{t_{n,1}-t_i}{h_i}\right)^{-\alpha}K_{n,1}(t_i+vh_i)((1-v)y_i+vy_{i+1})dv),$$

with $y_0=y(0)$ given by (6.4.9b).

For the full discretization of (6.4.18) we choose $\mu=2$, $d_1=0$, and $d_2=1$ *(product trapezoidal rule).* The resulting weights (6.4.12) are then found to be

$$w_{1,1}^{(n,i)}(\alpha)=b(\alpha)\{\left(\frac{t_{n,1}-t_{i+1}}{h_i}\right)^{2-\alpha}+\left(\frac{t_{n,1}-t_i}{h_i}\right)^{1-\alpha}\left(2-\alpha-\frac{t_{n,1}-t_i}{h_i}\right)\},$$

$$w_{1,2}^{(n,i)}(\alpha)=b(\alpha)\{\left(\frac{t_{n,1}-t_i}{h_i}\right)^{2-\alpha}-\left(\frac{t_{n,1}-t_{i+1}}{h_i}\right)^{1-\alpha}\left(2-\alpha+\frac{t_{n,1}-t_{i+1}}{h_i}\right)\}$$

$$(0\leqslant i\leqslant n-1),$$

and

$$w_{1,1}(\alpha) = c_1^{1-\alpha}/(2-\alpha), \quad w_{1,2}(\alpha) = c_1^{1-\alpha}/((1-\alpha)(2-\alpha)),$$

with $b(\alpha):=1/((1-\alpha)(2-\alpha))$. If $c_1=1$, then the fully discretized version of (6.4.18) becomes

$$b(\alpha)K(t_{n+1},t_{n+1})\hat{y}_{n+1}=-\{h_n^{\alpha-1}g(t_{n+1})+\frac{1}{2-\alpha}K(t_{n+1},t_n)\hat{y}_n \tag{6.4.19}$$

$$+\sum_{i=0}^{n-1}(h_i/h_n)^{1-\alpha}(w_{1,1}^{(n,i)}(\alpha)K(t_{n+1},t_i)\hat{y}_i+w_{1,2}^{(n,i)}(\alpha)K(t_{n+1},t_{i+1})\hat{y}_{i+1})\},$$

with $\hat{y}_{n+1} := \hat{u}_n(t_{n+1})$ $(=\hat{U}_{n,1})$, and where in the above expressions for the weights $w_{1,1}^{(n,i)}(\alpha)$, $t_{n,1}$ is to be replaced by $t_n + c_1 h_n = t_{n+1}$. For uniform meshes Π_N the method (6.4.19) is identical with the *product trapezoidal method* of R. Weiss [1972b] (compare also Eggermont [1981]). □

There is another way of discretizing the integrals in (6.4.18): instead of using the product trapezoidal rule (which employs two abscissas) we shall now use weighted *one-point Gauss* quadrature formulas. In this case the quadrature abscissas can no longer be characterized by a single parameter d_1 (corresponding to $\mu=1$ in (6.4.11)) since the weight functions $((t_{n,1}-t_i)/h_i - v)^{-\alpha}$ $(i \leqslant n)$ depend on n and on i. It follows from the elementary theory of weighted Gauss quadrature formulas (see, e.g., Isaacson & Keller [1966, pp. 331-334]) that, for $c_1 = 1$, we obtain formulas with degree of precision equal to one if the weights are given by

$$w_{1,1}^{(n,i)}(\alpha) = \int_0^1 \left[\frac{t_{n+1}-t_i}{h_i} - v\right]^{-\alpha} dv, \quad 0 \leqslant i \leqslant n-1,$$

and

$$w_{1,1}(\alpha) = 1/(1-\alpha);$$

the abscissas (which, according to the above observation, depend both on n and the interval σ_i) are of the form $t_i + d_1 h_i$, with

$$d_1 = d_1^{(n,i)}(\alpha) := \int_0^1 \left[\frac{t_{n+1}-t_i}{h_i} - v\right]^{-\alpha} v\,dv / w_{1,1}^{(n,i)}(\alpha), \quad i < n,$$

and $t_n + d_{1,1} h_n$, with $d_{1,1} = d_{1,1}(\alpha) := 1/(2-\alpha)$. The order of these one-point quadrature formulas equals that of the product trapezoidal rule used in (6.4.19), namely two. The resulting discretization of (6.4.18) (based on (6.4.11) with $\mu=1$, and with weights and abscissas as given above) coincides with a method derived by Branca [1976], [1978] for the nonlinear version of (6.4.1) (with $\alpha=1/2$).

6.4.2. *Convergence results and open problems*

In Theorem 5.5.1a we have shown that, under suitable smoothness conditions for $g(t)$ and $K(t,s)$, the collocation approximation $u \in S_{m-1}^{(-1)}(Z_N)$ to the solution of a first-kind integral equation with *bounded* kernel (i.e., equation (6.4.1) with $\alpha=0$) converges globally with order $p=m$; this holds in particular if the collocation parameters $\{c_j\}$ satisfy $0 < c_1 < \dots < c_m = 1$. More generally, if all collocation parameters lie in the interior of $(0,1]$, $0 < c_1 < \dots < c_m < 1$, then the same convergence result holds if the condition

$$\prod_{j=1}^{m} (1-c_j)/c_j < 1 \tag{6.4.20}$$

is satisfied. For the continuous case, $u \in S_m^{(0)}(Z_N)$, with $0 < c_1 < \dots < c_m = 1$, global convergence of order $p = m+1$ is obtained if, as in Theorem 5.5.1b,

$$\prod_{j=1}^{m-1}(1-c_j)/c_j<1. \tag{6.4.21}$$

For first-kind integral equations (6.4.1) with $0<\alpha<1$ results analogous to those of Theorems 5.5.1a and 5.5.1b are not yet known. (Compare the conjecture stated at the end of this section.) However, two special cases have been analyzed, namely, $u\in S_0^{(-1)}(Z_N)$ (with $c_1=1$), and $u\in S_1^{(0)}(Z_N)$ (also with $c_1=1$). Since the techniques used in the proofs of the corresponding convergence results (which are due to Weiss & Anderssen [1972], R. Weiss [1972a], [1972b], and Eggermont [1981]) do not seem to work for more general polynomial spline spaces, we shall refrain from presenting these proofs; the interested reader is referred to the original papers.

Theorem 6.4.1. *Assume that $g(t)$ and $K(t,s)$ in (6.4.1) satisfy $g\in C^3(I)$, $K\in C^3(S)$, with $g^{(k)}(0)=0$ $(k=0,1,2)$, and $K(t,t)\neq 0$ for all $t\in I$. Let $u\in S_0^{(-1)}(Z_N)$ and $\hat{u}\in S_0^{(-1)}(Z_N)$ denote, respectively, the collocation approximations defined by (6.4.15) and by its fully discretized counterpart (6.4.17), with $c_1=1$. If the underlying mesh sequence is quasi-uniform, then*

$$\|y-u\|_\infty = O(N^{-1}), \quad \text{and} \quad \|y-\hat{u}\|_\infty = O(N^{-1}).$$

Moreover,

$$\max_{0\leq n\leq N-1}|y(t_{n+1/2})-y_{n+1/2}| = O(N^{-(2-\alpha)}),$$

and

$$\max_{0\leq n\leq N-1}|y(t_{n+1/2})-\hat{y}_{n+1/2}| = O(N^{-(2-\alpha)});$$

i.e., we obtain local superconvergence of order $p^*(\alpha):=2-\alpha$ at the midpoints $t_{n+1/2}:=t_n+h_n/2$ of the subintervals σ_n. □

Note that by Theorem 1.3.17 the above hypotheses for $g(t)$ and $K(t,s)$ imply the existence of a (unique) solution $y\in C^2(I)$. In Weiss & Anderssen [1972] and in Eggermont [1979] the above local superconvergence results are established under slightly weaker conditions on $g(t)$ and $K(t,s)$.

Theorem 6.4.2. *Let $g(t)$ and $K(t,s)$ in (6.4.1) be subject to the conditions stated in Theorem 6.4.1, and assume that $u\in S_1^{(0)}(Z_N)$ and $\hat{u}\in S_1^{(0)}(Z_N)$ are the collocation approximations determined by (6.4.18) and (6.4.19), respectively, where $c_1=1$. Then for any quasi-uniform sequence of meshes these approximations satisfy*

$$\|y-u\|_\infty = O(N^{-2}), \quad \text{and} \quad \|y-\hat{u}\|_\infty = O(N^{-2}).$$

The same order of convergence is attained by $y_n:=u(t_n)$ and by $\hat{y}_n:=\hat{u}(t_n)$ for $t_n\in\bar{Z}_N$. □

R. Weiss [1972b] proved second-order convergence for the values $\{\hat{y}_n\}$ for all α satisfying $\alpha \in [\alpha_0, 1)$, with $\alpha_0 = 0.2117$; Eggermont [1981] then showed that the result holds for all $\alpha \in (0,1)$.

If the collocation equation (6.4.18) (with $c_1 = 1$) is discretized by one-point Gauss quadrature formulas, as described above, then the resulting approximation $\hat{u} \in S_1^{(0)}(Z_N)$ converges globally (on I) and locally (on $\bar{Z}_N$) with order $p=2$. This was shown by Branca [1976], [1978].

The convergence analysis for collocation methods in higher-degree polynomial spline spaces is, as indicated before, still an open problem. However, numerical calculations suggest the following result.

Conjecture 6.4.1. *Assume that $g(t)$ and $K(t,s)$ in (6.4.1) satisfy $g \in C^{m+1}(I)$, $K \in C^{m+1}(I)$, with $g^{(k)}(0) = 0$ $(k = 0, \ldots, m)$, and $K(t,t) \neq 0$ for all $t \in I$. Let the collocation parameters $\{c_j\}$ be such that $0 < c_1 < \ldots < c_m = 1$. If $u \in S_{m-1}^{(-1)}(Z_N)$ and $\hat{u} \in S_{m-1}^{(-1)}(Z_N)$ are, respectively, the collocation approximations determined by (6.4.6) and (6.4.13) (with discretization based on the m-point quadrature formulas (6.4.11), (6.4.12)), and if the underlying mesh sequence is quasi-uniform, then*

$$\|y - u\|_\infty = O(N^{-m}), \quad \text{and} \quad \|y - \hat{u}\|_\infty = O(N^{-m}). \quad \square$$

As for the case $0 < c_1 < \ldots < c_m < 1$, as well as for $u \in S_m^{(0)}(Z_N)$ with $0 < c_1 < \ldots < c_m = 1$, even less is known. (Compare, however, the numerical experiments given in Eggermont [1984b, pp. 441-443], and the analysis of product integration methods in Cameron & McKee [1985].)

6.4.3. First-kind Abel-type integral equations with nonexact data

So far it has been assumed that the function $g(t)$ occurring in a first-kind Volterra integral equation of the form (5.5.1) (regular kernel) or (6.4.1) (weakly singular kernel) is known exactly (i.e., $g(t)$ is given by some analytical expression). In practical applications, however, $g(t)$ is frequently given in the form of *discrete data* $\{\hat{g}(t_n)\colon n = 0, \ldots, N\}$, where $\hat{g}(t_n) := g(t_n) + \epsilon_n$; here , $g(t_n)$ represents the (unknown) exact value of $g(t)$ at $t = t_n$ (*signal*), and ϵ_n is some measurement error (*noise*). We are therefore faced with two questions: (i) can the values $\{\hat{g}(t_n)\}$ be directly used in a numerical method, and (ii) if instead of a given first-kind integral equation we solve a perturbed equation where $g(t)$ has been replaced by $\hat{g}(t)$, and if $g(t) - \hat{g}(t)$ is small in some sense (e.g., $\|g - \hat{g}\|_\infty < \epsilon$), will the difference between the solution of the exact equation and that of the perturbed equation also be small, and how will a given numerical method react to such a perturbation?

We first illustrate the second question by an example.

Example 6.4.3. Consider the integral equation

$$\int_0^t y(s)ds = g(t), \quad t \in I := [0,T], \ \text{with} \ g \in C^1(I), \ g(0)=0. \qquad (6.4.22a)$$

Its (unique) solution is $y(t)=g'(t)$. The solution of the perturbed equation,

$$\int_0^t y(s)ds = \hat{g}(t), \quad t \in I, \quad \text{with} \quad \hat{g}(t) := g(t) + \epsilon \sin(\omega t), \qquad (6.4.22b)$$

is given by $\hat{y}(t) = g'(t) + \delta(t)$, where $\delta(t) := \epsilon\omega\cos(\omega t)$. Hence,

$$\|y - \hat{y}\|_\infty = \|\delta\|_\infty \leqslant |\epsilon\omega|,$$

where we have equality for suitably chosen I. This means that, given some small $\epsilon > 0$, the norm $\|y - \hat{y}\|_\infty$ can be made arbitrarily large by choosing ω appropriately. In other words, the integral equation (6.4.22a) is *ill-posed*, in the sense that it is very sensitive to perturbations in $g(t)$: arbitrarily small perturbations of g may lead to large perturbations of the solution. The same holds for

$$\frac{1}{\Gamma(1-\alpha)} \int_0^t (t-s)^{-\alpha} y(s)ds = g(t), \quad t \in I, \quad 0 < \alpha < 1, \qquad (6.4.23a)$$

whose solution is given by the fractional derivative of $g(t)$,

$$y(t) = \frac{1}{\Gamma(\alpha)} \int_0^t (t-s)^{\alpha-1} g'(s)ds, \qquad (6.4.23b)$$

provided g satisfies $g \in C^1(I)$, $g(0)=0$ (cf. Theorem 1.3.6).

Suppose now that the solution of (6.4.23a) is approximated by applying fractional convolution quadrature (cf. Section 6.1.2) to the inversion formula (6.4.23b) with $t = t_n$ $(n \geq 0)$. We obtain the approximation

$$y_n = h^\alpha \sum_{j=0}^{n} w_{n,j}(\alpha) g'(t_j) \qquad (6.4.24a)$$

(here, we have assumed that the underlying mesh is uniform, with $h = TN^{-1}$). If the derivative $g'(t)$ is found by some numerical differentiation formula (e.g., a backward differentiation formula) we have

$$g'(t_j) \approx h^{-1} \sum_{l=j-k_0}^{j+k_1} b_{j,l}\, g(t_l), \qquad (6.4.24b)$$

where k_0 and k_1 are given nonnegative integers. Using this approximation for $g'(t_j)$ in (6.4.24a) we are led to the approximation

$$y_n = h^{\alpha-1} \sum_{j=0}^{n} \sum_{l=j-k_0}^{j+k_0} w_{n,j}(\alpha) b_{j,l}\, g(t_l). \qquad (6.4.24c)$$

If the value of $g(t)$ at $t = t_n$ is perturbed by ϵ_n, $\hat{g}(t_n) := g(t_n) + \epsilon_n$, while the remaining values remain unchanged, then there results a perturbed approximation $\hat{y}_n$ given by

$$\hat{y}_n = y_n + h^{\alpha-1}\beta_n\epsilon_n,$$

where the coefficient β_n depends on certain of the values $w_{n,j}(\alpha)$ and $b_{j,l}$. In other words, a single perturbation in the data $\{g(t_n)\}$ is propagated with an amplification factor which is proportional to $h^{\alpha-1}$. On the other hand, it follows from (6.4.24b) that a single perturbation in the data used in a numerical differentiation formula based on finite-difference techniques is propagated with an amplification factor proportional to h^{-1}. Thus, from a numerical point of view the inversion of the Abel equation (6.4.23a) is not as ill-posed as the inversion of (6.4.22a) by numerical differentiation.

In physical modeling processes the Abel-type integral equation

$$2\int_t^T s(s^2-t^2)^{-\frac{1}{2}} y(s)ds = g(t), \quad t \in I, \tag{6.4.25a}$$

plays a dominant role (see the extensive bibliographies in Anderssen [1977b] and in Anderssen & de Hoog [1982]; compare also the survey by Gorenflo & Vessella [1984b]). Here, if $g \in C^1(I)$, $g(T)=0$ the inversion formula has the form

$$y(t) = -\frac{1}{\pi}\int_t^T (s^2-t^2)^{-\frac{1}{2}} g'(s)ds, \tag{6.4.25b}$$

which involves again the derivative of $g(t)$ (consult also Sneddon [1972, pp. 207-211] for a discussion of inversion formulas for different types of first-kind Abel integral equations). It is easy to verify that if finite-difference techniques are employed to evaluate approximations to $y(t_n)$ by means of (6.4.25b), then perturbations in the data $\{g(t_n)\}$ are propagated with an amplification factor which is again proportional to $h^{-1/2}$ (as for the case (6.4.23b) with $\alpha = 1/2$).

These observations show that there will exist a minimum mesh spacing below which h must not be chosen if we want to obtain reliable numerical approximations. This, however, will put a limit on the attainable accuracy (i.e., we are faced with a loss of resolution in a physical problem). On the other hand, the fact that the given data $\{\hat{g}(t_n)\}$ is of observational nature means that, from a statistical point of view, it will be important to use as many abscissas $\{t_n\}$ as possible. Typically for an ill-posed problem, we are thus dealing with two competing claims.

We conclude from these results that, in general, it will not be feasible to approximate the solution of an Abel-type integral equation of the first-kind with nonexact data by applying numerical integration techniques to the corresponding inversion formula, or by treating the equation directly by one of the methods discussed so far. As has been observed by various authors (see, e.g., Anderssen [1973, p. 66]), if a numerical method is to yield reliable results, it must be able to cope with the following two completely different aspects of the problem:

(i) It must take into account the *structure of the given noisy data* $\{\hat{g}(t_n)=g(t_n)+\epsilon_n;\ n=0, \ldots, N\}$ which consists of the unknown exact values (signal) $\{g(t_n)\}$ (of which the derivative is required if the method is based on an inversion formula like (6.4.23b) or (6.4.25b)) and of the measurement error (noise) $\{\epsilon_n\}$. This means that the method will have to involve some preprocessing (smoothing, filtering, etc.) of the given data, in order to separate the signal from the noise.

(ii) It must also take into account the *inherent instability* associated with the *ill-posed nature* of the given first-kind integral equation. As has already been briefly pointed out in the Notes for Section 5.5, this will usually be done by using regularization techniques.

In other words, satisfactory methods for solving Abel-type integral equations like (6.4.25a) where the data is noisy can in general not be derived by appealing solely to theoretical numerical analysis: they will invariably involve tools from statistics.

In recent years the problem of devising suitable numerical methods for Abel-type integral equations of the form (6.4.23a) and, especially, (6.4.25a), has received a considerable amount of attention. Unfortunately, a more detailed description of the various approaches is beyond the scope of the present book since it would entail the introduction of a considerable body of results from, e.g., statistics (statistical fitting strategies, such as least-squares techniques) and stochastic processes and time series (used in methods based on Wiener filtering of the given data, to separate the unknown signal from the noise). Instead, we have to refer the reader to the relevant literature. Here, we should like to cite in particular the illuminating work of Anderssen [1973], [1976], [1977b], and, above all, of Anderssen & de Hoog [1982]; see also the Notes for this section where some additional papers are mentioned.

NOTES

Section 6.1.

The theory (and history) of fractional integration and differentiation is treated, e.g., in Davis [1927a], [1927b], Oldham & Spanier [1974], and Lavoie et al. [1976]; the last reference contains also a discussion of the generalization of finite-difference quotients to the case of fractional differentiation, as well as a summary of formulas and theorems on fractional derivatives.

The theory of numerical fractional integration and differentiation is due to Lubich [1983c], [1985b]. Based on this work Lubich [1985c], [1985b] introduced fractional linear multistep methods and gave convergence and stability results for second-kind integral equations with kernels $(t-s)^{\alpha} f(s,y)$. This analysis can clearly be extended to Volterra integro-differential equations with weakly singular kernels, by applying fractional convolution quadrature to the integral terms occurring in the linear multistep formula for such equations. However, no such convergence theory has been established as yet. The same is true for first-kind integral equation with weakly singular kernels (but compare the approaches suggested by Purga & Preem [1981], by Zheludev [1982], and by Cameron & McKee [1985]).

Section 6.2.

The methods of Huber [1939] and Wagner [1954] (employing, respectively, continuous piecewise polynomials of degree one and two) seem to represent the first attempts to solve numerically second-kind equations with weakly singular kernels. Oulès [1964] gave an example of a Runge-Kutta-type method for a special nonlinear equation with $\alpha=1/2$. High-order block and collocation methods for uniform meshes were derived, assuming the existence of a smooth solution on the (closed) interval I, by Linz [1969a], Garey [1974a], [1974b], El Tom [1974c], Brunner & Evans [1977], Westreich & Cahlon [1980], Cahlon [1981], Brunner & Nørsett [1981] (who showed that, if $0<\alpha<1$, local superconvergence of integer order on $\overline{Z}_N$ is not possible in polynomial spline collocation; see also Brunner [1981]), and Abdalkhani [1982], [1984]. Compare also Bownds [1979], Wood [1980], and Bownds & Wood [1979] (where kernel approximation techniques are combined with collocation and Galerkin methods).

Methods for second-kind integral equations with nonsmooth solutions were studied by de Hoog & Weiss [1973d], [1974] (product integration method of order $p=7/2$ for the case $\alpha=1/2$, based on Simpson's rule; see also Cameron & McKee [1984a] for an extension of this product integration approach), by Logan [1976] (using adaptive meshes; see also Chapter 8), by Lubich [1983b] (who presents the Runge-Kutta theory for weakly singular second-kind equation with nonsmooth solutions), by Kershaw [1982a] (collocation in $S_1^{(0)}(Z_N)$ with $c_1=0$, $c_2=1$); see also Brunner [1984c] for a survey and for additional references.

The theory of graded meshes in the approximation by polynomial splines goes back to Rice [1969]. For the complete theory and additional references (especially to the important work of Burchard and Burchard & Hale) we refer to de Boor [1978] and Schumaker [1981] (where notes on the history of this problem can be found). Graded meshes were then used in the numerical solution of second-kind Fredholm integral equations with weakly singular kernels (where nonsmooth behavior of the solution occurs at both endpoints of the interval of integration) by Chandler [1979], Graham [1980], [1982b] (Galerkin methods), Schneider [1981] (product integration methods), Vainikko & Uba [1981], and Vainikko, Pedas & Uba [1984] (collocation methods).

Theorems 6.2.6 and 6.2.7 may be found in Brunner [1985a]. In Brunner [1985b] these results are generalized to kernels of the form $(t-s)^{\nu}\ k(t,s,y)$, where $\nu=\rho-\alpha$, with $\rho\in\mathbb{N}_0$ and $0<\alpha<1$.

The papers of te Riele [1982] (who first employed nonpolynomial approximants) and of Brunner & te Riele [1984] contain extensive numerical experiments illustrating the performance of nonpolynomial spline collocation, especially when used in conjection with polynomial splines. We note in passing that nonpolynomial functions have recently been used by Doedel & Reddien [1984] to derive high-accuracy finite-difference methods for singular two-point boundary-value problems.

For a completely different class of singular Volterra integral equations, namely equations of the third kind (cf. (1.1.7)), there have been only a few attempts to devise numerical methods: we are aware of the papers by Anselone et al. [1963] and James [1968].

Among the numerous papers dealing with weakly singular second-kind Volterra integral equations occurring in practical applications we cite those of Mann & Wolf [1951], Levinson [1960], Handelsman & Olmstead [1972], Olmstead & Handelsman [1976a], [1976b], Olmstead [1977] (dealing with heat conduction problems and with gas absorption in liquids; see also the monograph by Cannon [1984] concerning the reduction of parabolic partial differential equations to integral equations for the boundary values), Franco, McKee & Dixon [1983] (here, Lighthill's equation is studied which contains a singularity of the form $(t^{3/2}-s^{3/2})^{-2/3}$ in its kernel), Groetsch [1982], Inselberg [1982], and O'Connor [1984]. Jakeman [1984] deals with the numerical inversion of a weakly singular second-kind equation (arising in stereology) whose data are nonexact. Additional references may be found in the bibliography at the end of this book.

Section 6.3.

The literature on the numerical solution of Volterra integro-differential equations with weakly singular kernels is comparatively small. Prosperetti [1977], [1982] introduces methods based on piecewise cubic Hermite interpolation for general second-order integro-differential equations, where care is taken that on a suitable initial interval the nonsmooth solution is approximated accurately. No convergence analysis is given, but these papers contain numerous references to such integro-differential equations occurring in practical applications. McKee [1979c] uses linear multistep methods with variable stepsizes to solve an integro-differential equation arising in a physical modeling process, while McKee & Stokes [1983] apply a related class of methods to equations of the form

$$y'(t) = f(t,y(t)) + \int_0^t (t-s)^{-\alpha} k(t) y'(s) ds.$$

These methods are based on product integration techniques, using backward difference interpolation formulas.

Makroglou [1977], [1981] has derived block methods which may be viewed as extensions of block methods for second-kind integral equations introduced by R. Weiss [1972a]; here, the solution is assumed to be smooth on I. For higher-order linear integro-differential equations Papatheodoru & Jesanis [1980] suggest numerical methods based on finite-element techniques and using piecewise cubic Hermite approximants on uniform meshes. Graded meshes are employed in Brunner [1985c]; here, generalizations of Theorems 6.3.1 and 6.3.2 to integro-differential equations with kernels $(t-s)^{\nu}\, k(t,s,y)$, $\nu = \rho - \alpha$, where $\rho \in \mathbb{N}_0$, $0 < \alpha < 1$, may be found.

Finally, collocation in suitable nonpolynomial spline spaces, analogous to the ones introduced in Definition 6.2.2, has been studied in Brunner [1982b], [1983]. The motivation for using such spaces for integro-differential equations with weakly singular kernels is given by Theorem 1.3.16 and, for rational α, by Corollary 1.3.5.

Section 6.4.

The use of collocation for solving first-kind integral equations with weakly singular kernels dates back to Huber [1939] (who used $S_1^{(0)}(Z_N)$ with $c_1 = 1$). The product trapezoidal and midpoint methods are due to R. Weiss [1972a], [1972b], Weiss & Anderssen [1972] (compare also Anderssen, de Hoog & Weiss [1973]), and Benson [1973]. Eggermont [1979], [1981] improved some of these results by using techniques which better reflect the analogy between the discrete and the given continuous problem. Other product integration methods have been proposed by Branca [1976], [1978] (methods of orders 2 and 3, based on one-point and two-point Gaussian quadrature), and by Cameron & McKee [1983] (who use backward difference interpolation to obtain methods of orders 3 and 4); see also Cameron & McKee [1984b], [1985] for further work on product integration methods.

Collocation methods (and their connection with methods based on product integration) in polynomial spline spaces have been considered in Brunner [1974a], [1977], [1978d], and in Hung [1979b] (who employed the differentiated version of (6.4.1) and collocation in $S_2^{(1)}(Z_N)$ to attain order $p = 3 - \alpha$ on $\bar{Z}_N$; he also showed that if $S_3^{(2)}(Z_N)$ is used as approximating space, then the method is divergent).

As for other types of methods, we mention the papers by Piessens & Verbaeten [1973] (approximation of $g(t)$ by Chebyshev series; inversion on the resulting equation by means of recurrence relations for hypergeometric functions), by Bownds [1976] (conversion to second-kind equation, with subsequent kernel approximation), Chawla & Kumar [1978] (Chebyshev series for $g(t)$ and $K(t,s)$, followed by collocation), and Favella & De Griffi [1981] (least-squares techniques). Lewanowicz [1984] discusses a modification of the method of Piessens & Verbaeten. A brief survey of convergence results for various methods may be found in McKee [1982b]; the paper by te Riele & Schroevers [1983] presents a detailed numerical comparison of the methods by Branca [1978], Chawla & Kumar [1978], and Anderssen, de Hoog & Weiss [1973]. It also introduces nonpolynomial spline collocation for first-kind equations with nonsmooth solutions. Methods dealing with the case $g(0)\neq 0$ (yielding solutions which are unbounded at $t=0$) are discussed in Fettis [1964] and Brunner [1975b].

Still remaining with first-kind equations where $g(t)$ is known exactly, we cite some contributions dealing with Abel-type equations whose kernels are either of the form $(t^2-s^2)^{-1/2}$ (with integration from 0 to t), or $s(s^2-t^2)^{-1/2}$ (with integration from t to T). Linz [1967], [1968] applied finite-difference techniques either to the inversion formula of the equation, or directly to the equation itself. The inversion formulas for the above (and for more general) types of Abel equations form also the basis of the methods introduced by Malinowski & Smarzewski [1978], Smarzewski & Malinowski [1978], [1979a], [1979b], [1983]: here, $g(t)$ is approximated by an interpolating or smoothing spline function of degree $m\geqslant 3$. Brunner [1973b], [1975a] uses collocation and uniform approximation techniques, while Atkinson [1974b] shows that application of the product trapezoidal method to the first of the above equations yields error bounds of the form $Ch^2\log(T/h)$. Hung [1979a] based collocation in $S_2^{(1)}(Z_N)$ on the differentiated form of the equation and attains convergence of order $p=5/2$ on $\overline{Z}_N$.

Turning to Abel-type integral equations with nonexact (noisy) data we once more point to the central and illuminating contributions of Anderssen [1973], [1976], [1977a], [1977b], [1980], and of Anderssen & de Hoog [1982]; here, the paper [1977b] contains not only an extensive list of references but classifies Abel-type integral equations arising in applications on the basis of the subject areas (pp. 26-37). Some further early references to Abel equations occurring in applications may be found in Brunner [1975a] and in Gorenflo & Vessella [1984a], [1984b]. Ugniewski [1977] presents a comparative study of various numerical methods; in addition see also, e.g., Piessens [1973], Kosarev [1973],

[1980], Anderssen & Jakeman [1974], [1975], Jakeman [1975], Jakeman & Anderssen [1975], and Gorenflo [1979] (who surveys numerical methods for Abel-type equations). Jakeman [1984] studied the numerical inversion of a second-kind equation (analogous to (6.4.25a), with $T=\infty$) whose data are noisy. Consult also the paper by Kanatani & Ishikawa [1985] for similar work. Finally, we mention a related problem which often arises when a physical model involving an Abel equation is considered: frequently, one is not directly interested in the solution of the equation but in some functionals containing the solution (e.g., certain moments). This may give rise to additional numerical difficulties. The reader interested in such problems is referred to Anderssen [1980] for a detailed introduction to the subject, and to Jakeman [1980] for a package of programs for problems of this kind (compare also Table 8.4.3).

CHAPTER 7

NUMERICAL STABILITY

In this chapter we shall deal with the problem of stability of numerical solutions (on unbounded intervals) to Volterra integral and integro-differential equations. Since the qualitative theory of general discrete Volterra equations is still in its infancy, most of the analysis of numerical stability will be carried out with respect to particular classes of test equations for which the discrete Volterra equations resulting from the application of the previously discussed numerical methods can be reduced to recurrence relations of (fixed) finite length. These test problems are related to ordinary differential equations; we thus first review a number of relevant concepts and results from the stability theory for ODEs. Even though there exist some recent results on the stability of the solutions of certain discrete Volterra equations not reducible to fixed-length recurrence relations (which will be presented, too), much work remains to be done towards a satisfactory and comprehensive theory of numerical stability for Volterra equations.

7.1. Stability of the exact solution

In the previous chapters we have analyzed the properties of solutions to Volterra equations and their discretizations on a bounded interval $[0,T]$; the resulting quantitative theory dealt primarily with sufficient conditions for the existence and uniqueness of solutions and with the question of the convergence of approximate solutions, as the mesh diameter h approaches zero. In applications this quantitative approach will usually have to be complemented by a study of the asymptotic behavior of these solutions as $T \to \infty$; here, approximate solutions are obtained on meshes with fixed diameters $h > 0$. This *qualitative theory* describing the sensitivity of the solution of a given Volterra equation (or of a corresponding discretization obtained by a numerical method) under perturbations of the data characterizing the equation is generally referred to as *stability theory.*

The aim of the subsequent sections will be to determine under which conditions and to what extent an approximate solution defined by a particular discretization of a Volterra integral or integro-differential equation exhibits the same perturbation sensitivity as the original (exact) solution. To this end we review a number of basic stability concepts, both for ordinary differential equations and for Volterra equations, which will be relevant in that analysis.

7.1.1. Definition of stability

Consider first a system of ODEs,

$$y'(t) = f(t, y(t)), \quad t_0 \leqslant t < \infty,$$

where $t_0 \geqslant 0$, $f:\mathbb{R}_+ \times D \to \mathbb{R}^m$, with $D := \{y \in \mathbb{R}^m : \|y\| < a\}$. Here, $a > 0$ is a suitable real number, and $\|\cdot\|$ denotes a norm in $\mathbb{R}^m$. Let $y(t) := y(t; t_0, y_0)$ be a solution of this system of ODEs which satisfies the initial condition $y(t_0) = y_0$.

Definition 7.1.1. (i) Let the solution $y(t) = y(t; t_0, y_0)$ be defined on $t_0 \leqslant t < \infty$; (ii) let the point $(t, y(t))$ belong to $\mathbb{R}_+ \times D$ for all $t \geqslant t_0$. Then $y(t)$ is said to be *stable* (over $[t_0, \infty)$) if

(a) there exists a $\gamma > 0$ such that every solution $y^*(t) = y^*(t; t_0, y_0^*)$ satisfies (i) and (ii) whenever $\|y_0 - y_0^*\| \leqslant \gamma$;

(b) given $\epsilon > 0$, there exists a $\delta > 0$, $0 < \delta < \gamma$, such that

$$\|y_0 - y_0^*\| \leqslant \delta \Rightarrow \|y(t; t_0, y_0) - y^*(t; t_0, y_0^*)\| \leqslant \epsilon$$

for $t_0 \leqslant t < \infty$.

A solution which is not stable is called *unstable*. □

In this definition δ will, in general, not only depend on ϵ but also on t_0: a solution $y^*(t)$ remains in an ϵ-neighborhood of $y(t)$ if it is sufficiently close to $y(t_0)$ at $t = t_0$; other solutions may enter and leave the "tube" corresponding to this ϵ-neighborhood at later times.

Definition 7.1.2. Let $y(t) = y(t; t_0, y_0)$ satisfy the conditions (i) and (ii) of Definition 7.1.1. Then $y(t)$ is said to be *uniformly stable* (over $[t_0, \infty)$), if, for given $\epsilon > 0$, there exists a $\delta > 0$ such that any solution $z(t)$ satisfying $\|y(t_1) - z(t_1)\| \leqslant \delta$ for some $t_1 \geqslant t_0$ exists and satisfies $\|y(t) - z(t)\| \leqslant \epsilon$ for all $t \geqslant t_1$. □

Note that now δ does not depend on t_1: once a solution enters the "tube" given by the ϵ-neighborhood of $y(t)$, it remains there for all $t \geqslant t_1$, in contrast to the case where $y(t)$ is only stable over $[t_0, \infty)$.

Definition 7.1.3. Let $y(t) = y(t; t_0, y_0)$ satisfy the conditions (i) and (ii) of Definition 7.1.1. Then $y(t)$ is said to be *asymptotically stable* (over $[t_0, \infty)$) if it is stable over $[t_0, \infty)$ and if there exists a $\delta' > 0$, $0 < \delta' < \gamma$, such that

$$\|y_0 - y_0^*\| \leqslant \delta' \Rightarrow \lim_{t \to \infty} \|y(t; t_0, y_0) - y^*(t; t_0, y_0)\| = 0. \quad \square$$

A detailed discussion of these and other stability concepts not mentioned here may be found, e.g., in Coppel [1965], Sanchez [1968], or Burton [1983b].

Before extending the above definitions to Volterra integral equations we note that if the function **f** characterizing the given system of ODEs is continuous and the solutions are uniquely determined by their initial values $\mathbf{y}_0$, then the solutions depend continuously on their initial values over any finite interval; stability over $[t_0,\infty)$ is then equivalent to stability over $[t_1,\infty)$ for any $t_1 \geq t_0$. Moreover, if the system of ODEs is linear,

$$\mathbf{y}'(t) = A(t)\mathbf{y}(t)+\mathbf{a}(t),$$

then a solution of this equation is (uniformly, asymptotically) stable if, and only if, the same is true for the null solution of the corresponding homogeneous equation. This holds since the above definitions involve only the difference between neighboring solutions. In this case the terminologies "stable solution" and "stable ODEs" are often used interchangeably.

The above initial-value problem is equivalent to the Volterra equation

$$\mathbf{y}(t) = \mathbf{y}_0 + \int_{t_0}^{t} \mathbf{f}(s,\mathbf{y}(s))ds, \quad t \geq t_0. \tag{7.1.1}$$

Since in a general Volterra integral equation the role of the initial value $\mathbf{y}_0$ is assumed by the function $\mathbf{g}_0$ in

$$\mathbf{y}(t) = \mathbf{g}_0(t) + \int_{t_0}^{t} \mathbf{k}(t,s,\mathbf{y}(s))ds, \quad t \geq t_0 \geq 0 \tag{7.1.2}$$

(in order to emphasize the analogy with (7.1.1) we have changed our earlier notation $\mathbf{g}$ to $\mathbf{g}_0$), we shall have to study the sensitivity of solutions to (7.1.2) under perturbations of $\mathbf{g}_0$ on the interval $[t_0,\infty)$. Hence, let

$$P(\mathbb{R}_+) := \{\mathbf{g} \in C(\mathbb{R}_+): \|\mathbf{g}\|_\infty := \sup_{t \in \mathbb{R}_+} \|\mathbf{g}(t)\|\}$$

be the normed space of *admissible perturbations* to $\mathbf{g}_0$, and let $\mathbf{k}$ in (7.1.2) be defined on $S \times D$, where $S := \{(t,s)\colon t_0 \leq s \leq t < \infty\}$, $D := \{\mathbf{y} \in \mathbb{R}^m\colon \|\mathbf{y}\| < a\}$ for some $a > 0$. Compare also Bownds & Cushing [1975a] for other choices of spaces of admissible perturbations and for the following definition.

Definition 7.1.4. (i) Let the solution $\mathbf{y}(t) = \mathbf{y}(t;\mathbf{g}_0)$ to (7.1.2) corresponding to the initial function $\mathbf{g}_0$ be defined for all $t \geq t_0$; (ii) let the point $(t,\mathbf{y}(t))$ belong to $\mathbb{R}_+ \times D$ for all $t \geq t_0$. Then $\mathbf{y}$ is said to be *stable on* $P(\mathbb{R}_+)$ (over $[t_0,\infty)$) if

(a) there exists a $\gamma > 0$ such that every solution $\mathbf{y}^*(t) = \mathbf{y}^*(t;\mathbf{g}_0^*)$ corresponding to $\mathbf{g}_0^* \in P(\mathbb{R}_+)$ satisfies (i) and (ii) whenever $\|\mathbf{g}_0 - \mathbf{g}_0^*\| \leq \gamma$; and

(b) given $\epsilon > 0$, there exists a $\delta > 0$, $0 < \delta < \gamma$, such that

$$\|\mathbf{g}_0 - \mathbf{g}_0^*\|_\infty \leq \delta \Rightarrow \|\mathbf{y} - \mathbf{y}^*\|_\infty < \epsilon.$$

This solution $\mathbf{y}(t)$ is called *uniformly stable on* $P(\mathbb{R}_+)$ (over $[t_0,\infty)$) if the above δ can be found independent of t_0. It is called *asymptotically stable on* $P(\mathbb{R}_+)$ (over $[t_0,\infty)$) if it is stable on $P(\mathbb{R}_+)$ and if, in addition, for each $t_0 \geq 0$ there exists a $\delta' > 0$, $0 < \delta' \leq \gamma$, such that

$$\|\mathbf{g}_0 - \mathbf{g}_0^*\|_\infty \leq \delta' \Rightarrow \lim_{t\to\infty} \|\mathbf{y}(t;\mathbf{g}_0) - \mathbf{y}^*(t;\mathbf{g}_0^*)\| = 0. \quad \square$$

In the case (7.1.1) corresponding to a system of ODEs the space of admissible perturbations for $\mathbf{y}_0$ is, according to Definitions 7.1.1 to 7.1.3, a special subspace of $P(\mathbb{R}_+)$, namely the space of all *constant* functions $\mathbf{g}(t)=\mathbf{c}$; here, $\|\mathbf{g}\|_\infty=\|\mathbf{g}\|=\|\mathbf{c}\|$.

The stability concepts introduced above are readily extended to *Volterra integro-differential* equations, either by modifying Definitions 7.1.1 to 7.1.3 in an obvious manner, or by rewriting the integro-differential equation as a system of Volterra integral equations of the second kind (cf. (3.4.1')). We refrain from formulating these definitions; compare also, e.g., Burton [1983b], or Burton & Mahfoud [1983].

It is beyond the scope of this book to present the general stability theory for Volterra equations; the interested reader is referred to the Notes at the end of this chapter for a short guide to the relevant literature. Instead, we shall discuss in some detail perturbation results for two important classes of Volterra equations which are reducible to certain systems of ODEs, namely, equations with polynomial convolution kernels and equations with finitely decomposable kernels. The motivation for choosing this approach lies in the fact that, at present, the qualitative theory and, especially, the stability theory for general (systems of) discrete Volterra equations still remains to be established; i.e., the tools for determining conditions under which a numerical method used to approximate the (uniformly, asymptotically) stable solution of a given general Volterra equation yields a solution which possesses the same stability property are still lacking. The exception to this are integral and integro-differential equations with convolution kernels integrable on $\mathbb{R}_+$ (compare Theorems 1.4.1 and 1.4.2): here, a qualitative theory for corresponding discrete equations has recently been presented by Lubich [1983a] and by Hairer & Lubich [1984] (see also Lubich [1985a]).

7.1.2. Differential equations with perturbed right-hand side

In many important cases the solution of the Volterra equation can be shown to be related to the solution of an initial-value problem for a system of ODEs, say

$$\mathbf{u}'(t) = \mathbf{f}(t,\mathbf{u}(t)), \quad \mathbf{u}(0) = \mathbf{u}_0, \tag{7.1.3}$$

where the right-hand side function depends on the initial function of the Volterra equation. In Section 7.1.4 and 7.1.5 we shall discuss the cases of *polynomial convolution kernels* and of *finitely decomposable kernels* (or *separable kernels*). In these cases the solution can either be expressed explicitly in terms of $\mathbf{u}(t)$ or it occurs as one of the components of $\mathbf{u}(t)$.

We shall investigate the condition that the solution $\mathbf{u}$ of (7.1.3) be *stable with respect to perturbations of the right-hand side function* $\mathbf{f}$. We observe that this requirement is much stronger than the usual stability condition in ODEs, where one requires that the solution be *stable with respect to perturbations of the initial vector* $\mathbf{u}_0$. In fact, it may well be that we do require too much. For instance, if y is just one of the components of $\mathbf{u}$, we need not insist on stability for all the components of $\mathbf{u}$, but only for the one that is identical to y (see Section 7.1.5.3). On the other hand, if the formula that expresses y in terms of $\mathbf{u}$ is ill-conditioned, then stability of $\mathbf{u}$ does not imply stability of y (see Section 7.1.5.1). In spite of these drawbacks of the ODE approach, we gain a lot of insight into the stability behavior of solutions of Volterra equations.

In Frank et al. [1978] (see also Dekker & Verwer [1984, p. 15]), a stability result is derived that precisely describes the effect of perturbing $\mathbf{f}$ in (7.1.3). In order to formulate this result we need the inner product $<\mathbf{u},\mathbf{v}>$ of the vectors $\mathbf{u}$ and $\mathbf{v}$, and the norm $\|\mathbf{u}\|_2 = \sqrt{<\mathbf{u},\mathbf{u}>}$. In a slightly modified form the stability result of Frank et al. [1978] is as follows.

Theorem 7.1.1. *Let* $\mathbf{f}$ *and* $\mathbf{f}^*$ *be continuous in the domain of definition D and let the following inequalities be satisfied in D:*

$$<\mathbf{f}(t,\mathbf{v}_1)-\mathbf{f}(t,\mathbf{v}_2),\mathbf{v}_1-\mathbf{v}_2> \leqslant M(t)\|\mathbf{v}_1-\mathbf{v}_2\|_2^2, \tag{7.1.4a}$$

$$\|\mathbf{f}(t,\mathbf{v})-\mathbf{f}^*(t,\mathbf{v})\|_2 \leqslant \delta(t). \tag{7.1.4b}$$

Then the exact solutions $\mathbf{u}$ *and* $\mathbf{u}^*$ *respectively corresponding to the right-hand side functions* $\mathbf{f}$ *and* $\mathbf{f}^*$, *and to the initial vectors* $\mathbf{u}_0$ *and* $\mathbf{u}_0^*$, *satisfy the inequality*

$$\|\mathbf{u}(t)-\mathbf{u}^*(t)\|_2 \leqslant \|\mathbf{u}_0-\mathbf{u}_0^*\|_2\exp(\int_0^t M(\tau)d\tau)+\int_0^t \delta(\tau)\exp(\int_\tau^t M(s)ds)d\tau. \tag{7.1.5}$$

Proof. For the solutions $\mathbf{u}$ and $\mathbf{u}^*$, condition (7.1.4a) yields

$$<\mathbf{f}(t,\mathbf{u}(t))-\mathbf{f}(t,\mathbf{u}^*(t)),\mathbf{u}(t)-\mathbf{u}^*(t)> \leqslant M(t)\|\mathbf{u}(t)-\mathbf{u}^*(t)\|_2^2.$$

Hence,

$$<\mathbf{u}'(t)-(\mathbf{u}^*(t))'-\boldsymbol{\delta}(t,\mathbf{u}^*(t)),\mathbf{u}(t)-\mathbf{u}^*(t)> \leqslant M(t)\|\mathbf{u}(t)-\mathbf{u}^*(t)\|_2^2, \tag{7.1.6}$$

where the prime denotes differentiation with respect to t, and where we have defined

$$\boldsymbol{\delta}(t,\mathbf{v}) := \mathbf{f}(t,\mathbf{v})-\mathbf{f}^*(t,\mathbf{v}).$$

Observing that

$$<\mathbf{v}'(t),\mathbf{v}(t)> \equiv \|\mathbf{v}(t)\|_2\frac{d}{dt}\|\mathbf{v}(t)\|_2,$$

inequality (7.1.6) can be written as

$$\|\mathbf{u}(t)-\mathbf{u}^*(t)\|_2\frac{d}{dt}\|\mathbf{u}(t)-\mathbf{u}^*(t)\|_2-<\boldsymbol{\delta}(t,\mathbf{u}^*(t)),\mathbf{u}(t)-\mathbf{u}^*(t)>$$
$$\leqslant M(t)\|\mathbf{u}(t)-\mathbf{u}^*(t)\|_2^2.$$

Using the Schwarz inequality (see, e.g., Lancaster [1969, p. 10]) we obtain, by (7.1.4),

$$\|\mathbf{u}(t)-\mathbf{u}^*(t)\|_2[\frac{d}{dt}\|\mathbf{u}(t)-\mathbf{u}^*(t)\|_2-\delta(t)-M(t)\|\mathbf{u}(t)-\mathbf{u}^*(t)\|_2]\leqslant 0,$$

that is,

$$\frac{d}{dt}\|\mathbf{u}(t)-\mathbf{u}^*(t)\|_2\leqslant M(t)\|\mathbf{u}(t)-\mathbf{u}^*(t)\|_2+\delta(t).$$

We compare the solution $\|\mathbf{u}(t)-\mathbf{u}^*(t)\|_2$ of this differential inequality with the solution of the differential equation

$$v'(t) = M(t)v(t)+\delta(t),$$

which is given by

$$v(t) = v(t_0)\exp(\int_0^t M(\tau)d\tau)+\int_0^t \delta(\tau)\exp(\int_\tau^t M(s)ds)d\tau.$$

Choosing $v(t_0)=\|\mathbf{u}_0-\mathbf{u}_0^*\|_2$ we conclude that $\|\mathbf{u}(t)-\mathbf{u}^*(t)\|_2\leqslant v(t)$ (cf. Coppel [1965, p. 29]), which yields the inequality (7.1.5). □

The condition (7.1.4a) is called the *one-sided Lipschitz condition.* It enables us to derive the rather sharp estimate (7.1.5) which is less pessimistic (since $M(t)$ can be negative) than the usual estimate based on the classical Lipschitz constant of the function $\mathbf{f}(t,\mathbf{u})$ (see, e.g., Kamke [1977]).

In the theorem, it is assumed that the exact solution curve $(t,\mathbf{u}(t))$ and the perturbed solution curve $(t,\mathbf{u}^*(t))$ both remain in the domain D. This implies that the space of perturbations, occurring in the stability theorems based on Theorem 7.1.1, is related to the domain of definition of the equation (7.1.3). At first sight, one might be inclined to choose the domain D as large as possible. On the other hand, however, if $\mathbf{f}$ and $\mathbf{f}^*$ are nonlinear in $\mathbf{u}$, the function $M(t)$ will usually assume smaller values as D is smaller. (Recall the analogous situation encountered in Theorem 1.3.9.)

Theorem 7.1.1 simplifies in the case of *linear* differential equations of the form

$$\mathbf{u}'(t) = A(t)\mathbf{u}(t)+\mathbf{a}(t), \tag{7.1.7}$$

where $A(t)$ is a matrix and $\mathbf{a}(t)$ a vector function. Let $Y(t)$ be the *fundamental matrix* for the homogeneous form of (7.1.7); that is, $Y(t)$ satisfies the equation $Y'(t)=A(t)Y(t)$ and the columns of $Y(t)$ form a fundamental system for (7.1.7). For instance, if A is a constant matrix, then $Y(t)=\exp(tA)$. Then, by *variation of constants,* the general solution of (7.1.7) is given by

$$\mathbf{u}(t) = Y(t)Y^{-1}(0)\mathbf{u}_0+Y(t)\int_0^t Y^{-1}(\tau)\mathbf{a}(\tau)d\tau. \tag{7.1.8}$$

From this explicit expression for the solution we derive the following theorem:

Theorem 7.1.2. *Let (7.1.3) be of the linear form (7.1.7), and let* $\boldsymbol{\delta}(t) := \mathbf{a}(t) - \mathbf{a}^*(t)$. *Then*

$$\mathbf{u}(t) - \mathbf{u}^*(t) = Y(t)Y^{-1}(0)(\mathbf{u}_0 - \mathbf{u}_0^*) + \int_0^t Y(t)Y^{-1}(\tau)\boldsymbol{\delta}(\tau)d\tau, \tag{7.1.9}$$

where $\mathbf{u}$ *and* $\mathbf{u}^*$ *correspond to* $\mathbf{a}$ *and* $\mathbf{a}^*$, *respectively.* □

It immediately follows from this theorem that *linear* differential equations, in which the matrix A is constant (so that $Y(t) = \exp(tA)$) and possesses eigenvalues in the *left half-plane,* have solutions that are stable with respect to *arbitrary, bounded* perturbations of the *inhomogeneous term* $\mathbf{a}(t)$. Furthermore, if $\boldsymbol{\delta}(t) \equiv 0$ for $\bar{t} \leqslant t < \infty$, that is, if we consider perturbations on a finite interval, then it follows from Theorem 7.1.2 that we have stability if $Y(t)$ is bounded as $t \to \infty$. Since most stability test equations used in the analysis of discrete Volterra equations can be associated with ODEs of the form (7.1.7), Theorem 7.1.2 forms a starting point in establishing stability criteria for these test equations.

7.1.3. The logarithmic norm

It follows from Theorem 7.1.1 that *general* differential equations which satisfy condition (7.1.3) with a *negative function* $M(t)$ for $t \geqslant t_1$ have solutions which are stable with respect to *arbitrary, bounded* perturbations of the *right-hand side function.* In order to determine the function $M(t)$, we introduce the following "measure" of a matrix A (see Lozinskiĭ [1958], Dahlquist [1959]):

Definition 7.1.5. Let $\|\cdot\|$ be some norm. Then

$$\mu(A) := \lim_{h \downarrow 0} \frac{\|I + hA\| - 1}{h}$$

is called the *logarithmic norm of the matrix A (with respect to the given norm).* □

The logarithmic norm can attain negative values and is therefore not a norm in the usual sense. In Table 7.1.1 a few important properties of $\mu(A)$ are listed (cf. Dekker & Verwer [1984, p. 31] and the references given there). Here, $\mu_p(A)$ corresponds to the logarithmic norms of A using the matrix norm $\|\cdot\|_p$. Furthermore, $\alpha(A)$ denotes an eigenvalue of $A = (a_{i,j})$.

Table 7.1.1. Properties of the logarithmic norm $\mu(A)$

(i)	$-\|A\| \leqslant -\mu(-A) \leqslant \mathrm{Re}(\alpha(A)) \leqslant \mu(A) \leqslant \|A\|$
(ii)	$\mu(aA) = a\mu(A),\ a \geqslant 0$
(iii)	$\mu(A+aI) = \mu(A)+a,\ a \in \mathbb{R}$
(iv)	$\max\{\mu(A)-\mu(-B), -\mu(-A)+\mu(B)\} \leqslant \mu(A+B) \leqslant \mu(A)+\mu(B)$
(v)	$\mu(aA+(1-a)B) \leqslant a\mu(A)+(1-a)\mu(B),\ a \in [0,1]$
(vi)	$\mu(A)-\mu(B) \leqslant \max\{\|\mu(A-B)\|, \|\mu(B-A)\|\} \leqslant \|A-B\|$
(vii)	$\|A\mathbf{v}\| \geqslant \max\{-\mu(-A), -\mu(A)\}\|\mathbf{v}\|$
(viii)	$\|A^{-1}\|^{-1} \geqslant \max\{-\mu(-A), -\mu(A)\}$, A nonsingular
(ix)	$\mu_1(A) = \max_j(a_{j,j} + \sum_{i \neq j} \|a_{i,j}\|)$
(x)	$\mu_2(A) = \frac{1}{2}\max \alpha(A+A^T)$
(xi)	$\mu_\infty(A) = \max_i(a_{i,i} + \sum_{j \neq i} \|a_{i,j}\|)$

Using the concept of the logarithmic norm, Frank et al. [1978] proved the following theorem.

Theorem 7.1.3. (a) *Let* $\mathbf{f}(t,\mathbf{u}) := A\mathbf{u}$, *where A is a constant matrix. Then* $\mathbf{f}$ *satisfies the one-sided Lipschitz condition (7.1.4a) with* $M(t) = \mu_2(A)$.
(b) *Let* $\mathbf{f}(t,\mathbf{u})$ *be continuously differentiable with respect to* $\mathbf{u}$ *for all* $(t,\mathbf{u}) \in D$. *Then* $\mathbf{f}$ *satisfies the one-sided Lipschitz condition (7.1.4a) in D with* $M(t) = \sup_D \mu_2(\partial \mathbf{f}(t,\mathbf{u})/\partial \mathbf{u})$.
(c) *The expressions for* $M(t)$ *in (a) and (b) are optimal with respect to the domain of definition D of* $\mathbf{f}$. □

Corollary 7.1.1. *Let* $\mathbf{f}(t,\mathbf{u})$ *be continuously differentiable with respect to* $\mathbf{u}$ *on D and let* $\mathbf{f}$ *satisfy condition (7.1.4b) on D with* $|\delta(t)| \leqslant \bar{\delta}$. *Then*

$$\|\mathbf{u}(t)-\mathbf{u}^*(t)\|_2 \leqslant e^{Lt}\|\mathbf{u}_0-\mathbf{u}_0^*\|_2 + \frac{\bar{\delta}}{L}(e^{Lt}-1),$$

where L is the one-sided Lipschitz constant defined by

$$L := \frac{1}{2}\sup_D \max_j \alpha_j\Big(\frac{\partial \mathbf{f}}{\partial \mathbf{u}}(t,\mathbf{u}) + \frac{\partial \mathbf{f}^T}{\partial \mathbf{u}}(t,\mathbf{u})\Big),$$

$\alpha_j(A)$ *denoting the j-th eigenvalue of A.*

Proof. It follows from Theorem 7.1.1 and from Theorem 7.1.3(b) that

$$\|\mathbf{u}(t)-\mathbf{u}(t^*)\|_2 \leqslant \|\mathbf{u}_0-\mathbf{u}_0^*\|_2 \exp\Big[\int_0^t L\,d\tau\Big] + \int_0^t \delta(\tau)\exp\Big[\int_\tau^t L\,ds\Big]d\tau$$

$$= e^{Lt}\|\mathbf{u}_0-\mathbf{u}_0^*\|_2 + \int_0^t e^{L(t-\tau)}\delta(\tau)\,d\tau$$

$$\leqslant e^{Lt}\|\mathbf{u}_0-\mathbf{u}_0^*\|_2+\bar{\delta}\int_0^t e^{L(t-\tau)}d\tau.$$

The estimate stated in the corollary follows now immediately. □

This corollary implies that we have stability with respect to *bounded perturbations* of both the initial vector $\mathbf{u}_0$ and of the right-hand side function, provided that the Lipschitz constant L is *negative.* If $L=0$, then we have stability with respect to perturbations of $\mathbf{f}$ of the form $\boldsymbol{\delta}(t)=\epsilon\mathbf{d}(t)$, where $\mathbf{d}\in L^1(\mathbb{R}_+)$ and where ϵ is the perturbation parameter. Such perturbations will be called *integrable perturbations.*

7.1.4. Polynomial convolution kernels

In the following we restrict our considerations to scalar equations; the extension to systems of Volterra equations is straightforward. Furthermore, we will write the Volterra integral equation in the form

$$\theta y(t) = g_0(t)+\int_0^t k(t,s,y(s))ds, \tag{7.1.10}$$

where $\theta=0$ and $\theta=1$ yield an equation of the first and second kind, respectively. The integro-differential equation will be written in the form

$$\begin{aligned} y'(t) &= f_0(t,y(t),g_0(t)+z(t)), \quad y(0) = y_0, \\ z(t) &= \int_0^t k(t,s,y(s))ds; \end{aligned} \tag{7.1.11}$$

the function $g_0(t)$ is inserted in order to indicate explicitly that the solution depends not only on the initial value y_0 but, just as the solution of (7.1.10), also on an initial function. We recall that (7.1.11) can also be represented by a system of two second-kind Volterra equations:

$$\begin{aligned} y(t) &= y_0+\int_0^t f_0(t,y(s),g_0(s)+z(s))ds \\ z(t) &= \int_0^t k(t,s,y(s))ds \end{aligned} \tag{7.1.11'}$$

An important class of Volterra equations, which can be reduced to initial-value problems for ODEs, has kernels of so-called *polynomial convolution* type (cf. (1.1.18)):

$$k(t,s,y) = \sum_{i=0}^{m}\frac{1}{i!}(t-s)^i\phi_i(s,y), \quad m\geqslant 1. \tag{7.1.12}$$

For instance, by (repeated) differentiation, the integral equation (7.1.10) reduces to:

$$\begin{aligned} \theta y'(t) &= g_0'(t)+\phi_0(t,y) \quad \text{if} \quad k(t,s,y) = \phi_0(s,y) \\ \theta y''(t) &= g_0''(t)+\phi_1(t,y) \quad \text{if} \quad k(t,s,y) = (t-s)\phi_1(s,y) \end{aligned}, \tag{7.1.13}$$

and, similarly, the integro-differential equation (7.1.11) reduces to:

$$\begin{aligned} y' &= f_0(t,y,z+g_0), \quad z' = \phi_0(t,y) \quad \text{if} \quad k(t,s,y) = \phi_0(s,y) \\ y' &= f_0(t,y,z+g_0), \quad z'' = \phi_1(t,y) \quad \text{if} \quad k(t,s,y) = (t-s)\phi_1(s,y) \end{aligned}. \tag{7.1.14}$$

7.1.4.1. *Linear equations*

Generally, an integral equation with a kernel of polynomial convolution form (7.1.12) will lead to an ODE of order m or $m+1$. In the case of *linear* polynomial convolution kernels the following result can be proved by using Theorem 7.1.2.

Theorem 7.1.4. *Let the kernel k be of the form (7.1.12) with* $\phi_i = i!c_i y$, *and let*

$$P(a;\zeta) := a\zeta^{m+1} - \sum_{i=0}^{m} i!c_i\zeta^{m-i}, \quad c_m \neq 0. \tag{7.1.15}$$

(a) *The solution of the integral equation (7.1.10) is stable with respect to bounded perturbations in* $C^{m+1}(\mathbb{R}_+)$ *if the polynomial* $P(\theta;\zeta)$ *has all its zeros in the left half-plane.*

(b) *The solution of the integro-differential equation (7.1.11) with* $f_0 = \xi y + \eta z + g_0$ *is stable with respect to bounded perturbations in* $C^{m+1}(\mathbb{R}_+)$ *if the polynomial* $\zeta^{m+2} + \eta P(-\xi/\eta;\zeta)$ *has all its zeros in the left half-plane.*

Proof. (a) Substitution of (7.1.12) with $\phi_i = i!c_i y$ into (7.1.10) and repeated differentiation yields the differential equation

$$\theta y^{(m+1)} - \sum_{i=0}^{m} i!c_i y^{(m-i)} = g_0^{(m+1)}, \quad \theta \text{ constant.}$$

Writing this equation in the form (7.1.7), the solution $\mathbf{u}(t)$ is given by (7.1.8) with $Y(t) = \exp(tA)$, where A is a constant matrix whose eigenvalues α satisfy the equation

$$\theta\alpha^{m+1} - \sum_{i=0}^{m} i!c_i\alpha^{m-i} = 0;$$

that is, the eigenvalues are the zeros of the polynomial $P(\theta;\zeta)$. Applying Theorem 7.1.2 we find that

$$\mathbf{u}(t) - \mathbf{u}^*(t) = e^{tA}(\mathbf{u}_0 - \mathbf{u}_0^*) + \int_0^t e^{(t-\tau)A}\delta(\tau)d\tau, \tag{7.1.16}$$

where $\boldsymbol{\delta}(t) := g_0^{(m+1)}(t) - g_0^{*(m+1)}(t)$. Part (a) of the theorem follows easily from (7.1.16).

(b) Substitution of (7.1.12) with $\phi_i = i!c_i y$ and of $f_0 = \xi y + \eta z + g_0$ into (7.1.11), and repeated differentiation yields

$$y' = \xi y + \eta z + g_0, \quad z^{(m+1)} - \sum_{i=0}^{m} i!c_i y^{(m-i)} = 0.$$

Elimination of the function z furnishes the differential equation

$$y^{(m+2)} - \xi y^{(m+1)} - \eta \sum_{i=0}^{m} i!c_i y^{(m-i)} = g_0.$$

Part (b) can now be proved along the lines of the proof of part (a). □

We remark that the conditions of Theorem 7.1.4 can be weakened in a few special cases. For instance, if the matrix A in (7.1.16) has a complete eigensystem with one or more zero eigenvalues, then we can still prove stability with respect to perturbations in $C^{m+1}(\mathbb{R}_+) \cap L^1(\mathbb{R}_+)$, provided that no eigenvalues are in the right half-plane.

The polynomials $P(\theta;\zeta)$ and $\zeta^{m+2} + \eta P(-\xi/\eta;\zeta)$ may be considered as characteristic polynomials associated with the Volterra equation. In order to decide whether the zeros of the characteristic polynomials are located in the left half-plane, we will use the well-known *Routh-Hurwitz criterion* (see, e.g., Coppel [1965], Lancaster [1969], Sanchez [1968]). For future reference, this criterion is given below:

Lemma 7.1.1. *Let the polynomial* $P(\zeta)$ *be defined by*

$$P(\zeta) = a_0\zeta^m + a_1\zeta^{m-1} + \ldots + a_m, \quad a_0 > 0,$$

and define the n by n Hurwitz matrices

$$Q_n := \begin{pmatrix} a_1 & a_3 & a_5 & \cdots & a_{2n-1} \\ a_0 & a_2 & a_4 & \cdots & a_{2n-2} \\ 0 & a_1 & a_3 & \cdots & a_{2n-3} \\ 0 & a_0 & a_2 & \cdots & a_{2n-4} \\ \vdots & & & & \vdots \\ 0 & 0 & & \cdots & a_n \end{pmatrix}. \tag{7.1.17}$$

Then the zeros of $P(\zeta)$ *lie all in the left half-plane, if, and only if,*

$$\det Q_n > 0, \quad n = 1, \ldots, m. \quad \square$$

It can be deduced from this criterion that, necessarily,

$$a_j > 0, \quad j = 0, \ldots, m. \tag{7.1.18a}$$

For $m > 2$, we have additional conditions: e.g.,

$$m=3:\ a_1a_2-a_3a_0>0,$$
$$m=4:\ a_1a_2a_3-a_0a_3^2-a_4a_1^2>0. \tag{7.1.18b}$$

Example 7.1.1. We start with the Volterra integral equation of the form (7.1.10) with

$$k(t,s,y) = [\xi+\eta(t-s)]y,$$

so that in (7.1.12): $\phi_0=\xi$, $\phi_1=\eta$, and $m=1$. The polynomial $P(\theta;\zeta)$ is given by

$$\theta\zeta^2-\xi\zeta-\eta, \quad \eta\neq 0.$$

Thus, if $\theta=1$ (second-kind equations) we have stability for $\xi<0$, $\eta<0$; if $\theta=0$ (first-kind equations) the stability condition is $\eta\xi>0$.

If $\eta=0$ then the polynomial $P(\theta;\zeta)$ reduces to $\theta\zeta-\xi$, so that, if $\theta=1$, we have also stability for $\eta=0$ and $\xi<0$; if $\theta=0$ we have stability along the whole ξ-axis. □

Example 7.1.2. Consider the renewal equation (3.2.8) of Feller with the convolution kernel

$$k(t,s,y) = \tfrac{1}{2}(t-s)^2e^{s-t}y.$$

Although this kernel is not of polynomial type, the integral equation can easily be reduced to a differential equation: multiplying both sides by e^t, we obtain

$$e^ty(t) = \tfrac{1}{2}t^2+\tfrac{1}{2}\int_0^t(t-s)^2e^sy(s)ds,$$

which is an integral equation for $z(t):=e^ty(t)$ with polynomial convolution kernel. Repeated differentiation yields the differential equation

$$y'''+3y''+3y' = 0.$$

Writing this equation in the form (7.1.7), it is easily seen that the eigenvalues α of the matrix A are the zeros of the polynomial

$$\alpha^3+3\alpha^2+3\alpha,$$

that is $\alpha=0$, $(-3\pm i\sqrt{3})/2$. Thus, the renewal equation of Feller is stable with respect to perturbations in $C^3(\mathbb{R}_+)\cap L^1(\mathbb{R}_+)$. □

Example 7.1.3. Consider the integro-differential equation

$$y'(t) = g_0(t)+\xi y+\eta\int_0^t[1+\mu(t-s)]y(s)ds, \quad \mu\neq 0.$$

In the notation of Theorem 7.1.4(b), this equation is characterized by $\phi_0=y$, $\phi_1=\mu y$ and $m=1$. Hence, the polynomial $P(a;\zeta)$ is given by $a\zeta^2-\zeta-\mu$ and the characteristic polynomial becomes

$$\zeta^3-\xi\zeta^2-\eta\zeta-\mu\eta.$$

From the Routh-Hurwitz criterion (7.1.18) we readily derive the stability conditions

$$\xi<0,\quad \eta<0,\quad \mu\eta<0,\quad (\xi+\mu)\eta>0,\quad \square$$

In the analysis of numerical methods, the linear convolution equation considered in Example 7.1.1 will play a central role. Therefore, a brief discussion of the behavior of its solution is justified. Following Brunner et al. [1980], we write the equation as the second-order differential equation

$$y''=\xi y'+\eta y+g'',\quad y(0)=y_0,\quad y'(0)=y_0'.$$

The solution of the corresponding homogeneous equation may be written in the form

$$y(t)=\frac{y_0}{r_1-r_2}[r_1e^{r_1t}-r_2e^{r_2t}],\quad r_1\neq r_2,$$

$$r_{1,2}:=\frac{1}{2}(\xi\pm\sqrt{D(\xi,\eta)}),\quad D(\xi,\eta):=\xi^2+4\eta.$$

If $D<0$ (with $\xi<0$, $\eta\leqslant 0$), then the exact solution is an exponentially damped oscillation with period $4\pi/\sqrt{-D}$; for $\xi=0$, $\eta<0$, we have the free oscillation $y(t)=y_0\cos(t\sqrt{-\eta})$. If $D>0$ (with $\xi<0$, $\eta\leqslant 0$), then r_1 and r_2 are both negative, so that $y(t)\to 0$ as $t\to\infty$.

7.1.4.2. Nonlinear equations

In the case of *nonlinear* polynomial convolution kernels, the integral equation leads to a nonlinear differential equation. Stability can then be investigated by using Theorem 7.1.3(b) and Corollary 7.1.1 of the preceding section. We will illustrate this by means of an example.

Example 7.1.4. Consider the first-kind equation

$$g_0(t)+\int_0^t[1+(t-s)y^2(s)]y(s)ds=0,\quad g_0\in C^2(\mathbb{R}_+)\cap L^1(\mathbb{R}_+).$$

Repeated differentiation leads to the differential equation

$$y'(t)=-g_0''(t)-y^3(t).$$

Since the right-hand side function of this equation is continuously differentiable with respect to y, it follows from Theorem 7.1.3(b) that the one-sided Lipschitz condition (7.1.4a) is satisfied with

$$M(t)=\mu_2(-3y^2(t))=-3y^2(t).$$

Hence, by Corollary 7.1.1, the solution of the differential equation is stable with respect to (integrable) perturbations of the right-hand side function. Consequently, the solution of the integral equation is stable with respect to perturbations $\in C^2(\mathbb{R}_+)\cap L^1(\mathbb{R}_+)$ of the initial function g_0. $\square$

The property that polynomial convolution kernels lead to differential equations, and the fact that the stability theory for differential equations is well developed, suggest using the kernels (7.1.12) as *test kernels* in the stability analysis of more general Volterra equations; that is, we approximate the given kernel by a kernel of the form (7.1.12), and we investigate the corresponding differential equation. It will, of course, be crucial that the test kernels be sufficiently close to the prescribed kernel. By the well-known theorem of Stone-Weierstrass it can be proved that the class of *continuous* polynomial convolution kernels is dense in the class of all continuous kernels. Hence, if k is not of the form (7.1.12), but is continuous in t,s and y, then it can be approximated by a function of the form (7.1.12) within any degree of accuracy (for an interesting discussion on this subject we refer to Bownds & Wood [1976]).

7.1.5. Finitely decomposable kernels

An alternative to polynomial convolution kernels are the *finitely decomposable* kernels, already introduced in Chapter 1 (see (1.1.8)). Here, we shall employ a different notation,

$$k(t,s,y) = \sum_{i=1}^{m} g_i(t) f_i(s,y), \tag{7.1.19}$$

in order to avoid confusion with notations to be used in the rest of this chapter. It will be assumed that g_i and f_i are real-valued functions. Notice that (7.1.12) is a special case of (7.1.19).

It is convenient to write (7.1.19) in the more compact form

$$k(t,s,y) = \langle \mathbf{g}(t), \mathbf{f}(s,y) \rangle, \tag{7.1.19'}$$

where $\langle\ ,\ \rangle$ denotes the inner product in $\mathbb{R}^m$ and $\mathbf{g}$ and $\mathbf{f}$ are vectors with components g_i and f_i, $i=1, \ldots ,m$. In this notation, the system of ODEs which can be associated with the integral equation assumes the form

$$\begin{aligned} \mathbf{u}'(t) &= \mathbf{f}(t,y(t)) \\ y(t) &= g_0(t) + \langle \mathbf{g}(t), \mathbf{u}(t) \rangle \end{aligned} \quad , \tag{7.1.20}$$

with initial condition $\mathbf{u}(0)=\mathbf{0}$.

For Volterra integro-differential equations of the form (7.1.11) we find, in a similar way, the associated system of ODEs:

$$\begin{aligned} y'(t) &= f_0(t,y(t),\ g_0(t) + \langle \mathbf{g}(t),\ \mathbf{v}(t) \rangle) \\ \mathbf{v}'(t) &= \mathbf{f}(t,y(t)) \end{aligned} \quad , \tag{7.1.21}$$

with initial conditions $y(0)=y_0$, $\mathbf{v}(0)=\mathbf{0}$.

As in the preceding section, the kernels (7.1.19), when continuous in t,s,y, are dense in the class of all continuous kernels and can, therefore, be used as *test kernels* in stability analysis.

7.1.5.1. Second-kind equations

We write (7.1.20) in the form

$$\mathbf{u}'(t) = \mathbf{f}(t, g_0(t) + <\mathbf{g}(t), \mathbf{u}(t)>). \tag{7.1.22}$$

The Jacobian matrix of this system is given by

$$A(t,\mathbf{u}) := \frac{\partial \mathbf{f}}{\partial y} \mathbf{g}^T. \tag{7.1.23}$$

Let us first consider the eigenvalues of A. Let $\mathbf{v}$ be an eigenvector with eigenvalue α. Then

$$A\mathbf{v} = \left[\frac{\partial \mathbf{f}}{\partial y} \mathbf{g}^T\right] \mathbf{v} = <\mathbf{g},\mathbf{v}> \frac{\partial \mathbf{f}}{\partial y} = \alpha \mathbf{v}. \tag{7.1.24}$$

It follows from this relation that there are $m-1$ zero eigenvalues corresponding to $m-1$ linearly independent eigenvectors which are orthogonal to $\mathbf{g}$ (i.e., $<\mathbf{g},\mathbf{v}>=0$). The m-th eigenvalue is obtained by forming the inner product of $\mathbf{g}$ and of the left and right-hand side of (7.1.24), to obtain

$$<A\mathbf{v}_m,\mathbf{g}> = <\frac{\partial \mathbf{f}}{\partial y},\mathbf{g}><\mathbf{g},\mathbf{v}_m> = \alpha_m <\mathbf{v}_m,\mathbf{g}>.$$

Since $\mathbf{g}$ is a real-valued vector we find, using (7.1.19),

$$\alpha_m = <\frac{\partial \mathbf{f}}{\partial y},\mathbf{g}> = \frac{\partial k}{\partial y}(t,t,y). \tag{7.1.25}$$

If $\partial k(t,t,y)/\partial y$ does not vanish, we have $\alpha_m \neq 0$, so that $\mathbf{v}_m$ is not orthogonal to $\mathbf{g}$. Thus, if

$$\frac{\partial k}{\partial y}(t,t,y) < 0, \tag{7.1.26}$$

then A possesses m linearly independent eigenvectors with nonpositive eigenvalues at the point (t,t,y). The condition (7.1.26) is a sufficient condition for stability in the *linear* case where A is constant (notice that (7.1.26) does not refer to a particular decomposition of the kernel k). However, in order to derive a sufficient condition for stability in the *nonlinear* case, we have to compute the one-sided Lipschitz constant L defined in Corollary 7.1.1. Similar to the derivation of (7.1.24), we have the eigenvector equation

$$(A+A^T)\mathbf{v} = \frac{\partial \mathbf{f}}{\partial y}<\mathbf{g},\mathbf{v}> + \mathbf{g}<\frac{\partial \mathbf{f}}{\partial y},\mathbf{v}> = \alpha \mathbf{v}. \tag{7.1.27}$$

Let us choose $m-2$ vectors $\mathbf{v}_j$, $j=1, \dots, m-2$, orthogonal to the space spanned by $\mathbf{g}$ and $\partial \mathbf{f}/\partial y$. These vectors satisfy (7.1.27) with $\alpha=0$. Hence, there are $m-2$ zero eigenvalues with eigenvectors $\mathbf{v}$ such that $<\mathbf{g},\mathbf{v}>= <\partial \mathbf{f}/\partial y,\mathbf{v}>=0$. The remaining eigenvectors $\mathbf{v}_{m-1}$ and $\mathbf{v}_m$ can be chosen in the space spanned by $\mathbf{g}$ and $\partial \mathbf{f}/\partial y$ and orthogonal to, respectively, $\mathbf{g}$ and $\partial \mathbf{f}/\partial y$.

The corresponding eigenvalues are equal and given by

$$\alpha_{m-1} = \alpha_m = <\mathbf{g}, \frac{\partial \mathbf{f}}{\partial y}> = \frac{\partial k}{\partial y}(t,t,y). \tag{7.1.28}$$

Thus, condition (7.1.26) is also a sufficient condition for stability in the nonlinear case. We can now easily prove the following theorem.

Theorem 7.1.5. *For integral equations of the second kind, with decomposable kernel of the form (7.1.19), the associated system of ODEs (7.1.20) has a stable solution* $\mathbf{u}(t)$ *with respect to integrable perturbations of the right-hand side function* $\mathbf{f}(t,g_0(t)+<\mathbf{g}(t),\mathbf{u}(t)>)$ *if*

(i) $\mathbf{f}$ *is continuously differentiable with respect to* $\mathbf{u}$ *in* D

(ii) $\sup_D \frac{\partial}{\partial y} k(t,t,y)<0$, *where* $y(t)=g_0(t)+<\mathbf{g}(t),\mathbf{u}(t)>$.

Proof. Since the one-sided Lipschitz constant L for the system (7.1.20) vanishes, we deduce from the proof of Corollary 7.1.1 that

$$\|\mathbf{u}(t)-\mathbf{u}^*(t)\|_2 \leqslant \|\mathbf{u}_0-\mathbf{u}_0^*\|_2+\int_0^t \delta(\tau)d\tau,$$

where $\delta(\tau)$ defines the perturbation of the right-hand side function according to (7.1.4b). Hence, we have stability if $\delta(t)=\epsilon d(t)$ with $d \in L^1(\mathbb{R}_+)$. $\square$

It follows from this theorem and the relation $y(t)=g_0(t)+<\mathbf{g}(t),\mathbf{u}(t)>$, that, for *bounded functions* $\mathbf{g}(t)$, the solution $y(t)$ of the integral equation is stable under the same conditions. However, if $\mathbf{g}(t)$ is not bounded as $t \to \infty$ (e.g., as in the equation (3.2.26) where $k=-t\cdot s\sqrt{y}$), then we are not justified to draw conclusions from the stability of $\mathbf{u}(t)$. In such cases, we introduce the new variable

$$\mathbf{v} = (v_i) := (g_i u_i), \tag{7.1.29}$$

so that

$$y(t) = g_0(t)+ <\mathbf{v}(t),\mathbf{e}>.$$

Stability of $\mathbf{v}$ now implies stability of y. It is easily verified that $\mathbf{v}$ satisfies the differential equation

$$\mathbf{v}'(t) = D_1(t)\mathbf{f}(t,g_0(t)+<\mathbf{v}(t),\mathbf{e}>)+D_2(t)\mathbf{v}(t), \tag{7.1.30}$$

where $\mathbf{g}(t)$ is assumed to be differentiable, and D_1 and D_2 are diagonal matrices defined by

$$D_1(t) := \mathrm{diag}(g_i(t)), \quad D_2(t) := \mathrm{diag}\left[\frac{g_i'(t)}{g_i(t)}\right].$$

The Jacobian matrix of (7.1.30) is given by

$$A = (a_{i,j}) := \left[g_i\frac{\partial f_i}{\partial y} + \frac{g_i'(t)}{g_i(t)}\delta_{i,j}\right], \tag{7.1.31}$$

where $\delta_{i,j}$ denotes the Kronecker symbol. The analysis of this matrix is much more complicated than that of (7.1.23) and we shall not give a detailed discussion. Instead, we consider the simple case where $m=1$ in the decomposition (7.1.19), and we derive a necessary condition for stability if $m>1$.

If $m=1$, then

$$\alpha(A+A^T) = 2\left[\frac{\partial k}{\partial y}(t,t,y)+\frac{g_1'(t)}{g_1(t)}\right],$$

so that, by Corollary 7.1.1, the estimate for $\|\mathbf{u}(t)-\mathbf{u}^*(t)\|_2$ holds with $L=\alpha(A+A^T)$. Thus, we have stability with respect to *integrable perturbations* of the right-hand side of (7.1.30) if

$$\frac{\partial k}{\partial y}(t,t,y)+\frac{g_1'(t)}{g_1(t)} \leqslant 0. \tag{7.1.32}$$

In the case of strict inequality we have stability with respect to *bounded perturbations.*

For $m>1$ the sum of the eigenvalues of $A+A^T$ is given by the trace of $A+A^T$, i.e., by

$$\sum_{i=1}^{m}\left[g_i\frac{\partial f_i}{\partial y}+\frac{g_i'}{g_i}\right];$$

a necessary condition to have nonpositive eigenvalues is therefore

$$\frac{\partial k(t,t,y)}{\partial y} \leqslant -\sum_{i=1}^{m}\frac{g_i'(t)}{g_i(t)}. \tag{7.1.33}$$

Thus, in order to have stability with respect to $L^1(\mathbb{R}_+)$ condition (7.1.33) must necessarily be satisfied.

We observe that the conditions (7.1.32) and (7.1.33) depend on the decomposition of the kernel, whereas the condition (ii) in Theorem 7.1.5 does not depend on the decomposition of k.

Summarizing, we can say that the conditions of Theorem 7.1.5 ensure a stable solution $\mathbf{u}(t)$, whereas conditions of the type (7.1.32) ensure stability of $g_i(t)u_i(t)$, $i=1, \ldots, m$, and therefore stability of $y(t)$. We have given this rather detailed discussion of the stability of equation (7.1.22) because an important practical method for solving second-kind integral equations actually determines an approximate solution $\mathbf{u}_n$ of the system (7.1.22) by a numerical ODE solver and then computes y_n by means of the formula

$$y_n := g_0(t_n)+ <\mathbf{g}(t_n), \mathbf{u}_n>.$$

In a number of papers Bownds has advocated the (approximate) decomposition of the kernel and the numerical solution of the system (7.1.22). The motivation is that the effort involved to find a decomposition of k is often amply compensated by the fact that solving ODEs is an $O(n)$ process whereas solving integral equations is an $O(n^2)$ process. A detailed discussion of this approach may be found in Bownds [1978b], [1981], [1982a], [1982b] (see also Chapter 8).

Example 7.1.5. Consider equation (3.2.26) where $k = -ts\sqrt{y}$. This kernel can be decomposed for $m=1$ with $g_1=-t$ and $f_1=s\sqrt{y}$. The associated differential equation (7.1.20) is given by

$$u'(t) = t\sqrt{g_0(t)-tu(t)};$$

this equation is stable with respect to integrable perturbations of $g_0(t)$ if conditions (i) and (ii) of Theorem 7.1.5 are satisfied; i.e., if

$$\sup_D \left[-\frac{t^2}{2\sqrt{y}}\right] < 0, \quad D := \{(t,u): g_0(t) > tu(t)\} = \{y : y > 0\},$$

where $y(t)=g_0(t)-tu(t)$. Thus, when u and the perturbed solution u^* are in D we have stability. However, this does not imply that y itself is also stable.

Let us now consider the differential equation (7.1.30):

$$v'(t) = -t^2\sqrt{g_0(t)+v(t)} + \frac{v(t)}{t};$$

this equation is stable if (7.1.32) is satisfied:

$$\sup_D \left[-\frac{t^2}{2\sqrt{y}} + \frac{1}{t}\right] \leqslant 0, \quad y = g_0(t)+v(t).$$

This condition is more restrictive than the condition above : it now implies stability of the integral equation solution y. □

7.1.5.2. *First-kind equations*

We confine our considerations to linear equations where in (7.1.19)

$$f_i(s,y) = \psi_i(s)y, \quad i=1, \ldots ,m,$$

and we assume that the first-kind equation can be written as a second-kind equation with kernel $k(t,t,y)/y = \langle \mathbf{g}(t), \boldsymbol{\psi}(t)\rangle \neq 0$. Then

$$y(t) = -\frac{g_0'(t)}{\langle \mathbf{g}(t), \boldsymbol{\psi}(t)\rangle} - \int_0^t \frac{\langle \mathbf{g}'(t), \boldsymbol{\psi}(s)\rangle}{\langle \mathbf{g}(t), \boldsymbol{\psi}(t)\rangle} y(s)ds$$

(cf. (3.3.8)), where it is assumed that g_0 and $\mathbf{g}$ are differentiable. Thus, the preceding analysis of second-kind equations applies to linear first-kind equations if we replace in Section 7.1.5.1 $g_0(t)$ by $-g_0'(t)/\langle \mathbf{g}(t), \boldsymbol{\psi}(t)\rangle$, $\mathbf{f}(s,y)$ by $\boldsymbol{\psi}(s)y$ and $\mathbf{g}(t)$ by $-\mathbf{g}'(t)/\langle \mathbf{g}(t), \boldsymbol{\psi}(t)\rangle$. For instance, the stability condition (7.1.32) in the case of one-term decompositions ($m=1$) assumes the form

$$\frac{g_1''(t)}{g_1'(t)} - 2\frac{g_1'(t)}{g_1(t)} - \frac{\psi_1'(t)}{\psi_1(t)} \leqslant 0. \tag{7.1.34}$$

Example 7.1.6. The equation (3.3.22) with kernel $k = ay\exp[a(t-s)]$ has a one-term decomposition with

$$g_1(t) = a\,e^{at}, \quad \psi_1(s) = e^{-as}.$$

On substitution into (7.1.34) we find that (7.1.34) is satisfied for all t, so that equation (3.3.22) is stable with respect to integrable perturbations of the initial function. □

7.1.5.3. Integro-differential equations

The Jacobian matrix of the system (7.1.21) is given by

$$A(t,y,\mathbf{v}) := \begin{bmatrix} \frac{\partial f_0}{\partial y} & \frac{\partial f_0}{\partial z} g_1 & \cdots & \frac{\partial f_0}{\partial z} g_m \\ \frac{\partial \mathbf{f}}{\partial y} & \mathbf{0} & \cdots & \mathbf{0} \end{bmatrix}. \tag{7.1.35}$$

Let $(w_0, \mathbf{w}^T)^T$ denote an eigenvector of $A + A^T$ with eigenvalue α. Then

$$\begin{aligned} &2\frac{\partial f_0}{\partial y} w_0 + \langle \mathbf{w}, \frac{\partial \mathbf{f}}{\partial y} + \mathbf{g}\frac{\partial f_0}{\partial z} \rangle = \alpha w_0 \\ &\left[\frac{\partial \mathbf{f}}{\partial y} + \mathbf{g}\frac{\partial f_0}{\partial z}\right] w_0 = \alpha \mathbf{w} \end{aligned} . \tag{7.1.36}$$

It follows from (7.1.36) that there are $m-1$ zero eigenvalues with eigenvectors $(0, \mathbf{w}^T)^T$ where $\mathbf{w}$ is orthogonal to $\partial \mathbf{f}/\partial y + \mathbf{g}\partial f_0/\partial z$. The remaining two eigenvectors have $w_0 \neq 0$; elimination of $\mathbf{w}$ from (7.1.36) and division by w_0 reveals that the corresponding eigenvalues satisfy the equation

$$\alpha^2 - 2\frac{\partial f_0}{\partial y}\alpha - \langle \frac{\partial \mathbf{f}}{\partial y} + \mathbf{g}\frac{\partial f_0}{\partial z}, \frac{\partial \mathbf{f}}{\partial y} + \mathbf{g}\frac{\partial f_0}{\partial z} \rangle = 0.$$

Evidently, these two eigenvalues cannot both be nonpositive. Therefore, the system (7.1.21) has, with respect to the inner product norm used, a *positive* one-sided Lipschitz constant L, so that we cannot draw conclusions from Corollary 7.1.1 on the stability of the solution $(y(t), \mathbf{v}(t))$ of (7.1.21).

As we have already observed, the condition $L \leqslant 0$ might be too severe, because we only want $y(t)$ to be stable, and not both $y(t)$ and $\mathbf{v}(t)$. In this connection, we remark that the *local*, *linearized* approximation to (7.1.21) is stable if the eigenvalues α of A (instead of $A + A^T$) satisfy $\mathrm{Re}\alpha \leqslant 0$. This *local* stability condition leads to the requirement that the zeros of the equation

$$\alpha^2 - \frac{\partial f_0}{\partial y}\alpha - \langle \frac{\partial \mathbf{f}}{\partial y}, \mathbf{g}\frac{\partial f_0}{\partial z} \rangle = 0$$

are in the left half-plane, i.e.,

$$\frac{\partial f_0}{\partial y}(t,y,z) < 0, \quad \frac{\partial f_0}{\partial z}(t,y,z)\frac{\partial k}{\partial y}(t,t,y) < 0. \tag{7.1.37}$$

Notice that the particular decomposition of the kernel k does not enter this stability condition, and therefore (7.1.37) can also be applied to nonseparable kernels (recall that the separable kernels are dense in the space of continuous kernels).

Example 7.1.7. The equation (3.4.11) can be characterized by

$$f_0(t,y,z) = g_0(t)+\frac{1}{y}\log\left[\frac{1+t}{1+\frac{1}{2}t}\right]+z$$

$$k(t,s,y) = \frac{1}{1+(1+t)y}.$$

It satisfies (7.1.37) for all $t>-1$ and is therefore at least locally stable. □

7.2. Stability of numerical methods

7.2.1. Weak and strong stability

We consider linear recursion of the form

$$\sum_{i=0}^{\kappa} M_i \mathbf{y}_{n-i} = \mathbf{w}_n, \quad n\geqslant\kappa \tag{7.2.1}$$

where the M_i are square matrices not depending on n, and where the $\mathbf{w}_n$ are given vectors. As we shall see, the application of any of the numerical methods discussed in Chapter 3 to 5 to a representative of the classes of test equations introduced in the following usually leads to a recursion of the above form; the numerical solution $\{y_n\}$ then appears in one or several components of the vector $\mathbf{y}_n$. In the special case where the M_i are scalars (and hence $\mathbf{y}_n=y_n$), we can directly appeal to the results of Theorems 2.6.8 and 2.6.9 to derive estimates for the difference $y_n-y_n^*$ due to a perturbed right-hand side $\mathbf{w}_n^*=w_n^*$ of (7.2.1). In the vector case a recursion containing only $\{y_n\}$ can be derived on the basis of the following lemma.

Lemma 7.2.1. *Let the sequence* $\{\mathbf{y}_n\}_0^\infty$ *satisfy the inhomogeneous difference equation*

$$G(E)\mathbf{y}_n = \mathbf{w}_{n+\kappa}, \quad n\geqslant 0,$$

where the entries of the constant matrix $G(z)$ *are polynomials in* z *of degree at most* κ *and where* E *is the shift operator defined in (2.6.5). Then each component of* $\mathbf{y}_n$ *satisfies again an inhomogeneous equation with characteristic polynomial* $\det[G(z)]$.

Proof. Let $H(z)$ be the classical adjoint of $G(z)$: $H(z):=\mathrm{adj}(G(z))$, i.e., the transpose of the matrix of cofactors of the elements of $G(z)$ (cf. e.g., Birkhoff & Mac Lane [1953, p. 305]). Since $H(z)G(z)=\det[G(z)]I$, the lemma easily follows. □

It should be remarked that $\det[G(z)]I$ and the entries of $H(z)$ may have common factors, so that the recursion obtained is of higher order than necessary, and consequently the corresponding characteristic polynomial has unwanted roots.

Thus, by applying this lemma to (7.2.1), we obtain the recursion

$$C(E)y_n = T_{inh}, \quad C(z) := \det\left[\sum_{i=0}^{\kappa} M_i z^{\kappa-i}\right];$$

here, T_{inh} denotes an inhomogeneous term consisting of linear combinations of the components of the vectors $\mathbf{w}_n$. Suppose that $\mathbf{w}_n$ itself depends linearly on the values of the initial function g_0 at a fixed number of mesh points $\{t_l\}$. We can then employ Theorem 2.6.8 and 2.6.9 to establish estimates for $y_n - y_n^*$ corresponding to a perturbed right-hand side g_0^* of (7.1.2). These observations give rise to the following stability definitions.

Definition 7.2.1. Let y_n be a component of the vector $\mathbf{y}_n$ satisfying the recursion (7.2.1), let $\mathbf{w}_n$ depend linearly on a fixed number of values of the initial function $g_0(t)$ at the mesh points $\{t_l\}$, and define the characteristic polynomial

$$C(z) := \det\left[\sum_{i=0}^{\kappa} M_i z^{\kappa-i}\right],$$

associated with (7.2.1) (or, equivalently, with the numerical method leading to (7.2.1)).

(a) $\{y_n\}_0^\infty$ is *weakly stable* if $C(z)$ satisfies the root condition.

(b) $\{y_n\}_0^\infty$ is *strongly stable* if $C(z)$ is a Schur polynomial. □

Recall that the concepts used in (a) and (b) of the above definition were introduced in Definition 2.6.2.

The characteristic polynomial $C(z)$ will of course depend not only on the numerical method but also on a set of parameters characterizing the kernel in the class of test equations under consideration. If, for example, we have $k(t,s,y) = \xi y$, then $C(z)$ will depend on $h\xi$ (which will remind the reader of numerical stability considerations associated with the ODE $y' = \xi y$). In view of practical applications it is convenient to compare numerical methods by means of their stability regions; i.e., the set of points in the parameter space (which will always involve the mesh parameter h) for which the methods exhibit a particular stability property.

Definition 7.2.2. Let the characteristic polynomial $C(z)$ associated with (7.2.1) depend on a parameter vector $\mathbf{p}$. Then the set of points in the (complex) $\mathbf{p}$-space for which $C(z)$ satisfies the root condition (is a Schur polynomial) is called the *region of weak (strong) stability* of the underlying numerical method (with respect to the given class of test kernels). □

In the subsequent study of stability results it will be useful to indicate explicitly if a stability region associated with a characteristic polynomial (and a given class of test equations) corresponds to a (weakly or strongly) stable method. We thus introduce the notion of the stability polynomial.

Definition 7.2.3. The characteristic polynomial $C(z)$ associated with (7.2.1) is a *stability polynomial* for the underlying numerical method if the method is weakly (strongly) stable for all $\mathbf{p}$ lying in the weak (strong) stability region. □

It thus follows from Definition 7.2.1 that a characteristic polynomial represents a stability polynomial if the inhomogeneous term in (7.2.1) is a linear combination of initial values on the given mesh.

7.2.2. Block-stability

The characteristic polynomial $C(z)$ does sometimes satisfy weaker conditions than those required in Definition 7.2.1. Consequently, weaker conditions of stability than those given in Definition 7.2.1 occur in the literature. To formulate these definitions it is convenient to introduce

$$\mathbf{Y}_n := (\mathbf{y}_n^T, \ldots, \mathbf{y}_{n+1-\kappa}^T), \quad \mathbf{W}_n := ((M_0^{-1}\mathbf{w}_n)^T, \mathbf{0}^T, \ldots, \mathbf{0}^T)^T \tag{7.2.2a}$$

(provided that M_0 is non-singular), and to write (7.2.1) as the one-step recursion

$$\mathbf{Y}_{n+1} = M\mathbf{Y}_n + \mathbf{W}_{n+1}, \quad M := \begin{bmatrix} -M_0^{-1}M_1 & \cdots & -M_0^{-1}M_\kappa \\ I & & \\ & \ddots & \\ & I & O \end{bmatrix}. \tag{7.2.2b}$$

Following Baker & Keech [1978a] we introduce

Definition 7.2.4. Let $\{\mathbf{Y}_n\}_0^\infty$ and $\{\mathbf{Y}_n^*\}_0^\infty$ denote the solution of (7.2.2) obtained for the inhomogeneous terms $\{\mathbf{W}_n\}_0^\infty$ and $\{\mathbf{W}_n^*\}_0^\infty$ respectively.
(i) The method is said to be *(block-) stable in the norm* $\|\cdot\|$ *with respect to an isolated (block) perturbation* if for any $r>0$ and $\rho\in[0,1]$

$$\mathbf{W}_n = \mathbf{W}_n^* \quad \text{for} \quad n>r \Rightarrow \|\mathbf{Y}_{n+r} - \mathbf{Y}_{n+r}^*\| \leqslant \rho\|\mathbf{Y}_r - \mathbf{Y}_r^*\|, \quad n>0. \tag{7.2.3}$$

The stability is called *strict* if $\rho\in[0,1)$, and the method is called *damped* if, for all $r>0$,

$$\|\mathbf{Y}_{r+n} - \mathbf{Y}_{r+n}^*\| \to 0 \quad \text{as} \quad n\to\infty \tag{7.2.4}$$

(ii) The method displays *error propagation which is bounded in norm* if

$$\max_{n\geqslant 0}\|\mathbf{W}_n - \mathbf{W}_n^*\| \leqslant \delta \Rightarrow \|\mathbf{Y}_n - \mathbf{Y}_n^*\| \leqslant \Omega\delta, \quad n=0,1,\ldots, \tag{7.2.5}$$

for some Ω. It displays *bounded propagation of isolated perturbations* if (7.2.5) holds for $\mathbf{W}_n = \mathbf{W}_n^*$, $n>r$. □

In the following we need a characterization of the eigenvalues of M:

Definition 7.2.5. The eigenvalues of a matrix M that are in the interior of the unit disk are called *strongly stable eigenvalues.* Eigenvalues on the unit circle corresponding to Jordan blocks of order 1 in the Jordan normal form of M are called *weakly stable eigenvalues.* Eigenvalues outside the unit circle are called *unstable.* □

Furthermore, we need the following lemmas (cf., e.g., Varga [1962], Ortega [1972]):

Lemma 7.2.2. *Let q^* be the largest order of all diagonal submatrices J_q of the Jordan normal form of M with $S(J_q)=S(M)$, where $S(\cdot)$ denotes the spectral radius. Then*

$$\|M^n\|_2 \sim \nu n^{q^*-1}[S(M)]^{n-q^*+1} \quad \text{as} \quad n\to\infty,$$

where ν is a positive constant which is bounded by the condition number of M. □

Lemma 7.2.3. *For any matrix M and any $\epsilon>0$, there exists some subordinate norm such that $\|M\|\leq S(M)+\epsilon$. If M has distinct eigenvalues, then there exists a norm such that $\|M\|=S(M)$.* □

By means of these lemmas we can characterize the various types of stability completely in terms of the eigenvalues of M. The following theorem summarizes the stability criteria given in Baker & Keech [1978a]).

Theorem 7.2.1. (a) *If M has only strongly stable eigenvalues, then the method is strictly stable (and therefore damped) with respect to isolated perturbations and displays error propagation which is bounded in norm.*
(b) *If M has both weakly and strongly stable eigenvalues, then the method displays error propagation of isolated perturbations which is bounded in norm.*
(c) *If all eigenvalues of M are simple and lie on the unit disk, then the method is stable with respect to isolated perturbations and displays error propagation which is at most linearly increasing in norm.*
(d) *If M has a complete system of eigenvectors and if the corresponding eigenvalues are all on the unit disk, then the method displays error propagation which is at most linearly increasing in norm.*

Proof. (a) Since $S(M)<1$ there exists a norm such that, by virtue of Lemma 7.2.3, $\|M\|<1$ (take $\epsilon=(1-S(M))/2$, cf. Baker & Keech [1978a]). If $\mathbf{W}_n=\mathbf{W}_n^*$ for $n>r$, then the method satisfies the inequality

$$\|\mathbf{Y}_{n+r}-\mathbf{Y}_{n+r}^*\| \leq \|M\| \; \|\mathbf{Y}_{n+r-1}-\mathbf{Y}_{n+r-1}^*\| < \|\mathbf{Y}_r-\mathbf{Y}_r^*\|,$$

so that the strict stability follows from (7.2.3). If $\|\mathbf{W}_n - \mathbf{W}_n^*\| \leqslant \delta$ for all n, then we have

$$\|\mathbf{Y}_n - \mathbf{Y}_n^*\| \leqslant \|M\| \, \|\mathbf{Y}_{n-1} - \mathbf{Y}_{n-1}^*\| + \delta$$
$$\leqslant \|M\|^{n-1}\delta + \|M\|^{n-2}\delta + \dots + \delta = \frac{1-\|M\|^n}{1-\|M\|}\delta,$$

so that (7.2.5) is satisfied with $\Omega = (1-\|M\|)^{-1}$.
(b) If $\mathbf{W}_n = \mathbf{W}_n^*$ for $n > r$, then the method satisfies the recurrence relation

$$\mathbf{Y}_n - \mathbf{Y}_n^* = M(\mathbf{Y}_{n-1} - \mathbf{Y}_{n-1}^*) = M^{n-r}(\mathbf{Y}_r - \mathbf{Y}_r^*), \quad n \geqslant r.$$

By means of Lemma 7.2.2 assertion (b) of the theorem is easily verified.
(c) It follows from Lemma 7.2.3 that there exists a norm such that $\|M\| = 1$. Proceeding as in the proof of part (a) of the theorem, we can easily show that (7.2.3) is satisfied with $\rho = 1$ and that (7.2.5) in satisfied with $\Omega = n$.
(d) In order to prove assertion (d) we introduce the (nonsingular) matrix N whose columns are just the eigenvectors of M, and we define the errors

$$\mathbf{e}_n := N^{-1}(\mathbf{Y}_n - \mathbf{Y}_n^*), \quad \mathbf{d}_n := N^{-1}(\mathbf{W}_n - \mathbf{W}_n^*).$$

The error equation

$$\mathbf{Y}_{n+1} - \mathbf{Y}_{n+1}^* = M(\mathbf{Y}_n - \mathbf{Y}_n^*) + \mathbf{W}_{n+1} - \mathbf{W}_{n+1}^*$$

can be "diagonalized" by premultiplying both sides with N^{-1}; i.e., we obtain

$$\mathbf{e}_{n+1} = N^{-1}MN\mathbf{e}_n + \mathbf{d}_{n+1},$$

where $N^{-1}MN$ is a diagonal matrix whose elements are the eigenvalues μ_j of M. Thus,

$$e_{n+1}^{(j)} = \mu_j e_n^{(j)} + d_{n+1}^{(j)},$$

where the upper index refers to the components of $\mathbf{e}_n$ and $\mathbf{d}_n$. The characteristic polynomial of this recursion is given by $z - \mu_j$. Since $|\mu_j| \leqslant 1$, it follows from Theorem 2.6.9 that for each j,

$$|e_n^{(j)}| \leqslant C_j \sum_{i=0}^{n} |d_i^{(j)}|,$$

where C_j is independent of n. Hence,

$$\|\mathbf{e}_n\| \leqslant C \sum_{i=0}^{n} \|\mathbf{d}_i\|,$$

so that

$$\|\mathbf{Y}_n - \mathbf{Y}_n^*\| = \|N\mathbf{e}_n\| \leqslant \|N\| C \sum_{i=0}^{n} \|\mathbf{d}_i\| \leqslant C\|N\| \, \|N^{-1}\| \sum_{i=0}^{n} \|\mathbf{W}_i - \mathbf{W}_i^*\|.$$

Thus (7.2.5) holds with $\Omega = C\|N\| \, \|N^{-1}\| n$. □

The above considerations are concerned with the stability of the error $\mathbf{Y}_n-\mathbf{Y}_n^*$, whereas our main interest is in the stability of the error $y_n-y_n^*$, which may be just one component of $\mathbf{Y}_n-\mathbf{Y}_n^*$. These quantities are related by the well-known norm equivalence theorem (see e.g., Ortega [1972, p. 18]):

$$\|\mathbf{Y}_n-\mathbf{Y}_n^*\|_\infty \leqslant c\|\mathbf{Y}_n-\mathbf{Y}_n^*\|, \quad c \text{ constant}$$

This relation shows that all components of $\mathbf{Y}_n-\mathbf{Y}_n^*$, in particular those which involve $y_n-y_n^*$, are uniformly bounded in n whenever $\|\mathbf{Y}_n-\mathbf{Y}_n^*\|$ is uniformly bounded in n.

We remark that, if the stability criteria of Theorem 7.2.1 in the cases (a) and (c) are satisfied, and if in addition the inhomogeneous term of the recursion (7.2.2) depends linearly on $g_0(t)$, then the criteria of Definition 7.2.1 (a) and (b) are respectively satisfied.

7.2.3. Local stability

In (7.2.1) we have explicitly assumed that the M_i are constant matrices. However, we sometimes encounter a situation where the matrices M_i depend on n. The stability conditions of the preceding subsection are then no longer sufficient conditions, but a number of them are *necessary conditions:*

Theorem 7.2.2. (a) *In a strictly stable method the matrix $M=M(n)$ in (7.2.2) has only strongly stable eigenvalues for $n \geqslant r$.*
(b) *If an isolated perturbation is propagated with bounded norm, then $M(n)$ has only strongly or weakly stable eigenvalues for $n \geqslant r$.* □

If $M(n)$ is slowly varying, then we expect that these conditions holding in a sequence of points $t_n, .. ,t_{n+m}$ imply stability in the range $[t_n, t_{n+m}]$. One may interpret them as *local* conditions and therefore we define the notion of local stability:

Definition 7.2.6. The method (7.2.2) is called *locally stable at t_n* $(n \geqslant r)$ if the matrix $M(n)$ has both strongly and weakly stable eigenvalues. It is called *locally stable in the strong sense* if $M(n)$ only possesses strongly stable eigenvalues. □

Thus, local stability is a necessary condition for bounded propagation of isolated perturbations.

In Bakke & Jackiewicz [1985] results are derived which present *sufficient* conditions for the boundedness of solutions of (homogeneous) recursions with nonconstant matrices M. The main tool in this work is the use of the theory of Poincaré-type difference equations. Unfortunately, the hypotheses of their theorems are difficult to verify, and so far explicit results have been obtained only for relatively simple cases.

7.3. VLM methods for integral equations

So far in the literature, the greater part of the stability investigations has been based on some (linear) test kernel for which one tries to reduce the numerical scheme to a *finite-term recurrence relation.* For a general kernel the numerical scheme is a recurrence relation, of which the number of terms is not fixed but grows with n, and therefore tends to infinity if $n \to \infty$ (for example, in the DQ method (3.2.4) all values of $y_0, y_1, \dots, y_n$ are involved yielding an $(n+1)$-term recurrence relation). Since LM formulas for ODEs yield finite recurrence relations, it is not surprising that the test kernels chosen in the literature stem from Volterra equations which can be reduced to a differential equation. Such kernels are *polynomial convolution kernels* (see Section 7.1.4) and *finitely decomposable kernels* (see Section 7.1.5) which also served as test kernels in the stability analysis of the exact solution.

The main result of our analysis of VLM methods is the local stability criterion stated in Theorem 7.3.20. It is derived by using decomposable kernels, but the theorem itself does not refer to any decomposition of the kernel. Extensive numerical experiments have shown that this local stability criterion is reliable in the sense that we never observed the development of instabilities when the theorem predicted otherwise. On the other hand, the theorem sometimes predicts instability whereas the numerical experiment did not show any sign of unstable behavior.

Recently, Lubich [1982b], [1983a], inspired by earlier work of Nevanlinna [1976a], [1976b], has given stability results which are not based on a reduction to a recurrence relation of finite length. In Section 7.3.6 we mention a few results obtained by Lubich for $(\tilde{\rho},\tilde{\sigma})$-reducible DQ methods when applied to second-kind equations with a *positive definite convolution kernel.* It is beyond the scope of this monograph to treat these developments in more detail.

7.3.1. Preliminaries

In order to save space, we shall discuss the VLM methods for first and for second-kind equations simultaneously. For that purpose we write the VLM formula (3.2.15) in the form

$$\theta \sum_{i=0}^{\kappa} \alpha_i y_{n-i} + \sum_{i=0}^{\kappa} \sum_{j=-\kappa}^{\kappa} \beta_{i,j} \tilde{F}_{n-i}(t_{n+j}) = h \sum_{i=0}^{\kappa} \sum_{j=-\kappa}^{\kappa} \gamma_{i,j} k_{n-i}(t_{n+j}), \quad (7.3.1)$$

so that, just as in (7.1.10), $\theta=0$ and $\theta=1$, correspond, respectively, to equations of the first and second kind.

In the results to be presented in the following subsections, it is convenient to associate with a polynomial $p(z)$ of degree κ in z the polynomial

$$p^{\#}(z) := \kappa p(z) - z\frac{dp(z)}{dz}. \quad (7.3.2)$$

Furthermore, we will frequently use a property of (ρ,σ)-reducible quadrature weights which we formulate in the following lemma:

Lemma 7.3.1. *Let the quadrature weights $w_{n,j}$ satisfy the (ρ,σ)-reducibility condition (2.6.7) and let $\{v_l\}$ be a sequence defined for $l \geqslant 0$. Then:*

(a) $\displaystyle\sum_{\nu=0}^{\kappa} a_\nu \sum_{l=0}^{n} w_{n-\nu,l} v_l = \sigma(E) v_{n-\kappa}, \quad n \geqslant \kappa.$

(b) $\displaystyle\sum_{\mu=0}^{\kappa} \sum_{\nu=0}^{\kappa} \nu a_\mu a_\nu \sum_{l=0}^{n} w_{n-\nu-\mu,l} v_l = \rho^{\#}(E)\sigma(E) v_n, \quad n \geqslant 2\kappa.$

(c) $\displaystyle\sum_{\nu=0}^{\kappa} a_\nu \sum_{l=0}^{n} (n-l) w_{n-\nu,l} v_l = \sigma^{\#}(E) v_{n-\kappa}, \quad n \geqslant \kappa.$

Proof. (a) From (2.6.7) it follows that, for $n \geqslant \kappa$,

$$\sum_{\nu=0}^{\kappa} a_\nu \sum_{l=0}^{n} w_{n-\nu,l} v_l = \sum_{l=n}^{n-\kappa} b_{n-l} v_l = \sum_{l=0}^{\kappa} b_l v_{n-l} = \sigma(E) v_{n-\kappa}.$$

(b) In a similar way we obtain, for $n \geqslant 2\kappa$,

$$\sum_{\mu=0}^{\kappa} \sum_{\nu=0}^{\kappa} \nu a_\mu a_\nu \sum_{l=0}^{n} w_{n-\nu-\mu,l} v_l = \sum_{\nu=0}^{\kappa} \sum_{l=0}^{\kappa} \nu a_\nu b_l v_{n-l-\nu}$$

$$= \sum_{\nu=0}^{\kappa} \nu a_\nu \sigma(E) v_{n-\nu-\kappa} = \sum_{\nu=0}^{\kappa} \nu a_\nu E^{\kappa-\nu} \sigma(E) v_{n-2\kappa}.$$

Since

$$\sum_{\nu=0}^{\kappa} \nu a_\nu z^{\kappa-\nu} = \kappa\rho(z) - z\frac{d\rho(z)}{dz} = \rho^{\#}(z),$$

part (b) is immediate,

(c) Part (c) can be proved in an analogous way. □

Finally, we shall frequently derive stability conditions which require that the zeros of a given polynomial $Q_m(z)$ of degree m in z be within the unit circle. These conditions can be expressed in terms of the coefficients of $Q_m(z)$ by applying the Routh-Hurwitz criterion (cf. Lemma 7.1.1) to the transformed polynomial $(1-\zeta)^m Q_m((1+\zeta)/(1-\zeta))$ (notice that the zeros ζ_j are in the left half-plane if the zeros z_j of $Q_m(z)$ are within the unit circle). For future reference we give these conditions for $m=2$ and $m=3$.

Lemma 7.3.2. *The zeros of the polynomial* $Q_m(z)=z^m+c_1 z^{m-1}+ \dots + c_{m-1}z+c_m$ *are within the unit circle if, and only if, for*

$m=2$: $|c_1|<c_2+1$; $c_2<1$;

$m=3$: $|c_1+c_3|<c_2+1$; $3c_3-c_2-c_1>3$; $c_3(c_1-c_3)-c_2>-1$. □

7.3.2. Stability results for the basic test equation

We start with the derivation of finite recursions for the VLM solution y_n and the lag term $\tilde{F}_n(t_n)$ for integral equations of the form

$$\theta y(t) = g_0(t)+\xi\int_0^t y(s)ds. \tag{7.3.3}$$

This equation is known in the literature as the *basic test equation.* This test equation was proposed by Mayers [1962], and since then it has been used extensively, in particular by Baker & Keech [1978a], for obtaining characteristic equations for various numerical methods.

For the VLM methods studied here, it is convenient to split the initial function g_0 and the y_n-dependent part in the lag term formula; we will write

$$\tilde{F}_n(t) = g_0(t)+\tilde{I}_n(t). \tag{7.3.4}$$

In the case (7.3.3) the term $\tilde{I}_n(t)$ does not depend on t: we set $\tilde{I}_n(t)=\tilde{I}_n$. Furthermore, in the following any fixed linear combination of $g_0(t)$-values will be denoted by T_{inh}.

The essential recursion which serves as a starting point for the stability results derived in this section is formulated in

Theorem 7.3.1. *If the VLM method* $\{(3.2.13); (7.3.1)\}$ *is applied to (7.3.3), then it reduces to the recursion*

$$\theta\alpha(E)y_n - h\xi\gamma(E)y_n + \beta(E)\tilde{I}_n = T_{inh}, \quad n\geqslant 0. \tag{7.3.5}$$

Proof. Substitution of $k=\xi y$ into (7.3.1) and using the notation α,β and γ as defined in (3.2.28) leads straightforwardly to (7.3.5). □

We immediately see that (7.3.5) reduces to a recursion only involving y_n, but not $\tilde{I}_n$, if $\beta(z)$ vanishes. Examples of VLM methods for which $\beta(z)$ vanishes are the MML and ILM methods. If $\theta=1$ and if we ignore the inhomogeneous term, then (7.3.5) is identical with the recursion obtained when the LM method $\{\alpha,\gamma\}$ is applied to the differential equation $y'=\xi y$ (notice that this equation is equivalent with (7.3.3)). This motivates

Definition 7.3.1. The VLM formula will be called (α,γ)-*reducible* if $\beta(z)\equiv 0$.□

Corollary 7.3.1. *Let the VLM formula be* (α,γ)*-reducible* $(\beta(z)\equiv 0)$ *and let* $k(t,s,y)=\xi y$. *Then, for any lag term formula, the characteristic polynomial is given by*

$$C_\theta(z) := \theta\alpha(z)-h\xi\gamma(z),$$

and it represents a stability polynomial for the VLM method. □

If the VLM method does not have a vanishing polynomial $\beta(z)$ (e.g., the DQ and ML method have $\beta(z)\not\equiv 0$), then we have to impose a condition on the weights $w_{n,l}$ in the lag term formula. Let us consider the case of $(\tilde{\rho},\tilde{\sigma})$-reducible lag terms (cf. Section 2.6.7). Here, $\tilde{I}_n$ satisfies the recursion

$$\tilde{\rho}(E)\tilde{I}_n = h\xi\tilde{\sigma}(E)y_n. \tag{7.3.6}$$

Applying Lemma 7.2.1 to the recursions (7.3.5) and (7.3.6) leads to the recursion

$$\theta\tilde{\rho}(E)\alpha(E)y_n - h\xi[\tilde{\rho}(E)\gamma(E)-\tilde{\sigma}(E)\beta(E)]y_n = T_{inh}. \tag{7.3.7}$$

Thus, we have a second corollary of Theorem 7.3.1:

Corollary 7.3.2. *Let the lag term formula be $(\tilde{\rho},\tilde{\sigma})$-reducible and let $k(t,s,y)=\xi y$. Then, for any VLM formula, the characteristic polynomial is given by*

$$C_\theta(z) := \theta\tilde{\rho}(z)\alpha(z) - h\xi[\tilde{\rho}(z)\gamma(z)-\tilde{\sigma}(z)\beta(z)],$$

and it represents a stability polynomial for the VLM method. □

Notice that the stability polynomials of methods with (α,γ)-reducible VLM formula *do not depend on the lag term formula.* This is a strong point in favour of (α,γ)-reducible formulas since, from a computational point of view, it is more convenient to adapt the VLM formula to the problem to be integrated, than to adapt the lag term formula (compare the discussion in Section 3.2.3).

The Corollaries 7.3.1 and 7.3.2 enable us to investigate the stability (with respect to the basic test kernel) of VLM methods with (α,γ)-reducible VLM formulas or $(\tilde{\rho},\tilde{\sigma})$-reducible lag terms. These methods include the MML, ILM, and $(\tilde{\rho},\tilde{\sigma})$-reducible ML and DQ methods (to be discussed in the subsections 7.3.2.1 (for second-kind equations) and 7.3.2.2 (for first-kind equations)). VLM methods which do not have (α,γ)-reducible VLM formulas or $(\tilde{\rho},\tilde{\sigma})$-reducible lag terms require a more complicated treatment. One such class is the VLM method with *block-reducible quadrature rules* for the lag term (see Definition 2.6.1b). Since the weights in the lag term (3.2.13) satisfy a more general relation, Lemma 7.3.1 becomes more complicated. However, the derivation of the corresponding characteristic polynomials is essentially the same as given above and only requires a more laborious notation.

The criteria derived on the basis of the basic test kernel $k=\xi y$ may also be indicative of the stability of numerical solutions of equations with more general kernels $k=k(t,s,y)$. In practice, one replaces ξ by $\partial k/\partial y(t_n,t_n,y_n)$, and one applies the stability criteria derived above (in Section 7.3.5.2 we shall see that the stability is determined mainly by kernel values on the diagonal (t_n,t_n,y_n)).

In addition to the notion of the stability region of a method as defined in Definition 7.2.2 (notice that in the present case $\mathbf{p}=h\xi$), we shall need in our discussion of stability the notion of the stability interval of a VLM method:

Definition 7.3.2. Let S_1 be the region of weak (strong) stability with respect to the kernel $k=\xi y$. Then the set $S_1 \cap \mathbb{R}$ is called the *interval of weak (strong) stability*. If the interval $(-H_0,0)$, with $-H_0:=\inf(S_1 \cap \mathbb{R})<0$, is contained in the interval of strong stability, then H_0 is called the *stability boundary*. □

In actual computation, one would wish that the VLM method has a stability region as large as that of the basic integral equation itself. It follows from Theorem 7.1.4 that this equation is stable with respect to bounded perturbations from $C^1(\mathbb{R}_+)$ if the polynomial $P(\theta;\zeta):=\zeta(\theta\zeta-\xi)$ has all its zeros in the left half-plane. For $\theta=1$ this means $\xi<0$, or $\mathrm{Re}\xi<0$ in the case of complex ξ-values. This motivates the following definition.

Definition 7.3.3. Let S_1 be the region of (strong) stability and let $\theta=1$. Then the method is said to be A_0*-stable* if $S_1 \cap \mathbb{R} \supset (-\infty,0)$, and A*-stable* if $S_1 \supset \{h\xi\text{: } \mathrm{Re}\xi<0\}$. □

7.3.2.1. Second-kind equations
$(\tilde{\rho},\tilde{\sigma})$-reducible lag term formulas

We recall (cf. Corollary 7.3.2) that $(\tilde{\rho},\tilde{\sigma})$-reducible lag term formulas lead to the stability polynomial

$$C_1(z) = \tilde{\rho}(z)\alpha(z)-h\xi[\tilde{\rho}(z)\gamma(z)-\tilde{\sigma}(z)\beta(z)], \quad \beta(z)\not\equiv 0. \tag{7.3.8}$$

The DQ methods are generated by $\{\alpha,\beta,\gamma\}\equiv\{1,-1,0\}$, defining the VLM formula, and by a consistent LM method for ODEs $\{\tilde{\rho},\tilde{\sigma}\}$, defining the lag term formula. The stability polynomial reduces to

$$C_1(z) = \tilde{\rho}(z)-h\xi\tilde{\sigma}(z),$$

which is identical with the stability polynomial of the LM method $\{\tilde{\rho},\tilde{\sigma}\}$. The stability boundaries H_0 can therefore be deduced from ODE stability theory:

Table 7.3.1. Stability boundaries for DQ ($\{\tilde{\rho},\tilde{\sigma}\}$) methods

$\{\tilde{\rho},\tilde{\sigma}\}$	G_2	G_3	G_4	G_5	G_6	BD_2	BD_3	BD_4	BD_5	BD_6
H_0	∞	6.0	3.0	1.8	1.2	∞	∞	∞	∞	∞

The ML methods are generated by two LM methods: $\{\rho,\sigma\}=\{\alpha+\beta,\gamma\}$, where $\alpha=a_0z^\kappa$ and $\beta=\Sigma_{i=0}^{\kappa}a_iz^{\kappa-i}$ (cf. (3.2.24)), and $\{\tilde{\rho},\tilde{\sigma}\}$, defining the lag term formula. The stability polynomial reduces to

$$C_1(z) = a_0z^\kappa[\tilde{\rho}(z)-h\xi\tilde{\sigma}(z)]-h\xi[\tilde{\rho}(z)\sigma(z)-\rho(z)\tilde{\sigma}(z)].$$

Table 7.3.2. Stability boundaries for ML $(\{\tilde{\rho},\tilde{\sigma}\}-\{\rho,\sigma\})$ methods

$\{\rho,\sigma\}\setminus\{\tilde{\rho},\tilde{\sigma}\}$	G_2	G_3	G_4	G_5	G_6	BD_2	BD_3	BD_4	BD_5	BD_6
AM_2	∞	∞	29.3	5.1	2.8	4.0	6.7	10.7	17.1	1.5
AM_3	3.0	6.0	∞	4.4	2.2	1.7	2.1	2.3	2.6	1.8
AM_4	1.5	2.0	3.0	6.9	2.3	1.1	1.2	1.3	1.4	1.4
AM_5	0.9	1.1	1.3	1.8	3.3	0.7	0.8	0.8	0.9	0.9
AM_6	0.6	0.7	0.7	0.9	1.2	0.5	0.5	0.6	0.6	0.6
BD_2	∞	9.3	4.3	3.2	2.6	∞	∞	∞	3.0	0.9
BD_3	9.1	2.8	1.8	1.3	1.0	8.8	∞	∞	2.9	0.7
BD_4	3.4	1.6	1.1	0.8	0.6	1.8	7.2	∞	17.7	0.7
BD_5	1.9	1.0	0.7	0.5	0.4	0.8	1.9	5.9	∞	1.0
BD_6	1.1	0.6	0.4	0.3	0.2	0.5	0.9	1.8	5.1	∞

In Table 7.3.2 stability boundaries corresponding to combinations of well-known LM formulas are listed. We conclude from these figures that, generally, ML methods have larger stability boundaries if the LM formulas $\{\rho,\sigma\}$ and $\{\tilde{\rho},\tilde{\sigma}\}$ are both chosen from the same class, e.g., Gregory lag terms should be combined with AM-based VLM formulas. Notice the surprising result that the non-A_0-stable G_4 and AM_3 formulas generate an A_0-stable ML method.

We observe that $C_1(z)$ reduces to the stability polynomial of DQ$(\{\tilde{\rho},\tilde{\sigma}\})$ if we set $\{\rho,\sigma\}=\{\tilde{\rho},\tilde{\sigma}\}$. Hence, we have the result:

Theorem 7.3.2. *If $\{\rho,\sigma\}$ defines an A_0-stable (or A-stable) LM method, then DQ $(\{\rho,\sigma\})$ and $ML(\{\rho,\sigma\}-\{\rho,\sigma\})$ are also A_0-stable (or A-stable).* □

(α,γ)-reducible VLM formulas

According to Corollary 7.3.1, an (α,γ)-reducible VLM formula gives rise to the stability polynomial

$$C_1(z) = \alpha(z) - h\xi\gamma(z). \tag{7.3.9}$$

The ILM and MML methods are (α,γ)-reducible, with $\{\alpha,\gamma\}$ corresponding to an LM method $\{\rho,\sigma\}$ for ODEs. Hence, ILM and MML methods have the same stability polynomial as the generating LM method. The stability intervals $S_1\cap\mathbb{R}$ are therefore identical with those listed in Table 7.3.1 if $\{\tilde{\rho},\tilde{\sigma}\}$ is replaced by $\{\rho,\sigma\}$ (we recall that the Gregory rules G_p are equivalent to the AM_p formulas). In particular, we have the following theorem.

Theorem 7.3.3. *If $\{\rho,\sigma\}$ defines an A_0-stable (or A-stable) LM method, then $ILM(\cdot-\{\rho,\sigma\})$ and $MML(\cdot-\{\rho,\sigma\})$ are also A_0-stable (or A-stable), irrespective of the choice of the lag term formula.* □

If no large stability regions are required (i.e., if $\partial k/\partial y$ is not negative with large modulus), then DQ methods based on Gregory rules are recommendable methods, because of their easy implementation. In the case of *stiff* equations ($\partial k/\partial y \ll 0$) we need large stability intervals and we are led to the DQ(BD), ILM($\cdot$ −BD) and MML($\cdot$ −BD) methods. Here, the ILM and MML methods are to be preferred: the DQ methods based on BD quadrature give rise to rather complicated implementations (cf. Section 3.2.3), whereas the ILM and MML methods with Gregory lag terms are much easier to implement. This observation has, in fact, been the reason for introducing the ILM methods and MML methods, and for developing the theory of VLM methods.

7.3.2.2. First-kind equations

It follows from Corollary 7.3.2 that for VLM methods with $(\tilde{\rho},\tilde{\sigma})$*-reducible lag terms* the polynomial

$$C_0(z) = [\tilde{\rho}(z)\gamma(z) - \tilde{\sigma}(z)\beta(z)] \tag{7.3.10}$$

is a stability polynomial. From this observation the following theorems can easily be proved.

Theorem 7.3.4. *Any convergent, $(\tilde{\rho},\tilde{\sigma})$-reducible DQ method is weakly stable for all h and for all ξ in the test kernel $k=\xi y$. It is strongly stable if $\tilde{\sigma}$ is a Schur polynomial.*

Proof. Since for DQ methods $\gamma(z)=0$ and $\beta(z)=-1$, (7.3.10) yields $C_0(z)=\tilde{\sigma}(z)$. From Theorem 3.3.4 we deduce that, in a convergent DQ method, $\tilde{\sigma}(z)$ necessarily satisfies the root condition so that the method is weakly stable; by Corollary 7.3.2 it is strongly stable if $\tilde{\sigma}$ is a Schur polynomial. □

Theorem 7.3.5. *Any convergent $ML(\{\rho,\sigma\}-\{\rho,\sigma\})$ method is weakly stable for all h and for all ξ in the test kernel $k=\xi y$. It is strongly stable if σ is a Schur polynomial.*

Proof. In Section 3.3.3 we have shown that ML methods with both the lag term and the VLM formula based on the same LM formula $\{\rho,\sigma\}$ (i.e., $\alpha+\beta=\rho$ and $\gamma=\sigma$) are algebraically equivalent to DQ($\{\rho,\sigma\}$) methods, provided, of course, that the starting values are identical. By applying Theorem 7.3.4 and by observing that stability includes stability with respect to the starting values, the assertion of the theorem is proved. □

Next we consider (α,γ)*-reducible* VLM methods. It follows from Corollary 7.3.1 that the stability polynomial

$$C_0(z) = -h\xi\gamma(z) \tag{7.3.11}$$

should satisfy the root condition.

Theorem 7.3.6. *$ILM(\cdot - \{\rho,\sigma\})$ and $MML(\cdot - \{\rho,\sigma\})$ methods are weakly stable for all h and for all ξ in the kernel $k = \xi y$, irrespective of the choice of the lag term, if σ satisfies the root condition. They are strongly stable if σ is a Schur polynomial.*

Proof. Since the ILM and MML methods are (α,γ)-reducible the stability polynomial is given by (7.3.11), and because $\gamma(z) \equiv \sigma(z)$, the stability polynomial satisfies the root condition and is a Schur polynomial if σ has the same properties. This proves the theorem. □

Theorem 7.3.7. *Local differentiation methods characterized in Theorem 3.3.8 are strongly stable for all h and for all ξ in the kernel $k = \xi y$, irrespective of the choice of the lag term.*

Proof. The local differentiation methods specified in Theorem 3.3.8 are (α,γ)-reducible with $\gamma(z) = \gamma_j z^{\kappa - j}$, $0 \leqslant j \leqslant \kappa$. Thus, γ is a Schur polynomial. The assertion of the theorem now follows from (7.3.11) and Corollary 7.3.1. □

Example 7.3.1. We consider ILM and MML methods with arbitrary lag term formulas and with VLM formulas based on the $\{\rho,\sigma\}$ formulas listed in Table 7.3.3 (cf. Lambert [1973, p. 41]). In addition, this table gives the order p of the LM formula $\{\rho,\sigma\}$ and indicates the stability behavior of the ILM and MML method.

Table 7.3.3. Stability of $ILM(\cdot - \{\rho,\sigma\})$ and $MML(\cdot - \{\rho,\sigma\})$ method

p	$\rho(z)$	$\sigma(z)$	stability behavior
1	$z-1$	1	stable
2	$z-1$	$\frac{1}{2}(z+1)$	stable
2	$z^2-(1+a)z+a$	$\frac{3-a}{2}z-\frac{1+a}{2}$	strongly stable if $a<1$
3	$z^2-(1+a)z+a$	$\frac{5+a}{12}z^2+2\frac{1-a}{3}z-\frac{1+5a}{12}$	stable if $a=1$
3	$z^3-(1+a)z^2+(a+b)z-b$	$\frac{23-5a-b}{12}z^2-\frac{4+2a-2b}{3}z+\frac{5+a+5b}{12}$	strongly stable if $a+b<3$, $-11<a-b<1$
κ	$\{\rho,\sigma\}$ corresponding to AM formulas (cf. Table 2.6.1)		unstable if $\kappa>1$
κ	$\{\rho,\sigma\}$ corresponding to BD formulas (cf. Table 2.6.2)		strongly stable

7.3.2.3. *Systems of integral equations*

We briefly discuss the case of the basic vector equation

$$\theta \mathbf{y}(t) = \mathbf{g}_0(t)+X\int_0^t \mathbf{y}(s)ds,$$

where X is a constant matrix. We shall assume that X has a complete eigensystem, so that this equation can be diagonalized. Let H be the matrix whose columns are the eigenvectors of X and define $\mathbf{g}_0^*:=H^{-1}\mathbf{g}_0$, $\mathbf{y}^*:=H^{-1}\mathbf{y}$, and $D:=H^{-1}XH$. Then

$$\theta \mathbf{y}^*(t) = \mathbf{g}_0^*(t)+D\int_0^t \mathbf{y}^*(s)ds.$$

Since D is the diagonal matrix whose diagonal elements are the eigenvalues ξ_j of X, this equation is a system of (uncoupled) basic test equations and all stability results derived in the preceding sections apply when ξ is replaced by ξ_j, $j=1, \ldots ,r$, with r denoting the order of the system.

7.3.3. *Stability results for the convolution test equation*

It has already been observed by Kershaw [1974, p. 159] that the use of the basic test equation "is obviously convenient, however its true relevance to the integral equation situation does not appear to have been thoroughly examined". Therefore, it may be of interest to consider the "next simple" equation, i.e., the convolution equation

$$y(t) = g_0(t)+\int_0^t [\xi+\eta(t-s)]y(s)ds, \tag{7.3.12}$$

and to see to what extent the stability criteria derived in the preceding section change.

We start with the analogue of Theorem 7.3.1; substitution of (7.3.12) into (7.3.1) yields:

Theorem 7.3.8. *If the VLM method* {(3.2.13; (7.3.1)} *is applied to* (7.3.12) *then it reduces to the recursion*

$$\theta\alpha(E)y_n - h\xi\gamma(E)y_n - h^2\eta[\gamma^*(E)+\gamma^{\#}(E)]y_n + E^{-\kappa}\sum_{i=0}^{\kappa}\sum_{j=-\kappa}^{\kappa}\beta_{i,j}\tilde{I}_{n-i}(t_{n+j}) = T_{inh}, \tag{7.3.13}$$

where $\gamma^{\#}(z)$ *is defined according to* (7.3.2) *and*

$$\gamma^*(z) := \sum_{i=0}^{\kappa}\gamma_i^* z^{\kappa-i}, \quad \gamma_i^* := \sum_{j=-\kappa}^{\kappa} j\gamma_{i,j}. \quad \square$$

In order to derive a recursion for the integral part $\tilde{I}_n(t)$ of the lag term $\tilde{F}_n(t)$ we shall from now on assume that $\tilde{F}_n(t)$ is $(\tilde{\rho},\tilde{\sigma})$-reducible. Then, by virtue of Lemma 7.3.1, we have

$$\tilde{\rho}(E)\tilde{I}_n(t_{n+j}) = h\xi\tilde{\sigma}(E)y_n + h^2\eta[j\tilde{\sigma}(E)+\tilde{\sigma}^{\#}(E)]y_n - h\eta\tilde{\rho}^{\#}(E)\tilde{J}_n, \tag{7.3.14}$$

$$\tilde{J}_n := h\sum_{l=0}^{n} w_{n,l}y_l.$$

Applying the operator $\tilde{\rho}(E)$ to both sides of (7.3.13) and substituting (7.3.14) yields

$$\begin{aligned}\theta\tilde{\rho}(E)\alpha(E)y_n - h\xi\tilde{\rho}(E)\gamma(E)y_n - h^2\eta\,\tilde{\rho}(E)[\gamma^*(E)+\gamma^{\#}(E)]y_n \\ + h\xi\beta(E)\tilde{\sigma}(E)y_n + h^2\eta[\beta^*(E)\tilde{\sigma}(E)+\beta(E)\tilde{\sigma}^{\#}(E)]y_n \\ - h\eta\beta(E)\tilde{\rho}^{\#}(E)\tilde{J}_n = T_{inh},\end{aligned} \tag{7.3.14'}$$

where $\beta^*(z)$ is defined in analogy to $\gamma^*(z)$.

If the VLM method is (α,γ)-reducible, then (7.3.14′) reduces to a recursion containing only y_n. Thus, since T_{inh} is a fixed linear combination of $g_0(t)$-values, we have the analogue of Corollary 7.3.1:

Corollary 7.3.3. *Let the VLM formula be (α,γ)-reducible, let the lag term be $(\tilde{\rho},\tilde{\sigma})$-reducible, and let $k(t,s,y)=[\xi+\eta(t-s)]y$. Then the characteristic polynomial is given by*

$$C_\theta(z) = \theta\tilde{\rho}(z)\alpha(z) - h\xi\tilde{\rho}(z)\gamma(z) - h^2\eta[\tilde{\rho}(z)\gamma^*(z)+\tilde{\rho}(z)\gamma^{\#}(z)-\beta^*(z)\tilde{\sigma}(z)],$$

and it represents a stability polynomial for the VLM method. □

If the VLM method is not (α,γ)-reducible, then we need an additional recursion for $\tilde{J}_n$. By means of Lemma 7.3.1 we derive

$$\tilde{\rho}(E)\tilde{J}_n - h\tilde{\sigma}(E)y_n = 0, \tag{7.3.15}$$

so that, by Lemma 7.2.1, the analogue of Corollary 7.3.2 becomes:

Corollary 7.3.4. *Let the lag term formula be $(\tilde{\rho},\tilde{\sigma})$-reducible and let $k(t,s,y)=[\xi+\eta(t-s)]y$. Then for any VLM formula, the characteristic polynomial is given by*

$$\begin{aligned}C_\theta(z) = {} & \theta\tilde{\rho}^2(z)\alpha(z) - h\xi\tilde{\rho}(z)[\tilde{\rho}(z)\gamma(z)-\tilde{\sigma}(z)\beta(z)] \\ & + h^2\eta\{\tilde{\rho}(z)[\beta^*(z)\tilde{\sigma}(z)-\gamma^*(z)\tilde{\rho}(z)-\gamma^{\#}(z)\tilde{\rho}(z)] \\ & + \tilde{\rho}(z)[\beta(z)\tilde{\sigma}^{\#}(z)+\beta^{\#}(z)\tilde{\sigma}(z)] - \tilde{\rho}^{\#}(z)\tilde{\sigma}(z)\beta(z)]\},\end{aligned}$$

and it represents a stability polynomial for the VLM method. □

Ideally, the VLM method should have the same region of strong stability as the convolution equation itself; that is, the region $\{(\xi,\eta)\colon \xi<0,\ \eta\leqslant 0\}$ if $\theta=1$, and $\{(\xi,\eta)\colon \xi\eta>0 \vee \eta=0\}$ for $\theta=0$ (see Example 7.1.1). In Brunner, Nørsett & Wolkenfelt [1980], such methods were called V_0-stable:

Definition 7.3.4. Let S_2 be the region of strong stability.
(i) Let $\theta=1$. Then the method is said to be *V_0-stable* if $S_2\cap\mathbb{R}^2\supset\{(h\xi,h^2\eta)\colon \xi<0,\ \eta\leqslant 0\}$.
(ii) Let $\theta=0$. Then the method is said to be *V_0-stable* if $S_2\cap\mathbb{R}^2\supset\{(h\xi,h^2\eta)\colon \xi\eta>0\vee\eta=0\}$. □

So far we have excluded the case $\xi=0$. In the case $\theta=1$, the resulting exact solution is an undamped oscillation (cf. the discussion at the end of Section 7.1.4.1), and consequently we should only look for regions of *weak* stability.

In other words, in order to simulate properly the correct qualitative behavior of the exact solution we should avoid choosing for $C_1(z)$ a Schur polynomial. In the following we relate VLM methods applied to kernels of the form $\eta(t-s)y$ to LM methods applied to the corresponding ODE. It follows from Corollary 7.3.4 that for $\theta=1$ and $\xi=0$ we may write the characteristic polynomial in the form

$$C_1(z) = P(z)-h^2\eta\Sigma(z).$$

In other words, the VLM method is algebraically equivalent to an LM method $\{\mathrm{P},\ \Sigma\}$ applied to the equation $y''=\eta y+T_{inh}$, where T_{inh} denotes some inhomogeneous term. This is, of course, not surprising, because our test equation is also equivalent to such an ODE if $\xi=0$.

In Lambert and Watson [1976], LM methods for this special test equation have been investigated in order to find criteria to avoid unwanted damping of the numerical solution. For that purpose they introduced the concept of the periodicity interval.

Definition 7.3.5. Let the LM method $\{\mathrm{P},\ \Sigma\}$ satisfy the following hypotheses: $\mathrm{P}(0)\neq 0$; P and Σ have no common factors; $\mathrm{P}(1)=\mathrm{P}'(1)=0$, $\mathrm{P}''(1)=2\Sigma(1)$; P has no roots outside the unit disk, with those on the unit circle being of multiplicity at most 2. Then the LM method is said to have *interval of periodicity* $(0,H_0^2)$ if, for all $-h^2\eta\in(0,H_0^2)$, $C_1(z)$ has distinct roots $z_1,\ \dots\ ,z_q$ satisfying

$$z_1=e^{i\phi(\eta h^2)},\quad z_2=e^{-i\phi(\eta h^2)},\quad |z_j|\leqslant 1,\quad j=3,\ \dots\ ,q,$$

where $\phi(\eta h^2)\in\mathbb{R}$. □

The main results of Lambert and Watson summarized in the following lemma answer the question of whether *a* given LM method has a non-vanishing interval of periodicity.

Lemma 7.3.3. *Let the LM method* $\{P,\Sigma\}$ *satisfy the hypotheses of Definition 7.3.5.*

(a) *If the method* $\{P,\Sigma\}$ *has interval of periodicity* $(0,H_0^2)$, $H_0^2>0$, *then* $\{P,\Sigma\}$ *is a symmetric method (i.e.,* $P(z)\equiv z^q P(1/z)$, $\Sigma(z)\equiv z^q\Sigma(1/z)$*), and, for all* $-h^2\eta\in(0,H_0^2)$, *all roots of* $C_1(z)$ *lie on the unit circle.*

(b) *If* $\{P,\Sigma\}$ *is a symmetric method and if* P *has only one double root on the unit circle, then the method* $\{P,\Sigma\}$ *has a non-vanishing interval of periodicity.* □

7.3.3.1. Second-kind equations

DQ $(\{\tilde\rho,\tilde\sigma\})$ *methods*

The stability polynomial is given by

$$C_1(z) = \tilde\rho^2(z)-h\xi\tilde\rho(z)\tilde\sigma(z)-h^2\eta z[\tilde\sigma(z)\tilde\rho'(z)-\tilde\rho(z)\tilde\sigma'(z)]. \qquad (7.3.16a)$$

Theorem 7.3.9. *A* $(\tilde\rho,\tilde\sigma)$*-reducible DQ method cannot be* V_0*-stable.* □

The proof as well as stability regions for the BD-reducible methods are given in Wolkenfelt [1981b].

In order to have a non-empty interval of periodicity we conclude from Lemma 7.3.3(a) that $\{\tilde\rho^2, z\tilde\sigma\tilde\rho'-z\tilde\rho\tilde\sigma'\}$ has to be a symmetric method for $-h^2\eta\in(0,H_0^2)$, $H_0^2>0$ (notice that the hypotheses of Definition 7.3.5 are satisfied, provided that $\{\tilde\rho,\tilde\sigma\}$ is consistent and $\tilde\rho(0)\neq 0$), and consequently $\tilde\rho$ *must have all its zeros on the unit circle.* This represents an easy criterion for deriving intervals of periodicity for DQ methods.

Example 7.3.2. (i) Consider the family of one-step methods $\{\tilde\rho,\tilde\sigma\}=\{z-1, az+1-a\}$. Then

$$P = \tilde\rho^2 = (z-1)^2, \quad \Sigma = z[\sigma\rho'-\rho\sigma'] = z.$$

Evidently, $\{P,\Sigma\}$ is symmetric and P has only one double root on the unit circle. Hence, by Lemma 7.3.3(b), the method DQ$(\{\tilde\rho,\tilde\sigma\})$ has a non-empty interval of periodicity for all a. Applying Definition 7.3.5 we find that this interval is given by (0,4) for all a.

(ii) Next, consider the family

$$\{\tilde\rho,\tilde\sigma\}=\{z^2-(1+a)z+a, \frac{5+a}{12}z^2+\frac{2(1-a)}{3}z-\frac{1+5a}{12}\}.$$

The polynomial $\tilde\rho$ has its zeros on the unit circle only for $a=\pm1$. Since $P=\tilde\rho^2$ should not have zeros with multiplicity greater than 2 we see that $a=-1$ is the only value for which $\{P,\Sigma\}$ may have a non-empty periodicity interval. However, since $\Sigma=4z(z^2+z+1)$, Σ is not symmetric so that, by Lemma 7.3.3(a), this method (two-step Milne-Simpson method) has no interval of periodicity. □

Theorem 7.3.10. *DQ methods have no interval of periodicity in the sense of Definition 7.3.5 if they are reducible to one of the following LM methods:*

(a) *Adams methods with* $\tilde{\kappa}>1$ ($\tilde{\rho}=z^{\tilde{\kappa}}-z^{\tilde{\kappa}-1}$).

(b) *Nyström methods with* $\tilde{\kappa}>2$ *or generalized Milne-Simpson methods* ($\tilde{\rho}=z^{\tilde{\kappa}}-z^{\tilde{\kappa}-2}$).

(c) *Backward differentiation methods with* $\tilde{\kappa}>1$ ($\tilde{\sigma}=b_0z^{\tilde{\kappa}}$).

Proof. (a) For Adams methods, $P=\tilde{\rho}^2=z^{2\tilde{\kappa}}-2z^{2\tilde{\kappa}-1}+z^{2\tilde{\kappa}-2}$. It follows from this expression that P is not symmetric if $\tilde{\kappa}>1$, so that by Lemma 7.3.3 there results an empty interval of periodicity. For $\tilde{\kappa}=1$ there is an interval of periodicity (0,4), as shown in Example 7.3.2(i).
(b) Since $P=z^{2\tilde{\kappa}}-2z^{2\tilde{\kappa}-2}+z^{2\tilde{\kappa}-4}$ the assertion of the theorem is clear for $\tilde{\kappa}>2$. The two-step Milne-Simpson method has been considered in Example 7.3.2 (ii). There remains the two-step Nyström method $\{z^2-1,2z\}$: although $\{P,\Sigma\}=\{(z^2-1)^2,\ 2z^3+2z\}$ is a symmetric method, the polynomial P has two double roots on the unit circle, so that Lemma 7.3.3(b) does not apply. Therefore, the 2-step Nyström method may possess an interval of periodicity.
(c) From $\tilde{\sigma}=b_0z^{\tilde{\kappa}}$ we deduce that

$$\Sigma=-b_0z^{\tilde{\kappa}}\sum_{j=0}^{\tilde{\kappa}} ja_jz^{\tilde{\kappa}-j},$$

which is not symmetric for $\tilde{\kappa}>1$. For $\tilde{\kappa}=1$, see Example 7.3.2 (i). □

MML ($\{\tilde{\rho},\tilde{\sigma}\}-\{\rho,\sigma\}$) *methods*

The stability polynomial is given by ($\beta^*=\rho^{\#},\gamma^*=0$)

$$C_1(z) = \tilde{\rho}(z)\rho(z)-h\xi\tilde{\rho}(z)\sigma(z)+h^2\eta[\rho^{\#}(z)\tilde{\sigma}(z)-\tilde{\rho}(z)\sigma^{\#}(z)]. \tag{7.3.16b}$$

Amini [1983] has proved

Theorem 7.3.11. *MML* ($\{\tilde{\rho},\tilde{\sigma}\}-\{\rho,\sigma\}$) *methods cannot be* V_0*-stable.* □

Stability regions for combinations of backward Euler and the trapezoidal rule are given in Figure 7.3.1. These regions are much larger than those of the corresponding DQ methods.

Also, intervals of periodicity can be expected to exist when an appropriate combination of lag term and MML formula is employed. For example, if ρ and $\tilde{\rho}$ each have distinct zeros on the unit circle which differ mutually except for the zero at $z=1$, then $P=\rho\tilde{\rho}$ is symmetric and has only one double root; this is one of the conditions in Lemma 7.3.3(b) for the existence of a nonvanishing interval of periodicity. In the DQ case, where $P=\tilde{\rho}^2$, it is this condition that often cannot be satisfied.

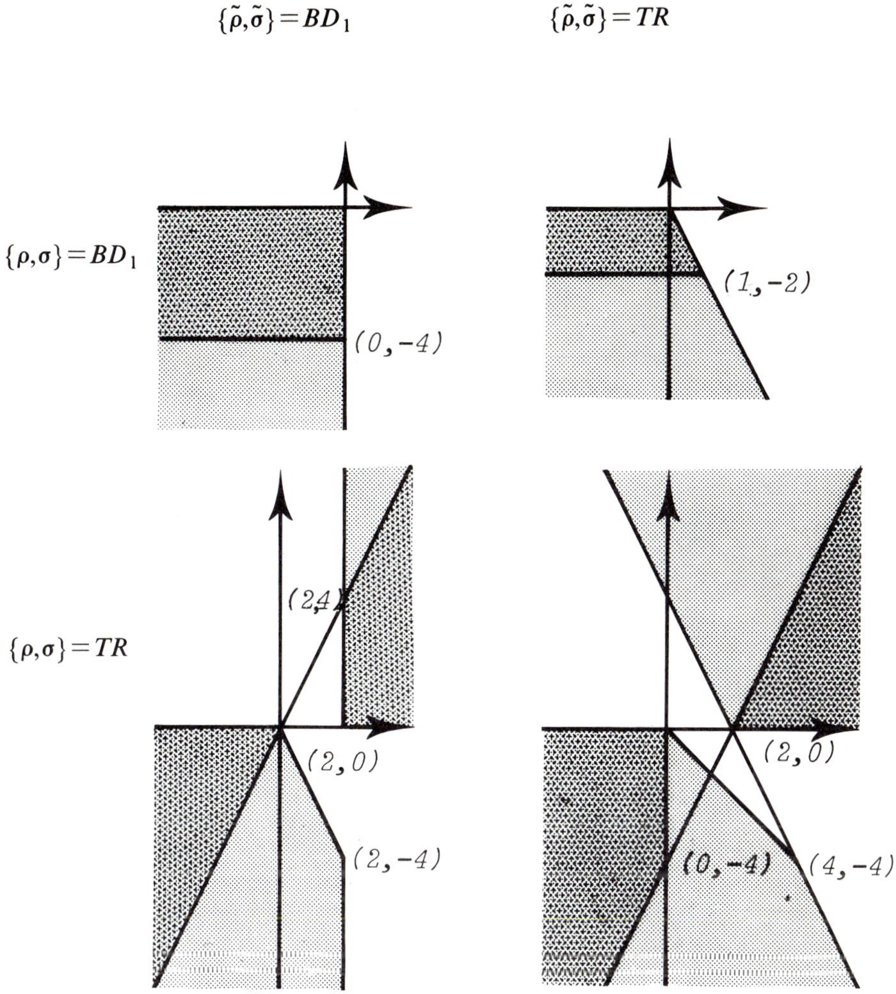

Figure 7.3.1. Stability regions for MML( ) and ILM( ) methods

ILM $(\{\tilde{\rho},\tilde{\sigma}\}-\{\rho,\sigma\})$ *methods*

With respect to V_0-stability a more positive result is offered by the ILM methods. The stability polynomial is given by ($\beta^* = -\sigma$, $\gamma^* = -\sigma^{\#}$)

$$C_1(z) = \tilde{\rho}(z)[\rho(z)-h\xi\sigma(z)]-h^2\eta\tilde{\sigma}(z)\sigma(z). \tag{7.3.16c}$$

This polynomial is equivalent to the stability polynomial obtained by Brunner & Lambert [1974] and by Matthys [1976] in their stability analysis of DQ methods for integro-differential equations (this is not surprising because ILM methods are constructed by differentiating the integral equation and by applying a numerical integration method to the resulting integro-differential equation (see Section 3.2.6)). Thus, by inspecting the stability regions given by Brunner and Lambert, we find that all four combinations, with $\{\tilde{\rho},\tilde{\sigma}\}$ and $\{\rho,\sigma\}$ defining either the trapezoidal rule (TR) or the backward Euler formula (BD_1), are V_0-stable methods. In Figure 7.3.1 the regions of stability are given together with those of the corresponding MML versions. Evidently, the ILM methods possess considerably larger stability regions. For $p=3, \ldots ,6$, stability regions of ILM (G_p-BD_p) and ILM (BD_p-BD_p) methods were given in van der Houwen, te Riele & Wolkenfelt [1978] and are reproduced in Figure 7.3.2. For $p=2$ these methods have the whole third quadrant of the $(h\xi,h^2\eta)$-plane in their stability region and are therefore V_0-stable.

If $\{\tilde{\rho},\tilde{\sigma}\}=\{\rho,\sigma\}$, then the stability region S_2 of the ILM method can be related to the stability region of the LM method $\{\rho,\sigma\}$ (cf. Brunner & Lambert [1974, p. 82]).

Theorem 7.3.12. *Let S^* be the stability region of the LM method $\{\rho,\sigma\}$ and let (λ,μ) be defined by $\lambda+\mu=\xi$, $\lambda\mu=-\eta$. Then the set $\{(h\xi,h^2\eta)\colon h\lambda\in S^*,\ h\mu\in S^*\}$ defines the region of strong stability S_2 of the ILM $(\{\rho,\sigma\}-\{\rho,\sigma\})$ method.*

Proof. For $\{\tilde{\rho},\tilde{\sigma}\}=\{\rho,\sigma\}$ the stability polynomial (7.3.16c) takes the form

$$C_1(z) = \rho^2(z)-h\xi\rho(z)\sigma(z)-h^2\eta\sigma^2(z) = [\rho(z)-h\lambda\sigma(z)][\rho(z)-h\mu\sigma(z)],$$

where λ and μ are defined in the theorem. Since $h\lambda$ and $h\mu$ are assumed to lie in S^*, all zeros of $\rho-h\lambda\sigma$ and of $\rho-h\mu\sigma$ are inside the unit disk, so that $C_1(z)$ is a Schur polynomial in the corresponding points $(h\xi,h^2\eta)$. According to Definition 7.2.2, these points form the stability region of the ILM method. □

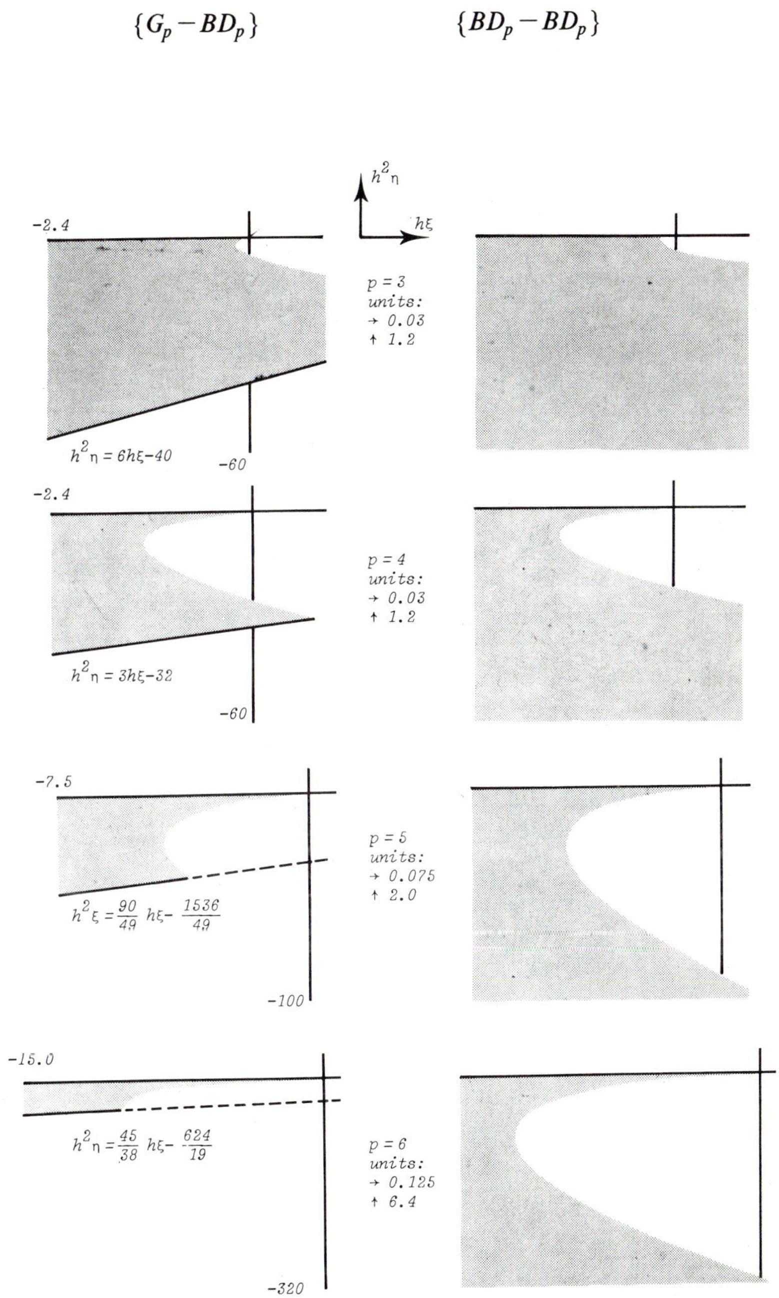

Figure 7.3.2a. Stability regions for ILM methods

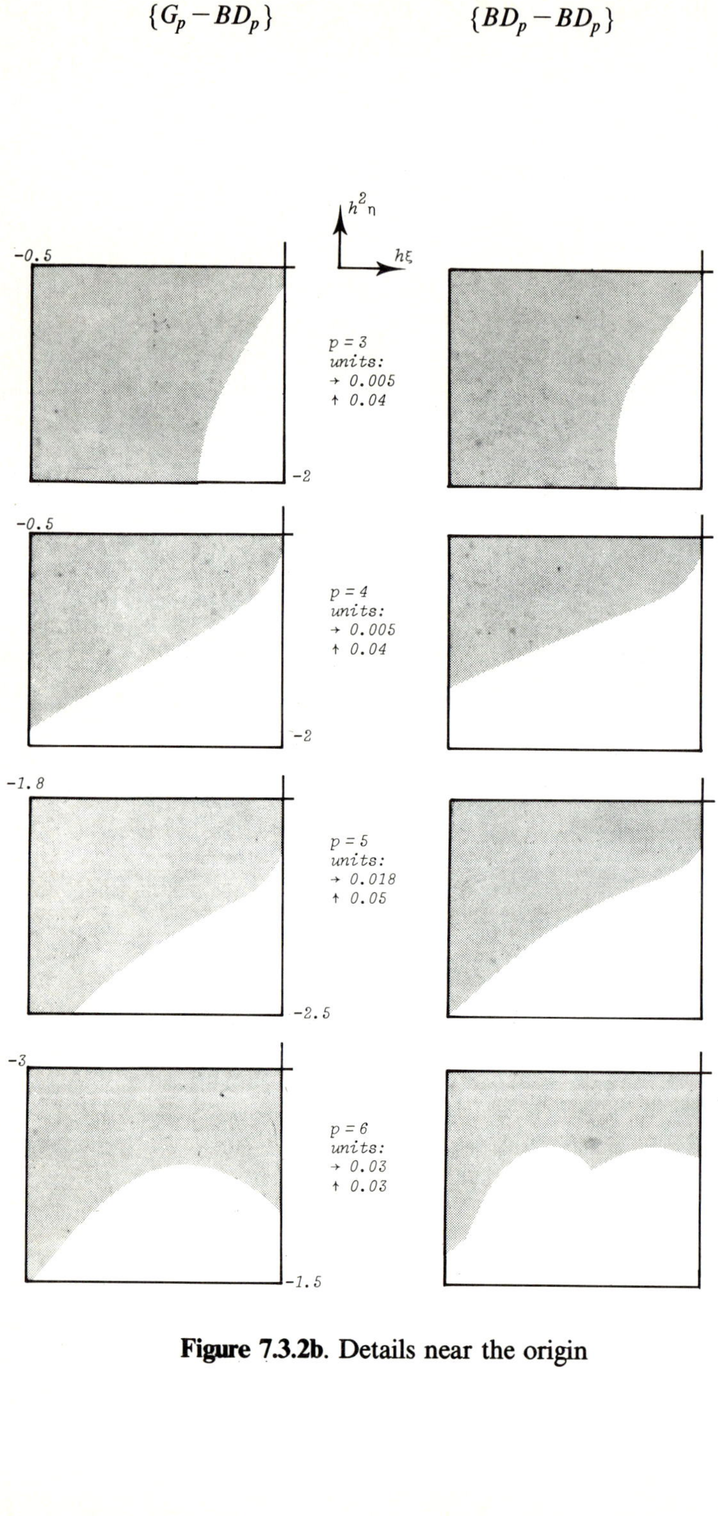

Figure 7.3.2b. Details near the origin

Corollary 7.3.5. *Let the LM method $\{\rho,\sigma\}$ be A-stable. Then the ILM $(\{\rho,\sigma\}-\{\rho,\sigma\})$ method is V_0-stable.*

Proof. By assumption, the stability region of $\{\rho,\sigma\}$ contains the left half-plane. It then follows from Theorem 7.3.12 that the stability region S_2 of the ILM method contains the points

$$\{(h\xi,h^2\eta)\colon \text{Re}\lambda<0, \quad \text{Re}\mu<0\}\supset\{(h\xi,h^2\eta)\colon \xi<0,\ \eta<0\}.$$

There remains to be shown that S_2 also contains the points $\{(h\xi,h^2\eta)\colon \xi<0,\ \eta=0\}$. Since for $\eta=0$ the convolution equation reduces to the basic test equation we can apply Theorem 7.3.3. Hence the strong stability region contains the left $h\xi$-axis. □

This result can be generalized by using a theorem of Matthys [1976, p. 90]; it follows from this result that the stability region S_2 of ILM $(\{\tilde{\rho},\tilde{\sigma}\}-\{\rho,\sigma\})$ contains the points $\{(h\xi,h^2\eta)\colon \xi<0,\ \eta<0\}$, provided that both $\{\tilde{\rho},\tilde{\sigma}\}$ and $\{\rho,\sigma\}$ are A-stable and that $\tilde{\rho}$ and σ do not have common zeros on the unit circle. It can be shown along the lines of the proof of Corollary 7.3.4 that S_2 contains the left $h\xi$-axis. Thus we have:

Theorem 7.3.13. *Let the LM methods $\{\tilde{\rho},\tilde{\sigma}\}$ and $\{\rho,\sigma\}$ be A-stable, and let $\tilde{\rho}$ and σ have no common zeros on the unit circle. Then the ILM $(\{\tilde{\rho},\tilde{\sigma}\}\ -\{\rho,\sigma\})$ method is V_0-stable.* □

Next, we turn to the special case where $\xi=0$. Then, by Lemma 7.3.3(b), the conditions which ensure that the ILM method has a non-empty interval of periodicity are: $\{\mathrm{P}=\rho\tilde{\rho},\ \Sigma=\sigma\tilde{\sigma}\}$ is symmetric and P has only one double root on the unit circle (notice that the hypotheses of Definition 7.3.5 are satisfied if both $\{\rho,\sigma\}$ and $\{\tilde{\rho},\tilde{\sigma}\}$ are consistent LM methods and if $\rho(0)\tilde{\rho}(0)\neq 0$).

Example 7.3.3. Consider the 3-step method

$$\{\rho,\sigma\} = \{(z-1)(z^2-az+1),\ \frac{1}{24}\ [(a+10)(z^3+1)+(14-13a)(z^2+z)]\},$$

where $|a|<2$. Evidently, ρ has three distinct zeros on the unit circle and σ is symmetric. In addition, we define $\{\tilde{\rho},\tilde{\sigma}\}$ in a similar way, with a replaced by $\tilde{a}$ where $\tilde{a}\neq a$ and $|\tilde{a}|<2$. It is easily verified that $\mathrm{P}=\rho\tilde{\rho}$ and $\Sigma=\sigma\tilde{\sigma}$ define a symmetric method and that P has only one double root. By Lemma 7.3.3(b) there is a non-vanishing interval of periodicity. □

7.3.3.2. First-kind equations

We confine our considerations to the (α,γ)-reducible ILM and MML methods.

Theorem 7.3.14. *Let S^* be the stability region of the LM method $\{\tilde{\rho},\tilde{\sigma}\}$, and let σ be a Schur polynomial. Then the set $\{(h\xi,h^2\eta)$: $-h\eta/\xi \in S^*\}$ defines the region of strong stability S_2 of the ILM $(\{\tilde{\rho},\tilde{\sigma}\}-\{\rho,\sigma\})$ method.*

Proof. The stability polynomial of ILM methods with $(\tilde{\rho},\tilde{\sigma})$-reducible lag term is given by

$$C_0(z) = -\sigma(z)[h\xi\tilde{\rho}(z)+h^2\eta\tilde{\sigma}(z)]$$

Since, by assumption, σ is a Schur polynomial, $C_0(z)$ is a Schur polynomial if $-h\eta/\xi$ lies in S^*. It then follows from Definition 7.2.2 that the points $(h\xi,h^2\eta)$ with $-h\eta/\xi \in S^*$ define S_2. □

Corollary 7.3.6. *Let the LM method $\{\tilde{\rho},\tilde{\sigma}\}$ be A_0-stable, and let σ be a Schur polynomial. Then the ILM $(\{\tilde{\rho},\tilde{\sigma}\}-\{\rho,\sigma\})$ method is V_0-stable.*

Proof. It follows from Theorem 7.3.14 that the set $\{(h\xi,h^2\eta)$: $\xi\eta>0\}$ lies in the stability region S_2 of the ILM method. According to Definition 7.3.4 (ii) there only remains to be shown that the whole ξ-axis lies also in S_2. This can be deduced from Theorem 7.3.6. □

Theorem 7.3.15. *Let S^* be the stability region of the LM method $\{\tilde{\rho}\sigma,\tilde{\rho}\sigma^\# -\rho^\#\tilde{\sigma}\}$. Then the set $\{(h\xi,h^2\eta)$: $-h\eta/\xi \in S^*\}$ defines the region of strong stability of the MML $(\{\tilde{\rho},\tilde{\sigma}\}-\{\rho,\sigma\})$ method.*

Proof. The stability polynomial of MML methods with $(\tilde{\rho},\tilde{\sigma})$-reducible lag term is given by

$$C_0(z) = -h\xi\tilde{\rho}(z)\sigma(z)+h^2\eta[\rho^\#(z)\tilde{\sigma}(z)-\tilde{\rho}(z)\sigma^\#(z)].$$

Proceeding as in the proof of Theorem 7.3.14 we readily arrive at the result of the present theorem. □

We remark that $\{\tilde{\rho}\sigma,\tilde{\rho}\sigma^\# -\rho^\#\tilde{\sigma}\}$ *formally* represents an LM method, that is, it does not need to be consistent or even zero-stable.

Example 7.3.4. Let $\{\rho,\sigma\}$ and $\{\tilde{\rho},\tilde{\sigma}\}$ respectively represent the BD_2 formula and the G_2 formula (trapezoidal rule). Then, σ is a Schur polynomial and $\{\tilde{\rho},\tilde{\sigma}\}$ is A_0-stable. Consequently, by virtue of Corollary 7.3.6, the corresponding ILM method is V_0-stable. For the MML version we have to consider the LM method

$$\{\tilde{\rho}\sigma,\tilde{\rho}\sigma^\# -\rho^\#\tilde{\sigma}\}=\{2z^2(z-1),\ (z+1)(2z+1)\}.$$

Applying Lemma 7.3.2 we find that the stability region S^* of this LM method contains the interval $(-2+\sqrt{2},0)$. Hence, the MML method has the (real) stability region

$$-2+\sqrt{2}<-h\frac{\eta}{\xi}<0.$$

Assuming that $\eta\xi>0$ this results in the stability condition

$$h<(2-\sqrt{2})\frac{\xi}{\eta}. \quad \square$$

7.3.4. *Stability results for polynomial convolution kernels*

The techniques used in the preceding section can, in principle, be applied to polynomial convolution kernels of the type (7.1.12). A motivation for considering such kernels might be that often the given kernel k can reasonably well be approximated by a truncated Taylor series:

$$k(t,s,y) \approx \sum_{i=0}^{m} \frac{1}{i!}(t-s)^i \frac{\partial^i k}{\partial t^i}(s,s,y). \tag{7.3.17}$$

This kernel is of the form (7.1.12) with $\phi_i(s,y) := (\partial^i k/\partial t^i)(s,s,y)$.

The derivation of characteristic polynomials associated with polynomial convolution kernels was carried out in Amini et al. [1983] for DQ methods and extended in Amini [1983] for ML and MML methods. The resulting characteristic polynomials, however, become increasingly complicated as m increases, and it is doubtful whether they are of practical use for, say, $m>2$. For the sake of completeness we give the basic recurrence relation for lag terms that are $(\tilde\rho,\tilde\sigma)$-reducible.

Theorem 7.3.16. *For polynomial convolution kernels of the form (7.1.12) the lag term (3.2.13) satisfies the relation*

$$\tilde\rho^m(E)\tilde F_{n-i}(t_{n+j}) = \tilde\rho^m(E)g_0(t_{n+j}) \tag{7.3.18}$$
$$+ \sum_{\mu=0}^{m} \frac{1}{\mu!} h^{\mu+1}\tilde\rho^{m-\mu}(E)[P_{i+j,\mu}(E)E^{-i}\phi_\mu(t_n,y_n)+Q_\mu(E)\tilde f_{n-i}^{(\mu)}],$$

$$\tilde f_{n-i}^{(\mu)} := \sum_{l=0}^{n-i} w_{n-i,l}\phi_\mu(t_l,y_l),$$

where the polynomials $P_{q,\mu}$ and Q_μ are defined recursively by

$$P_{q,0} = 0, \quad P_{q,\mu} = \tilde\rho^{\mu-1}\tilde\sigma_{q,\mu} - \sum_{r=0}^{\mu-1}\tilde\rho_{r,\mu}\tilde\rho^{\mu-1-r}P_{q,r},$$

$$Q_0 = 1, \quad Q_\mu = -\sum_{q=0}^{\mu-1}\tilde\rho_{q,\mu}\tilde\rho^{\mu-1-q}Q_q,$$

$$\tilde\rho_{q,\mu} := \sum_{r=0}^{\tilde\kappa}\tilde a_r\binom{\mu}{q}r^{\mu-q}z^{\tilde\kappa-r}, \quad \tilde\sigma_{q,\mu} := \sum_{r=0}^{\tilde\kappa}\tilde b_r(q+r)^\mu z^{\tilde\kappa-r}.$$

Proof. We write the lag term in the form

$$\tilde{F}_{n-i}(t_{n+j}) = g_0(t_{n+j}) + \sum_{\mu=0}^{m} \frac{1}{\mu!} h^{\mu+1} \tilde{F}_n^{(\mu,\mu)}, \tag{7.3.19}$$

where

$$\tilde{F}_n^{(\nu,\mu)} := \sum_{l \geq 0} w_{n-i,l}(n+j-l)^{\mu} \phi_{\nu}(t_l,y_l), \quad w_{n-i,l>n-i} := 0.$$

By means of the identity

$$(n+j-r-l)^{\mu} = (n+j-l)^{\mu} - \sum_{q=0}^{\mu-1} (n+j-r-l)^q r^{\mu-q},$$

a recurrence relation for $\tilde{F}_n^{(\nu,\mu)}$ can be derived. To that end we form a linear combination of $\{\tilde{F}_{n-r}^{(\nu,\mu)}\}$ with coefficients $\tilde{a}_r$, and we apply Lemma 7.3.1(a):

$$\begin{aligned}
\tilde{\rho}(E)\tilde{F}_{n-\tilde{\kappa}}^{(\nu,\mu)} &= \sum_{r=0}^{\tilde{\kappa}} \tilde{a}_r \sum_{l \geq 0} w_{n-r-i,l}(n-r+j-l)^{\mu} \phi_{\nu}(t_l,y_l) \\
&= \sum_{l \geq 0} [(n+j-l)^{\mu} \phi_{\nu}(t_l,y_l) \sum_{r=0}^{\tilde{\kappa}} \tilde{a}_r w_{n-r-i,l} \\
&\quad - \sum_{q=0}^{\mu-1} \sum_{r=0}^{\tilde{\kappa}} [\tilde{a}_r \binom{\mu}{q} r^{\mu-q} \sum_{l \geq 0} w_{n-r-i,l}(n-r+j-l)^q \phi_{\nu}(t_l,y_l)] \\
&= \sum_{l=n-i-\tilde{\kappa}}^{n-i} (n+j-l)^{\mu} \phi_{\nu}(t_l,y_l) \tilde{b}_{n-i-l} - \sum_{q=0}^{\mu-1} \sum_{r=0}^{\tilde{\kappa}} \tilde{a}_r \binom{\mu}{q} r^{\mu-q} \tilde{F}_{n-r}^{(\nu,q)}.
\end{aligned}$$

Introducing the polynomials $\tilde{\rho}_{q,\mu}$ and $\tilde{\sigma}_{q,\mu}$ this recurrence relation takes the form

$$\tilde{\rho}(E)\tilde{F}_n^{(\nu,\mu)} = \tilde{\sigma}_{i+j,\mu}(E)E^{-i}\phi_{\nu}(t_n,y_n) - \sum_{q=0}^{\mu-1} \tilde{\rho}_{q,\mu}(E)\tilde{F}_n^{(\nu,q)}.$$

Repeated application of this recursion finally leads to a relation of the form

$$\tilde{\rho}^{\mu}(E)\tilde{F}_n^{(\mu,\mu)} = P_{i+1,\mu}(E)E^{-1}\phi_{\mu}(t_n,y_n) + Q_{\mu}(E)\tilde{F}_n^{(\mu,0)},$$

and substitution into (7.3.19) yields the relation (7.3.18), where we have set $\tilde{f}_{n-i}^{(\mu)} := \tilde{F}_n^{(\mu,0)}$. □

By means of this theorem fixed-term recurrence relations can be derived for VLM methods. For (α,γ)-reducible methods we see that, by premultiplication of (7.3.1) by $\tilde{\rho}^m(E)$ and on substitution of (7.3.18), the quantities $\tilde{f}_{n-i}^{(\mu)}$ vanish, so that a recurrence relation of finite length is obtained. For non-(α,γ)-reducible methods we should premultiply by $\tilde{\rho}^{m+1}(E)$ and apply Lemma 7.3.1(a) to eliminate the quantities $\tilde{f}_{n-i}^{(\mu)}$.

7.3.5. Stability results for finitely decomposable kernels

The recurrence relations that can be derived for polynomial convolution kernels are rather complicated. This is due to the fact that they are actually discretizations of a *higher-order differential equation for* $y(t)$ (see in this connection Theorem 7.1.4). However, it has already been observed that for finitely decomposable kernels, such as (7.1.19), we can associate with a (second-kind) integral equations *a system of first-order differential equations for the vector function*

$$\mathbf{u}(t) := \int_{t_0}^{t} \mathbf{f}(s,y(s))ds \tag{7.3.20}$$

(see Section 7.1.5). We shall show that the recurrence relation for the numerical approximation $\mathbf{u}_n$ to $\mathbf{u}(t_n)$ is more suitable for the analysis of stability (recall that the class of decomposable kernels includes the polynomial convolution kernels).

In the following theorem we use the notation

$$\mathbf{f}_n = \mathbf{f}(t_n,y_n), \quad \mathbf{g}_n = \mathbf{g}(t_n). \tag{7.3.21}$$

Theorem 7.3.17. *Let k be defined by (7.1.19), and let the quadrature formula for the lag term be* $(\tilde{\rho},\tilde{\sigma})$*-reducible, with* $\tilde{\rho}(1)=0$*. Then the VLM method {(3.2.13); (7.3.1)} can be written in the form of the finite recursion*

$$\tilde{\rho}(E)\mathbf{u}_n = h\tilde{\sigma}(E)\mathbf{f}_n, \quad n \geqslant 0, \tag{7.3.22a}$$

$$\theta\alpha(E)y_{n-\kappa} = \sum_{i=0}^{\kappa} \sum_{j=-\kappa}^{\kappa} [\gamma_{i,j} h <\mathbf{g}_{n+j},\mathbf{f}_{n-i}> - \beta_{i,j}(g_0(t_{n+j})$$

$$+ <\mathbf{g}_{n+j},\mathbf{u}_{n-i}>)], \tag{7.3.22b}$$

where the vectors $\mathbf{u}_j$ *satisfy the starting conditions*

$$\mathbf{u}_j = h\sum_{l=0}^{\tilde{\kappa}-1} w_{n,l}\mathbf{f}_l, \quad j=0, \ldots ,\tilde{\kappa}-1. \tag{7.3.22c}$$

Proof. It follows from Lemma 7.3.1(a) and from $\tilde{\rho}(1)=0$ that

$$\tilde{\rho}(E)\tilde{F}_n(t) = h\tilde{\sigma}(E)k(t,t_n,y_n), \tag{7.3.23}$$

and from (7.1.19) and the definition of the lag term that

$$\tilde{F}_n(t) = g_0(t) + <\mathbf{g}(t), \mathbf{u}_n>, \quad \mathbf{u}_n := h\sum_{l=0}^{n} w_{n,l}\mathbf{f}_l. \tag{7.3.24}$$

Substitution of (7.3.24) into (7.3.23) yields

$$\tilde{\rho}(E)[g_0(t) + <\mathbf{g}(t), \mathbf{u}_n>] = h\tilde{\sigma}(E)<\mathbf{g}(t),\mathbf{f}_n>.$$

Using again $\tilde{\rho}(1)=0$, the above relation reduces to

$$<\mathbf{g}(t),\tilde{\rho}(E)\mathbf{u}_n - h\tilde{\sigma}(E)\mathbf{f}_n> \; = \; 0.$$

This relation can only be true if the vector $\mathbf{u}_n$ satisfies the equation (7.3.22a). The equation (7.3.22b) is immediate from substitution of (7.1.19) into the VLM formula (7.3.1) and using the representation (7.3.24) of the lag term. Finally, the starting conditions (7.3.22c) follow from (7.3.24). □

Unlike the recurrence relations derived for convolution kernels (see, for example, (7.3.23)), the relations (7.3.22) do not reduce to a linear difference equation with *constant coefficients when k is linearized.* Nevertheless, it provides some insight into the stability of the numerical schemes, as we shall see in the following subsections.

7.3.5.1. Connections with linear methods for ODEs

We will show that, for $\theta=1$, there exists a direct relation between DQ and ILM methods at the one hand, and linear methods applied to certain differential equations on the other hand.

Theorem 7.3.18. *Let the kernel be of the finitely decomposable form (7.1.19). Then the sequence* $\{y_n\}$, *obtained by a* $(\tilde{\rho},\tilde{\sigma})$*-reducible DQ method, is identical with the sequence* $\{g_0(t_n)+<\mathbf{g}_n,\mathbf{u}_n>\}$, *where* $\{\mathbf{u}_n\}$ *is the result of applying the LM method* $\{\tilde{\rho},\tilde{\sigma}\}$ *to the system of differential equations*

$$\mathbf{u}'(t) = \mathbf{f}(t,g_0(t)+<\mathbf{g}(t),\mathbf{u}(t)>), \tag{7.3.25}$$

with starting values (7.3.22c).

Proof. According to Theorem 7.3.17 the DQ method is algebraically equivalent to

$$\begin{aligned} \tilde{\rho}(E)\mathbf{u}_n &= h\tilde{\sigma}(E)\mathbf{f}(t_n,g_0(t_n)+<\mathbf{g}_n,\mathbf{u}_n>) \\ y_n &= g_0(t_n)+<\mathbf{g}_n,\mathbf{u}_n> \end{aligned} \quad , \tag{7.3.26}$$

with the starting values (7.3.22c). Evidently, these relations are also obtained if we apply the LM method $\{\tilde{\rho},\tilde{\sigma}\}$ to (7.3.25). □

This relationship with LM methods for ODEs suggests the application of the ODE stability theory to (7.3.26), in order to derive stability criteria for DQ methods. Since, in actual computation, the *linear* stability theory for ODEs has proved to be reliable we may have confidence in the resulting stability conditions.

The starting point of such a linear stability analysis is the possibility of diagonalizing the linearized version of the recurrence for $\mathbf{u}_n$ in (7.3.26). That implies that the Jacobian matrix $A:=\partial\mathbf{f}/\partial y\cdot\mathbf{g}^T$ must have m linearly independent eigenvectors. As we have shown in Section 7.1.5.1, A does indeed have m such eigenvectors $\{\mathbf{v}_j\}$. Therefore, we can apply the analysis developed for ODEs (see, e.g., Lambert [1973, p. 220]). The resulting stability criterion is that $\{\mathbf{u}_n\}$ is stable if the characteristic polynomial $\tilde{\rho}(z)-h\alpha_j\tilde{\sigma}(z)$ satisfies the root condition for all eigenvalues α_j of A; if the characteristic polynomial is a Schur polynomial for all α_j, then we have strong (or absolute) stability (cf. Definition 2.6.2 (iv)). Since the eigenvalues are given by

$$\alpha_j = 0, \quad j=1, \ldots ,m-1; \quad \alpha_m = \frac{\partial k}{\partial y}(t_n,t_n,y_n)$$

(see Section 7.1.5.1), we conclude that $\{\mathbf{u}_n\}$ is stable if $h\partial k/\partial y$ lies in the stability region of $\{\tilde{\rho},\tilde{\sigma}\}$. However, we want the sequence $\{y_n\}$ to be stable. By expanding $\mathbf{u}_n$ in terms of the eigenvectors $\mathbf{v}_j$ of A and by observing that the first $m-1$ eigenvectors are orthogonal to $\mathbf{g}_n$, we find

$$y_n = g_0(t_n)+<\mathbf{g}_n,\mathbf{u}_n> = g_0(t_n)+<\mathbf{g}_n, \sum_{j=1}^{m} c_j(n)\mathbf{v}_j>$$

$$= g_0(t_n)+c_m(n)<\mathbf{g}_n,\mathbf{v}_m>,$$

so that we only have to consider the stability of the component corresponding to α_m. On the other hand, we must now also require that $\mathbf{g}(t)$ be bounded as $t\to\infty$. Summarizing, we have the result that the linearized recursion (7.3.26) yields a *strongly stable solution* $\{y_n\}$ *if* $\mathbf{g}(t)$ *is bounded and if* $h\partial k/\partial y$ *lies in the stability region of* $\{\tilde{\rho},\tilde{\sigma}\}$.

The condition on $h\partial k/\partial y$ is similar to the stability condition obtained from the basic test equation; the condition on $\mathbf{g}(t)$, however, is new and impractical in actual computation, because we often do not know the decomposition of k, or it may not even exist (recall that we assumed the (approximate) decomposition of k only for the purpose of deriving practically useful stability conditions; the application of the DQ method itself is, of course, not dependent on this assumption).

Next we consider ILM methods.

Theorem 7.3.19. *Let the kernel be of the finitely decomposable type (7.1.19). Then the sequence* $\{y_n\}$, *obtained by the ILM* $(\{\tilde{\rho},\tilde{\sigma}\}-\{\rho,\sigma\})$ *method with* $\tilde{\kappa}=\kappa$, *is identical with the sequence* $\{y_n\}$ *obtained by applying the LM methods* $\{\tilde{\rho},\tilde{\sigma}\}$ *and* $\{\rho,\sigma\}$, *respectively, to the differential equations*

$$\mathbf{u}'(t) = \mathbf{f}(t,y(t)) \tag{7.3.27a}$$

$$y'(t) = k(t,t,y(t))+\frac{1}{h}\tau(E)g_0(t)+\frac{1}{h}<\tau(E)\mathbf{g}(t),\mathbf{u}(t)>. \tag{7.3.27b}$$

Here, h *is the integration step used by the ILM method and by the LM methods, and* $\tau(z):=-\sum_{i=0}^{\kappa}d_iz^{\kappa-i}$, *with* d_i *given in Table 3.2.3; furthermore, the starting values* $\mathbf{u}_j$, $j=0, \ldots ,\kappa-1$, *satisfy the starting conditions (7.3.22c).* □

The proof follows immediately from Theorem 7.3.17 and the definition of the ILM method.

In Söderlind [1979] the use of different LM methods, for integrating a compound system of ODEs like (7.3.27), was called a *linear multistep compound* (LMC) method. The linear stability theory for such methods proceeds as follows. Let

$$\mathbf{x}'(t) = A\mathbf{x}, \quad \mathbf{x} := [\mathbf{u},y]^T, \tag{7.3.28}$$

be a test equation with constant matrix A. Then the behavior of the solution is characterized by the zeros of the characteristic polynomial

$$\det[P(z) - hA\Sigma(z)],$$

where

$$P(z) := \begin{bmatrix} \tilde{\rho}(z)I_m & 0 \\ 0 & \rho(z) \end{bmatrix}, \quad \Sigma(z) := \begin{bmatrix} \tilde{\sigma}(z)I_m & 0 \\ 0 & \sigma(z) \end{bmatrix},$$

with I_m denoting the m by m identity matrix. If the characteristic polynomial is a Schur polynomial, then the LMC solution tends to 0, as $t_n \to \infty$, and is called *strongly (or absolutely) stable (with respect to (7.3.28))*.

In the case of (7.3.27) it is natural to choose for A the Jacobian matrix

$$A = \begin{bmatrix} 0 & \dfrac{\partial \mathbf{f}}{\partial y} \\ h^{-1}(\tau(E)\mathbf{g})^T & \dfrac{\partial k}{\partial y} \end{bmatrix}, \tag{7.3.29}$$

where $\mathbf{g}$, $\mathbf{f}$ and k are evaluated at some suitable point $(\bar{t},\bar{y})$. This matrix has $m-1$ zero eigenvalues and two eigenvalues satisfying the equation

$$\alpha^2 - \frac{\partial k}{\partial y}\alpha - h^{-1}\tau(E)\frac{\partial k}{\partial y} = 0,$$

where the shift operator E affects only the first argument of $\partial k/\partial y$. It is now easily verified that the characteristic polynomial associated with (7.3.29) reduces to

$$\tilde{\rho}(z)[\rho(z) - h\frac{\partial k}{\partial y}\sigma(z)] - h\tau(E)\frac{\partial k}{\partial y}\sigma(z)\tilde{\sigma}(z). \tag{7.3.30}$$

Notice the resemblance with the characteristic polynomial (7.3.16c) associated with the convolution equation. In particular, we have the result, analogous to Theorem 7.3.13: *if* $\{\tilde{\rho},\tilde{\sigma}\}$ *and* $\{\rho,\sigma\}$ *are A-stable, then the LMC* $(\{\tilde{\rho},\tilde{\sigma}\} - \{\rho,\sigma\})$ *method is strongly stable for all* $h\,\partial k/\partial y < 0$ *and* $h\tau(E)\partial k/\partial y \leq 0$. It may be expected that stability conditions derived from (7.3.30) for the LMC method also apply, in view of Theorem 7.3.18, to the ILM method.

7.3.5.2. A local stability result

In general, a VLM method cannot be viewed as an LM or LMC method for a system of ODEs. Therefore, we now consider the stability analysis of the general recursion (7.3.22). It has already been remarked that this recursion does not have constant coefficients and we shall, just as in the stability analysis of LM recursions for ODEs, "freeze" the coefficients in the recursion.

Firstly, we set $k(t,s,y)=K(t,s)y$, with $K(t,s)=\langle \mathbf{g}(t),\mathbf{f}(s)\rangle$, and we write (7.3.22) in the form

$$\sum_{i=0}^{\kappa^*} M_i(n)\mathbf{y}_{n-i} = \mathbf{w}_n, \quad \kappa^* := \max\{\kappa,\tilde{\kappa}\}, \tag{7.3.31}$$

where

$$\mathbf{y}_n := \begin{bmatrix} y_n \\ \mathbf{u}_n \end{bmatrix}, \quad M_i := \begin{bmatrix} \theta\alpha_i - h\sum_{j=-\kappa}^{\kappa}\gamma_{i,j}K(t_{n+j},t_{n-i}) & \sum_{j=-\kappa}^{\kappa}\beta_{i,j}\mathbf{g}_{n+j}^T \\ -\tilde{b}_i h\mathbf{f}_{n-i} & \tilde{a}_i I_m \end{bmatrix},$$

$$\mathbf{w}_n := \begin{bmatrix} -\sum_{i=0}^{\kappa}\sum_{j=-\kappa}^{\kappa}\beta_{i,j}g_0(t_{n+j}) \\ \mathbf{0} \end{bmatrix},$$

with the assumption that α_i, $\beta_{i,j}$, $\gamma_{i,j}$ and $\tilde{a}_i$, $\tilde{b}_i$ vanish if $i>\kappa$ and $i>\tilde{\kappa}$, respectively.

The easiest way to freeze the coefficients in (7.3.31) is to assume that K,$\mathbf{g}$ and $\mathbf{f}$ be constant (in fact, this is the approach in the analysis of LM recursions in ODEs). A less drastic approach is the assumption that $M_i(n)$ be a constant matrix, say $M_i(n)=M_i(\bar{n})$, where $\bar{n}$ is fixed. Such an assumption does not completely ignore variations of the kernel with t and s, and seems to be realistic if $K(t,s)$ is slowly varying. It may be considered as a *local* approach. According to Definition 7.2.6, the method is *locally stable* if the characteristic polynomial

$$C_\theta(z) := \det\left(\sum_{i=0}^{\kappa^*} M_i(n)z^{\kappa^*-i}\right) \tag{7.3.32}$$

has zeros which correspond either to strongly or to weakly stable eigenvalues of the companion matrix (i.e., the matrix M in the one-step recursion associated with (7.3.31)). In order to derive a more manageable stability condition, we need the following lemma:

Lemma 7.3.4. *Suppose that A, B, C_i and D_i, $i=1,2, \ldots ,m$, are square matrices of the same order. If the matrices D_i commute with B $(i=1, \ldots ,m)$, then*

$$\det\begin{pmatrix} A & C_1 & \cdots & C_m \\ D_1 & B & & O \\ \vdots & & \ddots & \\ D_m & O & & B \end{pmatrix} = \det(AB - \sum_{i=1}^{m} C_iD_i)[\det B]^{m-1}.$$

Proof. We first prove the lemma for regular matrices B. Since

$$\begin{pmatrix} A & C_1 & \cdots & C_m \\ D_1 & B & & O \\ \vdots & & \ddots & \\ D_m & O & & B \end{pmatrix} \begin{pmatrix} I & & & \\ -B^{-1}D_1 & B^{-1} & & O \\ \vdots & & \ddots & \\ -B^{-1}D_m & O & & B^{-1} \end{pmatrix}$$

$$= \begin{pmatrix} A - \sum_{i=0}^{m} C_iB^{-1}D_i & C_1B^{-1} & \cdots & C_mB^{-1} \\ & I & & O \\ & & \ddots & \\ O & & & I \end{pmatrix}$$

we find straightforwardly the result of the lemma. If B is singular we define $B^*(\epsilon):=B+\epsilon I$, which is a regular matrix for all sufficiently small scalars $|\epsilon|$ with $\epsilon \neq 0$. Furthermore,

$$B^*(\epsilon)D_m = (B+\epsilon I)D_m = D_mB+D_m\epsilon I = D_mB^*(\epsilon).$$

Hence, we can apply the lemma with B replaced by the regular matrix $B^*(\epsilon)$. By observing that both sides of this equation are continuous in ϵ, we may put $\epsilon=0$. Thus, it holds also for $B^*(0)$. However, $B^*(0)=B$, so that we have proved the lemma for singular matrices B, too. □

Theorem 7.3.20. *Let the lag term be $(\tilde{\rho},\tilde{\sigma})$-reducible. Then the VLM method is locally stable if the polynomial*

$$C_\theta^*(z) := \sum_{i=0}^{\kappa^*} \sum_{j=0}^{\kappa^*} [\theta\alpha_i\tilde{a}_j + h \sum_{l=-\kappa}^{\kappa} (\beta_{i,l}\tilde{b}_j - \gamma_{j,l}\tilde{a}_i)K(t_{n+l}, t_{n-j})]z^{2\kappa^* - i - j} \tag{7.3.33}$$

is a Schur polynomial. This condition is a necessary condition for bounded error propagation due to an isolated perturbation.

Proof. Substitution of the matrices M_i corresponding to (7.3.31) into (7.3.32) yields

$$C_\theta(z) = \det \begin{pmatrix} \sum_{i=0}^{\kappa^{\bullet}} [\theta\alpha_i - h \sum_{j=-\kappa}^{\kappa} \gamma_{i,j} K(t_{n+j}, t_{n-i})] z^{\kappa^{\bullet}-i} & \sum_{i=0}^{\kappa^{\bullet}} \sum_{j=-\kappa}^{\kappa} \beta_{i,j} \mathbf{g}_{n+j}^T z^{\kappa^{\bullet}-i} \\ -h \sum_{i=0}^{\kappa^{\bullet}} \tilde{b}_i \mathbf{f}_{n-i} z^{\kappa^{\bullet}-i} & \sum_{i=0}^{\kappa^{\bullet}} \tilde{a}_i z^{\kappa^{\bullet}-i} I_m \end{pmatrix}.$$

If we use Lemma 7.3.4 there results

$$C_\theta(z) = z^{(\kappa^{\bullet}-\tilde{\kappa})m-1} (\tilde{\rho}(z))^{m-1} C_\theta^*(z),$$

where $C_\theta^*(z)$ is given by (7.3.33).

According to Definition 7.2.3, all zeros of $C_\theta(z)$ should correspond to either strongly or weakly stable eigenvalues of the companion matrix M. Since $C_\theta^*(z)$ is a Schur polynomial by assumption, it remains to be shown that the roots of $\tilde{\rho}(z)$, which are on the unit circle, correspond to weakly stable eigenvalues of the companion matrix M. Let μ_0 be such a root; since $\tilde{\rho}$ satisfies the root condition, this root is simple and therefore it is a root of $C_\theta(z)$ with multiplicity $m-1$. It can be shown that there exist $m-1$ linearly independent eigenvectors $\mathbf{e}(\mu_0)$ of M corresponding to the eigenvalue μ_0: writing the companion matrix M in the form (7.2.8) and $\mathbf{e}(\mu_0)$ as the compound vector $(\mathbf{e}_1^T, \dots, \mathbf{e}_{\kappa^{\bullet}}^T)^T$, we derive from $M\mathbf{e}(\mu_0) = \mu_0 \mathbf{e}(\mu_0)$ the relations

$$\mathbf{e}_1 = \mu_0^{\kappa^{\bullet}-1} \mathbf{e}_{\kappa^{\bullet}}, \dots, \mathbf{e}_{\kappa^{\bullet}-1} = \mu_0 \mathbf{e}_{\kappa^{\bullet}}, \quad \left[\sum_{i=0}^{\kappa^{\bullet}} M_i \mu_0^{\kappa^{\bullet}-i} \right] \mathbf{e}_{\kappa^{\bullet}} = \mathbf{0}. \qquad (7.3.34)$$

Thus, the number of linearly independent eigenvectors $\mathbf{e}(\mu_0)$ equals the number of linearly independent vectors $\mathbf{e}_{\kappa^{\bullet}}$ satisfying (7.3.34), and the proof is completed if we can show that (7.3.34) has $m-1$ linearly independent solutions.

This last step of the proof uses the specific form of the matrices M_i and parallels the proof of a similar theorem given in van der Houwen & Wolkenfelt [1980], where the local stability of $(\tilde{\rho},\tilde{\sigma})$-reducible DQ methods was studied. Here, we refer to this reference for details. □

A convenient feature of the stability condition in Theorem 7.3.20 is that $C_\theta^*(z)$ *does not depend on the decomposition of* $K(t,s)$. Thus, only the *possibility* of separating $K(t,s)$ is used; the specific form of the decomposition, however, is not reflected into the characteristic equation. It has already been observed in our discussion of polynomial convolution kernels (see Section 7.1.4.2) that the use of test kernels is justified if they approximate sufficiently closely the actually given kernel. In the present case, the class of *continuous, separable kernels is dense,* by virtue of the Stone-Weierstrass theorem, *in the class of all continuous kernels.* This property of separable kernels, together with the fact that the decomposition is not needed to derive $C_\theta^*(z)$, makes Theorem 7.3.20 a useful starting point for establishing stability conditions.

The stability conditions based on $C_\theta^*(z)$ depend on the step number n, unless the kernel is of *convolution type,* i.e., $k(t,s,y)=K(t-s)y$. In such cases, Theorem 7.3.20 enables us to establish *a priori* stability conditions for a particular method.

If k is not of convolution type it will, in general, not be feasible to "check" $C_\theta^*(z)$ at every integration step. However, since $C_\theta^*(z)$ depends only on the behavior of $K(t,s)$ in the neighborhood of the diagonal $s=t$ in the (t,s)-plane, one can get a first indication of the stability of the method by writing

$$K(t_{n+l},t_{n-j}) = \xi+\eta lh-\zeta jh+O(h^2),$$
$$\xi := K(t_n,t_n), \quad \eta := K_t(t_n,t_n), \quad \zeta := K_s(t_n,t_n),$$

and by using a *linear approximation* to $K(t,s)$ in $C_\theta(z)$.

Substitution into (7.3.33) yields, for a $(\tilde\rho,\tilde\sigma)$-reducible VLM formula,

$$C_\theta^*(z) = z^{2\kappa^*-\kappa-\tilde\kappa}\{\theta\alpha(z)\tilde\rho(z)+h\xi[\beta(z)\tilde\sigma(z)-\tilde\rho(z)\gamma(z)]$$
$$+ h^2\eta[\beta^*(z)\tilde\sigma(z)-\tilde\rho(z)\gamma^*(z)]-h^2\zeta[\beta(z)\tilde\sigma^{\#}(z)-\tilde\rho(z)\gamma^{\#}(z)]\}, \tag{7.3.35}$$

where $\tilde\sigma^{\#},\gamma^{\#}$ and β^*,γ^* are defined in (7.3.2) and (7.3.13), respectively. We remark that, in the case of the basic test equation ($\eta=\zeta=0$), this polynomial reduces to the stability polynomials given in the Corollaries 7.3.1 and 7.3.2. For the convolution test equation ($\zeta=-\eta$) and (α,γ)-reducible VLM formulas ($\beta=0$), we are led back to the stability polynomial given in Corollary 7.3.3; however, if $\beta\not\equiv 0$, then (7.3.35) differs from the polynomial given in Corollary 7.3.4, demonstrating that $C_\theta^*(z)$ is not a stability polynomial in the sense of Definition 7.2.3.

For DQ methods we obtain the polynomial

$$C_\theta^*(z) = z^{\tilde\kappa-1}\{\theta\tilde\rho(z)-h\xi\tilde\sigma(z)+h^2\zeta\tilde\sigma^{\#}(z)\}, \tag{7.3.36}$$

and for ILM methods we find (using $\beta^*=-\sigma,\ \gamma^*=-\sigma^{\#}$)

$$C_\theta^*(z) = z^{2\kappa^*-\tilde\kappa-\kappa}\{\theta\rho(z)\tilde\rho(z)-h\xi\tilde\rho(z)\sigma(z)-\eta h^2[\sigma(z)\tilde\sigma(z)-\tilde\rho(z)\sigma^{\#}(z)]$$
$$+h^2\zeta\tilde\rho(z)\sigma^{\#}(z)\}. \tag{7.3.37}$$

From these polynomials practical stability conditions can be derived which furnish *necessary* conditions for a stable performance of the method.

If we are dealing with systems of integral equations, Theorem 7.3.20 remains valid, provided that we replace K by the Jacobian matrix of the kernel and if we change $C_\theta^*(z)$ to $\det(C_\theta^*(z))$.

Example 7.3.5. Consider the DQ(BD_p) methods. Then it follows from (7.3.36) that

$$C_\theta^*(z) = z^{p-1}\{\theta\tilde\rho(z)-h\xi\tilde\sigma(z)\},$$

where $\{\tilde\rho,\tilde\sigma\}$ generates the BD_p formula. For $\theta=1$, this polynomial is a Schur polynomial for all values of $h\xi$ in the stability region of BD_p; for $\theta=0$, we have local stability (for linear kernels) if $h\xi\in\mathbb{C}$. □

Example 7.3.6. Consider ILM $(\{\tilde{\rho},\tilde{\sigma}\}-\mathrm{BD}_p)$ methods. It follows from (7.3.37) that (since $\sigma^{\#}=0$)

$$C_\theta^*(z) = z^{2\kappa^*-\tilde{\kappa}-\kappa}\{\theta\rho(z)\tilde{\rho}(z)-h\xi\tilde{\rho}(z)\sigma(z)-h^2\eta\sigma(z)\tilde{\sigma}(z)\}.$$

This polynomial is identical with (7.3.16c) and consequently the analysis given for (7.3.16c) carries over to (7.3.37). □

7.3.6. Stability results for positive definite convolution kernels

In this section we summarize results obtained by Lubich [1983a] for second-kind Volterra equations with positive definite convolution kernels.

Lubich considered equations of the form

$$y(t) = g_0(t)+\lambda\int_0^t a(t-s)y(s)ds, \quad \mathrm{Re}\lambda<0, \tag{7.3.38}$$

where the continuous function $a(t)\in L^1(\mathbb{R}_+)$, $\{a(nh)\}_0^\infty\in l^1$, is a positive definite function.

Before presenting the results of Lubich, we first summarize a few relevant properties of positive definite functions (see, e.g., Edwards [1979, Chapter 9]).

Bochner was presumably the first who studied positive definite functions. His definition and its extension to L^1-functions is given in

Definition 7.3.6. (i) A continuous function a is *positive definite* if

$$\sum_{j,l=1}^{n} a(t_j-t_l)\bar{z}_j z_l\geqslant 0 \tag{7.3.39}$$

holds for any finite sequence $\{t_j\}_{j=1}^n$ of real numbers and any finite sequence $\{z_j\}_{j=1}^n$ of complex numbers.
(ii) A function $a\subset L^1$ is positive definite if $\int\int a(t-s)\overline{u(t)}u(s)dt\,6\,ds\geqslant 0$ for each continuous function u. □

Theorem 7.3.21. (a) *A function $a\in L^1$ is positive definite if, and only if, its Fourier transform is nonnegative.*
(b) *Positive definite functions are characterized by*

$$\mathrm{Re}\int_0^\infty a(t)e^{-zt}dt\geqslant 0 \quad \textit{for} \quad \mathrm{Re}z>0.$$

(c) *If a is continuous and positive definite, then $|a(t)|\leqslant a(0)$ for all t.*
(d) *Nonnegative, nonincreasing, convex functions on $[0,\infty)$ are positive definite.* □

A motivation for considering positive definite convolution kernels is

Theorem 7.3.22. *Positive definite L^1-functions form the largest class of L^1-kernels for which the solution of (7.3.38), for any λ with* $\mathrm{Re}\lambda<0$, *satisfies*

$$y(t)\to 0 \quad (\textit{is bounded}) \text{ whenever } g_0(t)\to 0 \quad (\textit{is bounded}) \text{ as } t\to\infty. \quad \Box \qquad (7.3.40)$$

The proof of this theorem follows from the Paley-Wiener theorem (cf. Theorem 1.4.1 and Bochner's Theorem given by 7.3.21(b)).

It would be desirable that the numerical solution satisfy the discrete analogue of (7.3.40). More precisely, we give the following definition.

Definition 7.3.7. A method is said to be *C-stable* if, for any convolution equation (7.3.38) with positive definite L^1-kernel and $\mathrm{Re}\lambda<0$, the numerical solution (computed with fixed stepsize h) satisfies

$$y_n\to 0 \quad \text{whenever} \quad g_0(t)\to 0, \quad \text{as} \quad t\to\infty. \quad \Box$$

The concept of C-stability for numerical methods applied to convolution equations was introduced in Hairer & Lubich [1984].

We state, without proof, the discrete analogue of Theorem 7.3.22 (cf. Lubich [1983a, p. 449]).

Theorem 7.3.23. *Let $a(t)$ in (7.3.38) be a positive definite L^1-kernel with $a(0)=1$, and let D be an open disk in the left half-plane, touching the origin and being contained in the stability region of the LM method $\{\tilde{\rho},\tilde{\sigma}\}$ generating the DQ method. Then, for $h\lambda\in D$, the numerical solution satisfies*

$$y_n\to 0 \quad (\textit{is bounded}) \text{ whenever } g_0(t)\to 0 \quad (\textit{is bounded}) \text{ as } t\to\infty. \quad \Box \qquad (7.3.41)$$

This theorem yields a sufficient condition for C-stability.

Corollary 7.3.7. *Let the LM method $\{\tilde{\rho},\tilde{\sigma}\}$ be A-stable. Then the DQ $(\{\tilde{\rho},\tilde{\sigma}\})$ method is C-stable.* $\Box$

If the generating LM method is $A(\alpha)$-stable, then there is at most a disk D with finite radius, so that (7.3.41) is satisfied only for restricted stepsizes.

7.3.7. Numerical examples

7.3.7.1. Second-kind equations

It has already been illustrated, by means of a numerical example in Section 3.2.9, that ILM methods have a favorable stability behavior when compared with DQ and (M)ML methods. This is confirmed by Theorem 7.3.12 in Section 7.3.3.1 (we note that, with respect to the basic test equations, the ILM and MML methods possess the same stability regions; see Theorem 7.3.3).

In this section, we shall provide more numerical evidence for the superior stability behavior of ILM methods when large Lipschitz constants of k with respect to y are present. In addition, we test the reliability of the local stability theorem of Section 7.3.5.2.

We start by considering again equation (3.2.37) with $\lambda=100$, but now we generate the VLM formula by a BD formula, instead of an AM formula as done in Table 3.2.9. The BD analogue of this table is presented by Table 7.3.3 (an asterisk indicates instability). A comparison with Table 3.2.9 reveals that, where the accuracy of the ILM method is increased, that of the (M)ML methods is decreased, in spite of the larger stability regions of BD formulas.

Table 7.3.3. Numbers of correct significant digits for problem (3.2.37) with $\lambda=100$ at $t=T=4$

h	$DQ(G_5)$	$ILM(G_5-BD_5)$	$ML(G_5-BD_4)$	$MML(G_5-BD_5)$
1/4	*	3.1	*	*
1/8	2.3	3.6	*	*
1/16	6.3	5.8	2.2	5.6

Our next test example is a nonlinear equation of Hammerstein type:

$$y(t) = -15t+17(e^t-1)-16\int_0^t (t-s+\frac{1}{16})e^{y(s)}ds, \qquad (7.3.42)$$

with exact solution $y(t)=t$ (cf. problem 25 in Table 8.2.3a). In Table 7.3.4, an "$*(t)$" indicates that at point t the Newton iteration process had not yet converged after 10 steps, and in all these cases the numerical process clearly showed an unstable behavior (for implementational details of the various methods we refer to Section 3.2.7).

Table 7.3.4. Numbers of correct significant digits for problem (7.3.42) at $t=T=8$

h	$DQ(G_5)$	$ILM(G_5-BD_5)$	$ML(G_5-BD_4)$	$MML(G_5-BD_5)$
1/2	*(2.5)	1.6	*(4.5)	*(5.0)
1/4	*(2.25)	3.9	*(3.25)	*(4.0)
1/8	*(3.375)	6.1	*(2.875)	*(3.625)
1/16	*(4.6875)	7.7	*(3.5625)	*(5.4375)

In the third example, we illustrate the insensitivity of the numerical solution to the stiffness of the equation:

$$y(t) = g_0(t)-\lambda\int_0^t \frac{1+t}{1+s}\, y^2(s)ds. \qquad (7.3.43a)$$

Here, $g_0(t)$ is chosen such that the exact solution is given by

$$y(t) = \sqrt{1+(1+t)\exp(-\mu t)}. \tag{7.3.43b}$$

Notice that $y(t)$ does not depend on λ. This equation resembles a frequently quoted test problem of de Hoog & Weiss [1975] which arises for $\lambda=\mu=10$ (cf. problem 6 of Table 8.2.3b). The results listed in Table 7.3.5 clearly illustrate the insensivity of the ILM and MML solutions for the stiffness of equation (7.3.43).

Table 7.3.5. Numbers of correct significant digits for problem (7.3.43) with $\mu=1$ at $t=T=192h$

h	λ	$DQ(G_4)$	$ILM(G_4-BD_4)$	$ML(G_4-BD_3)$	$MML(G_4-BD_4)$
	1	9.65	4.86	8.84	5.33
1/4	10	*(4.75)	4.89	*(8.50)	5.34
	1000	*(2.75)	4.90	*(5.75)	5.34
	1	9.60	6.34	8.81	6.26
1/16	10	9.71	6.40	8.92	6.29
	1000	*(1.0625)	6.40	*(1.7500)	6.29
	1	9.40	8.16	8.63	8.13
1/64	10	9.96	8.21	8.90	8.15
	1000	*(0.390625)	8.22	*(0.645625)	8.15

Finally, we present a test of the local stability Theorem 7.3.20. In addition to the number of correct significant digits, we have also computed the range of values assumed by $z_{max}:=\max|z_j|$, where $\{z_j\}$ are the zeros of the polynomial $C_1^*(z)$ in (7.3.33). If z_{max} does not exceed 1 during the integration, we expect stability; if it does then a necessary condition for stability is violated, so that the development of instabilities is possible. In this connection, we observe that the local stability criterion takes all components of $\mathbf{Y}_n$ into account, whereas we are only testing the stability of y_n. We have restricted our experiments to stiff equations since it is only for such equations that a large region of stability is an important feature of the integration method. (For an extensive test of both stiff and non-stiff problems we refer to van der Houwen & Blom [1985].) In the case of *linear* convolution kernels the range of z_{max}-values consists of just one value; in the case of *nonlinear* kernels, we replaced in (7.3.33) $K(t_i,t_j)$ by $\partial k/\partial y(t_i,t_j,y_j)$; since z_{max} now depends on the solution, we have an interval of z_{max}-values.

Table 7.3.6. Local stability test of ILM methods at $t=T=20$

Problem	T	h	G_4-AM_5 sd	G_4-AM_5 z_{max}	G_4-BD_4 sd	G_4-BD_4 z_{max}	BD_4-AM_5 sd	BD_4-AM_5 z_{max}	BD_4-BD_4 sd	BD_4-BD_4 z_{max}
(3.2.37)	20	1/10	4.7	1.11	4.7	1.00	5.9	1.56	5.9	1.00
$\lambda=10$		1/20	5.7	1.03	5.7	1.00	7.0	1.28	7.0	1.00
(3.2.37)	20	1/10	*	2.95	*	1.12	5.8	5.95	*	1.12
$\lambda=1000$		1/20	*	1.86	*	1.16	*	3.98	*	1.18
(7.3.42)		1/10	*	[0.58, 2.72]	5.9	[0.33, 0.95]	*	[0.79, 2.78]	5.7	[0.62, 0.95]
		1/20	*	[0.50, 2.84]	7.1	[0.45, 0.98]	*	[0.64, 2.86]	6.9	[0.62, 0.98]
(7.3.43)		1/10	0.3	[1.06, 1.29]	5.9	[0.90, 1.00]	0.3	[1.06, 1.32]	5.3	[0.90, 1.00]
$\lambda=10, \mu=1$		1/20	7.0	[0.95, 1.00]	7.0	[0.95, 1.00]	6.5	[0.95, 1.00]	6.5	[0.95, 1.00]
(7.3.43)		1/10	*	[0.86, 2.31]	1.8	[0.85, 1.00]	*	[0.83, 2.31]	1.8	[0.85, 1.00]
$\lambda=1, \mu=-5$		1/20	*	[0.92, 2.62]	3.2	[0.92, 1.00]	*	[0.92, 2.62]	3.2	[0.92, 1.00]
(7.3.44)	20	1/10	2.2	1.00	2.2	1.00	4.8	1.00	4.8	1.00
$\lambda=10$		1/20	3.7	1.00	3.7	1.00	5.1	1.00	5.1	1.00
(7.3.44)	20	1/10	*	1.96	*	1.14	*	1.10	*	1.12
$\lambda=1000$		1/20	*	1.22	*	1.16	*	1.14	*	1.18
(7.3.44)	20	1/10	*	2.22	*	1.05	2.8	.99	.5	1.03
$\lambda=1900$		1/20	*	1.57	*	1.18	*	1.15	*	1.18
(7.3.44)	20	1/10	*	2.35	2.2	.98	*	1.67	4.3	.95
$\lambda=3000$		1/20	*	1.82	*	1.16	*	1.12	*	1.15
(7.3.44)	20	1/10	*	2.57	*	1.98	*	2.58	4.3	.72
$\lambda=12000$		1/20	*	2.35	3.6	.97	*	1.79	5.1	.95

* unstable behavior of the numerical solution

In addition to the preceding three problems, we tested problem 16b of Table 8.2.3a,

$$y(t) = 1+\frac{1}{2}\lambda(1-e^{-t^2})-\lambda\int_0^t(t-s)e^{-(t-s)^2}y(s)ds, \tag{7.3.44}$$

whose exact solution is $y(t)=1$.

>From the results presented in Table 7.3.6 we conclude that:

(i) the numerical solution behaves stably if $z_{max}\leqslant 1$.

(ii) $z_{max}>1$ does not necessarily imply instability.

7.3.7.2. *First-kind equations*

We shall test the local stability Theorem 7.3.20 for ILM methods when applied to first-kind equations. In Table 7.3.7, results are given for linear convolution equations chosen from Table 8.4.4a, which do not have kernels that vanish on the diagonal $t=s$ (this is one of the conditions for convergence of the ILM method).

The first two problems were already used in convergence tests of VLM methods (cf. Section 3.3.6). The third problem is given by

$$\int_0^t (1+\sqrt{\epsilon+t-s})y(s)ds = t+\frac{2}{3}(t+\epsilon)^{3/2}-\frac{2}{3}\epsilon^{3/2}, \tag{7.3.45}$$

whose exact solution is given by $y(t)=1$ (cf. problem 4 of Table 8.4.4a).

Table 7.3.7. Local stability test of ILM methods at $t=T=20$

Problem	h	G_4-BD_4 sd	G_4-BD_4 $z_{\max}$	BD_4-BD_4 sd	BD_4-BD_4 $z_{\max}$
(3.3.21)	1/10	4.2	1.01	3.9	1.00
$a=1$	1/20	5.4	1.00	5.1	1.00
(3.3.21)	1/20	3.5	1.01	2.9	1.00
$a=-1^*$	1/20	4.6	1.00	4.0	1.00
(3.3.22)	1/10	3.6	.90	3.8	.90
$a=1^*$	1/20	3.5	.95	2.5	.95
(3.3.22)	1/10	5.4	1.10	3.7	1.11
$a=-1$	1/20	7.2	1.05	4.8	1.05
(7.3.45)	1/10	5.9	.98	8.1	.98
$\epsilon=1$	1/20	7.0	.99	9.0	.99
(7.3.45)	1/10	2.7	.82	7.1	.74
$\epsilon=10^{-2}$	1/20	3.0	.88	7.5	.84
(7.3.45)	1/10	2.4	.77	7.0	.62
$\epsilon=10^{-4}$	1/20	2.6	.83	7.5	.72

**sd* corresponds to absolute errors

The results in Table 7.3.7 show a picture similar to the one obtained for second-kind equations: stability if $z_{\max}\leqslant 1$, but not necessarily instability if $z_{\max}>1$.

7.4. VLM methods for integro-differential equations

The greater part of the stability investigations of methods for integro-differential equations have been concerned with the equation

$$y'(t) = g_0(t)+\xi y(t)+\eta\int_0^t y(s)ds, \quad \eta\neq 0. \tag{7.4.1}$$

This test equation was suggested by Brunner & Lambert [1974], and since then it has been used extensively for investigating stability properties of DQ and VDRK methods (cf. Matthys [1976], Baker et al. [1979], and Baker & Wilkinson [1980]). Equation (7.4.1) is called the *basic test equation.*

Recently, Lubich [1983a] has given an alternative approach based on discrete versions of the Paley-Wiener theorem. By writing the integro-differential equation as a system of second-kind integral equations and by applying the stability theory developed for second-kind equations (cf. Section 7.3.6 for further details), Lubich obtains stability results for DQ methods in the case of positive definite kernels.

In this section we shall confine our considerations to the stability of the VLM method {(3.2.13); (3.4.3)} when applied to equation (7.4.1).

7.4.1. Stability results for the basic test equation

The basic test equation is characterized by $f_0=g_0(t)+\xi y+\eta z$ and $k=y$ (cf. (7.1.11)). Substitution into (3.4.3) immediately leads to a finite recursion in terms of y_n, z_n and $\tilde{F}_n(t)=\tilde{F}_n$.

Theorem 7.4.1. *If the VLM method {(3.2.13); (3.4.3)} is applied to (7.4.1), then it reduces to the recursions*

$$\left.\begin{aligned}\hat{\alpha}(E)y_n - h\hat{\gamma}(E)[\xi y_n+\eta z_n] &= T_{inh}\\ \alpha(E)z_n - h\gamma(E)y_n+\beta(E)\tilde{F}_n &= 0\end{aligned}\right\}, \quad n\geqslant 0, \tag{7.4.2}$$

where T_{inh} is a fixed linear combination of $g_0(t)$-values. □

Proceeding as in Section 7.3.2 we obtain the following two corollaries:

Corollary 7.4.1. *Let the VLM formula be (α,γ)-reducible ($\beta(z)\equiv 0$), and let the method be applied to (7.4.1). Then, for any lag term formula, the characteristic polynomial is given by*

$$C(z) := \alpha(z)[\hat{\alpha}(z)-h\xi\hat{\gamma}(z)]-h^2\eta\gamma(z)\hat{\gamma}(z), \tag{7.4.3}$$

and it represents a stability polynomial for the VLM method. □

Corollary 7.4.2. *Let the lag term formula be $(\tilde{\rho},\tilde{\sigma})$-reducible, and let the method be applied to (7.4.1). Then, for any pair of VLM formulas, the characteristic equation is given by*

$$C(z) := \alpha(z)\tilde{\rho}(z)[\hat{\alpha}(z)-h\xi\hat{\gamma}(z]-h^2\eta\hat{\gamma}(z)[\tilde{\rho}(z)\gamma(z)-\tilde{\sigma}(z)\beta(z)] \qquad (7.4.4)$$

and it represents a stability polynomial for the VLM method. □

By means of Definition 7.2.2 we can define stability regions in the $(h\xi,h^2\eta)$-plane. Since the continuous equation (7.4.1) can be shown to be stable for $\xi<0$ and $\eta<0$ (e.g., by means of Theorem 7.1.4(b)), we are led to the following analogues of the definition of A_0-and A-stability for ODEs (cf. Lambert [1973, p. 233], Brunner & Lambert [1974] and Matthys [1976]).

Definition 7.4.1. Let S_2 be the region of strong stability. Then the method is said to be *A_0-stable* if $S_2\cap\mathbb{R}^2\supset\{(h\xi,h^2\eta):\xi<0,\ \eta<0\}$. It is said to be *A-stable* if $S_2\cap\mathbb{C}^2\supset\{(h\xi,h^2\eta):\xi=\lambda+\mu,\ \eta=-\lambda\mu,\ \mathrm{Re}\lambda<0,\ \mathrm{Re}\mu<0\}$. □

We shall now discuss the special cases of the (α,γ)-reducible VLM methods and the $(\tilde{\rho},\tilde{\sigma})$-reducible DQ methods. In the latter case, Corollary 7.4.2 shows that the stability polynomial reduces to (since $\alpha=1,\ \beta=-1,\ \gamma=0$)

$$C(z) = \tilde{\rho}(z)[\hat{\alpha}(z)-h\xi\hat{\gamma}(z)]-h^2\eta\hat{\gamma}(z)\tilde{\sigma}(z). \qquad (7.4.4a)$$

A comparison with (7.3.16c) reveals that (7.4.3) and (7.4.4a) have the same structure, so that stability results can be derived which are analogous to those obtained for the ILM $(\{\tilde{\rho},\tilde{\sigma}\}-\{\rho,\sigma\})$ methods when applied to the convolution test equation. Let us denote the (α,γ)-reducible VLM method and the $(\tilde{\rho},\tilde{\sigma}\}$-reducible DQ method by $VLM(\cdot-\{\alpha,\gamma\}-\{\hat{\alpha},\hat{\gamma}\})$ and by $DQ(\{\tilde{\rho},\tilde{\sigma}\}-\{\hat{\alpha},\hat{\gamma}\})$, respectively. Then the following stability results hold:

Theorem 7.4.2. *Let S^* be the stability region of the LM method $\{\hat{\alpha},\hat{\gamma}\}$, and let (λ,μ) be defined by $\lambda+\mu=\xi$, $\lambda\mu=-\eta$. Then the set $\{(h\xi,h^2\eta):h\lambda\in S^*,h\mu\in S^*\}$ defines the region of strong stability S_2 of the $DQ(\{\hat{\alpha},\hat{\gamma}\}-\{\hat{\alpha},\hat{\gamma}\})$ method and of the $VLM(\cdot-\{\hat{\alpha},\hat{\gamma}\}-\{\hat{\alpha},\hat{\gamma}\})$ method.* □

Theorem 7.4.3. *Let the LM methods $\{\hat{\alpha},\hat{\gamma}\}$ and $\{\tilde{\rho},\tilde{\sigma}\}$ be A-stable, and let $\tilde{\rho}$ and $\hat{\gamma}$ have no common zeros on the unit circle. Then the $DQ(\{\tilde{\rho},\tilde{\sigma}\}-\{\hat{\alpha},\hat{\gamma}\})$ method and the $VLM(\cdot-\{\tilde{\rho},\tilde{\sigma}\}-\{\hat{\alpha},\hat{\gamma}\})$ method are A_0-stable.* □

For DQ methods, these theorems were originally given by, respectively, Brunner & Lambert [1974] and Matthys [1976].

It obviously follows from Theorem 7.4.2 that any A-stable LM method $\{\hat{\alpha},\hat{\gamma}\}$ generates A_0-stable $DQ(\{\hat{\alpha},\hat{\gamma}\}-\{\hat{\alpha},\hat{\gamma}\})$ methods and A_0-stable $VLM(\cdot-\{\hat{\alpha},\hat{\gamma}\}-\{\hat{\alpha},\hat{\gamma}\})$ methods.

Notice that the stability results for the (α,γ)-reducible VLM methods hold (i) for *any lag term approximation* and (ii) *do not depend on the special structure* of the VLM formula; that is, they hold equally well for MML and ILM methods (this is not true for integral equations of the second kind; see Section 7.3.3.1).

In practical applications the equation to be solved is generally not of the form (7.4.1). We then apply the above stability results with ξ replaced by $(\partial f_0/\partial y)\ (t_n,y_n,z_n)$ and η replaced by $(\partial f_0/\partial z.\partial k/\partial y)\ (t_n,y_n,z_n)$. For a discussion of the stability in the case of systems we refer to Matthys [1976].

7.4.2. Numerical examples

According to Theorem 7.4.2, the methods DQ(BD_p-BD_p), MML($\cdot-BD_p-BD_p$) and ILM($\cdot-BD_p-BD_p$) all have the same region of stability, when applied to the basic test equation (7.4.1), and are A_0-stable for $p\leqslant 6$. This holds for any lag term used in the MML and ILM method; therefore, we choose the easily implementable Gregory rule G_p for evaluating the lag term. In contrast, the DQ method should employ BD quadrature, otherwise it loses its A_0-stability behavior. In order to illustrate this, we shall also apply the DQ(G_p-BD_p) method. In Table 7.4.1, results are shown for $p=4$. The test problem is the nonlinear equation

$$y'(t) = [g_0(t)-ay(t)-b\int_0^t(t+cs)^d y^3(s)ds]^3-1,\quad y(0)=1, \tag{7.4.5}$$

where

$$g_0(t) := 1+a+\frac{bt^{d+1}}{c(d+1)}\ [(1+c)^{d+1}-1]$$

(cf. problem 4 of Table 8.3.2c). For all values of the parameters a,b,c, and d the exact solution is $y(t)=1$.

Table 7.4.1. Numbers of correct significant digits for problem (7.4.5) with $a=40$, $b=15$, $c=2$, $d=3/2$ and $h=1/8$

t	DQ (BD_4-BD_4)	DQ (G_4-BD_4)	MML $(G_4-BD_4-BD_4)$	ILM $(G_4-BD_4-BD_4)$	
2	6.06	6.86	6.68	5.98	(6.73)
4	6.94	7.66	7.10	6.21	(6.63)
6	7.31	8.59	7.37	6.47	(6.86)
8	7.52	4.48	7.55	6.65	(7.04)
10	7.67	*	7.69	6.79	(7.18)
12	7.78	*	7.81	6.91	(7.29)
14	7.88	*	7.90	7.00	(7.39)
16	7.95	*	7.98	7.08	(7.47)

In our experiments two implementations of the various methods have been tested. In the first implementation the equation (7.4.5) was written in the form (7.1.11); in the second implementation we replaced (7.1.11) by the equivalent representation

$$y'(t) = f_0(t,y(t),z(t))$$
$$z(t) = g_0(t)+\int_0^t k(t,s,y(s))ds \qquad (7.4.6)$$

The VLM methods respectively applied to (7.1.11) and (7.4.6) are not algebraically equivalent (unless $g_0(t)$ is constant, and except for the DQ method). The MML method furnished similar results for both representations and the difference is not noticeable in the *sd*-values as listed in Table 7.4.1. However, the ILM method is more accurate when the representation (7.4.6) is used (the corresponding *sd*-values are listed in brackets). An heuristic explanation can be given by observing that the ILM($G_4-BD_4-BD_4$) method "differentiates" the function $g_0(t)$ when it is applied to (7.4.6). In the case (7.1.11), however, it only uses the value $g_0(t_n)$ in each step, so that less information on the forcing function $g_0(t)$ is employed during the integration process.

7.5. VRK methods for second-kind integral equations

As in the case of VLM methods, the greater part of the papers on the stability of VRK methods is based on test equations which are related to a differential equation, so that the VRK method can be reduced to a finite recursion. Recently, however, Hairer and Lubich [1984] have followed the non-separable kernel approach of Lubich [1983a] (see Section 7.3.6), to derive stability results for convolution equations with positive definite kernel, which are not reducible to differential equations. In Section 7.5.4 the main results of this approach will be presented.

In the test equation approach, the paper of Amini et al. [1983] presents rather general results: for the class of polynomial convolution kernels defined in (7.1.12), this paper gives the characteristic polynomial for VRK methods of extended, mixed, and γ-modified type (see Definition 4.1.1). In this section, we shall illustrate the derivation of characteristic polynomials by treating the *linear* case, i.e.,

$$k(t,s,y) = [\xi+\eta(t-s)]y, \qquad (7.5.1)$$

of which the basic test kernel ($\eta=0$) is a special case. For linear convolution kernels we shall derive explicit stability results that can be used in actual computation. In the case of general polynomial convolution kernels, it is doubtful whether feasible stability criteria can be obtained (see also our discussion in Section 7.3.4).

In the special case of the basic test equation we shall see that the recurrence relation

$$y_{n+1} = V(h\xi)\tilde{F}_n(t_n)+T_{inh} \tag{7.5.2}$$

is obtained. Here, T_{inh} denotes an inhomogeneous term consisting, as before, of a fixed linear combination of $g_0(t)$-values. Furthermore, $V(h\xi)$ is a rational function of $h\xi$ whose coefficients are functions of the VRK-parameters.

For extended Pouzet methods we have $y_n=\tilde{F}_n(t_n)$ (cf. (4.1.29)), so that

$$y_{n+1} = V(h\xi)y_n+T_{inh}, \tag{7.5.2$'$}$$

which is the recurrence relation of a Runge-Kutta method for ODEs. For mixed methods $y_n\neq\tilde{F}_n(t_n)$, and the stability behavior is influenced by the lag term.

In order to eliminate the effect of the lag term the γ-modified lag terms were introduced:

$$\tilde{F}^*_n(t) := \tilde{F}_n(t)+\gamma(t)[y_n-\tilde{F}_n(t_n)] \tag{7.5.3}$$

(cf. (4.1.6)), where $\gamma(t_n+\theta_jh)=:\gamma_j\in[0,1]$ and where $\tilde{F}_n(t)$ may be any approximation to $F_n(t)$. By replacing in the VRK method $\tilde{F}_n(t)$ by $\tilde{F}^*_n(t)$ with $\gamma(t)\equiv1$, the relation (7.5.2) changes to (7.5.2$'$), *irrespective of the choice of* $\tilde{F}_n(t)$. An additional advantage of the γ-modified lag term is that for VRK methods where one or more of the θ_js vanish, the choice $\gamma(t_n)=1$ and $\gamma(t)=0$ for $t\neq t_n$ yields VRK methods in which it is not necessary to evaluate $\tilde{F}_n(t)$ at $t=t_n$. The first examples of such methods can be found in Bel'tyukov [1965b], van der Houwen [1977b] and Amini et al. [1980]. In the latter reference this type of method was termed *economized version* of the Runge-Kutta method (note that $\gamma(t)\equiv0$ yields the unmodified VRK method). The term $\gamma(t)[y_n-\tilde{F}_n(t_n)]$ in $\tilde{F}^*_n(t)$ can be considered as a (weighted) residual correction of $\tilde{F}_n(t)$ because it measures the amount by which y_n fails to equal $\tilde{F}_n(t_n)$.

In the following section the basic recursion for a general VRK method with the linear kernel (7.5.1) is derived, and in subsequent sections the characteristic equations for the extended and mixed lag term formula are established.

7.5.1. Recurrence relations for the convolution test equation

Application of the γ-modified VRK method, as specified in Definition 4.1.1, with $\tilde{F}_n(t)$ of the form (4.1.22), yields, in the case of the convolution kernel (7.5.1),

$$\begin{aligned} Y_{n,j} &= \gamma_jy_n+(1-\gamma_j)\tilde{I}_n+g_0(t_n+\theta_jh)-\gamma_jg_0(t_n)+h\eta\theta_j\tilde{J}_n \\ &\quad +h\sum_{i=1}^{m}a_{j,i}[\xi+d_{j,i}h\eta-c_ih\eta]Y_{n,i}, \end{aligned} \tag{7.5.4}$$

where

$$\tilde{J}_n := h\sum_{l=-1}^{n-1}\sum_{j=1}^{m+1}w_{n,l,j}Y_{l,j}, \quad \tilde{I}_n(=\tilde{I}_n(t_n)) := \tilde{F}_n(t_n)-g_0(t_n).$$

More compactly, we may write (cf. the vector notation used in (5.2.16))

$$\mathbf{Y}_n = \gamma y_n + (\mathbf{e}-\gamma)\tilde{I}_n + h\eta\boldsymbol{\theta}\tilde{J}_n + [h\xi A + h^2\eta A_1]\mathbf{Y}_n + \mathbf{G}_n, \qquad (7.5.4')$$

where

$$\gamma := (\gamma_1, \dots, \gamma_m)^T, \quad \mathbf{e} := (1, \dots, 1)^T, \quad \boldsymbol{\theta} = (\theta_1, \dots, \theta_m)^T,$$

$$A := (a_{j,i}), \quad A_1 := (a_{j,i}(d_{j,i} - c_i)),$$

$$\mathbf{G}_n := (G_j) \quad \text{with} \quad G_j := g_0(t_n + \theta_j h) - \gamma_j g_0(t_n).$$

Introducing the matrix

$$M = [I - h\xi A - h^2\eta A_1], \qquad (7.5.5a)$$

and assuming that M is nonsingular, we find

$$\mathbf{Y}_n = M^{-1}[\gamma y_n + (\mathbf{e}-\gamma)\tilde{I}_n + h\eta\boldsymbol{\theta}\tilde{J}_n + \mathbf{G}_n]. \qquad (7.5.6)$$

We remark that in this section, **e** will always denote the unit vector defined above (**e** should not be confused with the coefficient vector $\mathbf{e} = (e_j)$ defined in Definition 4.1.1)).

The numerical solution y_{n+1}, given by (4.1.3), can be written as

$$y_{n+1} = \gamma_{m+1} y_n + (1 - \gamma_{m+1})\tilde{I}_n + g_0(t_{n+1}) - \gamma_{m+1} g_0(t_n) + h\eta\tilde{J}_n$$
$$+ h\xi < \mathbf{b}, \mathbf{Y}_n > + h^2\eta < \mathbf{b}_1, \mathbf{Y}_n > \qquad (7.5.7)$$

with

$$\gamma_{m+1} := \gamma(t_{n+1}), \quad \mathbf{b}_1 := (b_j(e_j - c_j)).$$

Introducing the vector

$$\mathbf{v} = \xi\mathbf{b} + h\eta\mathbf{b}_1 \qquad (7.5.5b)$$

and the functions

$$U = \gamma_{m+1} + h < \mathbf{v}, M^{-1}\gamma >,$$
$$V = 1 - \gamma_{m+1} + h < \mathbf{v}, M^{-1}(\mathbf{e}-\gamma) >, \qquad (7.5.5c)$$
$$W = 1 + h < \mathbf{v}, M^{-1}\boldsymbol{\theta} >,$$

we obtain, on substitution of (7.5.6) into (7.5.7),

$$y_{n+1} = U y_n + V\tilde{I}_n + h\eta W\tilde{J}_n + T_{inh}. \qquad (7.5.8)$$

Here, T_{inh} represents, as before, a linear combination of $g_0(t)$-values. The quantities U, V and W are functions of $(h\xi, h^2\eta)$, and are completely determined by the VRK parameters and the function $\gamma(t)$. In applications it is sometimes convenient to use a representation of the functions U, V and W, without the inverse of the matrix M. The following lemma may then be of help:

Lemma 7.5.1. *Let M be a nonsingular m by m matrix, and* $\mathbf{p}$ *and* $\mathbf{q}$ *m-dimensional vectors. Then*

$$1+<\mathbf{p},\ M^{-1}\mathbf{q}> \ = \ \frac{\det(M+\mathbf{q}.\mathbf{p}^T)}{\det(M)}\ ,$$

where $\mathbf{q}.\mathbf{p}^T$ *denotes the matrix product of the column vector* $\mathbf{q}$ *and the row vector* $\mathbf{p}^T$.

Proof. Let $\mathbf{x}:=M^{-1}\mathbf{q}$ and $x_{m+1}:=1+<\mathbf{p},\ \mathbf{x}>$. Then $(\mathbf{x}^T, x_{m+1})$ satisfies the system of $m+1$ equations

$$\begin{bmatrix} M_{1,1} & \cdots & M_{1,m} & 0 \\ \vdots & & \vdots & \vdots \\ M_{m,1} & \cdots & M_{m,m} & 0 \\ -p_1 & \cdots & -p_m & 1 \end{bmatrix} \begin{bmatrix} x_1 \\ \vdots \\ x_m \\ x_{m+1} \end{bmatrix} = \begin{bmatrix} q_1 \\ \vdots \\ q_m \\ 1 \end{bmatrix},$$

where the $M_{i,j}$ are the entries of M, and x_i, p_i, q_i the components of $\mathbf{x}$, $\mathbf{p}$, $\mathbf{q}$. By Cramer's rule we may write

$$x_{m+1} \ = \ 1+<\mathbf{p},\ M^{-1}\mathbf{q}> \ = \ \frac{\det(N)}{\det(M)}\ ,$$

with

$$N = \begin{bmatrix} M_{1,1} & \cdots & M_{1,m} & q_1 \\ \vdots & & \vdots & \vdots \\ M_{m,1} & \cdots & M_{m,m} & q_m \\ -p_1 & & -p_m & 1 \end{bmatrix}.$$

Subtracting the row vector $q_i(-\mathbf{p}^T,\ 1)$ from the i-th row vector of N leads to

$$\det(N) \ = \ \det(M+\mathbf{q}.\mathbf{p}^T),$$

which proves the lemma. □

Thus, applying this lemma to (7.5.5c) yields for U, V and W

$$U \ = \ \gamma_{m+1}\frac{\det(M+h\boldsymbol{\gamma}.(\mathbf{v}^T/\gamma_{m+1}))}{\det(M)}\ , \quad \gamma_{m+1}\neq 0,$$

$$V \ = \ (1-\gamma_{m+1})\frac{\det(M+h(\mathbf{e}-\boldsymbol{\gamma}).(\mathbf{v}^T/(1-\gamma_{m+1}))}{\det(M)}\ , \quad \gamma_{m+1}\neq 1, \qquad (7.5.5'\text{c})$$

$$W \ = \ \frac{\det(M+h\boldsymbol{\theta}.\mathbf{v}^T)}{\det(M)}\ .$$

The recurrence relation (7.5.8) specifies how the term $\tilde{I}_n$, the quadrature term $\tilde{J}_n$, and y_n (if $\gamma(t)\neq 0$) influence y_{n+1}. By imposing a special structure on the weights $w_{n,l,j}$, it is possible to derive also recurrence relations for $\tilde{I}_n$ and for $\tilde{J}_n$, so that, by eliminating $\tilde{I}_n$ and $\tilde{J}_n$ from (7.5.8) we obtain a recursion in terms of y_n-values only. This elimination step can be performed by using Lemma 7.2.1 (note that the recursions have constant coefficients).

7.5.2. Stability results for the basic test equation

In the case of the basic test kernel ($\eta=0$) the recursion (7.5.8) reduces to

$$\begin{cases} y_{n+1} = Uy_n + V\tilde{I}_n + T_{inh} \\ U := \gamma_{m+1} + h\xi<\mathbf{b}, (I-h\xi A)^{-1}\gamma> \\ V := 1-\gamma_{m+1} + h\xi<\mathbf{b}, (I-h\xi A)^{-1}(\mathbf{e}-\gamma)> \end{cases} . \qquad (7.5.9)$$

We observe that, for a given lag term approximation, this recursion is essentially determined by the matrix A and the vector $\mathbf{b}$. Thus, in the case of the basic test kernel, the difference of structure of PVRK and BVRK methods is not reflected into the stability analysis. For example, the PVRK formula specified by the fourth diagram in Example 4.1.1 and the BVRK formula specified by the first diagram in Example 4.1.8, possess identical vector $\mathbf{b}$ and matrix A and, therefore, have the same stability region when applied with the same lag term approximation.

Furthermore, all results derived in this section also apply to the basic vector equation

$$\mathbf{y}(t) = \mathbf{g}_0(t) + X\int_0^t \mathbf{y}(s)ds$$

provided that X is diagonalizable. We simply replace ξ by the eigenvalue ξ_j of X (see Section 7.3.2.3).

7.5.2.1. The case $\gamma(t)\equiv 1$

Theorem 7.5.1. *Let $\gamma(t)\equiv 1$ and apply the VRK method with $k=\xi y$. Then, for any choice of the weights $\{w_{n,j,l}\}$, the VRK method $\{(4.1.22), (4.1.4)\}$ reduces to the recursion*

$$y_{n+1} - R(h\xi)y_n = T_{inh}, \qquad (7.5.10a)$$

where

$$R(h\xi) := \frac{\det(I - h\xi A + h\xi\mathbf{e}.\mathbf{b}^T)}{\det(I - h\xi A)},$$

$$(7.5.10b)$$

$$T_{inh} = g_0(t_{n+1}) - g_0(t) + h\xi<\mathbf{b}, (I - h\xi A)^{-1}\mathbf{G}_n>.$$

Proof. Substituting $\gamma=\mathbf{e}$ and $\gamma_{m+1}=1$ into (7.5.9) and calling the resulting U function $R(h\xi)$ we obtain (7.5.10). The expression for R is obtained by means of Lemma 7.5.1. □

The recursion (7.5.10) is of a similar type, with the same function $R(h\xi)$, as obtained in ODE stability theory when the RK method

$$\begin{array}{c|c} \mathbf{c} & A \\ \hline & \mathbf{b}^T \end{array} \tag{7.5.11}$$

is applied to the test equation $y'=\xi y$. In fact, this was the motivation for introducing γ-modified VRK methods (see the discussion in Section 4.1.7). The function $R(h\xi)$ will play a central role in the subsequent analysis.

Corollary 7.5.1. *Let $\gamma(t)\equiv 1$ and let $k(t,s,y)=\xi y$. Then, for any choice of the weights $\{w_{n,l,j}\}$, the characteristic polynomial is given by*

$$C(z) = z-R(h\xi) \tag{7.5.12}$$

and it represents a stability polynomial for the VRK method. □

The function $R(h\xi)$ is identical with the so-called *stability function* of the RK method (7.5.11). Henceforth, we will call (7.5.11) the *associated RK method* (cf. (4.1.18)). The relation of the VRK method to its associated RK method is formulated in the next theorem.

Theorem 7.5.2. *Let the associated RK method possess the stability region S^*. Then, for any choice of the weights $\{w_{n,l,j}\}$ in (4.1.22), the γ-modified VRK method with $\gamma(t)\equiv 1$, possesses a region of strong stability equal to S^*.* □

7.5.2.2. *Extended methods*

We consider VRK methods of extended type, where the lag term is defined by (4.1.5a). It has already been observed that for PVRK methods $\tilde{F}_n(t_n)=y_n$, so that $y_n=\tilde{I}_n+g_n(t_n)$. In the case of the basic test kernel, this is true for any (extended) VRK method. Thus, we have

Theorem 7.5.3. *Let $k(t,s,y)=\xi y$. Then, for any function $\gamma(t)$, the extended VRK method reduces to the recursion (7.5.10).* □

Corollary 7.5.2. *Let the VRK method be of extended type and let $k(t,s,y)=\xi y$. Then, for any function $\gamma(t)$, the characteristic polynomial is given by*

$$C(z) = z-R(h\xi) \tag{7.5.13}$$

and it represents a stability polynomial for the VRK method. □

The analogue of Theorem 7.5.2 becomes:

Theorem 7.5.4. *Let the associated RK method possess the stability region S^*. Then, for any function $\gamma(t)$, the extended VRK method possess a region of strong stability equal to S^*.* □

7.5.2.3. Mixed methods

Finally, we consider the class of mixed VRK methods, where $\tilde{F}_n(t)$ is defined by (4.1.5b). In this case we have, in general, $y_n \neq \tilde{F}_n(t_n)$ so that we need again an additional recursion for $\tilde{I}_n$ in order to eliminate I_n from (7.5.9). We have already shown that for $(\tilde{\rho},\tilde{\sigma})$-reducible lag term formulas the recursion

$$\tilde{\rho}(E)\tilde{I}_n = h\xi\tilde{\sigma}(E)y_n \tag{7.5.14}$$

holds (see e.g., (7.3.6)). Hence, applying Lemma 7.2.1 to (7.5.9) and (7.5.14) we find that y_n satisfies an (inhomogeneous) recursion with characteristic polynomial

$$C(z) = \det\begin{pmatrix} \tilde{\rho}(z) & -h\xi\tilde{\sigma}(z) \\ -V & z-U \end{pmatrix} = (z-U)\tilde{\rho}(z) - h\xi V\tilde{\sigma}(z). \tag{7.5.15}$$

In particular we will consider the case where $\boldsymbol{\gamma}=\gamma\mathbf{e}$. From (7.5.5′c), (7.5.15) and Lemma 7.5.1 the following corollary can be derived:

Corollary 7.5.3. *Let $\boldsymbol{\gamma}=\gamma\mathbf{e}$ with $\gamma\in[0,1)$, and let $k(t,s,y)=\xi y$. Then the characteristic polynomial of the mixed VRK method, with $(\tilde{\rho},\tilde{\sigma})$-reducible $\tilde{F}_n(t)$, is given by*

$$C(z) = (z-\gamma R(h\xi))\tilde{\rho}(z) - (1-\gamma)h\xi R(h\xi)\tilde{\sigma}(z), \tag{7.5.15′}$$

and it represents a stability polynomial for the VRK method. □

The reason for considering VRK methods of mixed type with $\gamma(t)\not\equiv 1$, is the possibly reduced order of convergence if $\gamma(t_n)=1$, $n=0,1,\ldots$ (see Corollary 4.1.2). In van der Houwen, Wolkenfelt & Baker [1981] a few results are derived for $\boldsymbol{\gamma}=\gamma\mathbf{e}$ with $\gamma\in[0,1)$; these results will be stated in the theorem below. In this theorem the notation of weak A-stability plays a role.

Definition 7.5.1. The method is said to be *weakly A_0-stable* if $C(z)$ satisfies the root condition for all $h\xi<0$, and *weakly A-stable* if $C(z)$ satisfies the root condition for all $\mathrm{Re}(h\xi)<0$. □

Theorem 7.5.5. *Let the VRK method be of mixed type with $\boldsymbol{\gamma}=\gamma\mathbf{e}$, $\gamma\in[0,1)$, and with $(\tilde{\rho},\tilde{\sigma})$-reducible $\tilde{F}_n(t)$, and write the stability function $R(x)$ in the form $R(x)=P(x)/Q(x)$ with polynomials $P(x)$ and $Q(x)$ such that $P(0)=Q(0)=1$.*

(a) *If the method is weakly A_0-stable, then $R(x)$, $x \in \mathbb{R}$, satisfies*

(i) $|R(x)| = O(1/x)$ *as* $|x| \to \infty$.

(ii) $Q(x)$ *has no zeros for* $x \leqslant 0$.

(iii) $P(x)$ *does not change sign for* $x \leqslant 0$, *i.e.,* $P(x) \geqslant 0$.

(b) *If the method is weakly A-stable, then $R(x)$, $x \in \mathbb{C}$, satisfies*

(i) $|R(x)| = O(1/x)$ *as* $|x| \to \infty$.

(ii) $Q(x)$ *has no zeros in the half-plane* $\operatorname{Re} x \leqslant 0$.

(iii) $P(x)$ *has no zeros in the half-plane* $\operatorname{Re} x < 0$.

(c) *Let the associated RK method (7.5.11) be A-stable and let $R(x)$ be a p-th order approximation to* $\exp(x)$. *Then the method is not A-stable if $p \geqslant 3$.* □

A comparison with the stability results stated in the Theorems 7.5.2 and 7.5.4 reveals that the conditions to be imposed on $R(h\xi)$ are much stronger in the case of mixed VRK methods with $\gamma(t) \not\equiv 1$ than in the case of general VRK methods with $\gamma(t) \equiv 1$ or extended VRK methods with general $\gamma(t)$.

Example 7.5.1. Consider the PVRK formula defined in Example 4.1.5. It follows from (7.5.10b) that

$$R(x) = \frac{1+(1-2c)x+(1/2-2c+c^2)x^2}{(1-cx)^2}, \quad c = (3+\sqrt{3})/6.$$

The associated RK method is A-stable, so that the VRK method employing this PVRK formula is also A-stable if either $\gamma(t) \equiv 1$ or $\tilde{F}_n(t)$ is of extended type (4.1.5a). However, if $\tilde{F}_n(t)$ is of the mixed type (4.1.5b) and $\gamma = \gamma\mathbf{e}$ with $\gamma \in [0,1)$, then the PVRK method cannot be weakly A-stable, or even weakly A_0-stable, because $R(x) \to 1/2c^2 - 2/c + 1 \neq 0$ as $x \to \infty$. □

Although mixed methods with $\gamma(t) \not\equiv 1$ show a less favorable stability behavior than the extended methods or the $(\gamma(t) \equiv 1)$-modified methods, the introduction of the γ-modification improves the stability considerably when compared with the conventional mixed methods where $\gamma(t) \equiv 0$. This is demonstrated in the Tables 7.5.2 where stability intervals are listed for combinations of a number of VRK formulas and lag term approximations. The VRK formulas are specified in Table 7.5.1 and contain formulas which have been used in the numerical examples given in Chapter 4.

Table 7.5.1. Specification of VRK formulas for which stability boundaries will be given

PVRK_4	Standard explicit 4-stage formula (Example 4.1.3, first diagram)
BVRK_4	Explicit 5-stage formula of Hairer (Example 4.1.11 with $c=3-\sqrt{3})/8)$
PVRK_3	Diagonally-implicit 2-stage formula (Example 4.1.5)
GPVRK_4	2-stage Gauss-PVRK formula (Example 4.1.6)

In Table 7.5.2a stability boundaries for the conventional methods are listed, derived from (7.5.15′) by setting $\gamma=0$. Table 7.5.2b presents the stability intervals for the case $\gamma(t)\equiv 1$ derived from $C(z)=z-R(h\xi)$; since in this case the lag term is not involved in $C(z)$, we only mention the VRK formula. Finally, in Table 7.5.2c we consider mixed methods with $\gamma=\mathbf{e}_1:=[1,0,\ldots,0]^T$ and $\gamma_{m+1}=0$. The characteristic polynomial is given by (7.5.15); the function V follows directly from (7.5.5′c) and U is found from (7.5.5′c) by bringing γ_{m+1} inside the determinant sign and then substituting $\gamma=\mathbf{e}_1$ and $\gamma_{m+1}=0$.

Table 7.5.2a. Stability boundaries for mixed VRK methods with $\gamma(t)\equiv 0$

VRK formula $\{\tilde\rho,\tilde\sigma\}=$	G_2	G_3	G_4	G_5	G_6	BD_2	BD_3	BD_4	BD_5	BD_6
$PVRK_4$	2.6	2.5	2.4	2.4	2.4	3.0	3.0	2.5	2.3	2.1
$BVRK_4$	1.9	1.9	1.9	1.9	1.9	1.9	1.9	1.9	1.9	1.9
$PVRK_3$	2.2	2.2	2.2	2.2	2.2	2.2	2.2	2.2	2.2	2.2
$GPVRK_4$	8.3	7.5	7.2	7.2	7.2	1.1	1.2	7.7	6.5	5.7

Table 7.5.2b. Stability boundaries for VRK methods with $\gamma(t)\equiv 1$

VRK formula	$PVRK_4$	$BVRK_4$	$PVRK_3$	$GPVRK_4$
H_0	2.8	3.6	∞	∞

Table 7.5.2c. Stability boundaries for mixed VRK methods with $\gamma=\mathbf{e}_1$, $\gamma_{m+1}=0$

VRK formula $\{\tilde\rho,\tilde\sigma\}$	G_2	G_3	G_4	G_5	G_6	BD_2	BD_3	BD_4	BD_5	BD_6
$PVRK_4$	2.0	2.0			...				2.0	2.0
$BVRK_4$	1.1	1.1			...				1.1	1.1
$PVRK_3$	3.8	2.7	2.4	2.3	2.4	9.0	10.3	2.3	1.7	1.0
$GPVRK_4$	4.0	4.0			...				4.0	4.0

7.5.3. Stability results for the convolution test equation

We shall treat the stability of the two main classes of VRK methods, viz. the extended and mixed methods, when applied to the convolution equation (7.5.1).

7.5.3.1. Extended methods

Theorem 7.5.6. *Let the VRK method be of extended type. Then, in the case of the convolution equation, the method reduces (for any function $\gamma(t)$) to the finite recursion*

$$y_{n+1}-(R+W^*)y_n+(RW^*-h^2\eta R^*W)y_{n-1} = T_{inh}, \tag{7.5.16}$$

where $R:=1+h\langle \mathbf{v}, M^{-1}\mathbf{e}\rangle$, $R^*:=\langle \mathbf{b}, M^{-1}\mathbf{e}\rangle$, $W:=1+h\langle \mathbf{v}, M^{-1}\boldsymbol{\theta}\rangle$ *and* $W^*:=1+h^2\eta\langle \mathbf{b}, M^{-1}\boldsymbol{\theta}\rangle$.

Proof. The recursion (7.5.8) reads, upon substitution of $\tilde{I}_n=\tilde{F}_n(t_n)-g_0(t_n)$,

$$y_{n+1} = (U+V)y_n+h\eta W\tilde{J}_n+T_{inh}.$$

In order to eliminate $\tilde{J}_n$ we need a recursion for $\tilde{J}_n$. From (4.1.5a) and (4.1.22) it follows that

$$\tilde{J}_{n+1}-\tilde{J}_n = h\sum_{j=1}^{m} b_j Y_{n,j} = h\langle \mathbf{b}, \mathbf{Y}_n\rangle,$$

where, using (7.5.6),

$$\mathbf{Y}_n = M^{-1}[y_n\mathbf{e}+h\eta\boldsymbol{\theta}\tilde{J}_n+\mathbf{G}_n].$$

Thus

$$\tilde{J}_{n+1} = [1+h^2\eta\langle \mathbf{b}, M^{-1}\boldsymbol{\theta}\rangle]\tilde{J}_n+h\langle \mathbf{b}, M^{-1}\mathbf{e}\rangle y_n+h\langle \mathbf{b}, M^{-1}\mathbf{G}_n\rangle.$$

Introducing the functions $R=U+V$, R^* and W^* and applying Lemma 7.2.1 we obtain the recursion (7.5.16). □

The functions R, R^*, W and W^* can explicitly be expressed in terms of $h\xi$ and $h^2\eta$ by means of Lemma 7.5.1. Notice that R reduces to (7.5.10) for $\eta=0$.

Since T_{inh} is a fixed linear combination of $g_0(t)$-values we have

Corollary 7.5.4. *Let $k(t,s,y)=[\xi+\eta(t-s)]y$. Then, for any function $\gamma(t)$, the characteristic polynomial is given by*

$$C(z) = z^2-(R+W^*)z+RW^*-h^2\eta R^*W \tag{7.5.17}$$

and it represents a stability polynomial for the extended VRK method. □

V_0-stability can be defined for VRK methods in the same manner as for VLM methods (see Definition 7.3.4 (i)). So far, no general results on the existence of V_0-stable VRK methods have been published (notice that the stability polynomial (7.5.17) depends on all VRK parameters, and not only on **b** and A as in the case of the basic test kernel).

The following corollary of Theorem 7.5.6 presents a necessary condition for V_0-stability.

Corollary 7.5.5. *If the extended VRK method is V_0-stable, then the matrix $C:=A+aA_1$ has nonnegative eigenvalues for $a\geqslant 0$.*

Proof. If C has a negative eigenvalue α then the matrix $I-h\xi C$ is singular for $h\xi=1/\alpha$, and consequently the matrix M defined in (7.5.5a) is singular for $h\xi=1/\alpha$, $h^2\eta=a/\alpha$. Thus, the system for computing the vector $\mathbf{Y}_n$ of intermediate values is singular for at least one point $(h\xi, h^2\eta)$ in the region $h\xi<0$, $h^2\eta\leqslant 0$. This excludes V_0-stability. □

Example 7.5.2. Consider the class of two-stage, implicit PVRK methods. Then

$$A_1 = (\theta_2-\theta_1)\begin{bmatrix} 0 & -a_{1,2} \\ a_{2,1} & 0 \end{bmatrix}, \quad a_{1,2}\neq 0.$$

If $a_{2,1}$ and $a_{1,2}$ have different signs, then A_1 has a negative eigenvalue, and therefore C has a negative eigenvalue for sufficiently large values of a. Thus, it follows from Corollary 7.5.5 that two-stage, implicit VRK methods of extended type with $a_{1,2}a_{2,1}<0$ cannot be V_0-stable. In particular, the two-stage Gauss-PVRK method (cf. Example 4.1.6) is not V_0-stable. □

We shall now study the case $\xi=0$, $\eta\neq 0$ in more detail. In this special case, the test equation is the integrated form of the equation $y''=\eta y$ and one may wonder how our VRK method is related to the Runge-Kutta-Nyström (RKN) method for solving this differential equation. These RKN methods have been introduced in Section 5.2.2 (formula (5.2.9)) and are characterized by the vectors $\mathbf{b}$, $\bar{\mathbf{b}}$, $\mathbf{c}$, and the matrix $\bar{A}$. In vector form the method is defined by

$$\begin{aligned} y_{n+1} &= y_n+hy'_n+h^2\eta<\bar{\mathbf{b}}, \mathbf{Y}_n> \\ hy'_{n+1} &= hy'_n+h^2\eta<\mathbf{b}, \mathbf{Y}_n> \end{aligned}, \tag{7.5.18}$$

where

$$\mathbf{Y}_n = (I-h^2\eta\bar{A})^{-1}[y_n\mathbf{e}+hy'_n\mathbf{c}].$$

Application of Lemma 7.2.1 yields the characteristic polynomial

$$\bar{C}(z) = z^2-(\bar{R}+\bar{W}^*)z+\bar{R}\bar{W}^*-h^2\eta\bar{R}'^*\bar{W}, \tag{7.5.19}$$

where

$$\overline{R}:=1+h^2\eta\langle\overline{\mathbf{b}},\ \overline{M}^{-1}\mathbf{e}\rangle,\quad \overline{R}^*:=\langle\mathbf{b},\ \overline{M}^{-1}\mathbf{e}\rangle,$$
$$\overline{W}:=1+h^2\eta\langle\overline{\mathbf{b}},\ \overline{M}^{-1}\mathbf{c}\rangle,\quad \overline{W}^*:=1+h^2\eta\langle\mathbf{b},\ \overline{M}^{-1}\mathbf{c}\rangle,\quad \overline{M}:=I-h^2\eta\overline{A}.$$

A comparison of (7.5.17) and (7.5.19) reveals that the extended VRK method, when applied with kernel $k=\eta(t-s)y$, is identical with the RKN method (7.5.18) with

$$\overline{A}=A_1,\quad \overline{\mathbf{b}}=\mathbf{b}_1,\quad \mathbf{c}=\boldsymbol{\theta}, \tag{7.5.20}$$

where A_1, $\mathbf{b}_1$ and $\boldsymbol{\theta}$ correspond to the VRK method, with identical vectors $\mathbf{b}$ and $\mathbf{c}$ in both methods. The RKN method {(7.5.18), (7.5.20)} will be called the *associated RKN method.*

The following theorem is an immediate consequence of the preceding considerations.

Theorem 7.5.7. *Let the associated RKN method possess the stability region S^*. Then the extended VRK method possesses, with respect to the kernel $k=\eta(t-s)y$, a region of strong stability equal to S^*.* □

Example 7.5.3. Consider the class of diagonally implicit PVRK methods. From the definition of PVRK methods we deduce that

$$\overline{A}=A_1=(a_{i,j}(\theta_j-\theta_i)).$$

Thus, $\overline{A}$ is strictly lower triangular so that the associated RKN method is *explicit.* Hence, it follows from Theorem 7.5.7 that diagonally implicit PVRK methods cannot be V_0-stable. □

Example 7.5.4. Consider the family of one-stage BVRK methods with $\tilde{F}_n(t)$ defined by (4.1.5a), and with BVRK formula given by

$$\begin{array}{c|c|c} d & 1 & 1 \\ \hline & & 1 \end{array},$$

where d is a free parameter. Written explicitly, this method reads

$$\begin{aligned} y_{n+1} &= \tilde{F}_n(t_{n+1})+hk(t_n+dh,t_{n+1},y_{n+1}) \\ \tilde{F}_n(t) &= g_0(t)+h\sum_{l=0}^{n-1}k(t+(d-1)h,t_l+h,y_{l+1}) \end{aligned} \quad . \tag{7.5.21}$$

The method is a first-order consistent discretization for all d. For this example we have $M=1-h\xi-(d-1)h^2\eta$, $\mathbf{v}=\xi+(d-1)h\eta$ and $\boldsymbol{\theta}=\mathbf{b}=1$, so that

$$R=R^*=W=[1-h\xi-(d-1)h^2\eta]^{-1},$$
$$W^*=[1-h\xi-(d-2)h^2\eta]\,[1-h\xi-(d-1)h^2\eta]^{-1}.$$

The characteristic polynomial assumes the form

$$C(z) = z^2 - \frac{2-h\xi-(d-2)h^2\eta}{1-h\xi-(d-1)h^2\eta}\, z + \frac{1}{1-h\xi-(d-1)h^2\eta}\,.$$

Application of Lemma 7.3.2 with $m=2$ reveals that $C(z)$ has its roots within the unit circle if

$$h\xi<0, \quad h^2\eta\leqslant 0, \quad 2h\xi+(2d-3)h^2\eta<4 \quad (d\geqslant 1).$$

Thus, we have a V_0-stable method if $d\geqslant 3/2$. □

We remark that the PVRK method with the same **b** and A as in this example (i.e. $\mathbf{b}=1$, $A=1$), cannot be V_0-stable (cf. Example 7.5.2). Thus, here we have a case where the BVRK method is superior to the PVRK method (see also Brunner et al. [1980] for a related result; cf. Example 7.5.5 below).

7.5.3.2. Mixed methods

Here, we restrict our considerations to $(\tilde\rho,\tilde\sigma)$-reducible $\tilde F_n(t)$-approximations of the form (4.1.5b). We have already shown that $\tilde J_n$ satisfies the recursion (cf. (7.3.15))

$$\tilde\rho(E)\tilde J_n - h\tilde\sigma(E)y_n = 0,$$

and $\tilde I_n$ satisfies the recursion (cf. (7.3.14) with $j=0$)

$$\tilde\rho(E)\tilde I_n - h\xi\tilde\sigma(E)y_n - h^2\eta\tilde\sigma^{\#}(E)y_n + h\eta\tilde\rho^{\#}(E)\tilde J_n = 0.$$

Together with (7.5.8) we have three recursions in terms of y_n, $\tilde J_n$ and $\tilde I_n$. Using Lemma 7.2.1 we can eliminate $\tilde J_n$ and $\tilde F_n(t_n)$ to obtain a finite-term recursion for y_n, from which the following corollary can be derived:

Corollary 7.5.6. *Let $k=\xi+\eta(t-s)$, $\eta\neq 0$. Then the characteristic polynomial is given by*

$$C(z) = \begin{cases} (z-U)\tilde\rho^2(z) - h\xi V\tilde\rho(z)\tilde\sigma(z) - h^2\eta[V(\tilde\rho(z)\tilde\sigma^{\#}(z) - \tilde\rho^{\#}(z)\tilde\sigma(z)) \\ \qquad + W\tilde\rho(z)\tilde\sigma(z)], \quad \gamma(t)\neq 1, \\ (z-U)\tilde\rho(z) - h^2\eta W\tilde\sigma(z), \quad \gamma(t)\equiv 1, \end{cases} \tag{7.5.22}$$

and it represents a stability polynomial for the mixed VRK method. □

In van der Houwen, Wolkenfelt & Baker [1981] stability regions for a few combinations of VRK formulas and lag term formulas are shown, which indicate that choosing $\gamma(t)\equiv 1$ yields considerably larger stability regions than in the conventional case where $\gamma(t)\equiv 0$. (Also the *economized* version of the VRK method where $\gamma=(1,0,\ldots,0)^T$, $\gamma_{m+1}=0$, possesses larger stability regions than the conventional VRK method.) Therefore, we consider the polynomial $C(z)$ with $\gamma(t)\equiv 1$ in more detail, and we derive a necessary condition for stability in terms of conditions for the functions $U=U(h\xi,h^2\eta)$ and $W=W(h\xi,h^2\eta)$.

Theorem 7.5.8. *Let the VRK method be of mixed type, with $\gamma(t)\equiv 1$ and with $(\tilde{\rho},\tilde{\sigma})$-reducible $\tilde{F}_n(t)$. Let a be a nonnegative constant, and write $U(x,ax)\equiv P(x)/Q(x)$, $W(x,ax)\equiv P^*(x)/Q^*(x)$ with polynomials P, Q, P^* and Q^* which equal 1 at $x=0$ and which are such that P, Q and P^*, Q^* have no common zeros. Assume that the following conditions are satisfied:*

(i) $P(x)\neq Q(x)$ *on* $(-\infty,0)$.

(ii) *The method is weakly stable for* $h^2\eta=ah\xi$, $\xi\leqslant 0$.

Then:

(a) $P(x)/Q(x)=O(1)$ *and* $P^*(x)/Q^*(x)=O(1/x)$ *as* $x\to\infty$.

(b) $Q(x)$ *and* $Q^*(x)$ *do not vanish on* $(-\infty,0]$.

(c) $P^*(x)$ *does not change sign on* $(-\infty,0)$.

Proof. If $\gamma(t)\equiv 1$, then the coefficient of $z^{\tilde{\kappa}+1}$ in $C(z)$ is independent of $h\xi$ and $h^2\eta$. Thus, if $U(x,ax)$ and $xW(x,ax)$ are unbounded in the interval $(-\infty,0]$, then $C(z)$ has at least one unbounded coefficient (which cannot be removed by scaling) along the half-line $h^2\eta=ah\xi$, $\xi<0$, and therefore it has at least one unbounded zero. This proves (a) and (b).

Since $W(x,ax)$ is necessarily bounded, $Q^*(x)$ cannot have a zero in the interval $(-\infty,0)$, so that $Q^*(x)>0$ on $(-\infty,0)$. Suppose that $P^*(x)$ changes sign at $x=x_0$, $x_0<0$. Then $W(x_0,ax_0)=0$, so that for $h^2\eta=ah\xi=ax_0$ the characteristic polynomial reduces to $C(z)=[z-U(x_0,ax_0)]\tilde{\rho}(z)$, which has a zero at $z=1$. Hence, the characteristic equation

$$[z-U(x,ax)]\tilde{\rho}(z)-axW(x,ax)\tilde{\sigma}(z) = 0$$

has a zero $z=z_1(x)$ such that $z_1(x_0)=1$. We shall show that this zero is outside the unit circle for some x sufficiently close to x_0.

Let the zero x_0 of $W(x_0,ax_0)$ have multiplicity μ. Then

$$[z_1(x)-U(x,ax)]\tilde{\rho}(z_1(x)) = O((x-x_0)^\mu) \quad \text{as} \quad x\to x_0.$$

Since $z_1(x)\neq U(x,ax)$ in the neighborhood of x_0, we find, by means of the Mean-Value Theorem,

$$z_1(x)-z_1(x_0) = O((x-x_0)^\mu) \quad \text{as} \quad x\to x_0.$$

Evidently, μ is odd because $W(x,ax)$ changes sign at $x=x_0$. Thus, $z_1(x)-z_1(x_0)=z_1(x)-1$ changes sign at $x=x_0$ and therefore $z_1(x)>1$ for an x different from x_0 but sufficiently close to x_0. This proves (c) . □

We remark that for $a=0$ the result of this theorem is identical with that stated in Theorem 7.5.5(a). When a assumes all nonnegative values, we obtain necessary conditions for V_0-stability; the following corollary expresses these conditions directly in terms of the VRK parameters.

Corollary 7.5.7. *Consider the methods treated in Theorem 7.5.8, and define the matrices*

$$B \;:= A+aA_1-\mathbf{e}.(\mathbf{b}+a\mathbf{b}_1)^T,$$
$$B^* := A+aA_1-\boldsymbol{\theta}.(\mathbf{b}+a\mathbf{b}_1)^T,$$
$$C \;:= A+aA_1.$$

If $\det(I-xB)\neq\det(I-xC)$ *on* $(-\infty,0)$ *for all* $a\geq 0$, *then* V_0*-stability implies, for all* $a\geq 0$:

(a) $\det(I-xB)/\det(I-xC)$ *and* $x\det(I-xB^*)/\det(I-xC)$ *are bounded as* $x\to\infty$.

(b) C *does not have negative eigenvalues.*

(c) B^* *does not have negative eigenvalues of odd multiplicity.* □

Notice that the necessary condition (b) coincides with the necessary condition for V_0-stable extended VRK methods (see Corollary 7.5.5). Thus, it follows from the Examples 7.5.2 and 7.5.3 that diagonally implicit mixed PVRK methods and the two-stage mixed Gauss-PVRK method, γ-modified with $\gamma(t)\equiv 1$, are not V_0-stable. The following example offers a more positive result.

Example 7.5.5. Consider the family of one-stage, γ-modified, mixed BVRK methods with $\tilde{F}_n(t)$ reducible to the implicit Euler method $\{\tilde{\rho},\tilde{\sigma}\}=\{z-1,\, z\}$ and with the same BVRK formula as in Example 7.5.4. Thus,

$$\begin{aligned} y_{n+1} &= \tilde{F}_n(t_{n+1})+\gamma[y_n-\tilde{F}_n(t_n)]+hk(t_n+dh,t_{n+1},y_{n+1}) \\ \tilde{F}_n(t) &= g_0(t)+h\sum_{l=1}^{n}k(t,t_l,y_l) \end{aligned} \quad . \tag{7.5.23}$$

Like (7.5.21), this method is a first-order discretization of the continuous problem; its stability behavior was first studied by Brunner et al. [1980]. In terms of the notation introduced above we find

$$U=\gamma W,\quad V=(1-\gamma)W,\quad W=[1-h\xi-(d-1)h^2\eta]^{-1},$$

so that

$$C(z)=\begin{cases}(z-\gamma W)(z-1)^2-zW[(1-\gamma)h\xi(z-1)+h^2\eta(z-\gamma)], & \gamma\neq 1\\ (z-W)(z-1)-h^2\eta Wz, & \gamma=1\end{cases}.$$

We consider the cases $\gamma=0$ and $\gamma=1$. If $\gamma=1$, then $C(z)$ is identical with the characteristic polynomial derived in Example 7.5.4, so that we have V_0-stability for $d\geq 3/2$. If $\gamma=0$ we obtain

$$C(z)=z\Big[z^2-\frac{2-h\xi-(2d-3)h^2\eta}{1-h\xi-(d-1)h^2\eta}\,z+1+h\xi W\Big],$$

and the application of Lemma 7.3.2 leads to the stability region

$$h\xi<0, \quad h^2\eta\leqslant 0, \quad 2h\xi+(4d-5)h^2\eta<4 \quad (d\geqslant 1).$$

Hence, the method is V_0-stable for $d\geqslant 5/4$. □

7.5.4. Stability results for positive definite convolution kernels

In Hairer & Lubich [1984] it has been investigated to what extent the results of Lubich [1983a] (see Section 7.3.6) are valid for VRK methods. Here, we shall summarize the results obtained for PVRK methods.

As in the preceding sections on VRK methods, the associated RK method plays an important role, and stability properties are related to stability properties of this RK method. One such property is the algebraic stability introduced by Burrage & Butcher [1979] and Crouzeix [1979]:

Definition 7.5.2. An RK method is said to be *algebraically stable* if the matrices $B:=\text{diag}(\mathbf{b})$ and $M:=BA+A^TB-\mathbf{b}\mathbf{b}^T$ are nonnegative. □

In the following we shall call a VRK method algebraically stable if the associated RK method is algebraically stable.

Hairer and Lubich consider PVRK methods for convolution equations of the form (7.3.38) with positive definite L^1-kernel function $a(t)$, which is extended to the negative axis by setting $a(t)=\overline{a(-t)}$ for $t\leqslant 0$.

The following theorems represent the main results obtained in Hairer & Lubich [1984].

Theorem 7.5.9. *Every algebraically stable PVRK method is C-stable.* □

Theorem 7.5.10. *Consider a PVRK method with the properties that c_i-c_j is not integer if $i\neq j$ and $c_l+c_i-c_j$ is not integer if $i\neq j$ and $b_l\neq 0$. If this PVRK method is C-stable, then the matrices B and $D:=\text{diag}(M)$ are nonnegative.* □

Theorem 7.5.11. *Every C-stable PVRK method is A-stable.* □

Theorem 7.5.12. *An algebraically stable PVRK method satisfying the condition*

$$\sum_{\substack{i,j\\ c_i<c_j}} b_i a_{i,j}(c_i-c_j) = 0$$

has highest possible order 2. □

7.5.5. Numerical examples

We conclude the discussion of the stability of VRK methods with a numerical test of the diagonally implicit $PVRK_3$ formula (see Example 4.1.5) and the implicit $GPVRK_4$ formula (see Example 4.1.6). Both formulas are combined with an extended lag term formula and a mixed lag term formula. In the latter case, we respectively choose a third and fourth-order Gregory rule and apply, in addition to the conventional case, the γ-modification with $\gamma(t)\equiv 1$.

Table 7.5.3. Numbers of correct significant digits for problem (3.2.37) with $\lambda=100$ at $t=T=4$

Extended		mixed: $\gamma(t)\equiv 0$		mixed: $\gamma(t)\equiv 1$	
h	$PVRK_3$	h	G_3-PVRK_3	h	G_3-PVRK_3
4/7	*	4/9	*	4/9	*
4/13	*	4/18	*	4/18	*
4/26	*	4/36	-.6	4/36	0.4
h	$GPVRK_4$	h	$G_4-GPVRK_4$	h	$G_4-GPVRK_4$
4/7	*	4/9	*	4/9	-.6
4/13	*	4/18	*	4/18	2.8
4/26	3.6	4/36	4.2	4/36	4.8

Table 7.5.4. Numbers of correct significant digits for problem (3.2.42) at $t=T=8$

Extended		mixed: $\gamma(t)\equiv 0$		mixed: $\gamma(t)\equiv 1$	
h	$PVRK_3$	h	G_3-PVRK_3	h	G_3-PVRK_3
8/7	* (2.29)	8/9	* (1.78)	8/9	* (1.78)
8/13	* (3.69)	8/18	* (4.22)	8/18	* (5.33)
8/26	* (4.92)	8/36	-8.7	8/36	* (5.56)
8/52	* (5.38)	8/72	-4.0	8/72	-11
h	$GPVRK_4$	h	$G_4-GPVRK_4$	h	$G_4-GPVRK_4$
8/7	* (0.0)	8/9	* (2.67)	8/9	* (2.76)
8/13	* (1.85)	8/18	* (1.78)	8/18	* (2.67)
8/26	* (3.08)	8/36	* (2.89)	8/36	* (3.11)
8/52	* (5.08)	8/72	* (3.78)	8/72	* (4.44)

The examples are those already used in Section 7.3.7; the integration steps are chosen in such a way that the Tables 7.3.3-7.3.5 are comparable with the Tables 7.5.3-7.5.5 with respect to the computational effort involved.

In all tables, the more stable behavior of the modified methods ($\gamma(t)\equiv 1$) is clearly demonstrated. However, when compared with the ILM methods tested in Section 7.3.7.1, the γ-modified methods are considerably less satisfactory.

Table 7.5.5. Numbers of correct significant digits for problem (7.3.43) with $\mu=1$ at $t=T=192h$

h	λ	mixed: $\gamma(t)\equiv 0$ G_3-PVRK_3	mixed: $\gamma(t)\equiv 0$ G_4-PVRK_4	mixed: $\gamma(t)\equiv 1$ G_3-PVRK_3	mixed: $\gamma(t)\equiv 1$ $G_4-GPVRK_4$
	1	5.9	8.9	3.3	4.5
2/5	10	*	*	2.0	3.2
	1000	*	*	*	*
	1	7.0	9.6	4.4	6.1
1/10	10	*	8.0	2.8	4.9
	1000	*	*	-.3	0.9
	1	7.2	9.2	5.7	8.0
1/40	10	6.1	9.1	3.9	6.8
	1000	*	*	1.5	2.8

7.6. VDRK methods for integro-differential equations

We shall restrict our considerations to the stability of a VDRK method with respect to the basic test equation (7.4.1). The analysis closely follows the stability analysis given in Section 7.5 for VRK methods. Hence, many of the details will be omitted.

7.6.1. Recurrence relations for the basic test equation

Introducing the vectors

$$\mathbf{c} = (c_j) = (\sum_{i=1}^{m} a_{j,i}), \quad \mathbf{g}_n = (g_0(t_n+c_ih)), \quad \mathbf{e} = (1, \ldots, 1)^T,$$

the VDRK formula (4.2.9), when applied to (7.4.1), can be represented in the form

$$\mathbf{Y}_n = M^{-1}[\mathbf{e}y_n + h\eta\mathbf{c}\tilde{J}_n + hA\,\mathbf{g}_n], \tag{7.6.1}$$

where

$$M := I - AN, \quad N := h\xi + h^2\eta\bar{A},$$

and where $\tilde{J}_n$ is defined as in (7.5.4).

For y_{n+1}, defined by (4.2.8), we obtain

$$y_{n+1} = y_n + <\mathbf{b},\ N\mathbf{Y}_n> + h\eta<\mathbf{b},\mathbf{e}>\tilde{J}_n + h<\mathbf{b},\mathbf{g}_n>. \tag{7.6.2}$$

Substitution of (7.6.1) leads to the recursion

$$y_{n+1} = Uy_n + h\eta W\tilde{J}_n + T_{inh}, \tag{7.6.3a}$$

with

$$\begin{aligned} U &:= 1 + <\mathbf{b},\ NM^{-1}\mathbf{e}>, \\ W &:= <\mathbf{b},\ NM^{-1}\mathbf{c} + \mathbf{e}>. \end{aligned} \tag{7.6.3b}$$

Here, T_{inh} denotes some linear combination of $g_0(t)$-values.

We shall now derive stability results for *extended methods,* where the lag term is defined by (4.2.10a), and for *mixed methods,* where the lag term is defined by (4.2.10b). Finally, we shall compare these results with those obtained when a VRK method is applied to the integral equation representation of equation (7.4.1), i.e., the system

$$\begin{aligned} y(t) &= y_0 + \int_0^t g_0(s)ds + \int_0^t [\xi y(s) + \eta z(s)]ds \\ z(t) &= \int_0^t y(s)ds \end{aligned} \quad . \tag{7.6.4}$$

7.6.2. Extended methods

Theorem 7.6.1. *Let the VDRK method be of extended type. Then, in the case of the basic test equation, the method reduces to the finite recursion*

$$y_{n+1} - (U + U^*)y_n + (UU^* - h^2\eta WW^*)y_{n-1} = T_{inh}, \tag{7.6.5}$$

where $U^* = 1 + h^2\eta<\mathbf{b},M^{-1}\mathbf{c}>$ *and* $W^* = <\mathbf{b},M^{-1}\mathbf{e}>$.

Proof. It follows from the definition of $\tilde{J}_n$ and from (4.2.10a) that

$$\begin{aligned} \tilde{J}_{n+1} &= \tilde{J}_n + h<\mathbf{b},\mathbf{Y}_n> \\ &= [1 + h^2\eta<\mathbf{b},M^{-1}\mathbf{c}>]\tilde{J}_n + h<\mathbf{b},M^{-1}\mathbf{e}>y_n + T_{inh}. \end{aligned} \tag{7.6.6}$$

Application of Lemma 7.2.1 to (7.6.6) leads to (7.6.5). □

By means of Theorem 7.2.1 a stability polynomial in the sense of Definition 7.2.3 can be derived from the recursion (7.6.5). The functions U, U^*, W and W^*

occurring in this polynomial are completely determined by the VDRK parameters and can explicitly be expressed in terms of these parameters by using Lemma 7.5.1. It is then straightforward to compute stability regions in the $(h\xi,h^2\eta)$-plane. So far, no general results on A_0- and A-stable VDRK methods have been published.

Example 7.6.1. Consider the one-stage, second-order BVDRK formula specified in Example 4.2.5. For this formula we have

$$U = (1+\frac{1}{2}h\xi+\frac{1}{4}h^2\eta)M^{-1}, \quad U^* = (1-\frac{1}{2}h\xi+\frac{1}{4}h^2\eta)M^{-1},$$

$$W = \frac{1}{2}(1+U), \quad W^* = \frac{2}{h^2\eta}(U^*-1), \quad M = 1-\frac{1}{2}h\xi-\frac{1}{4}h^2\eta.$$

Hence, the stability polynomial is of the form

$$z^2-(U+U^*)z+1+U-U^*.$$

Its zeros lie inside the unit circle if

$$-1<U<U^*<1.$$

Substitution of U and U^* yields the stability conditions $\xi<0$ and $\eta<0$, so that the method is A_0-stable in the sense of Definition 7.4.1. □

7.6.3. Mixed methods

Theorem 7.6.2. *Let the VDRK method be of mixed type, and let the lag term formula be $(\tilde{\rho},\tilde{\sigma})$-reducible. Then, in the case of the basic test equation, the method reduces to the finite recursion*

$$[(E-U)\tilde{\rho}(E)-h^2\eta W\tilde{\sigma}(E)]y_n = T_{inh}. \tag{7.6.7}$$

Proof. Apply Lemma 7.2.1 to the recursions (7.6.3) and (7.3.15). □

Notice that the stability polynomial corresponding to (7.6.7) has the same structure as that of the γ-modified mixed VRK method when applied to the convolution test equation with $\gamma(t)\equiv 1$ (cf. (7.5.22)). As a consequence, Theorem 7.5.8 applies to mixed VDRK methods when U and W are replaced by the expressions given in (7.6.3).

7.6.4. Application of VRK methods

An alternative way to solve integro-differential equations by Runge-Kutta type methods is the application of a VRK method to the integral equation representation given by (7.1.11′). In the case of the basic test equation (7.4.1) we obtain the system (7.6.4). This system can be diagonalized to obtain a system of the form

$$\mathbf{y}^*(t) = \mathbf{g}_0^*(t)+D\int_0^t \mathbf{y}^*(s)ds, \text{ with } D := \begin{bmatrix} \xi_1 & 0 \\ 0 & \xi_2 \end{bmatrix},$$

where $\xi_{1,2}$ are the roots of the eigenvalue equation $x^2 - \xi x - \eta = 0$ (cf. Section 7.3.2.3). All stability results derived in Section 7.5.2 apply to this system of basic test equations if we substitute ξ_1 and ξ_2. For instance, the stability Theorems 7.5.2 and 7.5.4 lead to the following result.

Theorem 7.6.3. *Let the RK method associated to the VRK method be A-stable, and let the VRK method be either of extended type, or let $\gamma(t) \equiv 1$. Then the VRK method applied to (7.6.4) is A_0-stable in the sense of Definition 7.4.1.*

Proof. Let $\xi<0$ and $\eta<0$ in (7.6.4). It then follows from Lemma 7.1.1 that $\mathrm{Re}\xi_1<0$ and $\mathrm{Re}\xi_2<0$. Since the associated RK method is A-stable Theorems 7.5.2 and 7.5.4 imply that the region of strong stability of the VRK method contains the whole left-hand half-plane of the complex plane. Hence, the method is strongly stable for all ξ_1 and ξ_2 in the left half-plane, and therefore for all $\xi<0$ and $\eta<0$. Thus, we have A_0-stability in the sense of Definition 7.4.1. □

7.6.5. Numerical examples

We integrated the equation (7.4.5) using the implicit, fourth-order $PVDRK_4$ method specified in Example 4.2.3. We also applied the fourth-order method $GPVRK_4$ (see Example 4.1.6) to the integral equation representation of equation (7.4.5). Both methods were applied with extended and with mixed lag term formulas. In the latter case, fourth-order Gregory quadrature was used.

Table 7.6.1. Numbers of correct significant digits for problem (7.4.5) with $a=40$, $b=15$, $c=2$, $d=3/2$

t	Extended: $h=1/4$		mixed: $\gamma(t)\equiv 0$, $h=2/11$		mixed: $\gamma(t)\equiv 1$, $h=2/11$
	$PVDRK_4$	$GPVRK_4$	$G_4-PVDRK_4$	$G_4-GPVRK_4$	$G_4-GPVRK_4$
2	3.2	3.2	2.0	-3.1	4.0
4	3.4	3.4	*	*	4.2
6	3.5	3.5	*	*	4.3
8	3.6	3.6	*	*	4.4
10	3.6	3.6	*	*	4.4
12	3.7	3.7	*	*	4.5
14	3.7	3.7	*	*	4.5
16	3.8	3.8	*	*	4.5

NOTES

Section 7.1.

The last fifteen years have witnessed a steadily increasing activity in the research of the asymptotic behavior of solutions to Volterra equations. The following, necessarily limited, selection of references is intended as a first guide to this important domain.

The general study of questions on boundedness and asymptotic behavior of solutions to Volterra integral and integro-differential equations has its origin in the early 1960s; compare the early survey paper by Nohel [1964]. There are, however, a number of earlier investigations of the qualitative behavior of solutions to Volterra integral equation arising in heat transfer problems and in the theory of superfluidity; we mention the papers by Mann & Wolf [1951], Padmavally [1958], and Levinson [1960] (see also Nohel [1964] for more references, and compare Handelsman & Olmstead [1972], Keller & Olmstead [1972], Olmstead & Handelsman [1976a], [1976b], and Olmstead [1977] for related results). In addition, the reader is referred to the fundamental paper by Feller [1941] where asymptotic properties of solutions to linear convolution equations of the second kind (renewal equations) are studied; Karlin [1955], Bellman & Cooke [1963], and Wagner [1978] contain further results concerning such equations. See also Friedman [1963] and Ling [1978] for related results dealing with nonlinear convolution equations.

In connection with nonlinear perturbations of linear Volterra equations, a crucial question is that of the integrability of the corresponding resolvent kernel (recall also Theorem 1.3.7). Here, the first fundamental result is due to Paley & Wiener [1934] (cf. Theorem 1.4.1 in this book). For more recent results in this domain (including extensions to integro-differential equations and to Volterra equations with nonintegrable kernels) see, e.g., Miller [1968a], Nohel [1971], Corduneanu [1973, pp. 213-216], Grossman & Miller [1973], Bownds & Cushing [1975a, pp. 75-76], Shea & Wainger [1975], Londen [1984a], [1984b]; most of these papers list numerous additional references. Perturbation results based on the asymptotic behavior of the resolvent kernel may be found in Miller [1968a], [1968b], [1971], Miller, Nohel & Wong [1969], Grossman & Miller [1970], [1973], Levin [1977], Burton [1983b], Burton & Mahfoud [1983]. Asymptotic properties of resolvent kernels associated with Volterra integral and integro-differential equations have been studied by, e.g., Miller [1971], Levin [1977], Gripenberg [1980], [1982], Londen [1984a], [1984b]; further references may be found in the papers cited, as well as in the relevant contributions contained in the conference proceedings edited by Londen & Staffans [1980], Herdman et al. [1981], and Hannsgen et al. [1982]. Furthermore, we mention the review by Nohel [1976] where a detailed sketch of various important developments in the qualitative theory of Volterra equations has been given and which features a detailed list of references complementing those in the reviewed book, Corduneanu [1973]. Tsalyuk [1979] is an additional valuable source of references, particularly with regard to contributions by Soviet authors.

The concept of the one-sided Lipschitz constant was introduced by Frank, Schneid and Ueberhuber [1978] and has turned out to be a powerful tool for analyzing (nonlinear) stability of ODEs (see Frank, Schneid & Ueberhuber [1981a,b], [1985a,b]. The application of this approach to systems of ODEs associated with Volterra equations possessing separable kernels has not yet received much attention. A discussion of suitable ODE solvers for these rather peculiar systems was given in van der Houwen [1982] and Shampine [1985].

Section 7.2.

Much of the general quantitative and, especially, the qualitative theory of discrete Volterra equations remains to be developed. While some results on the existence and uniqueness of solutions to general discrete Volterra equations may be found in the thesis by Scott [1984], there are, at present, only a very few papers dealing with the asymptotic behavior of numerical solutions to general Volterra integral and integro-differential equations, as $h>0$ is fixed and N tends to infinity. Nevanlinna [1976a], [1976b] considers Volterra equations characterized by positive integral operators in a Hilbert space setting and analyzes direct quadrature methods for such equations where the underlying quadrature formulas preserve the positivity of the integral operator in the corresponding discrete setting. This framework allows an elegant study of the asymptotic behavior of numerical approximations obtained with fixed steplength. Compare also Nevanlinna [1982] for some extensions of this work.

Lubich [1982a], [1983a] established the discrete analogues of the Paley-Wiener theorem (cf. Theorem 1.4.1) and of the corresponding result of Grossman & Miller [1973] for integro-differential equations; on the basis of these results he developed a comprehensive theory of the asymptotic behavior of solutions to discrete Volterra equations corresponding to integral and integro-differential equations possessing integrable convolution kernels (see also Section 7.3.6).

The notion of block-stability and the related concepts discussed in Section 7.2.2 are due to Baker [1977, pp. 798-815] and to Baker & Keech [1978a] (compare also Baker [1980] for a discussion of the structure of recurrence relations arising in the discretization of special classes of Volterra equations by various numerical methods); these definitions are extended to integro-differential equations in Baker, Makroglou & Short [1979].

Section 7.3.

The so-called basic test equation (7.3.3) (related to the initial-value problem for $y'=\xi y$) was proposed in Mayers [1962]. Following the introduction of the notion of A-stability in the numerical analysis of ODEs by Dahlquist in 1963, this equation became the commonly used test equation in numerical methods for Volterra integral equations, in spite of its severe limitations, until about 1977. Among the numerous papers employing the basic test equation in the stability analysis of numerical methods for second-kind equations we only cite the rather comprehensive study by Baker & Keech [1978a]; some subsequent developments may be found in van der Houwen [1982]. The connection between this test equation and the so-called repetition factor of a DQ method (see Linz [1967], [1968], Noble [1969]) was investigated by McKee & Brunner [1980] and, especially, by Wolkenfelt [1983a].

Of the papers basing their stability analysis on more general test equations (i.e., equations with polynomial convolution kernels, or equations involving finitely decomposable kernels) we mention those by van der Houwen [1977b], van der Houwen & te Riele [1977/78], [1981], van der Houwen & Wolkenfelt [1980], Baker [1980], Wolkenfelt [1981a], [1981b], [1982], Jones & McKee [1982], Amini et al. [1983], Jones & McKee [1985], Jackiewicz & Kwapisz [1983b], and Bakke & Jackiewicz [1984], [1985].

The notion of V_0-stability was introduced by Wolkenfelt [1980], and Brunner, Nørsett & Wolkenfelt [1980].

As indicated above, the results of Section 7.3.6 are due to Lubich [1982a], [1982b], [1983a]. In addition, compare Kershaw [1982b]. An asymptotic theory of convolution quadratures for second-kind Volterra integral equations with weakly singular kernels (cf. Section 6.1), including the notion of A-stability, may be found in Lubich [1985a], [1985c]. Preliminary studies of numerical stability in product integration methods for certain integro-differential equations may be found in McKee [1979c] and McKee & Stokes [1983].

Section 7.4.

Brunner & Lambert [1974] introduced the basic test equation (7.4.1), with real parameters ξ and η, in order to define the notion of A_0-stability for LM methods for integro-differential equations. Matthys [1976] allowed ξ and η to assume complex values, thus extending the definition of A-stability, originally used for ODEs, to LM methods for integro-differential equations; in addition, he also considered A-stability for systems of such equations. Baker, Makroglou & Short [1979] gave a detailed study of the stability polynomials resulting from the application of various methods to the test equation (7.4.1). Compare also Wolkenfelt [1981a], [1982], where the stability properties of AM and BD methods are discussed.

Stability results for more general test equations can be derived by first rewriting the integro-differential equation as a system of two second-kind integral equations and by then resorting to the results of Section 7.3. Due to limitations of space details are omitted. For equations of the form

$$y'(t) = g(t)+\lambda y(t)-\int_0^t a(t-s)y(s)ds, \quad t\in\mathbb{R}_+,$$

with $a\in L^1(\mathbb{R}_+)$, Lubich [1983a] derives a number of results on the asymptotic behavior of solutions obtained by LM methods; these results are based on a discrete analogue of the Paley-Wiener-type theorem established by Grossman & Miller [1973].

Section 7.5.

The first stability analyses for VRK methods were given by Hung [1970], R. Weiss [1972a], de Hoog & Weiss [1975], El Tom [1974a], [1976], and Malina [1975a] (see also MR 52, #7178) who analyzed A-stability of spline, implicit PVRK, and block-type methods. Stability results for various PVRK and block-by-block methods with respect to the basic test equation may also be found in Baker & Keech [1978a], Amini [1980], Amini, Baker & Wilkinson [1980], and Baker & Wilkinson [1981]; see also Baker [1985] for remarks on A-, A_0-, and $A(\alpha)$-stability of certain PVRK methods.

Test equations of the second kind possessing more general kernels have been considered in van der Houwen [1977b], [1980a], [1980b] ($K(t,s)=A+Bt+Cs$; γ-modified VRK methods), van der Houwen & Blom [1978], van der Houwen, Wolkenfelt & Baker [1981] ($K(t,s)=\lambda+\mu(t-s)$; γ-modified VRK methods), and Amini et al. [1983] (linear equations with polynomial convolution kernels).

An example of a V_0-stable BVRK method was presented in Brunner, Nørsett & Wolkenfelt [1980]. See also Peluso & Piazza [1985] for a study of stability in the application of certain PVRK methods to a class of stiff integral equations.

The results on C-stable PVRK methods contained in Section 7.5.4 are due to Hairer & Lubich [1984]. A nice introduction to various notions of stability (particularly to algebraic stability) for RK methods may be found in Dekker & Verwer [1984].

As far as Volterra equations with weakly singular kernels are concerned, no analysis of the stability properties of the (discretized) collocation methods of Section 6.2 exists as yet.

Section 7.6.

Although there exist a few studies of stability properties of PVDRK methods vis-à-vis the basic test equations (7.4.1) (Makroglou [1977], Baker, Makroglou & Short [1979], Baker & Wilkinson [1980], [1982]), a more general stability theory for VDRK methods remains to be developed. Among the open problems are: (i) the construction of A-stable and A_0-stable VDRK methods of Pouzet and Bel'tyukov type; (ii) the extension of the C-stability theory of Hairer & Lubich [1984] to PVDRK and BVDRK methods (compare, however, the relevant remarks in this paper concerning the validity of some of its results for PVDRK methods); (iii) the stability analysis of (discretized) collocation methods for Volterra integro-differential equations whose kernels are weakly singular.

CHAPTER 8

SOFTWARE AND TEST EXAMPLES

In this chapter we describe various existing computer programs for Volterra integral and integro-differential equations; as will be seen, the development of automatic routines for such equations is in many ways still at an early stage, comparable to the situation in ODEs some twenty years ago. In addition, we also supply an extensive list of examples of Volterra equations (compiled from the literature) which may serve as test examples when assessing the performance of a given routine.

8.1. Introduction

During the last 20 years numerous authors have developed computer programs for the numerical solution of Volterra equations, most of them for integral equations of the second kind. While most of these routines are nonautomatic (i.e., they employ a fixed stepsize h, with the number N of subintervals supplied by the user; the routine may or may not return an error estimate), there have recently been attempts to produce automatic routines using variable stepsizes (with the user only prescribing the desired accuracy of the numerical approximation; based on some mechanism for controlling the local errors, the routine then selects the appropriate stepsizes).

The fact that almost all of these routines concern Volterra integral equations of the second kind is not too surprising: since many physical modeling processes lead to standard types of second-kind Volterra integral equations (see Table 8.2.3 and the sources given there), the provision of robust general routines for such equations seems both practicable and worthwhile. On the other hand, first-kind Volterra integral equations with regular or weakly singular kernels, and, to some extent, Volterra integro-differential equations are usually of a type as to call for special algorithms, tailored to the particular underlying problem (e.g., noisy data; information known from the physical background; stiffness of VIDEs). In such a case the user may be better served by a good method which he then can adapt (and program) to meet the requirements of his particular problem.

8.2. Routines for second-kind Volterra integral equations

In this section (Tables 8.2.1 and 8.2.2), we present brief descriptions of existing (nonautomatic and automatic) codes for second-kind Volterra integral equations. Most of the automatic codes using variable stepsizes are based on local error control techniques designed originally for ODEs. As we have observed before, an initial-value problem for a first-order ODE,

$$y'(t) = f(t,y(t)), \quad y(0) = y_0, \quad t \in I := [0,T], \tag{8.2.1}$$

is equivalent to a special second-kind Volterra equation,

$$y(t) = F_n + \int_{t_n}^{t} f(s,y(s))ds, \quad t \in [t_n, t_{n+1}], \tag{8.2.2}$$

where the lag term,

$$F_n := y_0 + \int_0^{t_n} f(s,y(s))ds \quad (=y(t_n)),$$

is independent of t. In other words, the solution of the special integral equation (8.2.2) is directly related to the kernel (with the kernel being the derivative of the solution), and so a numerical method for (8.2.2) will not involve a lag term approximation.

For a general Volterra integral equation of the second kind,

$$y(t) = g(t) + \int_0^t k(t,s,y(s))ds, \quad t \in I := [0,T], \tag{8.2.3}$$

we have

$$y(t) = F_n(t) + h_n \Phi_n(t), \quad t \in [t_n, t_{n+1}], \tag{8.2.4a}$$

with increment function $\Phi_n(t)$ defined by

$$h_n \Phi_n(t) := \int_{t_n}^{t} k(t,s,y(s))ds, \tag{8.2.4b}$$

and with lag term

$$F_n(t) := g(t) + \int_0^{t_n} k(t,s,y(s))ds \tag{8.2.4c}$$

(cf. (4.1.1) and (4.1.2)).

While $F_n(t_n)=y(t_n)$ we now have, in general, $F_n(t) \neq y(t_n)$ for $t > t_n$. hence, the solution of (8.2.3) is no longer directly related to the kernel, as it was in the ODE case (8.2.1) (where $F_n(t)$ does not depend on t). Consequently, a change in t will imply a change in $F_n(t)$, and hence necessitates the evaluation of a new lag term approximation. This in turn means that a reliable and efficient automatic code for (8.2.3) will have to base its stepsize strategy not only on the behavior of the increment function $\Phi_n(t)$ but also on that of the lag term $F_n(t)$. To illustrate this, consider two examples (both for the case where $k(t,s,y) = K(t,s)y$ in (8.2.3)); the solution is assumed to be a slowly varying function.

(i) $$K(t,s) = \exp(\gamma_0(s-t))/(1+\gamma_1(t-s)), \quad \text{with} \quad \gamma_i >> 0.$$

Here, the kernel is varying rapidly near the line $t=s$ (where $K(t,t)=1, t \in I$). Once we move away from $t=s$, then $K(t,s)$ tends to be almost constant. If we base our stepsize choice only on $\Phi_n(t)$ we shall have to work with a fine mesh; this will imply an inefficient lag term approximation (i.e., an excessive number of kernel evaluations in $\tilde{F}_n(t_{n+1})$ for an almost constant integrand). Although the code will be able to meet the prescribed error tolerance, it will be achieved at high cost.

(ii) $$K(t,s) = (t-s)^2\exp(-ts).$$

While this kernel is varying slowly near $t=s$ (with $K(t,t)=0$ for all $t \in I$) it possesses a peak at $(t,s)=(T,0)$ ($K(T,0)=T^2$). If stepsize control is based on the behavior if $\Phi_n(t)$ alone, the resulting large stepsize will lead to inaccurate approximations for the lag term $F_n(t)$, thus preventing the approximation of the exact solution for small error tolerances.

We note that the code ORION (cf. Table 8.2.1) is based on these observations; in addition, we refer to the investigations of Arndt [1984], [1985], where a related adaptive stepsize control is suggested.

In the following tables the references contain not only the source of a particular code (this is indicated by an asterisk), but often also papers and books providing details on related work (such as underlying ODE techniques).

8.2.1. *Summary of published routines*

Table 8.2.1. Routines for equations with regular kernels

- RUNGEKUTTA, VOLTERRASEPARES, VOLTERRALIES, etc. (ALGOL) - Reformulation of integral equation as an initial-value problem. Application of explicit 4-stage RK method of order $p=4$ (extended or mixed mode); application of Adams-type method of order $p=5$. Fixed step size (chosen by user); no error estimates. - Pouzet [1963], [1964], [1967]*. Baker & Keech [1978b]: stable modification of mixed method. Phillips [1977a], [1977b]: modification using fewer kernel evaluations.
- VIESUB (FORTRAN) - Method based on Simpson's rule which is used in one of the following modes: (i) LM method using Simpson's rule and the 3/8-rule. (ii) Block-by-block Simpson rule. User chooses one of the two options; both have order $p=4$. Variable stepsize and local error control based on increment function. - Logan [1976]*.

Table 8.2.1 cont'd. Regular kernels

- VOLTS (FORTRAN)
- Approximation of solution in $S_{2m-1}^{(m-1)}(Z_N)$, $m=1,2,3$; Hermite collocation at knots (multiple collocation points): order $p=2m$; full discretization of collocation equation by m-point Gauss quadrature.
  Stepsize sequence given by user; no local or global error estimates.
- Esser [1976]*, [1978].

- V2EXT, V2MIXT (ALGOL)
- Extended and mixed explicit PVRK methods of order $p=3$ (mixed method uses Gregory formula (2.6.23) with $q=3$).
  Fixed stepsize (chosen by user); global error estimate by Fehlberg embedding technique.
- Lomakovich & Ishchuk [1974], Baker & Keech [1978c]*.

- VOLTEX (FORTRAN)
- Midpoint method with extrapolation.
  Estimation of accuracy of elements in extrapolation tableau.
  Stepsize control strategy so that overall computational effort is minimized.
- Hock [1977], [1979b], [1980], [1981].

- RK (FORTRAN)
- Extended explicit PVRK methods of orders 4 and 5 (cf. Example 4.1.4).
  Global error estimation by using embedded forms of RK formulas.
  Fixed stepsize (chosen by user).
- Tanfulla & Ribighini [1981]*, Lapidus & Seinfeld [1971, pp. 70-73], van der Houwen [1977a, pp. 164-171].

- VE1 (FORTRAN)
- Approximation of (convolution) kernel by Fourier-Chebyshev series; conversion of resulting integral equation into initial-value problem for system of ODEs.
  Number of terms in series chosen by user or by program.
  ODE solvers are either Adams methods or RK-Fehlberg method with local error control.
- Bownds [1978b], [1981], [1982a], [1982b], Bownds & Appelbaum [1985]*, Golberg [1978], Shampine & Gordon [1975], Forsythe, Malcolm & Moler [1977].
  See also Shampine [1985] for a modification in case of stiffness.

Table 8.2.1 cont'd. Regular kernels

- ORION (FORTRAN) - Predictor-corrector method with variable stepsize and variable order based on two independent meshes for I: (i) Control of local error for increment function on $[t_n, t_{n+1}]$ ("t-grid"), using ODE technique of Sedgwick [1973]. (ii) Control of lag term error on $[0, t_n]$ by evaluating lag term on independently generated "s-grid". Error tolerance prescribed by user. - Kunkel [1982]*, Bader & Kunkel [1984]*.
- RKVIEP, RKSOLV (FORTRAN) - Extension of Pouzet's fixed-step algorithm into variable-step method, but using 6-stage explicit RK formula of order 5 embedded into 8-stage RK formula of order 6. Local error estimator; variable stepsize; global error estimator. - Duncan [1982]*, Verner [1978].
- VOLCON (FORTRAN) - Extended explicit PVRK method (cf. Example 4.1.3, first tableau) for equations with convolution kernels. Lag term approximation evaluated by discrete FFT techniques. Fixed stepsize; extrapolation with global error estimates. - Hairer, Lubich & Schlichte [1985]*, Henrici [1979].
- IVRKX, IVRKXC (FORTRAN) - Implicit method using collocation in $S_{m-1}^{(-1)}(Z_N)$; fully discretized collocation equation based on (5.3.43); collocation parameters (and quadrature abscissas) are Lobatto points. Fixed stepsize; extrapolation with global error estimates; tolerance prescibed by user. IVRKXC exploits convolution structure of kernel. - Schlichte [1984]*, Weiss [1972a], de Hoog & Weiss [1975].
- INTSOL (FORTRAN) - Adams-type predictor-corrector method with variable stepsize and variable order. Control of error in increment function on $[t_n, t_{n+1}]$; estimation of lag term error (program recognizes problems in lag term approximations: if error too large, then it relaxes tolerance and gives a message to user). Error tolerance specified by user. - Jones & McKee [1985]*, Jones [1986]*.

Table 8.2.1 cont'd. Regular kernels

- COLVI2 (FORTRAN)
- Implicit method using collocation in $S_{m-1}^{(-1)}(Z_N)$, together with corresponding iterated collocation.
  Full discretization of collocation equation, by m-point interpolatory quadrature formulas.
  Collocation parameters: Gauss, Radau II or Lobatto chosen by user.
  Global error estimate, variable stepsize; error tolerance prescribed by user.
- Blom & Brunner [1985], [1986]*.

- CONFRK (FORTRAN)
- Extended explicit Fehlberg PVRK method for equations with nonlinear convolution kernels (approach along the lines of VOLCON).
  Fehlberg parameters supplied by the user (example routine provided).
  Fixed stepsize; convolution structure, vanishing b_s, repetition of c_i, all exploited. FFT version uses NAG routines.
- Baker & Derakhshan [1985]*; Hairer, Lubich & Schlichte [1985].

Table 8.2.2. Routines for equations with weakly singular kernels

- VIESIN (FORTRAN)
- Conversion of integral equation (1.3.44), with $\alpha = 1/2$, into system (1.3.46); solution of system by block-by-block Simpson rule; order $p = 7/2$.
  Variable stepsize, local error control using increment function on $[t_n, t_{n+1}]$.
- Logan [1976]*, de Hoog & Weiss [1974].

- MATE (FORTRAN)
- Solution of thin-section equation from stereology with non-exact data (given on non-uniform grid); kernel singularity of form $(s^2 - t^2)^{-1/2}$, interval of integration $[t, \infty)$.
  Evaluation of various linear functionals of solution. See references below for details on statistical tools and product--integration techniques underlying the algorithm.
  Option for analogous first-kind equation (cf. Table 8.4.3).
- Jakeman [1980], Anderssen [1973], Anderssen & Jakeman [1974], Jakeman & Anderssen [1975].

8.2.2. *Test problems*

In the literature numerous Volterra equations have been used for testing the performance of numerical methods. In order to facilitate future testing of existing or new methods, we have collected many of these equations in the tables below. In this collection we have confined ourselves to those problems which are, in our opinion, suitable for testing the performance of a numerical method. Thus, the kernels, the initial functions, etc., should be available in closed, analytical form. Also, in most problems, the exact solution is explicitly known. Otherwise, an approximate value at some reference point is provided.

The references quoted in the tables of test problems are papers which, to the best of our knowledge, first mention the particular test equation or class of test equations.

We distinguish between nonsingular convolution kernels, decomposable kernels, general nonsingular kernels, weakly singular kernels, and vector kernels (i.e. systems of equations). All equations are written in the standard form (8.2.3). In the various tables we list k, g and the exact solution y. We do not specify the value of T because it may vary in the quoted references. In a number of cases the problem will depend on one or more parameters. These will not be specified either, for the same reason.

Table 8.2.3a. Nonsingular convolution kernels

Equation (8.2.3)	References
1. $k=2(t-2s)y$; $g=2t\sqrt{\pi}$; $y=\mathrm{erf}(t)$	Young [1954b]
2. $k=(t-s)y$; $g=\sin(t)$; $y=\frac{1}{2}[\sin(t)+\sinh(t)]$	Baker [1977, p.785]
3. $k=-[2(t-s)+3]y$; $g=2t+1$; $y=e^{-t}$	Baker et al. [1980]
4. $k=[\xi+\eta(t-s)]y$; $y=\sin(t)$ $g=-(\xi+\eta t)+(1+\eta)\sin(t)+\xi\cos(t)$;	Wolkenfelt [1981a, p. 22]
5. $k=-2y/(t-s+2)^2$; $g=2-2/(t+2)$; $y=1$	Friedlander [1941]
6. $k=-2y/(t-s+2)^2$; $g=1/(t+2)^2$; $y(2)\approx .027\,326\,910\,025\,016\,7$	Friedlander [1941]
7. $k=-y/[2(t-s+2)^3]^{1/2}$; $g=1/(t+2)^{3/2}$; $y(40)\approx .001\,004\,031\,976\,4994$	Friedlander [1941]

Table 8.2.3a. cont'd. Nonsingular convolution kernels

	Equation (8.2.3)	References
8.	$k=\cos(t-s)+\sin(t-s)y;\ g=0;\ y=t$	Pouzet [1967]
9.	$k=-\cos(t-s)y;\ g=1+t-\cos(t);\ y=t$	de Hoog & Weiss [1975]
10.	$k=\sin(t-s)y;\ g=t;\ y=t+t^3/6$	Baker [1977, p. 785]
11a.	$k=10\cos(t-s)y;\ g=(1-5t)\sin(t);\ y=\sin(t)$	Fox & Goodwin [1953]
11b.	$k=2\cos(t-s)y;\ g=e^t;\ y=e^t(1+t)^2$	Baker [1977, p. 785]
12.	$k=-(t-s)\cos(t-s)y;\ g=\cos(t);$ $y=\frac{2}{3}\cos(\sqrt{3}t+1)$	Baker [1977, p. 784]
13.	$k=\lambda e^{\mu(t-s)}y;\ g=1;$ $y=\begin{cases}1+\mu t, & \text{if } \mu=\lambda\\ e^{(\lambda-\mu)t}(1+\frac{\mu}{\mu-\lambda}(e^{(\mu-\lambda)t}-1)), & \text{if } \mu\neq\lambda\end{cases}$	Krylov [1949] ($\lambda=-\mu=1$) Wolkenfelt[1981a,p.120], Bownds[1982a]
14.	$k=-\cosh(t-s)y;\ g=\sinh(t);$ $y=\frac{2}{\sqrt{5}}e^{-t/2}\sinh(\frac{\sqrt{5}}{2}t)$	Baker [1977, p. 785]
15.	$k=-\lambda\log(1+t-s)y;\ y=1-t;$ $g=\frac{1}{2}\lambda(1-t^2)\log(1+t)+\frac{3}{4}\lambda t^2-(\frac{1}{2}\lambda+1)t+1;$	Garey [1975a]* ($\lambda=4$) Section 3.2.9, 3.2.11
16a.	$k=\lambda^2(t-s)e^{-\lambda(t-s)}y,\ \lambda>0;$ $g=1-e^{-\lambda t}(1+\lambda t);\ y=\frac{1}{4}(2\lambda t-1+e^{-2\lambda t})$	Deligonoul & Bilgen [1984] ($\lambda=\frac{1}{2}$, $\lambda=1$)

* In this paper the kernel is erroneously given as $-\lambda\log(1+t+s)$

Table 8.2.3a. cont'd. Nonsingular convolution kernels

Equation (8.2.3)	References
16b. $k=-\lambda(t-s)e^{-(t-s)^2}y$; $g=1+\frac{1}{2}\lambda(1-e^{-t^2})$; $y=1$	Section 7.3.7.1.
17. $k=\frac{1}{2}(t-s)^2e^{s-t}y$; $g=\frac{1}{2}t^2e^{-t}$; $y=\frac{1}{3}[1-e^{-3t/2}(\cos(\frac{1}{2}t\sqrt{3})$ $+\sqrt{3}\sin(\frac{1}{2}t\sqrt{3}))]$	Feller [1941] Example 1.4.3 Section 3.2.2, 3.2.9, 3.2.11
18. $k=-(1+t-s)y^2$; $g=1+t+\frac{1}{2}t^2$; $y=1$	Baker [1977, p. 882]
19. $k=-3\sin(t-s)y^2$; $g=1+\sin^2(t)$; $y=\cos(t)$	Pouzet [1962, p. 33]
20. $k=3\cos(t-s)y^2$; $g=\cos(t)-\sin(t)-\sin(2t)$; $y=\cos(t)$	Lomakovich& Ishchuk [1974]
21. $k=e^{t-s}y^2$; $g=2-e^t$; $y=1$	Baker [1977, p. 832]
22. $k=-2y(1+y^2)/(t-s+2)^2$; $g=\cos(t)$; $y(0)\approx-.4718905296$, $y(40)\approx-.650\,131\,101\,333\,44$	Friedlander [1941]
23. $k=(t-s)^3(4-t+s)e^{s-t}y^4/(1+2y^2+2y^4)$; $g=1$; $y(10)\approx1.259\,955\,823\,372\,3$	an der Heiden [1980]
24. $k=-(16+t-s)e^y$; $g=17(e^t-1)$; $y=t$	van der Houwen et al. [1981]
25. $k=-(1+16(t-s))e^y$; $g=-15t+17(e^t-1)$; $y=t$	van der Houwen & te Riele [1981] Section 7.3.7.1
26. $k=e^{s-t}(y+e^{-y})$; $g=e^{-t}$; $y=\log(1+e)$	Golberg [1978] Section 3.2.9, 3.2.11

Table 8.2.3b. Decomposable kernels

	Equation (8.2.3)	References
1.	$k=y;\ g=\cos(t);\ y=\frac{1}{2}[\sin(t)+\cos(t)+e^t]$	Baker [1977, p. 785]
2.	$k=\lambda y\colon g=\lambda t;\ y=e^{\lambda t}-1$	Baker & Keech [1978b]
3.	$k=2(t-2s)y;\ g=2t\sqrt{\pi};\ y=\mathrm{erf}(t)$	Young [1954b]
4.	$k=(-a+bt+cs)y;\ y=\sin(t);$ $g=a-bt+(1-c)\sin(t)-(a-(b+c)t)\cos(t)$	Wolkenfelt [1981a, p. 22]
5.	$k=-(t^2-y)^2;\ g=t^2(g+\frac{56}{5}t^3);\ y=9t^2$	van der Houwen & Schilder [1977]
6.	$k=-\lambda(1+t)y^2/(1+s);\ y=[1+(1+t)e^{-\mu t}]^{1/2};$ $g=y(t)+\lambda(1+t)[1-e^{\mu t}+\mu\ln(1+t)/\mu$	de Hoog & Weiss [1975] Section 7.3.7.1
7.	$k=e^{-t}y^2;\ g=\frac{1}{2}e^{-t}(1+e^{-2t});\ y=e^{-t};$	Baker [1977, p. 882]
8.	$k=-t^2y^3;\ g=t^2+t^9/7;\ y=t^2$	van der Houwen & Schilder [1977]
9.	$k=-ay^{2m+1};\ y=[\sin(b\pi t)+\frac{1}{c}e^{-ct}]^{\frac{1}{2m+1}};$ $g=y(t)+a[\frac{1-\cos(b\pi t)}{b\pi}+\frac{1-e^{-ct}}{c^2}]$	Bownds [1981]
10.	$k=e^{-2t}/y;\ g=e^{-2t};\ y=e^{-t}$	Baker et al. [1980]
11a.	$k=-ts\sqrt{y};\ g=e^{t^2}-t+te^{t^2/2};\ y=e^{t^2}$	Garey [1971, p. 101] Section 3.2.7, 3.2.9, 3.2.11
11b.	$k=-ts\sqrt{y};\ g=t(1+t^{5/2}/5);\ y=1$	Section 3.2.7
11c.	$k=-ts\sqrt{y};\ y=(e^{-t}+1)^2$ $g=(e^{-t}+1)^2-t(t+1)e^{-t}+t+\frac{1}{2}t^3;$	Section 3.2.9, 3.2.11

Table 8.2.3b cont'd. Decomposable kernels

Equation (8.2.3)	References
12. $k = = -\sin(t-y);\ g = 1+t-\cos(t);\ y = t$	Garey [1971, p. 99]
13. $k = -[100+e^{-t}\cos(sy)]y;\ y = 1$ $g = 1+100t+e^{-t}\sin(t)$	van der Houwen et al. [1981]
14. $k = -[16+t-s]\,[1-\frac{1}{100}e^{-t}\cos(sy)]y;\ y = 1;$ $g = 1+16t+\frac{1}{2}t^2-\frac{1}{100}e^{-t}[1-\cos(t)+16\sin(t)]$	van der Houwen et al. [1981]
15. $k = -\log((t+1)y);\ y = e^t$ $g = e^t+\frac{1}{2}t^2+t\log(t+1);$	Garey & Harris [1977]
16. $k = -\exp(-\alpha(y-1)^2),\ \alpha \geqslant 1;\ y = t;$ $g = t+\frac{1}{\sqrt{\alpha}}\operatorname{erf}(\sqrt{\alpha})+\frac{1}{\sqrt{\alpha}}\begin{cases}-\operatorname{erf}(\sqrt{\alpha}(1-t)) & \text{if } t \leqslant 1\\ +\operatorname{erf}(\sqrt{\alpha}(t-1)) & \text{if } t>1\end{cases}$	Logan [1976, p. 85]

Table 8.2.3c. General nonsingular kernels

Equation (8.2.3)	References
1. $k = t^2 e^{-ts} y;\ g = t+(1+t^2)e^{-t^2}-1;\ y = t$	Pouzet [1960]
2. $k = [te^{s(t-2s)}+e^{+2s^2}]y^2;\ g-1-t;\ y = e^{t^2}$	Laudet & Oulès [1960]
3. $k = e^{-ts}y^2;\ g = 1;\ y(4) \approx 1.53694399227063$	Pouzet [1962]
4. $k = -\sin(ty-s);\ g = 2-\cos(t);\ y = 1$	Garey [1971, p. 100]
5. $k = \frac{1}{4}ts\exp(t^2s^2y/4);\ g = 0;$ $y = \frac{1}{8}t^3+\frac{1}{896}t^{10}+\frac{73}{4816896}t^{17}+\ldots$	Bel'tyukov [1965b]

Table 8.2.3d. Weakly singular kernels

Equation (8.2.3) with $k=K/\sqrt{t-s}$	References
1. $K=-\frac{1}{4}y;\ g=y(t)+\frac{\pi}{8}+\frac{1}{4}\arcsin\left[\frac{1-t}{1+t}\right];$ $y=\frac{1}{\sqrt{1+t}}$	Linz [1969a]
2. $K=-y;\ g=y(t)+2\sqrt{t}M(1;\frac{3}{2};-t)^{*};\ y=e^{-t}$	Brunner & Nørsett [1981]
3. $K=-y;\ g=1+2\sqrt{t};\ y=1$	Miller & Feldstein [1971]
4. $K=-y;\ g=\frac{1}{2}\pi t+\sqrt{t};\ y=\sqrt{t}$	te Riele [1982]
5. $K=-y;\ g=\frac{4}{3}t\sqrt{t}+t;\ y=t$	te Riele [1982]
6. $K=-y;\ g=\frac{3}{8}\pi t^2+t^{3/2};\ y=t\sqrt{t}$	te Riele [1982]
7. $K=-y;\ g=\frac{16}{15}t^2\sqrt{t}+t^2;\ y=t^2$	te Riele [1982]
8. $K=y;\ g=-\frac{1}{2}\pi t+\sqrt{t};\ y=\sqrt{t}$	Logan [1976]
9. $K=-y;\ g=2\sqrt{t};\ y=1-e^{\pi t}\operatorname{erfc}\sqrt{\pi t}^{**}$	Noble [1964]
10. $K=-y;\ g=1;\ y=e^{\pi t}\operatorname{erfc}\sqrt{\pi t}^{**}$	Miller & Feldstein [1971]
11. $K=-\frac{y}{\sqrt{\pi}};\ g=\frac{2\sqrt{t}}{\sqrt{\pi}};\ y=\sum_{k=1}^{\infty}\frac{(-1)^{k-1}t^{k/2}}{\Gamma(\frac{k}{2}+1)}$	Oulès [1964]
12. $K=-y;\ y=\sqrt{t}e^{-\alpha t}$ $g=\sqrt{t}e^{-\alpha t}+\frac{1}{2}\pi te^{-\frac{1}{2}\alpha t}[I_0(-\frac{1}{2}\alpha t)+I_1(-\frac{1}{2}\alpha t)]^{***};$	Logan [1976]

* M: Kummer's function (see Abramowitz & Stegun [1970, p. 504]

** erfc: complementary error function (see Abramowitz & Stegun [1970, p. 297])

*** I_ν: modified Bessel function (ibid p. 375).

Table 8.2.3d cont'd. Weakly singular kernels

Equation (8.2.3) with $k=K/\sqrt{t-s}$	References
13. $K=-\frac{ry}{\sqrt{\pi}}$; $g=r(1-e^{-t})$; $y=\frac{r}{1+r^2}[e^{r^2t}\mathrm{erfc}(r\sqrt{t})$ $-e^{-t}+2r\sqrt{\pi}F(\sqrt{t})]$; $F(t)=e^{-t^2}\int_0^t e^{u^2}du$	Ghez & Lew [1973]
14. $K=-y^2$; $g=\sqrt{t}+\frac{4}{3}t\sqrt{t}$; $y=\sqrt{t}$	te Riele [1982]
15. $K=-y^3$; $g=\sqrt{t}+\frac{3}{8}\pi t^2$; $y=\sqrt{t}$	te Riele [1982]
16. $K=-y^4$; $g\sqrt{t}+\frac{15}{16}t^2\sqrt{t}$; $y=\sqrt{t}$	te Riele [1982]
17. $K=-\exp[s(1-s)^2-y^2]$; $g=(3-t)\sqrt{t}$; $y=(1-t)\sqrt{t}$	Logan [1976]
18. $K=-\frac{(y-\sin(s))^3}{\sqrt{\pi}}$; $g=0$; $y=O(t^3\sqrt{t})$ as $t\to 0$	Padmavally [1958]
19. $K=-\frac{y^4}{\sqrt{\pi}}$; $g=2\sqrt{\frac{t}{\pi}}$; $y=O(\sqrt{t})$ as $t\to 0$	Chambré [1959]

Table 8.2.3e. Systems with nonsingular kernels

Equation (8.2.3)	References
1. $k=A\begin{bmatrix}3y_1(1-y_1-y_2)\\ 1-y_1-y_2\end{bmatrix}$; $g=\frac{1}{100}\begin{bmatrix}b^{21}\\ 1+(10-b^{20})b\end{bmatrix}$; $A:=\begin{bmatrix}a^{21} & 0\\ (1-a^{20})a & a/1000\end{bmatrix}$, $a:=\exp\left[\frac{s-t}{20}\right]$, $b:=\exp\left[\frac{-t}{20}\right]$; $y(50)=\begin{bmatrix}.031716689391939\\ .62784627209779\end{bmatrix}$	Hethcote & Tudor [1980]
2. $k=\begin{bmatrix}0 & 1\\ -25 & 0\end{bmatrix}y$; $g=\begin{bmatrix}0\\ \frac{259}{9}+24\sin(t)+\frac{16}{3}\sin(3t)\end{bmatrix}$; $y=\begin{bmatrix}\sin(t)+\frac{1}{3}\sin(3t)+\frac{1}{5}\sin(5t)\\ \cos(t)+\cos(3t)+\cos(5t)\end{bmatrix}$	Hock [1981]

8.3. Routines for Volterra integro-differential equations

There are, in comparison with second-kind Volterra integral equations, very few codes available for Volterra integro-differential equations. Recall, however, that any integro-differential equation can be written as a system of second-kind equations to which the codes mentioned in Table 8.2.1 apply.

8.3.1. Summary of published routines

Table 8.3.1. Routines for integro-differential equations

- RUNGEKUTTA (ALGOL60)
- Explicit PVDRK method, based on the same Runge-Kutta formulas as in the code RUNGEKUTTA for second-kind integral equations (cf. Table 8.2.1).
  Fixed stepsize (chosen by user); no error estimates.
- Pouzet [1962], [1967]*.

- DMRODE (FORTRAN)
- Explicit Runge-Kutta-Merson formula of order $p=4$, adapted to Volterra functional differential equations of the form

$$y'(t) = f(t,y(t), y(\alpha(t,y(t)))), \quad \alpha(t,y(t))\leq t,$$

  of which the integro-differential equations form one particular subclass; uses extrapolation; local error estimates and automatic change of stepsize.
- Neves [1975a], [1975b]*, Lapidus & Seinfeld [1971, p. 68-69], Lambert [1973, p. 131-132], van der Houwen [1977a, p. 168].

- VIDFRK (FORTRAN)
- Extended explicit Fehlberg PVDRK method for nonlinear integro-differential equations involving a nonlinear convolution Volterra operator.
  Direct extension of techniques of CONFRK (see Table 8.2.1).
- Derakhshan & Baker [1985]*.

8.3.2. Test problems

We write the integro-differential equations in the form (3.4.1); i.e.,

$$y'(t) = f(t,y,(t),z(t)), \quad 0 \leq t \leq T,$$
$$z(t) = \int_0^t k(t,s,y(s))ds. \tag{8.3.1}$$

In the Tables 8.3.2 the functions k,f and y are listed, where we distinguish between equations with convolution kernels, decomposable kernels, general nonsingular kernels and weakly singular kernels.

Table 8.3.2a. Nonsingular convolution kernels

Equation (8.3.1)	References
1. $k=(t+cs)^d y^3$; $f=[g_0(t)-ay-bz]^3-1$; $g_0:=1+a+\frac{bt^{d+1}}{c(d+1)}[(1+c)^{d+1}-1]$; $y=1$	Wolkenfelt [1977] $(c=-1)$

Table 8.3.2b. Decomposable kernels

Equation (8.3.1)	References
1. $k=\cos(s)$; $f=z$; $y=-\cos(t)$	Makroglou [1982]
2. $k=ts\cos(s^2)$; $f=z$; $y=-\frac{1}{4}\cos(t^2)$	Feldstein & Sopka [1974]
3. $k=-2tse^{-y^2}$; $f=1-te^{-t^2}+y+z$; $y=t$	Linz [1969d] Section 3.4.4
4. $k=ts\exp(y)$; $f=\frac{5}{2}t-\frac{1}{2}\exp(t^2)+z$; $y=t^2$	Feldstein & Sopka [1974]
5. $k=\frac{s^2}{t}\exp(y)$; $f=-\frac{1+3t^4}{12t}+\exp(-y)+z$; $t\geq 1$; $y=\log(t)$	Feldstein & Sopka [1974]

Table 8.3.2c. General nonsingular kernels

Equation (8.3.1)	References
1. $k=\dfrac{1}{1+(1+t)y};\ y=\dfrac{1}{1+t}$ $f=-t-\dfrac{1}{(1+t)^2}+z+\dfrac{1}{y}\log\left[\dfrac{1+t}{1+t/2}\right];$	Brunner & Lambert [1974] Section 3.4.4
2. $k=-t(1-2t)\exp(s(s-t))y;\ f=1-2t+z-y;$ $y=\exp(-t^2)$	Pouzet [1960] Section 3.4.4
3. $k=t^2s\cos(t^2y);\ f=2t-\frac{1}{2}\sin(t^4)+z;\ y=t^2$	Feldstein & Sopka [1974]
4. $k=(t+cs)^d y^3;\ f=[g_0(t)-ay-bz]^3-1;$ $g_0(t):=1+a+\dfrac{bt^{d+1}}{c(d+1)}\ [(1+c)^{d+1}-1];\ y=1$	Wolkenfelt [1977] Section 7.4.2, 7.6.5

Table 8.3.2d, Weakly singular kernels

Equation (8.3.1) with $k=K/(t-s)^\alpha$
1. $K=cy,\ 0<\alpha<1$, c constant; $f=p(t)y+z$ $+(2+\nu)t^{1+\nu}-p(t)(1+t^{2+\nu})-c[\dfrac{t^{1-\alpha}}{1-\alpha}+B(1-\alpha,\ 3+\nu)t^{3+\nu-\alpha}]$ $y=1+t^{2+\nu},\ \nu=\rho-\alpha,\ \rho\in\mathbb{N}_0,\ p(t)$ smooth

8.4. Routines for first-kind Volterra integral equations

In the following we present a list of published computer codes for Volterra and Abel-type integral equations of the first kind and a number of test problems (for equations with exact data). We shall be referring to these first-kind integral equations as follows:

$$g(t)+\int_0^t K(t,s)y(s)ds = 0, \tag{8.4.1a}$$

with $K(t,s)$ bounded;

$$g(t)+\int_0^t k(t,s,y(s))ds = 0, \tag{8.4.1b}$$

with $k(t,s,y)$ bounded (for each finite y);

$$g(t)+\int_0^t (t-s)^{-\alpha} K(t,s)y(s)ds = 0, \tag{8.4.2a}$$

with $0<\alpha<1$ and $K(t,s)$ bounded;

$$g(t)+\int_0^t (t-s)^{-\alpha} k(t,s,y(s))ds = 0, \tag{8.4.2b}$$

with $0<\alpha<1$ and $k(t,s,y)$ bounded (for each finite y);

$$g(t)+\int_0^t (t^2-s^2)^{-1/2} y(s)ds = 0; \tag{8.4.3a}$$

and

$$g(t)+\int_t^T (s^2-t^2)^{-1/2} sy(s)ds = 0. \tag{8.4.3b}$$

(where $T=\infty$ is also encountered, see, e.g., Anderssen [1977b], Anderssen & de Hoog [1982]).

8.4.1. Summary of published routines

Table 8.4.1. Routines for equation (8.4.1a)

- VLIMP, VLISHP, VLIGM (FORTRAN) - Respectively based on: Midpoint method with one (optimal) Richardson extrapolation step using stepsizes h and $h/3$, Midpoint method with end correction and using VLIMP to obtain the starting values, Product midpoint method with $K(t,s)$ as weight function (indefinite integral of $K(t,s)$ with respect to s must be given by user), using one (optimal) extrapolation. All methods have order $p=3$, fixed stepsizes and no error estimates. - Squire [1969]*, [1970].
- REGULAR (ALGOL) - Equation (8.4.1a) with exact and non-exact data; implementation of regularization method of Philips and Tikhonov; additional option for integrating first-kind Fredholm equations. - te Riele [1977b]*.

Table 8.4.2. Routines for equation (8.4.2b)

- NLSI2, NLSI3 (FORTRAN)
- Integrate equation (8.4.2b) with $\alpha = 1/2$; collocation in $S_1^{(0)}(Z_N)$ and $S_2^{(0)}(Z_N)$; discretization of collocation equation by means of weighted one-point Gauss or two-point Radau quadrature; orders are $p = 2$ and $p = 3$; fixed stepsize, no error estimates.
- Branca [1976]*, [1978].

Table 8.4.3. Routines for equation (8.4.3b)

- ABELA (FORTRAN)
- Integrates equation (8.4.3b) with non-exact data.
  Least-square polynomial (using Chebyshev polynomials) is fitted to observed data for $g(t)$, with degree $\leqslant 15$ determined automatically by program; exact evaluation of inversion formula by Gauss-Jacobi quadrature.
- Piessens [1973]*.

- ABEL (FORTRAN)
- Integrates equation (8.4.3b) with exact data; interpolation of discrete data for $g(t)$ by cubic splines; solution of integral equation by product midpoint method (adaptation of routine VLIGM, cf. Table 8.4.1).
- Fan & Squire [1975]*.

- INTEQABEL (ALGOL)
- Integrates equation (8.4.3b) with exact data.
  Interpolation of discrete data for $g(t)$ by polynomial splines of degree $m = 2l - l$ $(1 < l \leqslant n)$; approximation for y based on inversion formula.
- Smarzewski & Malinowski [1979a], [1979b]*.

- MATE (FORTRAN)
- Solution of plane-section equation from stereology with non-exact data (given on non-uniform grid); interval of integration $[t, \infty)$ $(T = \infty)$.
- Evaluation of various linear functionals of solution. See references below for details on statistical tools and product-integration techniques underlying the algorithm.
  Option for analogous second-kind equation (cf. Table 8.2.2).
- Jakeman [1980], Anderssen [1973], Jakeman & Anderssen [1975].

8.4.2. Test problems

In the Tables 8.4.4 we distinguish between the families of convolution kernels, of weakly singular kernels, and kernels not belonging to these families.

All equations are written in the form (8.4.1b). The functions k and g, and the references where numerical experiments are reported, are listed in the Tables 8.4.4. In addition, the exact solution y is provided in most cases.

Table 8.4.4a. Nonsingular convolution kernels

Equation (8.4.1)	References
1. $k=[1+\eta(t-s)]y$; $g=-\dfrac{t}{\eta}-\dfrac{1}{\eta^2}(e^{-\eta t}-1)$; $y=te^{-\eta t}$	Brunner [1978c] ($\eta=1$: Kobayasi [1967])
2. $k=ae^{a(t-s)}y$; $g=-\sinh(at)$; $y=e^{-at}$	Gladwin [1975] Section 3.3.6, 7.3.7.2
3. $k=(a^2+1)\cos(t-s)y$; $g=a\cos(t)-\sin(t)-ae^{at}$; $y=e^{at}$	Linz [1967] Section 3.3.6, 7.3.7.2
4. $k=1+\sqrt{\epsilon+t-s}$; $g=-t-\frac{2}{3}(t+\epsilon)^{3/2}+\frac{2}{3}\epsilon^{3/2}$; $y=1$	Section 7.3.7.2
5. $k=\sin(2\pi(t-s))y$; $y=e^{-t}$ $g=[-\sin(2\pi t)+2\pi\cos(2\pi t)-e^{-t}]/(1+4\pi^2)$;	Knopp et al. [1976]
6. $k=\begin{cases}\sin[2\pi(t-s)]y, & t-s\leq\frac{1}{2}\\ 0 & t-s>\frac{1}{2}\end{cases}$; $y=e^{-t}+\delta(t)^*$ $g=\dfrac{1}{1+4\pi^2}\begin{cases}[-\sin(2\pi t)+2\pi\cos(2\pi t)-e^{-t}], & t\leq\frac{1}{2}\\ 2\pi(1+e^{1/2})e^{-t}, & t>\frac{1}{2}\end{cases}$	Knopp et al. [1976]
7. $k=e^{s-t}[y+e^{-y}]$; $g=e^{-t}-\log(t+e)$; $y=\log(t+e)$	Section 3.3.6

* $\delta(t)$: Dirac δ function

Table 8.4.4b. General nonsingular kernels

Equation (8.4.1)	References
1. $k = y$; $g = -te^{at}(a<0)$; $y = -(1+at)e^{at}$	Brunner [1978c] ($a = -1$)
2. $k = (t+s)y$; $g = 1-2t\sin(t)-\cos(t)$; $y = \cos(t)$	Keech [1977b]

Table 8.4.4c. Weakly singular kernels

Equation (8.4.1b) with $k = K/(t-s)^{\alpha}$	References
1. $K = y$, $0<\alpha<1$; $g = -\dfrac{t^{2-\alpha}}{(1-\alpha)(2-\alpha)}$; $y = t$	Brunner [1978c]
2. $K = y$, $0<\alpha<1$; $g = \dfrac{6t^{4-\alpha}}{(1-\alpha)(2-\alpha)(3-\alpha)(4-\alpha)}$; $y = t^3$	Hung [1979b]
3. $K = y$, $\alpha = \frac{1}{2}$; $g = 1-e^t$; $y = \dfrac{1}{\sqrt{\pi}} e^t \operatorname{erf}(\sqrt{t})$	Piessens & Verbaeten [1973]
4. $K = y$, $\alpha = \frac{1}{2}$; $g = -\dfrac{2\log(\sqrt{1+t}+\sqrt{t})}{\sqrt{1+t}}$; $y = \dfrac{1}{1+t}$	Hung [1979b]
5. $K = y$, $\alpha = \frac{1}{2}$; $g = -\frac{3}{8}\pi t^2$; $y = t^{3/2}$	te Riele & Schroevers [1986]
6. $K = (1+ts)y$, $\alpha = \frac{1}{2}$; $g = -t(1+\frac{3}{4}t^2)$; $y = 2t^{1/2}/\pi$	Branca [1976]
7. $K = \dfrac{3+s-t}{1+y} - \dfrac{t(1-y)}{3+s}$, $\alpha = \frac{1}{2}$; $g = -(t+4)(\dfrac{\pi}{2} - \arctan(\dfrac{2}{\sqrt{t}}))$; $y = (t+4)^{1/2}$	Branca [1976]

NOTES

Section 8.2

The earliest published code for (systems of) second-kind Volterra integral equations seems to be the one by Rumyantsev [1965]. As for later (mostly non-automatic) codes see, e.g., Squire [1970] and Verlan & Sisikov [1978]. Papers describing the state of the art in integral equation codes are due to Miller [1974a], [1974b], Delves [1978], Riddell & Delves [1980]; compare also Noble [1977, pp. 957-958] and Elliott [1980, pp. 12-13] for remarks on aims in the construction of computer programs and on the availability of codes in program libraries (at the time of writing, neither the IMSL nor the NAG library contained routines for Volterra equations; on the other hand, compare Riddell & Delves [1980] for Fredholm routines contained in the NAG library).

Stoutemyer [1977] used a computer algebra language (REDUCE) to construct automatically closed-form and series solutions for Volterra and Fredholm integral equations; see also the remarks by Bogen [1979] on this approach.

As for tests of the performance of many of the routines listed in Table 8.2.1 (including, in some cases, comparisons with other such methods), we refer the reader to Baker, Riddell et al. [1980], Duncan [1982], Schlichte [1984], Hairer, Lubich & Schlichte [1985], Jones & McKee [1985], Blom & Brunner [1985], and Bader & Kunkel [1985].

The book by Krasnov et al. [1977] contains many examples of linear Volterra integral equations of the second kind with closed-form solutions. Sets of test examples were given by te Riele (1977) (personal communication) and, subsequently, by Baker, Riddell et al. [1980], Duncan [1982], Bownds [1982a] (this paper contains an analysis of the salient features of the test equations which include most of te Riele's problems), Schlichte [1984], and Jones & McKee [1985].

Section 8.3.

The paper by Jackiewicz [1984] extends techniques used for ODEs (method of defect correction) to estimate the global error of Euler's method when applied to nonlinear VIDEs. Goldfine [1977] discusses the computer implementation of Taylor series methods for VIDEs (and for second-kind Volterra equations); here, the principal feature is the PL/1-FORMAC program for the symbolic computation of the required derivatives of f and k in (8.3.1). So far no automatic codes exist for VIDEs with weakly singular kernels.

Section 8.4.

The paper by Anderssen [1976] contains, in addition to a description of three algorithms for the solution of (8.4.3b) with noisy data, an analysis of the shortcomings of classical methods like those proposed by Minerbo & Levy [1969] and Piessens & Verbaeten [1973]. Compare also Anderssen [1977b] and Anderssen & de Hoog [1982] for remarks on strategies to be used when devising algorithms for such first-kind equations.

Finally, we are grateful to Dr. A.J. Jakeman (Canberra) who provided us with a complete listing of his package MATE (cf. Tables 8.2.2 and 8.4.3).

REFERENCES

Abdalkhani, J. [1982], *Collocation and Runge-Kutta methods for Volterra integral equations with weakly singular kernels,* PhD thesis, Dalhousie University, Halifax, Nova Scotia. *403*

Abdalkhani, J. [1984], *Collocation and Runge-Kutta methods for Volterra integral equations of the second kind with weakly singular kernels,* Congr. Numer., 42, 87-100. *403*

Abd-Elal, L.F. [1979], *An asymptotic expansion for a regularization technique for numerical singular integrals and its application to Volterra integral equations,* J. Inst. Math. Appl., 23, 291-309.

Abramowitz, M. & Stegun, I.A. [1970], *Handbook of Mathematical Functions,* National Bureau of Standards, Applied Mathematical Series, 55, Washington (Dover, New York). *56, 59, 60, 62, 91, 108*

Abel, N.H. [1823]. *Solution de quelques problèmes à l'aide d'intégrales définies,* Oeuvres Complètes, I, Grøndahl & Søn, 1881, 11-27. *1, 7*

Abel, N.H. [1826], *Auflösung einer mechanischen Aufgabe,* J. Reine Angew, Math., 1, 153-157 (also in Oeuvres Complètes, I, Grøndahl & Søn, Christiania, 97-101). *7, 46*

Agachev, Yu.R. [1981], *Convergence of the spline-subregion method for integral equations (Russian),* Izv. Vyssh. Uchebn. Zaved. Mat., 6 (229), 3-10 (see also Soviet Math., 25, 3-10).

Agachev, Yu. R. & Leonov, A.I. [1979], *On the approximate solution of integral equations with a weak singularity (Russian),* Izv. Vyssh. Uchebn. Zaved. Mat., 7 (206), 69-72 (see also Soviet Math., 23, 73-77).

Agarwal, R.P. & Thandapani, E. [1981], *Remarks on generalizations of Gronwall's inequality,* Chinese J. Math., 9, 1-22. *49*

Albrecht, J. & Collatz, L. (eds.) [1980], *Numerical Treatment of Integral Equations,* Internat. Ser. Numer. Math., 53, Birkhauser Verlag, Basel-Boston-Stuttgart. *45, 163*

Alexander, R. [1977], *Diagonally implicit Runge-Kutta methods for stiff ODE's,* SIAM J. Numer. Anal., 14, 1006-1021. *180, 231, 232*

Alliney, S. & Sgallari, F. [1984], *An 'ill-conditioned' Volterra integral equation related to the reconstruction of images from projections,* SIAM J. Appl. Math., 44, 627-645.

Amini, S. [1980], *An analysis of stability behaviour in the numerical treatment of certain Volterra integral equations,* PhD thesis, University of Manchester. *496*

Amini, S. [1982], *On the stability of numerical methods for Volterra integral equations of the second kind,* in: Baker & Miller [1982], 43-46.

Amini, S. [1983], *Stability analysis of methods employing reducible rules for Volterra integral equations,* BIT, 23, 322-328. *453*

Amini, S., Baker, C.T.H. & Wilkinson, J.C. [1980], *Basic analysis of Runge-Kutta methods for Volterra integral equations of the second kind,* Report No. 46, Dept. of Math., University of Manchester. *496*

Amini, S., Baker, C.T.H., Houwen, P.J. van der & Wolkenfelt, P.H.M. [1983], *Stability analysis of numerical methods for Volterra integral equations with polynomial convolution kernels,* J. Integral Equations, 5, 73-92. *453, 473, 495, 496*

Anderssen, R.S. [1973], *Computing with noisy data with an application to Abel's equation,* in: *Error, Approximation, and Accuracy* (F.R. de Hoog & C.L. Jarvis, eds.), University of Queensland Press, St. Lucia, 61-79. *400, 401, 504, 516*

Anderssen, R.S. [1976], *Stable procedures for the inversion of Abel's equations,* J. Inst. Math. Appl., 17, 329-342. *401, 406, 519*

Anderssen, R.S. [1977a], *Some numerical aspects of improperly posed problems or why regularization works and when not to use it,* Techn. Report No. 32, Computer Centre, Australian National University, Canberra. *406*

Anderssen, R.S. [1977b], *Application and numerical solution of Abel-type integral equations,* MRC Techn. Summary Report No. 1787, University of Wisconsin, Madison. *45, 400, 401, 515, 520*

Anderssen, R.S. [1980], *On the use of linear functionals for Abel-type integral equations in applications,* in: Anderssen, de Hoog & Lukas [1980], 195-221. *406, 407*

Anderssen, R.S. & Bloomfield, P. [1974], *Numerical differentiation procedures for nonexact data,* Numer. Math., 22, 157-182. *321*

Anderssen, R.S. & Hoog, F.R. de [1982], *Application and numerical solution of Abel-type integral equations,* Research Report No. 7-1982, Dept. of Math., Australian National University, Canberra. *45, 400, 401, 406, 515, 520*

Anderssen, R.S. & Hoog, F.R. de [1984], *Finite difference methods for the numerical differentiation of non-exact data,* Computing, 33, 259-267. *321*

Anderssen, R.S., Hoog, F.R. de & Lukas, M.A. (eds.) [1980], *Application and Numerical Solution of Integral Equations,* Sijthoff & Noordhoff, Alphen aan de Rijn. *163*

Anderssen, R.S., Hoog, F.R. de & Weiss, R. [1973], *On the numerical solution of Brownian motion processes,* J. Appl. Probab., 10, 409-418. *405, 406*

Anderssen, R.S. & Jakeman, A.J. [1974], *On computational stereology,* in: *Proc. 6th Austral. Comput. Confer,* Vol. 2, Sydney, 353-362. *407, 504*

Anderssen, R.S. & Jakeman, A.J. [1975], *Abel-type integral equations in stereology II: Computational methods of solution and the random sphere approximation*, J. Microscopy, 105, 135-153. *407*

Anderssen, R.S. & White, E.T. [1971], *Improved numerical methods for Volterra integral equations of the first kind*, Comput. J., 14, 442-443. *320*

Andrade, C., Franco, N.B. & McKee, S. [1981], *Convergence of linear multistep methods for Volterra first kind equations with $k(t,t) \neq 0$*, Computing, 27, 189-204. *165*

Andrade, C. & McKee, S. [1979], *On optimal high accuracy linear multistep methods for first kind Volterra integral equations*, BIT, 19, 1-11. *165*

Andrade, C. & McKee, S. [1983], *Optimal stability functionals and their applications to discretization merthods for Volterra integro-differential equations*, Z. Angew. Math. Mech., 63, 453-455.

Andreoli, G. [1914], *Sulle equazioni integrali*, Rend. Palermo, 37, 76-112. *45*

Anselone, P.M. (ed) [1964], *Nonlinear Integral Equations*, University of Wisconsin Press, Madison.

Anselone, P.M., Bueckner, H.F. & Greenspan, D. [1963], *Solution of an integral equation of the third kind by successive approximations*, MRC Techn. Summary Report No. 345, University of Wisconsin, Madison. *404*

Aparcin, A.S. [1973], *The application of various quadrature formulae to the approximate solution of Volterra integral equations of the first kind by the method of quadrature sums (Russian)*, in: *Differential and Integral Equations*, No. 2, Irkutsk. Gos. Univ., 107-116, 309-310.

Aparcin, A.S. [1976], *On the numerical solution of systems of Volterra integral equations of the first kind by a quadrature method (Russian)*, Optimization Methods and Operations Research, Work Collect., Irkutsk, 79-88.

Aparcin, A.S. [1979], *On the numerical solution of Volterra integral equarions of the first kind by the regularized quadrature method*, Optimization Methods and Their Applications, 9, Work Collect., Irkutsk, 99-107. *165*

Aparo, E. [1959], *Sulla resoluzione numerica delle equazioni integrali di Volterra di seconda specie*, Atti Acad. Naz. Lincei Rend. Cl. Sci. Fis. Mat. Natur., 26, 183-188. *177*

Arndt, H. [1984], *On stepsize control for Volterra integral equations*, in: *Numerical Methods of Approximation Theory*, (L. Collatz, G. Meinardus & H. Werner, eds.), Internat. Ser. Numer. Math., 67, Birkhäuser Verlag, Basel-Boston-Stuttgart, 9-17. *501*

Arndt, H. [1985], *An adaptive step size control for Volterra integral equations*, in: Hämmerlin & Hoffmann, [1985], 44-52. *501*

Artin, E. [1964], *The Gamma Function*, Holt, Rinehart and Winston, New York. *15*

Ash, J.M. & Jones, R.L. [1981], *Optimal numerical differentiation using three function evaluations*, Math. Comp., 37, 159-167. *305*

Atdaev, S. [1976], *On the theory of solutions of nonlinear multidimensional*

Volterra operator equations (Russian), Izv. Akad. Nauk Turkmen. SSR Ser. Fiz.-Tehn. Him. Geol. Nauk, 6, 9-16.

Atdaev, S. & Asirov, S. [1976], *On convergence of some iterative processes in Banach spaces with a cone (Russian),* Izv. Akad. Nauk Turkmen. SSR Ser. Fiz.-Tehn. Him. Geol. Nauk, 5, 7-15.

Atkinson, K.E. [1974a], *An existence theorem for Abel integral equations,* SIAM J. Math. Anal., 5, 729-736. *48*

Atkinson, K.E. [1974b], *The numerical solution of an Abel integral equation by a product trapezoidal method,* SIAM J. Numer. Anal., 11, 97-101. *406*

Atkinson, A.E. [1976], *A Survey of Numerical Methods for the Solution of Fredholm Integral Equations of the Second Kind,* Society for Industrial and Applied Mathematics, Philadelphia. *47*

Axelsson, O. [1969], *A class of A-stable methods,* BIT, 9, 185-199. *317*

Bakaev, N.Y. & Tarasov, R.P. [1980], *Stability of solutions of a class of first-order convolution integral equations,* Differential Equations, 15, 1593-1600.

Bader, G. & Kunkel, P. [1985], *An adaptive multistep method for the solution of second-kind Volterra integral equations* (to appear). *503, 519*

Baker, C.T.H. [1974a], *Methods for Volterra equations of the first kind,* in: Delves & Walsh [1974], 162-174. *163*

Baker, C.T.H. [1974b], *Methods for integro-differential equations,* in: Delves & Walsh [1974], 189-206. *163, 233*

Baker, C.T.H. [1976], *Initial value problems for Volterra integro-differential equations,* in: Hall & Watt [1976], 296-307. *96, 163, 233*

Baker, C.T.H. [1977], *The Numerical Treatment of Integral Equations,* Clarendon Press, Oxford. *91, 92, 163, 167, 177, 494, 505, 506, 507, 508*

Baker, C.T.H. [1978], *Runge-Kutta methods for Volterra equations of the second kind,* in: *Numerical Analysis* (G.A. Watson, ed.), Lecture Notes in Math., 630, Springer-Verlag, Berlin-Heidelberg-New York, 1-13.

Baker, C.T.H. [1980], *Structure of recurrence relations in the study of stability in the numerical treatment of Volterra equations,* J. Integral Equations, 2, 11-29. *494, 495*

Baker, C.T.H. [1982a], *An introduction to the numerical treatment of Volterra and Abel-type integral equations,* in: *Topics in Numerical Analysis,* Lecture Notes in Math., 965, Springer-Verlag, Berlin-Heidelberg-New York, 1-38. *96, 163*

Baker, C.T.H. [1982b], *Families of quadrature rules and some ramifications of reducibility,* in: *Numerical Integration* (G. Hämmerlin, ed.), Internat. Ser. Numer. Math.,57, Birkhäuser Verlag, Basel-Boston-Stuttgart, 12-24.

Baker, C.T.H. [1985], *Concerning $A(\alpha)$-stable mixed Volterra-Runge-Kutta methods,* in: Hämmerlin & Hoffmann [1985], 53-67. *496*

Baker, C.T.H. & Derakhshan, M.S. [1985], *Preliminary report on a fast algorithm for the numerical solution (with error estimates) of a Volterra integral equation of the 2nd kind having a convolution kernel,* Report No. 106, Dept. of Math., University of Manchester. *504*

Baker, C.T.H. & Keech, M.S. [1978a], *Stability regions in the numerical treatment of Volterra integral equations,* SIAM J. Numer. Anal., 15, 394-417. *430, 431, 436, 494, 495, 496*

Baker, C.T.H. & Keech, M.S. [1978b], *Stability analysis of certain Runge-Kutta procedures for Volterra integral equations,* ACM Trans. Math. Software, 4, 305-315. *501, 508*

Baker, C.T.H. & Keech, M.S. [1978c], *Methods for two-sided Runge-Kutta approximations to the solution of Volterra integral equations of the second kind,* Report No. 27, Dept. of Math., University of Manchester. *502*

Baker, C.T.H., Makroglou, A. & Short, E. [1979], *Regions of stability in the numerical treatment of Volterra integro-differential equations,* SIAM J. Numer. Anal., 16, 890-910. *469, 494, 495, 496*

Baker, C.T.H. & Miller, G.F. (eds.) [1982], *Treatment of Integral Equations by Numerical Methods,* Academic Press, London. *163*

Baker, C.T.H., Riddell, I.J., Keech, M.S. & Amini, S. [1980], *Runge-Kutta methods with error estimates for Volterra integral equations of the second kind,* in: Albrecht & Collatz [1980], 24-42. *505, 508, 519*

Baker, C.T.H. & Wilkinson, J.C. [1980], *Basic stability analysis of Runge-Kutta methods for Volterra integro-differential equations,* Report No. 50, Dept. of Math., University of Manchester. *469, 496*

Baker, C.T.H. & Wilkinson, J.C. [1981], *Stability analysis of Runge-Kutta methods applied to a basic Volterra integral equation,* J. Austral. Math. Soc. Ser. B, 22, 515-538. *496*

Baker, C.T.H. & Wilkinson, J.C. [1982], *On the construction of stability polynomials for modified R-K methods for Volterra integro-differential equations,* in: Baker & Miller [1982], 33-42. *496*

Bakke, V.L. & Jackiewicz, Z. [1984], *Boundedness of solutions of difference equations and application to numerical solution of Volterra integral equations of the second kind,* Numer. Anal. Techn. Report 65R20-3, University of Arkansas, Fayetteville. *495*

Bakke, V.L. & Jackiewicz, Z. [1985], *Stability of reducible quadrature methods for Volterra integral equations of the second kind,* Numer. Math., 47, 159-173. *433, 495*

Balasubramanian, R., Norrie, D.H. & Vries, G. de [1979], *The application of the least squares finite element method to Abel's integral equation,* Internat. J. Numer. Methods Engrg., 14, 201-209.

Bateman, H. [1910], *Report on the history and present state of the theory of integral equations,* Report to the British Association for the

Advancement of Science (Sheffield), 345-424. *46*

Beer, A. [1865], *Einleitung in die Elektrostatik, die Lehre vom Magnetismus und die Elektrodynamik,* F. Vieweg & Sohn, Braunschweig. *10*

Beesack, P.R. [1969], *Comparison theorems and integral inequalities for Volterra integral equations,* Proc. Amer. Math. Soc., 20, 61-66. *40, 48*

Beesack, P.R. [1975], *Gronwall Inequalities,* Carleton Math. Lecture Notes, 11, Carleton University, Ottawa. *39, 40, 49*

Beesack, P.R. [1984a], *More generalized disxrete Gronwall inequalities,* Preprint. *49*

Beesack, P.R. [1984b], *On some Gronwall-type integral inequalities in* n *independent variables,* J. Math. Anal. Appl., 100, 393-408. *48*

Beesack, P.R. [1984c], *Systems of multidimensional Volterra integral equations and inequalities,* Carleton Math. Series, No. 209, Dept. of Math. and Statis. Carleton University, Ottawa. *48*

Bellen, A [1985], *Constrained mesh methods for functional differential equations,* in: *Delay Equations, Approximation and Application* (G. Meinardus & G. Nürnberger, eds.), Internat. Ser. Numer. Math., 74, Birkhäuser Verlag, Basel-Boston, 52-70.

Bellman, R. & Cooke, K.L. [1963], *Differential-Difference Equations,* Academic Press, New York. *16, 38, 47, 48, 493*

Bel'tyukov, B.A. [1961], *A method of approximate solution of Volterra integral equations (Russian),* Sibirsk. Mat. Zh., 2, 789-791. *318*

Bel'tyukov, B.A. [1965a], *An application of nonclosed quadrature formulae to the numerical solution of nonlinear integral equations of Volterra type (Russian),* Trudy Naucn. Ob"ed. Prepodav. Fiz.-Mat. Fak. Ped. Inst. Dal'n. Vstok., 5, 10-20. *164*

Bel'tyukov, B.A. [1965b], *Analogue of the Runge-Kutta method for the solution of nonlinear integral equations of Volterra type,* Differential Equations, 1, 417-433. 174, 177, 192, 194, 195, 473, 509

Bel'tyukov, B.A. & Kuznechikhina, L.N. [1976], *A Runge-Kutta method for the solution of two-dimensional nonlinear Volterra integral equations,* Differential Equations, 12, 1169-1173. *177*

Benson, M. [1973], *Errors in numerical quadrature for certain singular integrands and the numerical solution of Abel integral equations,* PhD thesis, University of Wisconsin, Madison. *89, 92, 405*

Berikelashvili, G.K. & Pavlenishvili, G.D. [1980], *On the approximate solution of some systems of nonlinear Volterra integral equations of the second kind (Russian),* Soobshch. Akad. Nauk Gruz. in SSR, 99, 312-316.

Bernfeld, S.R. & Lord, M.E. [1978], *A nonlinear variation of constants method for integro-differential and integral equations,* Appl. Math. Comput., 4, 1-14. *47*

Bernier, J. [1945], *Les principales méthodes de résolution numérique des équations intégrales de Fredholm et de Volterra,* Ann. Radioelectr., 1, 311-318. *163*

Bharucha-Reid, A.T. [1972], *Random Integral Equations,* Academic Press, New York. *46, 47*

Birkhoff, G. (ed.) [1973], *A Source Book in Classical Analysis,* Harvard University Press, Cambridge (Mass.), Ch. 13, 435-465. *46*

Birkhoff, G. & Mac Lane, S. [1953], *A Survey of Modern Algebra,* Mac Millan, New York. *428*

Blom, J.G. & Brunner, H. [1985], *The numerical solution of nonlinear Volterra integral equations of the second kind by collocation and iterated collocation,* Report NM-R8522, Centre for Mathematics and Computer Science, Amsterdam. *504, 519*

Blom, J.G., & Brunner, H. [1986], Algorithm XXX: *Discretized collocation and iterated collocattion for nonlinear Volterra integral equations of the second kind* (to appear). *504*

Bochner, M. [1913], *An Introduction to the Study of Integral Equations* (2nd ed.), Cambridge Tracts in Mathematics and Mathematical Physics, No. 10, Cambridge University Press, Cambridge (Mass.). *47*

Bock, T.L. [1977], *Numerical inversion of the Abel integral function, using least squares splines,* Ann. Physik (7), 34, 335-340.

Bogen, R.A. [1979], *Addendum to "Analytically solving integral equations by using computer algebra",* ACM Trans. Math. Software, 5, 234-237. *519*

Bois-Reymond, P. du [1888], *Bemerkungen über* $\Delta z = \partial^2 z / \partial x^2 + \partial^2 z / \partial y^2 = 0$, J. Reine Angew. Math., 103, 204-229. *9, 45*

Bona, K. & Garey, L. [1973], *Methods for solving second-kind Volterra integral equations using generalized quadrature formulae,* in: Proc. Third Manitoba Confer. Numer. Math. (R.S.D. Thomas & H.C. Williams, eds.), University of Manitoba, Winnipeg, 183-193.

Boor, C. de [1978], *A Practical Guide to Splines,* Springer-Verlag, New-York-Heidelberg-Berlin. *52, 239, 317, 348, 350, 404*

Bownds, J.M. [1976], *On solving weakly singular Volterra equations of the first kind with Galerkin approximations,* Math. Comp., 30, 747-757 (corrigendum: Math. Comp., 31, 808). *406*

Bownds, J.M. [1978a], *A modified Galerkin approximation method for Volterra equations with smooth kernels,* Appl. Math. Comput., 4, 67-79. *319*

Bownds, J.M. [1978b], *On an initial value method for quickly solving Volterra integral equations: a review,* J. Optim. Theory Appl., 24, 133-151. *319, 426, 502*

Bownds, J.M. [1979], *A combined recursive collocation and kernel approximation technique for certain singular Volterra integral equations,* J. Integral Equations, 1, 153-164. *319, 403*

Bownds, J.M. [1981], *A subroutine for solving Volterra integral equations,* Part I: *Theory and performance;* Part II (with Appelbaum, L.): *The code and its documentation,* Preprint, University of Arizona, Tucson. *319, 426, 502, 508*

Bownds, J.M. [1982a], *Theory and performance of a subroutine for solving Volterra integral equations,* Computing, 28, 317-332. *319, 502, 506, 519*

Bownds, J.M. [1982b], *Comments on the performance of a FORTRAN subroutine for certain Volterra equations,* in: Baker & Miller [1982], 163-167. *426, 502*

Bownds, J.M. & Appelbaum, L. [1985], *Algorithm 627: A FORTRAN subroutine for solving Volterra integral equations,* ACM Trans. Math. Software, 11, 58-65. *502*

Bownds, J.M. & Cushing, J.M. [1972], *Some stability criteria for linear systems of Volterra integral equations,* Funkcial. Ekvac., 15, 101-117.

Bownds, J.M. Cushing, J.M. [1975a], *Some stability theorems for systems of Volterra equations,* Applicable Anal., 5, 65-77. *13, 14, 48, 493*

Bownds, J.M. & Cushing, J.M. [1975b], *On preserving stability of Volterra integral equations under a general class of perturbations,* Math. Systems Theory, 9, 117-131.

Bownds, J.M. & Wood, B. [1976], *On numerically solving nonlinear Volterra integral equations with fewer computations,* SIAM J. Numer. Anal., 13, 705-719.

Bownds, J.M. Wood, B. [1977], *A note on solving Volterra integral equations with convolution kernels,* Appl. Math. Comput., 3, 307-315. *319*

Bownds, J.M. & Wood, B. [1979], *A smoothed projection method for singular, nonlinear Volterra equations,* J. Approx. Theory, 25, 120-141. *403*

Branca, H.-W. [1976], *Die nichtlineare Volterra Integralgleichung vom Abel'schen Typ und ihre numerische Behandlung,* Dissertation, Universität Köln. 47, 396, 398, 405, 516, 518

Branca, H.-W. [1978], *The nonlinear Volterra equation of Abel's kind and its numerical treatment,* Computing, 20, 307-324. *27, 47, 396, 398, 405, 406, 516*

Brass, H. [1977], *Quadraturverfahren,* Vandenhoek & Ruprecht, Göttingen-Zürich. *58, 91*

Brauer, F. [1972], *A nonlinear variation of constants formula for Volterra equations,* Math. Systems Theory, 6, 226-234. *47*

Brauer, F. [1975], *On a nonlinear integral equation for population growth problems,* SIAM J. Math. Anal., 6. 312-317.

Brauer, F. [1976], *Constant rate harvesting of populations governed by Volterra integral equations,* J. Math. Anal. Appl., 56, 18-27.

Brunner, H. [1971], *The solution of nonlinear Volterra integral equations by piecewise polynomials,* in: *Proc. First Manitoba Confer, Numer. Math.* (R.S.D. Thomas & H.C. Williams, eds.), University of Manitoba, Winnipeg, 65-78. *318, 320*

Brunner, H. [1973a], *On the numerical solution of nonlinear Volterra integro-differential equations,* BIT, 13, 381-390. *319, 320*

Brunner, H. [1973b], *The numerical solution of a class of Abel integral equations by piecewise polynomials,* J. Comput. Phys., 12, 412-416. *406*

Brunner, H. [1973c], *The solution of Volterra integral equations of the first kind by piecewise polynomials,* J. Inst. Math. Appl., 12, 295-302.

Brunner, H. [1974a], *Global solution of the generalized Abel integral equation by implicit interpolation,* Math. Comp., 28, 61-67. *405*

Brunner, H. [1974b], *On the approximate solution of first-kind integral equations,* Computing, 13, 67-79. *321*

Brunner,. H. [1975a], *Projection methods for the approximate solution of integral equations of the first kind,* in: *Proc. Fifth Manitoba Confer. Numer. Math.* (B.C. Hartnell & H.C. Williams, eds.), University of Manitoba, Winnipeg, 3-23. *406*

Brunner, H. [1975b], *On the approximate solution of the Abel equation with discontinuous solution,* BIT, 15, 136-143. *406*

Brunner, H. [1976], *The approximate solution of linear and nonlinear first-kind integral equations of Volterra type,* in: *Numerical Analysis,* (G.A. Watson, ed.), Lecture Notes in Math., 506, Springer-Verlag, Berlin-Heidelberg-New York. *321*

Brunner, H. [1977], *Discretization of Volterra integral equations of the first kind,* Math. Comp., 31, 708-716. *232, 318, 320, 405*

Brunner, H. [1978a], *Discretization of Volterra integral equations of the first kind (II),* Numer. Math., 30, 117-136. *232, 299, 302, 305, 320*

Brunner, H. [1978b], *Superconvergence of collocation methods for Volterra integral equations of the first kind,* Computing, 21, 151-157. *308, 320*

Brunner, H. [1978c], *The approximate solution of integral equations by projection methods based on collocation,* Lecture Notes, Report No. 1/78, Dept. of Math., University of Trondheim. *163, 318, 517, 518*

Brunner, H. [1978d], *On superconvergence in collocation methods for Abel integral equations,* in: *Proc. Eighth Manitoba Confer. Numer. Math.* (D. McCarthy & H.C. Williams, eds.), University of Manitoba, Winnipeg, 117-128. *405*

Brunner, H. [1979], *A note on collocation methods for Volterra integral equations of the first kind,* Computing, 23, 179-187. *308, 320*

Brunner, H. [1980], *Superconvergence in collocation and implicit Runge-Kutta methods for Volterra-type integral equations of the second kind,* in: Albrecht & Collatz [1980], 54-72. *231, 318*

Brunner, H. [1981], *The application of the variation of constants formulas in the numerical analysis of integral and integro-differential equations,* Utilitas Math., 19, 255-290. *318, 319, 403*

Brunner, H. [1982a], *A survey of recent advances in the numerical treatment of Volterra integral and integro-differential equations,* J. Comput. Appl. Math., 8, 213-229. *96, 163*

Brunner, H. [1982b], *On collocation approximations for Volterra equations with weakly singular kernels,* in: Baker & Miller [1982], 409-420. *373, 403*

Brunner, H. [1983], *Nonpolynomial spline collocation for Volterra equations with weakly singular kernels,* SIAM J. Numer. Anal., 20, 1106-1119. *48, 373, 405*

Brunner, H. [1984a], *Implicit Runge-Kutta methods of optimal order for Volterra integro-differential equations,* Math. Comp., 42, 95-109. *233, 319, 320*

Brunner, H. [1984b], *Iterated collocation methods and their discretiations for Volterra integral equations,* SIAM J. Numer. Anal., 20, 1132-1145. *318*

Brunner, H. [1984c], *The numerical solution of integral equations with weakly singular kernels,* in: *Numerical Analysis,* (D.F. Griffiths, ed.), Lecture Notes in Math., 1066, Springer-Verlag, Berlin-Heidelberg-New York, 50-71. *378, 403*

Brunner, H. [1985a], *The numerical solution of weakly singular Volterra equations by collocation on graded meshes,* Math. Comp., 45, 417-437. *404*

Brunner, H. [1985b], *The approximate solution of Volterra equations with nonsmooth solutions,* Utilitas Math., 27, 57-95. *404*

Brunner, H. [1985c], *On the numerical solution by collocation of Volterra integro-differential equations with nonsmooth solutions,* in: Hämmerlin & Hoffmann [1985], 74-92. *405*

Brunner, H. & Evans, M.D. [1977], *Piecewise polynomial collocation for Volterra-type integral equations of the second kind,* J. Inst. Math. Appl., 20, 415-423. *403*

Brunner, H., Hairer, E. & Nørsett, S.P. [1982], *Runge-Kutta theory for Volterra integral equations of the second kind,* Math. Comp., 39, 147-163. 175, 177, 181, 186, 190, 231, 268

Brunner, H. & Lambert, J.D. [1974], *Stability of numerical methods for Volterra integro-differential equations,* Computing, 12, 75-89. *91, 167, 319, 448, 469, 470, 495, 514*

Brunner, H. & Nørsett, S.P. [1980], *Runge-Kutta theory for Volterra integral equations of the second kind,* Report No. 1/80, Dept. of Math., University of Trondheim. *177, 181, 231*

Brunner, H. & Nørsett, S.P. [1981], *Superconvergence of collocation methods for Volterra and Abel equations of the second kind,* Numer. Math., 36, 347-358. *204, 318, 403, 510*

Brunner, H., Nørsett, S.P. & Wolkenfelt, P.H.M. [1980], *On V_0-stability of numerical methods for Volterra integral equations of the second kind,* Report NW 84/80, Mathematisch Centrum Amsterdam. *444, 484, 486, 495, 496*

Brunner, H. & Riele, H.J.J. te [1984], *Volterra-type integral equations of the second kind with nonsmooth solutions: high order methods based on collocation techniques,* J. Integral Equations, 6, 187-203. *363, 374, 404*

Burrage, K. & Butcher, J.C. [1979], *Stability criteria for implicit Runge-Kutta methods,* SIAM J. Numer. Anal., 19, 245-258. *487*

Burton, T.A. [1983a], *Structure of solutions of Volterra equations,* SIAM Review, 25, 343-364. *47, 48*

Burton, T.A. [1983b], *Volterra Integral and Differential Equations,* Academic Press, New York. 25, 27, 38, 47, 48, 410, 412, 493

Burton, T.A. & Mahfoud, W.E. [1983], *Stability criteria for Volterra equations,* Trans. Amer. Math. Soc. 279, 143-174. *48, 412, 493*

Butcher, J.C. [1963], *Coefficients for the study of Runge-Kutta integration processes,* J. Austral. Math. Soc., 3, 185-201. *179, 231*

Butcher, J.C. [1964a], *On Runge-Kutta processes of high order,* J. Austral. Math. Soc., 4, 179-194. *179, 231*

Butcher, J.C. [1964b], *Implicit Runge-Kutta processes,* Math. Comp., 18, 50-64. *178, 231*

Butcher, J.C. [1965], *On the attainable order of Runge-Kutta methods,* Math. Comp., 19, 408-417. *179, 231*

Butcher, J.C. [1972], *An algebraic theory of integration methods,* Math. Comp., 26, 79-106. *177, 231*

Butcher, J.C. [1976], *Implicit Runge-Kutta and related methods,* in: Hall & Watt [1976], 136-151. *179, 199, 231*

Butcher, J.C. [1985], *The non-existence of ten stage eighth order explicit Runge Kutta methods,* BIT, 25, 521-540. *179, 231*

Cahlon, B. [1981], *Numerical solution of non-linear Volterra integral equations,* J. Comput. Appl. Math., 7, 121-128. *403*

Callender, E.D. [1971], *Single step methods and low order splines for solutions of ordinary differential equations,* SIAM J. Numer. Anal., 8, 61-66. *317*

Cameron, R.F. [1981], *Direct solution of applicable Volterra integral equations,* D. Phil. thesis, University of Oxford.

Cameron, R.F. & McKee, S. [1981], *The direct numerical solution of a Volterra integral equation arising out of viscoelastic stress in materials,* Comput. Methods Appl. Mech. Engrg., 29, 219-232.

Cameron, R.F. & McKee, S. [1983], *High-accuracy product integration methods for the Abel equation, and their application to a problem in scattering theory,* Internat. J. Numer. Methods Engrg., 19, 1527-1536. *405*

Cameron, R.F. & McKee, S. [1984a], *Product integration methods for second-kind Abel integral equations,* J. Comput. Appl. Math., 11, 1-10. *403*

Cameron, R.F. & McKee, S. [1984b], *High accuracy convergent product integration methods for the generalized Abel equation,* J. Integral Equations, 7, 103-125. *405*

Cameron, R.F. & McKee, S. [1985], *The analysis of product integration methods for Abel's equation using discrete fractional differentiation,* IMA J. Numer. Anal., 5, 339-353. *398, 403, 405*

Campbell, G.M. [1970], *On the numerical solution of Volterra integral equations,* PhD thesis, Pennsylvania State University, University Park. *167, 319*

Campbell, G.M. & Day, J.T. [1970], *The numerical solution of nonlinear integral equations,* BIT, 10, 10-19. *162, 167, 319*

Campbell, G.M. & Day, J.T. [1971], *A block by block method for numerical solution of Volterra integral equations,* BIT, 11, 120-124. *167*

Cannon, J.R. [1984], *The One-Dimensional Heat Equation,* Addison-Wesley (Encyclopedia Math, Appl. 23), Reading (Mass.). *46, 404*

Caqué, J. [1864], *Méthode nouvelle pour l'intégration des équations différentielles linéaires ne contenant qu'une variable indépendante,* J. Math. Pures Appl. (2), 9, 185-222. *10*

Carson, J.R. [1926], *Electric Circuit Theory and the Operational Calculus,* Chelsea Publ. Co. (Reprint, 1953), New York. *163*

Chamayou, J.M.F. [1973a], *L'équation de Volterra de la théorie statistique de la création de paires de Frenkel dans un solide soumis à une irradiation neutronique,* Thèse, Université Paul Sabatier, Toulouse.

Chamayou, J.M.F. [1973b], *Volterra's functional integral equations of the statistical theory of damage,* J. Comput. Phys., 13, 70-93.

Chamayou, J.M.F. & El Tom, M.E.A. [1975], *On the approximate solution of the delay integral equation of the statistical theory of radiation damage,* Comput. Phys. Comm., 9, 131-140.

Chambré, P.L. [1959], *Nonlinear heat transfer problem,* J. Appl. Phys., 30, 1683-1688. *511*

Chang, S.H. [1982], *On certain extrapolation methods for the numerical solution of integro-differential equations,* Math. Comp., 39, 165-171. *166*

Chandler, G.A. [1979], *Superconvergence of numerical methods to second-kind integral equations,* PhD thesis, Australian National University, Canberra. *318, 404*

Chashchin, O.N. [1981], *Regularization of Volterra operator equations of the first kind in a scale of Banach spaces (Russian),* in: *Approximate Methods of Solutions and Questions of the Correctness of Inverse Problems,* Collect. Sci. Works, Novosibirsk, 132-144.

Chatelin, F. [1983], *Spectral Approximation of Linear Operators,* Academic Press, New York. *269, 318*

Chatelin, F. & Lebbar, R. [1981], *The iterated projection solution for the Fredholm integral equation of the second kind,* J. Austral. Math. Soc. Ser. B, 22, 439-451. *269, 271, 318*

Chawla, M.M. & Kumar, S. [1978], *Chebyshev series method for the approximate solution of Volterra integral equations with singular kernels,* J. Math. Phys. Sci., 12, 473-481. *406*

Chen, G. & Grimmer, R. [1982], *Asymptotic expansions of a penalty method for computing a regulator problem governed by Volterra equations,* in: Hannsgen, Herdman, Stech & Wheeler [1982], 217-231.

Chester, C.R. [1971], *Techniques in Partial Differential Equations,* McGraw-Hill, New York.

Chirikalov, V.A. [1981], *Solution of Volterra integral equations by means of quadrature formulas of increasing accuracy (Russian),* Vychisl. Prikl. Mat. (Kiev), 43, 147-153.

Chu, S.C. & Metcalf, F.T. [1967], *On Gronwall's inequality,* Proc. Amer. Math. Soc., 18, 439-440. *39, 48*

Clarysse, J.H. [1979], *Rational predictor-corrector methods for nonlinear Volterra integral equations of the second kind,* in: *Padé Approximations and its Applications* (L. Wuytack, ed.), Lecture Notes in Math., 765, Springer-Verlag Berlin-Heidelberg-New York, 278-294.

Cochran, J.A. [1972], *Analysis of Linear Integral Equations,* McGraw-Hill, New York. *40, 47*

Collatz, L. [1966], *The Numerical Treatment of Differential Equations,* 3rd edition, Springer-Verlag, New York: 512-518. *163*

Cooke, K.L. & Yorke, J.A. [1973], *Some equations modeling growth processes and epidemiccs,* Math. Biosci., 16, 75-101.

Coppel, W.A. [1965], *Stability and Asymptotic Behavior of Differential Equations,* Heath, Boston. *48, 410, 414, 419*

Corduncanu, A. [1979], *Uniform asymptotic stability for systems of Volterra integral equations,* Bul. Inst. Politehn. Iasi Sect. I, 15, 27-34.

Corduneanu, A. [1982], *A note on the Gronwall inequality in two independent variables,* J. Integral Equations, 4, 271-276. *48*

Corduneanu, A. [1983], *An integral inequality in two independent variables,* An Stiint. Univ. "Al. Cuza" Iasi Sect. I a Mat. (N.S.), 29, 69-74. *48*

Corduneanu, C. [1973], *Integral Equations and Stability of Feedback Systems,* Academic Press, New York. *38, 48, 493, 494*

Coroian, I. [1974], *On the Runge-Kutta-Fehlberg method regarding the nonlinear integral equation of the Volterra type,* Stud. Cerc. Mat., 26, 505-511. *177, 232*

Cremers, C.J. & Birkebak, R.C. [1966], *Application of the Abel integral equation to spectroscopic data,* Appl. Opt., 5, 1057-1064.

Crouzeix, M. [1979], *Sur la B-stabilité des méthodes de Runge-Kutta,* Numer. Math., 32, 75-82. *487*

Cryer, C.W. [1972], *Numerical methods for functional differential equations,* in: *Delay and Functional Differential Equations and Their Applications* (K. Schmitt, ed.), Academic Press, New York, 17-101. *163, 232, 233, 319*

Cullum, J. [1971], *Numerical differentiation and regularization,* SIAM J. Numer. Anal., 8, 254-265. *321*

Curle, S.N. [1978], *Solution of an integral equation of Lighthill,* Proc. Roy. Soc. London Ser. A, 364, 435-441.

Cushing, J.M. [1977], *Integro-differential Equations and Delay Models in Population Dynamics,* Lecture Notes in Biomath., 20, Springer-Verlag, Berlin-Heidelberg-New York. *45*

Dahlquist, G. [1959], *Stability and error bounds in the numerical integration of ordinary differential equations,* Report No. 130, Kungl. Tekn. Hogsk. Handl., Stockholm. *91, 164, 415*

Davis, H.T. [1926], *The Present Status of Integral Equations,* Indiana University Studies, Vol. XIII (Study No. 70), Bloomington. *10, 45, 163*

Davis, H.T. [1927a], *A Survey of Methods for the Inversion of Integrals of Volterra Type,* Indiana University Studies, XIV (Studies No. 76,77), Bloomington. *8, 21, 47, 403*

Davis, H.T. [1927b], *The application of fractional operators to functional equations,* Amer. J. Math., 49, 123-142. *46, 403*

Davis, H.T. [1930], *The Theory of Volterra Integral Equations of the Second kind,* Indiana University Studies, Vol. XVII (Studies No. 88,89,90), Bloomington. *46*

Davis, H.T. [1962], *Introduction to Nonlinear Differential and Integral Equations,* Dover, New York. *47*

Davis, J. & Jacoby, J.J. [1969], *Numerical integration of linear integral equations with mild discontinuity,* Numer. Math., 13, 357-361. *92*

Davis, P.J. & Rabinowitz, P. [1984], *Methods of Numerical Integration,* 2nd ed., Academic Press, New York-London. *51, 59, 60, 85, 91, 92, 250*

Day, J.T. [1967a], *A starting method for solving nonlinear Volterra integral equations,* Math. Comp., 21, 179-188. *160, 167, 319*

Day, J.T. [1967b], *Note on the numerical solution of integro-differential equations,* Comput. J. 9, 394-395. *167, 319*

Day, J.T. [1968], *On the numerical solution of Volterra integral equations,* BIT, 8, 134-137. *161, 167, 319*

Day, J.T. [1969], *Numerical solution of the convolution integral equation,* BIT, 9,

81-82. *167*

Day, J.T. [1970], *On the numerical solution of integro-differential equations,* BIT, 10, 511-514. *162, 166, 167, 319*

Day, J.T. [1977], *On a starting method for the numerical solution of linear Volterra integral equations,* Z. Angew. Math. Mech., 57, T417-T418. *319*

Dekker, K. & Verwer, J.G. [1984], *Stability of Runge-Kutta Methods for Stiff Nonlinear Differential Equations,* CWI Monographs, 2, North-Holland, Amsterdam - New York - Oxford. *231, 413, 415, 496*

Deligonoul, Z.S. & Bilgen, S. [1984], *Solution of the Volterra equation of renewal theory with the Galerkin technique using cubic splines,* J. Statist. Comput. Simulation, 20, 37-45. *506*

Derakhshan, M.S. & Baker, C.T.H. [1985], *A fast algorithm with error estimates for a convolution integro-differential equation,* Report No 107, Dept. of Math., University of Manchester. *512*

Delves, L.M. [1978], *Numerical software for integral equations,* in: *Numerical Software* (D.A.H. Jacobs, ed.), Academic Press, London-New York, 303-323. *163, 519*

Delves, L.M. & Walsh, J. (eds.) [1974], *Numerical Solution of Integral Equations,* Clarendon Press, Oxford.

Diekmann, O. [1979], *Integral equations and population dynamics,* in: te Riele [1977a], 115-149. *45*

Dieudonné, J. [1981], *History of Functional Analysis,* North-Holland, Amsterdam. *10, 46*

Dirichlet, P.G.L. [1837], *Sur les series dont le terme général dépend de deux angles, et qui servent à exprimer des fonctions arbitraires entre deux limites données,* J. Reine Angew. Math., 17, 36-56. *46*

Dixon, J. & McKee, S. [1983], *Singular Gronwall inequalities,* Numer. Anal. Report NA/83/44, Hertford College, University of Oxford. *44, 48, 49*

Dixon, J. & McKee, S. [1984], *Repeated integral inequalities,* IMA J. Numer. Anal., 4, 99-107. *48*

Dixon, J. & McKee, S. [1985], *A unified approach to convergence analysis of discretization methods for Volterra-type equations,* IMA J. Numer. Anal., 5, 41-57. *164*

Dixon, J.A., McKee, S. & Jelthsch, R. [1985], *Convergence analysis of discretization methods for nonlinear first-kind Volterra integral equations,* to appear. *47*

Doedel, E.J. & Reddien, G.W. [1984], *Finite difference methods for singular two-point boundary-value problems,* SIAM J. Numer. Anal., 21, 300-313. *404*

Donelson, J. & Hansen, E. [1971], *Cyclic composite multistep predictor-corrector methods,* SIAM J. Numer. Anal., 8, 137-157. *65*

Duncan, R.P. [1982], *A Runge-Kutta method using variable stepsizes for Volterra integral equations of the 2nd kind,* Techn. Report 157/82, Dept. of Computer Science, University of Toronto. *231, 503, 519*

Durbin, J. [1971], *Boundary crossing probabilities for the Brownian motion and Poison processes and techniques for computing the power of the Kolmogorov-Smirnov-test,* J. Appl. Probab., 8, 431-453.

Edwards, R.E. [1979], *Fourier Series. A modern Introduction, I (2nd ed.),* Springer -Verlag, New-York - Heidelberg - Berlin. *463*

Eggermont, P.P.B. [1979], *A new analysis of the Euler, midpoint- and trapezoidal-discretization methods for the numerical solution of Abel-type integral equations,* MIPG Techn. Report No. MIPG 34, Dept. of Computer Science, SUNY, Buffalo. *395, 405, 397*

Eggermont, P.P.B. [1980], *Special discretizatyion methods for the equation of image reconstruction and for Abel-type integral equations,* MIPG Techn. Report No. MIPG 50, Dept. of Computer Science, SUNY, Buffalo.

Eggermont, P.P.B. [1981], *A new analysis of the trapezoidal discretization method for the numerical solution of Abel-type integral equations,* J. Integral Equations, 3, 317-332. *396, 398, 405*

Eggermont, P.P.B. [1982], *Collocation as a projection method and superconvergence for Volterra integral equations of the first kind,* in: Baker & Miller [1982], 131-138. *320*

Eggermont, P.P.B. [1983], *Collocation for Volterra integral equations of the first kind with iterated kernel,* SIAM J. Numer. Anal., 20, 1032-1048. *320*

Eggermont, P.P.B. [1984a], *Approximation properties of quadrature methods for Volterra integral equations of the first kind,* Math. Comp., 43, 455-471.

Eggermont, P.P.B. [1984b], *Stability and robustness of collocation methods for Abel-type integral equations,* Numer. Math., 45, 431-445. *166, 398*

Eggermont, P.P.B. [1985], *Beyond superconvergence of collocation methods for Volterra integral equations of the first kind,* in: Hämmelin & Hoffmann [1985], 110-119. *320*

Elden, L. [1983], *The numerical solution of a non-characteristic Cauchy problem for a parabolic equation,* in: *Numerical Treatment of Inverse Problems in Differential and Integral Equations* (P. Deuflhard & E. Hairer, eds.), Birkhäuser Verlag, Basel-Boston-Stuttgart, 246-268.

El-Gendi, S.E. [1969], *Chebychev solution of differential, integral, and integro-differential equations,* Comput. J., 12, 282-287. *164, 319*

Elliott, C.M. & McKee, S. [1981], *On the numerical solution of an integro-differential equation arising from wave-power hydraulics,* BIT. 21, 318-325.

Elliott, D. [1979], *Lagrange interpolation - decline and fall?* Internat. J. Math. Ed. Sci. Tech., 10, 1-12. *317*

Elliott, D. [1980], *Integral equations - ninety years on,* in: Anderssen, de Hoog

& Lukas [1980], 1-20. *10, 519*

El Tom, M.E.A. [1971], *Application of spline functions to Volterra integral equations,* J. Inst. Math. Appl., 8, 354-357. *318*

El Tom, M.E.A. [1973], *Numerical solution of Volterra integral equations by spline functions,* BIT, 13, 1-7. *318*

El Tom, M.E.A. [1974a], *On the numerical stability of spline function approximations to the solution of Volterra integral equations of the second kind,* BIT, 14, 136-143. *496*

El. Tom, M.E.A. [1974b], *On spline function approximations to the solution of Volterra integral equations of the first kind,* BIT, 14, 288-297. *320*

El Tom, M.E.A. [1974c], *Spline function approximations to the solution of singular Volterra integral equations of the second kind,* J. Inst. Math. Appl., 14, 303-309. *403*

El. Tom, M.E.A. [1975], *Efficient algorithms for Volterra integral equations of the second kind,* Computing, 14, 153-166. *318*

El Tom, M.E.A. [1976], *Application of spline functions to systems of Volterra integral equations of the first and second kinds,* J. Inst. Math. Appl., 17, 295-310. *318, 320, 496*

Engels, H. [1980], *Numerical Quadrature and Cubature,* Academic Press, London. *91*

Erdélyi, A. [1953], *Higher Transcendental Functions,* Vol. I, McGraw-Hill, New York.

Erdélyi, A. [1955], *Higher Trancendental Functions,* Vol. III, McGraw-Hill, New York.

Esser, R. [1976], *Numerische Lösung einer verallgemeinerten Volterraschen Integralgleichung zweiter Art,* Dissertation, Universität Köln. *318, 502*

Esser, R. [1978], *Numerische Behandlung einer Volterraschen Integralgleichung,* Computing, 19, 269-284. *318, 502*

Evans, G.A., Hyslop, J. & Morgan, A.P.G. [1981], *Iterative solution of Volterra integral equations using Clenshaw-Curtiss quadrature,* J. Comput. Phys., 40, 64-76. *164*

Evans, G.C. [1910/1911], *Volterra's integral equation of the second kind with discontinuous kernel,* Trans. Amer. Math. Soc., 11, 393-413; 12, 429-472. *46*

Fan, L.S. & Squire, W. [1975], *Inversion of Abel's integral equation by a direct method,* Comput. Phys. Comm., 10, 98-103. *516*

Favella, L.F. & De Griffi, E.M. [1981], *On a weakly singular Volterra integral equation,* Calcolo, 18, 153-195. *406*

Fedotkin, I.M., Aizen, A.M. & Goloshchuk, I.A. [1975], *Application of integral equations to heat conduction problems in which the heat coefficient varies,*

J. Engrg. Phys., 28, 392-395.

Feldstein, A. & Miller, R.K. [1971], *Error bounds for compound quadrature of weakly singular integrals,* Math. Comp., 25, 505-520. *87, 92*

Feldstein, A. & Sopka, J.R. [1974], *Numerical methods for nonlinear Volterra integro-differential equations,* SIAM J. Numer. Anal., 11, 826-846. *166, 232, 513, 514*

Feller, W. [1941], *On the integral equation of renewal theory,* Ann. Math. Statist., 12, 243-267. *47, 126, 493, 507*

Feller, W. [1966], *An Introduction to Probability Theory and Its Applications,* Vol II, Wiley, New York (pp. 346-372: renewal theory; pp. 441-445: theory and applications of the renewal equation). *38, 47*

Fenyö, S. & Stolle, H.W. [1982/1983/1984a/1984b], *Theorie und Praxis der linearen Integralgleichungen,* Bd. I[1982], Bd II[1983], Bd. III[1984a], Bd IV[1984b], VEB Deutscher Verlag der Wiss., Berlin/Birkhäuser Verlag, Basel-Boston-Stuttgart. *46, 47, 92, 163*

Fettis, H.E. [1964], *On the numerical solution of the Abel equation,* Math. Comp., 18, 491-496. *406*

Filippi, S. & Schoedon, P.M. [1971], *Explizite klassische und Hermite'sche Runge-Kutta-Verfahren zur numerischen Lösung von nichtlinearen Volterra-Integralgleichungen,* Mitt. Math. Sem. Giessen, 93, 64-81. *177, 232*

Fleurier, C. & Chapelle, C. [1974], *Inversion of Abel's integral equation-application to plasma spectroscopy,* Comput. Phys. Comm., 7, 200-206.

Forsythe, G.E. Malcolm, M.A. & Moler, C.R. [1977], *Computer Methods for Mathematical Computations,* Prentice-Hall, Englewood Cliffs (N.J.). *502*

Fox, L. & Goodwin, E.T. [1953], *The numerical solution of nonsingular linear integral equations,* Phil. Trans. Roy. Soc. London Ser. A, 245, 501-534. *95, 163, 506*

Franco, N.B., McKee, S. & Dixon, J. [1983], *A numerical solution of Lighthill's integral equation for the surface temperature distribution of a projectile,* Mat. Apl. Comput., 2, 257-271. *404*

Frank, R., Schneid, J. & Ueberhuber, C.W. [1978], *Einseitige Lipschitz-bedingungen für gewöhnliche Differentialgleichungen,* Bericht No. 33/78, Institut für Numerische Mathematik, TU-Wien. *413, 416, 494*

Frank, R., Schneid, J. & Ueberhuber, C.W. [1981a], *The concept of B-convergence,* SIAM J. Numer. Anal., 18, 753-780. *494*

Frank, R., Schneid, J. & Ueberhuber, C.W. [1981b], *B-convergence of Runge-Kutta methods,* Bericht No. 48/81, Institut für Numerische Mathematik, TU-Wien. *494*

Frank, R., Schneid, J. & Ueberhuber, C.W. [1985a], *Stability properties of implicit Runge-Kutta methods,* SIAM J. Numer. Anal., 22, 497-514. *494*

Frank, R., Scheneid, J. & Ueberhuber, C.W. [1985b], *Order results for implicit Runge-Kutta methods applied to stiff systems,* SIAM J. Numer. Anal., 22, 514-534. *494*

Friedlander, F.G. [1941], *The reflexion of sound pulses by convex parabolic reflectors,* Proc. Cambr. Phil. Soc., 37, 134-149. *6, 505, 507*

Friedman, A. [1963], *On integral equations of Volterra type,* J. Analyse Math., 11, 381-413. *17, 47, 48, 493*

Gabdulkhaev, B.G. [1979], *Optimization of collocation methods,* Soviet Math. Dokl., 20, 823-827.

Gaifullina, L.P. [1980], *On the spline method for solving integral equations (Russian),* Izv. Vyssh. Uchebn. Zaved. Mat., 1, 68-71 (see also Soviet Math., 24, 86-90).

Gamondi, G. & Italiani, M. [1965], *Sull'analisi numerica delle equazioni integrali di Volterra,* Calcolo, 2, 249-266. *177*

Garey, L. [1971], *Numerical methods of solution for Volterra integral equations,* PhD thesis, Dalhousie University, Halifax, Nova Scotia. *164, 319, 508, 509*

Garey, L. [1972a], *Predictor-corrector methods for nonlinear Volterra integral equations of the second kind,* BIT, 12, 325-333. *164*

Garey, L. [1972b], *Implicit methods for Volterra integral equations of the second kind,* in: *Proc. Second Manitoba Confer. Numer. Math.* (D. McCarthy & H.C. Williams, eds.), Univerrsity of Manitoba, Winnipeg, 167-177. *319*

Garey, L. [1974a], *Numerical methods for second kind Volterra equations with singular kernels,* in: *Proc. Fourth Manitoba Confer. Numer. Math.* (B.L. Hartnell & H.C. Williams, eds.), University of Manitoba, Winnipeg, 253-263. *403*

Garey, L. [1974b], *The numerical solution of Volterra integral equations with singular kernels,* BIT, 14, 33-39. *403*

Garey, L. [1975a], *Solving nonlinear second-kind Volterra equations by modified increment methods,* SIAM J. Numer. Anal., 12, 501-508. *319, 506*

Garey, L. [1975b], *Block methods for nonlinear Volterra integral equations,* BIT, 15, 401-408. *319*

Garey, L. [1975c], *Step by step methods for the numerical solution of Volterra integro-differential equations,* in: *Proc. Fifth Manitoba Confer. Numer. Math.* (B.C. Hartnell & H.C. Williams, eds.), University of Manitoba, Winnipeg, 319-329. *320*

Garey, L. [1982], *Finding a zero of the solution of an integral equation,* Utilitas Math., 22, 149-159.

Garey, L. & Harris, J. [1977], *An iteration method for nonlinear Volterra integral equations,* in: *Proc. Seventh Manitoba Confer. Numer. Math.* (D.

McCarthy & H.C. Williams, eds.), University of Manitoba, Winnipeg, 399-405. *164, 509*

Gautschi, W. [1981], *A survey of Gauss-Christoffel quadrature formulae,* in: *E.B. Christoffel: The Influence of His Work on Mathematics and the Physical Sciences* (P.L. Butzer & Fehér, eds.), Birkhäuser Verlag, Basel, 72-147. *91*

Gear, C.W. [1971], *Numerical initial value problems in ordinary differential equations,* Prentice Hall, Englewood Cliffs (N.J.). *91*

Ghermanesco, M. [1959], *Equations intégrales aux deux limites variables,* C.R. Acad. Sci. Paris, 249, 1606-1607. *45*

Ghermanesco, M. [1961], *Equations intégrales aux deux limites variables,* Ann. Mat. Pura Appl. (4), 54, 33-56. *45*

Ghez, R. & Lew, J.S. [1973], *Interface kinetics and crystal growth under conditions of constant cooling rate I: Constant diffusion coefficient,* J. Crystal Growth, 20, 273-282. *511*

Ghizzetti, A. & Ossicini, A. [1970], *Quadrature Formulae,* 2nd ed., McGraw-Hill, New York. *91, 307*

Gladwin, C.J. [1972], *Methods of high order for the numericall solution of first kind Volterra integral equations,* in: *Proc. Second Manitoba Confer. Numer. Math.* (R.S.D. Thomas & H.C. Williams, eds.), University of Manitoba, Winnipeg, 179-193. *165*

Gladwin, C.J. [1973], *Some remarks on the numerical solution of first kind Volterra integral equations,* in: *Proc. Third Manitoba Confer. Numer. Math.* (R.S.D. Thomas & H.C. Williams, eds.), University of Manitoba, Winnipeg, 223-237. *165*

Gladwin, C.J. [1975], *Numerical solution of Volterra integral equations of the first kind,* PhD thesis, Dalhousie University, Halifax, Nova Scotia. *165, 517*

Gladwin, C.J. [1979], *Quadrature methods for Volterra integral equations of the first kind,* Math. Comp., 33, 705-716. *165*

Gladwin, C.J. [1982], *On optimal integration methods for Volterra integral equations of the first kind,* Math. Comp., 39, 511-518. *165*

Gladwin, C.J. [1985], *A stability problem in the numerical solution of Volterra integral equations of the first kind,* (to appear in Math. Comp.). *165*

Gladwin, C.J. & Jeltsch, R. [1974], *Stability of quadrature rule methods for Volterra integral equations of the first kind,* BIT, 14, 144-151. *142, 165*

Glasmacher, W. & Sommer, D. [1966], *Implizite Runge-Kutta-Formeln,* Forschungsberichte des Landes Nordrhein-Westfalen, Westdeutscher Verlag, Köln. *91, 231*

Golberg, M.A. [1978], *On a method of Bownds for solving Volterra integral equations,* J. Optim. Theory Appl., 24, 221-232. *319, 502, 507*

Golberg, M.A. (ed.) [1979a], *Solution Methods for Integral Equations: Theory and Applications,* Plenum Press, New York. *163*

Golberg, M.A. [1979b], *A survey of numerical methods for integral equations,* in: Golberg [1979a], 1-58.

Goldfine, A. [1973], *The numerical solution of Volterra integral and integro-differential equations,* PhD thesis, Pennsylvania State University, University Park. *232, 319, 320*

Goldfine, A. [1977], *Taylor series methods for the solution of Volterra integral and integro-differential equations,* Math. Comp., 31, 691-707. *232, 319, 320, 519*

Goldman, A. & Visscher, W. [1978], *Applications of integral equations in particle-size statistics,* J. Optim. Theory Appl., 24, 207-220 (also in: Golberg [1979a, 169-182]).

Goldstine, H.H. [1977], A History of Numerical Analysis from the 16th through the 19th Century, Springer-Verlag, New York-Heidelberg-Berlin. *91*

Gorenflo, R. [1979], *Numerical treatment of Abel integral equations,* in: Inverse and Improperly Posed Problems in Differential Equations (G. Anger, ed.), Akademie-Verlag, Berlin, 125-133. *407*

Gorenflo, R. & Vessella, S. [1984a], *Abel integral equations, a historico - bibliographical survey,* Istituto di Analisi Globale e Applicazioni, Firenze. *46, 406*

Gorenflo, R. & Vessella, S. [1984b], *Basic theory and some applications of Abel integral equations,* Istituto di Analisi Globale e Applicazioni, Firenze. *45, 400, 406*

Graham, I.G. [1980], *The numerical solution of Fredholm integral equations of the second kind,* PhD thesis, Univerrsity of New South Wales, Kensington. *318, 404*

Graham, I.G. [1982a], *Singularity expansions for the solutions of second kind Fredholm integral equations with weakly singular convolution kernels,* J. Integral Equations, 4, 1-30.

Graham, I.G. [1982b], *Galerkin methods for second kind integral equations with singularities,* Math. Comp., 39, 519-533. *318, 404*

Grigorovich, G.A. & Nikoolajshvili, S.S. [1978], *On the numerical solution of Volterra integral equations by use of spline approximations (Russian),* Dokl. Semin. Inst. Prikl. Mat., 12-13, 67-70.

Gripenberg, G. [1980], *On the resolvent of nonconvolution Volterra kernels,* Funkcial. Ekvac., 23, 83-95. *493*

Gripenberg, G. [1982], *Decay estimates for resolvents of Volterra equations,* J. Math. Anal. Appl., 85, 473-487. *48, 493*

Groetsch, C.W. [1982], *Numerical treatment of a nonlinear problem arising in heat transfer,* in: *Nonlinear Analysis and Applications* (S.P.Singh and J.H.Burry, eds.) Lecture Notes Pure Appl. Math., 80, Marcel Dekker, New York-Basel, 357-361. *404*

Groetsch, C.W. [1984], *The Theory of Tikhonov Regularization for Fredholm Equations of the First Kind,* Res. Notes in Math., 105, Pitman, Boston. *321*

Gronwall, T.H. [1918/19], *Note on the derivatives with respect to a parameter of the solutions of a system of differential equations,* Ann. of Math.(2), 20, 292-296. *38*

Grossman, S.I. & Miller, R.K. [1970], *Perturbation theory for Volterra integro-differential systems,* J. Differential Equations, 8, 457-474. *19, 38, 47, 493*

Grossman, S.I. & Miller, R.K. [1973], *Nonlinear Volterra integro-differential systems with L^1-kernels,* J. Differential Equations, 13, 551-566. *37, 38, 47, 48, 493, 494, 496*

Grundy, R.E. [1978], *The solution of Volterra integral equations of the convolution type using two-point rational approximants,* J. Inst. Math. Appl., 22, 147-158.

Guillou, A. & Soulé, J.L. [1969], *La résolution numérique des problèmes differentiels aux conditions initiales par des méthodes de collocation,* RAIRO Anal. Numér., 3, 17-44. *247, 317*

Guzek, J.A. & Kemper, G.A. [1973], *A new error analysis for a cubic spline approximate solution of a class of Volterra integro-differential equations,* Math. Comp., 27, 563-570. *319*

Hacia, L. [1974], *On approximate solution of integral equations of the Fredholm-Volterra type,* Fasc. Math., 7, 45-51.

Hacia, L. [1981], *On solving of Fredholm-Volterra equations, Fasc.* Math., 13, 21-30.

Hairer, E. [1981], *Order conditions for numerical methods for partitioned ordinary differential equations,* Numer. Math., 36, 431-445. *231*

Hairer, E. [1982], *Extended Volterra-Runge-Kutta methods,* in: Baker & Miller [1982], 221-231. *195, 231*

Hairer, E. & Lubich, Ch. [1984], *On the stability of Volterra-Runge-Kutta methods,* SIAM J. Numer. Anal., 21, 123-135. 412, 464, 472, 487, 496, 497

Hairer, E., Lubich, Ch. & Nørsett, S.P. [1983], *Order of convergence of one-step methods for Volterra integral equations of the second kind,* SIAM J. Numer. Anal., 20, 569-579. 177, 197, 202, 204, 205, 206, 207

Hairer, E., Lubich, Ch. & Schlichte, M. [1985], *Fast numerical solution of nonlinear Volterra convolution equations,* SIAM J. Sci. Statist. Comput., 6, 532-541. *503, 504, 519*

Hairer, E. & Maass, P. [1985], *Fractional multistep methods for the nonlinear*

Basset equation, Preprint No. 313, Sonderforschungsbereich 123, University of Heidelberg.

Hairer, E., Nørsett, S.P. & Wanner, G. [1986], *Numerical of Ordinary Differential Equations,* in preparation. *178, 179, 231*

Hairer, E. & Wanner, G. [1983], *On the instability of the BDF formulas,* SIAM J. Numer. Anal., 20, 1206-1209. *92*

Hall, G. & Watt, J.M. (eds.) [1976], *Modern Numerical Methods for Ordinary Differential Equations,* Clarendon Press, Oxford. *91*

Hämmerlin, G. & Hoffmann, K.-H. (eds.) [1985], *Constructive Methods for the Practical Treatment of Integral Equations,* Internat. Ser. Numer. Math., 73, Birkhäuser Verlag, Basel-Boston-Stuttgart. *163*

Handelsman, R.A. & Olmstead, W.E. [1972], *Asymptotic solution to a class of nonlinear Volterra integral equations,* SIAM J. Appl. Math., 22, 373-384. *404, 493*

Hannsgen, K.B., Herdman, T.L., Stech, H.W. & Wheeler, R.L. (eds.) [1982], *Volterra and Functional Differential Equations,* Lecture Notes Pure Appl. Math., 81, Marcel Dekker, New York-Basel. 11, 38, 45, 47, 493

Harvey, D. [1971], *Global approximations of initial value problems for special second order ordinary differential equations,* M.A. thesis, Dalhousie University, Halifax, Nova Scotia. *317*

Heiden, U. van der [1980], *Analysis of Neutral Networks,* Lecture Notes in Biomath., 35, Springer-Verlag, Berlin-Heidelberg-New York. *507*

Hellinger, E. & Toeplitz, O. [1927], *Integralgleichungen und Gleichungen mit unendlich-vielen Unbekannten,* Encyklopadie der Math. Wiss., 11.3, 1335-1601 (Reprint: Chelsea, New York, 1953). *10, 45, 46*

Hendry, W.L. [1976], *A Volterra integral equation of the first kind,* J. Math. Anal. Appl., 54, 266-278.

Henrici, P. [1962], *Discrete Variable Methods in Ordinary Differential Equations,* Wiley, New York. *64, 67, 76, 77, 91, 92, 164, 205, 232, 330*

Henrici, P. [1974], *Applied and Computational Complex Analysis,* Vol 1, Wiley, New York. *92, 332, 341*

Henrici, P. [1977], *Applied and Computational Complex Analysis,* Vol. 2, Wiley, New York. *15, 16, 47*

Henrici, P. [1979], *Fast Fourier methods in computational complex analysis,* SIAM Review, 21, 481-527. *340, 341, 503*

Herdman, T.L., Rankin, S.M. & Stech, H.W. (eds.) [1981], *Integral and Functional Differential Equations,* Lecture Notes Pure Appl. Math., 67, Marcel Dekker, New York-Basel. *11, 38, 47, 493*

Hethcote, H.W. & Tudor, D.W. [1980], *Integral equation models for endemic infectious diseases,* J. Math. Biol., 9, 37-47. *511*

Heun, K. [1900], *Neue Methode zur approximativen Integration der Differentialgleichungen einer unabhängigen Veranderrlichen,* Z. Math. Phys., 45, 23-38. *95, 178*

Hilbert, D. [1912], *Grundzuge einer allgemeinen Theorie der linearen Integralgleichungen,* Teubner, Leipzig.

Hildebrand, F.B. [1974], *Introduction to Numerical Analysis,* 2nd ed., McGraw-Hill, New York. *61, 73, 91, 92*

Hock, W. [1977], *Asymptotische Entwicklungen bei Mehrschrittverfahren zur numerischen Behandlung von Volterra-Integralgleichungen,* Dissertation Universität Würzburg. *164, 502*

Hock, W. [1979a], *Asymptotic expansions for multistep methods applied to nonlinear Volterra integral equations of the second kind,* Numer. Math., 33, 77-100. *164*

Hock, W. [1979b], *An extrapolation method with stepsize control for nonlinear second-kind Volterra integral equations,* Preprint No. 56, Universität Würzburg. *164, 502*

Hock, W. [1980], *Ein Extrapolationsverfahren fur Volterra-Integral-gleichungen,* in: Albrecht & Collatz [1980], 102-114. *164, 502*

Hock, W. [1981], *An extrapolation method with stepsize control for nonlinear Volterra integral equations,* Numer. Math., 38, 155-178. *164, 502, 511*

Hoffer, E. [1976], *Partially implicit methods for large stiff systems of ODEs with only a few equations introducing small time constants,* SIAM J. Numer. Anal., 13, 645-663.

Holyhead, P.A.W. [1976], *Direct methods for the numerical solution of Volterra integral equations of the first kind,* PhD thesis, University of Southampton. *165*

Holyhead, P.A.W., & McKee, S. [1976], *Stability and convergence of multistep methods for linear Volterra integral equations of the first kind,* SIAM J. Numer. Anal., 13, 269-292. *147, 165*

Holyhead, P.A.W., McKee, S. & Taylor, P.J. [1975], *Multistep methods for solving linear Volterra integral equations of the first kind,* SIAM J. Numer. Anal., 12, 698-711. *147, 165*

Hönig, Ch.S. [1975], *Volterra-Stieltjes Integral Equations,* North-Holland, Amsterdam. *47*

Hoog, F.R. de [1973], *Asymptotic errors for finite difference schemes,* in: *Error, Approximation, and Accuracy* (F.R. de Hoog & C.L. Jarvis, eds.), University of Queensland Press, St Lucia, 159-172. *165*

Hoog, F.R. de & Weiss, R. [1973a], *On the solution of Volterra integral equations of the first kind,* Numer. Math., 21, 22-32. *320*

Hoog, F.R. de & Weiss, R. [1973b], *High order methods for Volterra integral equations of the first kind,* SIAM J. Numer. Anal. 10, 647-664. *303, 320*

Hoog, F.R. de & Weiss, R. [1973c], *Asymptotic expansions for product integration,* Math. Comp., 27, 295-306. *89, 92*

Hoog, F.R. de & Weiss, R. [1973d], *On the solution of a Volterra integral equation with a weakly singular kernel,* SIAM J. Math. Anal., 4, 561-573. *403*

Hoog, F.R. de & Weiss, R. [1974], *High order methods for a class of Volterra integral equations with weakly singular kernels,* SIAM J. Numer. Anal., 11, 1166-1180. *403, 504*

Hoog, F.R. de & Weiss, R. [1975], *Implicit Runge-Kutta methods for second kind Volterra integral equations,* Numer. Math., 23, 199-213. 134, 177, 204, 231, 267, 319, 466, 496, 503, 506, 508

Houwen, P.J. van der [1977a], *Construction of Integration Formulas for Initial Value Problems,* North-Holland, Amsterdam. *187, 502*

Houwen, P.J. van der [1977b], *On the numerical solution of Volterra integral equations of the second kind I: Stability,* Report NW 42/77, Mathematisch Centrum, Amsterdam. 177, 231, 473, 494, 496

Houwen, P.J. van der [1977c], *Volterra integral equations of the second kind (Dutch),* in: te Riele [1977a], 124-144. *163*

Houwen, P.J. van der [1980a], *Convergence and stability analysis in Runge-Kutta type methods for Volterra integral equations of the second kind,* Report NW 83/80, Mathematisch Centrum, Amsterdam. *205, 231, 496*

Houwen P.J. van der [1980b], *Convergence and stability results in Runge-Kutta type methods for Volterra integral equations of the second kind,* BIT, 20, 375-377. *231, 496*

Houwen, P.J. van der [1981], *A-stable Runge-Kutta methods for Volterra integral equations of the second kind,* in: *Numerical Methods for Stiff Initial Value Problems,* (G. Dahlquist & R. Jeltsch, eds.), Bericht No. 9, TH Aachen, 122-126. *177*

Houwen, P.J. van der [1982], *Algebraically equivalent linear multistep solutions of Volterra integral equations and certain systems of ODEs,* Report NW 144/82, Mathematisch Centrum, Amsterdam. *164, 494, 495*

Houwen, P.J. van der [1985], *Stability results for discrete Volterra equations,* in: *Mathematical Analysis* (J.M. Rassias, ed.), Teubner-Texte zur Mathematik, 79, BSB B.G. Teubner Verlaggesellschaft, Leipzig.

Houwen, P.J. van der & Blom, J.G. [1978], *On the numerical solution of Volterra integral equations of the second kind II: Runge-Kutta methods,* Report NW 61/78, Mathematisch Centrum, Amsterdam. *177, 231, 496*

Houwen, P.J. van der & Blom, J.G. [1985], *Stability results for discrete Volterra equations: Numerical experiments,* in: Hämmerlin & Hoffmann [1985], 166-178. *466*

Houwen, P.J. van der & Riele, H.J.J. te [1977/78], *Backward differentiation formulas for Volterra integral equations of the second kind, I: Convergence and stability,* Report NW 48/77; *II: Numerical experiments,* Report 57/78, Mathematisch Centrum, Amsterdam. *164, 495*

Houwen, P.J. van der & Riele, H.J.J. te [1981], *Backward differentiation type formulas for Volterra integral equations of the second kind,* Numer. Math., 37, 205-217. *101, 164, 495, 507*

Houwen, P.J. van der & Riele, H.J.J. te [1982], *Linear multistep methods for Volterra integral equations of the second kind,* in: Baker & Miller [1982], 79-94. *164*

Houwen P.J. van der, & Riele, H.J.J. te [1985], *Linear multistep methods for Volterra integral and integro-differential equations,* Math. Comp., 45, 439-461. *164, 166*

Houwen, P.J. van der Riele, H.J.J. te & Wolkenfelt, P.H.M. [1978], *On the stability of multistep formulas for systems of Volterra integro-differential equations,* Report NW 63/78, Mathematisch Centrum, Amsterdam. *448*

Houwen, P.J. van der & Schilder, J.N. [1977], *Numerical experiments with Runge-Kutta type methods for Volterra integral equations of the second kind,* Note 12/77, Mathematisch Centrum, Amsterdam. *508*

Houwen, P.J. van der & Wolkenfelt, P.H.M. [1980], *On the stability of multistep methods for Volterra integral equations of the second kind,* Computing, 24, 341-347. *461*

Houwen, P.J. van der, Wolkenfelt, P.H.M. & Baker, C.T.H. [1981], *Convergence and stability analysis for modified Runge-Kutta methods in the numerical treatment of second kind Volterra integral equations,* IMA J. Numer. Anal., 1, 303-328. *202, 207, 208, 478, 484, 495, 496, 507*

Hrusa, W.J. & Nohel, j.A. [1984], *Global existence and asymptotics in one-dimensional nonlinear viscoelasticity,* in: *Trends and Applications of Pure Mathematics to Mechanics* (P.G. Ciarlet & M. Roseau, eds.), Lecture Notes in Phys., 195, Springer-Verlag, Berlin-Heidelberg-New York, 165-187. *45, 48*

Huber, A. [1939], *Eine Näherungsmethode zur Auflösung Volterrascher Integralgleichungen,* Monatsh. Math. Phys., 47, 240-246. *95, 163, 318, 320, 403, 405*

Hung, H.-S. [1968], *Application of linear spline functions to the numerical solution of Volterra integral equations of the second kind,* Comp. Sci. Techn. Report No. 27, University of Wisconsin, Madison. *318*

Hung, H.-S. [1970], *The numerical solution of differential and integral equations by spline functions,* MRC Techn. Summary Report No. 1053, University of Wisconsin, Madison. *317, 318, 319, 496*

Hung, H.-S. [1978], *Error analysis of a linear spline method for solving an Abel integral equation,* MRC Techn. Summary Report No. 1904, University of Wisconsin, Madison.

Hung, H.-S. [1979a], *A higher order global approximation method for solving an Abel integral equation by quadratic splines,* MRC Techn. Summary Report No. 1921, University of Wisconsin, Madison. *406*

Hung, H.-S. [1979b], *Spline approximation to the solution of a class of Abel integral equations,* MRC Techn. Summary Report No. 1921, University of Wisconsin, Madison. *405, 518*

Hurwitz, W.A. [1908], *Note on certain iterated and multiple integrals,* Ann. of Math., 9, 183-192. *46*

Iguchi, K. [1972], *A starting method for solving nonlinear Volterra integral equations of the second kind,* Comm. ACM, 15, 460-461. *161, 167*

Imomnazarov, B. [1979], *Regularization of equations of the first kind with abstract Volterra operator,* Soviet Math. Dokl., 20, 642-645.

Inselberg, A. [1982], *An integral equation arising in a convective heat (mass) transfer problem through a boundary layer,* J. Math. Anal. Appl., 90, 440-455. *404*

Isaacson, E. & Keller, H.B. [1966], *The Analysis of Numerical Methods,* Wiley, New York. *58, 91, 92, 396*

Ishchuk, V.A. [1975], *Two-sided methods of Runge-Kutta form for a nonlinear Volterra integral equation (Ukrainian; Engl./Russian summary),* Dopovidi Akad. Nauk Ukrain. RSR Ser. A, 14-17, 91. *232*

Ishchuk, V.A. & Lutsev, E.M. [1981], *Two-sided Runge-Kutta method for the solution of a system of nonlinear integral equations of Volterra type,* Soviet Math., 25, 34-42. *232*

Jackiewicz, Z. [1979], *Convergence of multistep methods for Volterra functional differential equations,* Numer. Math., 32, 307-332. *166*

Jackiewicz, Z. [1984], *Global error estimation in the numerical solution of intergro differential equations by Euler's method,* Zastos. Mat., 8, 487-501. *166, 519*

Jackiewicz, Z. & Kwapisz. M. [1983a], *Convergence of multistep methods for Volterra integro-differential equations,* Ann. Polon. Math., 43, 121-139. *166*

Jackiewicz, Z. & Kwapisz, M. [1983b], *A note on the stability of θ- methods for Volterra integral equations of the second kind,* Numer. Anal. Techn. Report 65R20-1, University of Arkansas, Fayetteville. *495*

Jain, M.K. & Sharma, K.D. [1967], *Numerical solution of linear differential equations and Volterra integral equations using Lobatto quadrature formulas,* Comput. J., 10, 101-107. *164*

Jakeman, A.J. [1975], *Numerical inversion of Abel type integral equations in stereology,* PhD thesis, Australian National University, Canberra. *407*

Jakeman, A.J. [1980], *How, when and why to use Mate - A package of programs containing methods for Abel type equations and their functionals,* Mikroskopie, 37(Suppl.), 458-465. *407, 504, 516*

Jakeman, A.J. [1984], *Numerical inversion of a second kind singular Volterra equation - the thin section equation of stereology,* Utilitas Math., 26, 193-213. *404, 407*

Jakeman, A.J. & Anderssen, R.S. [1975], *Abel type integral equations in stereology I: General discussion,* J. Microscopy, 105, 121-133. *407, 504, 516*

Jakeman, A.J. & Young, P.C. [1980], *System identification and estimation for convolution integral equations,* in: Anderssen, de Hoog & Lukas [1980], 235-255. *321*

Jakeman, A.J. & Young, P.C. [1984], *Recursive filtering and smoothing procedures for the inversion of ill-posed causal problems,* Utilitas Math., 25, 351-376. *321*

James, R.C. [1968], *The numerical solution of singular Volterra integral equations,* SIAM J. Numer. Anal., 5, 352-362. *404*

Janikowski, J. [1962], *Equation intégrale non linéaire d'Abel,* Bull. Soc. Sci. Lettres Lodz, 13, No. 11 (MR 29, 440). *27, 47*

Jeltsch, R. & Nevanlinna, O. [1983], *Stability of semidiscretizations of hyperbolic problems,* SIAM J. Numer. Anal., 20, 1210-1218.

Joe, S. [1985a], *Collocation methods using piecewise polynomials for second kind integral equations,* J. Comput. Appl. Math., 12 & 13, 391-400. *269, 318*

Joe, S. [1985b], *Discrete collocation methods for second kind Fredholm integral equations,* SIAM J. Numer. Anal., 22, 1167-1177. *269, 318*

Joe, S. [1985c], *The numerical solution of second-kind Fredholm integral equations,* PhD thesis, University of New South Wales, Kensington. *269, 318*

Jones, J.G. [1961], *On the numerical solution of convolution integral equations and systems of such equations,* Math. Comp., 15, 131-142. *142, 163, 164*

Jones, G.S. [1964], *Fundamental inequalities for discrete and discontinuous functional equations,* J. Soc. Indust. Appl. Math. 12, 43-57. *40, 41, 49*

Jones, H.M. [1986], *A variable step variable order package for solving Volterra integral equations of the second kind* (to appear). *503*

Jones, H.M. & McKee, S. [1982], *The numerical stability of multistep methods for convolution integral equations,* J. Comput. Math. Appl., 8, 291-303. *495*

Jones, H.M. & McKee, S. [1985], *Variable step size predictor-corrector schemes for second kind Volterra integral equations,* Math. Comp., 44, 391-404. *164, 495, 503, 519*

Jordan, G.S. & Wheeler, R.L. [1976], *A generalization of the Wiener-Lévy theorem applicable to some Volterra equations,* Proc. Amer. Math. Soc., 57, 109-114. *48*

Jörgens, K. [1982], *Linear Integral Operators,* Pitman, Boston-London. *47*

Kadner, H. [1967], *Die numerische Behandlung von Integralgleichungen nach der*

Kollokationsmethode, Numer. Math., 10, 241-260.

Kagiwada, H.H. & Kalaba, R. [1974], *Integral Equations via Imbedding Methods,* Addison-Wesley, Reading (Mass.).

Kalaïda, A.F. [1982], *Some schemes of higher accuracy for the quadrature method of solving integral equations of Volterra type (Russian),* Vychisl. Prikl. Mat. (Kiev), 47, 3-9.

Kalla, S.L., Viloria, L. & Conde, S. [1978], *On an integral equation associated with a production problem,* Rev. Técn. Fac. Ingr. Univ. Zulia, 1, 45-54.

Kamke, E. [1977], *Differentialgleichungen, Lösungsmethoden und Lösungen, Teil I,* Teubner, Stuttgart. *414*

Kanatani, K.-I. & Ishikawa, O. [1985], *Error analysis for the stereological estimation of sphere size distribution: Abel type integral equation,* J. Comput. Phys., 57, 229-250. *407*

Karlin, S. [1955], *On the renewal equation,* Pacific J. Math., 5, 229-357. *47, 493*

Kasperczuk, A. et. al. [1978], *Numerical solution to Abel integral equation in optical diagnostics of axisymmetrical objects,* J. Techn. Phys., 19, 137-150.

Keech, M.S. [1976], *Semi-explicit, one-step methods for the numerical solution of nonsingular Volterra integral equations of the first kind,* Numer. Anal. Report No. 19, University of Manchester. 167-232

Keech, M.S. [1977a], *Stability in the numerical solution of intial value problems in integral equations,* PhD thesis, University of Manchester.

Keech, M.S. [1977b], *A third order, semi-explicit method in the numerical solution of first kind Volterra integral equations,* BIT, 17, 312-320. *147, 167, 232, 518*

Keilson, J. & Nunn, W.R. [1979], *Laguerre transformation as a tool for the numerical solution of integral equations of convolution type,* Appl. Math. Comput., 5, 313-359.

Keller, G. [1982], *Numerical solution of initial-value problems by collocation methods using generalized piecewise functions,* Computing, 28, 199-211. *317*

Keller, J.B. & Olmstead, W.E. [1972], *Temperature of a nonlinearly radiating semi-infinite rod,* Quart. Appl. Math. 29, 559-566. *7, 493*

Kershaw, D. [1970], *The determination of the density distribution of a gas flowing in a pipe from mean density measurements,* J. Inst. Math. Appl., 6, 111-114.

Kershaw, D. [1974], *Volterra equations of the second kind,* in: Delves & Walsh [1979], 140-161. *96, 163, 442*

Kershaw, D. [1982a], *Some results for Abel-Volterra integral equations of the second kind,* in: Baker & Miller [1982], 273-282. *403*

Kershaw, D. [1982b], *The stability of a numerical method for a second-kind Abel equation,* in: Baker & Miller [1982], 459-461. *495*

Knirk, D.L. [1976], *Accurate and efficient quadrature for Volterra integral equations,* J. Comput. Phys., 21, 371-399.

Knopp, T.J. et al. [1976], *Transcoronary intravascular transport functions obtained via a stable deconvolution technique,* Ann. Biomedical Engrg., 4, 44-59. *517*

Kobayasi, M. [1966], *On numerical solution of the Volterra integral equations of the second kind by linear multistep methods,* Rep. Statist. Appl. Res. Un. Japan. Sci. Engrs., 13, 1-21. *122, 164*

Kobayasi, M. [1967], *On numerical solution of the Volterra integral equations of the first kind by trapezoidal rule,* Rep. Statist. Appl. Res. Un. Japan. Sci. Engrs., 14, 1-14. *164, 517*

Kosarev, E.L. [1973], *The numerical solution of Abel's integral equation,* U.S.S.R. Comput. Math. and Math. Phys., 13, 271-277. *406*

Kosarev, E.L. [1980], *Applications of integral equations of the first kind in experimental physics,* Comput. Phys. Comm., 20, 69-75. *407*

Kowalewski, G. [1930], *Integralgleichungen,* de Gruyter & Co, Berlin. *47*

Kramarz, L. [1978], *Global approximations to solutions of initial value problems,* Math. Comp., 32, 35-59. *317*

Krasnov, M.L., Kisselev, A.I. & Makarenko, G.I. [1977], *Equations Intégrales: Problèmes et Exercices,* Editions Mir, Moscow. *47, 519*

Krylov, V.I. [1949], *Application of the Euler-Laplace formula to approximate solutions of integral equations of Volterra type (Russian),* Trudy Mat. Inst. Steklov, 28, 33-72 (Compare also the detailed review in Zentralblatt, 41, p. 79). *91, 95, 163, 165, 167, 177, 506*

Krylov, V.I. [1962], *Approximate Calculation of Integrals,* Macmillan, London. *58, 91, 92*

Kumar, S. [1979], *On a method of Noble for second kind Volterra integral equations,* BIT, 19, 482-488. *167, 319*

Kumar, S. [1980a], *On modified increment methods of Garey for nonlinear second kind Volterra integral equations,* BIT, 20, 346-350. *319*

Kumar, S. [1980b], *A recurrence relation for solution of singular Volterra integral equations using Chebyshev polynomials,* BIT, 21, 123-125. *319*

Kumar, S. [1981], *Numerical methods for Volterra and singular integral equations,* PhD thesis, India Institute of Technology, New Delhi. *167, 319*

Kung, F.C. & Chen, S.Y. [1978], *Solution of integral equations using a set of block pulse functions,* J. Franklin Inst., 305, 283-291.

Kunkel, P. [1982], *Ein adaptives Verfahren zur Lösung von Volterra'schen Integralgleichungen zweiter Art,* Diplomarbeit, Universität Heidelberg. *503*

Kurihara, M. [1978], *On approximate solutions of nonlinear Volterra integro-differential equations in Chebyshev series,* Mem. Numer. Math., 5, 33-72.

Kutta, W. [1901], *Beitrag zur näherungsweisen Integration totaler Differentialgleichungen,* Z. Math. Phys., 46, 435-453. *95, 177, 178, 187*

Lakshmikantham, V. & Leela, S. [1969], *Differential and Integral Inequalities,* Vol. I, Academic Press, New York. *48*

Lalesco, T. [1908], *Sur l'équation de Volterra,* J. Math. Pures Appl. (6), 4, 125-202. *9, 10, 21*

Lalesco, T. [1912], *Introduction à la Théorie des Equations Intégrales,* Hermann & Fils, Paris. *21, 46, 47*

Lambert, J.D. [1973], *Computational Methods in Ordinary Differential Equations,* Wiley, London. *67, 91, 92, 106, 164, 181, 441, 456, 470, 512*

Lambert, J.D. & Watson, I.A. [1976], *Symmetric multistep methods for periodic initial value problems,* J. Inst. Math. Appl., 18, 189-202. *444, 446*

Lancaster, P. [1969], *Theory of Matrices,* Academic Press, New York. *419*

Lapidus, L. & Seinfeld, J.H. [1971], *Numerical Solution of Ordinary Differential Equations,* Academic Press, New York. *187, 231, 233, 502, 512*

Laudet, M. & Oulés, H. [1960], *Sur l'intégration numérique des équations intégrales du type de Volterra,* in: Symposium Rome [1960], 117-121. *163, 509*

Lauricella, G. [1908], *Sulle equazioni integrali,* Ann. Mat. Pura Appl. (3), 15, 21-45. *46*

Lavoie, J.L., Osler, T.J. & Tremblay, R. [1976], *Fractional derivatives and special functions,* SIAM Rev. 18, 240-268. *403*

Lavrent'ev, N.M. [1977], *On the regularization of operator equations of Volterra type (Russian),* in: *Problems of Mathematical Physics and Numerical Mathematics,* Work Collect., Moscow 1977, 199-205.

Lax, M.D. [1981], *Solving random linear Volterra integral equations using the methods of moments,* J. Integral Equations, 3, 357-363.

Lee, A.C.H. & Padgett, W.J. [1980], *A random nonlinear integral equation in population growth problems,* J. Integral Equations, 2, 1-9.

Lee, D.E. [1979], *Computational considerations for some regularizable ill-posed problems in engineering,* in: *Proc. Int. Symp. Mathematical Theory of Networks and Systems, Delft 1979,* Western Periodicals, North Hollywood, 3, 495-498.

Le Roux, J. [1895], *Sur les intégrales des équations linéaires aux dérivées partielles du second ordre à deux variables indépendantes,* Ann. Sci. Ecole Normale Supér. (3), 12, 227-316. *9*

Levic, V.G., Podgaeckii, E.M. & Filinovskii, V.J. [1972], *A successive approximation method for solving a nonlinear Volterra integral equation of the second kind,* U.S.S.R. Comput. Math. and Math. Phys., 10, 138-145.

Levin, J.J. [1977], *Resolvents and bounds for linear and nonlinear Volterra equations,* Trans. Amer. Math. Soc., 228, 207-222. *47, 48, 493*

Levinson, N. [1960], *A nonlinear Volterra equation arising in the theory of superfluidity,* J. Math Anal. Appl., 1, 1-11. *404, 493*

Lewanowicz, S. [1984], *Application of modified moments in the numerical solution of the Abel integral equation,* Zastos. Mat., 12, 311-317. *405*

Lihtarnikov, L.M. [1968], *An application of the Runge-Kutta method to the solution of Volterra type integro-differential equations,* Trudy Naucn. Ob"ed. Prepodav. Fiz.-Mat. Fak. Ped. Inst. Dal'n. Vostok., 9, 27-33. *232*

Ling, R. [1978], *Integral equations of Volterra type,* J. Math. Anal. Appl., 64, 381-397. *48, 493*

Linkovskii, G.B. & Krapivin, V.F. [1961], *Numerical solution of an integro-differential equation with a quasi-linear operator and a generalized Volterra operator (Russian),* Sibirsk. Mat. Zh., 2, 797-800.

Linz, P. [1967], *The numerical solution of Volterra integral equations by finite difference methods,* MRC Techn. Summary Report No. 825, University of Wisconsin, Madison. *134, 135, 142, 163, 165, 166, 167, 319, 406, 495, 517*

Linz, P. [1968], *Numerical methods for Volterra integral equations with applications to certain boundary value problems,* PhD thesis, University of Wisconsin, Madison. *122, 163, 165, 166, 167, 406, 495*

Linz, P. [1969a], *Numerical methods for Volterra integral equations with singular kernels,* SIAM J. Numer. Anal., 6, 365-374. *403, 510*

Linz, P. [1969b], *A method for solving nonlinear Volterra integral equations of the second kind,* Math. Comp., 23, 595-600. *134, 167, 319*

Linz, P. [1969c], *Numerical methods for Volterra integral equations of the first kind,* Comput. J., 12, 393-397. *135, 165, 166*

Linz, P. [1969d], *Linear multistep methods for Volterra integro-differential equations,* J. Assoc. Comput. Mach., 16, 295-301. *151, 156, 166, 167, 319, 513*

Linz, P. [1971], *Product integration methods for Volterra integral equations of the first kind,* BIT, 11, 413-421. *166, 320*

Linz, P. [1974], *A simple approximation method for solving Volterra integro-differential equations of the first kind,* J. Inst. Math. Appl., 14, 211-215.

Linz, P. [1975], *Convergence of a discretization method for integro-differential equations,* Numer. Math., 25, 103-107.

Linz, P. [1979], *Theoretical Numerical Analysis: An Introduction to Advanced Techniques,* Wiley, New York. *47*

Linz, P. [1980], *A survey of methods for the solution of Volterra integral equations of the first kind*, in: Anderssen, de Hoog & Lukas [1980], 183-194. *96, 163, 166*

Linz, P. [1982], *The solution of Volterra equations of the first kind in the presence of large incertaintes*, in: Baker & Miller [1982], 123-130. *321*

Liouville, J. [1832], *Sur quelques questions de géometrie et de mécanique, et sur un nouveau genre de calcul pour résoudre ces questions*, J. Ecole Polytechn., 21, 1-69. *8*

Liouville, J. [1837], *Sur le développement des fonctions ou parties de fonctions en séries dont les divers termes sont assujettis à satifaire á une méme équation différentielle du second ordre contenant un paramètre variable*, J. Math. Pures Appl. (1), 2, 16-35; 418-436. *9*

Logan, J.E. [1976], *The approximate solution of Volterra integral equations of the second kind, PhD thesis*, University of Iowa, Iowa City. *48, 167, 378, 403, 501, 504, 509, 510, 511*

Lomakovich, A.N. [1969a], *Solutions of Volterra type integral equations by the Runge-Kutta-Fehlberg method (Ukrainian. Engl./Russian summaries)*, Dopovidi Akad. Nauk Ukrain. RSR Ser. A 1969, 307-310. 369. *177, 232*

Lomakovich, A.N. [1969b], *Solution of integro-differential equations of Volterra type by the Runge-Kutta-Fehlberg method (Russian. French summary)*, Vychisl. Prikl. Mat. (Kiev) Vyp., 7, 64-76. *177, 232*

Lomakovich, A.N. & Ischuk, V.A. [1974], *The approximate solution of certain nonlinear Volterra type integral equations by a bilateral method of Runge-Kutta-Fehlberg form (Russian)*, Vychisl. Prikl. Mat. (Kiev), 23, 29-40. (A translation by C.T.H. Baker, M.S. Keech and P. Sermer appeared as Numer. Anal. Report No. 26, Dept. of Math. University of Manchester, 1978). *177, 232, 502, 507*

Lomakovich, A.N. & Vasilenko A.P. [1968], *The solution of Volterra type integral equations by a method of the Runge-Kutta form*, Differential Equations, 4, 1083-1087. *177, 232*

Londen, S.-O. [1984a], *Asymptotic estimates for resolvents of some integral equations*, in: Infinite-dimensional Systems (F. Kappel & W. Schappacher, eds.), Lecture Notes in Math., 1076, Springer-Verlag, Berlin-Heidelberg-New York. 139-146. *493*

Londen, S.-O. [1984b], *Asymptotic estimates for resolvents of some nonintegrable Volterra Kernels*, MRC Techn. Summary Report No. 2654, University of Wisconsin, Madison. *493*

Londen, S.-O. & Staffans, O.J. (eds.) [1979], *Volterra Equations*, Lecture Notes in Math., 737, Springer-Verlag, Berlin-Heidelberg-New York. *11, 38, 45, 47, 493*

Lonseth, A.T. [1977], *Sources and applicaations of integral equations*, SIAM Rev., 19, 241-278.

Loscalzo, F.R. [1969], *An introduction to the application of spline functions to initial value problems,* in: Theory and Applications of Spline Functions (T.N.E. Greville, ed.), Academic Press, New York, 37-64. *317*

Loscalzo, F.R. & Talbot, T.D. [1967], *Spline function approximations for solutions of ordinary differential equations,* SIAM J. Numer. Anal., 4, 433-445. *317*

Lozinskiĭ, S.M. [1958], *Error estimate for the numerical integration of ordinary differential equations. I (Russian),* Izv. Vyssh. Uchebn. Zaved. Mat., 5, 52-90. *415*

Lubich, Ch. [1981], *Numerische Behandlung Volterra'scher Integrodifferentialgleichungen,* Diplomarbeit, Universität Innsbruck. *219, 222, 228, 233*

Lubich [1982a], *Zur Stabilität linearer Mehrschrittverfahren für Volterra-Gleichungen,* Dissertation, Universität Innsbruck. *494, 495*

Lubich, Ch. [1982b], *On the stability of linear multistep methods for Volterra integral equations of the second kind,* in: Baker & Miller [1982], 233-238. *48, 434, 495*

Lubich, Ch. [1982c], *Runge-Kutta theory for Volterra integrodifferential equations,* Numer. Math. 40, 119-135. *219, 220, 222, 225, 226, 228, 233*

Lubich, Ch. [1983a], *On the stability of linear multistep methods for Volterra convolution equations,* IMA J. Numer. Anal. 3, 439-465. *48, 412, 434, 463, 464, 469, 472, 487, 494, 495, 496*

Lubich, Ch. [1983b], *Runge-Kutta theory for Volterra and Abel integral equations of the second kind,* Math. Comp., 41, 87-102. *29, 30, 48, 403*

Lubich, Ch. [1983c], *Numerical fractional calculus,* Preprint Nr. 230, Sonderforschungsbereich 123, Universität Heidelberg. *323, 325, 403*

Lubich, Ch. [1985a], *A stability analysis of convolution quadratures for Abel-Volterra integral equations,* IMA J. Numer. Anal., 5. *412, 495*

Lubich, Ch. [1985b], *Discretized fractional calculus,* SIAM J. Math. Anal. *323, 325, 328, 330, 332, 334, 335, 336, 403*

Lubich, Ch. [1985c], *Fractional linear multistep methods for Abel-Volterra integral equations of the second kind,* Math. Comp., 45, 463-469. *344, 403, 495*

Lutzen, J. [1982], *Joseph Liouville's contribution to the theory of integral equations,* Historia Math., 9, 373-391. *46*

Lyness, J.N. [1977] *Has numerical differentiation a future?,* in: Proc. Seventh Manitoba Confer. Numer. Math. (D. McCarthy & H.C. Williams, eds.), University of Manitoba, Winnipeg, 107-129. *321*

Lyness, J.N. & Ninham, B.W. [1967], *Numerical quadrature and asymptotic expansions,* Math. Comp., 21, 162-178. *92*

MacCamy, R.C. & Weiss, P. [1979a], *Numerical solution of Volterra integral equations,* Nonlinear Anal., 3, 677-695. *319*

MacCamy, R.C. & Weiss, P. [1979b], *Numerical approximations for Volterra integral equations,* in: Londen & Staffans [1979], 173-191. *319*

Magnitskii, N.A. [1979], *Volterra linear integral equations of the first and third kinds,* U.S.S.R. Comput. Math. and Math. Phys., 19, 182-200. *47*

Makroglou, A. [1977], *Numerical solution of Volterra integro-differential equations,* PhD thesis, University of Manchester. *319, 405, 496*

Makroglou, A. [1980], *Convergence of a block-by-block method for nonlinear Volterra integro-differential equations,* Math. Comp., 35, 783-796. *167, 319*

Makroglou, A. [1981], *A block-by-block method for Volterra integro-differential equations with weakly singular kernel,* Math. Comp., 37, 95-99. *166, 405*

Makroglou, A. [1982], *Hybrid methods in the numerical solution of Volterra integro-differential equations,* IMA J. Numer. Anal., 2, 21-35 (Errata: IMA J. Numer. Anal., 3, 381-382. *166, 513*

Malina, L. [1975a], *A-stable methods of high order for Volterra integral equations,* Apl. Mat., 20, 336-344. *318, 319, 496*

Malina, L. [1975b], *A note on the convergence for the numerical scheme from the Micula's paper,* Mathematica (Cluj), 17 (40), 187-190. *318*

Malinowski, H. & Smarzewki, R. [1978], *A numerical method for solving the Abel integral equation,* Zastos. Mat., 16, 275-281. *406*

Mann, W.R. & Wolf, F. [1951], *Heat transfer between solids and gases under nonlinear boundary conditions,* Quart. Appl. Math., 9, 163-184. *404, 493*

Marchuk, G.I. & Shaidurov, V.V. [1983], *Difference Methods and Their Extrapolation,* Springer-Verlag, Berlin-Heidelberg-New York. *129, 164, 166*

Markowich, P. & Renardy, M. [1983], *A nonlinear Volterra integro-differential equation describing the stretching of polymeric liquids,* SIAM J. Math. Anal., 14, 66-97.

Markowich, P. & Renardy, M. [1983], *The numerical solution of a class of quasilinear parabolic Volterra integral equations arising in polymer rheology,* SIAM J. Math. Anal., 20, 890-908.

Marti, J.T. [1980], *On the convergence of an algorithm computing minimum norm solutions of ill-posed problems,* Math. Comp., 34, 521-527.

Matthys, J [1976], *A-stable linear multistep methods for Volterra integro-differential equations,* Numer. Math., 27, 85-94. *65, 80, 91, 164, 166, 325, 448, 451, 469, 470, 495*

Mayers, D.F. [1962], *Equations of Volterra type,* in: Numerical Solution of Ordinary and Partial Differential Equations (L. Fox, ed.), Pergamon Press, Oxford, 165-173. *163, 436, 495*

Mazilu, P,. & Paltineanu, G. [1979], *Semi-discretization method for the numerical solving of Volterra integral equations,* in: Proc. Int. Symp.

Applications of Mathematics in System Theory, Brasov (Romania) 1978, 245-253.

Mazur, J. [1971], *Numerical solution of integral equations of the first kind applied to slit correction in small-angle X-ray scattering,* J. Res. Nat. Bur. Standards, 75B, 173-187.

McKee, S. [1979a], *Best convergence rates of linear multistep methods for Volterra first-kind equations,* Computing, 21, 343-358. *165*

McKee, S. [1979b], *Cyclic multistep methods for solving Volterra integro-differential equations,* SIAM J. Numer. Anal., 16, 106-114. *166*

McKee, S. [1979c], *The analysis of variable step, variable coefficient linear multistep method for solving a singular integro-differential equation arising from the diffusion of discrete particles in a turbulent fluid,* J. Inst. Math. Appl., 23, 373-388. *405, 495*

McKee, S. [1982a], *Generalized discrete Gronwall lemmas,* Z. Angew. Math. Mech., 62, 429-434. *44, 49*

McKee, S. [1982b], *A review of linear multistep methods and product integration methods and their convergence analysis for first-kind Volterra integral equations,* in: Baker & Miller [1982], 153-161. *163, 406*

McKee, S. [1984], *On the convergence of Euler's method,* Bull. Inst. Math. Appl., 20, 115-120. *165*

McKee, S. & Brunner, H. [1980], *The repetition factor and numerical stability of Volterra integral equations,* Comput. Math. Appl., 6, 339-347. *495*

McKee, S. & Stokes, A. [1983], *Product integration methods for the nonlinear Basset equation,* SIAM J. Numer. Anal., 20, 143-160. *405, 495*

Mey, G. de [1976], *An integral equation method for the numerical calculation of ion drift and diffusion in evaporated dielectrics,* Computing, 17, 169-176.

Micula, G. [1973], *Approximate solution of the differential equation $y''=f(x,y)$ with spline functions,* Math. Comp., 27, 807-816 (Corrigendum: Math. Comp., 29 (1975), 673-674). *317*

Micula, G. [1975], *The numerical solution of Volterra integro-differential equations by spline functions,* Rev. Roumaine Math. Pures Appl., 20, 349-358. *319*

Micula, G. [1976], *Bemerkungen zur numerischen Behandlung von nichtlinearen Volterraschen Integralgleichungen mit Splines,* Z. Angew. Math. Mech., 56, T302-304.

Micula, G. [1978], *Functii spline si aplicatii (Romanian),* Editura Technica, Bucarest. *317, 318, 319*

Micula, G. [1979], *Numerischen Behandlung der Volterra-Integralgleichungen mit Splines,* Studia Univ. Babes-Bolyai, Math., 24, 46-54. *318*

Micula, M. [1973], *Procédés de type Runge-Kutta-Fehlberg pour l'approximation de la solution de l'équation integrodifférentielle de premier ordre de type*

Volterra, Studia Univ. Babes-Bolyai Math. 18, 61-68. *232*

Micula, M. [1979], *Numerische Behandlung der Volterra-Integralgleichungen mit splines,* Studia Univ. Babes-Bolyai Math., 24, 46-54. *318*

Micula, M. [1981], *The Runge-Kutta-Fehlberg procedures for the numerical solution of Volterra integral equations,* Studia Univ. Babes-Bolyai Math., 26, 56-61. *177, 232*

Micula, M. & Micula, G. [1973], *Sur la résolution numérique des équations integrales du type de Volterra de seconde espéce à l'aide des fonctions splines,* Studia Univ. Babes-Bolyai Math. 18, 65-68. *318*

Mikeladze, Sh. E. [1935], *De la résolution numérique des équations intégrales (Russian),* Bull. Acad. Sci. URSS, VII, Nr. 2, 255-297 (French summary: 297-300). *49, 95, 163*

Mikeladze, Sh. E. [1951], *On approximate solution of integral equations of Volterra type (Russian),* Akad.Nauk Gruzin. SSR Trudy Mat. Inst. Rasmadze, 18, 315-325. *95, 163, 164*

Miller, G.F. [1974a], *Algorithms for integral equations,* in: Software for Numerical Mathematics (D.J. Evans, ed.), Academic Press, London, 139-147. *519*

Miller, G.F. [1974b], *Provision of library programs for the numerical solution of integral equations,* in: Delves & Walsh [1974], 247-256. *519*

Miller, R.K. [1968a], *On Volterra integral equations with nonnegative integrable resolvents,* J. Math. Anal. Appl., 22, 319-340. *493*

Miller, R.K. [1968b], *On the linearization of Volterra integral equations,* J. Math. Anal. Appl., 23, 198-208. *47, 493*

Miller, R.K. [1971], *Nonlinear Volterra Integral Equations,* Benjamin, Menlo Park (Cal.). *11, 23, 24, 25, 27, 38, 40, 45, 47, 48, 493*

Miller, R.K. & Feldstein, A. [1971], *Smoothness of solutions of Volterra integral equations with weakly singular kernels,* SIAM J. Math. Anal., 2, 242-258. *28, 510*

Miller, R.K., Nohel, J.A. & Wong, J.S.W. [1969], *Perturbations of Volterra integral equations,* J. Math. Anal. Appl., 25, 676-691. *493*

Minerbo, G.N. & Levy, M.E. [1969], *Inversion of Abel's integral equation by means of orthogonal polynomials,* SIAM J. Numer. Anal., 6, 596-616. *520*

Mocarsky, W.L. [1971], *Convergence of step-by-step methods for nonlinear integro-differential equations,* J. Inst. Math. Appl., 8, 235-239. *156, 232, 320*

Monna, A.F. [1973], *Functional Analysis in Historical Perspective,* Oosthoek Publ. Co., Utrecht. *10*

Moore, R.E. [1982], *A generalization of the method of upper and lower solutions for integral equations,* Nonlinear Anal., 6, 829-831.

Morozov, V.A. [1984], *Methods for Incorrectly Posed Problems,* Springer-Verlag, Berlin-Heidelberg-New York. *321*

Multhei, H.N. [1980], *Numerische Lösung gewöhnlicher Differentialgleichungen mit Splinefunktionen,* Computing, 25, 317-335. *317*

NAG [1984], *FORTRAN library manual-mark 11,* Numerical Algorithms Group, Oxford. *5*

Nazarenko, T.I. [1963], *On the numerical solution of the Cauchy problem for a class of linear integro-differential equations of Volterra type (Russian),* Sibirsk. Mat. Zh., 4, 303-308.

Netravali, A.N. [1973], *Spline approximation to the solution of the Volterra integral equation of the second kind,* Math Comp., 27, 99-106. *318*

Neumann, C. [1877], *Untersuchungen über das logarithmische und Newton'sche Potential,* Teubner, Leipzig. *10*

Nevalinna, O. [1976a], *Positive quadratures for Volterra equations,* Computing, 16, 349-357. *434, 494*

Nevanlinna, O. [1976b], *On the numerical solution of some Volterra equations on infinite intervals,* Report No.2, Institut Mittag-Leffler (See also: Anal. Numér. Théor. Approx., 5, 31-57. *434, 494*

Nevanlinna, O. [1982], *On the stability of discrete Volterra equations,* in: Baker & Miller [1982], 139-147. *494*

Neves, K.W. [1975a], *Automatic integration of functional differential equations: an approach,* ACM Trans. Math. Software, 1, 357-368. *233, 512*

Neves, K.W. [1975b], Algorithm 497: *Automatic integration of functional differential equations,* ACM Trans. Math. Software, 1, 369-371. *233, 512*

Nicholson, R.S. [1965], *Some examples of the numerical solution of nonlinear integral equations,* Anal. Chem., 37, 667-671.

Nikolaishvili, Sh. & Grigorovich, G. [1976], *A new method for the numerical solution of Volterra integral equations (Russian),* Sem. Inst. Prikl. Mat. Dokl., 11, 67-72.

Noble, B. [1964], *The numerical solution of nonlinear integral equations and related topics,* in: Anselone [1964], 215-318. *6, 96, 147, 163, 164, 493, 510*

Noble, B. [1969], *Instability when solving Volterra integral equations of the second by multistep methods,* in: Proc. Confer. Numer. Solution Diff. Equations (J.L. Morris, ed.), Lecture Notes in Math., 109, Springer-Verlag, Berlin-Heidelberg-New York, 23-39. *122, 163, 164, 495*

Noble, B. [1971], *A bibliography on 'Methods for solving integral equations',* MRC Techn. Summary Report No. 1176 (Author listing), MRC Techn. Summary Report No. 1177 (Subject listing), University of Wisconsin, Madison. *45, 163*

Noble, B. [1977], *The numerical solution of integral equations,* in: The State of the Art in Numerical Analysis (D. A.H. Jacobs, ed.), Academic Press, London-New York, 915-966. *163, 519*

Nohel, J.A. [1964], *Problems in the qualitative behavior of solutions of nonlinear Volterra equations,* in: Anselone [1964], 191-214. *38, 493*

Nohel, J.A. [1971], *Perturbations of Volterra equations and admissibility,* in: *Japan-United States Seminar on Ordinary and Functional Equations* (M. Urabe, ed.), Lecture Notes in Math., 243, Springer-Verlag, Berlin-Heidelberg-New York, 40-53. *38, 46, 493*

Nohel, J.A. [1976], *Review of 'Integral Equations and Stability of Feedback Systems'* (C. Corduneanu), SIAM Rev., 12, 520-526. *48, 493*

Nørsett, S.P. [1974], *Semi-explicit Runge-Kutta methods,* Techn. Rep. No. 6-74, Dept. of Mathematics, University of Trondheim. *180, 188, 225, 231*

Nørsett, S.P. [1980], *Collocation and perturbed collocation methods,* in: *Numerical Analysis* (G.A. Watson, ed.), Lecture Notes in Math., 773, Springer-Verlag, Berlin-Heidelberg-New York, 119-132. *317*

Nørsett, S.P. & Wanner, G. [1981], *Perturbed collocation and Runge-Kutta methods,* Numer. Math., 38, 193-208. *247, 317*

Nurimov, T. [1971], *Untersuchung der Lösung einer zweidimensionalen nichtlinearen Volterraschen Integralgleichung (Russian),* Doklady Akad. Nauk. USSR, 11, 6-8. *247, 317*

Nychka, D. [1984], *The solution of Abel-type integral equations and their application in stereology,* PhD thesis, Dept. of Statistics, University of Wisconsin, Madison.

O'Connor, A.J. [1984], *The nonlinear cooling of a semi-infinite solid. Padé approximation methods,* in: *Proc. CTAC-83 Conference,* North-Holland, Amsterdam, 881-892. *404*

O'Neill, J.C. [1967], *Some numerical solutions of Volterra's integral equations,* PhD thesis, University of Pittsburgh. *163, 319*

O'Neill, J.C. & Byrne, G.D. [1968], *A starting method for the numerical solution of Volterra's integral equation of the second kind,* BIT, 8, 43-47. *163, 319*

Oldham, K.B. & Spanier, J. [1974], *The Fractional Calculus,* Academic Press, New York. *8, 15, 403*

Olmstead, W.E. [1977], *A nonlinear integral equation associated with gas absorption in a liquid,* Z. Angew. Math. Phys., 28, 513-523. *404, 493*

Olmstead, W.E. & Handelsman, R.A. [1976a], *Diffusion in a semi-infinite region with nonlinear surface dissipation,* SIAM Rev. 18, 275-291. *404, 493*

Olmstead, W.E. & Handelsman, R.A. [1976b], *Asymptotic solution to a class of nonlinear Volterra integral equations. II,* SIAM J. Appl. Math., 30, 180-189. *404, 493*

Ortega, J.M. [1972], *Numerical Analysis: A Second Course,* Academic Press, New York-London. *431, 433*

Oule'O's, H. [1960a], *Une méthode de résolution numérique de l'équation intégrale du type de Volterra,* Rev. Française Traitement Information Chiffres, 3, 137-142. *164, 177*

Oulés, H. [1960b], *Une méthode de résolution numérique de l'équation intégrodifférentielle linéaire de Volterra,* C.R. Acad. Sci. Paris, 250, 3937-3939. *166*

Oulés, H. [1964], *Résolution numérique d'une équation intégrale singulière,* Rev. FranUaise Traitement Information Chiffres, 7, 117-111. *403, 510*

Padmavally, K. [1958], *On a nonlinear integral equation,* J. Math. Mech., 7, 533-555. *493, 511*

Paley, R.E.A.C. & Wiener, N. [1934], *Fourier Transforms in the Complex Domain,* American Mathematical Society (Colloquium Publications, XIX), New York. *11, 36, 48, 493*

Papatheodorou, T.S. & Jesanis, M.E. [1980], *Collocation methods for Volterra integro-differential equations with singular Kernels,* J. Comput. Appl. Math., 6, 3-8. *405*

Peano, G. [1913], *Resto nelle formule di quadratura, espresso con un integrale definito,* Atti Accad. Naz. Lincei Rend. Cl. Sci. Math. Natur. (5), 22, 562-569. *91*

Pelekh, Ya.N. [1984], *Nonlinear methods of higher order accuracy for the solution of Volterra integral equations (Russian),* Mat. Metody i Fiz.-Mekh. Polya, 20, 11-16. *232*

Peluso, R. & Piazza, G. [1985], *Metodi di tipo Runge-Kutta per la risoluzione di equazioni integrali di Volterra di seconda specie fortemente stiff,* Boll. Un. Mat. Ital. B (6), 4, 155-165. *496*

Pennachi, R. [1960], *Su un metodo numerico per la risoluzione della equazione integrale di Volterra,* in: Symposium Rome, [1960], 330-350.

Petsoulas, A.G. [1975], *The approximate solution of Volterra integral equations,* J. Approx. Theory, 14, 152-159.

Phillips, C. [1977a], *Numerical solution of Volterra integral equations of the second kind,* PhD thesis, University of Liverpool. *501*

Phillips, C. [1977b], *An improved algorithm for the solution of Volterra integral equations of the second kind by Runge-Kutta processes,* Preprint, Dept. of Computational and Statistical Science, University of Liverpool. *501*

Phillips, G.M. [1970], *Analysis of numerical interative methods for solving integral and integro-differential equations,* Comput. J., 13, 297-300. *164*

Phillips, G.M. [1971], *An error estimate for Volterra integral equations,* BIT, 11, 181-186. *163*

Picard, E. [1890], *Mémoire sur la théorie des équations aux dérivées partielles et la méthode des approximations successives,* J. Math. Pures Appl. (4), 6, 145-210 (see particularly: 197-200). *46*

Piessens, R. [1973], *Calculation of the radial distribution of emitters in a cylindrical source,* Comput. Phys. Comm., 5, 294-298. *406, 516*

Piessens, R. & Verbaeten, P. [1973], *Numerical solution of the Abel integral equation,* BIT, 13, 451-457. *406, 518, 520*

Piessens, R. et al. [1983], *QUADPACK. A Subroutine Package for Automatic Integration,* Springer-Verlag, Berlin-Heidelberg-New York. *91, 250*

Pitt, H.R. [1958], *Tauberian Theorems,* Oxford University Press, London. *38*

Pogorzelski, W. [1966], *Integral Equations and Their Applications,* Pergamon Press, Oxford. *47*

Porath, G. [1974], *Lineare Volterrasche Integralgleichungen zweiter Art mit Kernen vom allgemeinen Typ,* Beitrage Numer. Math., 2, 147-162.

Porath, G. [1975a], *Das Ersatzoperatorenverfahren fur linearen Volterrasche Integralgleichungen zweiter Art,* Beitrage Numer. Math., 3, 115-130. *319*

Porath, G. [1975b], *Uber die numerische Behandlung Volterrascher Integralgleichungen zweiter Art,* Acta Univ. Carolinae, Math. Phys., 15, 121-123.

Porath, G. [1981], *Ein Approximations-Iterationsverfahren für nichtlineare Volterrasche Integralgleichungen zweiter Art,* Abh. Akad. Wiss. DDR, Abt. Math. Naturwiss. Tech., 2N, 379-383.

Porath, G. & Tabbert, E. [1978], *Ueber ein verfahrensorientiertes Programmsystem für Volterrasche Integralgleichungen zweiter Art,* in: *Anwendung numerischer und Stochastischer Methoden zur Lösung technischer und technologischer Probleme,* Numerik, 1, wiss. Kolloq. Wismar 1977, 42-47.

Porath, G. & Tabbert, E. [1979], *Das Tschebyscheffsche Iterationsverfahren für lineare Volterrasche Integralgleichungen zweiter Art,* Beitrage Numer. Math., 7, 83-96. *319*

Postnikov, A.G. [1980], *Tauberian Theory and Its Applications,* Proc. Steklov Institute, No. 144, American Math. Society, Providence, R.I. *38*

Pouzet, P. [1960], *Méthode d'intégration numérique des équations intégrales et integro-différentielles du type Volterra de seconde espéce. Formules de Runge-Kutta,* in: Rome Symposium [1960], 362-368. *174, 177, 218, 232, 509, 514*

Pouzet, P. [1961], *Evaluation des erreurs de troncature pour les formules de Runge-Kutta relatives au traitement numérique, des équations intégrales de type Volterra,* C.R. Acad. Sci. Paris, 252, 1901-1903. *177*

Pouzet, P. [1962], *Etude, en vue de leur approximation numérique, des solutions d'équations intégrales et intégrodifférentielles de type Volterra pour des problèmes de conditions initiales,* Thése, Université de Strasbourg. 163, *174, 177, 218, 228, 231, 232, 507, 509*

Pouzet, P. [1963], *Etude en vue de leur traitement numérique des équations intégrales de type Volterra,* Rev. Française Traitement Information Chiffres, 6, 79-112. *163, 174, 186, 204, 218, 222, 501*

Pouzet, P. [1964], *Algorithme de résolution des équations intégrales de type Volterra par des méthodes par pas,* Rev. Française Traitement Information Chiffres, 7, 169-173. *501*

Pouzet, P. [1967], *Systèmés différentiels, équations intégrales et intégro-différentielles,* in: *Procédures ALGOL en analyse numérique, Tome I,* Editions du Centre National de la Recherche Scientifique, Paris, 193-203. *501, 506*

Powell, M.J.D. [1981], *Approximation Theory and Methods,* Cambridge University Press, Cambridge. *350*

Prasad, G. [1924], *On the numerical solution of integral equations,* Proc. Edinburgh Math. Soc., 42, 46-59. *95, 163, 167, 177, 187*

Prenter, P.M. [1975], *Splines and Variational Methods,* Wiley, New York. *317*

Prosperetti, A. [1977], *A numerical method for the solution of certain classes of nonlinear Volterra integro-differential and integral equations,* Internat. J. Numer. Methods Engrg., 11, 431-438. *404*

Prosperetti, A. [1982], *A method for the solution of singular Volterra integro-differential equations,* J. Comput. Phys., 46, 462-468. *404*

Purga, A. & Preem, R. [1981], *On the solution of Abel integral equation by the method of fractional integration,* Eesti NSV Tead. Akad. Toimetised Fuus.-Mat., 30, 9-13. *403*

Rabinowitz, P. & Sloan, I.H. [1984], *Product integration in the presence of a singularity,* SIAM J. Numer. Anal., 21, 149-166.

Radziuk, J. [1977], *The numerical solution from measurement data of linear integral equations of the first kind,* Internat. J. Numer. Meth. Engrng., 11, 729-740. *321*

Rakotch, E. [1973], *Numerical solution of Volterra integral equations,* Numer. Math., 20, 271-279. *163*

Rall, R.B. [1978], *Resolvent kernels of Green's function kernels and other finite-rank modifications of Fredholm and Volterra kernels,* J. Optim. Theory Appl., 24, 59-88.

Ramm, A.G. [1978], *Investigation of some classes of integral equations and their applications,* in: *Abel Inversion and its Generalizations,* Collect. Artic., Novosibirsk, 120-179.

Reynolds, D. [1984], *On linear singular Volterra integral equations of the second kind,* J. Math. Anal. Appl., 103, 230-262. *46*

Rice, J.R. [1969], *On the degree of convergence of nonlinear spline approximation,* in: *Approximation with Special Emphasis on Spline Functions* (I.J. Schoenberg, ed.), Academic Press, New York, 349-365. *404*

Richtmyer, R.D. & Morton, K.W. [1967], *Difference Methods for Initial Value Problems,* Interscience, New York. *117*

Riddell, I. & Delves, L.M. [1980], *The numerical testing of integral equation routines,* in: *Production and Assessment of Numerical Software* (I. Riddell & L.M. Delves, eds.), Academic Press, London, 41-56. *519*

Riele, H.J.J. te (ed.) [1977a], *Colloquium Numerical Software. Part 2 (Dutch),* MC Syllabus 29.2, Mathematisch Centrum, Amsterdam.

Riele, H.J.J. te [1977b], *Regularization methods for first-kind integral equations (Dutch),* in: te Riele [1977a], 147-176. *515*

Riele, H.J.J. te (ed.) [1979a], *Colloquium Numerical Treatment of Integral Equations,* MC Syllabus 41, Mathematisch Centrum, Amsterdam. *163*

Riele, H.J.J. te [1979b], *Introduction and global survey of numerical methods for integral equations,* in: te Riele [1979a], 1-25. *96, 163*

Riele, H.J.J. te [1982], *Collocation methods for weakly singular second kind Volterra integral equations with non-smooth solution,* IMA J. Numer. Anal., 2, 437-449. *363, 374, 404, 510, 511*

Riele, H.J.J. te & Schroevers [1986], *A comparative survey of numerical methods for the linear generalized Abel integral equation,* Z. Angew. Math. Mech., 66.. *406, 518*

Riemann, B. [1847], *Versuch einer allgemeinen Auffassung der Integration und Differentiation,* Ges. Math. Werke, 2nd ed., 1892 (H. Weber, Ed.), Teubner, Leipzig, 353-366. *8*

Roberts, J.H. & Mann, W.R. [1951], *A nonlinear integral equation of Volterra type,* Pacific J. Math., 1, 431-445.

Roemisch, W. & Schulze, R. [1979], *Kennwertmethoden fur stochastische Volterrasche Integralgleichungen,* Wiss. Z. Humboldt-Univ. Berlin Math. Natur. Reihe, 28, 523-533.

Rosse, E.W., Jr. [1981], *On Volterra integral equations of pulse-convolution type,* Trans. 26th Conf. Army Math., Hanover/N.H., ARO Rep. 81-1, 153-166.

Rumyantsev, I.A. [1965], *Program for solving a system of Volterra integral equations of the second kind,* U.S.S.R. Comput. Math. and Math. Phys., 5, 218-224. *519*

Runge, C. [1895], *Über die numerische Auflösung von Differentialgleichungen,* Math. Anal., 46, 167-178. *95, 178*

Saaty, T.L. [1967], *Modern Nonlinear Equations,* McGraw-Hill, New York. *47*

Salzer, H.E. [1960], *Optimal points for numerical differentiation,* Numer. Math., 2, 214-227. *305*

Sato, T. [1953], *Sur l'équation intégrale non linéaire de Volterra,* Compositio Math., 11, 271-290.

Sanchez, D.A. [1968], *Ordinary Differential Equations and Stability Theory: An Introduction,* Freeman, San Francisco. *410, 419*

Sathiaraj, D.D. & Sankar, R. [1982], *The method of successive updated iterated defect correction and its application to second kind Volterra integral equations,* BIT, 22, 390-394. *318*

Schaum, H.-J. [1977], *Problemorientierte Verfahren zur numerischen Integration nichtlinearer Integrodifferentialgleichungen Volterraschen Typs,* Dissertation, TH Aachen.

Schlichte, M. [1984], *Anwendung eines impliziten Runge-Kutta-Verfahrens auf Volterra'sche Integralgleichungen zweiter Art mit Faltungskern,* Diplomarbeit, Institut fur Angewandte Mathematik, Universität Heidelberg. *319, 503, 519*

Schmaedeke, W.W. [1968], *Approximate solutions for Volterra integral equations the first kind,* J. Math. Anal. Appl., 23, 604-613.

Schmeidler, W. [1950], *Integralgleichungen mit Anwendungen in Physik und Technik,* Akad. Verlagsges. Geest & Portig, Leipzig. *47*

Schmeisser, G. & Schirmeier, H. [1976], *Praktische Mathematik,* de Gruyter, Berlin. *51, 73, 91, 92*

Schmidt, J.W. [1976], *Über lineare Ungleichungen vom Gronwallschen Typ,* Beitrage Numer. Math., 5, 171-189. *39, 40, 41, 48, 49*

Schneider, C. [1980], *Produktintegration mit nicht-aquidistanten Stutzstellen,* Numer. Math., 35, 35-43. *89, 92, 378*

Schneider, C. [1981], *Product integration for weakly singular integral equations,* Math. Comp., 36, 207-213. *378, 404*

Schoedon, P.M. [1970], *Beitrage zur numerischen Lösung nichtlinearer Volterrascher Integralgleichungen und Integrodifferentialgleichungen mit den Runge-Kutta-Verfahren,* Dissertation, TH Aachen. *177, 232*

Schumaker, L.L. [1981], *Spline Functions: Basic Theory,* Wiley, New York. *239, 317, 348, 350, 404*

Scott, J.A. [1984], *A unified analysis of discretization methods,* D.Phil. thesis, University of Oxford. *48, 49, 164, 165, 494*

Scott, J.A. [1985], *A unified analysis of discretization methods for Volterra-type equations,* in: Hammerlin & Hoffmann [1985], 244-255. *164*

Sedgwick, A. [1973], *An effective variable order variable step Adams method,* Techn. Report No. 53, Dept. of Comput. Science, University of Toronto.

Shampine, L.F. [1985], *Solving Volterra integral equations with ODE codes,* Sandia Report, Sand. 85-0678, Albuquerque (to appear in: IMA J. Numer. Anal.). *494, 502*

Shampine, L.F. & Gordon, M.K. [1975], *Computer Solution of Ordinary Differential Equations,* Freeman, San Francisco, 5. *502*

Shea, D.F. & Wainger, S. [1975], *Variants of the Wiener-Levy theorem, with applications to stability problems for some Volterra integral equations,* Amer. J. Math., 97, 312-343. *48, 493*

Singh, P. [1976], *A note on the solution of two-dimensional Volterra integral equations by splines,* Indian J. Math., 18, 61-64. *318*

Skerstena, A.Ja. [1974], *Numerical solution of the Volterra integral equation of the second kind (Russian),* Latvii. Math. Ezegodnic, 15, 155-168.

Sloan, I.H. [1976], *Improvement by iteration for compact operator equations,* Math. Comp., 30, 758-764. *269, 318*

Sloan, I.H. [1980], *A review of numerical methods for Fredholm equations of the second kind,* in: Anderssen, de Hoog & Lukas [1980], 51-74. *269*

Sloan, I.H. [1984], *Four variants of the Galerkin method for integral equations of the second kind,* IMA J. Numer. Anal., 21, 123-125. *269, 318*

Small, J.C. & Booker, J.R. [1977], *Methods for the numerical solution of the equations of viscoelasticity,* Internat. J. Numer. Analyt. Methods Geomech., 1, 139-150.

Smarzewski, R. [1976], *A method for solving the Volterra integral equation of the first kind,* Zastos. Mat., 15, 117-123. *320*

Smarzewski, R. & Malinowski, H. [1978], *Numerical solution of a class of Abel integral equations,* J. Inst. Math. Appl., 22, 159-170. *406*

Smarzewski, R. & Malinowski, H. [1979a], *Determination of the solution of an Abel integral equation (Algorithm 65),* Zastos. Mat., 16, 505-511. *406, 516*

Smarzewski, R. & Malinowski, H. [1979b], *On the numerical solution of an Abel integral equation,* Zastos. Mat., 16, 497-503. *406, 516*

Smarzewski, R. & Malinowski, H. [1983], *Numerical solution of generalized Abel integral equations by spline functions,* Zastos. Mat., 17, 677-687. *406*

Smit, J.H. [1976], *Two-sided error bounds in the numerical solution of generalized Volterra integro-differential equations,* Doctoral thesis, University of Leiden. *166*

Sneddon, I.H. [1972], *The Use of Integral Transforms,* McGraw-Hill, New York. *47, 400*

Söderlind, G. [1979], *Some stability properties of linear multistep compound discretizations of partitioned differential systems,* Report TRITA-NA-7910, The Royal Institute of Technology, Stockholm. *458*

Sonine, N. [1884], *Sur la généralisation d'une formule d'Abel,* Acta Math., 3, 171-176. *8, 95*

Soni, K. & Soni, R.P. [1981], *On the asymptotic behavior of the solution of a nonlinear Volterra integral equation,* in: *Proc. Analytical and Numerical Approaches to Asymptotic Problems in Analysis,* North-Holland Math. Stud., 47, 369-368.

Spohn, D. [1965], *Sur les formules à pas liés dans l'intégration numérique des équations intégrales du type de Volterra,* in: *Proc. 4ème Congr. Calcul et de Traitement de l'Information AFIRO* Dunod, Paris, pp. 349-356. *164*

Squire, W. [1969], *Numerical solution of linear Volterra equations of the first kind,* Aerospace Engineering Report TR15, West Virginia University, Morgantown. *515*

Squire, W. [1970], *Integration for Engineers and Scientists,* American Elsevier Publ. Co., New York. *515, 519*

Srivastava, H.M. & Buschman, R.G. [1977], *Convolution Integral Equations,* Wiley Eastern Ltd., New Delhi. *47*

Steinberg, J. [1972], *Numerical solution of Volterra integral equations,* Numer. Math., 19, 212-217. *73, 91, 163*

Stetter, H.J. [1973], *Analysis of Discretization Methods for Ordinary Differential Equations,* Springer-Verlag, New York-Heidelberg-Berlin. 67, *78, 91, 164, 177, 231*

Stoutemyer, R.D. [1977], *Analytically solving integral equations by using computer algebra,* ACM Trans. Math. Software, 3, 128-146. *519*

Ström, T. [1975], *On logarithmic norms,* SIAM J. Numer. Anal., 12, 741-753.

Stroud, A.H. [1974], *Numerical Quadrature and Solution of Ordinary Differential Equations,* Springer-Verlag, Berlin-Heidelberg-New York. *53, 59, 60, 91*

Stroud, A.H. & Secrest, D. [1966], *Gaussian Quadrature Formulas,* Prentice Hall, Englewood Cliffs (N.J.). *91*

Suyama, Y. & Nakamori, K. [1952], *Pri numera solvada de la integrale ekvacio de Volterra-tipo (Esperanto) (=On numerical solution of the integral equation of Volterra type),* Mem. Fac. Sci. Kyushu Univ. Ser. A, 6, 121-129. *177*

Symposium Rome [1960], *Symposium on the Numerical Treatment of Ordinary Differential Equations, Integral and Integro-Differential Equations,* Birkhäuser Verlag, Basel.

Szegö, G. [1939], *Orthogonal Polynomials,* American Mathematical Society Colloquium Publications, 23, New York. *307*

Szopa, J. [1976], *Application of Volterra stochastic integral equations of the second kind to the analysis of dynamic systems of variable inertia,* J. Tech. Physics, 17, 423-433.

Szufla, S. [1984], *On the existence of L^p-solutions of Volterra integral equations in Banach spaces,* Funkcial. Ekvac., 27, 157-172. *47*

Tabbert, E. [1979], *Struktur und Organisation eines Programmiersystems zur Lösung von Volterraschen Integralgleichungen,* in: *Problemsemin. Mathematische Verfahrenstechnik,* Weissig (DDR), 107-118.

Takeda, T. [1976], *Numerical method for solution of the integral equation of the first kind - application to analysis of plasma density profile,* J. Comput. Phys., 21, 305-318.

Tanfulla, M. & Ribighini, G. [1981], *Procedure di stima dell' errore nella risoluzione di equazioni integrali di tipo Volterra,* Riv Mat. Univ. Parma (4),

7, 423-487. *188, 232, 502*

Tavernini, L. [1971], *One-step methods for the numerical solution of Volterra functional differential equations,* SIAM J. Numer. Anal., 8, 786-795. *232, 320*

Tavernini, L. [1973], *Linear multistep methods for the numerical solution of Volterra functional differential equations,* Applicable Anal., 1, 169-185. *166*

Taylor, P.J. [1976], *The solution of Volterra integral equations of the first kind using inverted differentiation formulas,* BIT, 16, 416-425. *147, 165*

Taylor, P.J. [1982], *Applications of results of Vainikko to Volterra integral equations,* in: Baker & Miller [1982], 185-196. *166*

Ten, Men Yan [1979a], *Block method for the solution of two-dimensional Volterra equations of the first kind,* Differential Equations, 15, 794-798. *320*

Ten Men Yan [1979b], *On unimprovable estimates of solutions of some difference inequalities (Russian),* in: *Optimization Methods and their Applications,* 9, Work Collect., Irkutsk, 150-159.

Tesei, A. [1980], *Approximation results for Volterra integro-partial differential equations,* in: Albrecht & Collatz [1980], 213-227. *45*

Tikhonov, A.N. & Arsenin, V.Y. [1977], *Solution of Ill-Posed Problems,* Wiley, New York. *321*

Timan, A.F. [1963], *Theory of Approximation of Functions of a Real Variable,* Pergamon Press, Oxford. *348*

Tivonchuk, V.I. [1972], *Solution of nonlinear integral equations of mixed type by the method of averaging functional corrections,* Differential Equations, 5, 453-457.

Tivonchuk, V.I. [1981], *Solution of linear Volterra integral equations by the method of averaging functional corrections in conjunction with splines,* Ukrainian Math. J., 32, 281-287.

Tivonchuk, V.I. & Shlepakov, L.N. [1978], *Spline iteration method for solving Volterra linear integral equations (Russian),* in: *Boundary value Problems in Mathematical Physics,* Akad. Nauk Ukr. SSR, Inst. Mat., Kiev, 159-166, 182.

Tivonchuk, V.I. & Shlepakov, L.N. [1981], *Numerical-analytical method for solving systems of linear Volterra integral equations,* Soviet Math., 25, 101-107.

Tricomi, F. [1957], *Integral Equations,* Wiley-Interscience, New York. *5, 47*

Tsalyuk, Z.B. [1979], *Volterra integral equations,* J. Soviet Math., 12, 715-758. *38, 46, 48, 96, 163, 494*

Tsokos, C.P. & Padgett, W.J. [1974], *Random Integral Equations with Applications to Life Sciences and Engineering,* Academic Press, New York. *46, 47*

Tsuruta, K. & Ohmori, K. [1976], *A posteriori error estimates for Volterra integro-differential equations,* Mem. Numer. Math., 3, 33-47.

Ugniewski, S. [1977], *Solving the Abel integral equation by interpolation methods,* Zastos. Mat., 16, 91-109. *406*

Vainikko, G. & Uba, P. [1981], *A piecewise polynomial approximation to the solution of an integral equation with weakly singular kernel,* J. Austral. Mat. Soc. Ser. B, 22, 431-438. *404*

Vainikko, G., Pedas, A. & Uba, P. [1984], *Methods for Solving Weakly Singular integral Equations (Russian),* Tartu. Gos. Univ., Tartu. *404*

Varga, R.S. [1962], *Matrix Iterative Analysis,* Prentice-Hall, Englewood Cliffs (N.J.), 431.

Vata, P. [1976], *Intégration numérique d'une équation intégrale non linéaire Volterra,* in: *Proc. Inst. Math. Iasi* (1974), Editura Acad. R.S.R., Bucharest, 185-191.

Verlan, A.F. [1975], *Error equations for the solution of integral equations of Volterra type (Russian),* Tocn. Nadezn. Kibernet. Sist., 3, 3-10.

Verlan, A.F. & Sizikov, V.S. [1978], *Methods for the Solution of Integral Equations using Computer Programs (Russian),* Izdat. 'Naukova Dumka', Kiev. (MR 58, # 25047). *519*

Verner, J.H. [1978], *Families of imbedded Runge-Kutta methods,* Queen's Mathematical Preprints No. 1978-19, Queen's University, Kingston, Ontario. *503*

Verner, J.H. [1979], *John Butcher's algebraic theory: motivation for selecting simplifying conditions,* in: *Proc. Ninth Manitoba Confer. Numer. Math. Computing* (G.H.J. van Rees & H.C. Williams, eds.), University of Manitoba, Winnipeg, 125-155. *179*

Vessella S. [1985], *Stability results for Abel Equation,* in: Hämmerlin & Hoffmann [1985], 273-279.

Vivanti, G. [1929], *Elemente der Theorie der linearen Integralgleichungen,* Helwingsche Verlagsbuchhandlung, Hannover. *10, 46, 47*

Volterra, V. [1884], *Sopra un problema di elettrostatica,* Rend. Accad. Lincei (3), 315-318.

Volterra, V. [1896], *Sulla inversione degli integrali definiti,* Atti Accad. Sci. Torino, 31, 311-323 (Nota I), 400-408 (Nota II), 557-567 (Nota III), 693-708 (Nota IV); Rend. Accad. Lincei (5), 5, 177-185. (see also: Opere Matematiche, II, Accademia Nazionale dei Lincei, Roma (1956), 216-262.) *8, 9, 10, 46*

Volterra, V. [1897], *Sopra alcune questioni di inversione di integrali definiti,* Ann. Mat. Pura Appl. (2), 25, 139-178. *10, 21, 46, 94*

Volterra, V. [1913], *Lečons sur les équations intégrales et intégro-différentielles,* Gauthier-Villars, Paris. *94*

Volterra, V. [1959], *Theory of Functionals and of Integral and Integro-Differential Equations,* Dover, New York.

Wagner, C. [1954], *On the numerical solution of Volterra integral equations,* J. Math. Phys., 32, 289-301. *96, 318, 403*

Wagner, E. [1978], *Ueber die Asymptotik der Lośungen linearer Volterrascher Integralgleichungen 2. Art von Faltungstyp,* Beitrage Anal., 11, 165-183. *38, 48, 493*

Wahr, H. [1977], *Kollokationsverfahren bei Integrodifferentialgleichungen,* Dissertation, Universität Karlsruhe. *319*

Weiss, P. [1978], *Numerical solutions of Volterra integral equations,* PhD thesis, Carnegie-Mellon University, Pittsburgh. *319*

Weiss, R. [1972a], *Numerical procedures for Volterra integral equations,* PhD thesis, Australian National University, Canberra. *177, 204, 319, 397, 403, 405, 496, 503*

Weiss, R. [1972b], *Product integration for the generalized Abel equation,* Math. Comp., 26, 177-190. *396, 397, 398, 405*

Weiss, R. & Anderssen, R.S. [1972], *A product integration method for a class of singular Volterra equations,* Numer. Math., 18, 442-456. *397, 405*

Westreich, D. & Cahlon, B. [1980], *Numerical solution of Volterra integral equations with continuous or discontinuous terms,* J. Inst. Math. Appl., 26, 175-186. *403*

Whittaker, E.T. [1918], *On the numerical solution of integral equations,* Proc. Roy. Soc. London Ser. A, 94, 367-383. *95, 163*

Widder, D.V. [1941], *The Laplace Transform,* Princeton University Press, Princeton. *38, 47*

Williams, H.M. [1980], *Variable step-size predictor-corrector schemes for Volterra second kind integral equations,* PhD thesis, University of Oxford. *164*

Wolfe, M.A. [1969], *The numerical solution of non-singular integral and integro-differential equations by iteration with Chebyshev series,* Comput. J., 12, 193-196. *164*

Wolfe, M.A. & Phillips, G.M. [1968], *Some methods for the solution of non-singular Volterra integro-differential equations,* Comput. J., 11, 334-336. *166, 232*

Wolkenfelt, P.H.M. [1977], *Backward differentiation formulas for Volterra integro-differential equations,* Report NW 53/77, Mathematisch Centrum, Amsterdam. *166, 513, 514*

Wolkenfelt, P.H.M. [1979], *Linear multistep methods and the construction of quadrature formulas for Volterra integral and integro-differential equations,* Report NW 76/79, Mathematisch Centrum, Amsterdam. *68, 77, 91, 164, 166*

Wolkenfelt, P.H.M. [1980], *Stability analysis of reducible quadrature methods for*

Volterra integral equations of the seond kind, Report NW 79/80, Mathematisch Centrum, Amsterdam. *495*

Wolkenfelt, P.H.M. [1981a], *The numerical analysis of reducible quadrature methods for Volterra integral and integro-differential equations,* Doctoral thesis, Mathematisch Centrum, Amsterdam. *65, 68, 80, 84, 91, 101, 104, 110, 122, 140, 143, 144, 152, 156, 164, 165, 166, 325, 495, 505, 506, 508*

Wolkenfelt, P.H.M. [1981b], *On the numerical stability of reducible quadrature methods for second kind Volterra integral equations,* Z. Angew. Math. Mech., 61, 399-401. *122, 149*

Wolkenfelt, P.H.M. [1981c], *Reducible quadrature methods for Volterra integral equations of the first kind,* BIT, 21, 232-241. *143, 165*

Wolkenfelt, P.H.M. [1982], *The construction of reducible quadrature rules for Volterra integral and integro-differential equations,* IMA J. Numer. Anal., 2, 131-152. *68, 84, 91, 164, 166, 325, 495*

Wolkenfelt, P.H.M. [1983a], *On the relation between the repetition factor and numerical stability of direct quadrature methods for second kind Volterra integral equations,* SIAM J. Numer. Anal., 20, 1049-1061. *495*

Wolkenfelt, P.H.M. [1983b], *Modified multilag methods for Volterra functional equations,* Math. Comp., 40, 301-316. *101, 104, 110, 122, 140, 144, 146, 152, 156, 164, 166*

Wolkenfelt, P.H.M., Houwen, P.J. van der & Baker, C.T.H. [1981], *Analysis of numerical methods for second kind Volterra equations by imbedding techniques,* J. Integral Equations, 3, 61-82. *65, 110, 164*

Wood, B. [1980], *Observations on the combined recursive collocation and kernel approximation technique for solving Volterra integral equations,* in: *Approximation Theory III* (E.W. Cheney, ed.), Academic Press, New York, 919-925. *403*

Wouk, A. [1964], *Direct integration, existence and uniqueness,* in: Anselone [1964], 3-33.

Wright, K. [1970], *Some relationships between implicit Runge-Kutta, collocation and Lanczos τ methods, and their stability properties,* BIT, 10, 217-227. *317*

Yatsenko, Yu. P. [1983], *Asymptotically optimally exact algorithm for solution of a linear Volterra equation (Russian),* Vychisl. Prikl. Mat. (Kiev), 50, 122-127.

Young, A. [1954a], *Approximate product-integration,* Proc. Roy. Soc. London Ser. A, 224, 552-561. *92, 318*

Young, A. [1954b], *The application of product-integration to the numerical solution of integral equations,* Proc. Roy. Soc. London Ser. A, 224, 561-573. *92, 318, 505, 508*

Zabreyko, P.P. et al. [1975], *Integral Equations - A Reference Text,* Noordhoff, Leiden.

Zheludev, V.A. [1975], *The approximate solution of a class of equations in convolutions by means of splines,* U.S.S.R. Comput. Math. and Math. Phys., 15, 26-45. *321*

Zheludev, V.A. [1981], *A stable solution of a class of convolution equations (Russian),* Izv. Vyssh. Uchebn. Zaved. mat., 3, 35-45 (see also Soviet Math., 25, 35-46). *321*

Zheludev, V.A. [1982], *Derivatives of fractional order and the numerical solution of convolution equations,* Differential Equations, 18, 1404-1413. *403*

NOTATION

Relations

$:=$	means "equal by definition".
$=$	means "identically equal". Hence, $g(t) \equiv 0$ indicates that the function g vanishes for all t in the domain of definition.
$\approx$	means "approximately equal".
$f(t) = O(g(t))$ as $t \to a$	means "$f(t)$ is of order $g(t)$ as t approaches a"; that is, $f(t)/g(t)$ remains bounded as $t \to a$. The symbol O is usually referred to as Landau's O symbol.
$f(t) \sim g(t)$ as $t \to a$	means "$f(t)$ behaves asymptotically like $g(t)$ as t approaches a": the error between f and g goes to zero as $t \to a$ (or, equivalently, $\lim_{t \to a} (f(t)/g(t)) = 1$).

Sets and spaces

$\{c_j : j = 1, \ldots, m\}$	denotes the set consisting of the elements $c_j, \ldots, c_m$.
$\{t \in A : P(t)\}$	is the set of all elements t in A possessing the property $P(t)$.
$A \cup B$, $A \cup B$	union, intersection of the sets A and B.
$A \times B$	denotes the set of ordered pairs (a,b) with $a \in A$ and $b \in B$.
$\mathbb{N}$	$:= \{1, 2, \ldots\}$ (the set of natural numbers); $\mathbb{N}_0 := \mathbb{N} \cup \{0\}$.
$\mathbb{Z}$	$:= \{\ldots, -1, 0, 1, \ldots\}$ (the set of integers).
$\mathbb{R}$	denotes the set of real numbers.
$\mathbb{R}_+$	denotes the set of positive real numbers.
$\mathbb{R}^m$	denotes the real m-dimensional Euclidean space; $\mathbb{R}^1 = \mathbb{R}$
$\mathbb{C}$	denotes the set of complex numbers
$[a,b]$	$:= \{x \in \mathbb{R} : a \leqslant x \leqslant b\}$ (closed interval).
(a,b)	$:= x \in \mathbb{R} : a < x < b\}$ (open interval).
$(a,b]$, $[a,b)$	left-open, right-open interval.
I	$:= [0,T]$ ($t > 0$, T finite).
S	$:= \{(t,s) \in \mathbb{R} \times \mathbb{R} : 0 \leqslant s \leqslant t \leqslant T\}$.
π_m	denotes the set of real polynomials of degree less than, or equal to, m.

$C[a,b]$, $C(I)$ — is the space of real-valued functions which are continuous on $[a,b]$ or on I.

$C^r[a,b]$, $C^r(I)$ — denotes the space of real-valued functions which are r times continuously differentiable on $[a,b]$ or on I. We frequently use the terminology "smooth function on I": by this we mean that the function is in $C^r(I)$, for some sufficiently large $r>0$.

$L^p[a,b]$ — denotes the space of real-valued functions f for which

$$(\int_a^b |f(t)|^p dt)^{1/p}$$

is finite. Here, $1\leqslant p<\infty$. The terminology "f is integrable on $[a,b]$" means that $f\in L^1[a,b]$.

$\mathrm{Lip}_\beta(I)$ — is the space of real-valued, Hölder-continuous functions on I, where $0<\beta\leqslant 1$: for each such f there exists a finite constant L such that $|f(t)-f(t')|\leqslant L|t-t'|^\beta$ for all t, $t'\in I$. For $\beta=1$ we obtain the space of Lipschitz-continuous functions.

Matrices and vectors

$A=(a_{i,j})$ — denotes a square matrix; i is the row index, j the column index of its element $a_{i,j}$.

A^T — denotes the transpose of A : $A^T:=(a_{j,i})$.

$\det(A)$ — is the determinant of A.

$S(A)$ — $:=\max\{|\alpha_j|:\alpha_j$ eigenvalues of $A\}$ denotes the spectral radius of A.

I — denotes the identity matrix.

O — denotes the zero matrix.

e — $:= (1, \dots ,1)^T$ denotes the identity vector (unless otherwise stated).

Norms, etc.

$\|A\|_p$ — denotes the subordinate matrix norm (or: l_p-norm) induced by the l_p-norm in $\mathbb{R}^m$.

$\|A\|_1$ — $:=\max\{\sum_{i=1}^m |a_{i,j}|: j=1, \dots ,m\}$.

$\|A\|_\infty$:= $\max\{\sum_{j=1}^{m} |a_{i,j}|: j=1, \ldots ,m\}$ is also called the maximum norm of A.

$\|A\|_2$:= $\sqrt{S(AA^T)}$ is also called the spectral norm of A.

$\mu(A)$:= $\lim_{h\downarrow 0} \frac{\|I+hA\|-1}{h}$ denotes the logarithmic norm of A corresponding to $\|A\|$.

$\|f\|_p$:= $\left[\int_a^b |f(t)|^p\right]^{1/p}$ denotes the l_p-norm for $f\in L^p[a,b]$, $1\leqslant p<\infty$.

$\|f\|_\infty$:= $\max\{f(t): t\in[a,b]\}$ denotes the maximun (or: l_∞-)norm for $f\in C[a,b]$.

$\|\mathbf{x}\|_p$:= $\left[\sum_{i=1}^{m} |x_i|^p\right]^{1/p}$ denotes the l_p-norm for $\mathbf{x}\in\mathbb{R}^m$, $1\leqslant p<\infty$.

$\|\mathbf{x}\|_\infty$:= $\max\{|x_i|: i=1, \ldots ,m\}$ denotes the maximum norm (or: l_∞-) norm for $\mathbf{x}:=(x_1, \ldots ,x_m)^T\in\mathbb{R}^m$.

$<\mathbf{u}, \mathbf{v}>$:= $\sum_{i=1}^{m} u_i\bar{v}_i$ denotes the Euclidean inner product of **u** and **v**.

Functions and polynomials

$\exp(t)$ denotes the value of the exponential function for t.

$\log(t)$ is the natural logarithm of t.

$\log_{10}(t)$ is the logarithm of t to base 10.

$\lfloor t\rfloor$:= $\max\{n\in\mathbb{Z};\ n\leqslant t\}$ ("floor function " or " integer function").

$(x-c)_+^m$:= $\begin{cases} 0\ , & \text{if } x-c<0,\\ (x-c)^m, & \text{if } x-c\geqslant 0\end{cases}$ ("truncation function").

$\Gamma(t)$:= $\int_0^\infty \exp(-s)s^{t-1}ds$ denotes the gamma function (Euler's second integral).

$B(\lambda,\mu)$ $:=\int_0^1 s^{\lambda-1}(1-s)^{\mu-1}ds$ $(\lambda,\mu>-1)$ denotes the beta function (Euler's first integral). It is related to the gamma function by:

$$B(\lambda,\mu)=\frac{\Gamma(\lambda)\Gamma(\mu)}{\Gamma(\lambda+\mu)}$$ u.$

$P_m(t)$ denotes the Legendre polynomial of degree m (with respect to the interval $[-1,1]$).

Operators

Δ, ∇ denote, respectively, the forward and the backward difference operators, defined on uniform meshes:

$\Delta f(t):=f(t+h)-f(t)$, $\nabla f(t):=f(t)-f(t-h)$;

$\Delta^r f(t):=\Delta^{r-1}(\Delta f(t))$, $\nabla^r f(t):=\nabla^{r-1}(\nabla f(t))$.

E denotes the forward shift operator defined on the mesh Π_N: $Ef(t_n):=f(t_{n+1})$.

$f|_\sigma$ denotes the restriction of the function f to the interval σ.

$\sum_{j=k}^m a_j$ $:=\begin{cases}0, & \text{if } m<k \\ a_k+\ldots+a_m, & \text{if } k\geqslant m.\end{cases}$

$\prod_{j=k}^m a_j$ $:=\begin{cases}1, & \text{if } m<k \\ a_k\ldots a_m, & \text{if } k\geqslant m.\end{cases}$

Meshes

Π_N denotes a partition (or mesh) for the interval I of the form $0=t_0<t_1<\ldots<t_N=T$ (here, the index N indicating the dependence of the mesh points $t_n=t_n^{(N)}$ on N will usually be suppressed).

h $:=\max\{h_n: n=0,\ldots,N-1\}$, with $h_n:=t_{n+1}-t_n$, denotes the diameter of the mesh Π_N (again, h will depend on N). The mesh Π_N will be called uniform if $h_n=h$ for all n.

$h\downarrow 0$ $(h\uparrow 0)$ means that h approaches 0 through positive (negative) real numbers.

INDEX

Abbreviations

ODE	ordinary differential equation.
PDE	partial differential equation.
VIDE	Volterra integro-differential equation.
DQ method	direct quadrature method.
VLM method	Volterra linear multistep method.
IBD method	indirect backward differentiation method.
ILM method	indirect linear multistep method.
(M)ML method	(modified) multilag method.
LM method	linear multistep method.
LMC method	linear multistep compund method.
AM method	Adams-Moulton method.
AM_p method	AM method of order p.
BD method	backward differentiation method.
BD_p method	BD method of order p.
G_p rule	Gregory rule of order p.
ETR method	extrapolated trapezoidal method
RK method	Runge-Kutta method (for ODE).
RKN method	Runge-Kutta-Nyström method.
VRK method	Volterra Runge-Kutta method (for second-kind Volterra equation).
PVRK method	Pouzet VRK method.
BVRK method	Bel'tyukov-VRK method.
VDRK method	Runge-Kutta method (for VIDE).
PVDRK method	Pouzet-VDRK method.
BVDRK method	Bel'tyukov-VDRK method.
FLM method	fractional linear multistep method.

Author index

See the list of references for page numbers where authors are cited.

Subject index